AF411170

Cell Signaling in Model Plants 2.0

Cell Signaling in Model Plants 2.0

Editors

Jen-Tsung Chen
Parviz Heidari

MDPI • Basel • Beijing • Wuhan • Barcelona • Belgrade • Manchester • Tokyo • Cluj • Tianjin

Editors
Jen-Tsung Chen
Department of Life Sciences
National University of Kaohsiung
Taiwan

Parviz Heidari
Faculty of Agriculture
Shahrood University of Technology
Iran

Editorial Office
MDPI
St. Alban-Anlage 66
4052 Basel, Switzerland

This is a reprint of articles from the Special Issue published online in the open access journal *International Journal of Molecular Sciences* (ISSN 1422-0067) (available at: www.mdpi.com/journal/ijms/special_issues/CSMP2).

For citation purposes, cite each article independently as indicated on the article page online and as indicated below:

LastName, A.A.; LastName, B.B.; LastName, C.C. Article Title. *Journal Name* **Year**, *Volume Number*, Page Range.

ISBN 978-3-0365-1939-5 (Hbk)
ISBN 978-3-0365-1938-8 (PDF)

Contents

International Journal of
Molecular Sciences

Cell Signaling in Model Plants 2.0

Jen-Tsung Chen [1],* and Parviz Heidari [2]

[1] Department of Life Sciences, National University of Kaohsiung, Kaohsiung 81148, Taiwan
[2] Faculty of Agriculture, Shahrood University of Technology, Shahrood 3619995161, Iran; heidar-ip@shahroodut.ac.ir
* Correspondence: jentsung@nuk.edu.tw

Citation: Chen, J.-T.; Heidari, P. Cell Signaling in Model Plants 2.0. *Int. J. Mol. Sci.* **2021**, *22*, 8007. https://doi.org/10.3390/ijms22158007

Received: 5 July 2021
Accepted: 14 July 2021
Published: 27 July 2021

Publisher's Note: MDPI stays neutral with regard to jurisdictional claims in published maps and institutional affiliations.

Plant cell signaling is an intensive research topic in which reductionist can be achieved when we investigate the systems of model plants [1]. To continue our previous Special Issue "Cell Signaling in Model Plants", the second volume explores more insights into the regulatory mechanisms of plant growth and development and eventually collects 18 publications that consist of ten original research articles and eight literature reviews.

Among the model plants, *Arabidopsis thaliana* is the most utilized system for revealing the in-depth mechanisms of cell signaling in this Special Issue. Krogman et al. investigated the concentration of cytosolic calcium ion ($[Ca^{2+}]_{cyt}$) in different root cells of *A. thaliana* that express *GCaMP3* for producing a modified version of a genetically encoded calcium indicator [2]. According to the results, the authors confirmed that GCaMP3 cell lines could be a reliable system for studying the connections between an environmental stimulus and root development that communicate through calcium signaling. Mohammad-Sidik et al. studied the response and underlying mechanisms of $[Ca^{2+}]_{cyt}$ that are elevated by extracellular ATP (eATP) and ADP (eADP) in the root of *A. thaliana* [3]. The authors identified AtANN1 (annexin 1) that mediates the increase of $[Ca^{2+}]_{cyt}$ in root cells in response to both eATP and eADP, which might be a reactive oxygen species (ROS)-activated Ca^{2+} channel in the plasma membrane. The mechanism that regulates the size of *Arabidopsis* root meristem was investigated by Hashem et al. [4] with an emphasis on the involvement of hormone crosstalk and ROS. Finally, the authors suggest that polyamines modulate the size of root meristem and that it is through the interaction with auxin and cytokinin signaling and the accumulation of ROS is achieved. In the report by Villacampa et al. [5], it was demonstrated that the growth and proliferation of meristematic cells could be affected by microgravity and partial gravity together with red light photostimulation. The authors proposed that the resulting data might be valuable in the future for designing bioregenerative life support systems and space farming. The protein family of mitochondrial transcription termination factor (mTERF) has been identified as a key player in plant development, abiotic stress tolerance, etc. Jiang et al. contributed a report studying the function of *Arabidopsis* mTERF27 when faced with salt stress [6]. It was confirmed that mTERF27 interacted with multiple organellar RNA editing factor 8 (MORF8) directly, which may be crucial for regulating mitochondrial gene expression and thus affecting the development of mitochondria under salt stress. Cyclic nucleotide-gated channels (CNGCs) are a group of cation channels, and 20 subunits of CNGCs have been identified in *A. thaliana*. They have been found to participate in the development, defense, and stress response in plants. Jarratt-Barnham et al. indicated that the heterotetrameric complexes of CNGC subunits act differently from the homotetramers and that cyclic nucleotide-gated channel-like proteins ma bey involved in the regulation of the complex formation [7].

In plant research, the exploration of cell signaling pathways is crucial for breeding stress-tolerant crops. Min et al. studied the role of clade A Type 2C protein phosphatases in rice (OsPP2C09) with an emphasis on the relationship with abscisic acid (ABA) signaling under osmotic stress [8]. The authors concluded that OsPP2C09 regulates an

1

ABA-independent signaling pathway through activating promoters that contain a drought-responsive element. Under stress conditions, lipid second messengers including phosphatidic acid, phosphoinositide, lysophospholipids, and sphingolipids, could modulate the physiological responses of plants. Rodas-Junco et al. summarized the involvement of lipid second messengers in plant cell signaling under osmotic stress and the knowledge is critical for the development of strategies to generate stress-tolerant crops [9]. Lyu et al. investigated the roles of the gene family, *CitGF14s*, that codes for phosphorylated proteins in *Citrus sinensis* in response to biotic and abiotic stresses [10]. The authors found certain isoforms of CitGF14s displayed tissue-specific expression patterns and unraveled the protein–protein interaction network. Finally, they confirmed that several candidates may play an important role in regulating the development and stress responses in *C. sinensis*. A receptor for brassinosteroid signaling, SlBRI1, was modified in a phosphorylation site of serine 1040 and then tested for its functions in tomatoes [11]. Wang et al. found that the modified SlBRI1 exhibited a positive response in term of both growth and yield under heat stress. It was thus confirmed that the phosphorylation site of ser-1040 in SlBRI1 affects heat tolerance and might be applied in protecting growth and yield from high-temperature stress in crops. In response to biotic and abiotic stresses, a significant reprogramming was found to occur in the level of plant transcriptome. Chong et al. contributed a review article for refining the knowledge of signal integration by mediators, particularly the cyclin-dependent kinase 8 (CDK8) module [12]. The CDK8 module was thought to predominantly act as a transcriptional repressor, however, it may play a contrasting regulatory role that depends on the type of biotic and abiotic stress.

In cell signaling, the knowledge obtained from model plants could be applied in crops to accelerate the progress of research and breeding programs. FLAVIN-BINDING, KELCH REPEAT, F-BOX 1 (FKF1) has been recognized as a blue-light receptor that plays a role in the promotion of flowering in *A. thaliana*. Shibuya et al. identified an FKF1 homolog SlFKF1 in tomatoes and confirmed that the expression of *SlFKF1* at a low level was associated with late flowering, increased leaflets, and low concentrations of lycopene [13]. A mitogen-activated protein kinase (MAPK) is a type of protein kinase that has been well-proved in *A. thaliana* to play a role in defense responses. Additionally, a MAPK in *Lotus japonicus*, LjMPK6, was known to be involved in regulating symbiosis between legume and rhizobia. Yan et al. identified a type 2C protein phosphatase, LjPP2C, and confirmed that it is required for dephosphorylating LjMPK6 to regulate the nodule development in *L. japonicas* [14]. Chaulagain and Frugoli contributed a review article on the advances in the symbiosis between legumes and rhizobia with an emphasis on the regulation of nodule number [15]. The signaling pathways for initiation and organogenesis of the nodule, nitrate-dependent signaling, and autoregulation of nodulation were comprehensively discussed, and it set up the direction of future research on the fine-tuning of the plant's response to rhizobia.

Magnesium ions (Mg^{2+}) are the second most abundant cation in living cells and they serve as a signal of the adenylate status. Kleczkowski and Igamberdiev refined the signaling pathway of Mg^{2+}, particularly the adenylate kinase (AK) equilibrium and its involvement in the adenylate status [16]. It was concluded that AK plays a central role in Mg^{2+} signaling and is an allosteric effector in cellular metabolism for energy homeostasis in cells.

ROS are a universal regulatory element in plant cells and participate in an array of signaling pathways. Breygina and Klimenko reviewed the role of ROS and the underlying mechanism involving ion transport systems during plant sexual reproduction [17]. They found that ROS are involved at most stages of the life cycle in the male gametophyte, and it chiefly acts through ion currents by a feedback loop.

Parasitic plants absorb water and nutrients from hosts and could cause substantial losses of yield. Hu et al. summarized the virulence mechanisms in obligate root parasites of the genera *Orobanche* and *Striga* focusing on the activities of proteins with nucleotide-binding and leucine-rich repeat domains, NLR proteins, that encoded by resistance genes of the host [18]. This present review presents recent advances in this field achieved by

CRISPR genome editing and RNAi silencing and indicates that the new findings might contribute to establishing strategies for controlling parasitic weeds.

The communication between the nucleus, organelles, and cellular compartments is a critical system for plant cells to face changing environments, including stresses. Mielecki et al. contributed a review for a better understanding of retrograde signaling in stress responses, such as its involvement in the induction of cell death and in the biogenesis of organelles [19]. The authors stated that the knowledge may be important for developing novel strategies for improving the adaptability of plants in changing environments.

Theoretically, cell signaling affects virtually every aspect of plant cell structure and function, and thus to scientists, unveiling the global network and the underlying machinery is always a great challenge. This Special Issue only presents a snapshot of this intensive field of plant biology and we believe that with the rapid development of advanced technology such as integrative multi-omics and CRISPR genome editing, more and more critical networks of plant cell signaling will be unraveled in the near future.

Funding: This research received no external funding.

Conflicts of Interest: The author declares no conflict of interest.

References

1. Chen, J.-T.; Heidari, P. Cell Signaling in Model Plants. *Int. J. Mol. Sci.* **2020**, *21*, 6062. [CrossRef] [PubMed]
2. Krogman, W.; Sparks, J.A.; Blancaflor, E.B. Cell type-specific imaging of calcium signaling in *Arabidopsis thaliana* seedling roots using GCaMP3. *Int. J. Mol. Sci.* **2020**, *21*, 6385. [CrossRef] [PubMed]
3. Mohammad-Sidik, A.; Sun, J.; Shin, R.; Song, Z.; Ning, Y.; Matthus, E.; Wilkins, K.A.; Davies, J.M. Annexin 1 is a component of eATP-induced cytosolic calcium elevation in *Arabidopsis thaliana* roots. *Int. J. Mol. Sci.* **2021**, *22*, 494. [CrossRef]
4. Hashem, A.M.; Moore, S.; Chen, S.; Hu, C.; Zhao, Q.; Elesawi, I.E.; Feng, Y.; Topping, J.F.; Liu, J.; Lindsey, K. Putrescine Depletion Affects Arabidopsis Root Meristem Size by Modulating Auxin and Cytokinin Signaling and ROS Accumulation. *Int. J. Mol. Sci.* **2021**, *22*, 4094. [CrossRef] [PubMed]
5. Villacampa, A.; Ciska, M.; Manzano, A.; Vandenbrink, J.P.; Kiss, J.Z.; Herranz, R.; Medina, F.J. From Spaceflight to Mars g-Levels: Adaptive Response of A. Thaliana Seedlings in a Reduced Gravity Environment Is Enhanced by Red-Light Photostimulation. *Int. J. Mol. Sci.* **2021**, *22*, 899. [CrossRef] [PubMed]
6. Jiang, D.; Chen, J.; Zhang, Z.; Hou, X. Mitochondrial Transcription Termination Factor 27 Is Required for Salt Tolerance in *Arabidopsis thaliana*. *Int. J. Mol. Sci.* **2021**, *22*, 1466. [CrossRef] [PubMed]
7. Jarratt-Barnham, E.; Wang, L.; Ning, Y.; Davies, J.M. The Complex Story of Plant Cyclic Nucleotide-Gated Channels. *Int. J. Mol. Sci.* **2021**, *22*, 874. [CrossRef] [PubMed]
8. Min, M.K.; Kim, R.; Hong, W.-J.; Jung, K.-H.; Lee, J.-Y.; Kim, B.-G. OsPP2C09 Is a Bifunctional Regulator in Both ABA-Dependent and Independent Abiotic Stress Signaling Pathways. *Int. J. Mol. Sci.* **2021**, *22*, 393. [CrossRef] [PubMed]
9. Rodas-Junco, B.A.; Racagni-Di-Palma, G.E.; Canul-Chan, M.; Usorach, J.; Hernández-Sotomayor, S.M. Link between Lipid Second Messengers and Osmotic Stress in Plants. *Int. J. Mol. Sci.* **2021**, *22*, 2658. [CrossRef] [PubMed]
10. Lyu, S.; Chen, G.; Pan, D.; Chen, J.; She, W. Molecular Analysis of 14-3-3 Genes in *Citrus sinensis* and Their Responses to Different Stresses. *Int. J. Mol. Sci.* **2021**, *22*, 568. [CrossRef] [PubMed]
11. Wang, S.; Hu, T.; Tian, A.; Luo, B.; Du, C.; Zhang, S.; Huang, S.; Zhang, F.; Wang, X. Modification of Serine 1040 of SlBRI1 Increases Fruit Yield by Enhancing Tolerance to Heat Stress in Tomato. *Int. J. Mol. Sci.* **2020**, *21*, 7681. [CrossRef] [PubMed]
12. Chong, L.; Shi, X.; Zhu, Y. Signal integration by cyclin-dependent kinase 8 (CDK8) module and other mediator subunits in biotic and abiotic stress responses. *Int. J. Mol. Sci.* **2021**, *22*, 354. [CrossRef] [PubMed]
13. Shibuya, T.; Nishiyama, M.; Kato, K.; Kanayama, Y. Characterization of the FLAVIN-BINDING, KELCH REPEAT, F-BOX 1 Homolog SlFKF1 in Tomato as a Model for Plants with Fleshy Fruit. *Int. J. Mol. Sci.* **2021**, *22*, 1735. [CrossRef] [PubMed]
14. Yan, Z.; Cao, J.; Fan, Q.; Chao, H.; Guan, X.; Zhang, Z.; Duanmu, D. Dephosphorylation of LjMPK6 by Phosphatase LjPP2C is Involved in Regulating Nodule Organogenesis in *Lotus japonicus*. *Int. J. Mol. Sci.* **2020**, *21*, 5565. [CrossRef] [PubMed]
15. Chaulagain, D.; Frugoli, J. The Regulation of Nodule Number in Legumes Is a Balance of Three Signal Transduction Pathways. *Int. J. Mol. Sci.* **2021**, *22*, 1117. [CrossRef]
16. Kleczkowski, L.A.; Igamberdiev, A.U. Magnesium signaling in plants. *Int. J. Mol. Sci.* **2021**, *22*, 1159. [CrossRef] [PubMed]
17. Breygina, M.; Klimenko, E. ROS and ions in cell signaling during sexual plant reproduction. *Int. J. Mol. Sci.* **2020**, *21*, 9476. [CrossRef]
18. Hu, L.; Wang, J.; Yang, C.; Islam, F.; Bouwmeester, H.J.; Muños, S.; Zhou, W. The Effect of Virulence and Resistance Mechanisms on the Interactions between Parasitic Plants and Their Hosts. *Int. J. Mol. Sci.* **2020**, *21*, 9013. [CrossRef]
19. Mielecki, J.; Gawroński, P.; Karpiński, S. Retrograde signaling: Understanding the communication between organelles. *Int. J. Mol. Sci.* **2020**, *21*, 6173. [CrossRef]

International Journal of *Molecular Sciences*

Article

Putrescine Depletion Affects Arabidopsis Root Meristem Size by Modulating Auxin and Cytokinin Signaling and ROS Accumulation

Ahmed M. Hashem [1,2,3,†], Simon Moore [1,4,†], Shangjian Chen [1], Chenchen Hu [1], Qing Zhao [1], Ibrahim Eid Elesawi [1,2,5], Yanni Feng [1], Jennifer F. Topping [4], Junli Liu [4], Keith Lindsey [4] and Chunli Chen [1,2,*]

[1] College of Life Science and Technology, Huazhong Agricultural University, Wuhan 430070, China; ahmedhashem@webmail.hzau.edu.cn (A.M.H.); simon.moore@durham.ac.uk (S.M.); chenshangjian@webmail.hzau.edu.cn (S.C.); xccb456@webmail.hzau.edu.cn (C.H.); qing.zhao@webmail.hzau.edu.cn (Q.Z.); ibrahimeid@webmail.hzau.edu.cn (I.E.E.); feng@mail.hzau.edu.cn (Y.F.)

[2] Key Laboratory of Plant Resource Conservation and Germplasm Innovation in Mountainous Region (Ministry of Education), Institute of Agro-Bioengineering, College of Life Science, Guizhou University, Guiyang 550025, China

[3] Biotechnology Department, Faculty of Agriculture, Al-Azhar University, Cairo 11651, Egypt

[4] Department of Biosciences, Durham University, South Road, Durham DH1 3LE, UK; j.f.topping@durham.ac.uk (J.F.T.); junli.liu@durham.ac.uk (J.L.); keith.lindsey@durham.ac.uk (K.L.)

[5] Agricultural Biochemistry Department, Faculty of Agriculture, Zagazig University, Zagazig 44511, Egypt

* Correspondence: chenchunli@mail.hzau.edu.cn

† These authors contributed equally to this work.

Citation: Hashem, A.M.; Moore, S.; Chen, S.; Hu, C.; Zhao, Q.; Elesawi, I.E.; Feng, Y.; Topping, J.F.; Liu, J.; Lindsey, K.; et al. Putrescine Depletion Affects Arabidopsis Root Meristem Size by Modulating Auxin and Cytokinin Signaling and ROS Accumulation. *Int. J. Mol. Sci.* **2021**, *22*, 4094. https://doi.org/10.3390/ijms22084094

Academic Editor: Jen-Tsung Chen

Received: 5 March 2021
Accepted: 12 April 2021
Published: 15 April 2021

Publisher's Note: MDPI stays neutral with regard to jurisdictional claims in published maps and institutional affiliations.

Abstract: Polyamines (PAs) dramatically affect root architecture and development, mainly by unknown mechanisms; however, accumulating evidence points to hormone signaling and reactive oxygen species (ROS) as candidate mechanisms. To test this hypothesis, PA levels were modified by progressively reducing ADC1/2 activity and Put levels, and then changes in root meristematic zone (MZ) size, ROS, and auxin and cytokinin (CK) signaling were investigated. Decreasing putrescine resulted in an interesting inverted-U-trend in primary root growth and a similar trend in MZ size, and differential changes in putrescine (Put), spermidine (Spd), and combined spermine (Spm) plus thermospermine (Tspm) levels. At low Put concentrations, ROS accumulation increased coincidently with decreasing MZ size, and treatment with ROS scavenger KI partially rescued this phenotype. Analysis of double *AtrbohD/F* loss-of-function mutants indicated that NADPH oxidases were not involved in H_2O_2 accumulation and that elevated ROS levels were due to changes in PA back-conversion, terminal catabolism, PA ROS scavenging, or another pathway. Decreasing Put resulted in a non-linear trend in auxin signaling, whereas CK signaling decreased, re-balancing auxin and CK signaling. Different levels of Put modulated the expression of PIN1 and PIN2 auxin transporters, indicating changes to auxin distribution. These data strongly suggest that PAs modulate MZ size through both hormone signaling and ROS accumulation in *Arabidopsis*.

Keywords: polyamine; root meristem; hormone signaling; ROS; auxin response; PIN transporter; cytokinin response

1. Introduction

Polyamines (PAs) are small polycationic compounds found in all living organisms. Due to their characteristic positive charges, they can interact with negatively charged molecules such as DNA, RNA, proteins, and phospholipids, and therefore influence their activity [1]. Putrescine (Put), spermidine (Spd), spermine (Spm), and thermospermine (Tspm) are the most common polyamines found in plants [2,3]. PAs have been shown to be implicated in the regulation of several plant physiological processes, including flower development, embryogenesis, organogenesis, senescence, and fruit maturation

5

and development [4], and are also involved in *Arabidopsis* meristem development [5] and stress responses [6–8]. Recent studies on plant polyamines have been reviewed [9]. Put is the central diamine substrate compound for the biosynthesis of higher triamine and tetraamine polyamines. Unlike many other plants, *Arabidopsis thaliana* has only one Put biosynthesis pathway, through arginine decarboxylase [10]. In *Arabidopsis*, arginine decarboxylase 1 (ADC1) and ADC2 are key rate-limiting enzymes in Put synthesis, which catalyze L-arginine conversion into Put [10]. It was recently reported that the *ADC1* gene is involved in *N*-acetylputrescine biosynthesis from $N\delta$-acetylornithine [11].

Cellular PA homeostasis is controlled by a combination of mechanisms, including transcriptional and upstream open reading frames (uORFs), translational regulation of PA biosynthesis genes, conjugation, back-conversion, and terminal catabolism [12–14]. The intricacies of PA biosynthesis are exhibited by several examples—treatment with D-Arg, the ADC-specific competitive inhibitor, results in a reduction in Put and increased Spd and Spm content in *Pringlea antiscorbutica* [15]; *adc2* mutants with low Put content show no change in Spd and Spm levels [16]; overexpression of *ADC* in plants generally results in high Put accumulation but in many cases exhibits relatively small changes in Spd and Spm [17]; and silencing of *ADC* genes significantly reduces Put, Spd, and Spm + Tspm in *Arabidopsis* ecotype Wassilewskija (Ws) [18]. This indicates that PA levels are under strict regulation [19] and emphasizes the complexity of the PA metabolic pathway.

Modifications in PA content can differentially affect root growth and development. For example, treatment with D-Arginine (D-Arg), a specific ADC1/2 competitive inhibitor, leads to a reduction in Put and a longer primary root length in *Pringlea antiscorbutica* [15]; perturbation of Put biosynthesis in *adc1* and *adc2* single T-DNA mutants shows no root phenotype, whereas the double mutant is lethal [20]; mutation in *BUD2* gene, which encodes S-adenosylmethionine decarboxylase 4 (SAMDC4), a key enzyme required for PA biosynthesis in *A. thaliana*, resulted in Put accumulation and altered root architecture [21]; silencing both *ADC1* and *ADC2* shows a significant reduction in primary root length [18], and high levels of Put treatment exhibit a similar phenotype in *Arabidopsis* [22]. Exogenous application of low levels of Put has no effect on root growth in *Arabidopsis* [23,24], but treatment with 1 mM Put increases root length in strawberries [25]. Treatment with an inhibitor of Tspm biosynthesis increased primary root growth, whereas treatment with exogenous Tspm inhibited root growth [26]; and inhibition or induction of *Arabidopsis* polyamine oxidase 5 (AtPAO5) significantly affected root length and development [27]. Such findings indicate complex regulatory effects of PAs on root development.

The PA back-conversion and terminal catabolism processes are mediated by two classes of amine oxidases, copper-containing amine oxidases (CuAOs) and FAD-dependent polyamine oxidases (PAOs). These enzymatic reactions lead to the production of H_2O_2 [8,28–30], suggesting another mechanism by which PAs can affect root growth and development [28], in addition to their role as ROS scavengers [31], acting to reduce stress-induced ROS accumulation in leaves and roots [32,33].

ROS homeostasis plays a vital role in root growth and development. A study of how changes in ROS accumulation affect root development [34] reported that the balance between hydrogen peroxide (H_2O_2) and superoxide (O_2^-) affects root growth and meristem structure by influencing the transition from cell division to cell differentiation. It was also demonstrated that perturbation of polyamine catabolism ZmPAO1 strongly influences root development and xylem differentiation in *Zea mays*, mediated by H_2O_2 production [35]. Such studies suggest that altering ROS homeostasis is one mechanism by which PAs could affect root development.

Several studies indicate that PAs also modify hormone response and crosstalk under specific physiological and developmental processes [36,37]; for example, transcriptome studies revealed that changes in endogenous PA content altered the expression levels of genes associated with biosynthesis and signaling of several plant hormones such as auxin, ethylene, and gibberellins [38]; it is necessary for interactions between Tspm, auxin, and cytokinin to be tightly controlled for proper xylem development and plant growth [27]; PAs and auxin affect root growth and development in two sweet orange cultivars [39]; and several studies have further explored the relationships between PAs and auxin response [40]. The above data provide evidence to suggest that the adjustment of the hormone response is an additional mechanism by which PAs affect root development.

The above studies demonstrate that changes in PA levels can influence plant phenotypes in multiple species and tissues; however, mechanisms linking PA levels to phenotype are not well understood. In this work, we focus on the root meristem to identify PA-driven mechanisms that regulate the meristem phenotype under conditions of Put depletion in *Arabidopsis*.

2. Results

Since *adc1* and *adc2* single T-DNA mutants exhibit no root phenotype (Figure 1A,C) and the double mutant is lethal, it was decided to adjust PA levels by simultaneously reducing the activity of ADC1 and ADC2, which are key enzymes in the Put biosynthesis pathway. Several previous studies [41–46] have used D-Arg treatment to inhibit ADC1/2 and deplete Put levels. We progressively reduced Put biosynthesis, the substrate of Spd, Spm, and Tspm, by inhibiting ADC1/2 enzyme activity using the competitive inhibitor D-Arg, which specifically inhibits ADC1/2 activity by blocking binding of the Put substrate L-Arg. Given the specificity of D-Arg and its mode of action, this method was thought to be equivalent to using ADC genetic knockdowns, with the additional advantage of allowing improved flexibility in the control over ADC activity and Put biosynthesis.

ADC2::GUS results confirmed the differential localization of ADC2 expression to the root (Figure 1D). The results also suggested that ADC2 activity is higher in the root than that of ADC1, since while D-Arg treatment inhibited root length in WT, *adc1*, and *adc2*, D-Arg inhibition of ADC1 activity in the *adc2* mutant resulted in a shorter root than inhibition of ADC2 did in the *adc1* mutant, indicating that the *adc2* mutant root phenotype is more sensitive to D-Arg than *adc1* (Figure 1B,C).

2.1. D-Arg Application Promotes or Inhibits Root Growth and Meristem Size in a Concentration-Dependent Manner

To explore how Put depletion modulates root growth and development, we treated 5-day-old *Arabidopsis* wild-type Col-0 seedlings for a further 3 days with a range of concentrations from 0.01 mM to 1 mM D-Arg, a competitive inhibitor of ADC1/2 activity (Figure 2A,B). To determine the effect of D-Arg on polyamine content, seedlings were treated with 0.01, 0.05, 0.1, and 0.6 mM of D-Arg for three days. Whole seedlings were used for HPLC analysis. Data revealed a significant reduction in Put of approximately 30–35% upon 0.01, 0.05, and 0.1 mM treatment, and then a significant decrease of approximately 60% upon 0.6 mM treatment. The Spm plus Tspm concentration was unchanged with 0.01 mM D-Arg treatment, and it increased by 78% at 0.05 mM, by 100% at 0.1 mM, and then only by 35% at 0.6 mM treatment, relative to controls (Figure 2A). Interestingly, Spd remained stable until a considerable decrease in Put occurred at 0.6 mM D-Arg, when Spd declined by approximately 50% (Figure 2A). Spd levels appeared to be more tightly controlled than for Put and Spm + Tspm.

After 3 days of treatment, we observed an inverted-U trend in root growth as D-Arg concentrations increased (Figure 2B,C). Compared to controls, root growth was unchanged at 0.01 mM and 0.05 mM treatments, enhanced at 0.1 mM, unchanged at 0.3 mM, and then root growth gradually decreased at treatment concentrations of 0.6 mM and greater.

During the 3-day D-Arg treatment period, we observed interesting changes in daily root growth as demonstrated by the new root growth-rate curve (Figure 2C). We noticed no significant change in root length at concentrations from 0.01 to 0.6 after one day of treatment, whereas 0.8 and 1 mM showed a significant decrease. However, the root growth rate increased at days two and three at 0.1 mM and then started to decline at 0.8 and 1 mM. The root reduction on day three became highly significant, at 0.6 mM and more, compared to untreated controls.

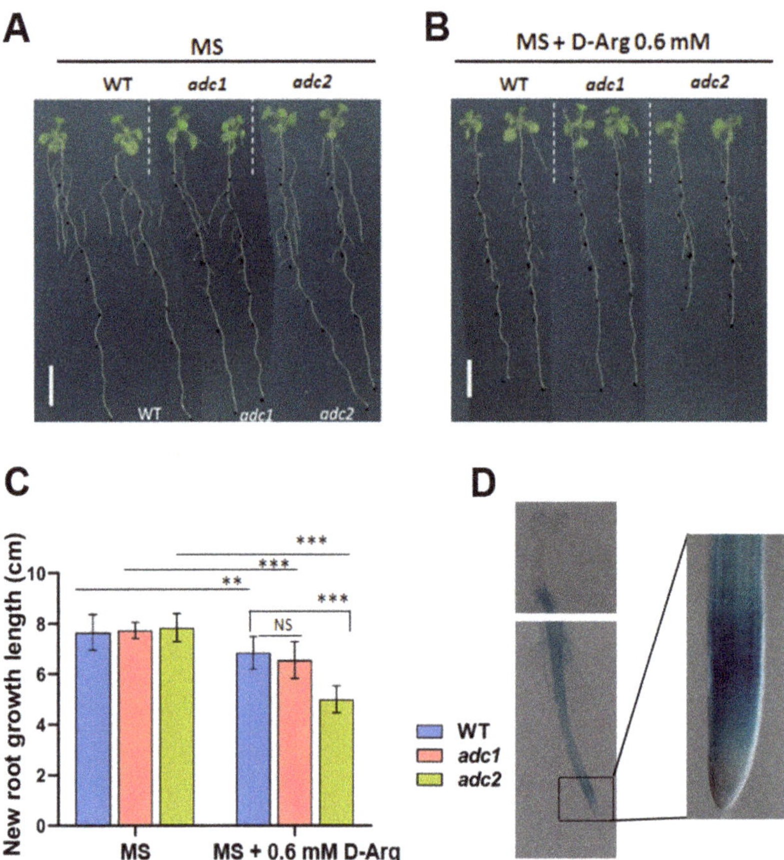

Figure 1. *adc1* and *adc2* single mutants exhibit no root phenotype and the results indicate that ADC2 localizes mainly to the root and *adc2* is more sensitive to D-Arg than *adc1*. For (**A**), (**B**) and (**C**), 5-day-old seedlings at the same developmental stage were transferred to new media containing water (Ctrl) or 0.6 mM D-Arg and were treated for 8 days. (**A**) Untreated wild-type (WT), *adc1*, and *adc2* root phenotypes are the same. Scale bar = 1 cm. (**B**) D-Arg treatment inhibits root length of WT, *adc1*, and *adc2*. Scale bar = 1 cm. (**C**) D-Arg treatment exhibits no difference between WT and *adc1* root length; however, the *adc2* root is significantly shorter that of WT and *adc1* phenotypes. (**D**) Histochemical localization of GUS activity in 2-day-old transgenic seedling harboring promoter ADC2::GUS (*pAtADC2::GUS*). Data shown are averages ± SD (n > 30). Asterisks denote significant differences (** $p < 0.01$, *** $p < 0.001$; Student's *t*-test).

Root growth is the result of both cell division in the meristem zone (MZ) and cell expansion in the elongation zone [47]. We chose to more closely examine the meristem structure under D-Arg treatments with a series concentration (Figure 2D) and measured the meristem length and number of cells in the MZ cortical layer. Root cells were analyzed by means of differential interference contrast (DIC) microscopy and compared to the untreated control. A medium D-Arg concentration of 0.1 mM enlarged the meristem by approximately 18%, and MZ size decreased by 20% at 0.6 mM D-Arg and by 25% at 0.8 mM D-Arg (Figure 2E). The MZ cell number slightly increased, but not significantly, at 0.1 mM and then decreased significantly at 0.6 and 0.8 mM D-Arg (Figure 2F).

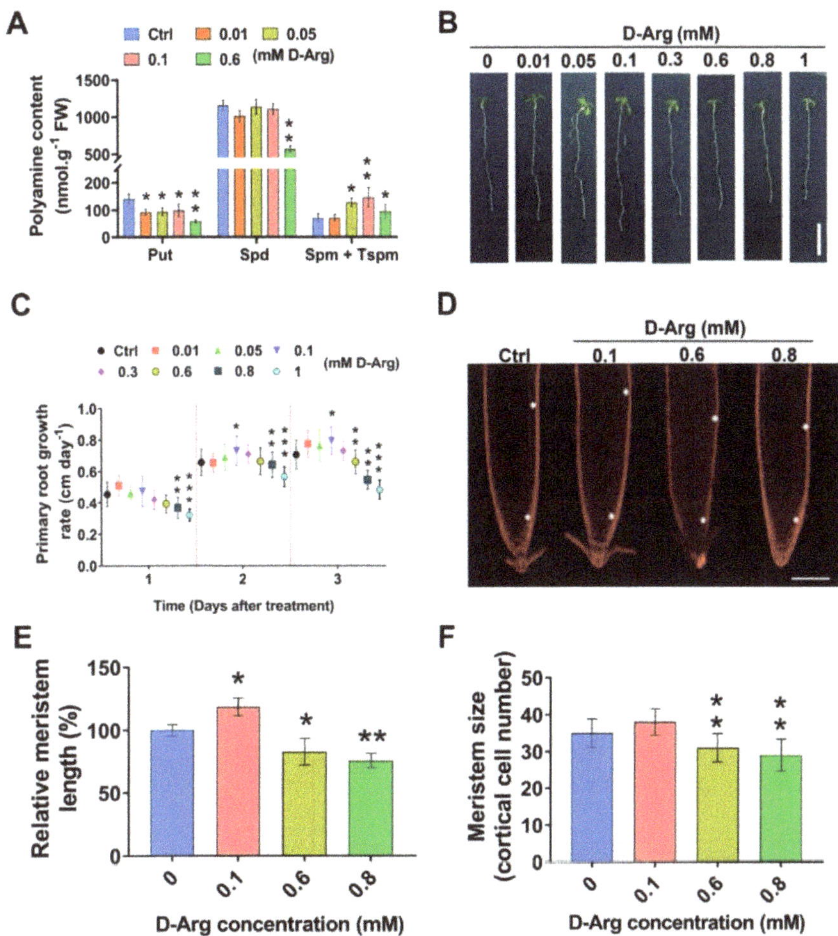

Figure 2. ADC1/2 inhibition differentially modifies polyamine (PA) content and affects root growth and meristematic zone (MZ) size in a non-linear concentration-dependent manner. Five-day-old seedlings at the same developmental stage were transferred to new media containing water (Ctrl) or D-Arg and were treated for 3 days. (**A**) Differential effect of D-Arg treatment on polyamine content. Putrescine (Put), spermidine (Spd), and spermine plus thermo-spermine (Spm + Tspm) levels were measured using HPLC. (**B**) Whole seedling phenotype shows a non-linear effect of increased D-Arg treatment. Scale bar = 1 cm. (**C**) Daily new root growth length varies for different D-Arg concentrations. (**D**) Root meristem imaging, using PI staining and confocal microscopy, exhibits the non-linear effects of D-Arg on MZ size. Two white asterisks on each root indicate the quiescent center (QC) (bottom) and first elongated cell (top). Scale bar = 100 μm. (**E**) Meristem cell length normalized to control. (**F**) Cortical cell number in the meristem zone. Data shown are averages ± SD (n > 30). Asterisks denote significant differences compared with the control (* $p < 0.05$, ** $p < 0.01$; Student's *t*-test).

2.2. KI Application Indicates That MZ Size Inhibition at High D-Arg Concentration Is Partially Due to H_2O_2 Accumulation

To investigate whether treatment with different D-Arg concentrations alters ROS accumulation in WT seedlings, we used 3,3′-diaminobenzidine (DAB) (Figure 3A) and nitrotetrazolium blue (NBT) (Figure 3C) staining to detect the presence of H_2O_2 and O_2^- respectively in vivo. DAB staining indicated no difference in ROS accumulations at low and medium D-Arg treatment, but at high D-Arg treatment of 0.6 mM and 0.8 mM, ROS increased significantly (Figure 3B). Furthermore, we did not detect significant changes in NBT staining at any D-Arg concentration (Figure 3D).

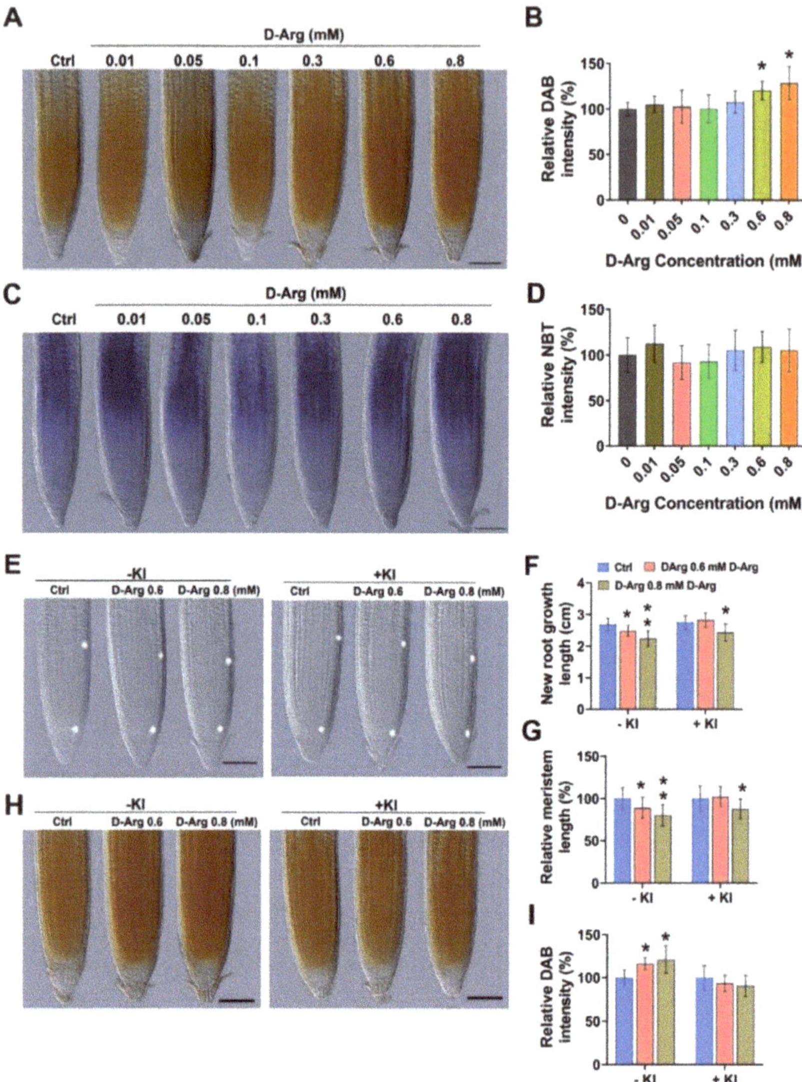

Figure 3. D-Arg-induced H_2O_2 accumulation, root length, and MZ size reduction is partially rescued by ROS scavenging with 1 mM KI. Five-day-old seedlings at the same developmental stage were transferred to new media containing water (Ctrl) or D-Arg and treated for 3 days. MZ imaging by differential interference contrast (DIC) microscopy. (**A,B**) 3,3′-diaminobenzidine (DAB) staining and relative stain intensity indicate that D-Arg treatment increases H_2O_2 accumulation in root meristem. (**C,D**) Nitrotetrazolium blue (NBT) staining and relative stain intensity indicates that D-Arg treatment does not change O_2^- accumulation in the root meristem. (**E–G**) Treatment with and without D-Arg and 1 mM KI shows partial rescue of root meristem and root length by KI. (**H,I**) DAB staining of root meristem and relative stain intensity, with or without D-Arg and KI, show successful H_2O_2 scavenging by KI. All scale bars = 100 μm. Data shown are averages ± SD (n > 30). Asterisks indicate significant differences compared with the control (* $p < 0.05$, ** $p < 0.01$; Student's *t*-test).

To determine whether the reduction in root length and root meristem at higher D-Arg concentrations was attributable to H_2O_2 accumulation, we exposed control roots and roots treated with 0.6 mM and 0.8 mM D-Arg to the ROS scavenger potassium iodide (KI, 1 mM), as described previously [34]. KI application successfully scavenged H_2O_2 (Figure 3H,I) and completely rescued the shorter root phenotype caused by 0.6 mM of D-Arg treatment, whereas at 0.8 mM D-Arg, root length and meristem size were only partially recovered (Figure 3E–G), suggesting that at higher D-Arg levels, reduced root length is not completely ROS-dependent and that other factors are involved.

2.3. H_2O_2 Accumulation Is Not Caused by NADPH Oxidases

During plant growth and development, ROS are predominantly generated by the NADPH oxidases RBOHD and RBOHF class III peroxidases [48,49] and polyamine amine oxidases (PAO) [8]. Some reports show a link between PAs and NADPH oxidases (RBOHD/F) in tobacco and *Arabidopsis* [50,51]. To investigate the involvement of NADPH oxidases in H_2O_2 accumulation at high D-Arg treatments, we examined the phenotype of seedlings carrying double *atrbohD/F* loss-of-function mutants. In this set of experiments, the D-Arg treatment period was extended from 3 to 6 days and new root growth length was observed to ensure that NADPH oxidases were not a significant source of ROS accumulation under conditions of Put depletion. The results showed that, similarly to the WT results, exogenous high D-Arg application inhibited the new root growth of *atrbohD/F* (Figure 4A,B).

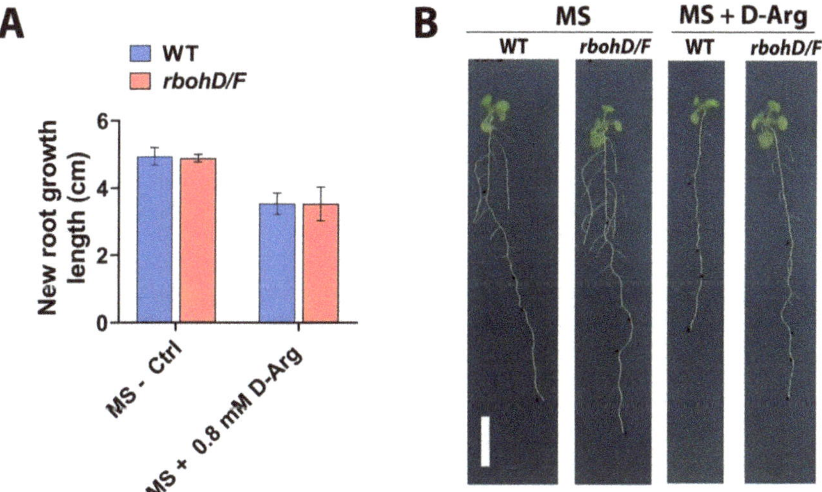

Figure 4. Effect of D-Arg treatment on WT and *AtrbohD/F* root growth indicates that H_2O_2 accumulation is not caused by NADPH oxidases. Five-day-old seedlings at the same developmental stage were transferred to new media containing water (Ctrl) or 0.8 mM D-Arg and treated for 6 days. (**A**) New root growth length of WT and *AtrbohD/F*. (**B**) Seedling phenotype. Scale bar = 1 cm. Data shown are averages ± SD (n > 15).

2.4. Auxin Response Exhibits a Non-Linear Trend as ADC1/2 Activity Decreases

Auxin signaling promotes cell division and is involved in regulating root meristem size [47]. We analyzed the effects of Put depletion on the auxin response using the reporter line *DR5::GFP*. Interestingly, increasing levels of exogenous D-Arg had a non-linear effect on *DR5*-dependent GFP fluorescence (Figure 5A). We observed that 0.1 mM D-Arg decreased *DR5* activity by 15%, but in contrast, higher D-Arg at 0.6 and 0.8 mM significantly enhanced it by 20–25%, respectively (Figure 5B).

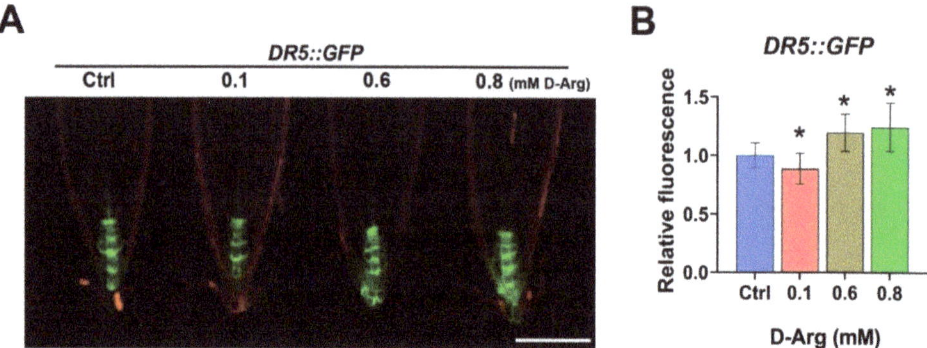

Figure 5. D-Arg treatment has a non-linear effect on auxin response. Five-day-old seedlings at the same developmental stage were transferred to new media containing water (Ctrl) or D-Arg and treated for 3 days. (**A**) Confocal images of *DR5::GFP* auxin response reporter. (**B**) *DR5::GFP* normalized fluorescence indicates a non-linear auxin response trend. Data are averages ± SD (n > 15). Scale bar = 100 μm. Asterisks denote significant differences compared with the control (* $p < 0.05$; Student's *t*-test).

2.5. CK Response Gradually Decreases as D-Arg Increases

Cytokinin (CK) signaling promotes cell differentiation and negatively regulates root meristem size [47]. To investigate whether reduced Put affects CK signaling, we used the reporter line *proARR5::GFP*. Arabidopsis response regulator 5 (ARR5) is a type-A negative regulator of cytokinin response and is strongly induced by CK [52]. Under different levels of D-Arg treatment, we observed that *proARR5::GFP* fluorescence gradually decreased (Figure 6A). The CK response was reduced significantly at 0.1 mM D-Arg by 25% and at higher D-Arg levels of 0.6 mM and 0.8 mM by 33% to 35% compared to the untreated control (Figure 6B). *ARR5* expression was measured using qRT-PCR at different concentrations of applied D-Arg. At low (0.01 mM) D-Arg, we observed a slight but not significant reduction in the ARR5 transcript level (Figure 6C). Similarly to the *proARR5::GFP* results, medium and high levels of D-Arg, at 0.1 mM and 0.6 mM, significantly downregulated *ARR5* transcription by approximately 50%, indicating that D-Arg treatment and Put depletion progressively reduces the CK response in the *Arabidopsis* root. Cytokinin activity is regulated by the balance between biosynthesis and degradation [53,54]. Cytokinin oxidase/dehydrogenases (CKXs) play a key role in CK degradation [54]. It has been reported that CKX family genes play critical roles in determining root architecture in several plant species [55–59]. In *Arabidopsis*, AtCKX1, AtCKX4, and AtCKX7 have been shown to either localize to the root or their overexpression has been shown to affect root phenotype [53,56,60,61]; however, this does not preclude other members of the CKX family playing a significant role in root development. Our results reveal that ADC1/2 inhibition significantly induced cytokinin oxidase1 (*CKX1*) expression levels and slightly induced *CKX7* in the root, whereas *CKX4* was unchanged (Figure 6D).

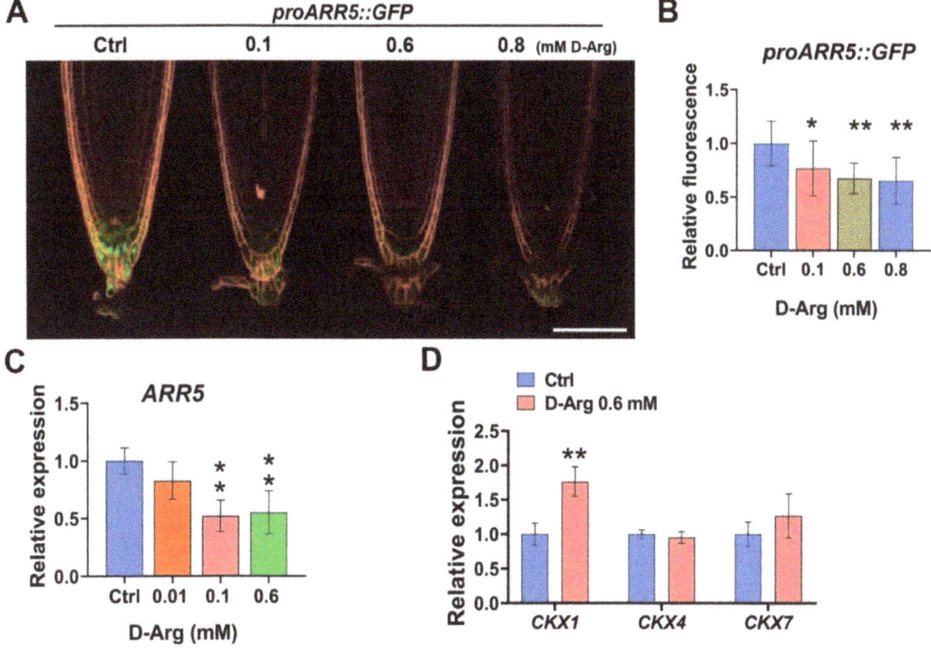

Figure 6. D-Arg treatment reduces the cytokinin response and increases *CKX1* expression in the root meristem. Five-day-old seedlings at the same developmental stage were transferred to new media containing water (Ctrl) or D-Arg and treated for 3 days. (**A**,**B**) Confocal imaging indicates that increasing D-Arg gradually reduces cytokinin response *proARR5::GFP* and relative *proARR5::GFP* fluorescence. Scale bar = 100 μm. (**C**,**D**) qPCR data indicate that D-Arg treatment modulates the relative expression of *ARR5* and cytokinin oxidase (*CKX*) genes. Data shown are averages ± SD (n > 15). Asterisks indicate significant differences compared with the control (* $p < 0.05$, ** $p < 0.01$; Student's *t*-test).

2.6. Differential Modulation of PIN1 and PIN2 Protein at Low and High D-Arg Treatment

We explored the level of protein and localization of auxin efflux transporters PIN1 and PIN2 using two reporter lines, *proPIN1::PIN1:GFP* and *proPIN2::PIN2:GFP*. The auxin efflux transporter PIN1 is localized to the vascular tissue membrane of the root meristem and is involved in auxin transport from the stele to the quiescent center and columella initials [62]; PIN2 is localized mainly in the outer cell layers and is an essential component for basipetal auxin transport [63], and both PIN1 and PIN2 are important for correct auxin distribution [63]. Confocal imaging at low levels of D-Arg revealed significant increases in *PIN1:GFP* by 23% (Figure 7A,B) and *PIN2:GFP* by 35% (Figure 7E,F) compared to untreated controls. At high D-Arg, the results showed that the level of *PIN1:GFP* was significantly increased by 30% compared to the untreated control (Figure 7C,D). In contrast, *PIN2:GFP* at the same D-Arg concentration was significantly decreased by 20% (Figure 7G,H). Our results therefore suggest that Put depletion alters PIN1 and PIN2 levels, cellular auxin efflux, and auxin distribution.

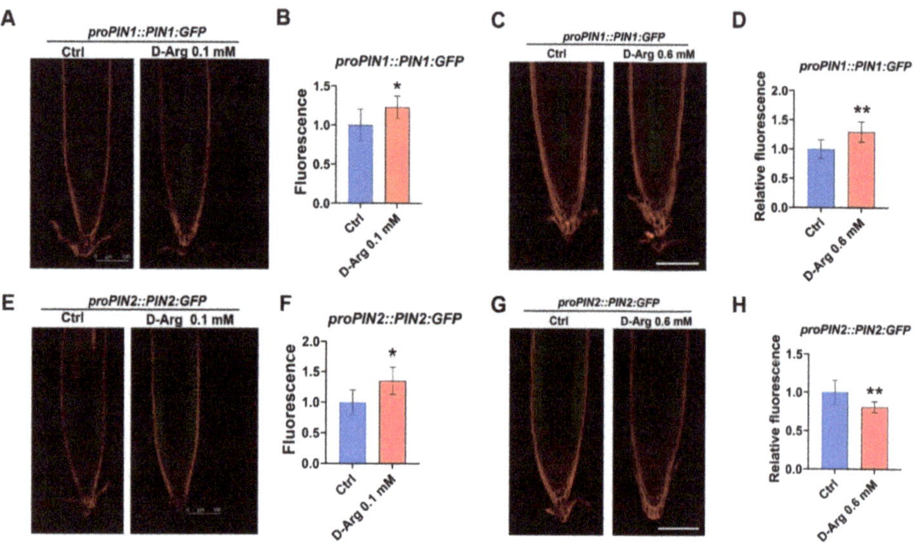

Figure 7. D-Arg treatment differentially affects PIN1 and PIN2 auxin efflux carriers. Five-day-old seedlings at the same developmental stage were transferred to new media containing water (Ctrl) or 0.6 mM D-Arg and treated for 3 days. (**A,B**) Confocal imaging of *proPIN1::PIN1:GFP* reporter line and relative *proPIN1::PIN1:GFP* fluorescence indicate that low D-Arg treatment increases PIN1 protein. (**C,D**) *proPIN1::PIN1:GFP* reporter line and relative *proPIN1::PIN1:GFP* fluorescence indicate that high D-Arg treatment increases PIN1. (**E,F**) *proPIN2::PIN2:GFP* reporter line and relative *proPIN2::PIN2:GFP* fluorescence indicate that low D-Arg treatment increases PIN2 protein. (**G,H**) *proPIN2::PIN2:GFP* reporter line and relative *proPIN2::PIN2:GFP* fluorescence indicate that high D-Arg treatment reduces PIN2 protein. All scale bars =100 μm. Data shown are averages ± SD (n > 15). Asterisks indicate significant differences compared with the control (* $p < 0.05$, ** $p < 0.01$; Student's *t*-test).

3. Discussion

The goal of our study was to investigate the effect of PAs on root growth and development in *Arabidopsis* and identify possible mechanisms by which PAs affect root phenotype under conditions of ADC1/2 inhibition. Under normal conditions, *ADC1* expression is localized to the shoot and *ADC2* expression to the root [64]. Our *ADC2::GUS* results (Figure 1D) confirm that *ADC2* expression is localized to the root. Given that neither *adc1* or *adc2* single mutants exhibited a root phenotype (Figure 1A), it is reasonable to assume that Put levels, directly or indirectly, affect the localization and/or level of *ADC1/2* expression. This assumption is supported by the literature, which provides evidence of *ADC1/2* regulation by both PAs and hormones [37,65–69]. To identify links between Put depletion and MZ size, we therefore used D-Arg application to simultaneously inhibit both ADC1 and ADC2 to ensure a reduction in Put levels, rather than developing specific *ADC2* knockdowns. Although *ADC1/2* regulation, localization, or redundancy are important components of this complex biological system, it was not part of this initial study but could be addressed in future research.

PA levels were modified by using the specific ADC1/2 enzyme inhibitor D-Arg to perturb PA biosynthesis instead of using PA treatment, since exogenous application, for example of Spd, could artificially increase Spd levels, affect PA balance, and generate ROS due to reverse biosynthesis and terminal catabolism [8,28,29]. D-Arg application has been used to deplete Put in several research papers [41–46]; however, none of these papers addressed the issue of whether D-Arg application could produce any significant additional effects by, for instance, regulation of other enzyme activities or by increasing the availability of the ADC1/2 substrate, L-Arg. This is an outstanding issue for future research, which is discussed further in the conclusion.

Inhibition of ADC1/2 reduced Put levels and resulted in unpredictable changes in the levels of higher PAs. As Put levels decreased, Spd remained stable until high D-Arg application, when levels dropped significantly, and Spm + Tspm levels initially increased and then decreased at high D-Arg. These unexpected and differing trends in PA levels under conditions of progressive ADC1/2 inhibition demonstrate the complexity of the PA forward and reverse biosynthesis pathways that re-balance PA levels when Put is depleted. A more detailed examination of how PA levels change under different conditions of enzyme inhibition, PA treatment, or under stress conditions could prove useful in investigating the regulation of PA balancing.

Put depletion resulted in non-intuitive phenotype outcomes, with increased MZ size observed at low levels of D-Arg application and progressively decreasing MZ size at higher D-Arg. This inverted-U trend in MZ size suggests that more than one mechanism regulates MZ size: for example, with one mechanism promoting MZ size at low D-Arg concentrations and another antagonistic mechanism inhibiting MZ size at higher D-Arg levels. At low D-Arg levels the first mechanism acts to increase MZ size, but as D-Arg treatment increases, the second mechanism becomes dominant in order to reduce MZ size.

It has been shown that root MZ size is determined by the balance between cell division and differentiation and that this process is regulated by two mechanisms—auxin and CK signaling balance [47], and changes in ROS homeostasis [34] that are independent of auxin and CK signaling. We therefore examined these two candidate mechanisms. Since auxin promotes cell division, whereas CK promotes differentiation, the auxin-to-CK signaling balance modulates MZ size [47], rather than individual auxin or CK signaling trends. Auxin signaling displayed a U-shaped trend as D-Arg increased, whereas cytokinin progressively decreased. At low D-Arg application, no change in ROS accumulation was observed, but the auxin:CK ratio shifted in favor of auxin to enlarge the MZ, consistent with the literature [47]. However, at higher D-Arg levels, we observed increasing ROS accumulation, which is known to inhibit MZ size [34,70–72]. The results for DAB and NBT staining (Figure 3D) suggest that high Put depletion led to the accumulation of H_2O_2 but not of O_2^- and that that H_2O_2 accumulation is caused by PAO activity, rather than by NADPH oxidases (Figure 4). The ROS effect was confirmed with phenotype rescue by ROS scavenging using KI; however, the partial rescue at high D-Arg also suggests that additional mechanism(s) could be involved in MZ size and root growth regulation.

We concluded that the hormone effect promoted MZ size at low D-Arg but that at higher D-Arg the antagonistic ROS effect dominated to reduce the MZ. The underlying mechanisms whereby Put depletion modulates ROS accumulation and auxin and cytokinin signaling have yet to be unraveled; however, the literature points to several promising lines of investigation: (1) previous studies report that PAs play two opposing roles in the regulation of ROS levels as both ROS scavengers under stress [32,33] and as a source of ROS generated by terminal and back conversion pathways [8,28,29], (2) Spd and Spm have been shown to regulate CKX [37,38,73] and auxin conjugation [37], and (3) Tspm inhibits the final steps in the CK biosynthesis pathway [3,74].

The results of our work indicate that Put depletion affects root phenotype through the antagonistic actions of hormone signaling and ROS accumulation; however, the question remains as to whether this is due to direct or indirect effects of changing Put levels. As noted earlier, PA levels are closely linked by forward and reverse biosynthesis pathways. Furthermore, terminal catabolism of excess PAs and back-conversion to lower-level PAs can produce bioactive products. These mechanisms constitute an extremely complex process by which perturbation of a single PA can result in changes in other PAs and in the generation of bioactive products. This is in part illustrated by our HPLC experiments, which produced non-intuitive results in regard to the changes in higher PAs when Put was depleted. It is therefore likely that perturbation of any single PA results in PA re-balancing and effects phenotype outcomes by multiple direct and indirect pathways. Several studies also indicate that the overall PA balance seems to be more important than the specific effect of individual PAs [15,75].

Previous work [47] showed that changes in relative auxin to CK signaling modulate MZ size through the regulation of auxin transporters and modified auxin distribution. In our work, at low D-Arg the *proPIN1::PIN1:GFP* protein reporter showed an increase in PIN1 and the *proPIN2::PIN2:GFP* showed an increase in PIN2, whereas upon high D-Arg treatment, PIN1 increased and PIN2 decreased (Figure 7), indicating that ADC1/2 inhibition differentially modulates PIN1 and PIN2 protein levels to modify auxin transport and distribution.

The above data demonstrate that hormone signaling and ROS accumulation are two mechanisms by which PAs regulate MZ size (Figure 8).

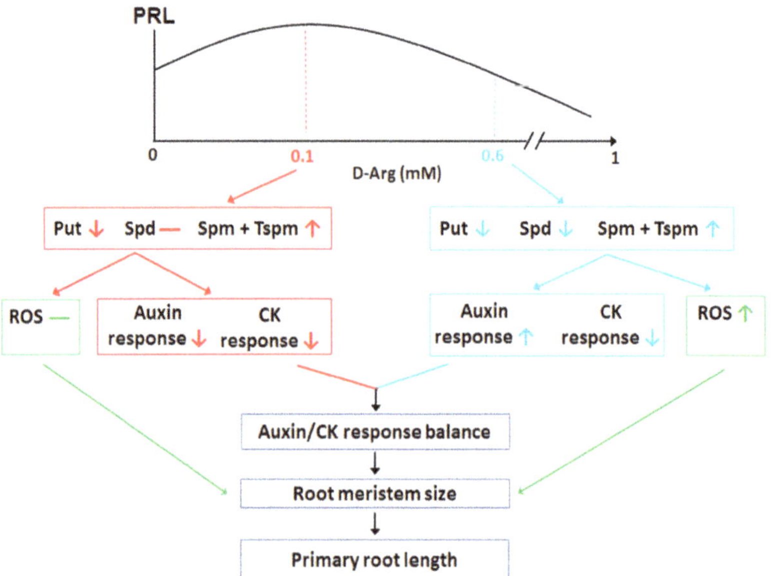

Figure 8. ADC1/2 inhibition differentially changes PA levels and has a non-linear effect on meristem size by modulating relative auxin/CK signaling and ROS accumulation. Exogenous application of increasing D-Arg concentrations differentially modified Put, Spd, and Spm + Tspm levels and resulted in an inverted-U trend in primary root growth. Medium D-Arg at 0.1 mM (red boxes) promoted root growth, decreased the auxin and CK response, and had no effect on ROS levels (green boxes). High D-Arg 0.6 mM (blue boxes) reduced root growth, differentially affected auxin and CK response, and accumulated ROS. PRL, primary root length. Vertical arrows denote a statistically significant increase↑ or decrease↓, whereas a horizontal line indicates no significant change.

4. Materials and Methods

4.1. Plant Materials and Growth Conditions

All experiments were performed using *Arabidopsis thaliana* ecotype Columbia (Col-0). Transgenic marker lines are in Col-0 backgrounds as follows: *DR5::GFP* for auxin-response; *proPIN1::PIN1:GFP* and *proPIN2::PIN2:GFP* for PIN1/2 auxin efflux carrier proteins; *proARR5::GFP* for cytokinin response; and *proADC2::GUS* for localization of *ADC2* expression. Genetic materials, T-DNA insertion mutant *adc1* and *adc2* for polyamine biosynthesis pathway and *atrbohD/F* for ROS experiments were used.

Seeds were surfaced-sterilized for 2 min in 75% ethanol and then 10 min in 1% sodium hypochlorite. Seeds were washed five times with sterilized deionized water. For simultaneous germination, all seeds were stratified for 2–3 days at 4 °C before germination.

Seeds were grown on 120 × 120 × 17 mm square plates containing half-strength Murashige and Skoog (MS) medium 2.2 g L^{-1} (Duchefa Biochemie) including vitamins, MES 0.5 g L^{-1} (Duchefa Biochemie), and 1% (*w/v*) sucrose and solidified with 1% agar

(Duchefa Biochemie). PH was adjusted to 5.8 with KOH. Plated were sealed with Micropore tape.

To maintain the root for molecular assay and physical root measurements, seedlings were grown and placed vertically in the growth incubator at 22 °C with light cycles of 16 h light/8 h dark (light intensity: ~120 µmol/s/m^2, and relative humidity of 60%). For putrescine depletion, 5-day-old seedlings were transferred to MS medium containing 0, 0.01, 0.05, 0.1, 0.3, 0.6, 0.8, and 1 mM of D-Arginine (Sangon Biotech, Shanghai, China) for a further 3 days. For experiments involving double *atrbohD/F* loss-of-function mutants, the D-Arg treatment period was extended from 3 to 6 days and new root growth length was observed to ensure that NADPH oxidases were not a significant source of ROS accumulation under conditions of Put depletion.

4.2. Root Length, Root Meristem Size, and Cell Length Measurements

To track changes in root growth, five-day-old seedlings with the same root length were transferred to a fresh medium containing either water as control or D-Arginine. The position of the root tip was marked on the back of plates at the time of transfer and new root growth length was measured as root growth extension during the treatment period, using images taken with a Nikon D300s digital camera (Nikon Corp., Tokyo, Japan).

For root meristem measurements, after 3 days D-Arg treatment seedlings were cleared with chloral hydrate solution [76]. Root meristem size was determined as the length from the quiescent center (QC) to where the cell was double the size of the previous cell along the cortex cell layer [77,78]. Root meristems were analyzed and imaged using an Olympus BX61 microscope (Olympus, Tokyo, Japan) equipped with differential interference contrast (DIC) optics, 20x UPlanSApo objective, and a CCD camera (Olympus DP74). For Photo acquisition, Olympus cellSens software was used.

4.3. Confocal Laser Scanning Microscopy

All confocal root images were taken with a Leica SP8 confocal laser scanning microscope (https://www.leica-microsystems.com) after three days of D-Arg treatment. Ten micrograms per milliliter of propidium iodide (PI) solution (Sangon Biotech, Shanghai, China) were used to visualize cell walls. Stained roots were visualized with the excitation wavelength set at 548 nm for PI and at 488 nm for green fluorescent protein (GFP).

4.4. RNA Extraction, cDNA Synthesis, and qRT-PCR

Whole roots of Col-0 wild type non-treated (Ctrl) and treated with D-Arginine (D-Arg) were used. Total RNA extraction was performed using TransZol Up reagent (TransGen, Beijing, China). RNA concentration and quality were determined with a Nanodrop 2000 Spectrophotometer (Thermo Fisher Scientific). cDNA was synthesized from 1 µg of RNA using a TransScript One-Step gDNA Removal and cDNA Synthesis SuperMix kit (TransGen, Beijing, China), following the manufacturer's instructions. cDNA was diluted 1:10 for quantitative real-time PCR (qPCR).

Quantitative real-time PCR analyses were performed using TransStart Tip Green qPCR SuperMix (TransGen, Beijing, China) with a Roche LightCycler 480 thermal cycler instrument, 384-well (Roche). Relative expression values were calculated using the $2^{-\Delta\Delta Ct}$ method [79], and ACTIN2 was used as a reference gene. Primers are listed in Table S1. Three biological replicates were performed for each sample and each biological replicate was represented by three technical replicates.

4.5. Quantification of Free Polyamine

Free PA quantification was performed using high-performance liquid chromatography (HPLC). Five-day-old seedlings at the same developmental stage were selected and transferred to the treatment media containing D-Arg for another 3 days. Zero point three five grams of fresh weight of seedlings were harvested for analysis. The extraction and quantification methods were performed as described in [80]. 1,6-hexanediamine was used as an internal standard.

4.6. Histochemical GUS Staining

Transgenic Arabidopsis containing ADC2 promoter::GUS fusions were stained for GUS according to the method described in [32].

4.7. Determination of O_2^- and H_2O_2 by NBT and DAB Staining

3,3′-Diaminobenzidine (DAB, 1 mg mL^{-1}) staining (Sigma Aldrich) was used to detect H_2O_2 levels in roots, and Nitrotetrazolium blue (NBT, 1 mg mL^{-1}) (Sangon Biotech, Shanghai, China) was used for O_2^- detection. DAB and NBT staining were performed as described by [32]. At least 30 root meristems for each staining were analyzed and imaged using an Olympus BX61 microscope (Olympus, Tokyo, Japan) equipped with differential interference contrast (DIC) optics, 20× UPlanSApo objective, and a CCD camera (Olympus DP74). For photo acquisition, Olympus cellSens software was used.

4.8. Image Analysis

Root length, meristem size, cell counts, DAB, and NBT images were investigated using ImageJ (http://www.imagej.nih.gov/ij/). DAB, and NBT stained images were quantified as described in [32]. Mean relative fluorescence for confocal images was calculated with ImageJ. In quantifying fluorescence, at least 15 seedlings per line were used.

4.9. Statistical Analysis

All experiments were performed at least three times. For statistical comparisons, we used Student's t-test. Data shown are averages ± SD. Asterisks indicate significant differences compared with the control (*, $p < 0.05$, **, $p < 0.01$, ***, $p < 0.001$).

5. Conclusions

In this study, we investigated the effect of PAs on root development in *Arabidopsis* with the goal of identifying mechanisms by which PAs affect root growth. We explored changes in ROS accumulation and hormone signaling under conditions of ADC1/2 inhibition and Put depletion, and the results demonstrated that the effects of PAs on root phenotype are mediated by ROS and hormone signaling. Furthermore, the data generated by this study indicate that PA regulation of root phenotype is extremely complex, involving (1) intricate forward and reverse PA biosynthesis, (2) changes in ROS accumulation generated by PA catabolism and reverse biosynthesis and potentially by PA-mediated H_2O_2 scavenging, and (3) the complexities of hormonal crosstalk, resulting in a modified auxin distribution and CK response and changes to the ratio of auxin to CK signaling. This complexity is further evidenced by the contrasting changes in PA levels observed as ADC1/2 inhibition increased; by the non-linear trends in root growth, MZ size, and auxin response; and also by the differential regulation of PIN1 and PIN2 transporters.

Having identified candidate mechanisms linking PAs to root phenotype, future work requires more detailed investigations of these pathways under conditions of Put depletion. As noted earlier, although D-Arg treatment has been used in several studies to reduce Put, questions remain about the possible side-effects of D-Arg treatment. Therefore, the first step in future research will be to address this issue by developing ADC1/2 knockdowns and making comparisons with D-Arg treatment results for gene expression, hormone signaling, ROS accumulation, and phenotype. It would be ideal to be able to calibrate each knockdown with a specific D-Arg treatment level and, provided there is no significant difference between results for the knockdowns and D-Arg application, the two methods can be combined to deplete Put, utilizing the convenience of D-Arg that also allows fine-tuning of Put levels, which is difficult to achieve with knockdowns alone. The next step is to investigate mechanisms by which PAs modulate hormone signaling. The literature suggests links between PAs and auxin conjugation [37], CK biosynthesis [3,74], and CK degradation [37,38], and therefore the initial focus should be on components of these signaling pathways using genetic material. During this phase, possible ROS effects will be removed through the application of KI. A similar approach can be adopted for analyzing links between PAs and ROS.

Our initial results and the literature indicate that PA levels are closely inter-related and that it is difficult to predict how the perturbation of one PA will affect other PAs, hormone signaling, ROS accumulation, and phenotype. Therefore, another interesting research area would be a more detailed examination of PA balancing and downstream effects by treating WT and *adc1* and *adc2* single mutants and knockdowns with D-Arg and PAs, and observing them under stress conditions. Since ADC1/2 enzymes appear to have some level of redundancy given that single mutants do not exhibit phenotypes, the investigation of enzyme expression, localization, and redundancy under different conditions could also be carried out.

Supplementary Materials: Supplementary Materials can be found at https://www.mdpi.com/article/10.3390/ijms22084094/s1.

Author Contributions: A.M.H., S.C., C.H., Q.Z., I.E.E. and Y.F. performed the experiments with support from S.M. and J.F.T., J.L., and K.L. C.C. conceived the project. C.C. and A.M.H. designed the experiments. A.M.H., S.M., and C.C. analyzed the data and wrote the manuscript. All authors read and approved of this content. All authors have read and agreed to the published version of the manuscript.

Funding: This research was supported by the National Natural Science Foundation of China (No.31971520), Advanced Foreign Experts Project (G20200017071, G20190017014), and Fundamental Research Funds for the Central Universities (2662018PY099) from the Chinese government.

Institutional Review Board Statement: Not applicable.

Informed Consent Statement: Not applicable.

Data Availability Statement: Not applicable.

Conflicts of Interest: The authors declare no conflict of interest.

Abbreviations

PAs	Polyamines
ADC	Arginine decarboxylase
CK	Cytokinin
MZ	Meristem zone
Put	Putrescine
Spd	Spermidine
Spm	Spermine
Tspm	Thermospermine
uORFs	Upstream open reading frames
D-Arg	D-Arginine
CuAOs	Copper-containing amine oxidases
PAOs	Polyamine oxidase
RBOH	Respiratory burst oxidase homolog
ROS	Reactive oxygen species
CKX	Cytokinin oxidase
ARR	*Arabidopsis* response regulator
QC	Quiescent center

References

1. Igarashi, K.; Kashiwagi, K. Modulation of cellular function by polyamines. *Int. J. Biochem. Cell Biol.* **2010**, *42*, 39–51. [CrossRef] [PubMed]
2. Imai, A.; Matsuyama, T.; Hanzawa, Y.; Akiyama, T.; Tamaoki, M.; Saji, H.; Shirano, Y.; Kato, T.; Hayashi, H.; Shibata, D.; et al. Spermidine synthase genes are essential for survival of *Arabidopsis*. *Plant Physiol.* **2004**, *135*, 1565–1573. [CrossRef] [PubMed]

3. Takano, A.; Kakehi, J.I.; Takahashi, T. Thermospermine is not a minor polyamine in the plant kingdom. *Plant Cell Physiol.* **2012**, *53*, 606–616. [CrossRef] [PubMed]

4. Chen, D.; Shao, Q.; Yin, L.; Younis, A.; Zheng, B. Polyamine function in plants: Metabolism, regulation on development, and roles in abiotic stress responses. *Front. Plant Sci.* **2018**, *9*, 1945. [CrossRef]

5. Zhang, Y.; Wu, R.; Qin, G.; Chen, Z.; Gu, H.; Qu, L.J. Over-expression of WOX1 leads to defects in meristem development and polyamine homeostasis in arabidopsisf. *J. Integr. Plant Biol.* **2011**, *53*, 493–506. [CrossRef]

6. Alcázar, R.; Altabella, T.; Marco, F.; Bortolotti, C.; Reymond, M.; Koncz, C.; Carrasco, P.; Tiburcio, A.F. Polyamines: Molecules with regulatory functions in plant abiotic stress tolerance. *Planta* **2010**, *231*, 1237–1249. [CrossRef]

7. Shi, H.; Chan, Z. Improvement of plant abiotic stress tolerance through modulation of the polyamine pathway. *J. Integr. Plant Biol.* **2014**, *56*, 114–121. [CrossRef]

8. Tiburcio, A.F.; Altabella, T.; Bitrián, M.; Alcázar, R. The roles of polyamines during the lifespan of plants: From development to stress. *Planta* **2014**, *240*, 1–18. [CrossRef]

9. Takahashi, T. Plant Polyamines. *Plants* **2020**, *9*, 511. [CrossRef]

10. Hanfrey, C.; Sommer, S.; Mayer, M.J.; Burtin, D.; Michael, A.J. Arabidopsis polyamine biosynthesis: Absence of ornithine decarboxylase and the mechanism of arginine decarboxylase activity. *Plant J.* **2001**, *27*, 551–560. [CrossRef]

11. Lou, Y.-R.; Ahmed, S.; Yan, J.; Adio, A.M.; Powell, H.M.; Morris, P.F.; Jander, G. Arabidopsis ADC1 functions as an Nδ-acetylornithine decarboxylase. *J. Integr. Plant Biol.* **2020**, *62*, 601–613. [CrossRef]

12. Chang, K.S.; Lee, S.H.; Hwang, S.B.; Park, K.Y. Characterization and translational regulation of the arginine decarboxylase gene in carnation (*Dianthus caryophyllus* L.). *Plant J.* **2000**, *24*, 45–56. [CrossRef]

13. Baron, K.; Stasolla, C. The role of polyamines during in vivo and in vitro development. *Vitr. Cell. Dev. Biol. Plant* **2008**, *44*, 384–395. [CrossRef]

14. Guerrero-González, M.L.; Rodríguez-Kessler, M.; Jiménez-Bremont, J.F. uORF, a regulatory mechanism of the Arabidopsis polyamine oxidase 2. *Mol. Biol. Rep.* **2014**, *41*, 2427–2443. [CrossRef]

15. Hummel, I.; Couée, I.; El Amrani, A.; Martin-Tanguy, J.; Hennion, F. Involvement of polyamines in root development at low temperature in the subantarctic cruciferous species *Pringlea antiscorbutica*. *J. Exp. Bot.* **2002**, *53*, 1463–1473. [CrossRef]

16. Urano, K.; Yoshiba, Y.; Nanjo, T.; Ito, T.; Yamaguchi-Shinozaki, K.; Shinozaki, K. *Arabidopsis* stress-inducible gene for arginine decarboxylase AtADC2 is required for accumulation of putrescine in salt tolerance. *Biochem. Biophys. Res. Commun.* **2004**, *313*, 369–375. [CrossRef]

17. Alcázar, R.; García-Martínez, J.L.; Cuevas, J.C.; Tiburcio, A.F.; Altabella, T. Overexpression of ADC2 in Arabidopsis induces dwarfism and late-flowering through GA deficiency. *Plant J.* **2005**, *43*, 425–436. [CrossRef]

18. Sánchez-Rangel, D.; Chávez-Martínez, A.I.; Rodríguez-Hernández, A.A.; Maruri-López, I.; Urano, K.; Shinozaki, K.; Jiménez-Bremont, J.F. Simultaneous Silencing of two arginine decarboxylase genes alters development in *Arabidopsis*. *Front. Plant Sci.* **2016**, *7*, 300. [CrossRef]

19. Bhatnagar, P.; Minocha, R.; Minocha, S.C. Genetic manipulation of the metabolism of polyamines in poplar cells. The regulation of putrescine catabolism. *Plant Physiol.* **2002**, *128*, 1455–1469. [CrossRef]

20. Urano, K.; Hobo, T.; Shinozaki, K. Arabidopsis ADC genes involved in polyamine biosynthesis areb essential for seed development. *FEBS Lett.* **2005**, *579*, 1557–1564. [CrossRef]

21. Cui, X.; Ge, C.; Wang, R.; Wang, H.; Chen, W.; Fu, Z.; Jiang, X.; Li, J.; Wang, Y. The BUD2 mutation affects plant architecture through altering cytokinin and auxin responses in Arabidopsis. *Cell Res.* **2010**, *20*, 576–586. [CrossRef] [PubMed]

22. Mirza, J.I.; Bagni, N. Effects of exogenous polyamines and difluoromethylornithine on seed germination and root growth of Arabidopsis thaliana. *Plant Growth Regul.* **1991**, *10*, 163–168. [CrossRef]

23. Ghuge, S.A.; Carucci, A.; Rodrigues-Pousada, R.A.; Tisi, A.; Franchi, S.; Tavladoraki, P.; Angelini, R.; Cona, A. The apoplastic copper AMINE OXIDASE1 mediates jasmonic acid-induced protoxylem differentiation in *Arabidopsis* roots. *Plant Physiol.* **2015**, *168*, 690–707. [CrossRef] [PubMed]

24. Fraudentali, I.; Rodrigues-Pousada, R.A.; Tavladoraki, P.; Angelini, R.; Cona, A. Leaf-wounding long-distance signaling targets AtCuAOβ leading to root phenotypic plasticity. *Plants* **2020**, *9*, 249. [CrossRef]

25. Tarenghi, E.; Carré, M.; Martin-Tanguy, J. Effects of inhibitors of polyamine biosynthesis and of polyamines on strawberry microcutting growth and development. *Plant Cell. Tissue Organ Cult.* **1995**, *42*, 47–55. [CrossRef]

26. Yoshimoto, K.; Takamura, H.; Kadota, I.; Motose, H.; Takahashi, T. Chemical control of xylem differentiation by thermospermine, xylemin, and auxin. *Sci. Rep.* **2016**, *6*, 21487. [CrossRef]

27. Alabdallah, O.; Ahou, A.; Mancuso, N.; Pompili, V.; Macone, A.; Pashkoulov, D.; Stano, P.; Cona, A.; Angelini, R.; Tavladoraki, P. The *Arabidopsis* polyamine oxidase/dehydrogenase 5 interferes with cytokinin and auxin signaling pathways to control xylem differentiation. *J. Exp. Bot.* **2017**, *68*, 997–1012. [CrossRef]

28. Cona, A.; Rea, G.; Angelini, R.; Federico, R.; Tavladoraki, P. Functions of amine oxidases in plant development and defence. *Trends Plant Sci.* **2006**, *11*, 80–88. [CrossRef]

29. Angelini, R.; Cona, A.; Federico, R.; Fincato, P.; Tavladoraki, P.; Tisi, A. Plant amine oxidases "on the move": An update. *Plant Physiol. Biochem. PPB* **2010**, *48*, 560–564. [CrossRef]

30. Corpas, F.J.; del Río, L.A.; Palma, J.M. Plant peroxisomes at the crossroad of NO and H2O2 metabolism. *J. Integr. Plant Biol.* **2019**, *61*, 803–816. [CrossRef]

31. Takahashi, T.; Kakehi, J.I. Polyamines: Ubiquitous polycations with unique roles in growth and stress responses. *Ann. Bot.* **2010**, *105*, 1–6. [CrossRef]
32. Fu, Y.; Guo, C.; Wu, H.; Chen, C. Arginine decarboxylase ADC2 enhances salt tolerance through increasing ROS scavenging enzyme activity in *Arabidopsis thaliana*. *Plant Growth Regul.* **2017**, *83*, 253–263. [CrossRef]
33. López-Gómez, M.; Hidalgo-Castellanos, J.; Muñoz-Sánchez, J.R.; Marín-Peña, A.J.; Lluch, C.; Herrera-Cervera, J.A. Polyamines contribute to salinity tolerance in the symbiosis Medicago truncatula-Sinorhizobium meliloti by preventing oxidative damage. *Plant Physiol. Biochem.* **2017**, *116*, 9–17. [CrossRef]
34. Tsukagoshi, H.; Busch, W.; Benfey, P.N. Transcriptional Regulation of ROS Controls Transition from Proliferation to Differentiation in the root. *Cell* **2010**, *143*, 606–616. [CrossRef]
35. Tisi, A.; Federico, R.; Moreno, S.; Lucretti, S.; Moschou, P.N.; Roubelakis-Angelakis, K.A.; Angelini, R.; Cona, A. Perturbation of polyamine catabolism can strongly affect root development and xylem differentiation. *Plant Physiol.* **2011**, *157*, 200–215. [CrossRef]
36. Bitrián, M.; Zarza, X.; Altabella, T.; Tiburcio, A.F.; Alcázar, R. Polyamines under abiotic stress: Metabolic crossroads and hormonal crosstalks in plants. *Metabolites* **2012**, *2*, 516–528. [CrossRef]
37. Anwar, R.; Mattoo, A.K.; Handa, A.K. Polyamine interactions with plant hormones: Crosstalk at several levels. In *Polyamines: A Universal Molecular Nexus for Growth, Survival, and Specialized Metabolism*; Springer: Tokyo, Japan, 2015; pp. 267–302.
38. Marco, F.; Alcázar, R.; Tiburcio, A.F.; Carrasco, P. Interactions between polyamines and abiotic stress pathway responses unraveled by transcriptome analysis of polyamine overproducers. *OMICS* **2011**, *15*, 775–781. [CrossRef]
39. Mendes, A.F.S.; Cidade, L.C.; Otoni, W.C.; Soares-Filho, W.S.; Costa, M.G.C. Role of auxins, polyamines and ethylene in root formation and growth in sweet orange. *Biol. Plant.* **2011**, *55*, 375. [CrossRef]
40. Saini, S.; Sharma, I.; Kaur, N.; Pati, P.K. Auxin: A master regulator in plant root development. *Plant Cell Rep.* **2013**, *32*, 741–757. [CrossRef]
41. Roberts, D.R.; Walker, M.A.; Thompson, J.E.; Dumbroff, E.B. The Effects of inhibitors of polyamine and ethylene biosynthesis on senescence, ethylene production and polyamine levels in cut carnation flowers. *Plant Cell Physiol.* **1984**, *25*, 315–322. [CrossRef]
42. Tiburcio, A.F.; Kaur-Sawhney, R.; Galston, A.W. Effect of polyamine biosynthetic inhibitors on alkaloids and organogenesis in tobacco callus cultures. *Plant Cell. Tissue Organ Cult.* **1987**, *9*, 111–120. [CrossRef] [PubMed]
43. Liu, J.-H.; Nada, K.; Honda, C.; Kitashiba, H.; Wen, X.-P.; Pang, X.-M.; Moriguchi, T. Polyamine biosynthesis of apple callus under salt stress: Importance of the arginine decarboxylase pathway in stress response. *J. Exp. Bot.* **2006**, *57*, 2589–2599. [CrossRef] [PubMed]
44. Wang, J.; Sun, P.P.; Chen, C.L.; Wang, Y.; Fu, X.Z.; Liu, J.H. An arginine decarboxylase gene PtADC from Poncirus trifoliata confers abiotic stress tolerance and promotes primary root growth in *Arabidopsis*. *J. Exp. Bot.* **2011**, *62*, 2899–2914. [CrossRef] [PubMed]
45. Wu, H.; Fu, B.; Sun, P.; Xiao, C.; Liu, J.H. A NAC transcription factor represses putrescine biosynthesis and affects drought tolerance. *Plant Physiol.* **2016**, *172*, 1532–1547. [CrossRef]
46. Gao, C.; Sheteiwy, M.S.; Han, J.; Dong, Z.; Pan, R.; Guan, Y.; Hamoud, Y.A.; Hu, J. Polyamine biosynthetic pathways and their relation with the cold tolerance of maize (*Zea mays* L.) seedlings. *Plant Signal. Behav.* **2020**, *15*, 7722. [CrossRef]
47. Dello Ioio, R.; Nakamura, K.; Moubayidin, L.; Perilli, S.; Taniguchi, M.; Morita, M.T.; Aoyama, T.; Costantino, P.; Sabatini, S. A genetic framework for the control of cell division and differentiation in the root meristem. *Science* **2008**, *322*, 1380–1384. [CrossRef]
48. Daudi, A.; Cheng, Z.; O'Brien, J.A.; Mammarella, N.; Khan, S.; Ausubel, F.M.; Bolwell, G.P. The Apoplastic Oxidative Burst Peroxidase in *Arabidopsis* is a Major Component of Pattern-Triggered Immunity. *Plant Cell* **2012**, *24*, 275–287. [CrossRef]
49. Kadota, Y.; Shirasu, K.; Zipfel, C. Regulation of the NADPH oxidase RBOHD during plant immunity. *Plant Cell Physiol.* **2015**, *56*, 1472–1480. [CrossRef]
50. Gémes, K.; Kim, Y.J.; Park, K.Y.; Moschou, P.N.; Andronis, E.; Valassaki, C.; Roussis, A.; Roubelakis-Angelakis, K.A. An NADPH-oxidase/polyamine oxidase feedback loop controls oxidative burst under salinity. *Plant Physiol.* **2016**, *172*, 1418–1431. [CrossRef]
51. Liu, C.; Atanasov, K.E.; Tiburcio, A.F.; Alcázar, R. The polyamine putrescine contributes to H2O2 and RbohD/F-dependent positive feedback loop in *Arabidopsis* PAMP-triggered immunity. *Front. Plant Sci.* **2019**, *10*, 894. [CrossRef]
52. Brandstatter, I.; Kieber, J.J. Two genes with similarity to bacterial response regulators are rapidly and specifically induced by cytokinin in *Arabidopsis*. *Plant Cell* **1998**, *10*, 1009–1019. [CrossRef]
53. Werner, T.; Motyka, V.; Laucou, V.; Smets, R.; Van Onckelen, H.; Schmülling, T. Cytokinin-deficient transgenic *Arabidopsis* plants show multiple developmental alterations indicating opposite functions of cytokinins in the regulation of shoot and root meristem activity. *Plant Cell* **2003**, *15*, 2532–2550. [CrossRef]
54. Werner, T.; Köllmer, I.; Bartrina, I.; Holst, K.; Schmülling, T. New insights into the biology of cytokinin degradation. *Plant Biol.* **2006**, *8*, 371–381. [CrossRef]
55. Werner, T.; Motyka, V.; Strnad, M.; Schmülling, T. Regulation of plant growth by cytokinin. *Proc. Natl. Acad. Sci. USA* **2001**, *98*, 10487–10492. [CrossRef]
56. Werner, T.; Hanus, J.; Holub, J.; Schmülling, T.; Van Onckelen, H.; Strnad, M. New cytokinin metabolites in IPT transgenic *Arabidopsis thaliana* plants. *Physiol. Plant* **2003**, *118*, 127–137. [CrossRef]
57. Laplaze, L.; Benkova, E.; Casimiro, I.; Maes, L.; Vanneste, S.; Swarup, R.; Weijers, D.; Calvo, V.; Parizot, B.; Herrera-Rodriguez, M.B.; et al. Cytokinins act directly on lateral root founder cells to inhibit root initiation. *Plant Cell* **2007**, *19*, 3889–3900. [CrossRef]

58. Gao, S.; Fang, J.; Xu, F.; Wang, W.; Sun, X.; Chu, J.; Cai, B.; Feng, Y.; Chu, C. Cytokinin oxidase/dehydrogenase4 Integrates cytokinin and auxin signaling to control rice crown root formation. *Plant Physiol.* **2014**, *165*, 1035–1046. [CrossRef]

59. Reid, D.E.; Heckmann, A.B.; Novák, O.; Kelly, S.; Stougaard, J. Cytokinin oxidase/dehydrogenase3 maintains cytokinin homeostasis during root and nodule development in *Lotus japonicus*. *Plant Physiol.* **2016**, *170*, 1060–1074. [CrossRef]

60. Dello Ioio, R.; Linhares, F.S.; Scacchi, E.; Casamitjana-Martinez, E.; Heidstra, R.; Costantino, P.; Sabatini, S. Cytokinins determine *Arabidopsis* root-meristem size by controlling cell differentiation. *Curr. Biol.* **2007**, *17*, 678–682. [CrossRef]

61. Köllmer, I.; Novák, O.; Strnad, M.; Schmülling, T.; Werner, T. Overexpression of the cytosolic cytokinin oxidase/dehydrogenase (CKX7) from *Arabidopsis* causes specific changes in root growth and xylem differentiation. *Plant J.* **2014**, *78*, 359–371. [CrossRef]

62. Gälweiler, L.; Guan, C.; Müller, A.; Wisman, E.; Mendgen, K.; Yephremov, A.; Palme, K. Regulation of polar auxin transport by AtPIN1 in Arabidopsis vascular tissue. *Science* **1998**, *282*, 2226–2230. [CrossRef] [PubMed]

63. Blilou, I.; Xu, J.; Wildwater, M.; Willemsen, V.; Paponov, I.; Friml, J.; Heidstra, R.; Aida, M.; Palme, K.; Scheres, B. The PIN auxin efflux facilitator network controls growth and patterning in *Arabidopsis* roots. *Nature* **2005**, *433*, 39–44. [CrossRef] [PubMed]

64. Hummel, I.; Bourdais, G.; Gouesbet, G.; Couée, I.; Malmberg, R.L.; El Amrani, A. Differential gene expression of arginine decarboxylase ADC1 and ADC2 in *Arabidopsis thaliana*: Characterization of transcriptional regulation during seed germination and seedling development. *N. Phytol.* **2004**, *163*, 519–531. [CrossRef]

65. Chang, S.C.; Kaufman, P.B.; Kang, B.G. Changes in Endogenous levels of free polyamines during petiole elongation in the semiaquatic plant *Ranunculus sceleratus*. *Int. J. Plant Sci.* **1999**, *160*, 691–697. [CrossRef]

66. Locke, J.M.; Bryce, J.H.; Morris, P.C. Contrasting effects of ethylene perception and biosynthesis inhibitors on germination and seedling growth of barley (*Hordeum vulgare* L.). *J. Exp. Bot.* **2000**, *51*, 1843–1849. [CrossRef]

67. Ha, B.H.; Cho, K.J.; Choi, Y.J.; Park, K.Y.; Kim, K.H. Characterization of arginine decarboxylase from *Dianthus caryophyllus*. *Plant Physiol. Biochem.* **2004**, *42*, 307–311. [CrossRef]

68. Ziosi, V.; Bregoli, A.M.; Bonghi, C.; Fossati, T.; Biondi, S.; Costa, G.; Torrigiani, P. Transcription of ethylene perception and biosynthesis genes is altered by putrescine, spermidine and amino ethoxyvinyl glycine (AVG) during ripening in peach fruit (*Prunus persica*). *N. Phytol.* **2006**, *172*, 229–238. [CrossRef]

69. Stes, E.; Biondi, S.; Holsters, M.; Vereecke, D. Bacterial and plant signal integration via D3-type cyclins enhances symptom development in the *Arabidopsis-Rhodococcus fascians* interaction. *Plant Physiol.* **2011**, *156*, 712–725. [CrossRef]

70. Tsukagoshi, H. Defective root growth triggered by oxidative stress is controlled through the expression of cell cycle-related genes. *Plant Sci.* **2012**, *197*, 30–39. [CrossRef]

71. Tsukagoshi, H. Control of root growth and development by reactive oxygen species. *Curr. Opin. Plant Biol.* **2016**, *29*, 57–63. [CrossRef]

72. Yamada, M.; Han, X.; Benfey, P.N. RGF1 controls root meristem size through ROS signalling. *Nature* **2020**, *577*, 85–88. [CrossRef]

73. Gonzalez, M.E.; Marco, F.; Minguet, E.G.; Carrasco-Sorli, P.; Blázquez, M.A.; Carbonell, J.; Ruiz, O.A.; Pieckenstain, F.L. Perturbation of spermine synthase gene expression and transcript profiling provide new insights on the role of the tetraamine spermine in Arabidopsis defense against Pseudomonas viridiflava. *Plant Physiol.* **2011**, *156*, 2266–2277. [CrossRef] [PubMed]

74. Kakehi, J.; Kuwashiro, Y.; Niitsu, M.; Takahashi, T. Thermospermine is required for stem elongation in *Arabidopsis thaliana*. *Plant Cell Physiol.* **2008**, *49*, 1342–1349. [CrossRef]

75. Couée, I.; Hummel, I.; Sulmon, C.; Gouesbet, G.; El Amrani, A. Involvement of polyamines in root development. *Plant Cell. Tissue Organ Cult.* **2004**, *76*, 1–10. [CrossRef]

76. Perilli, S.; Sabatini, S. Analysis of root meristem size development. In *Plant Developmental Biology: Methods and Protocols*; Hennig, L., Köhler, C., Eds.; Humana Press: Totowa, NJ, USA, 2010; pp. 177–187.

77. Ivanov, V.B.; Dubrovsky, J.G. Longitudinal zonation pattern in plant roots: Conflicts and solutions. *Trends Plant Sci.* **2013**, *18*, 237–243. [CrossRef] [PubMed]

78. Napsucialy-Mendivil, S.; Alvarez-Venegas, R.; Shishkova, S.; Dubrovsky, J.G. *Arabidopsis* homolog of trithorax1 (ATX1) is required for cell production, patterning, and morphogenesis in root development. *J. Exp. Bot.* **2014**, *65*, 6373–6384. [CrossRef]

79. Schmittgen, T.D.; Livak, K.J. Analyzing real-time PCR data by the comparative C(T) method. *Nat. Protoc.* **2008**, *3*, 1101–1108. [CrossRef]

80. Gong, X.; Liu, J.-H. Detection of free polyamines in plants subjected to abiotic stresses by high-performance liquid chromatography (HPLC). In *Plant Stress Tolerance: Methods and Protocols*; Sunkar, R., Ed.; Springer: New York, NY, USA, 2017; pp. 305–311.

International Journal of
Molecular Sciences

MDPI

Review

Link between Lipid Second Messengers and Osmotic Stress in Plants

Beatriz A. Rodas-Junco [1,*], Graciela E. Racagni-Di-Palma [2], Michel Canul-Chan [3], Javier Usorach [4] and S. M. Teresa Hernández-Sotomayor [4,*]

[1] CONACYT—Facultad de Ingeniería Química, Campus de Ciencias Exactas e Ingenierías, Universidad Autónoma de Yucatán (UADY), Periférico Norte Kilómetro 33.5, Tablaje Catastral 13615 Chuburná de Hidalgo Inn, C.P. 97203 Mérida, Mexico

[2] Departamento de Biología Molecular, Universidad Nacional de Río Cuarto, C.P. 5800 Río Cuarto, Argentina; g_ragacni@hotmail.com

[3] Facultad de Ciencias Químicas, Universidad Veracruzana, Prolongación de Avenida Oriente 6 Num. 1009, Rafael Alvarado, C.P. 94340 Orizaba, Mexico; mcanul@uv.mx

[4] Unidad de Bioquímica y Biología Molecular de Plantas, Centro de Investigación Científica de Yucatán (CICY), Calle 43 No. 130, Col. Chuburná de Hidalgo, C.P. 97205 Mérida, Mexico; jiusorach@unvime.edu.ar

[*] Correspondence: beatriz.rodas@correo.uady.mx (B.A.R.-J.); ths@cicy.mx (S.M.T.H.-S.)

Abstract: Plants are subject to different types of stress, which consequently affect their growth and development. They have developed mechanisms for recognizing and processing an extracellular signal. Second messengers are transient molecules that modulate the physiological responses in plant cells under stress conditions. In this sense, it has been shown in various plant models that membrane lipids are substrates for the generation of second lipid messengers such as phosphoinositide, phosphatidic acid, sphingolipids, and lysophospholipids. In recent years, research on lipid second messengers has been moving toward using genetic and molecular approaches to reveal the molecular setting in which these molecules act in response to osmotic stress. In this sense, these studies have established that second messengers can transiently recruit target proteins to the membrane and, therefore, affect protein conformation, activity, and gene expression. This review summarizes recent advances in responses related to the link between lipid second messengers and osmotic stress in plant cells.

Keywords: lipid messengers; phosphatidic acid; phospholipase C; phospholipase D; sphingholipids; lysophospholipids

Citation: Rodas-Junco, B.A.; Racagni-Di-Palma, G.E.; Canul-Chan, M.; Usorach, J.; Hernández-Sotomayor, S.M.T. Link between Lipid Second Messengers and Osmotic Stress in Plants. *Int. J. Mol. Sci.* **2021**, *22*, 2658. https://doi.org/10.3390/ijms22052658

Academic Editor: Jen-Tsung Chen

Received: 19 January 2021
Accepted: 2 March 2021
Published: 6 March 2021

1. Introduction

Plants use complex signal transduction networks to orchestrate biochemical, genetic, and physiological responses under different stress conditions. Among the components involved in that response are molecules called second messengers. These molecules are "master regulators" since they generate a high degree of amplification via signal transduction and modulate key downstream molecular regulatory components involved in the response to stress. Lipids are major components of biological membranes that serve as platforms for important signaling functions [1,2]. Lipid-second messengers may be formed from membrane structural lipids by hydrolytic activity of phospholipases such as phospholipase D (PLD), phospholipase C (PLC), and phospholipase A2 (PLA2). In the context of stress in plants, salt, and drought represent osmotic factors that limit crop productivity [3]. In this context, salt or drought are different types of stress that result in a series of different changes at the cellular or plant level, generating specific changes at the biochemical, molecular, and physiological levels in plants. Understanding the molecular mechanism by which plants respond to osmotic stress signals is pivotal for the development of biotechnological tools for the generation of tolerant plants. This review will focus on assessing the current knowledge of lipids second messengers (phosphoinositides,

Int. J. Mol. Sci. **2021**, *22*, 2658. https://doi.org/10.3390/ijms22052658

https://www.mdpi.com/journal/ijms

phosphatidic acid, sphingolipids, and lysophospholipids), which have been shown to be key regulators of osmotic stress responses in plant cells.

2. Lipid-Derived Second Messengers in Plant Cells

Phospholipids are important components in all membranes in eukaryotes and play a role in signaling mechanisms in plant cells. Enzymes as phospholipases, lipid kinases or phosphatases modify membrane lipids to generate signaling molecules known as lipid-derived second messengers. Important lipid second messengers include phosphatidylinositols, diacylglycerols, phosphatidic acid, sphingolipids, and lysophospholipids [4,5] (Figure 1). Several research groups have reported that lipid second messengers activate or recruit proteins to membranes, which leads to the activation of downstream signaling pathways that result in cellular events and physiological responses. In this review, we will attempt to highlight some of the recent studies on the of the functional mechanism of lipid-derived second messengers, with an emphasis on their regulation, particularly in response to osmotic in plant cells.

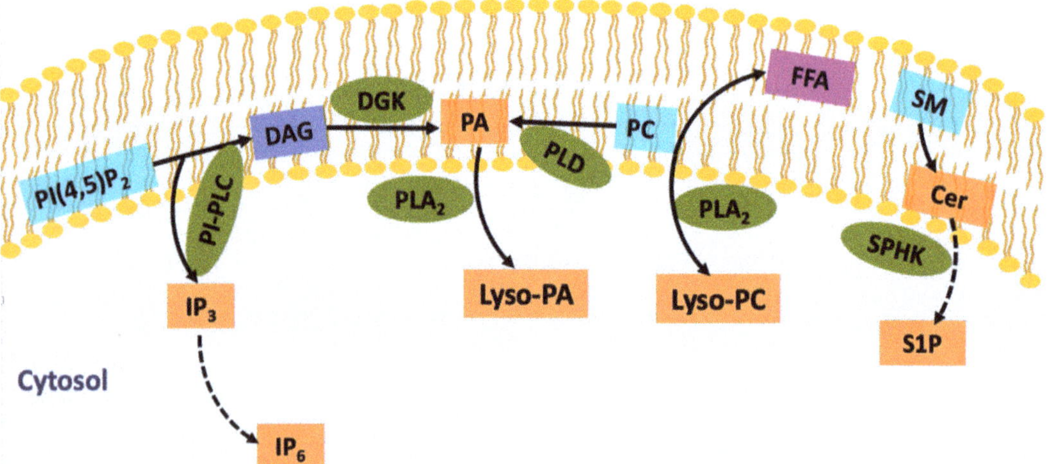

Figure 1. Generation of lipid second messengers in plants. Phospholipid precursors (blue box) involved in the production of intracellular second messengers (orange box). PI-PLC leads to the cleavage of PIP$_2$ into DAG and IP$_3$. DGK converts DAG to PA, which is a second messenger on its own right. PA, which can also be generated by PC hydrolysis by PLD. DAG can also be synthesized from PC via PLD. IP$_3$ diffuses into the cytosol, where it is converted to IP$_6$. Fatty acids of phospholipids are liberated by PLA2s and converted to eicosanoids. Lysophospholipids are also precursors of a different class of lipid mediators, including Lyso-PC or Lyso-PA. Sphingomyelin is a precursor of ceramide that can then be phosphorylated to generate ceramide 1-phosphate and to form sphingosine, which is phosphorylated to generate sphingosine 1-phosphate. PIP$_2$, phosphatidylinositol (4,5)-bisphosphate; PC, phosphatidylcholine; PI-PLC, phosphoinositide-phospholipase C; IP$_3$, inositol (1,4,5)-trisphosphate; IP$_6$, myo-inositol-1,2,3,4,5,6 hexaskisphosphate; DAG, diacylglycerol; PLD, phospholipase D; PA, phosphatidic acid; PLA2, phospholipase A2; Lyso-PA, lyso-phosphatidic acid; Lyso-PC, lyso-phosphatidylcholine; FFA, Free Fatty Acid; SM, sphingomyelin; Cer, ceramide; SPHK, sphingosine kinase; S1P, sphingosine-1-phosphate.

3. Phosphoinositide Signaling

Phosphoinositides (PI) are a class of inositol-containing phospholipids present in the plasma membrane. In plants, the inositol ring is sequentially phosphorylated at several different positions, generating five isomers: phosphatidylinositol (PI), PI-3 phosphate (PI3P), PI-4 phosphate (PI4P or PIP), PI-5 phosphate, PI-3-5- bisphosphate (PI-3,5-P$_2$), and PI-4,5-bisphosphate (PI-4,5-P$_2$ or PIP$_2$) [6,7].

Unlike the majority of membrane lipids, PIs show only a minor abundance, and their dynamic formation occurs a set of specific kinases and phosphatases, and is maintained via constant turnover [8]. Additionally, they modulate fundamental cellular processes, such as membrane trafficking, cytoskeleton organization, polar tip growth, and stress responses [9]. At the poles across kingdoms, phosphoinositide is involved in polar tip growth [10]. PIs can work as ligands for different proteins called PI "modulins" and regulate their subcellular distribution or activity via interactions. PI binding takes place through the inositol polyphosphate head groups and PI binding domains of phosphoinositide, such as pleckstrin homology (PH) domains, Phox homology (POX) domains, and Fab1-YOTB-Vac1-EEA1 (FYVE) domains [8]. Examples of PI modulin activities include the regulation of ion channels [9], ATPase activity, and hormonal and stress signaling [9]. In Arabidopsis and rice, the presence of proteins with FVYE domain has been reported in response to abiotic stress tolerance [11]. In phosphoinositide signaling, the generation of a second messenger occurs through the activation of phospholipases. PI-phospholipase C (PLC) catalyzes the hydrolysis of PIP_2 to generate the soluble second messenger's inositol 1,4,5-trisphosphate (IP_3) and diacylglycerol (DAG). In plants, DAG is converted into phosphatidic acid (PA), while IP_3 may be further phosphorylated to form inositol hexakisphosphate (IP_6). PA may also be generated by hydrolysis of structural phospholipids such as phosphatidylcholine (PC) and phosphatidylethanolamine (PE) by phospholipase D (PLD) enzymes [12]. Although the study of these lipid second messengers has provided evidence of their importance in plant defense response under stress, many questions still need to be answered. As a continuation, the roles of IP_3, IP_6, PA, and other lipid second messengers in plants are described below.

3.1. IP_3 as a Second Messenger in Plant Cells

PIs constitute a class of membrane phospholipids that are substrates for phosphoinositide-specific phospholipase C (PI-PLC). PI-PLC catalyzes the hydrolysis of PIP_2 to generate two important second messengers, IP_3 and DAG [12]. In plant systems, the role of IP_3 in releasing Ca^{2+} from cellular stores has been widely reported [13]. However, a critical component that is still unknown in plant cells is a putative IP_3 receptor (IP_3-R). The search for an IP_3-R has been underway for many years. Various authors have approached the search for an IP_3 receptor through in silico and in vivo analyses, and an interesting approach that has been taken is the search for homologous gene(s) that encode the IP_3 receptor in plants. Sequencing of the green algae Chlamydomonas sp. genome, which does possess such a receptor, has made it possible to generate valuable genetic information to explain that this gene has been discarded during the evolution of plants. Additionally, at the protein level this does not clarify whether plants express an IP_3-R, as it indicates only that there is no plant protein that has an IP_3 receptor RIH domain [ryanodine, (RYR) and IP_3 homology] in animals in structural homolog databases [14,15].

On the other hand, an interesting aspect of IP_3 as a second messenger that is well documented is the rapid intracellular changes that this molecule shows under biotic or abiotic stimulation. For example, Monteiro et al. [16] reported that IP_3 caused an influx of Ca^{2+} in the growing pollen tube of *Agapanthus umbellatus* under osmotic shock treatment. Additionally, biphasic changes in IP_3 were detected in response to gravity in *Arabidopsis* inflorescence stems [17] and *Avena sativa* [18] or cold exposure in Arabidopsis suspension cells [19]. The release of IP_3 has often been linked to the activation of PI-PLC [13,14,20]. For example, in Arabidopsis, an increase in IP_3 via PI-PLC activation in response to blue light induces the release of Ca^{2+} [21]. Legendre et al. [22] hypothesize that the activation of PI-PLC and increase in IP_3 could be a way by which polygalacturonic acid triggers an oxidative burst in soybean cell suspensions. Recently, Ren et al. [23] showed that the increase in IP_3 after heat shock in Arabidopsis plants is partially dependent on the activity of AtPLC3 (Arabidopsis thaliana Phosphoinositide-Specific Phospholipase C Isoform 3). Collectively, these examples indicate that the increase in IP_3 as a consequence of PI-PLC activity, could be dependent on an increase in the substrate PIP_2 levels, as observed in response to abiotic stress in plants [13].

3.2. Inositol 1,2,3,4,5,6-Hexakisphosphate as a Putative Signaling Mediator

Myo-inositol-1,2,3,4,5,6 hexaskisphosphate (IP_6 or phytic acid) is a component of plant cells that regulates many cellular functions. In plants, IP_3 might be phosphorylated into IP_6 by two inositol kinases, inositol polyphophate multikinase 6/3 (IPK2), and inositol polyphosphate (IPK1). IP_6 accumulates in large amounts in seeds, pollen, and other storage tissues, where it serves as a source for Pi, inositol, and minerals [3,24]. As a signaling molecule, IP_6 has received attention in recent years. Some authors point out that IP_6 is the central signaling molecule rather than IP_3 [25–28]; however, it is also clear that there is an important contribution of IP_3 as a precursor for IP_6 generation. In contrast, there are reports showing that IP_6 controls cellular reactions through the mobilization of intracellular Ca^{2+} deposits. For example, Lemtiri-Chlieh et al. [25] suggested a signaling role of IP_6 in abscisic acid (ABA)-regulated Ca^{2+} release in guard cells in which the vacuole may contribute to the release of Ca^{2+} in response to IP_6. In this way, it is necessary to determine whether these molecules send different signals in plants, and it would be interesting to undertake studies that allow evaluation of the impact of IP_3 and IP_6 on the same cellular response.

3.3. Phosphatidic Acid

Phosphatidic acid (PA) may be formed from structural membrane lipids such as (PC and PE by phospholipase D, mainly to produce PA species such as PA 18:3/18:2 and PA 18:2/18:2. Additionally, the combined action of PI-PLC and diacylglycerol kinase (DGK) generates the PA species 16:0/18:2 and 16:0/18:3 [29]. Therefore, lipidomic tools have allowed research to reveal which metabolic pathway is activated in response to stress. Differential ^{32}P radiolabeling and chromatography technique has been most commonly used to reveal the signaling mechanisms that are involved in hormonal signaling, cytoskeleton, and vesicle trafficking [30–34]. One limitation biochemistry methodologies have faced is that cellular levels of PA are highly dynamic in response to stimuli and to the various enzymatic reactions that modulate its production and degradation.

The role of PA, as a second messenger, has been established by identifying PA-binding domains (PABD) within PA effectors in different plant cell processes. This suggests the importance of this molecule as a central messenger in phospholipid-mediated signaling. Recently, an increasing number of PABDs fused with fluorescent proteins have been used as probes to obtain images of the spatiotemporal distribution of PA in plant cells [35,36]. For instance, the PABD-derived probe Spo20p (Spo20p-PABD) was fused with YFP to monitor PA in growing pollen tubes in tobacco [35]. This biosensor allowed us to detect that the different distribution of PA in the subapical zone is important in the regulation of endocytosis and in the actin dynamics for growth of the pollen tube. Using an optogenetic biosensor, Li et al. [36] development a probe with NADPH oxidase PA-binding domain (RBOHD-PABD) based on Förster resonance energy transfer (FRET) and found that biosensor can monitor the dynamic changes in PA in the plasma membrane in Arabidopsis cells in response to saline and hormonal stress. These findings have contributed to understanding the dynamics of PA in cells under specific environmental conditions, however there is still the challenge of delving into the subcellular distribution of PABD when expressed as PA sensors fused with XFP in response to stress.

Another aspect that has been addressed for the study of PA is through the enzyme PLD. Genetically modified plants have also been used to address the role of some PLD isoforms in the production of PA in response to abiotic stress [37,38]. The results showed that the cellular response derived from the activation of the PI-PLC pathway is functionally different from that resulting from PLD, although both enzymes can generate PA.

For a thorough understanding of the molecular mechanism by which PA regulates different developmental processes in plants, the reader is referred to many excellent reviews on this subject [39–42].

3.4. Other Lipid Second Messengers

The roles of other lipid classes in plant cells during abiotic stress, such as sphingolipids and lysophospholipids, have recently been discovered. The term sphingolipids covers a class of lipids composed of the following three blocks: the long chain base (LCB), the amide-linked fatty acyl chain to the LCB, and the polar head group. LCB is considered the simplest functional sphingolipid and may be linked to a very-long-chain fatty acid via an amide bond to form a ceramide [43]. LCB esterification with a phosphate group at C1 occurs to form phosphorylated LCBs (LCB-P). In plants, the different classes of sphingolipids and LCB-Ps allow these molecules to function both as bioactive lipid components to regulate diverse cellular processes, including signaling, and as structural components in the membrane in plant cells [43,44]. Although the first evidence of the role of LCBs as second messengers was reported for stomatal closure [45,46], its identity remains unclear. For this reason, several research groups have focused on genetic analysis with mutants to establish whether a particular LCB-P is a mediator of signaling. Michaelson et al. [47] analyzed a mutant with a T-DNA insertion in the *4-desaturase* gene in Arabidopsis and exposed it to ABA. Their results showed that phosphorylated 4E-sphingenine (SPH-P) was not involved in stomatal closure in Arabidopsis. In contrast, complex sphingolipids such as glucosyl ceramide (GlcCer) and glucosyl inositol phosphoryl ceramides (GIPC) have also been reported in plant tissues; however, they have not yet been assigned a role as signaling molecules in plants. Thus, an interesting question to be investigated is whether plants possess an enzymatic degradation pathway for structural and complex sphingolipids such as GIPCs to generate signaling molecules involved in the response to stress in plants. For more details on sphingolipid biosynthesis, see the recent reviews by Huby et al. [43] and Cassim et al. [48].

Lysophospholipids (LPLs) are phospholipids that harbor one fatty acyl chain and are generally produced from a large pool of glycerol- and sphingosine-based phospholipids in the membrane lipid bilayer by phospholipase A [1]. Examples of these are lysophosphatidic acid (LPA), lysophostatidylcholine (LPC), sphingosylphosphorylcholine (SPC), and sphingosine 1-phosphate (S1P). The signal functions of LPLs are much less well documented than those of phospholipids. For instance, LPA has been suggested to participate in osmotic signaling in algae [49]. LPC and S1P, have also been proposed as second messengers in plant cells [50,51]. In 2007, Drissner and coworkers reported that LPC is an important signal in arbuscular mycorrhizal symbiosis in *Solanum tuberosum* L.

These findings infer that LPLs exhibit a wide range of biological activities. It is therefore necessary to elucidate the underlying mechanisms by which the LPLs signal is transduced in plant cells. One aspect that has been addressed is the identification of receptors. Although in animal cells it has been established that the effect of LPLs is mediated by G protein-coupled receptors (GPCRs), this in plants is still controversial. Coursol et al. [52] reported that heterotrimeric G proteins have been identified as molecular elements in S1P signaling during ABA regulation in Arabidopsis guard cells. In contrast, Wielandt et al. [53] reported that plasma membrane $^+$H-ATPase (PM $^+$H-ATPase) as a lysophospholipid receptor evidenced the participation of LPLs as important plant signaling molecules in the regulation of electrochemical gradients in Arabidopsis.

4. Link between Lipid Second Messengers and Osmotic Stress

4.1. Osmotic Stress-Induced Lipid Second Messengers

Osmotic stress is one of the most important abiotic stresses for crop productivity. Plant cells experience osmotic stress when the solute concentrations in their apoplast change and respond with compensatory adaptations to reestablish osmotic equilibrium [4]. To survive osmotic stress, such as high salinity or dehydration, plant cells activate signaling pathways that lead to a wide range of responses in gene expression and metabolism. Although the importance of salinity and drought has been recognized for a long time, the identity of the molecular components involved in signaling tolerance in plants has been gradually established. Evidence has shown the importance of lipid-mediated reorganization of cell membranes, as well as its role in signaling to respond to changes in osmotic stress in plant cells [54–56]. However, more work is needed to fully describe the impact that lipid second messengers have on the molecular landscape during osmotic stress in plant.

4.2. IP$_3$ and IP$_6$ upon Osmotic Stress

A worldwide problem in the cultivation of plants is caused by high salinity in soils, which causes cells to lose water and experience reduced turgor pressure [57]. Osmotic stress imposed by NaCl or KCl generates a rapid increase in IP$_3$ and mobilization of Ca^{2+} in different models of plants, such as Arabidopsis [58,59], *Daucus carota* L. [56,60], and *Nicotiana tabacum* [15]. Previous work has reported that osmotic stress activates the PI-PLC pathway [61]. For example, Hirayama et al. [62] reported a PLC gene, AtPLC, in Arabidopsis that is induced by salt and drought stress. Another study [61] analyzed the expression patterns of TaPLC1 under drought and high salinity stress (200 mM NaCl or 20% (*w*/*v*) PEG) in wheat plants. Their data showed that the expression of TAPLC1 was low in the seedling stage and was strongly induced under osmotic stress conditions. Additionally, in our group, Usorach (2016) (unpublished data) observed a 20% increase in the in vitro activity of PI-PLC by ^{3}[H]-IP$_3$ formation in barley coleoptiles under conditions of saline stress (NaCl: 50200 nM), which was contrary to that observed by osmotic stress with mannitol and sorbitol (100–400 nM). Interestingly, in barley roots, PI-PLC activity increased by 50% under both saline and osmotic stress (unpublished data).

These results suggest that PI-PLC activity is different for each plant tissue that is subjected to osmotic stress, though it must also be take considered that enzymatic activities are affected in plants by the type of stress.

Additionally, the use of pharmacological approaches, such as PI-PLC inhibitors, has provided a molecular view of the link between the PI-PLC pathway and IP$_3$ under osmotic stress. This strategy has made it possible to observe how the calcium signal is affected by inhibiting the production of IP$_3$ and blocking metabolite biosynthesis induced by water stress. In this context, Parre et al. [63] reported that the inhibition of PI-PLC by U73122 decreased IP$_3$ levels and in the Ca^{2+} signaling. These results showed that Ca^{2+}/PI-PLC signaling is a committed step in the biosynthesis of proline (an osmolyte) in response to water stress. Recently, a connection between phosphoinositides and osmotic stress gene expression was also demonstrated. Takahashi et al. [59] reported that hyperosmotic stress induces a rapid and transient elevation in IP$_3$ levels in Arabidopsis T87 cells due to PI-PLC activation. However, when the cells were treated with neomycin and U73122, not only the levels of IP$_3$ but also the expression of hyperosmotic stress-inducible genes decreased under hyperosmolality.

The involvement of IP$_3$ as a lipid second messenger is still controversial because the increase in the levels of IP$_3$ contrasts with the relatively high levels of IP$_6$, which consequently generates a potent release of Ca^{2+} compared to IP$_3$. The two explanations for this could be: (1) IP$_6$ is also an important form of phosphate storage (e.g., seeds), so tissue specificity is an important factor for that response; and (2) the constant breakdown of inositol polyphosphates (IPPs) causes a flux from IP$_6$ and consequently Ca^{2+} release. However, there is still a long way to go to clarify the IP$_6$ signaling mechanism in plants [3,64].

Guard cells, as an experimental model, have made it possible to study the role of IP$_6$ in the ABA (drought stress hormone) response, which induces stomatal closure, conserving water and ensuring plant survival [65]. In an interesting work, Lemtiri-Chlieh et al. [25] demonstrated by laser scanning confocal microscopy in dye-loaded patch-clamped guard cell protoplasts that the detected increase in cytoplasmic Ca^{2+} was due to its release from endomembrane stores triggered by IP$_6$.

In contrast, signaling PIs are terminated through the action of PI phosphatases and inositol polyphosphate phosphatases (PTases). In the case of IP$_3$, 5TPases have the ability to hydrolyze it to prevent its accumulation and consequently alter the oscillations of Ca^{2+} in stress-related pathways. In this sense, strategies such as mutation or overexpression of inositol type I 5PTase genes have been used to establish the importance of IP$_3$ in saline signaling. For example, Golani and coworkers [66] reported that T-DNA insertion mutants of At5PTase9 increase salt sensitivity and that overexpression of this gene increased salt tolerance. Another example is Arabidopsis SAL1 [also known as FIERY1 (FRY1)], a gene encoding an inositol polyphosphate-1 phosphatase in Arabidopsis that enhances salt tolerance.

Multiple laboratories have developed mutants to evaluate the importance of FRY1 in of IP_3 metabolism [55,67,68]. It has been shown that *fry1*-mutant plants treated with ABA induces a sustained increase in IP_3 levels (not transient levels), which improves stress responses. For instance, Xiong et al. [55] showed that loss-of-function mutations in FRY1 enhanced the induction of stress-responsive genes such as RD29A, KIN1, and COR15A upon drought, salt and ABA treatments. However, overexpression or ectopic expression of Arabidopsis SAL1 could not enhance salt tolerance [69]. These findings are very interesting and have allowed us to raise the possibility that specific genes could be regulated through a different pathway.

4.3. Involvement of PLD-Derived PA in Osmotic Stress

PA plays an important and complex role in plant drought and salt stress tolerance in plants [70]. PA reportedly has the ability to act as a docking site for proteins that play an important role in salinity or drought conditions. Likewise, putative proteins and PA binding motifs have been identified, making it possible to know the identity of the signaling components involved in the response to osmotic stress [41,57]. In this context, McLoughlin et al. [41] identified eight putative PA-binding proteins recruited to membranes in response to salt stress in Arabidopsis roots through a proteomic approach. Among these were clathrin heavy chain (CHC) isoforms and glyceraldehyde 3-phosphate dehydrogenase (GAPDH), which were recruited towards the membrane for their interaction with PA in response to saline stress. Other examples are proteins of the sucrose nonfermenting-1-related protein kinase 2 (SnRK2) family [29,41,71]. Julkowska et al. [57] performed an in planta study to characterize the interaction of the PABD in SnRK2 upon saline stress. Their results showed that PABD/domain 1 in SnRK2.4 plays a role in the response to saline stress in Arabidopsis. An interesting approach was taken by Yu et al. [72] to investigate the relationship between PA and MAPK (mitogen-activated protein kinase) signaling in response to salt stress in Arabidopsis. The authors reported that salt stress induces a transient increase in the amount of PA and its binding to mitogen protein kinase 6 (MPK6) and stimulates its kinase activity, which phosphorylated salt overly sensitive 1 (SOS1 Na^+/H^+ antiporter) [72]. However, knockout of PLDα1-derived PA resulted in the generation of less PA and reduced MPK6 activity, leading to the accumulation of more Na^+ in leaves and increased sensitivity to NaCl stress.

In contrast, some reports have that the molecular species of PA (i.e., PAs with different fatty acyl chains) may exhibit different affinities towards their target proteins [73]. For example, the PA molecular species 16:0/18:2 has the highest affinity for MAPK6. Together, these results indicate that the regulation of PA towards its target proteins under stress conditions is extremely complex due to (1) the fatty acid composition of PA formed by the different contributions of the PI-PLC and PLD pathways that would be active, (2) different PA species interacting with the different target proteins, and (3) different isoforms of PLD and their preference for different substrates (i.e., PC, PE, or PG). This raises the possibility of specificity in signaling, which consequently allows interaction with different effectors. PA and cytoskeletal dynamics are intimately interconnected in plant cells to adapt to saline concentrations. During the response to salt stress, plant cells undergo microtubule depolymerization and reorganization, and both processes are believed to be essential for plant survival under salt stress [64,74]. However, what are the molecular mechanisms that mediate the changes in actin or tubulin dynamics by PA?

Currently, it is known that both the enzyme PLD and its PA product are important regulators of the behavior of actin filaments through the regulation of actin capping proteins (CPs) [75] or the arrangement of cortical microtubules [76,77]. In this context, it has been reported that PLD may be a linker that connects microtubules with the plasma membrane. Gardiner et al. [78] reported a microtubule-binding protein (MAP) with PLD activity in Arabidopsis. Later, Lee et al. [79] demonstrated for the first time that PLD is involved in the regulation of the actin cytoskeleton in soybean cell culture, since the exogenous addition of PA induced actin polymerization. Through proteomic analysis, it has been observed that tubulin is also a target protein for PA [29] and that the binding might not be direct but might occur through a MAP called AtMAP65-1, since the increase in PA by PLDα action recruits AtMAP65-1 to the membrane and induces stabilization of the microtubules, which confers survival against saline stress [80]. Our understanding of the role of PA formation in the osmotic stress response has greatly increased through traditional model systems. However, it is necessary to explore the potential mechanisms by which PA causes downstream effects in emergent models that are of agronomic importance. For example, barley (*Hordeum vulgare*) crops are severely affected by the high salinity of soils and hyperosmotic stress, which makes them excellent experimental models to study the role of PA and its relationship with tubulin in the cytoskeleton. Probing this hypothesis, we explored by confocal microscopy whether microtubule organization was affected by osmotic stress, when the coleoptiles and barley roots were treated with NaCl and mannitol (Figure 2A–F). The distribution of the microtubules was heterogeneous in the cytoplasm of the coleoptiles cells subjected to saline stress (Figure 2B) while in the roots, it became evident that the organization of the microtubules was interrupted by the increase in intracellular compartments (Figure 2E) compared to the control. In relation to mannitol, no differences were observed in the distribution of microtubules in coleoptiles and roots (Figure 2C,F). These results indicate that the activation of PLD under saline stress is important for the reorganization of microtubules in coleoptiles and barley roots, but whether PLD interacts directly or indirectly via PA needs to be determinate.

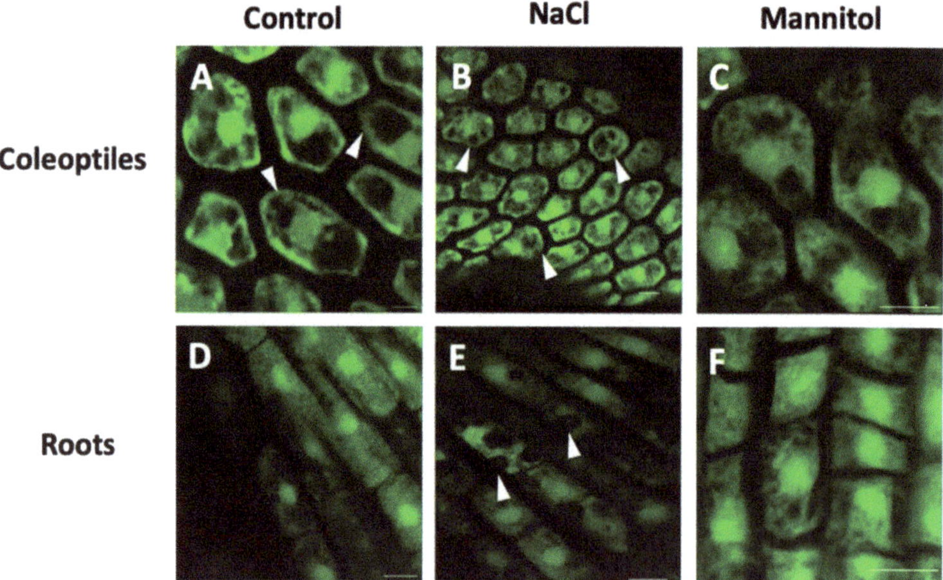

Figure 2. Organization of microtubules in coleoptiles and barley roots under osmotic stress. The images (**A,D**) show the distribution of the microtubules in the central plane of the cells of the coleoptile apex and the radical apex in roots without treatment. The distribution of the microtubules was disrupted when cells of coleoptiles and root were treated with NaCl (100 mM, images **B,E**) or mannitol (200 mM, images **C,F**). Fluorescence-labeled microtubules were visualized with a confocal laser microscope (Nikon Eclipse Ti). Scale bar = 20 μm.

4.4. Other Second Messengers Involved in Osmotic Stress

In comparison to the large body of work related to sphingolipids in the mammalian system, there is a paucity of published studies analyzing bioactive sphingolipids in plants [81]. However, an understanding of the roles of sphingolipids in the response to osmotic stress has been facilitated by mutants in plants. Experiments by Wu et al. [82] showed that At-ACER (an *Arabidposis thaliana* alkaline ceramidase) mutants were more sensitive to salinity stress and displayed increased ceramides and reduced LCBs, which suggests that ceramides are an important component in the response to salinity. In another work, Zhang et al. [80] explored the participation of the rice S1P lyase gene (OsPL1) in transgenic tobacco plants under saline stress through a functional analysis. Another bioactive component that has generated interest is sphingosine kinase (SPHK), which phosphorylates phytosphingosine to generate phyto-S1P. This is due to its interaction with PA as a component in transduction and the ABA effect in stomatal closure. Guo et al. [83] reported that SPHK1 and SPHK2 are molecular targets of PA and are part of the signaling networks in Arabidopsis. This could suggest that the interaction between PA and sphingolipids is a critical point to coordinate the response to stress in plants. Another compound analyzed is phosphoryl ceramide (GIPC). Jian et al. [84], using a mutant, identified the importance of plant-specific GIPC sphingolipids in the modulation of salt-associated ionic stress in the plasma membrane. Recently, Yang et al. reported that NaCl (300 nM) inhibited sphingolipid accumulation in a ceramide kinase-deficient mutant. These observations suggest that these compounds may also fulfill important signaling roles. Although there is no direct evidence linking sphingolipids and salt stress, sphingolipidomic analysis could yet reveal a link.

In relation to lysophospholipids, it has been suggested that LPC could be a candidate second messenger since it regulates different protein kinases, phosphatases and other signaling molecules. For example, MPK6 has been reported to be a target protein for lysophospholipids derived from pPLAIIIγ. Studies suggest that the activation of MAPK6 causes the phosphorylation of the antiporter Na^+/H^+ SOS1, which contributes to reducing Na^+ levels in plants [85]. In contrast, analysis of a pPLAIIIγa knockout mutant in Arabidopsis showed that the plants were sensitive, while overexpression improved the tolerance of the plants to saline stress [86]. Future studies should be carried out to establish whether pPLAIIIγ responds by modulating other independent pathways to SOS during osmotic stress.

5. Conclusions and Perspectives

Plants constantly face different types of abiotic stresses and their response involves the generation of second messengers. In this review, we summarize the second messengers derived from lipids and the molecular scenarios of their involvement in the response to osmotic stress in plants (Figure 3). Interestingly, multiple studies indicate that these second messengers drive downstream responses involving protein-protein interactions. Although research using omics studies has contributed to the understanding of the mechanism that these signaling molecules carry out, it is necessary to further exploit the field of genetic manipulation. In this sense, it would be interesting to use editing technologies and genetic approaches such as knockout lines, to learn more about the function of IP_6, lysophospholipids and sphingolipids in planta in other experimental models. Another aspect to be addressed is the identification of more molecular targets of lysophospholipids and sphingolipids that allow to explain the effects of osmotic stress in different plant cells. Therefore, in the future, efforts should be devoted to conducting new studies that combine genetic and molecular approaches that could contribute to the understanding of osmotic signaling in plant cells. In conclusion, lipid second messengers are important players in osmotic signaling in plant cells, and there are still potential studies that need to be conducted to clarify the molecular mechanism. This will allow to development of strategies to generate crops with least negative impacts on normal physiology due to osmotic stress.

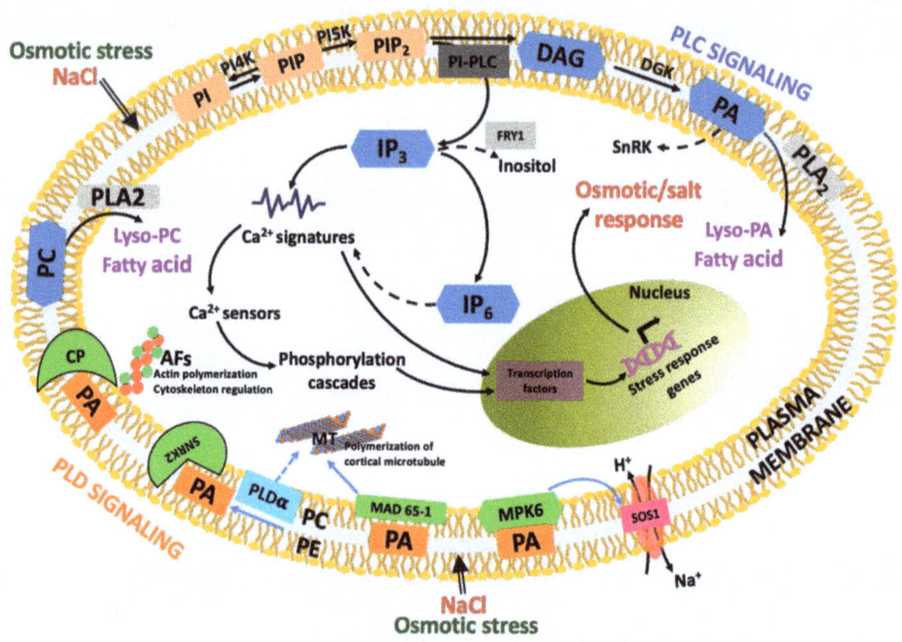

Figure 3. Proposed model for lipid-derived second messengers under osmotic and salt stress in plant cells. Osmotic and salt stress is perceived at the cell membrane, which activates PI-PLC, PLD and sphingolipid signaling to produce lipid second messengers that trigger the release of calcium from different sources, directly or indirectly. The changes in calcium concentration are sensed by calcium sensor proteins (e.g., CaM calmodulin, CML calmodulin-like protein sensors). In this response, PI-PLC and PLD signaling promotes a chain of reactions that includes IP$_3$, IP$_6$, and PA. PA has numerous targets, such as SNRK2 (snf1-related protein kinase2), MAD 65-1 (microtubule-associated protein MAD 65-1), and MPK6, that produce diverse cellular effects, such as actin and cortical microtubule polymerization. Finally, lysophosphatidic acid (lyso-PA) or lysophosphatidylcholine (LPC) can also stimulate many cellular processes.

Author Contributions: Conceptualization, B.A.R.-J.; writing-Original preparation, B.A.R.-J., S.M.T.H.-S. and J.U.; writing—review and editing, B.A.R.-J., G.E.R.-D.P. and M.C.-C.; design of the images, M.C.-C. All authors have read and agreed to the published version of the manuscript.

Funding: This research received no external funding.

Acknowledgments: This work was supported by Facultad de Ingeniería Química, Universidad Autónoma de Yucatán and Centro de Investigación Científica de Yucatán. The authors would like to thank M.C. Angela Ku González for her technical support in obtaining the images with the confocal microscope.

Conflicts of Interest: The authors declare no conflict of interest.

References

1. Hou, Q.; Ufer, G.; Bartels, D. Lipid signalling in plant responses to abiotic stress. *Plant Cell Environ.* **2016**, *39*, 1029–1048. [CrossRef]
2. Okazaki, Y.; Saito, K. Roles of lipids as signaling molecules and mitigators during stress response in plants. *Plant J.* **2014**, *79*, 584–596. [CrossRef]
3. Munnik, T.; Vermeer, J.E.M. Osmotic stress-induced phosphoinositide and inositol phosphate signalling in plants. *Plant Cell Environ.* **2010**, *33*, 655–669. [CrossRef]
4. Munnik, T.; Meijer, H.J. Osmotic stress activates distinct lipid and MAPK signalling pathways in plants. *FEBS Lett.* **2001**, *498*, 172–178. [CrossRef]
5. Xue, H.-W.; Chen, X.; Mei, Y. Function and regulation of phospholipid signalling in plants. *Biochem. J.* **2009**, *421*, 145–156. [CrossRef] [PubMed]

6. Balla, T. Phosphoinositides: Tiny lipids with giant impact on cell regulation. *Physiol. Rev.* **2013**, *93*, 1019–1137. [CrossRef]
7. Tuteja, N.; Sopory, S.K. Chemical signaling under abiotic stress environment in plants. *Plant Signal. Behav.* **2008**, *3*, 525–536. [CrossRef]
8. Heilmann, I. Phosphoinositide signaling in plant development. *Development* **2016**, *143*, 2044–2055. [CrossRef] [PubMed]
9. Heilmann, I. Using genetic tools to understand plant phosphoinositide signalling. *Trends Plant Sci.* **2009**, *14*, 171–179. [CrossRef] [PubMed]
10. Boss, W.F.; Im, Y.J. Phosphoinositide signaling. *Annu. Rev. Plant Biol.* **2012**, *63*, 409–429. [CrossRef]
11. Xiao, S.; Shao, M.; Wang, N.; Li, W.; Liu, F. Identification and evolution of FYVE domain-containing proteins and their expression patterns in response to abiotic stresses in rice. *Plant Mol. Biol. Rep.* **2016**, *34*, 1064–1082. [CrossRef]
12. Rodas-Junco, B.A.; Nic-Can, G.I.; Muñoz-Sánchez, A.; Hernández-Sotomayor, S.M. Phospholipid signaling is a component of the salicylic acid response in plant cell suspension cultures. *Int. J. Mol. Sci.* **2020**, *21*, 5285. [CrossRef]
13. Abd-El-Haliem, A.M.; Joosten, M.H. Plant phosphatidylinositol-specific phospholipase C at the center of plant innate immunity. *J. Integr. Plant Biol.* **2017**, *59*, 164–179. [CrossRef]
14. Krinke, O.; Ruelland, E.; Valentová, O.; Vergnolle, C.; Renou, J.-P.; Taconnat, L.; Flemr, M.; Burketová, L.; Zachowski, A. Phosphatidylinositol 4-kinase activation is an early response to salicylic acid in Arabidopsis suspension cells. *Plant Physiol.* **2007**, *144*, 1347–1359. [CrossRef]
15. Im, Y.J.; Phillippy, B.Q.; Perera, I.Y. InsP 3 in plant cells. In *Lipid Signaling in Plants*; Springer: Berlin/Heidelberg, Germany, 2010; pp. 145–160.
16. Monteiro, D.; Liu, Q.; Lisboa, S.; Scherer, G.E.F.; Quader, H.; Malhó, R. Phosphoinositides and phosphatidic acid regulate pollen tube growth and reorientation through modulation of $[Ca^{2+}]_c$ and membrane secretion. *J. Exp. Bot.* **2005**, *56*, 1665–1674. [CrossRef] [PubMed]
17. Perera, I.Y.; Hung, C.-Y.; Brady, S.; Muday, G.K.; Boss, W.F. A universal role for inositol 1,4,5-trisphosphate-mediated signaling in plant gravitropism. *Plant Physiol.* **2005**, *140*, 746–760. [CrossRef] [PubMed]
18. Yun, H.S.; Joo, S.-H.; Kaufman, P.B.; Kim, T.-W.; Kirakosyan, A.; Philosoph-Hadas, S.; Kim, S.-K.; Chang, S.C. Changes in starch and inositol 1,4,5-trisphosphate levels and auxin transport are interrelated in graviresponding oat (*Avena sativa*) shoots. *Plant Cell Environ.* **2006**, *29*, 2100–2111. [CrossRef] [PubMed]
19. Ruelland, E.; Cantrel, C.; Gawer, M.; Kader, J.-C.; Zachowski, A. Activation of phospholipases C and D is an early response to a cold exposure in arabidopsis suspension cells. *Plant Physiol.* **2002**, *130*, 999–1007. [CrossRef]
20. Ruelland, E.; Pokotylo, I.; Djafi, N.; Cantrel, C.; Repellin, A.; Zachowski, A. Salicylic acid modulates levels of phosphoinositide dependent-phospholipase C substrates and products to remodel the Arabidopsis suspension cell transcriptome. *Front Plant Sci.* **2014**, *5*, 608. [CrossRef] [PubMed]
21. Harada, A.; Sakai, T.; Okada, K. Phot1 and phot2 mediate blue light-induced transient increases in cytosolic Ca^{2+} differently in Arabidopsis leaves. *Proc. Natl. Acad. Sci. USA* **2003**, *100*, 8583–8588. [CrossRef] [PubMed]
22. Legendre, L.; Yueh, Y.G.; Crain, R.; Haddock, N.; Heinstein, P.F.; Low, P.S. Phospholipase C activation during elicitation of the oxidative burst in cultured plant cells. *J. Biol. Chem.* **1993**, *268*, 24559–24563. [CrossRef]
23. Ren, H.; Gao, K.; Liu, Y.; Sun, D.; Zheng, S. The role of AtPLC3 and AtPLC9 in thermotolerance in Arabidopsis. *Plant Signal. Behav.* **2017**, *12*, e1162368. [CrossRef]
24. Belgaroui, N.; Lacombe, B.; Rouached, H.; Hanin, M. Phytase overexpression in Arabidopsis improves plant growth under osmotic stress and in combination with phosphate deficiency. *Sci. Rep.* **2018**, *8*, 1–12. [CrossRef]
25. Lemtiri-Chlieh, F.; Macrobbie, E.A.C.; Webb, A.A.R.; Manison, N.F.; Brownlee, C.; Skepper, J.N.; Chen, J.; Prestwich, G.D.; Brearley, C.A. Inositol hexakisphosphate mobilizes an endomembrane store of calcium in guard cells. *Proc. Natl. Acad. Sci. USA* **2003**, *100*, 10091–10095. [CrossRef]
26. Arisz, S.A.; Testerink, C.; Munnik, T. Plant PA signaling via diacylglycerol kinase. *Biochim. Biophys. Acta Mol. Cell Biol. Lipids* **2009**, *1791*, 869–875. [CrossRef]
27. Tsui, M.M.; York, J.D. Roles of inositol phosphates and inositol pyrophosphates in development, cell signaling and nuclear processes. *Adv. Enzym. Regul.* **2010**, *50*, 324–337. [CrossRef] [PubMed]
28. Williams, S.P.; Gillaspy, G.E.; Perera, I.Y. Biosynthesis and possible functions of inositol pyrophosphates in plants. *Front. Plant Sci.* **2015**, *6*, 67. [CrossRef] [PubMed]
29. Testerink, C.; Munnik, T. Molecular, cellular, and physiological responses to phosphatidic acid formation in plants. *J. Exp. Bot.* **2011**, *62*, 2349–2361. [CrossRef]
30. Ramos-Díaz, A.; Hérnandez-Sotomayor, S.M.T. Does aluminum generate a bonafide phospholipd signal cascade? *Plant Signal. Behav.* **2007**, *2*, 263–264. [CrossRef]
31. Junco, B.R.; Muñoz-Sánchez, J.; Vazquez-Flota, F.; Hernández-Sotomayor, S. Salicylic-acid elicited phospholipase D responses in Capsicum chinense cell cultures. *Plant Physiol. Biochem.* **2015**, *90*, 32–37. [CrossRef] [PubMed]
32. Van Leeuwen, W.; Vermeer, J.E.; Gadella, T.W.; Munnik, T. Visualization of phosphatidylinositol 4,5-bisphosphate in the plasma membrane of suspension-cultured tobacco BY-2 cells and whole Arabidopsis seedlings. *Plant J.* **2007**, *52*, 1014–1026. [CrossRef]
33. Monreal, J.A.; López-Baena, F.J.; Vidal, J.; Echevarría, C.; García-Mauriño, S. Involvement of phospholipase D and phosphatidic acid in the light-dependent up-regulation of sorghum leaf phosphoenolpyruvate carboxylase-kinase. *J. Exp. Bot.* **2010**, *61*, 2819–2827. [CrossRef] [PubMed]

34. Van Der Luit, A.H.; Piatti, T.; van Doorn, A.; Musgrave, A.; Felix, G.; Boller, T.; Munnik, T. Elicitation of suspension-cultured tomato cells triggers the formation of phosphatidic acid and diacylglycerol pyrophosphate. *Plant Physiol.* **2000**, *123*, 1507–1516. [CrossRef] [PubMed]

35. Potocký, M.; Pleskot, R.; Pejchar, P.; Vitale, N.; Kost, B.; Žárský, V. Live-cell imaging of phosphatidic acid dynamics in pollen tubes visualized by Spo20p-derived biosensor. *New Phytol.* **2014**, *203*, 483–494. [CrossRef] [PubMed]

36. Li, W.; Song, T.; Wallrad, L.; Kudla, J.; Wang, X.; Zhang, W. Tissue-specific accumulation of pH-sensing phosphatidic acid determines plant stress tolerance. *Nat. Plants* **2019**, *5*, 1012–1021. [CrossRef]

37. Distéfano, A.M.; Valiñas, M.A.; Scuffi, D.; la Mattina, L.; Have, A.T.; García-Mata, C.; Laxalt, A.M. Phospholipase D δ knock-out mutants are tolerant to severe drought stress. *Plant Signal. Behav.* **2015**, *10*, e1089371. [CrossRef]

38. Lu, S.; Fadlalla, T.; Tang, S.; Li, L.; Ali, U.; Li, Q.; Guo, L. Genome-wide analysis of phospholipase d gene family and profiling of phospholipids under abiotic stresses in *Brassica napus*. *Plant Cell Physiol.* **2019**, *60*, 1556–1566. [CrossRef] [PubMed]

39. Yao, H.-Y.; Xue, H.-W. Phosphatidic acid plays key roles regulating plant development and stress responses. *J. Integr. Plant Biol.* **2018**, *60*, 851–863. [CrossRef] [PubMed]

40. Tanguy, E.; Kassas, N.; Vitale, N. Protein–phospholipid interaction motifs: A focus on phosphatidic acid. *Biomolecules* **2018**, *8*, 20. [CrossRef]

41. McLoughlin, F.; Arisz, S.A.; Dekker, H.L.; Kramer, G.; de Koster, C.G.; Haring, M.A.; Munnik, T.; Testerink, C. Identification of novel candidate phosphatidic acid-binding proteins involved in the salt-stress response of *Arabidopsis thaliana* roots. *Biochem. J.* **2013**, *450*, 573–581. [CrossRef]

42. Liu, Y.; Su, Y.; Wang, X. Phosphatidic acid-mediated signaling. *Results Probl. Cell Differ.* **2013**, *991*, 159–176. [CrossRef]

43. Huby, E.; Napier, J.A.; Baillieul, F.; Michaelson, L.V.; Dhondt-Cordelier, S. Sphingolipids: Towards an integrated view of metabolism during the plant stress response. *New Phytol.* **2020**, *225*, 659–670. [CrossRef] [PubMed]

44. Ali, U.; Li, H.; Wang, X.; Guo, L. Emerging roles of sphingolipid signaling in plant response to biotic and abiotic stresses. *Mol. Plant.* **2018**, *11*, 1328–1343. [CrossRef] [PubMed]

45. Coursol, S.; le Stunff, H.; Lynch, D.V.; Gilroy, S.; Assmann, S.M.; Spiegel, S. Arabidopsis sphingosine kinase and the effects of phytosphingosine-1-phosphate on stomatal aperture. *Plant Physiol.* **2005**, *137*, 724–737. [CrossRef] [PubMed]

46. Ng, C.K.Y.; Carr, K.; McAinsh, M.R.; Powell, B.; Hetherington, A.M. Drought-induced guard cell signal transduction involves sphingosine-1-phosphate. *Nature* **2001**, *410*, 596–599. [CrossRef]

47. Michaelson, L.V.; Zäuner, S.; Markham, J.E.; Haslam, R.P.; Desikan, R.; Mugford, S.G.; Albrecht, S.; Warnecke, D.; Sperling, P.; Heinz, E.; et al. Functional characterization of a higher plant sphingolipid Δ4-desaturase: Defining the role of sphingosine and sphingosine-1-phosphate in arabidopsis. *Plant Physiol.* **2008**, *149*, 487–498. [CrossRef]

48. Cassim, A.M.; Gouguet, P.; Gronnier, J.; Laurent, N.; Germain, V.; Grison, M.; Boutté, Y.; Gerbeau-Pissot, P.; Simon-Plas, F.; Mongrand, S. Plant lipids: Key players of plasma membrane organization and function. *Prog. Lipid Res.* **2019**, *73*, 1–27. [CrossRef]

49. Meijer, H.J.G.; Berrie, C.P.; Iurisci, C.; Divecha, N.; Musgrave, A.; Munnik, T. Identification of a new polyphosphoinositide in plants, phosphatidylinositol 5-monophosphate (PtdIns5P), and its accumulation upon osmotic stress. *Bioch. J.* **2001**, *360*, 491–498. [CrossRef]

50. Viehweger, K.; Dordschbal, B.; Roos, W. Elicitor-activated phospholipase A2 generates lysophosphati-dylcholines that mobilize the vacuolar H$^+$ Pool for pH signaling via the activation of Na$^+$-dependent proton fluxes. *Plant Cell* **2002**, *14*, 1509–1525. [CrossRef]

51. Ryu, S.B. Phospholipid-derived signaling mediated by phospholipase A in plants. *Trends Plant Sci.* **2004**, *9*, 229–235. [CrossRef] [PubMed]

52. Coursol, S.; Fan, L.-M.; le Stunff, H.; Spiegel, S.; Gilroy, S.; Assmann, S.M. Sphingolipid signalling in Arabidopsis guard cells involves heterotrimeric G proteins. *Nat. Cell Biol.* **2003**, *423*, 651–654. [CrossRef]

53. Wielandt, A.; Pedersen, J.; Falhof, J.; Kemmer, G.C.; Lund, A.; Ekberg, K.; Fuglsang, A.; Günther Pomorski, T.; Buch-Pedersen, M.; Palmgren, M. Specific activation of the plant P-type plasma membrane H$^+$-ATPase by lysophospholipids depends on the autoinhibitory N- and C-terminal domains. *J. Biol. Chem.* **2015**, *290*, 16281–16291. [CrossRef]

54. Golldack, D.; Li, C.; Mohan, H.; Probst, N. Tolerance to drought and salt stress in plants: Unraveling the signaling networks. *Front Plant Sci.* **2014**, *5*, 151. [CrossRef] [PubMed]

55. Xiong, L.; Lee, B.-H.; Ishitani, M.; Lee, H.; Zhang, C.; Zhu, J.-K. FIERY1 encoding an inositol polyphosphate 1-phosphatase is a negative regulator of abscisic acid and stress signaling in Arabidopsis. *Genes Dev.* **2001**, *15*, 1971–1984. [CrossRef] [PubMed]

56. Dewald, D.B.; Torabinejad, J.; Jones, C.A.; Shope, J.C.; Cangelosi, A.R.; Thompson, J.E.; Prestwich, G.D.; Hama, H. Rapid accumulation of phosphatidylinositol 4,5-bisphosphate and inositol 1,4,5-trisphosphate correlates with calcium mobilization in salt-stressed Arabidopsis. *Plant. Physiol.* **2001**, *126*, 759–769. [CrossRef]

57. Julkowska, M.M.; McLoughlin, F.; Rankenberg, J.M.; Kawa, D.; Klimecka, M.; Haring, M.A.; Munnik, T.; Kooijman, E.E.; Testerink, C. Identification and functional characterization of the A rabidopsis Snf 1-related protein kinase SnRK 2.4 phosphatidic acid-binding domain. *Plant Cell Environ.* **2015**, *38*, 614–624. [CrossRef]

58. König, S.; Mosblech, A.; Heilmann, I. Stress-inducible and constitutive phosphoinositide pools have distinctive fatty acid patterns in Arabidopsis thaliana. *FASEB J.* **2007**, *21*, 1958–1967. [CrossRef]

59. Takahashi, S.; Katagiri, T.; Hirayama, T.; Yamaguchi-Shinozaki, K.; Shinozaki, K. Hyperosmotic stress induces a rapid and transient increase in inositol 1,4,5-trisphosphate independent of abscisic acid in arabidopsis cell culture. *Plant Cell Physiol.* **2001**, *42*, 214–222. [CrossRef] [PubMed]

60. Drobak, B.K.; Watkins, P.A. Inositol(1,4,5)trisphosphate production in plant cells: An early response to salinity and hyperosmotic stress. *FEBS Lett* **2000**, *481*, 240–244. [CrossRef]
61. Zhang, K.; Jin, C.; Wu, L.; Hou, M.; Dou, S.; Pan, Y. Expression analysis of a stress-related phosphoinositide-specific phospholipase C gene in wheat (*Triticum aestivum* L.). *PLoS ONE* **2014**, *9*, e105061. [CrossRef] [PubMed]
62. Hirayama, T.; Ohto, C.; Mizoguchi, T.; Shinozaki, K. A gene encoding a phosphatidylinositol-specific phospholipase C is induced by dehydration and salt stress in Arabidopsis thaliana. *Proc. Natl. Acad. Sci. USA* **1995**, *92*, 3903–3907. [CrossRef] [PubMed]
63. Parre, E.; Ghars, M.A.; Leprince, A.-S.; Thiery, L.; Lefebvre, D.; Bordenave, M.; Richard, L.; Mazars, C.; Abdelly, C.; Savouré, A. Calcium signaling via phospholipase C is essential for proline accumulation upon ionic but not nonionic hyperosmotic stresses in Arabidopsis. *Plant Physiol.* **2007**, *144*, 503–512. [CrossRef]
64. Wang, X.; Zhang, W.; Li, W.; Mishra, G. Phospholipid signaling in plant response to drought and salt stress. In *Advances in Molecular Breeding Toward Drought and Salt Tolerant Crops*; Jenks, M.A., Hasegawa, P.M., Jain, S.M., Eds.; Springer: Dordrecht, The Netherlands, 2007; pp. 183–192.
65. Lemtiri-Chlieh, F.; Macrobbie, E.A.C.; Brearley, C.A. Inositol hexakisphosphate is a physiological signal regulating the K$^+$-inward rectifying conductance in guard cells. *Proc. Natl. Acad. Sci. USA* **2000**, *97*, 8687–8692. [CrossRef]
66. Golani, Y.; Kaye, Y.; Gilhar, O.; Ercetin, M.; Gillaspy, G.; Levine, A. Inositol polyphosphate phosphatidylinositol 5-phosphatase9 (At5PTase9) controls plant salt tolerance by regulating endocytosis. *Mol. Plant* **2013**, *6*, 1781–1794. [CrossRef] [PubMed]
67. Xiong, L.; Zhu, J.-K. Molecular and genetic aspects of plant responses to osmotic stress. *Plant Cell Environ.* **2002**, *25*, 131–139. [CrossRef] [PubMed]
68. Kim, C.; Kim, B. Characterization of drought tolerance in Arabidopsis mutant fry1–6. *J. Emer. Investig.* **2019**, *1*, 1–6.
69. Chen, G.; Snyder, C.L.; Greer, M.S.; Weselake, R.J. Biology and biochemistry of plant phospholipases. *Crit. Rev. Plant. Sci.* **2011**, *30*, 239–258. [CrossRef]
70. Bargmann, B.O.R.; Laxalt, A.M.; Ter-Riet, B.; van Schooten, B.; Merquiol, E.; Testerink, C.; Haring, M.A.; Bartels, D.; Munnik, T. Multiple PLDs required for high salinity and water deficit tolerance in plants. *Plant Cell Physiol.* **2008**, *50*, 78–89. [CrossRef]
71. McLoughlin, F.; Testerink, C. Phosphatidic acid, a versatile water-stress signal in roots. *Front. Plant Sci.* **2013**, *4*, 525. [CrossRef]
72. Yu, L.; Nie, J.; Cao, C.; Jin, Y.; Yan, M.; Wang, F.; Liu, J.; Xiao, Y.; Liang, Y.; Zhang, W. Phosphatidic acid mediates salt stress response by regulation of MPK6 in Arabidopsis thaliana. *New Phytol.* **2010**, *188*, 762–773. [CrossRef] [PubMed]
73. Ufer, G.; Gertzmann, A.; Gasulla, F.; Röhrig, H.; Bartels, D. Identification and characterization of the phosphatidic acid-binding A. thaliana phosphoprotein PLDrp1 that is regulated by PLDα1 in a stress-dependent manner. *Plant J.* **2017**, *92*, 276–290. [CrossRef] [PubMed]
74. Wang, S.; Kurepa, J.; Hashimoto, T.; Smalle, J.A. Salt stress–induced disassembly of Arabidopsis cortical microtubule arrays involves 26S proteasome–dependent degradation of SPIRAL1. *Plant Cell* **2011**, *23*, 3412–3427. [CrossRef]
75. Li, J.; Cao, L.; Staiger, C.J. Capping protein modulates actin remodeling in response to reactive oxygen species during plant innate immunity. *Plant Physiol.* **2017**, *173*, 1125–1136. [CrossRef] [PubMed]
76. Pleskot, R.; Pejchar, P.; Staiger, C.J.; Potocký, M. When fat is not bad: The regulation of actin dynamics by phospholipid signaling molecules. *Front. Plant Sci.* **2014**, *5*, 5. [CrossRef] [PubMed]
77. Zhang, Q.; Lin, F.; Mao, T.; Nie, J.; Yan, M.; Yuan, M.; Zhang, W. Phosphatidic acid regulates microtubule organization by interacting with MAP65-1 in response to salt stress in Arabidopsis. *Plant Cell* **2012**, *24*, 4555–4576. [CrossRef]
78. Gardiner, J.; Harper, J.; Weerakoon, N.; Collings, D.; Ritchie, S.; Gilroy, S.; Cyr, R.; Marc, J. A 90-kD phospholipase D from tobacco binds to microtubules and the plasma membrane. *Plant Cell* **2001**, *13*, 2143–2158. [CrossRef]
79. Lee, S.; Park, J.; Lee, Y. Phosphatidic acid induces actin polymerization by activating protein kinases in soybean cells. *Mol. Cells* **2003**, *15*, 313–319.
80. Zhang, H.; Zhai, J.; Mo, J.; Li, D.; Song, F. Overexpression of rice sphingosine-1-phoshpate lyase gene OsSPL1 in transgenic tobacco reduces salt and oxidative stress tolerance. *J. Integr. Plant Biol.* **2012**, *54*, 652–662. [CrossRef]
81. Lynch, D.V.; Dunn, T.M. An introduction to plant sphingolipids and a review of recent advances in understanding their metabolism and function. *New Phytol.* **2004**, *161*, 677–702. [CrossRef]
82. Wu, J.X.; Li, J.; Liu, Z.; Yin, J.; Chang, Z.Y.; Rong, C.; Wu, J.L.; Bi, F.C.; Yao, N. The Arabidopsis ceramidase AtACER functions in disease resistance and salt tolerance. *Plant J.* **2015**, *81*, 767–780. [CrossRef] [PubMed]
83. Guo, L.; Mishra, G.; Taylor, K.; Wang, X. Phosphatidic acid binds and stimulates Arabidopsis sphingosine kinases. *J. Biol. Chem.* **2011**, *286*, 13336–13345. [CrossRef] [PubMed]
84. Jiang, Z.; Zhou, X.; Tao, M.; Yuan, F.; Liu, L.; Wu, F.; Wu, X.; Xiang, Y.; Niu, Y.; Liu, F.; et al. Plant cell-surface GIPC sphingolipids sense salt to trigger Ca^{2+} influx. *Nat. Cell Biol.* **2019**, *572*, 341–346. [CrossRef] [PubMed]
85. Yu, D.; Boughton, B.A.; Hill, C.B.; Feussner, I.; Roessner, U.; Rupasinghe, T.W.T. Insights into oxidized lipid modification in barley roots as an adaptation mechanism to salinity stress. *Front. Plant Sci.* **2020**, *11*, 1. [CrossRef] [PubMed]
86. Li, J.; Li, M.; Yao, S.; Cai, G.; Wang, X. Patatin-related phospholipase pPLAIIIγ involved in osmotic and salt tolerance in Arabidopsis. *Plants* **2020**, *9*, 650. [CrossRef]

International Journal of
Molecular Sciences

Article

Characterization of the FLAVIN-BINDING, KELCH REPEAT, F-BOX 1 Homolog *SlFKF1* in Tomato as a Model for Plants with Fleshy Fruit

Tomoki Shibuya [1], Manabu Nishiyama [2], Kazuhisa Kato [2,*] and Yoshinori Kanayama [2,*]

[1] Faculty of Life and Environmental Science, Shimane University, Matsue 690-8504, Japan; tomoki.s.t.f@gmail.com

[2] Graduate School of Agricultural Science, Tohoku University, Aoba-ku, Sendai 980-8572, Japan; manabu.nishiyama.c3@tohoku.ac.jp

* Correspondence: kazuhisa.kato.d8@tohoku.ac.jp (K.K.); yoshinori.kanayama.a7@tohoku.ac.jp (Y.K.)

Abstract: FLAVIN-BINDING, KELCH REPEAT, F-BOX 1 (FKF1) is a blue-light receptor whose function is related to flowering promotion under long-day conditions in *Arabidopsis thaliana*. However, information about the physiological role of FKF1 in day-neutral plants and even the physiological role other than photoperiodic flowering is lacking. Thus, the FKF1 homolog SlFKF1 was investigated in tomato, a day-neutral plant and a useful model for plants with fleshy fruit. It was confirmed that SlFKF1 belongs to the FKF1 group by phylogenetic tree analysis. The high sequence identity with *A. thaliana* FKF1, the conserved amino acids essential for function, and the similarity in the diurnal change in expression suggested that SlFKF1 may have similar functions to *A. thaliana* FKF1. CONSTANS (CO) is a transcription factor regulated by FKF1 and is responsible for the transcription of genes downstream of *CO*. *cis*-Regulatory elements targeted by CO were found in the promoter region of *SINGLE FLOWER TRUSS (SFT)* and *RIN*, which are involved in the regulation of flowering and fruit ripening, respectively. The blue-light effects on *SlFKF1* expression, flowering, and fruit lycopene concentration have been observed in this study and previous studies. It was confirmed in RNA interference lines that the low expression of *SlFKF1* is associated with late flowering with increased leaflets and low lycopene concentrations. This study sheds light on the various physiological roles of FKF1 in plants.

Keywords: *Solanum lycopersicum*; Solanaceae; FLAVIN-BINDING; KELCH REPEAT; F-BOX 1; blue light; flowering; ripening; lycopene

Citation: Shibuya, T.; Nishiyama, M.; Kato, K.; Kanayama, Y. Characterization of the FLAVIN-BINDING, KELCH REPEAT, F-BOX 1 Homolog *SlFKF1* in Tomato as a Model for Plants with Fleshy Fruit. *Int. J. Mol. Sci.* **2021**, 22, 1735. https://doi.org/10.3390/ijms22041735

Academic Editors: Jen-Tsyng Chen, Parviz Heidari and Sixue Chen

Received: 11 December 2020
Accepted: 4 February 2021
Published: 9 February 2021

1. Introduction

Cryptochrome (CRY) and phototropin are blue-light receptors in plants [1]. Studies using *Arabidopsis thaliana* have shown that these photoreceptors control plant responses, including de-etiolation, hypocotyl elongation, and photoperiodic flowering by CRY [2–4] and phototropism, stomatal opening, and chloroplast localization by phototropin [5–8], indicating that their functions are diverse. Even in long-day plants other than *A. thaliana*, flowering promotion by blue light, which is supposed to involve CRY, has been reported in *Petunia* and *Eustoma* [9,10].

There are other blue-light receptors, such as FLAVIN-BINDING, KELCH REPEAT, F-BOX 1 (FKF1), and ZEITLUPE (ZTL), whose function is reportedly related to flowering promotion under long-day conditions in *A. thaliana*. FKF1 and ZTL proteins control the function of CONSTANS (CO) antagonistically [11], and here, we focus on FKF1. FKF1 interacts with GIGANTEA (GI) in a blue-light-dependent manner in *A. thaliana* and induces the degradation of CYCLING DOF FACTOR 1 (CDF1), which suppresses *CO* transcription, and its family proteins [12–15]. FKF1 also stabilizes the CO protein by suppressing its degradation by CONSTITUTIVE PHOTOMORPHOGENIC1 (COP1) and SUPPRESSOR

OF PHYA-105 (SPA) [16–19]. Furthermore, FKF1 promotes *FLOWERING LOCUS T* (*FT*) transcription by inducing the degradation of CDF1, which is the transcriptional repressor of *FT* [16]. In this way, FKF1 plays an essential role in flowering promotion by blue light. As FKF1 in long-day plants, it is suggested that the *Gypsophila* FKF1 homolog is involved in flowering promotion, besides *A. thaliana* [20,21]. Reportedly, OsFKF1 promotes flowering in rice, a short-day plant, regardless of the photoperiod condition [22]. The FKF1 homolog is also considered to play an important role in the developmental phase transition of liverwort [23]. In addition, FKF1 and ZTL regulate the clock period by ubiquitination [24].

Although FKF1 has been reported as a negative regulator of cellulose synthesis [25], there are still few reports on its physiological roles other than flowering promotion. *FKF1* in *A. thaliana* is expressed in the vascular bundle sheath of leaves in relation to flowering control. However, its expression is also found in other tissues, including cotyledons, leaves, guard cells, and root tips, and their physiological importance remains unknown [26,27]. Additionally, FKF1-like sequences are conserved on the genomes of many plant species regardless of photoperiod responsiveness; short-day plants rice and soybean have phylogenetically FKF1-orthologous genes [22,28]. A few reports have suggested the possible functions of FKF1, such as stem and root growth and potassium response [29–31]. From these findings, it is expected that FKF1 plays various roles in plants.

Tomato is generally considered as one of the day-neutral plants, whose FKF1 has not been investigated well, and can also be used as a useful model for plants with fleshy fruit, which has accumulated data for bioinformatics. Hence, in this study, the sequence and physiological roles of the tomato *FKF1* homolog *SlFKF1* were analyzed.

2. Results

2.1. SlFKF cDNA Sequence

Using a BLAST search for a sequence orthologous to the amino acid sequence of *A. thaliana* FKF1, only one *FKF1*-like gene (XM_004228691.3) was found to be present in the tomato genome. This gene was labeled *SlFKF1*. The cDNA of *SlFKF1* was prepared from cv. Micro-Tom by reverse transcription–polymerase chain reaction (RT-PCR) and sequenced. Consequently, its sequence was the same as that on the database. On the alignment based on the amino acid sequence, SlFKF1 was 75.1% identical to *A. thaliana* FKF1, and the amino acids essential for the function of FKF1 were also conserved in SlFKF1 [15,16,32] (Figure 1A). As a result of a phylogenetic tree analysis based on alignment, including ZTL groups having the same domain structure as FKF1 but different functions, it was deduced that SlFKF1 belongs to the FKF1 group (Figure 1B).

2.2. Expression Analysis of SlFKF1

SlFKF1 expression in the wild-type cv. Micro-Tom tomato showed clear diurnal change under a 16 h day length (Figure 2A). The expression was very low from the dark period to the first 6 h of the light period and then increased. The expression peak of *SlFKF1* was shown at Zeitgeber time (ZT) 9. This expression pattern was similar to that of *A. thaliana* FKF1 [13]. *SlFKF1* expression was observed in all organs tested in Figure 2B and was higher in mature leaves than in other organs. Furthermore, the effect of light quality on *SlFKF1* expression was investigated in leaves and fruit (Figure 2C). High expression levels were found in the fruit, similar to those in the leaves. The expression levels were lower under blue light in both organs.

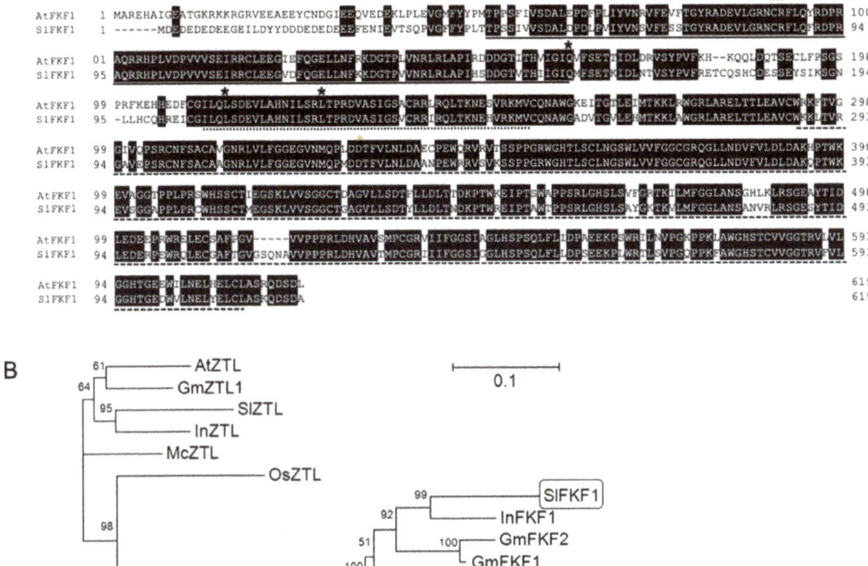

Figure 1. (**A**) Amino acid sequence alignment of SlFKF1 and *A. thaliana* FLAVIN-BINDING, KELCH REPEAT, F-BOX 1 (FKF1). Identical amino acids are shown in black boxes. Asterisks indicate the amino acids essential for the function of FKF1 [15,16,32]. (**B**) A phylogenetic tree based on the amino acid sequences of FKF1 and ZEITLUPE (ZTL) homologs from various species. LOV/PAS, F-box, and KELCH-repeat domain are shown on continuous, dotted, and broken lines, respectively, according to InterPro. The tree was constructed by the neighbor-joining method after sequence alignment using the Clustal W program. Branch numbers refer to the percentage of replicates that support the branch using the bootstrap method (1000 replicates). The scale bar corresponds to 0.1 amino acid substitutions per residue. Table S2 shows the accession numbers of the proteins used to construct the phylogenetic tree.

2.3. Effect of Light Quality on Lycopene Concentration

The effect of blue light on lycopene concentration, a major pigment in tomato fruit and a functional component [33], was investigated. Consequently, the lycopene concentration was lower under blue light than under white light, although it was not significantly different between white light and red light (Figure 3).

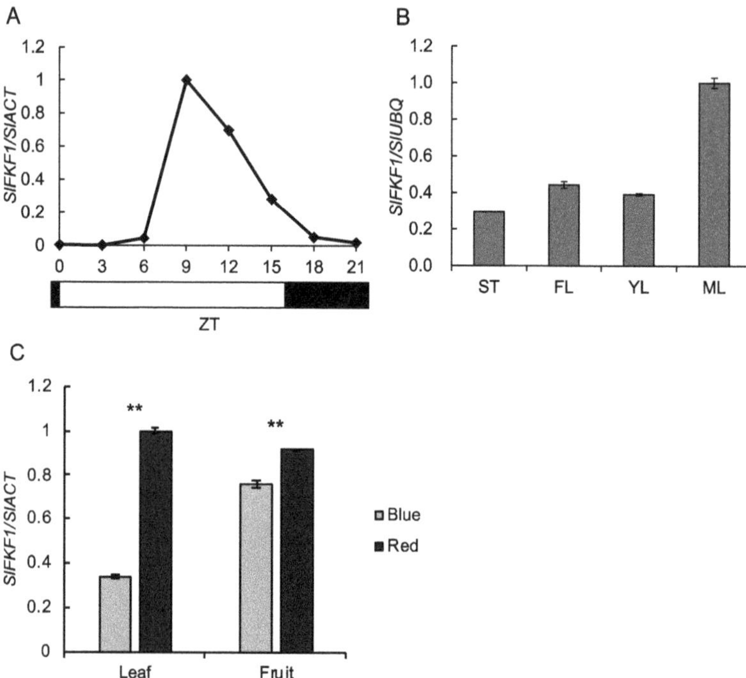

Figure 2. Diurnal change in the expression level of *SlFKF1* in leaves (**A**); expression level of *SlFKF1* in stems (ST), flowers (FL), young leaves (YL), and mature leaves (ML) (**B**); and expression level of *SlFKF1* in leaves and fruit under blue and red light (**C**). Total RNA was prepared from the wild-type plants of cv. Micro-Tom tomato. The relative expression levels were normalized against *SlUBQ* or *SlACT* with standard errors ($n = 3$), and the maximum level of the transcripts was set at 1.0. Values with ** are significantly different between blue and red light in each organ, according to Welch's *t*-test (**C**).

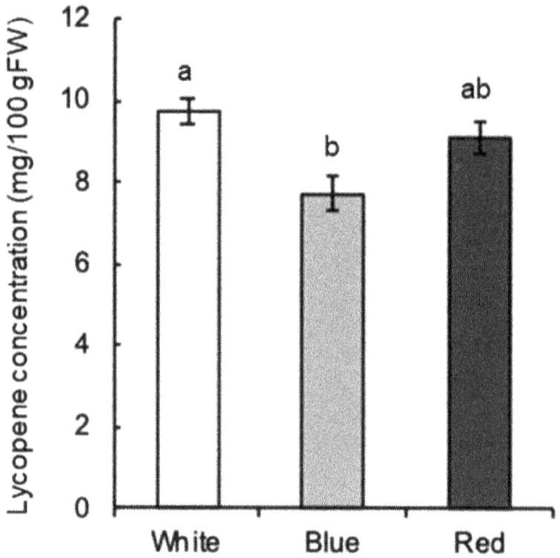

Figure 3. Lycopene concentration in fruit under white (W), blue (B), and red (R) light. Lycopene was extracted from the wild-type plants of cv. Micro-Tom tomato. Values indicate means with standard errors ($n = 3$). $p < 0.01$, values with different letters between treatments, according to Tukey's test.

2.4. Analysis of SlFKF1 Promoter Sequences

Because FKF1 regulates downstream genes via the transcription factor CO [1,16], the possibility of regulating important factors related to flowering, ripening, and pigment synthesis was investigated. Putative CO-binding motifs were searched in the promoter regions of *FT*, *RIN*, and *PSY* tomato homologs, which are key factors for flowering promotion, fruit ripening control, and lycopene synthesis, respectively (Table 1). These genes were analyzed because *SINGLE FLOWER TRUSS* (*SFT*) is the tomato homolog of *FT* [34], and *RIN* and *PSY* are well-known important traits of tomato fruit development as factors governing the ripening process and carotenoid accumulation, respectively [35]. The motifs were found in *SFT*, *PSY1*, *PSY2*, and *RIN* promoters, whereas they were not found in the promoter of *PSY3* in the tomato genome.

Table 1. CONSTANS (CO)-responsive elements on the promoters of *SFT*, *RIN*, and *PSY* homologs.

Gene	Locus [1]	Putative CONSTANS Responsive Element [2]	Strand	Position of 1st C from ATG
SFT	Solyc03g063100	TTT**CCACA**AAA	Top	−379
PSY1	Solyc03g031860	TT**CCCACAC**TG	Bottom	−554
	Solyc03g031860	AAA**TGTGGT**GT	Bottom	−269
	Solyc03g031860	GTC**TGTGGT**CT	Bottom	−186
PSY2	Solyc02g081330	TT**GTGTGGT**CA	Bottom	−274
PSY3	Solyc01g005940	not found		
RIN	Solyc05g012020	CTA**CCACA**AGG	Top	−1049
	Solyc05g012020	AT**GTGTGGC**TA	Bottom	−701

[1] Locus number in the Sol Genomics Network (SGN). [2] Bold letters indicate the core motif.

2.5. Transformation Experiments

SlFKF1 RNA interference (RNAi)-suppressed tomato plans were prepared to confirm the promoter analysis results above. The RNAi lines were differentiated from independent transformation events and produced normal seeds for phenotypic observations. The expression of each transgenic line at the expression peak of *SlFKF1* (ZT9) was investigated, and it was confirmed that expression-suppressed lines with RNAi have lower expression than the wild type (Figure 4A). The flowering in RNAi lines was investigated, and the number of days and leaves until flowering increased in RNAi lines compared to the wild type, suggesting that flowering was delayed by the suppression of *SlFKF1* expression (Figure 4D,F). Another interesting phenotype of RNAi lines was the increased number of leaflets (Figure 4E). The main stem length increased (Figure 4B), and this increase was not accompanied by an increase in internode length but the number of leaves (Figure 4C,F). In fact, because of the long main shoot length and a large number of leaves and leaflets, the RNAi line appeared to have a large plant volume (Figure S1). Regarding fruit coloration, the number of days from flowering to the breaker stage increased, and the degree of coloring was lower in RNAi lines than in the wild type 10 days after the beaker stage (Figure 5A,B).

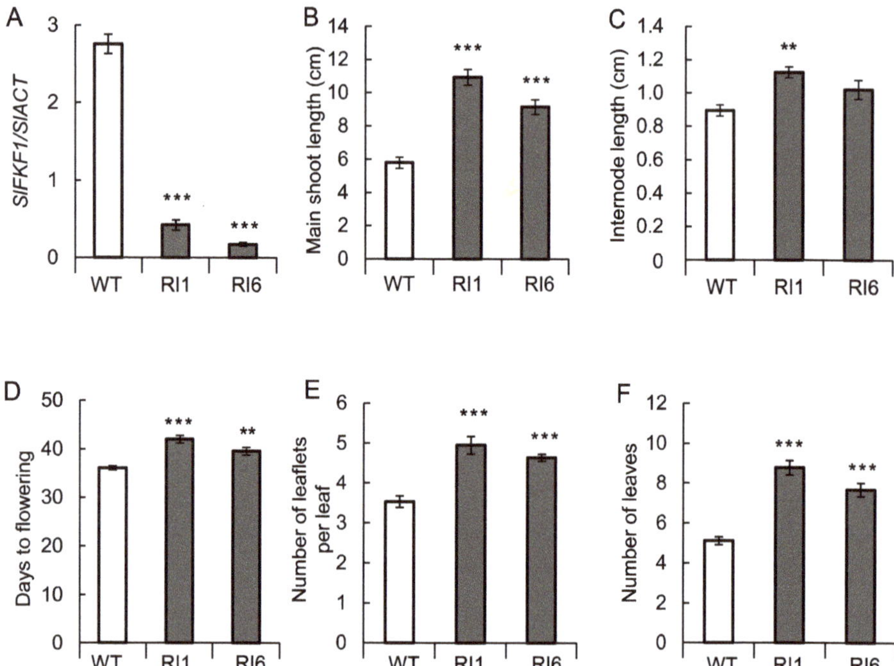

Figure 4. SlFKF1 mRNA levels and growth of SlFKF1 RNA interference (RNAi) lines (RI) and wild type (WT). SlFKF1 mRNA levels (**A**) were measured with the main shoot length (**B**) and internode length (**C**). Days to flowering indicate the number of days to first inflorescences from sowing (**D**). The number of leaflets per leaf was also measured (**E**). The number of leaves indicates the number of leaves to first flowers (**F**). Values indicate means with standard errors ($n = 3$ in A and $n = 9$ in **B**–**F**). *** $p < 0.001$ and ** $p < 0.01$, significantly different from WT, according to Dunnett's test.

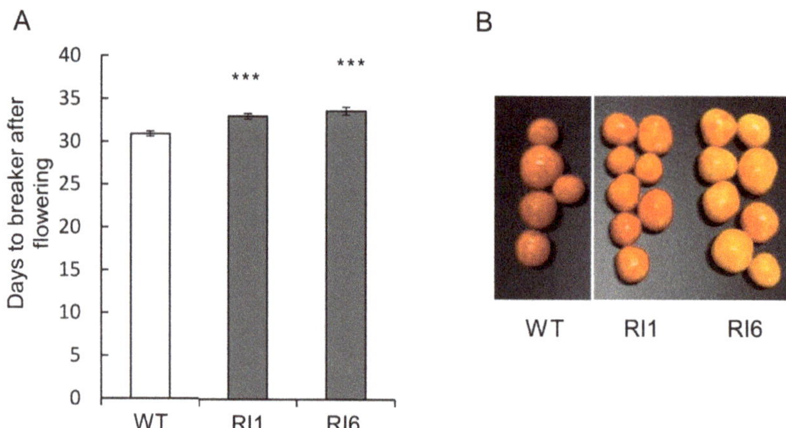

Figure 5. Number of days from flowering to the breaker stage (**A**) and fruit coloration 10 days after the breaker stage (**B**) in SlFKF1 RNAi lines (RI) and wild type (WT). Values indicate means with standard errors ($n = 12$–26). *** $p < 0.01$, significantly different from WT, according to Dunnett's test.

3. Discussion

FKF1 has three domains: LOV, F-box, and KELCH repeat. In *A. thaliana*, it forms a gene family with ZTL [20]. ZTL has been reported to have a different function from FKF1 and is involved in the decomposition of TIMING OF CAB EXPRESSION 1 (TOC1), one of the components of the circadian clock [36–38]. The phylogenetic tree analysis results, including FKF1 and ZTL homologs, confirmed that SlFKF1 belongs to the FKF1 group. The high sequence identity with *A. thaliana* FKF1, the conserved amino acids essential for function, and the similarity in the diurnal change in expression suggested that SlFKF1 may have similar functions to *A. thaliana* FKF1. SlFKF1 was expressed not only in leaves, which are the photoreceptive organs for photoperiodic flowering control, but also in various organs, and its expression level was also high in fruit. Because the flowering of tomato is not affected by the photoperiod and the role of FKF1 in fruit has not been reported so far, in tomato, the physiological role of FKF1 other than in photoperiodic flowering should be examined.

CO is a transcription factor regulated by FKF1 and is responsible for the transcription of genes downstream of *CO* [1,16]. A CO-responsive element targeted by CO was reported by Tiwari et al. [39], and its CCACA core motif was identified by Gnesutta et al. [40]. SFT is the tomato homolog of FT involved in flowering promotion as a florigen [34], and RIN and PSY play important roles in governing the ripening process and carotenoid accumulation [35]. Therefore, it is possible that the FKF1-CO-mediated pathway regulates the expression of these genes and affects these developmental processes in tomato.

The heterozygotes of wild-type and mutant alleles of *SFT* in a determinant cultivar showed an approximately twofold increase in yield [41]. Thus, the physiological and agricultural importance of the *FT*-related flowering pathway in day-neutral plants is something of interest. A CO *cis*-element was found in the promoter region of *SFT*. Late flowering and increased leaflets were commonly observed in the *sft* mutant [42,43] and SlFKF1 RNAi lines, suggesting the relationship between *SlFKF1* and *SFT*. In *A. thaliana*, FKF1 is considered a blue-light receptor responsible for promoting flowering under long-day conditions, whereas in rice, a short-day plant, the FKF1 homolog promotes flowering regardless of day length [22]. The results suggested that an FKF1 homolog may also function in the flowering pathway of day-neutral plants. *A. thaliana* FKF1 promotes flowering by suppressing the function of the COP1/SPA system that degrades the CO protein [17–19], and tomato likely has a similar mechanism. In contrast, it has been proposed in *A. thaliana* that FKF1 positively regulates the gibberellin (GA) signal through the degradation of the DELLA protein and activates the GA-dependent flowering promotion pathway [44]. However, in tomato, because GA rather suppresses flowering [45], it would not be possible to apply this *A. thaliana* model to tomato.

Because SlFKF1 expression was suppressed by blue light, the effect of blue light on lycopene concentration was investigated for comparison. In this study, the effect of blue-light irradiation on lycopene concentration in fruit was measured during cultivation. Although the effects of blue-light irradiation on lycopene concentration have been investigated in previous reports [46,47], their experimental conditions were different from this study. The effects of supplemental blue light over a long period after flowering in a greenhouse [46] and blue-light irradiation after harvest [47] have been investigated in previous reports. In addition, the effects of blue light on lycopene concentration are different between these reports. In this study, to limit the effect on vegetative growth, after cultivation under white light, the effect was investigated by irradiating blue light from the breaker only for 10 days. As a result, this study and a previous report [47] showed low fruit lycopene concentrations under blue light. Collectively, it is likely that blue light alone for a relatively short period negatively affects lycopene accumulation in tomato fruit.

The fruit lycopene concentration was low under blue light, as described above, and there are *cis*-regulatory elements targeted by CO in the promoter regions of RIN and PSY. Additionally, fruit *SlFKF1* expression was low under blue light. Because there are reports that light quality regulates growth and environmental response through the regulation of photoreceptor gene expression [48,49], low *SlFKF1* expression might be related to low lycopene concentration under blue light as one possible mechanism. It was confirmed in RNAi lines that low expression of *SlFKF1* is associated with low concentrations of lycopene. So far, the effect of blue light on tomato fruit color has assumed the involvement of CRY [50–52], and there have been few reports on FKF1. A similar control mechanism is possible for flowering because blue light delays flowering in tomato [53,54], and leaf *SlFKF1* expression is also low under blue light. It is interesting to note that blue light promotes flowering in *A. thaliana* and some other long-day plants in contrast to tomato [10,21,55].

A previous report suggested a relationship between auxin and FKF1 in the adventitious rooting of longan [30]. Auxin plays an important role in the growth and ripening of tomato fruit [56]. Another report suggested that potassium levels, which are important in fruit growth and quality, affect *FKF1* expression in banana roots [31]. Therefore, the function of SlFKF1 may be related to auxin signaling and potassium nutrition in fruit. Because other photoreceptors such as CRY and PHY are considered important agronomic traits [57], the assumed role of FKF1 in fruit crops is promising based on this study.

FKF1 is present in the genomes of a wide variety of species, whereas so far, knowledge about its role has been limited. Its role in flowering in day-neutral plants and even the physiological role other than flowering is discussed in this study using tomato as a model plant, while preliminary transcriptome analysis has shown that blue light induces the expression of several transcriptional regulation- and signal transduction-related genes during tomato fruit ripening [58]. Starting with this study, the varying physiological roles of FKF1, which have been comparatively unknown compared to phytochrome and CRY, will be elucidated.

4. Materials and Methods

4.1. Plant Materials

Tomato cv. Micro-Tom wild type was cultivated for cloning and expression analysis in a growth chamber (LH240SP; Nihon Ika Co., Ltd., Osaka, Japan) at 25 °C with a white fluorescent lamp in a 16-h photoperiod. Photosynthetic photon flux density (PPFD) was 100 µmol m^{-2} s^{-1}, and Sumisoil N150 (Sumika Agro-tech Co., Ltd., Osaka, Japan) was used as cultivation soil. The plants were supplemented with nutrient solution (Hyponex Japan) every week. The cultivation was conducted with reference to Tsunoda et al. [59].

4.2. Sequence Analysis of Tomato FKF1 Homolog SlFKF1 cDNA

A BLAST search was performed to search for a candidate protein of SlFKF1 and an open reading frame (ORF), encoding it based on the amino acid sequence of FKF1 (AT1G68050) of *A. thaliana* FKF1 in Tomato Genome CDS (ITAG release 2.40) of the Sol Genomics Network (SGN; https://solgenomics.net accessed on 8 February 2021) and NW_004194292.1 and XM_004228691.1 of the National Center for Biotechnology Information (NCBI; https://www.ncbi.nlm.nih.gov accessed on 8 February 2021). Using a primer set designed based on the candidate sequence obtained from the database, its ORF sequence was cloned by PCR with the cDNA obtained by reverse transcription of RNA extracted from cv. Micro-Tom tomato leaves and confirmed by sequencing.

4.3. Expression Analysis

For a diurnal change in expression, mature leaves that were fully developed and not senesced were randomly collected every 3 h from the start of the light period (ZT0) to ZT21 at approximately 30 days after germination. For the expression analysis in stems, flowers, immature leaves, and mature leaves, samples were collected on ZT8 approximately 50 days after germination. For the expression analysis in leaves under blue and red light, the plants at 2 months after sowing were cultivated for 1 day under a 16 h photoperiod with red or blue light emitting diodes (LEDs) (CCS Inc., Kyoto, Japan), and leaves were sampled for RNA extraction. For the expression analysis in fruit under blue and red light, the plants at the fruit breaker stage were cultivated for 10 days under a 16 h photoperiod with red or blue

LEDs, and the pericarp was sampled for RNA extraction. PPFD was 100 μmol m^{-2} s^{-1}, and the samples were collected on ZT10. Red and blue LEDs had peaks at 655 and 470 nm, respectively. RNA was prepared from these samples using the RNeasy Plant Mini Kit (Qiagen) for real-time PCR. The removal of genomic DNA and reverse transcription were performed using the Quantiscript Reverse Transcription Kit (Qiagen) and the ReverTra Ace qPCR RT Master Mix (Toyobo). Real-time PCR was performed using the QuantiTect SYBR Green PCR Kit (Qiagen) and the THUNDERBIRD SYBR qPCR Mix (Toyobo) according to Ikeda et al. [60]. *SlACT* and *SlUBQ* were used as reference genes [61,62], and Table S1 shows the nucleotide sequences for the primers.

4.4. Alignment, Phylogenetic Analysis, and Promoter Analysis

Using the amino acid sequences of AtFKF1 and AtZTL as queries, a BLAST search was performed for non-redundant protein sequences (nr) of each plant species in the NCBI to search for homologs. For tomato FKF1, a BLAST search was also performed on tomato genome protein sequences (ITAG release 2.40) on the SGN to obtain locus information about matching proteins, and the gene model was confirmed by referring to genomic detail. A phylogenetic tree was created using the neighbor-joining method for the alignment obtained by executing Clustal W2 on Genetyx version 10 for FKF1 and ZTL amino acid sequences. Bootstrap probabilities were calculated by 1000 trials.

For promoter analysis, each gene was searched by a keyword and BLAST on the SGN, and locus names were confirmed. Next, the locus was searched, and an upstream 3000-base sequence was obtained from the genomic sequence of genomic detail using the function of Get flanking sequences on SL2.50ch07. Additionally, the position of the start codon was confirmed from the cDNA sequence and protein sequence of the genomic detail, and the start codon upstream 1500-base sequence was determined. The CO-binding sequence CCACA was searched from the obtained 1500-base sequence by the Text Search function on Genetyx version 14.

4.5. Determination of Lycopene Concentration

The plants at the fruit breaker stage were cultivated for 10 days under a 16 h photoperiod with red LEDs, blue LEDs, or a white fluorescent lamp, and the pericarp was sampled for the determination of lycopene concentration. PPFD was 100 μmol m^{-2} s^{-1}, and the samples were collected on ZT10. Red and blue LEDs (CCS Inc.) had peaks at 655 and 470 nm, respectively. According to Ito and Horie [63], lycopene was extracted using dimethyl ether/methanol (7:3), filtrated with DSMIC JP 13 (Advantec), and quantified by measuring the absorbance at 505 nm.

4.6. Transformation Experiment

RNAi was used for expression suppression. A partial fragment (341 bp) of *SlFKF1* was amplified by RT-PCR using the cDNA of cv. Micro-Tom as a template and introduced into pBI-RNAi-GW (Inplanta Innovations, Inc., Yokohama, Japan), a vector for preparing an RNAi construct with the CaMV *35S* promoter. Table S1 shows the nucleotide sequences for the primers used to prepare these vectors. The transformation was outsourced to Inplanta Innovations. The cultivation of transformed tomato plants and expression analysis were performed as described above.

Supplementary Materials: Can be found at https://www.mdpi.com/1422-0067/22/4/1735/s1. Table S1: List of primer sequences, Table S2: Accession numbers of proteins used for the phylogenetic tree analysis, Figure S1: Wild-type (WT) and SlFKF1 RNAi line (RI6) plants with ripe fruit.

Author Contributions: Conceptualization, K.K. and Y.K.; methodology, Y.K. and T.S.; investigation, T.S.; formal analysis, M.N. and T.S.; resources, T.S.; data curation, K.K.; writing—original draft preparation, T.S.; writing—review and editing, M.N. and Y.K.; visualization, T.S.; supervision, M.N. and Y.K.; project administration, Y.K.; funding acquisition, K.K. and Y.K. All authors have read and agreed to the published version of the manuscript.

Funding: This research was supported by Grants-in-Aid for Scientific Research (16H02534, 19K06012).

Conflicts of Interest: The authors declare no conflict of interest.

References

1. Shibuya, T.; Kanayama, Y. Flowering response to blue light and its molecular mechanisms in *Arabidopsis* and horticultural plants. *Adv. Hort. Sci.* **2014**, *28*, 179–183.
2. Ahmad, M.; Cashmore, A.R. HY4 gene of *A. thaliana* encodes a protein with characteristics of a blue-light photoreceptor. *Nature* **1993**, *366*, 162–166. [CrossRef] [PubMed]
3. Guo, H.; Yang, H.; Mockler, T.C.; Lin, C. Regulation of flowering time by *Arabidopsis* photoreceptors. *Science* **1998**, *279*, 1360–1363. [CrossRef] [PubMed]
4. El-Assal, S.; Alonso-Blanco, C.; Peeters, A.; Raz, V.; Koornneef, M. A QTL for flowering time in *Arabidopsis* reveals a novel allele of *CRY2*. *Nat. Genet.* **2001**, *29*, 435–440. [CrossRef]
5. Kagawa, T.; Wada, M. Blue light-induced chloroplast relocation in *Arabidopsis thaliana* as analyzed by microbeam irradiation. *Plant Cell Physiol.* **2000**, *41*, 84–93. [CrossRef]
6. Sakai, T.; Kagawa, T.; Kasahara, M.; Swartz, T.E.; Christie, J.M.; Briggs, W.R.; Wada, M. *Arabidopsis* nph1 and npl1: Blue light receptors that mediate both phototropism and chloroplast relocation. *Proc. Natl. Acad. Sci. USA* **2001**, *98*, 6969–6974. [CrossRef]
7. Kinoshita, T.; Doi, M.; Suetsugu, N.; Kagawa, T.; Wada, M.; Shimazaki, K. Phot1 and phot2 mediate blue light regulation of stomatal opening. *Nature* **2001**, *414*, 656–660. [CrossRef] [PubMed]
8. Wada, M.; Kagawa, T.; Sato, Y. Chloroplast movement. *Annu. Rev. Plant Biol.* **2003**, *54*, 455–468. [CrossRef] [PubMed]
9. Fukuda, N.; Ishii, Y.; Ezura, H.; Olsen, J.E. Effects of light quality under red and blue light emitting diodes on growth and expression of *FBP28* in petunia. *Acta Hort.* **2011**, *907*, 361–366. [CrossRef]
10. Shibuya, T.; Takahashi, T.; Hashimoto, S.; Nishiyama, M.; Kanayama, Y. Effects of overnight radiation with monochromatic far-red and blue light on flower budding and expression of flowering-related and light quality-responsive genes in *Eustoma grandiflorum*. *J. Agric. Meteorol.* **2019**, *75*, 160–165. [CrossRef]
11. Hwang, D.Y.; Park, S.; Lee, S.; Lee, S.S.; Imaizumi, T.; Song, Y.H. GIGANTEA regulates the timing stabilization of CONSTANS by altering the interaction between FKF1 and ZEITLUPE. *Mol. Cells* **2019**, *42*, 693–701.
12. Fornara, F.; Panigrahi, K.C.S.; Gissot, L.; Sauerbrunn, N.; Rühl, M.; Jarillo, J.A.; Coupland, G. *Arabidopsis* DOF transcription factors act redundantly to reduce CONSTANS expression and are essential for a photoperiodic flowering response. *Dev. Cell* **2009**, *17*, 75–86. [CrossRef] [PubMed]
13. Imaizumi, T.; Tran, H.G.; Swartz, T.E.; Briggs, W.R.; Kay, S.A. FKF1 is essential for photoperiodic-specific light signaling in *Arabidopsis*. *Nature* **2003**, *426*, 302–306. [CrossRef] [PubMed]
14. Imaizumi, T.; Schultz, T.F.; Harmon, F.G.; Ho, L.A.; Kay, S.A. FKF1 F-box protein mediates cyclic degradation of a repressor of CONSTANS in *Arabidopsis*. *Science* **2005**, *309*, 293–297. [CrossRef]
15. Sawa, M.; Nusinow, D.A.; Kay, S.A.; Imaizumi, T. FKF1 and GIGANTEA complex formation is required for day-length measurement in *Arabidopsis*. *Science* **2007**, *318*, 261–265. [CrossRef]
16. Song, Y.H.; Smith, R.W.; To, B.J.; Millar, A.J.; Imaizumi, T. FKF1 conveys timing information for CONSTANS stabilization in photoperiodic flowering. *Science* **2012**, *336*, 1045–1049. [CrossRef] [PubMed]
17. Ponnu, J. Molecular mechanisms suppressing COP1/SPA E3 ubiquitin ligase activity in blue light. *Physiol. Plant.* **2020**, *169*, 418–429. [CrossRef] [PubMed]
18. Lee, B.-D.; Cha, J.-Y.; Kim, M.R.; Shin, G.-I.; Paek, N.-C.; Kim, W.-Y. Light-dependent suppression of COP1 multimeric complex formation is determined by the blue-light receptor FKF1 in *Arabidopsis*. *Biochem. Biophys. Res. Commun.* **2019**, *508*, 191–197. [CrossRef] [PubMed]
19. Lee, B.-D.; Kim, M.R.; Kang, M.-Y.; Cha, J.-Y.; Han, S.-H.; Nawkar, G.M.; Sakuraba, Y.; Lee, S.Y.; Imaizumi, T.; McClung, C.R.; et al. The F-box protein FKF1 inhibits dimerization of COP1 in the control of photoperiodic flowering. *Nat. Commun.* **2017**, *8*, 2259. [CrossRef] [PubMed]
20. Shibuya, T.; Murakawa, Y.; Nishidate, K.; Nishiyama, M.; Kanayama, Y. Characterization of flowering-related genes and flowering response in relation to blue light in *Gypsophila paniculata*. *Hort. J.* **2017**, *86*, 94–104. [CrossRef]
21. Hori, Y.; Nishidate, K.; Nishiyama, M.; Kanahama, K.; Kanayama, Y. Flowering and expression of flowering-related genes under long-day conditions with light-emitting diodes. *Planta* **2011**, *234*, 321–330. [CrossRef]
22. Han, S.-H.; Yoo, S.-C.; Lee, B.-D.; An, G.; Paek, N.-C. Rice FLAVIN-BINDING, KELCH REPEAT, F-BOX 1 (OsFKF1) promotes flowering independent of photoperiod. *Plant Cell Environ.* **2015**, *38*, 2527–2540. [CrossRef] [PubMed]
23. Kubota, A.; Kita, S.; Ishizaki, K.; Nishihama, R.; Yamato, K.T.; Kohchi, T. Co-option of a photoperiodic growth-phase transition system during land plant evolution. *Nat. Commun.* **2014**, *5*, 3668. [CrossRef] [PubMed]
24. Lee, C.-M.; Feke, A.; Li, M.-W.; Adamchek, C.; Webb, K.; Pruneda-Paz, J.; Bennett, E.J.; Kay, S.A.; Gendron, J.M. Decoys untangle complicated redundancy and reveal targets of circadian clock F-BOX proteins. *Plant Physiol.* **2018**, *177*, 1170–1186. [CrossRef] [PubMed]
25. Yuan, N.; Balasubramanian, V.K.; Chopra, R.; Mendu, V. The photoperiodic flowering time regulator fkf1 negatively regulates cellulose biosynthesis. *Plant Physiol.* **2019**, *180*, 2240–2253. [CrossRef] [PubMed]

26. Nelson, D.C.; Lasswell, J.; Rogg, L.; Cohen, M.A.; Bartel, B. FKF1, a Clock-controlled gene that regulates the transition to flowering in *Arabidopsis*. *Cell* **2000**, *101*, 331–340. [CrossRef]

27. Obulareddy, N.; Panchal, S.; Melotto, M. Guard cell purification and RNA isolation suitable for high-throughput transcriptional analysis of cell-type responses to biotic stresses. *Mol. Plant Microbe Interact.* **2013**, *26*, 844–849. [CrossRef]

28. Li, F.; Zhang, X.; Hu, R.; Wu, F.; Ma, J.; Meng, Y.; Fu, Y. Identification and molecular characterization of *FKF1* and *GI* homologous genes in soybean. *PLoS ONE* **2013**, *8*, e79036. [CrossRef]

29. Zeng, J.; Mo, Y.; Chen, J.; Li, C.; Zhao, L.; Liu, Y. Expression and interaction proteins analysis of BjuFKF1 in stem mustard. *Sci. Hortic.* **2020**, *269*, 109430. [CrossRef]

30. Huang, F.; Fu, Z.; Zeng, L.; Morley-Bunker, M. Isolation and characterization of *GI* and *FKF1* homologous genes in the subtropical fruit tree. *Dimocarpus longan. Mol. Breed.* **2017**, *37*, 90. [CrossRef]

31. He, Y.; Li, R.; Lin, F.; Xiong, Y.; Wang, L.; Wang, B.; Guo, J.; Hu, C. Transcriptome changes induced by different potassium levels in banana roots. *Plants* **2020**, *9*, 11. [CrossRef]

32. Han, L.; Mason, M.; Risseeuw, E.P.; Crosby, W.L.; Somers, D.E. Formation of an SCFZTL complex is required for proper regulation of circadian timing. *Plant J.* **2004**, *40*, 291–301. [CrossRef] [PubMed]

33. Pesaresi, P.; Mizzotti, C.; Colombo, M.; Masiero, S. Genetic regulation and structural changes during tomato fruit development and ripening. *Front. Plant Sci.* **2014**, *5*, 124. [CrossRef] [PubMed]

34. Lifschitz, E.; Eviatar, T.; Rozman, A.; Shalit, A.; Goldshmidt, A.; Amsellem, Z.; Alvarez, J.; Eshed, Y. The tomato FT ortholog triggers systemic signals that regulate growth and flowering and substitute for diverse environmental stimuli. *Proc. Natl. Acad. Sci. USA* **2006**, *103*, 6398–6403. [CrossRef]

35. Liu, R.; How-Kit, A.; Stammittia, L.; Teyssier, E.; Rolin, D.; Mortain-Bertrand, A.; Halle, S.; Liu, M.; Kong, J.; Wu, C.; et al. A DEMETER-like DNA demethylase governs tomato fruit ripening. *Proc. Natl. Acad. Sci. USA* **2015**, *112*, 10804–10809. [CrossRef]

36. Más, P.; Kim, W.-Y.; Somers, D.E.; Kay, S.A. Targeted degradation of TOC1 by ZTL modulates circadian function in *Arabidopsis thaliana. Nature* **2003**, *426*, 02163. [CrossRef]

37. Kiba, T.; Henriques, R.; Sakakibara, H.; Chua, N.-H. Targeted degradation of PSEUDO-RESPONSE REGULATOR5 by an SCFZTL complex regulates clock function and photomorphogenesis in *Arabidopsis thaliana. Plant Cell* **2007**, *19*, 2516–2530. [CrossRef]

38. Fujiwara, S.; Wang, L.; Han, L.; Suh, S.-S.; Salomé, P.A.; McClung, R.C.; Somers, D.E. Post-translational regulation of the *Arabidopsis* circadian clock through selective proteolysis and phosphorylation of pseudo-response regulator proteins. *J. Biol. Chem.* **2008**, *283*, 23073–23083. [CrossRef] [PubMed]

39. Tiwari, S.B.; Shen, Y.; Chang, H.-C.; Hou, Y.; Harris, A.; Ma, S.F.; McPartland, M.; Hymus, G.J.; Adam, L.; Marion, C.; et al. The flowering time regulator CONSTANS is recruited to the *FLOWERING LOCUS T* promoter via a unique *cis*-element. *New Phytol.* **2010**, *187*, 57–66. [CrossRef]

40. Gnesutta, N.; Kumimoto, R.W.; Swain, S.; Chiara, M.; Siriwardana, C.S.; Horner, D.; Ben, F.; Holt, B.F.; Mantovania, R. CONSTANS imparts DNA sequence specificity to the histone fold NF-YB/NF-YC dimer. *Plant Cell* **2017**, *29*, 1516–1532. [CrossRef] [PubMed]

41. Krieger, U.; Lippman, Z.B.; Zamir, D. The flowering gene *SINGLE FLOWER TRUSS* drives heterosis for yield in tomato. *Nat. Genet.* **2010**, *42*, 459–463. [CrossRef]

42. Shalit, A.; Rozman, A.; Goldshmidt, A.; Alvarez, J.P.; Bowman, J.; Eshed, Y.; Lifschitz, E. The flowering hormone florigen functions as a general systemic regulator of growth and termination. *Proc. Natl. Acad. Sci. USA* **2009**, *106*, 8392–8397. [CrossRef] [PubMed]

43. Molinero-Rosales, N.; Latorre, A.; Jamilena, M.; Lozano, R. SINGLE FLOWER TRUSS regulates the transition and maintenance of flowering in tomato. *Planta* **2004**, *218*, 427–434. [CrossRef] [PubMed]

44. Yan, J.; Li, X.; Zeng, B.; Zhong, M.; Yang, J.; Yang, P.; Li, X.; He, C.; Lin, J.; Liu, X.; et al. FKF1 F-box protein promotes flowering in part by negatively regulating DELLA protein stability under long-day photoperiod in *Arabidopsis. J. Integr. Plant Biol.* **2020**, *62*, 1717–1740. [CrossRef]

45. Silva, G.F.F.; Silva, E.M.; Correa, J.P.O.; Vicente, M.H.; Jiang, N.; Notini, M.M.; Junior, A.C.; De Jesus, F.A.; Castilho, P.; Carrera, E.; et al. Tomato floral induction and flower development are orchestrated by the interplay between gibberellin and two unrelated microRNA-controlled modules. *New Phytol.* **2019**, *221*, 1328–1344. [CrossRef] [PubMed]

46. Xie, B.-X.; Wei, J.-J.; Zhang, Y.-T.; Song, S.-W.; Su, W.; Sun, G.-W.; Hao, Y.-W.; Liu, H.-C. Supplemental blue and red light promote lycopene synthesis in tomato fruits. *J. Integr. Agric.* **2019**, *18*, 590–598. [CrossRef]

47. Dhakal, R.; Baek, K.-H. Short period irradiation of single blue wavelength light extends the storage period of mature green tomatoes. *Postharvest Biol. Technol.* **2014**, *90*, 73–77. [CrossRef]

48. Lin, C.; Yang, H.; Guo, H.; Mockler, T.; Chen, J.; Cashmore, A.R. Enhancement of blue-light sensitivity of Arabidopsis seedlings by a blue light receptor cryptochrome 2. *Proc. Natl. Acad. Sci. USA* **1998**, *95*, 2686–2690. [CrossRef]

49. Novák, A.; Boldizsár, Á.; Ádám, É.; Kozma-Bognár, L.; Majláth, I.; Båga, M.; Tóth, B.; Chibbar, R.; Galiba, G. Light-quality and temperature-dependent CBF14 gene expression modulates freezing tolerance in cereals. *J. Exp. Bot.* **2016**, *67*, 1285–1295. [CrossRef]

50. Kim, M.-J.; Kim, P.; Chen, Y.; Chen, B.; Yang, J.; Liu, X.; Kawabata, S.; Wang, Y.; Li, Y. Blue and UV-B light synergistically induce anthocyanin accumulation by co-activating nitrate reductase gene expression in Anthocyanin fruit (*Aft*) tomato. *Plant Biol.* **2020**. [CrossRef]

51. Giliberto, L.; Perrotta, G.; Pallara, P.; Weller, J.; Fraser, P.D.; Bramley, P.M.; Fiore, A.; Tavazza, M.; Giuliano, G. Manipulation of the blue light photoreceptor cryptochrome 2 in tomato affects vegetative development, flowering time, and fruit antioxidant content. *Plant Physiol.* **2005**, *137*, 199–208. [CrossRef]

52. Liu, C.-C.; Ahammed, G.J.; Wang, G.-T.; Xu, C.-J.; Chen, K.-S.; Zhou, Y.-H.; Yu, J.-Q. Tomato CRY1a plays a critical role in the regulation of phytohormone homeostasis, plant development, and carotenoid metabolism in fruits. *Plant Cell Environ.* **2018**, *41*, 354–366. [CrossRef] [PubMed]

53. Nanya, K.; Ishigami, Y.; Hikosaka, S.; Goto, E. Effects of blue and red light on stem elongation and flowering of tomato seedlings. *Acta Hort.* **2012**, *956*, 261–266. [CrossRef]

54. Kim, H.M.; Hwang, S.J. The growth and development of 'Mini Chal' tomato plug seedlings grown under monochromatic or combined red and blue light-emitting diodes. *Hort. Sci. Technol.* **2019**, *37*, 190–205.

55. Yoshida, H.; Mizuta, D.; Fukuda, N.; Hikosaka, S.; Goto, E. Effects of varying light quality from single-peak blue and red light-emitting diodes during nursery period on flowering, photosynthesis, growth, and fruit yield of everbearing strawberry. *Plant Biotechnol.* **2016**, *33*, 267–276. [CrossRef] [PubMed]

56. Kanayama, Y. Sugar metabolism and fruit development in tomato. *Hort. J.* **2017**, *86*, 417–425. [CrossRef]

57. Mawphlang, O.I.L.; Kharshiing, E.V. Photoreceptor mediated plant growth responses: Implications for photoreceptor engineering toward improved performance in crops. *Front. Plant Sci.* **2017**, *8*, 1181. [CrossRef] [PubMed]

58. Kawabata, S.; Miyamoto, K.; Li, Y. cDNA microarray analysis of differential gene expression in tomato fruits exposed to blue, UV-A, and UV-A+UV-B. *Acta Hort.* **2011**, *907*, 371–374. [CrossRef]

59. Tsunoda, Y.; Hano, S.; Imoto, N.; Shibuya, T.; Ikeda, H.; Amagaya, K.; Kato, K.; Shirakawa, H.; Aso, H.; Kanayama, Y. Physiological roles of tryptophan decarboxylase revealed by overexpression of *SlTDC1* in tomato. *Sci. Hortic.* **2021**, *275*, 109672. [CrossRef]

60. Ikeda, H.; Shibuya, T.; Imanishi, S.; Aso, H.; Nishiyama, M.; Kanayama, Y. Dynamic metabolic regulation by a chromosome segment from a wild relative during fruit development in a tomato introgression line, IL8-3. *Plant Cell Physiol.* **2016**, *57*, 1257–1270. [CrossRef] [PubMed]

61. Mohammed, S.A.; Nishio, S.; Takahashi, H.; Shiratake, K.; Ikeda, H.; Kanahama, K.; Kanayama, Y. Role of vacuolar H$^+$-inorganic pyrophosphatase in tomato fruit development. *J. Exp. Bot.* **2012**, *63*, 5613–5621. [CrossRef]

62. Miura, K.; Shiba, H.; Ohta, M.; Kang, S.; Sato, A.; Yuasa, T.; Iwava-Inoue, M.; Kamada, H.; Ezura, H. SlICE1 encoding a MYC-type transcription factor controls cold tolerance in tomato, *Solanum lycopersicum*. *Plant Biotechnol.* **2012**, *29*, 253–260. [CrossRef]

63. Ito, H.; Horie, H. Proper solvent selection for lycopene extraction in tomatoes and application to a rapid determination. *Bull. Natl. Inst. Veg. Tea Sci.* **2009**, *8*, 165–173.

International Journal of
Molecular Sciences

Article

Mitochondrial Transcription Termination Factor 27 Is Required for Salt Tolerance in *Arabidopsis thaliana*

Deyuan Jiang, Jian Chen, Zhihong Zhang and Xin Hou *

State Key Laboratory of Hybrid Rice, College of Life Sciences, Wuhan University, Wuhan 430070, China; dyjiang@whu.edu.cn (D.J.); 2015202040070@whu.edu.cn (J.C.); zzh@whu.edu.cn (Z.Z.)
* Correspondence: xinhou@whu.edu.cn

Abstract: In plants, mTERF proteins are primarily found in mitochondria and chloroplasts. Studies have identified several mTERF proteins that affect plant development, respond to abiotic stresses, and regulate organellar gene expression, but the functions and underlying mechanisms of plant mTERF proteins remain largely unknown. Here, we investigated the function of *Arabidopsis* mTERF27 using molecular genetic, cytological, and biochemical approaches. *Arabidopsis* mTERF27 had four mTERF motifs and was evolutionarily conserved from moss to higher plants. The phenotype of the mTERF27-knockout mutant *mterf27* did not differ obviously from that of the wild-type under normal growth conditions but was hypersensitive to salt stress. mTERF27 was localized to the mitochondria, and the transcript levels of some mitochondrion-encoded genes were reduced in the *mterf27* mutant. Importantly, loss of mTERF27 function led to developmental defects in the mitochondria under salt stress. Furthermore, mTERF27 formed homomers and directly interacted with multiple organellar RNA editing factor 8 (MORF8). Thus, our results indicated that mTERF27 is likely crucial for mitochondrial development under salt stress, and that this protein may be a member of the protein interaction network regulating mitochondrial gene expression.

Keywords: *Arabidopsis*; mTERF27; MORF8; salt stress; mitochondrial morphology

Citation: Jiang, D.; Chen, J.; Zhang, Z.; Hou, X. Mitochondrial Transcription Termination Factor 27 Is Required for Salt Tolerance in *Arabidopsis thaliana*. *Int. J. Mol. Sci.* **2021**, *22*, 1466. https://doi.org/10.3390/ijms22031466

Academic Editor: Jen-Tsung Chen

Received: 14 December 2020
Accepted: 26 January 2021
Published: 2 February 2021

Publisher's Note: MDPI stays neutral with regard to jurisdictional claims in published maps and institutional affiliations.

1. Introduction

Mitochondria, which originated through the endosymbiosis of an α-proteobacterial ancestor, are considered the "power house" of the cell, providing the necessary energy for cellular function. Mitochondria have their own intrinsic genomes, RNA, and ribosomes. The regulation of mitochondrial genome expression is vital to the coordination of energy demands during particular growth and developmental stages in plants [1]. Plant mitochondria have unique and complex RNA metabolism mechanisms, combining the characteristics of their prokaryotic ancestors with the new features of evolution in eukaryotic hosts [2]. Compared to animals, the mitochondrial genomes of plants are relatively larger [3]. Post-transcriptional mechanisms, including RNA editing, RNA splicing, maturation of transcriptional ends, RNA degradation, and other processing steps play a controlling role in gene expression pattern of mitochondria [4–7]. During the integration of the endosymbiont into the present-day mitochondrial genome, most mitochondrial genetic content was transferred to the nucleus of the host cell [8]. While plant mitochondria are larger than animal genomes, they retain only a minor portion of their ancestral genomes. In *Arabidopsis thaliana*, the mitochondrial genome consists of 57 mitochondrial genes encoding for subunits of the respiratory chain, and the cytochrome maturation complexes, 40 ribosomal proteins, tRNAs, and rRNAs have been reported [2,9,10]. Thus, thousands of originally mitochondrial genes are now expressed under central control of the nucleus, and their protein products are subsequently imported into the mitochondria. As a result, mitochondrial biogenesis relies on the coordinated expression of organellar and nuclear genomes [2,9,11].

49

The mitochondrial transcription termination factor (mTERF) protein family is a key player affecting gene expression in plastid and mitochondrial genomes [12]. mTERF proteins are characterized by a modular architecture consisting of tandem repeats of a conserved 30-amino acid sequence, known as the mTERF motif [12]. In animals, the mTERF family has only four members, mTERF1–4, which are all localized to the mitochondria [13]. By contrast, more than 30 different mTERF proteins are widely distributed across plant nuclear genomes [14].

Thirty-five mTERF proteins have been identified in *Arabidopsis*, and mutations in several of these proteins have previously been associated with defects in development or stress responses [14,15]. For example, deficiencies in mTERF1/SOLDAT10, mTERF4/BSM/RUG2, mTERF5/MDA1, mTERF6, mTERF9/TWIRT1, mTERF15, or mTERF18/SHOT1 block development [16–21], while deficiencies in mTERF5/MDA1, mTERF6, mTERF9/TWIRT1, mTERF10, mTERF11, and mTERF18/SHOT1 affect responses to various abiotic stresses [18,22–24]. Most mTERF proteins in plants target the mitochondria and chloroplasts, playing an active part in organellar gene expression and RNA transcription [14,25]. For example, mTERF15 participates in mitochondrial intron splicing, mTERF22 affects the expression of many mitochondrial genes [20,26], and mTERF18/SHOT1 influences the steady-state abundance of various mitochondrial transcripts [18]. In *Chlamydomonas reindhardtii*, the mTERF-like protein MOC1 promotes the termination of anti-sense mitochondrial transcription [27]. However, even though plants have more mTERFs than mammals, the functional network of mTERFs in plants remains little known.

In addition to mTERFs, many other nuclear-encoded protein families, such as pentatricopeptide repeat (PPR) proteins and multiple organellar RNA editing factor (MORF) proteins, also play vital roles in the transcriptional and post-transcriptional regulation of organellar gene expression [9]. Like mTERFs, PPRs, which are characterized by tandem repeats of a degenerate 35-amino acid motif, are a large group of eukaryote-specific nucleic acid binding proteins encoded by the nucleus; PPRs function as RNA-binding proteins and regulate the processing of chloroplastic and mitochondrial RNA [28]. PPRs have also recently been shown to participate in the tolerance of certain stressors, such as salt and other abiotic factors [29]. The *Arabidopsis* genome encodes hundreds of PPR proteins [28].

The *Arabidopsis* genome encodes 10 MORF proteins (also termed RNA-editing factor interacting proteins, RIPs) [30,31]: MORF2, MORF9, and MORF10 are located in plastids; MORF1, MORF3, MORF4, MORF6, and MORF7 are located in mitochondria; and MORF5 and MORF8 are found in both organelles [32]. MORF deficiencies were shown to affect plant development and RNA editing at multiple sites in both mitochondria and plastids, many of which are associated with different individual PPR proteins [30–32]. MORFs have also been shown to interact with various PPR proteins, and they can form both homo- and heteromers [30–32]. Some MORFs may participate in the response to salt and other abiotic stresses, as well as the development of mitochondria and chloroplasts [33].

In this study, we identify and functionally characterize mTERF27 in *Arabidopsis thaliana*. Cell fluorescence imaging analyses showed that mTERF27 was localized to the mitochondria. Loss of mTERF27 decreased salt tolerance. Quantitative real-time PCR (qRT-PCR) and transmission electron microscope (TEM) analyses showed that defects in mTERF27 compromised mitochondrial gene expression and development under salt stress. Finally, we explored the direct interaction between mTERF27 and MORF8, and showed that mTERF27 interacted with itself to form homomers.

2. Results

2.1. mTERF27 Is a Mitochondria-Localized mTERF Protein

Land plant genomes have considerably larger numbers of mTERF proteins than other eukaryotes; *Arabidopsis* encodes at least 35 mTERF proteins [12,15]. Nonetheless, few *Arabidopsis* mTERF genes have been characterized in detail. *AT1G21150* (*mTERF27* [15]) is one of the previously unreported mTERF genes. Domain architecture analysis using SMART (http://smart.embl-heidelberg.de) indicated that mTERF27 carried four mTERF motifs

(Figure 1A). The homologous features of the *mTERF27* gene sequence were identified using Dicots PLAZA 4.0 (https://bioinformatics.psb.ugent.be/plaza/versions/plaza_v4_dicots/). From genomes of 55 species contained in Dicots PLAZA 4.0 database, 52 BHI (Best-Hits-and-Inparalogs) orthologs of mTERF27 were found (Figure S1A). The phylogenetic tree, constructed based on the 52 orthologs of *mTERF27*, indicated that *mTERF27* were relatively well-conserved across the plant kingdom, from *Physcomitrella patens* to higher plants (Figure S1B).

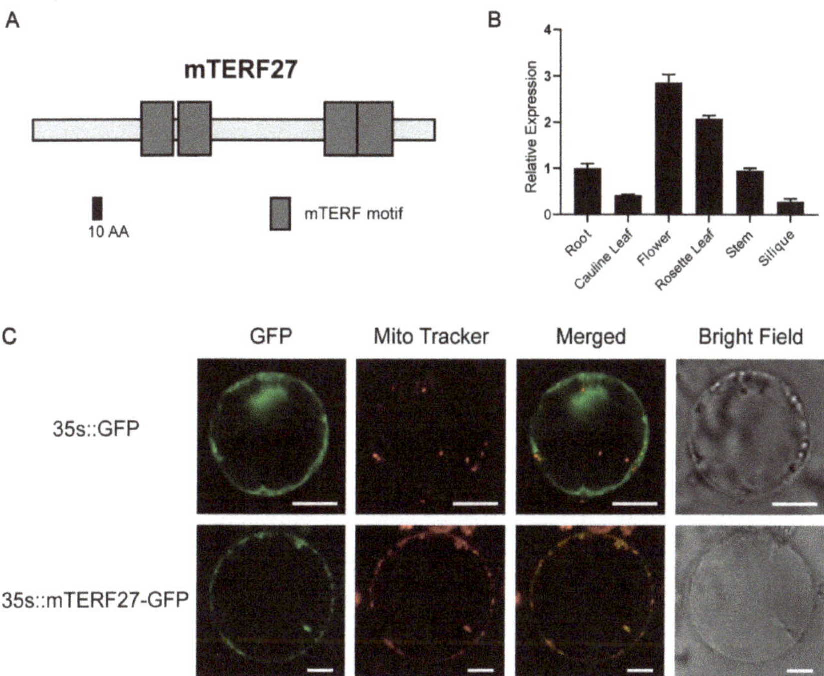

Figure 1. Modular architecture, expression pattern, and subcellular localization of mTERF27. (**A**) Schematic representation of the mTERF27 protein, drawn using SMART. mTERF motifs are shown as gray boxes. (**B**) Tissue-specific expression patterns of mTERF27, determined using qRT-PCR. Data are shown as means ± SD, n = 3. *ACTIN2* was used as an internal control. (**C**) Transient expression of the vectors *35s::GFP* and *35s::mTERF27-GFP* in *Arabidopsis* protoplasts. Mitochondrial locations are shown using MitoTracker Red (Invitrogen). Scale bars = 10 μm.

To investigate the gene expression patterns of *mTERF27* in *Arabidopsis*, we assessed the expression of *mTERF27* in various plant organs using qRT-PCR. The *mTERF27* gene was constitutively expressed in all tissues and organs examined, and the expression level was relatively higher in rosette leaves and flower (Figure 1B).

Most *Arabidopsis* mTERF proteins are found in mitochondria and/or chloroplasts; the mitochondrial localization of the mTERF27 had been preliminarily detected in the guard cells of *Arabidopsis*. [17]. To verify this result, we constructed a vector containing the full length CDS of *mTERF27* fused to GFP. Then, this vector (*35s::mTERF27-GFP*) and the empty control vector (*35s::GFP*) were transiently expressed in separate *Arabidopsis* protoplasts. Florescent localization analysis indicated that mTERF27-GFP signals overlapped well with the MitoTracker signal corresponding to mitochondria (Figure 1C). These results indicated that *mTERF27* encodes a mitochondria localized mTERF protein.

2.2. Disruption of mTERF27 Reduced Arabidopsis Resistance to Salt Stress

To explore the cis-acting regulatory elements of *mTERF27*, we used PlantCARE (http://bioinformatics.psb.ugent.be/webtools/plantcare/html/) to analyze the promoter region 2000 bp upstream of the *mTERF27* start codon. As shown in Figure 2A, the existence of cis-acting regulatory elements such as the abscisic acid responsive element, MeJA responsive element, auxin responsive element, salicylic acid responsive element, light responsive element, and a drought-responsive element implied that *mTERF27* might be involved in abiotic stress response. Furthermore, recent studies have suggested that plant mTERFs may play a role in the response to various abiotic stresses, such as salt [33].

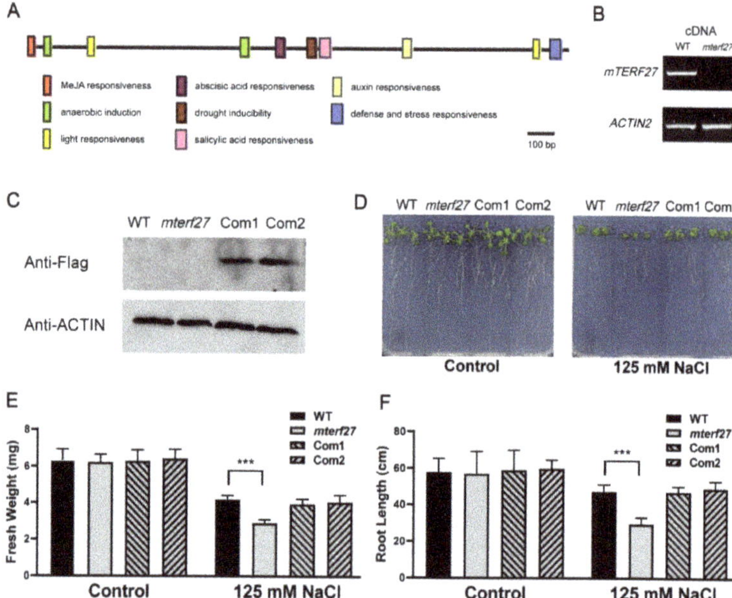

Figure 2. Identification of the *Arabidopsis mterf27* mutant and phenotypes of WT and *mterf27* seedlings under salt stress treatment. (**A**) Cis-acting regulatory elements of mTERF27 analyzed by PlantCARE. The bar represents 100 bp of nucleic acids. (**B**) PCR amplifications showing *mTERF27* expression in WT and *mterf27* plants. *ACTIN2* was used as an internal control. cDNA, complementary DNA. (**C**) Western blots verifying mTERF27-Flag expression in *mTERF27*-complemented lines (Com1 and Com2) based on total proteins extracted using the anti-Flag antibody. Anti-β-actin was used as internal control. (**D**) Fourteen-day-old WT, *mterf27*, Com1, and Com2 plants grown for 10 days in either 1/2 MS medium (Control) or 1/2 MS medium supplemented with 125 mM NaCl. (**E**) Fresh weights and (**F**) root lengths of the WT, *mterf27*, Com1, and Com2 plants shown in Figure 2D. Data shown are means ± SD of three independent experiments. Asterisks show significant differences compared to the WT: ***, $p < 0.001$ (Student's *t* test).

To investigate the phenotype of the loss-of-function mutant of mTERF27, we obtained one transfer DNA (T-DNA) insertion line from the Arabidopsis Biological Resource Center (https://abrc.osu.edu): SAIL_902, which putatively carries a T-DNA insertion in the *AT1G21150* (*mTERF27*) gene. PCR and sequencing analyses confirmed the T-DNA insertion site of the SAIL_902 mutant (here referred to as the *mterf27* mutant) (Figure S2A,B). The 1009 bp T-DNA was inserted in the exon of *AT1G21150* (position 7407135 in chromosome 1) in the mutant line (Figure S2C). RT-PCR analyses showed that *mTERF27* transcripts were absent in the homozygous mutant lines but present in the wild-type lines (Figure 2B).

The homozygous mutant plants displayed a wild-type (WT)-like phenotype under control growth conditions (Figure 2D, left), and loss of *mTERF27* did not affect photosynthetic activity (Figure S3). To generate lines complementing the *mTERF27* mutant, the coding region of *AT1G21150* was fused with the Flag tag, and the resulting construct (*35s::mTERF27-Flag*) was introduced into homozygous *mterf27* plants. Two independent T1 transgenic plants were identified as complemented lines (referred to as Com1 and Com2). Western-blot analyses indicated that the mTERF27-Flag protein was expressed in both complement lines (Figure 2C).

To clarify the function of mTERF27 in the response of *Arabidopsis* to salt stress, WT, *mterf27*, Com1, and Com2 plants were cultivated on 1/2 MS medium with or without 125 mM NaCl supplementation. Interestingly, when grown in salt-stressed conditions, the *mterf27* mutant displayed reduced growth compared to the WT (Figure 2D, right). To see the details, chlorophyll fluorescence which represents photosynthetic efficiency was examined. The results showed that loss of *mTERF27* did not affect photosynthetic activity even under salt-stressed conditions (Figure S3). In addition, the fresh weight and root length of WT, *mterf27*, Com1, and Com2 seedlings with or without salt stress were measured. While the fresh weights of WT, *mterf27*, Com1, and Com2 seedlings grown under control conditions did not differ significantly, the fresh weight of the *mterf27* seedlings was significantly lower than that of all other seedlings (Figure 2E). Similarly, the root lengths of WT, *mterf27*, Com1, and Com2 seedlings grown under control conditions did not differ significantly, but the root length of the *mterf27* seedlings was significantly lower than the root lengths of all other seedlings (Figure 2F). In these comparisons, Com1 and Com2 seedlings generally displayed phenotypes similar to that of the WT. Thus, we hypothesized mTERF27 participated in salt stress tolerance in *Arabidopsis*.

2.3. Disruption of mTERF27 Affected Mitochondrial Gene Transcription and Altered Mitochondrial Morphology

mTERF27 is a mitochondria-localized mTERF family protein, which may affect mito-chondrial gene transcription. To figure out if salt stress affected the accumulation of certain mitochondrial transcripts in *mterf27* plants, total RNA was extracted from WT and *mterf27* plants grown in control and salt-stressed conditions. RT-qPCR results showed that without salt stress, the transcript levels of *atp4* (subunit of complex V), *cob* (subunit of complex III), *cox1* (subunit of complex IV), *nad9* (subunit of complex I) and *rps12* (Ribosomal protein S12) in the *mterf27* mutant were lower than those in the WT plants, indicating that loss of mTERF27 led to a certain deficiency in the expression of some mitochondrial genes. (Figure 3A). After salt stress, the expression of *atp4*, *atp9* (subunit of complex V), *cox1*, *rps12* and *rrn26* was reduced in the WT plants, while these reductions were pronounced in the *mterf27* mutant plants (Figure 3A). These suggested that mTERF27 affected the expression of some mitochondrial genes in *Arabidopsis*, especially under salt stress.

Recent reports have indicated that deficiencies in many mTERF proteins influence the development of chloroplasts or mitochondria [29,33]. To evaluate the effects of *mTERF27* mutation on mitochondrial biogenesis, we examined the morphologies of mitochondria from the leaves of WT and *mterf27* plants grown in control or salt-stressed conditions using TEM. Under control growth conditions, TEM images showed that both WT and *mterf27* plants had normally structured mitochondrial cristae with small inner spaces (Figure 3B). In contrast, when WT and *mterf27* plants were grown under salt stress, the *mterf27* mito-chondria lacked cristae and had large internal space, while the WT mitochondria showed regular cristae (Figure 3B). This suggested that mTERF27 is required for mitochondrial development under salt stress.

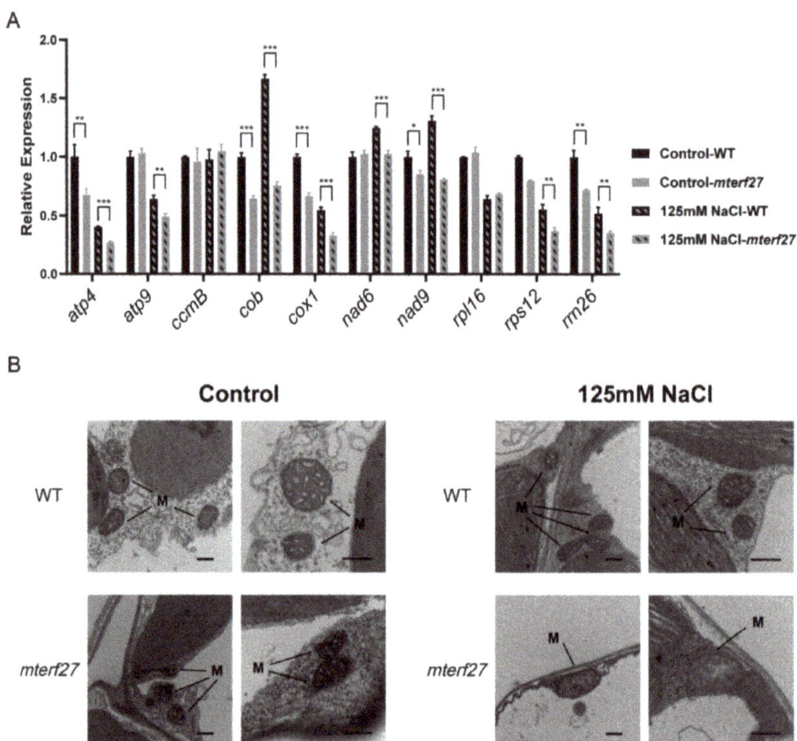

Figure 3. Mitochondrial gene expression levels and mitochondrial morphology in WT and *mterf27* plants. (**A**) Relative expression of mitochondrial genes in 14-day-old WT and *mterf27* plants were grown for 10 days in either 1/2 MS medium (Control) or 1/2 MS medium supplemented with 125 mM NaCl. Relative expression was measured using qRT-PCR, with *ACTIN2* as the internal control. Data represent means ± SD of three independent experiments. Asterisks show significant differences compared to the WT: ***, $p < 0.001$; **, $p < 0.01$; *, $p < 0.05$ (Student's *t* test). (**B**) TEM images of leaves from plants shown in (**A**). M, mitochondria. Scale bars = 0.5 μm.

2.4. mTERF27 Forms Homomers and Interacts with MORF8

Similar to mTERFs, PPRs are a large group of eukaryotic-specific nucleic acid binding proteins encoded by the nucleus, and *Arabidopsis* genome encodes hundreds of PPR proteins. Many PPRs and MORFs are known to interact [30–32], so we tested whether mTERF27 interacted with any MORFs in *Arabidopsis*. To investigate this, yeast two-hybrid assays were used to detect interactions between mTERF27 and mitochondria-localized MORFs (i.e., MORF1, MORF3, MORF4, MORF5, MORF6, MORF7 and MORF8). We also used yeast two-hybrid assays to determine whether mTERF27 formed homomers. Growth analyses in selective media (SD-T/-L/-H and SD-T/-L/-H/-A) showed that mTERF27 interacted with MORF8, but not with any other MORFs; mTERF27 also interacted with itself (Figure 4A and Figure S4). These interactions were validated in planta using firefly luciferase complementation imaging assays (Figure 4B). Yeast growth on selective medium and bioluminescence signals produced by the catalysis of luciferin partly reflected that mTERF27 has weaker interaction with MORF8, comparing with its homomer interaction.

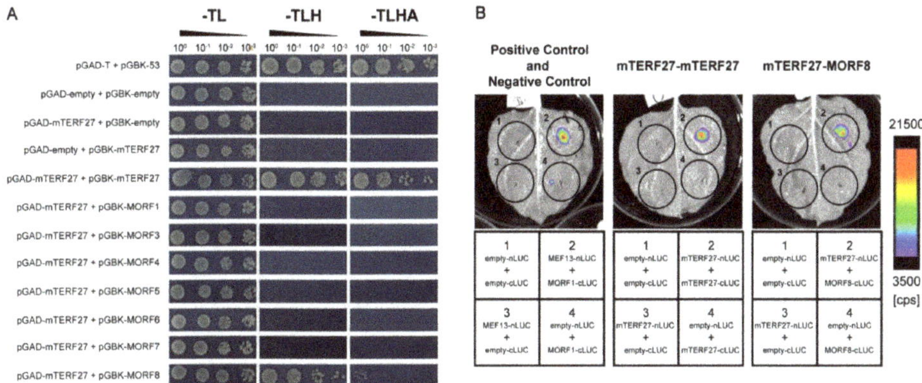

Figure 4. mTERF27 directly interacts with mTERF27 and MORF8. (**A**) Yeast two-hybrid assays showing interactions between mTERF27 and MORF8. pGAD, pGADT7 with the GAL4 activation domain; pGBK, pGBKT7 with the GAL4 DNA binding domain; -TL, SD/-Trp-Leu dropout medium; -TLH, SD/-Trp-Leu-His dropout medium; -TLHA, SD/-Trp-Leu-His-Ade dropout medium. The 53-T interaction were used as positive controls. (**B**) Firefly luciferase complementation assays showing mTERF27-mTERF27 and mTERF27-MORF8 interactions in planta. MEF13-MORF1 interaction were used as positive controls [34]. Color scale represents the luminescent signal intensity measured by cps (counts per second).

3. Discussion

Our results showed that the *mterf27* mutants were hypersensitive to salt stress. Under salt stress, the loss of mitochondria-localized mTERF27 disrupted mitochondrial development and caused defects in mitochondrial gene expression. Furthermore, mTERF27 directly interacted with MORF8 and possibly formed homomers.

The mTERF protein family shares several features with the PPR and MORF protein families. For example, PPR and mTERF proteins both harbor tandem repeats of a conserved domain [28], and mutations in both types of proteins have been linked to developmental defects or inhibited stress responses [33]. In addition, PPRs participate in organellar RNA metabolism and function as specific RNA-binding proteins [9]. Similarly, it has been reported that some mTERF proteins bind mitochondrial and chloroplastic nucleic acids in *Arabidopsis* [20,35–37].

MORFs may provide an ordered spatial connection between PPRs and other proteins [9,32]. Indeed, recent studies have indicated that the MORF8 protein, which is located in both the mitochondria and the chloroplasts [32], directly interacts with multiple PPRs in *Arabidopsis* [30,34,38]. MORF8 is also involved in the establishment and/or mediation of a direct or indirect connection between MEF13 (a PPR protein) and MORF3 [34]. Here, our results suggested that mTERF27 might interact with MORF8 to participate in mitochondrial gene expression and RNA metabolism (Figure 4A,B). Curiously, the interaction between mTERF27 and MORF8 is weaker than mTERF27 homomer interaction (Figure 4A,B). Similar to the model of MEF13-MORF1-MOEF3 interaction [34], other factors may affect the interaction between mTERF27 and MORF8 in mitochondria. However, there is no evidence to show the relation between the mTERF homomer interaction and mTERF-MORF interaction so far. The molecular mechanism remains unclear.

In plants, mutations in some mTERF genes lead to paleness, significant retardations in growth and development, and even arrested embryogenesis [17,20,21]. However, mutant lines defective in some other mTERFs exhibit less severe growth and developmental defects or display hypersensitivity to abiotic stress [22,26,37]. Previous studies showed that deficiency of mTERF5/MDA1, mTERF6, mTERF9/TWIRT1, mTERF10, and/or mTERF11 resulted in altered response to salt stress and/or ABA treatment in the mutants [19,22,23,39]. A recent study reported that mTERF9 and mTERF5 are negative regulators of salt tolerance, and have contributions to plastid gene expression and retrograde signaling in *Arabidopsis thaliana* [24]. Our results showed that similar to mTERF9 and mTERF5, mTERF27 is also involved in plant salt response and mitochondrial gene expression.

Although the expression level of some mitochondrial genes was lower in *mterf27* mutants compared with WT, there was no visible defects of *mterf27* mutant under normal growth conditions. Previous studies have indicated that mTERF7, mTERF22, and mTERF27 have close phylogenetic relationships [26]. Therefore, under normal growth conditions, mTERF7 and mTERF22 might complement the loss function of mTERF27 partly. However, mTERF27 may play a critical role in plant salt response. Under salt stress conditions, *mterf27* mutant displayed retarded growth phenotype. mTERFs may play diverse roles in organelles. Various pairs of transcription factors, including mTERFs, may have redundant or complementary function networks in plant mitochondria and chloroplasts. Indeed, some mTERF proteins in plants are involved in the transcription termination of chloroplast genes [35,37]. However, recent reports showed that plant mTERFs may have a more complicated mechanism in organelle gene expression [36,40–42].

The characterization of mTERF27 in this study helps to clarify plant organellar gene expression in response to salt and other abiotic stress. Our work also provides a basis for further analyses of the mTERF−MORF protein−protein interaction network and investigations of its functional relevance.

4. Materials and Methods

4.1. Plant Materials and Growth Conditions

Arabidopsis thaliana, ecotype Columbia-0 (Col-0), was grown in a growth chamber under 16 h of light at 22 °C and 8 h of darkness at 20 °C. To grow seedlings on agar plates, surface-sterilized seeds were planted on 1/2 MS medium containing 1.0% (*w/v*) sucrose and 0.8% (*w/v*) agar, cold-treated for 2 days, and transferred to a growth chamber. To test seedling salt tolerance, seeds were planted on normal 1/2 MS medium, cold treated for 2 days, grown for 2 days under normal conditions, transferred to either unmodified 1/2 MS medium (control growth conditions) or 1/2 MS medium supplemented with 125 mM NaCl (salt-stressed growth conditions), and then grown for 10 days in the growth chamber. To grow seedlings in soil, sown seeds were cold-treated for 2 days and then transferred to a green room under same growth conditions (16 h of light at 22 °C and 8 h of darkness at 20 °C). *Nicotiana benthamiana* was cultured in autoclaved vermiculite in a green room under a 16 h light/8 h dark photoperiod at 25 °C; 4–5-week-old plants were used for transient expression analysis.

4.2. Plant Transformation

The coding region of the *mTERF27* gene was amplified from total RNA using reverse-transcriptase PCR (RT-PCR). The resulting cDNA was cloned into vector pCAMBIA1300 to produce a construct expressing the Flag-tagged mTERF27 protein. *Agrobacterium tumefaciens* strain GV3101 was used for transformation. The constructs were transferred into *mterf27* mutants using the floral dip method, and transgenic plants were identified using hygromycin resistance analysis, PCR genotyping, and western blots.

4.3. Chlorophyll Fluorescence Measurements

Chlorophyll fluorescence imaging and analysis were performed using a chlorophyll imaging system (FluorCam FC 800-C/1010, PSI), with photosynthetic parameters determined as described previously [43]. Before each measurement, plants were dark-adapted for 20 min. The F_v/F_m ratio was defined as $(F_m − F_o)/Fm$. The nonphotochemical quenching (NPQ) was calculated as $(F_m − F'm)/F'_m$, where F_m is the maximum fluorescence value in the dark-adapted state; F'_m is the maximum fluorescence value in any light-adapted state; and F_o is the minimal fluorescence value in the dark-adapted state.

4.4. RNA Isolation and Quantitative Reverse-Transcriptase PCR (RT-qPCR)

Total RNA was isolated with the leaves of WT, *mterf27*, Com1, and Com2 plants using a RNeasy Plant Mini Kit (Qiagen). cDNA was synthesized using a PrimeScript RT reagent Kit with gDNA Eraser (Takara). qPCRs were performed with the 7300Plus real-time PCR system (ABI) using TB Green *Premix Ex Taq* II (Tli RNaseH Plus) (Takara). The *ACTIN2* gene was used as an endogenous control. Primers used to detect the mitochondrial transcripts were designed as previously described [26].

4.5. Protein Preparation and Western Blots

Total protein samples were prepared and western blots were performed following a previous study [43]. For immunoblotting analysis, we separated equal amounts of protein sample on 10% SDS PAGE gels and transferred them to nitrocellulose membranes. After blocking nonspecific binding with 5% milk, we subsequently incubated the blot with specific primary antibodies generated against the indicated proteins and secondary horseradish peroxidase conjugated antibodies (Abbkine). Signals were detected using the SuperSignal™ West Pico PLUS Chemiluminescent Substrate (Thermo Scientific) according to the manufacturer's protocol. The primary antibodies used were Anti-Flag (Sigma-Aldrich, #F3165) and Anti-β-actin (Abbkine, #A01050-2).

4.6. Microscopy

To transiently express *mTERF27* in *Arabidopsis* protoplasts, the full-length CDS of *mTERF27* was cloned into pHBT-sGFP plasmids as previously described [44]. Mesophyll protoplasts were extracted from 4-week-old darkness-treated *Arabidopsis* leaves and transformed with the GFP plasmids as previously described [44]. MitoTracker Red (Invitrogen) was used to specifically dye the mitochondria. The organelle and GFP signals were detected with a confocal microscope (TCS SP8, Leica). The excitation and emission wavelengths were as follows: GFP, excitation at 488 nm and emission at 510–540 nm; MitoTracker, excitation at 644 nm and emission at 650–680 nm.

For TEM analysis, leaves from WT and *mterf27* seedlings were prepared as described previously [45]. Leaves were observed and imaged using an HT7800 Compact-Digital TEM system (Hitachi).

4.7. Protein Interaction Assays

Yeast two-hybrid assays were performed using the yeast strain Y2H Gold (Clontech), following the manufacturer's instructions. For construction of the Gateway entry clones, PCR products were inserted into pDONR207 via BP reactions (Gateway BP clonase enzyme mix; Invitrogen), then cloned into the expression vectors (pGBKT7 or pGADT7) which contain the attR1-CmR-ccdB-attR2 fragment via LR reactions (Gateway LR clonase enzyme mix; Invitrogen) as previously described [46]. These vectors were transformed into Y2H Gold yeast (Clontech). The transformants were grown on SD/-Trp-Leu, SD/-Trp-Leu-His, and SD/-Trp-Leu-His-Ade dropout selective culture-media.

To perform firefly luciferase complementation imaging assays, the coding regions of the target genes were fused with either nLUC or cLUC and cloned into the pCAMBIA1300 vector as previously described [47]. These vectors were transformed into *A. tumefaciens*. The positive clones were injected into *N. benthamiana* as previously described [47], the bioluminescent signals were detected by NightSHADE LB985 system (Berthold).

Supplementary Materials: The following are available online at https://www.mdpi.com/1422-0067/22/3/1466/s1, Figure S1. Amino acid sequences alignment and Neighbor-joining phylogenetic tree of mTERF27 ortholog genes in plants; Figure S2. Sequences flanking the transfer DNA (T-DNA) insertion point in the *mterf27* mutant; Figure S3. Chlorophyll fluorescence analysis of the wild-type (WT) and *mterf27* plants shown in Figure 2D; Figure S4. Negative controls for the yeast two-hybrid assay used to determine protein interactions; Table S1. Primers used in this study.

Author Contributions: D.J. and X.H. designed the research; D.J. and J.C. performed the research; D.J., J.C., and X.H. analyzed the data; D.J., Z.Z., and X.H. wrote the manuscript. All authors have read and agreed to the published version of the manuscript.

Funding: This work was supported by the National Natural Science Foundation of China (31570238).

Institutional Review Board Statement: Not applicable.

Informed Consent Statement: Not applicable.

Data Availability Statement: Data are available on request to the corresponding author.

Conflicts of Interest: The authors declare no conflict of interest.

References

1. Millar, A.H.; Whelan, J.; Soole, K.L.; Day, D.A. Organization and regulation of mitochondrial respiration in plants. *Annu. Rev. Plant. Biol.* **2011**, *62*, 79–104. [CrossRef] [PubMed]
2. Hammani, K.; Giege, P. RNA metabolism in plant mitochondria. *Trends Plant. Sci* **2014**, *19*, 380–389. [CrossRef] [PubMed]
3. Gualberto, J.M.; Mileshina, D.; Wallet, C.; Niazi, A.K.; Weber-Lotfi, F.; Dietrich, A. The plant mitochondrial genome: Dynamics and maintenance. *Biochimie* **2014**, *100*, 107–120. [CrossRef] [PubMed]
4. Binder, S.; Brennicke, A. Gene expression in plant mitochondria: Transcriptional and post-transcriptional control. *Philos. Trans. R Soc. Lond. B Biol. Sci.* **2003**, *358*, 181–188. [CrossRef]
5. Brown, G.G.; Colas des Francs-Small, C.; Ostersetzer-Biran, O. Group II intron splicing factors in plant mitochondria. *Front. Plant. Sci.* **2014**, *5*, 35. [CrossRef]
6. Small, I.D.; Schallenberg-Rudinger, M.; Takenaka, M.; Mireau, H.; Ostersetzer-Biran, O. Plant organellar RNA editing: What 30 years of research has revealed. *Plant. J.* **2020**, *101*, 1040–1056. [CrossRef]
7. Marchetti, F.; Cainzos, M.; Shevtsov, S.; Cordoba, J.P.; Sultan, L.D.; Brennicke, A.; Takenaka, M.; Pagnussat, G.; Ostersetzer-Biran, O.; Zabaleta, E. Mitochondrial Pentatricopeptide Repeat Protein, EMB2794, Plays a Pivotal Role in NADH Dehydrogenase Subunit nad2 mRNA Maturation in Arabidopsis thaliana. *Plant. Cell Physiol.* **2020**, *61*, 1080–1094. [CrossRef]
8. Gray, M.W.; Burger, G.; Lang, B.F. Mitochondrial evolution. *Science* **1999**, *283*, 1476–1481. [CrossRef]
9. Zmudjak, M.; Ostersetzer Biran, O. RNA metabolism and transcript regulation. *Annu. Plant. Rev. Online* **2018**, 143–184. [CrossRef]
10. Ostersetzer-Biran, O. Respiratory complex I and embryo development. *J. Exp. Bot* **2016**, *67*, 1205–1207. [CrossRef]
11. Best, C.; Mizrahi, R.; Ostersetzer-Biran, O. Why so Complex? The Intricacy of Genome Structure and Gene Expression, Associated with Angiosperm Mitochondria, May Relate to the Regulation of Embryo Quiescence or Dormancy-Intrinsic Blocks to Early Plant Life. *Plants* **2020**, *9*, 598. [CrossRef] [PubMed]
12. Linder, T.; Park, C.B.; Asin-Cayuela, J.; Pellegrini, M.; Larsson, N.G.; Falkenberg, M.; Samuelsson, T.; Gustafsson, C.M. A family of putative transcription termination factors shared amongst metazoans and plants. *Curr. Genet.* **2005**, *48*, 265–269. [CrossRef] [PubMed]
13. Roberti, M.; Polosa, P.L.; Bruni, F.; Deceglie, S.; Gadaleta, M.N.; Cantatore, P. MTERF factors: A multifunction protein family. *Biomol. Concepts* **2010**, *1*, 215–224. [CrossRef] [PubMed]
14. Kleine, T.; Leister, D. Emerging functions of mammalian and plant mTERFs. *Biochim. Biophys. Acta* **2015**, *1847*, 786–797. [CrossRef] [PubMed]
15. Kleine, T. Arabidopsis thaliana mTERF proteins: Evolution and functional classification. *Front. Plant. Sci* **2012**, *3*, 233. [CrossRef]
16. Meskauskiene, R.; Wuersch, M.; Laloi, C.; Vidi, P.-A.; Coll, N.S.; Kessler, F.; Baruah, A.; Kim, C.; Apel, K. A mutation in the Arabidopsis mTERF-related plastid protein SOLDAT10 activates retrograde signaling and suppresses 1O(2)-induced cell death. *Plant. J.* **2009**, *60*, 399–410. [CrossRef]
17. Babiychuk, E.; Vandepoele, K.; Wissing, J.; Garcia-Diaz, M.; De Rycke, R.; Akbari, H.; Joubes, J.; Beeckman, T.; Jaensch, L.; Frentzen, M.; et al. Plastid gene expression and plant development require a plastidic protein of the mitochondrial transcription termination factor family. *Proc. Natl. Acad. Sci. USA* **2011**, *108*, 6674–6679. [CrossRef]
18. Kim, M.; Lee, U.; Small, I.; des Francs-Small, C.C.; Vierling, E. Mutations in an Arabidopsis mitochondrial transcription termination factor-related protein enhance thermotolerance in the absence of the major molecular chaperone HSP101. *Plant. Cell* **2012**, *24*, 3349–3365. [CrossRef]
19. Robles, P.; Micol, J.L.; Quesada, V. Arabidopsis MDA1, a Nuclear-Encoded Protein, Functions in Chloroplast Development and Abiotic Stress Responses. *PLoS ONE* **2012**, *7*. [CrossRef]
20. Hsu, Y.-W.; Wang, H.-J.; Hsieh, M.-H.; Hsieh, H.-L.; Jauh, G.-Y. Arabidopsis mTERF15 Is Required for Mitochondrial nad2 Intron 3 Splicing and Functional Complex I Activity. *PLoS ONE* **2014**, *9*. [CrossRef]
21. Romani, I.; Manavski, N.; Morosetti, A.; Tadini, L.; Maier, S.; Kuhn, K.; Ruwe, H.; Schmitz-Linneweber, C.; Wanner, G.; Leister, D.; et al. A Member of the Arabidopsis Mitochondrial Transcription Termination Factor Family Is Required for Maturation of Chloroplast Transfer RNAIle(GAU). *Plant. Physiol.* **2015**. [CrossRef] [PubMed]
22. Xu, D.; Leister, D.; Kleine, T. Arabidopsis thaliana mTERF10 and mTERF11, but Not mTERF12, Are Involved in the Response to Salt Stress. *Front. Plant. Sci.* **2017**, *8*, 1213. [CrossRef] [PubMed]

23. Robles, P.; Navarro-Cartagena, S.; Ferrandez-Ayela, A.; Nunez-Delegido, E.; Quesada, V. The Characterization of Arabidopsis mterf6 Mutants Reveals a New Role for mTERF6 in Tolerance to Abiotic Stress. *Int. J. Mol. Sci.* **2018**, *19*, 2388. [CrossRef] [PubMed]

24. Nunez-Delegido, E.; Robles, P.; Ferrandez-Ayela, A.; Quesada, V. Functional analysis of mTERF5 and mTERF9 contribution to salt tolerance, plastid gene expression and retrograde signalling in Arabidopsis thaliana. *Plant. Biol.* **2020**, *22*, 459–471. [CrossRef]

25. Robles, P.; Micol, J.L.; Quesada, V. Unveiling Plant mTERF Functions. *Molecular Plant.* **2012**, *5*, 294–296. [CrossRef]

26. Shevtsov, S.; Nevo-Dinur, K.; Faigon, L.; Sultan, L.D.; Zmudjak, M.; Markovits, M.; Ostersetzer-Biran, O. Control of organelle gene expression by the mitochondrial transcription termination factor mTERF22 in Arabidopsis thaliana plants. *PLoS ONE* **2018**, *13*, e0201631. [CrossRef]

27. Wobbe, L.; Nixon, P.J. The mTERF protein MOC1 terminates mitochondrial DNA transcription in the unicellular green alga Chlamydomonas reinhardtii. *Nucleic Acids Res.* **2013**, *41*, 6553–6567. [CrossRef]

28. Barkan, A.; Small, I. Pentatricopeptide repeat proteins in plants. *Annu. Rev. Plant. Biol* **2014**, *65*, 415–442. [CrossRef]

29. Robles, P.; Quesada, V. Transcriptional and Post-transcriptional Regulation of Organellar Gene Expression (OGE) and Its Roles in Plant Salt Tolerance. *Int. J. Mol. Sci.* **2019**, *20*, 1056. [CrossRef]

30. Bentolila, S.; Heller, W.P.; Sun, T.; Babina, A.M.; Friso, G.; van Wijk, K.J.; Hanson, M.R. RIP1, a member of an Arabidopsis protein family, interacts with the protein RARE1 and broadly affects RNA editing. *Proc. Natl. Acad. Sci. USA* **2012**, *109*, E1453–E1461. [CrossRef]

31. Takenaka, M.; Zehrmann, A.; Verbitskiy, D.; Kugelmann, M.; Hartel, B.; Brennicke, A. Multiple organellar RNA editing factor (MORF) family proteins are required for RNA editing in mitochondria and plastids of plants. *Proc. Natl. Acad. Sci. USA* **2012**, *109*, 5104–5109. [CrossRef] [PubMed]

32. Zehrmann, A.; Hartel, B.; Glass, F.; Bayer-Csaszar, E.; Obata, T.; Meyer, E.; Brennicke, A.; Takenaka, M. Selective homo- and heteromer interactions between the multiple organellar RNA editing factor (MORF) proteins in Arabidopsis thaliana. *J. Biol. Chem.* **2015**, *290*, 6445–6456. [CrossRef] [PubMed]

33. Quesada, V. The roles of mitochondrial transcription termination factors (MTERFs) in plants. *Physiol Plant.* **2016**, *157*, 389–399. [CrossRef] [PubMed]

34. Glass, F.; Haertel, B.; Zehrmann, A.; Verbitskiy, D.; Takenaka, M. MEF13 Requires MORF3 and MORF8 for RNA Editing at Eight Targets in Mitochondrial mRNAs in Arabidopsis thaliana. *Mol. Plant.* **2015**, *8*, 1466–1477. [CrossRef] [PubMed]

35. Zhang, Y.; Cui, Y.L.; Zhang, X.L.; Yu, Q.B.; Wang, X.; Yuan, X.B.; Qin, X.M.; He, X.F.; Huang, C.; Yang, Z.N. A nuclear-encoded protein, mTERF6, mediates transcription termination of rpoA polycistron for plastid-encoded RNA polymerase-dependent chloroplast gene expression and chloroplast development. *Sci. Rep.* **2018**, *8*, 11929. [CrossRef] [PubMed]

36. Ding, S.; Zhang, Y.; Hu, Z.; Huang, X.; Zhang, B.; Lu, Q.; Wen, X.; Wang, Y.; Lu, C. mTERF5 Acts as a Transcriptional Pausing Factor to Positively Regulate Transcription of Chloroplast psbEFLJ. *Mol. Plant.* **2019**, *12*, 1259–1277. [CrossRef] [PubMed]

37. Xiong, H.B.; Wang, J.; Huang, C.; Rochaix, J.D.; Lin, F.M.; Zhang, J.X.; Ye, L.S.; Shi, X.H.; Yu, Q.B.; Yang, Z.N. mTERF8, a Member of the Mitochondrial Transcription Termination Factor Family, Is Involved in the Transcription Termination of Chloroplast Gene psbJ. *Plant. Physiol.* **2020**, *182*, 408–423. [CrossRef]

38. Brehme, N.; Bayer-Császár, E.; Glass, F.; Takenaka, M. The DYW subgroup PPR protein MEF35 targets RNA editing sites in the mitochondrial rpl16, nad4 and cob mRNAs in Arabidopsis thaliana. *PLoS ONE* **2015**, *10*, e0140680. [CrossRef]

39. Robles, P.; Micol, J.L.; Quesada, V. Mutations in the plant-conserved MTERF9 alter chloroplast gene expression, development and tolerance to abiotic stress in Arabidopsis thaliana. *Physiol. Plant.* **2015**, *154*, 297–313. [CrossRef]

40. Alamdari, K.; Fisher, K.E.; Sinson, A.B.; Chory, J.; Woodson, J.D. Roles for the chloroplast-localized pentatricopeptide repeat protein 30 and the 'mitochondrial' transcription termination factor 9 in chloroplast quality control. *Plant. J.* **2020**, *104*, 735–751. [CrossRef]

41. Méteignier, L.V.; Ghandour, R.; Meierhoff, K.; Zimmerman, A.; Chicher, J.; Baumberger, N.; Alioua, A.; Meurer, J.; Zoschke, R.; Hammani, K. The Arabidopsis mTERF-repeat MDA1 protein plays a dual function in transcription and stabilization of specific chloroplast transcripts within the psbE and ndhH operons. *New Phytol.* **2020**, *227*, 1376–1391. [CrossRef] [PubMed]

42. Meteignier, L.V.; Ghandour, R.; Zimmerman, A.; Kuhn, L.; Meurer, J.; Zoschke, R.; Hammani, K. Arabidopsis mTERF9 protein promotes chloroplast ribosomal assembly and translation by establishing ribonucleoprotein interactions in vivo. *Nucleic Acids Res.* **2021**. [CrossRef] [PubMed]

43. Hou, X.; Fu, A.; Garcia, V.J.; Buchanan, B.B.; Luan, S. PSB27: A thylakoid protein enabling Arabidopsis to adapt to changing light intensity. *Proc. Natl. Acad. Sci. USA* **2015**, *112*, 1613–1618. [CrossRef] [PubMed]

44. Zhang, Q.; Xu, Y.; Huang, J.; Zhang, K.; Xiao, H.; Qin, X.; Zhu, L.; Zhu, Y.; Hu, J. The Rice Pentatricopeptide Repeat Protein PPR756 Is Involved in Pollen Development by Affecting Multiple RNA Editing in Mitochondria. *Front. Plant. Sci.* **2020**, *11*, 749. [CrossRef]

45. Yi, B.; Zeng, F.; Lei, S.; Chen, Y.; Yao, X.; Zhu, Y.; Wen, J.; Shen, J.; Ma, C.; Tu, J.; et al. Two duplicate CYP704B1-homologous genes BnMs1 and BnMs2 are required for pollen exine formation and tapetal development in Brassica napus. *Plant. J.* **2010**, *63*, 925–938. [CrossRef]

46. Jiang, D.; Tang, R.; Shi, Y.; Ke, X.; Wang, Y.; Che, Y.; Luan, S.; Hou, X. Arabidopsis Seedling Lethal 1 Interacting With Plastid-Encoded RNA Polymerase Complex Proteins Is Essential for Chloroplast Development. *Front. Plant. Sci.* **2020**, *11*, 602782. [CrossRef]
47. Qiao, J.; Li, J.; Chu, W.; Luo, M. PRDA1, a novel chloroplast nucleoid protein, is required for early chloroplast development and is involved in the regulation of plastid gene expression in Arabidopsis. *Plant. Cell Physiol.* **2013**, *54*, 2071–2084. [CrossRef]

International Journal of
Molecular Sciences

Review

Magnesium Signaling in Plants

Leszek A. Kleczkowski [1,*] and **Abir U. Igamberdiev** [2]

[1] Department of Plant Physiology, Umeå Plant Science Centre, University of Umeå, 901 87 Umeå, Sweden
[2] Department of Biology, Memorial University of Newfoundland, St. John's, NL A1B3X9, Canada; igamberdiev@mun.ca
* Correspondence: leszek.kleczkowski@umu.se

Abstract: Free magnesium (Mg^{2+}) is a signal of the adenylate (ATP+ADP+AMP) status in the cells. It results from the equilibrium of adenylate kinase (AK), which uses Mg-chelated and Mg-free adenylates as substrates in both directions of its reaction. The AK-mediated primary control of intracellular $[Mg^{2+}]$ is finely interwoven with the operation of membrane-bound adenylate- and Mg^{2+}-translocators, which in a given compartment control the supply of free adenylates and Mg^{2+} for the AK-mediated equilibration. As a result, $[Mg^{2+}]$ itself varies both between and within the compartments, depending on their energetic status and environmental clues. Other key nucleotide-utilizing/producing enzymes (e.g., nucleoside diphosphate kinase) may also be involved in fine-tuning of the intracellular $[Mg^{2+}]$. Changes in $[Mg^{2+}]$ regulate activities of myriads of Mg-utilizing/requiring enzymes, affecting metabolism under both normal and stress conditions, and impacting photosynthetic performance, respiration, phloem loading and other processes. In compartments controlled by AK equilibrium (cytosol, chloroplasts, mitochondria, nucleus), the intracellular $[Mg^{2+}]$ can be calculated from total adenylate contents, based on the dependence of the apparent equilibrium constant of AK on $[Mg^{2+}]$. Magnesium signaling, reflecting cellular adenylate status, is likely widespread in all eukaryotic and prokaryotic organisms, due simply to the omnipresent nature of AK and to its involvement in adenylate equilibration.

Keywords: adenylate energy charge; adenylate kinase; cellular magnesium; free magnesium; nucleoside diphosphate kinase; thermodynamic buffering

Citation: Kleczkowski, L.A.; Igamberdiev, A.U. Magnesium Signaling in Plants. *Int. J. Mol. Sci.* **2021**, *22*, 1159. https://doi.org/10.3390/ijms22031159

Academic Editors: Jen-Tsung Chen and Parviz Heidari
Received: 13 December 2020
Accepted: 19 January 2021
Published: 25 January 2021

Publisher's Note: MDPI stays neutral with regard to jurisdictional claims in published maps and institutional affiliations.

1. Introduction

Magnesium (Mg) is one of the most abundant cations in living cells, second only to potassium [1]. Total Mg concentration in plant cells is in the range of 15–25 mM, and most of it is stored in vacuoles, away from metabolism [2]. A substantial pool of total cellular Mg is required to synthesize chlorophyll in photosynthetic tissues, and the rest is used for ribosome bridging during translation and for chelation with nucleotides, nucleic acids and other phosphate-containing compounds. Normally, as much as 20% of total Mg is in chloroplasts, but it may increase to 50% under low light conditions or during Mg deficiency [3]. Many enzymes require the binding of Mg for activity and/or regulation [4,5]. Processes such as phloem loading [6–8], leaf senescence [9], stomata opening and ionic balance of the cell [10,11] are only a few of many examples that illustrate the requirement for adequate Mg homeostasis.

Chelation of nucleotides by Mg is an essential feature of cell metabolism. Among nucleotides, adenylates (ATP, ADP, and AMP) are the most abundant, with ATP being produced both during the reductive (photosynthetic light reactions) and oxidative (respiration) phosphorylation [12]. Most, if not all, enzymes which require ATP in fact use it in its chelated form (MgATP). Assuming that, in leaves, the concentration of free Mg (Mg^{2+}) is in the order of 1–5 mM [13], 95–99% of ATP is predicted to be complexed as MgATP. A similar situation occurs with other nucleoside triphosphates (NTPs) [14]. In some compartments (e.g., cytosol) and under specific physiological conditions, the internal $[Mg^{2+}]$ may decrease well below 1 mM, e.g., to 0.2 mM in chloroplasts during the induction phase of photosynthesis (see below). Under these conditions, the proportion of Mg-chelated NTPs decreases. The same rule applies to ADP and other nucleoside diphosphates (NDPs), and to a certain extent to AMP and other nucleoside monophosphates (NMPs).

In many cases, it is MgADP rather than free ADP which is a substrate/product in enzymatic reactions, whereas AMP binds Mg very weakly, and thus is used in metabolism predominantly as free AMP (Figure 1A). The binding of ATP and Mg is very tight, with stability constant (K_{MgATP}) of 73 mM^{-1}, which implies that a 1:1 (molar ratio) mixture of Mg and ATP will result in nearly all of Mg and ATP tied up as MgATP. The binding of ADP (or other NDPs) by Mg is less strong, with stability constant (K_{MgADP}) of 4 mM^{-1}. This implies that, in physiological conditions, roughly 30% (or less) of ADP exists in the cell as free ADP, and 70% (or more) as a complex with Mg (MgADP) (Figure 1A) [15]. The uncomplexed forms of both ATP and ADP may cause appreciable inhibition of many enzymes, especially the kinase-type phosphotransferases [16]. The binding of Mg to inorganic pyrophosphate (PP$_\mathrm{i}$), another high energy storing compound, is also highly dependent on [Mg^{2+}] (Figure 1B).

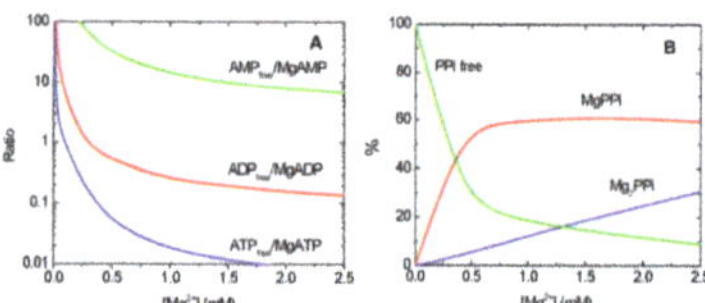

Figure 1. Effects of Mg^{2+} on Mg-chelation with adenylates and PP$_\mathrm{i}$. (**A**) The ratios of free and Mg-bound adenylates, depending on [Mg^{2+}]. (**B**) The percentage of free and Mg-chelated PP$_\mathrm{i}$, depending on [Mg^{2+}]. All lines were drawn according to the values of stability constants for chelation of adenylates and PP$_\mathrm{i}$ with Mg [15], using Origin software (OriginLab Corporation, Northampton, MA, USA).

Both Mg-complexed and free adenylates are equilibrated by adenylate kinase (AK), a ubiquitous enzyme present in all organisms. The result of AK-maintained equilibrium, [Mg^{2+}], is set as a signal from total adenylate pool in a given cellular compartment. The [Mg^{2+}], in turn, depends on metabolic status of the cell, and on rates of transport of Mg^{2+} and adenylates on specific transporters (see below). AK in different cell compartments establishes the concentration of [Mg^{2+}] via equilibration of adenylate species, as well as membrane potential of cell organelles [17], and the equilibrium of AK determines [Mg^{2+}] in a controlled way depending on the rates of ATP production and consumption and, in turn, optimizing these rates. This [Mg^{2+}] directly regulates multiple enzymes and translocators, thus representing a powerful feedback signal from the energy level of the cell and its compartments expressed in the concentrations of adenylate species.

Mg^{2+} has previously been suggested as a signal in human T cell immunodeficiency, with the plasma membrane-bound Mg^{2+} transporter identified as a major player [18]. We argue here that the Mg^{2+} signaling is in fact widespread in plants and, probably, in other organisms as well, and that it is a simple consequence of the status of Mg-free and Mg-complexed adenylates in a given cell or tissue, which is maintained via AK. AK has previously been implicated as a central hub in the concept of "adenylate charge" theory [19], but also as a regulator of AMP and ATP signaling in plants and animals [20,21].

In this review/opinion paper, we have focused on magnesium and cell energetics in plants, but we also refer to other organisms, when applicable.

2. AK and NDPK in Cell Energetics

In plant cells, AK activity is widely distributed through different compartments, namely plastids, mitochondria, nuclei, and cytosol [22–26]. Different isozymes of leaf AK were purified from various species using affinity chromatography [27]. Arabidopsis contains a total of eight genes of AK. Of those, at least two genes code for plastidial AK isozymes [23,25], at least three genes code for mitochondrial AKs [25,28], and one gene codes for nuclear AK [26]. The remaining two genes code for AKs which may be dually targeted between mitochondria and plastids, and mitochondria and cytosol, respectively [25] (Table 1). Mitochondrial AKs are believed to be located in the space between outer and inner membranes (the intermembrane space; IMS), possibly membrane-bound, but not in the mitochondrial matrix [25]. The pool of free and Mg-bound adenylates in the matrix, despite the lack of matrix-own AK, is affected by AKs from mitochondrial IMS via the involvement of adenylate translocators in the inner mitochondrial membrane [12,14]. Contrary to plants, human mitochondria do contain an AK isozyme in the matrix [29].

Table 1. Subcellular location and roles of Arabidopsis adenylate kinases (AKs). Classification of AK1-8 generally follows that of Lange et al. [25], with the exception of nuclear isozyme AK6, which replaced what is presented here as AK8. Mitochondrial location most likely refers to the presence of AK in mitochondrial intermembrane space (IMS), but not the matrix. C, cytosol; M, mitochondria; N, nucleus; P, plastids.

AK Name and Gene	Location	Function	Reference
AK1 (*At2g37250*)	M [a], P [b]	control of growth	[24,25,30]
AK2 (*At5g47840*)	P	plastid development	[24,25]
AK3 (*At5g50370*)	M [c], C [d]	unknown	[25,28]
AK4 (*At5g63400*)	M	unknown	[28]
AK5 (*At5g35170*)	P	no phenotype for the knockout	[25]
AK6 (*At5g60340*)	N	control of stem growth;	[26]
		control of root growth; ribosome maturation	[31]
AK7 (*At3g01820*)	M	unknown	[25]
AK8 [e] (*At2g39270*)	M	unknown	[25]

[a] [25,30]; [b] [24]; [c] [28]; [d] [25]; [e] referred to as AMK6 in ref [25].

Plant AKs react almost exclusively with AMP and ATP as substrates [22,26,32]. Human AKs, however, while more or less specific for AMP, can react with a variety of NTPs, depending on the AK isozyme. Thus, erythrocyte and serum AKs react only with ATP; muscle AK preferentially reacts with ATP and, to some extent, other NTPs; and liver AK reacts with UTP and GTP [33].

The roles of plant AKs, as determined using knockout mutants, are listed in Table 1. Whereas the disturbance of adenylate equilibrium in a given compartment in such mutants and the resulting effects on cell energetics are most likely the major causes of a given phenotype, a role of AK protein itself as a regulator cannot be discounted. One example for this is the reported complexation of AK with chloroplast glyceraldehyde-3-phosphate dehydrogenase, with the complex proposed to optimize photosynthesis during rapid fluctuation in environmental resources [34]. Another example concerns a plant nucleus-associated AK6 isozyme, which is homologous to human nuclear AK6 [26,31]. The plant protein has AK activity and was found to be essential for stem growth in Arabidopsis [26], but also contributing to Arabidopsis root growth control [31]. Plants lacking AK6 over-accumulated 80S ribosomes relative to polysome levels, consistent with the AK6 role in ribosome maturation [31]. In yeast, an orthologue of AK6 encodes the well-characterized ribosome assembly factor Fap7 [35], which has been reported to mediate cleavage of 20S pre-rRNA by directly interacting with an Rps protein. Using pulldown and two-hybrid system, the Arabidopsis AK6 was found to interact physically with Arabidopsis' own Rps14 [26]. Interestingly, it has been earlier observed that different plant AK isozymes can bind RNA and that, at least in plants, RNA-binding by AK may be related to regulatory mechanisms [36].

Aside from equilibrating adenylates, in some tissues AK was also reported to act more unidirectionally toward either ATP formation or utilization. For instance, under some conditions, e.g., in drying seeds where tissue dehydration leads to a decline in mitochondrial energy production, AK may become the main ATP (and AMP) producing mechanism [37]. Following imbibition, however, cellular adenylate balance is rapidly restored from AMP by both AK and oxidative phosphorylation in mitochondria. Another study involved antisense inhibition of the expression of plastidial AK in potato tubers, which led to dramatic effects on the overall metabolism and tuber yield [38]. In field trials, the transgenic plants had up to a 10-fold increase in ADP-glucose (key precursor to starch synthesis [39]), a 2–4-fold increase in some amino acids, and an almost two-fold increase in starch, with tuber yield nearly doubled when compared to WT plants. This suggested that, under in vivo conditions, the amyloplastic AK acts in the ATP-consuming direction, and competes for ATP both with ADP-glucose pyrophosphorylase, which produces ADP-glucose, and with plastidial pathways of amino acid biosynthesis [38].

Whereas ATP is the only NTP which arises via photophosphorylation (in chloroplasts) and oxidative phosphorylation (in mitochondria) [40], other NTPs in plants are formed mainly via nucleoside diphosphate kinase (NDPK) activity [14,41,42]. Its reaction can be described as: ATP + NDP $\leftrightarrow$ ADP + NTP. In animals and bacteria, AK can apparently substitute for NDPK, due to the apparent bifunctionality of AKs in those organisms [43]. In plants, however, where AK is specific for adenylates, these two activities are frequently metabolically "coupled" together [42,44,45], which may involve physical interaction [46]. Both AK and NDPK have been linked to stress perception [47], and they are major components of the so-called cell thermodynamical buffering system [14,48,49], which has been proposed to operate during photosynthesis and respiration [12,50,51], and, arguably, during starch synthesis in plastids and during cell wall polysaccharide formation in the plasma membrane and endoplasmic reticulum (ER) [42].

Both AK and NDPK are functionally coupled to photophosphorylation and oxidative phosphorylation, establishing that concentrations of nucleoside phosphates depend on the rates of ATP synthesis and consumption, and optimizing the operation of ATP synthases [12]. Plants contain several genes for NDPK, e.g., five genes in Arabidopsis and rice, coding for different isozymes located in cytosol, plastids, mitochondria and, possibly, ER [41]. In potato roots, cytosolic NDPK activity is believed to supply UTP for the reaction of UDP-glucose pyrophosphorylase (UGPase), to produce UDP-glucose, a key precursor to sucrose and cell wall polysaccharides [52]. Additionally, in cereal seeds, cytosolic NDPK may provide UTP which, indirectly, is used for starch synthesis [44].

3. AK-Mediated Adenylate Equilibrium and Mg^{2+} Signaling

The AK reaction has frequently been presented as: 2 ADP $\leftrightarrow$ ATP + AMP [19]. However, the true substrates/products of AK are: MgADP + ADP $\leftrightarrow$ MgATP + AMP [53,54]. An important consequence of this is that apparent equilibrium constant (K_{app}) of the first reaction (with total adenylates), where K_{app} = [ATP][AMP]/[ADP]^2, depends on free magnesium (Mg^{2+}) concentration in the reaction mixture and can be described as a bell-shaped curve, peaking at 1.5 at ca. 0.2 mM Mg^{2+} (Figure 2). At any other value of [Mg^{2+}], the K_{app} corresponds to two values of [Mg^{2+}], one below and one above 0.2 mM. In studies with plant material, the right side of the bell-shaped curve is most relevant, because internal [Mg^{2+}] in plants is usually above 0.2 mM [55]. The AK equilibrium-linked [Mg^{2+}] values can be easily computed from the following Equation (1)

$$[Mg^{2+}] = [0.7 - 0.25K_{app} \pm 0.57(1.5 - K_{app})^{1/2}]/(K_{app} - 0.1) \tag{1}$$

where a given K_{app} of AK can be calculated from the experimentally determined total of each of the adenylates taking part in the reaction (AMP, ADP, and ATP) [55]. In contrast to its K_{app}, the true equilibrium constant of AK, defined as K_{true} = [MgATP][AMP]/[MgADP][ADP], is not dependent on [Mg^{2+}] and has a fixed value of ca. 5.5 (Figure 2). The non-linear relationship between K_{app} and [Mg^{2+}] was observed in several studies with purified AKs [16,53,54,56,57], and similar principles most likely also apply under in vivo conditions [14,49].

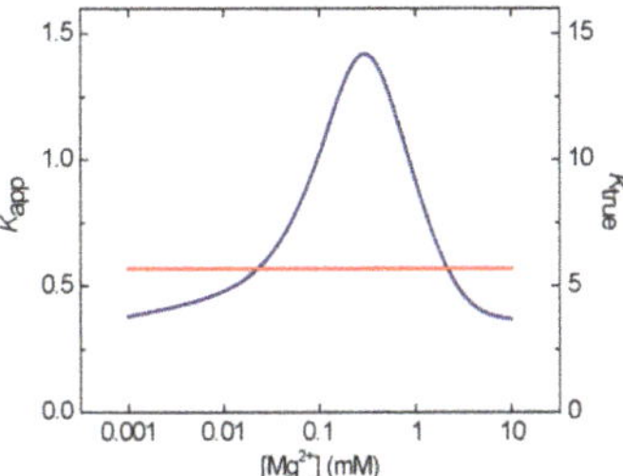

Figure 2. Effects of [Mg^{2+}] on K_{app} and K_{true} of AK. The K_{app} peaks at ca. 0.2 mM Mg^{2+}. Please note that the scale on X-axis is logarithmic. The lines for K_{app} (blue) and K_{true} (red) were computed as described in ref 55, using Origin software (OriginLab Corporation, Northampton, MA, USA).

As evident from Equation (1), the K_{app} of AK, although different from the K_{true}, can be a useful parameter linking concentrations of all adenylates to [Mg^{2+}]. Thus, knowing the contents of total ATP, ADP, and AMP in a given biological preparation, and assuming that they are under equilibrium governed by AK, can provide information about intracellular [Mg^{2+}]. Similarly, when only [Mg^{2+}] is known, this can be linked with a given K_{app} of AK (Figure 2). A similar computational set can be established for Mg^{2+} and other nucleotides (guanylates, cytidylates, and uridylates) via the corresponding buffering equilibria of NDPK and UMP/CMP kinase. The latter reacts reversibly with UMP and CMP rather than AMP, and uses ATP as a second substrate [25,58]. It is unknown whether the UMP/CMP kinase reaction requires a combination of Mg-chelated and free nucleotides, as is the case for AK, but magnesium is apparently required for the reaction [58].

When considering the Mg requirement for NDPK, its true reaction can be described either as MgATP + NDP ↔ ADP + MgNTP, or MgATP + MgNDP ↔ MgADP + MgNTP. To the best of our knowledge, it is unknown whether nucleoside diphosphates (including ADP) used by NDPK are reactive as Mg-chelated or Mg-free species. This could, at least theoretically, make significant difference in terms of the relationship between the K_{app} of NDPK and [Mg^{2+}]. With respect to UMP/CMP kinase, the K_{app} of its reaction (UMP/CMP + ATP ↔ UDP/CDP + ADP) is likely to depend on [Mg^{2+}] in the same way as AK. Additionally, regardless of what true substrates are for NDPK and the other kinase, cellular pools of non-adenylate nucleotides are generally much smaller than those of adenylates [42]. Thus, intracellular [Mg^{2+}] would still respond more strongly to AK-mediated equilibrium than that of NDPK and UMP/CMK kinase.

4. Magnesium and the Adenylate Energy Charge Theory

Equilibrium of adenylates maintained by AK is at the core of the adenylate energy charge (AEC) theory, developed and popularized by Atkinson [19]. The theory assumes that AK uses total adenylates as substrates and that the adenylate concentrations at AK equilibrium account for the energy status in metabolism. This can be defined as AEC = ([ATP] + $\frac{1}{2}$ [ADP])/([ATP] + [ADP] + [AMP]), with the concentrations of adenylates equilibrated by AK reaction. The theory has been criticized [16,59,60] on the grounds that it does not take into account a crucial role of magnesium for the AK reaction. Most importantly, mass action (K_{app}) of AK is very much dependent on [Mg^{2+}] (Figure 2), and is not constant, as assumed for AEC. As a consequence of that, at low [Mg^{2+}], as it is in the cytosol, free ATP may actually inhibit MgATP-utilizing enzymes rather than serving as a substrate. The same concerns glycolytic kinases involved in ATP formation, which use MgADP rather than free ADP as a substrate, with [Mg^{2+}] being a key player in controlling MgADP availability [53]. These and other arguments against AEC as a key parameter controlling energy status of cellular processes have been summarized by Purich and Fromm [16,59] and Pradet and Raymond [60], and we will not cover them here. However, it is important, in our opinion, to emphasize that Mg signaling as controlled by AK is not compatible with AEC theory. This is simply because the AEC does not take into account true substrates (Mg-bound and Mg-free) of AK and the crucial role of [Mg^{2+}] in making these substrates available for AK.

5. Magnesium Status in Cells

With a total cellular concentration of magnesium at 15–25 mM with 15–20% bound to chlorophyll [61] and free magnesium concentration frequently at a less than millimolar level [62,63], this implies that most of magnesium is complexed, and only a small fraction exists as Mg^{2+}. It has been reported that up to 90% of the cytosolic pool of nucleotides is bound to Mg [64]. The same applies to chloroplasts [13], and probably to all other compartments which contain metabolically active nucleotides. For instance, in chloroplast stroma, free Mg^{2+} ranges from ca. 0.2 to 5 mM (Table 2), with the lower values characteristic for darkened leaves, and the higher values for illuminated leaves [55,65,66]. This constitutes less than 10% of total Mg in chloroplasts. The rest is confined mostly to chlorophyll in thylakoids, but also chelates stromal pools of phosphorylated compounds (e.g., ATP) and dicarboxylic acids [65].

Table 2. [Mg^{2+}] in cellular compartments and methods used to measure [Mg^{2+}].

Compartment	[Mg^{2+}], mM	Method	Reference
Cytosol	0.25	^{31}P-NMR	[67]
	0.40	^{31}P-NMR	[64]
	0.9	Ionophore	[68]
	0.2–0.4	From K_{app} of AK	[55,62]
Mitochondria	2.4	^{31}P-NMR	[67]
	1.0–3.0	From K_{app} of AK	[55]
Chloroplasts	0.5–2.0	Ionophore	[66]
	1.0–3.0	Ionophore	[65]
	0.2–5.0	From K_{app} of AK	[55]
Vacuole	5–80 [a]	X-ray analysis	[2,69]
ER lumen	Unknown		
Peroxisomes	Unknown		

[a] The value of 80 mM was obtained by feeding leaves with high Mg–sap solutions.

A major role in Mg^{2+} homeostasis in plants belongs to the vacuole [11]. Vacuoles buffer and balance fluctuating concentrations of external nutrients, but they can also alleviate the effects of excessive concentrations of such compounds. In Arabidopsis leaves fed with high-Mg–sap solutions, vacuoles may accumulate up to 80 mM Mg [69]. This concentration is one order of magnitude higher than vacuolar [Mg] under normal conditions [2]. Upon withdrawal of Mg from the nutrient solution, a typical first symptom of Mg deficiency is the higher accumulation of starch and sucrose in the leaves, followed by leaf chlorosis, which may lead to a lower photosynthetic rate [1]. It has been suggested that Mg deficiency limits the carbohydrate transport from source organs to the sink by affecting the loading of sucrose to the phloem, which requires an adequate Mg concentration [6–8]. It has been reported that the process of Mg translocation can be hampered under severe Mg-deficiency, having effects both on photoassimilate partitioning and root growth [70]. The disruption of Mg transport in young plants could result in reduced growth of the plant at a later growth stage.

Cells are usually quite resistant to external Mg-deficient conditions. As pointed out by Gout et al. [67], it takes 14 days to decrease cellular magnesium content by five-fold, when placing sycamore cells into Mg-free media. They also found that during first 10 days of Mg-deficient conditions, the cytosolic [Mg^{2+}] did not change at all due to a release of Mg^{2+} from the vacuole. Only after 10 days was there a decrease in cytosolic [Mg^{2+}], accompanied by a cessation of cell growth and a decrease in respiration.

To the best of our knowledge, there are no data on [Mg^{2+}] in the nucleus, ER lumen, and peroxisomes. For the nucleus, we can only assume that its [Mg^{2+}] is similar to that in the cytosol, given the porous structure of the nuclear envelope. In addition, the nucleus contains its own AK [26], which probably can access the same adenylate pool (given the pores in nuclear membrane) as in the cytosol. For the ER, the major obstacle for the determination of Mg^{2+} has been the presence of high (millimolar) concentration of Ca^{2+}, preventing reliable Mg^{2+} detection there with the use of ionophores [71]. However, the ER has at least one Mg^{2+} transporter (see below), and metabolism within ER strongly depends on ATP supply [72,73], suggesting an important role for Mg^{2+} in this organelle. Peroxisomes probably have low [Mg^{2+}], because of the lack of an Mg^{2+} transporter in their membrane, but they also constitute only a tiny fraction of cell volume, thus are unlikely to contribute significantly to overall cellular Mg^{2+} homeostasis. Both ER and peroxisomes do not have their own AK, and adenylate metabolism there must be independent of AK equilibrium.

6. Feasibility of the Estimations of [Mg^{2+}] Based on Adenylate Measurements

Assays of intracellular [Mg^{2+}] in plants most frequently have been performed with Mg-binding ionophores, using fluorescence spectrophotometry to detect the Mg–ionophore complex, sometimes also with help of fluorescence microscopy [66,74,75]. However, many of the Mg^{2+} fluorescent probes proved unsatisfactory, due to their lack of specificity or low affinity for Mg^{2+} [74]. Additionally, the ionophores need to be loaded into cells before the assays, which may perturb normal metabolism. The use of ^{31}P-NMR was more successful, and it permitted non-invasive in vivo studies, allowing simultaneous identification and quantification of free and Mg-complexed nucleotides as well as [Mg^{2+}] in whole cells, the cytosol, and organelles [63,76]. Those and other methods of Mg determination in biological samples, e.g., electron probe X-ray microanalysis (XRMA) or ^{13}C-NMR citrate/isocitrate ratio, have been discussed by Romani and Scarpa [77], and they all require specialized scientific tools and expertise. On the other hand, adenylates can be easily quantified by a variety of methods, using standard laboratory equipment. Although an indirect measure, the calculated [Mg^{2+}] values that were derived from adenylate contents are comparable to those obtained by other methods (Table 2).

Scientific literature abounds with measurements of adenylate species in whole organs/tissues and, to a lesser extent, in fractionated preparations containing purified organelles [55,78]. These data can be recalculated for estimations of [Mg^{2+}], especially in organelles. Based on its subcellular localization, the AK-mediated equilibrium of adenylates and the resulting Mg^{2+}-signaling encompasses chloroplasts (both stroma and IMS), cytosol, nucleus and the IMS of mitochondria (Table 1)). The outer membranes of chloroplasts and mitochondria are permeable to small compounds, e.g., adenylates, and thus the IMS in both organelles is under AK equilibrium, which extends through the permeable outer membranes to the cytosol [62,67]. Adenylate data collected for any of these compartments should tightly correlate with an internal [Mg^{2+}] there; as indeed is the case when compared to other methods of [Mg^{2+}] determination (Table 2).

A different situation occurs if the adenylate data are collected for whole tissues, cells, or protoplasts, where mixing up of various adenylate pools occurs, and the final result reflects the adenylate status in whole tissue/cells, but not in given compartments. The calculated K_{app} of AK for such a system will be a mean K_{app} of all cellular AKs and should apply only to compartments where AK equilibrium is established. This excludes peroxisomes, ER, and vacuoles, which lack AK isozymes (Table 1). Among those organelles, vacuoles do not have any adenylate translocators and, besides, it has already been reported that potato tuber vacuoles contain no adenylates [79]. However, both peroxisomes and ER do have adenylate translocators (see below), and thus adenylates present in these organelles may affect the K_{app} of AK based on total contents of adenylates from whole plant cells/tissues. The contribution of peroxisomal adenylates is probably close to negligible (small size of peroxisomes), whereas the pool of adenylates in the ER might be significant, given that the ER represents a continuous membrane system, often quite abundant in the cytosol.

Thus, in summary, when using the concept of the K_{app} of AK to derive $[Mg^{2+}]$ from total contents of adenylates in tissues or whole cells, the main drawback of this method is that it yields an average $[Mg^{2+}]$ for all compartments where AK equilibrium applies (thus excluding vacuoles, ER, and peroxisomes). This approach is obviously only an approximation (e.g., possible error from the contribution of ER's adenylates), but it could be useful, especially when studying a process known to be confined to a single compartment, where substantial changes in $[Mg^{2+}]$ have already been observed by other methods, e.g., during the light-induction phase of CO_2 fixation in chloroplasts [55] or during anoxia in the cytosol [50].

In earlier studies [55], based on published data for adenylate contents in plant tissues and organs, we have found that levels of Mg^{2+} in leaves depend on the developmental stage (young leaves having lower $[Mg^{2+}]$ than old ones), light conditions (darkened leaves have lower $[Mg^{2+}]$ than in illuminated ones), and salt stress conditions (stressed plants have lower $[Mg^{2+}]$). For instance, based on data from Nieman et al. [80], $[Mg^{2+}]$ in young and mature pepper leaves was 3 and 6 mM, respectively, whereas salt stress decreased $[Mg^{2+}]$ in safflower buds (from 10 to 4 mM). When applying the same approach to adenylate data from one study on the effects of salt (NaCl) stress on energetics of cyanobacteria [81], the calculations again suggest strong effects of salt on internal $[Mg^{2+}]$. When stressed, the cyanobacteria maintained low $[Mg^{2+}]$ levels of 0.24 mM, which markedly increased to 2.2 mM upon salt withdrawal. This probably reflects the observed two-fold higher photosynthesis rate of these microorganisms under normal conditions, resulting in more ATP produced and thus requiring more magnesium. Cyanobacteria are prokaryotes and do not have any organelles; therefore, aside from thylakoid-like membrane system, the calculated $[Mg^{2+}]$ might be representative of $[Mg^{2+}]$ anywhere inside of their cells.

7. Magnesium Translocators

Soils are usually low in Mg content, because Mg binds soil only weakly and can be easily leached out by rains. To adapt to such conditions, plants have evolved a highly efficient system for Mg acquisition from the soil; its transport via xylem; and its distribution to targeted tissues/cells. This system has been comprehensively reviewed [2,4,82], and we will not cover it in this paper. Instead, we will focus on Mg^{2+} traffic into and within a cell. Mg^{2+} is taken up first by specific translocators in the plasma membrane, and then it is distributed to several membrane-surrounded compartments, each using its own specific set of Mg^{2+} transporters (Figure 3).

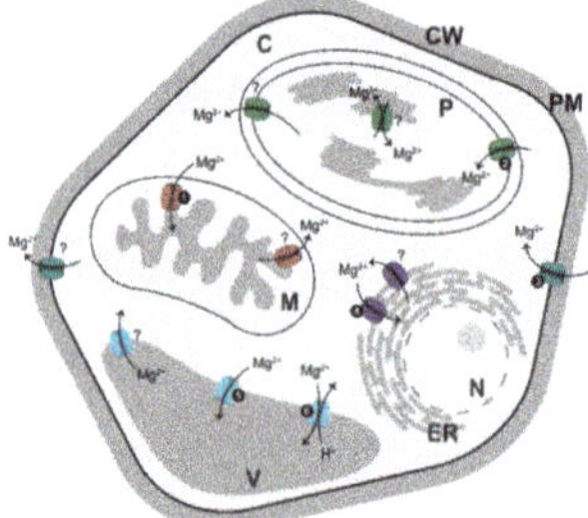

Figure 3. Distribution of Mg^{2+} transporters in membranes in plants. All transporters shown here have been identified in Arabidopsis [83]. The light-grey area corresponds to compartments where AK equilibrium is established. Numbers correspond to: (1) MGT5; (2) MGT10; (3) MGT1, MGT5, MGT6, MGT9; (4) MGT4, MGT7; (5) MGT2, MGT3; and (6) MHX. Question marks refer to transporters involved in Mg^{2+} export from a given compartment/cell; their nature remains unclear. Abbreviations: C, cytosol; CW, cell wall; ER, endoplasmic reticulum; M, mitochondrion; N, nucleus; P, plastid; PM, plasma membrane; V, vacuole.

Most Mg^{2+} transporters belong to a single family of proteins, which in turn belongs to the CorA protein superfamily [84]. In plants, this family was first described in Arabidopsis by two groups, which named it *At*MRS2 [85] and *At*MGT [84]. For simplicity, we will refer to those transporters as belonging to the MGT family. In Arabidopsis, there are 10 genes for MGT [86], whereas in rice nine genes have been identified [87]. Most of the Mg^{2+} transporters are responsible for Mg^{2+} import into a given compartment, and the rest are involved in Mg^{2+} export. The importers are relatively well described, whereas the nature of Mg^{2+} exporters is less clear [83]. Some of the importers may become exporters, depending on [Mg^{2+}], as is the case for *At*MGT5, which has a dual role as an Mg-importer at micromolar levels, and an exporter at a millimolar range [86]. Besides Mg^{2+}, some MGT members may also transport other cations, including Zn^{2+} and Cu^{2+}. There are also other carriers predominantly transporting K^+ or Ca^{2+}, but which are also permeable to Mg^{2+}. Non-selective cation channels are the other candidates for Mg^{2+} transport [1,88]. In addition, Mg^{2+} availability may affect the activities of plasma membrane transporters for Ca^{2+}, K^+, and H^+ [88].

Under conditions of Mg excess, as in so called serpentine soils [89], plants deploy an elaborate system to avoid Mg toxicity. This system is composed of plasma membrane- and tonoplast-localized calcineurin B-like proteins (CBLs) and their downstream components, CBL-interacting protein kinases (CIPKs) [90]. At high external Mg^{2+}, there is an interaction between the plasma membrane-associated CBL and CIPK components which modulates the activity of a number of ion channels/transporters, facilitating the uptake or exclusion of Mg^{2+}. In the next step, high cytosolic [Mg^{2+}] triggers changes in internal Ca^{2+}, which are then sensed by tonoplast CBLs. This, in turn, triggers tonoplast CIPKs to activate Mg^{2+} transporters or channels to detoxify the cytosol from Mg^{2+} [90]. The CBL/CIPK system may also have a similar role in protecting against the toxicity of other ions, including excess of Na^+ ([90–93]. Some studies, based on knockout mutants, have identified plasma membrane-bound MGT6 and ER-associated MGT7 as likely candidates involved in the detoxification of Mg^{2+} [93]. MGT6 was also required for plant adaptation to a low [Mg^{2+}] [94].

Arabidopsis contains four MGT proteins in the plasma membrane, whereas vacuolar tonoplast contains two MGT transporters (MGT2 and MGT3) [69] and the so-called MHX transporter [95] (Figure 3). MHX is structurally distinct from MGT/MRS2 transporters and shows the highest similarity to mammalian Na^+/Ca^{2+} exchangers, which are part of the Ca^{2+}/cation (CaCA) exchanger superfamily [96]. The MHX protein exchanges vacuolar protons for cytosolic Mg^{2+} and Zn^{2+} [95,97]. In Arabidopsis, MHX co-localizes with a major chromosomal quantitative trait locus (QTL), affecting seed Mg content [98].

Besides being located in the plasma membrane and tonoplast, Mg^{2+} transporters are also elsewhere in the cell, i.e., in plastids, mitochondria and ER (Figure 3). Some of the transporters are tissue-specific, e.g., Arabidopsis mitochondrial MGT5 which is exclusively expressed in anthers at early stages of flower development, underlying its role in pollen development and male fertility [86]. MGT4 in the ER and MGT9 in the plasma membrane are also essential for pollen development [82,83].

Exact subcellular location needs to be reevaluated for some Mg^{2+} translocators, because several studies have reported discrepant results, especially concerning putative ER-location. As pointed out by Yan et al. [93], membrane proteins can be mis-targeted to ER, especially when overexpressed in a transient expression system. It would also be interesting to see how plants that have adapted to growth on serpentine soils and deal with excess Mg^{2+} are managing their AK-mediated energy metabolism.

8. Adenylate Translocators

In Arabidopsis, at least 16 distinct genes for adenylate carriers have been identified [99], which code for proteins distributed in the plasma membrane, plastids, mitochondria, peroxisomes, and ER (Figure 4). Together, they represent an efficient system of energy partitioning between different cell compartments. Most of them are antiporters, transferring one adenylate in exchange for another adenylate species (or inorganic phosphate, P_i, as is the case for some mitochondrial carriers). In most cases, ATP is exchanged for ADP, but there are also uniporters for ATP transport (at the plasma membrane) or for the transport of all adenylates (ATP, ADP, and AMP) in plastids and mitochondria. Importantly, mitochondria have an antiporter transporting ATP in exchange for AMP and, to some extent, ADP [99]. It is believed that adenylate translocators, with the exception of MgATP/P_i exchangers [99], use free adenylates for transport across a given membrane [62,100], implying a major role for [Mg^{2+}] in regulating a supply of free adenylates to the translocators.

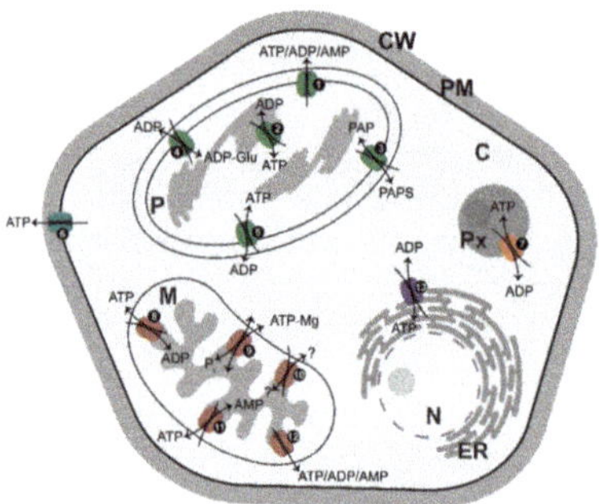

Figure 4. Distribution of major adenylate carriers in membranes in plants. The names of the transporters and major transported molecules are as they are given by da Fonseca-Pereira et al. [99]. The light-grey area corresponds to compartments where AK equilibrium is established. Numbers correspond to: (1) AtBT1, *Arabidopsis thaliana* ATP/ADP/AMP carrier; (2) ATP/ADP carrier; (3) TAAC/PAPST1; (4) ZmBT1, maize (*Zea mays*) plastid ADP-Glucose/ADP carrier; (5) NTT1-2, ATP/ADP carriers; (6) PM-ANT1; (7) PNC1-2, ATP/ADP carrier; (8) AAC1-3, ATP/ADP carriers; (9) APC1-3, MgATP/P_i carriers; (10) ZmBT1, maize mitochondrial transporter, the substrate and transport mode of which are unclear; (11) ADNT1, AMP/ATP carrier; (12) AtBT1, ATP/ADP/AMP carrier; and (13) ER-ANT1, ATP/ADP carrier. Abbreviations: C, cytosol; CW, cell wall; ER, endoplasmic reticulum; M, mitochondrion; N, nucleus; P, plastid; PAP, 3′-phosphoadenosine 5′-phosphate; PAPS, 3′-phosphoadenosine 5′-phosphosulfate; PM, plasma membrane; Px, peroxisome.

Adenylate translocators have variable organ/tissue-specific expression patterns under different environmental conditions and at different developmental stages, suggesting specific non-redundant functions for each of the translocators [99]. For instance, the chloroplast ATP/ADP antiporter has been identified as one of several membrane-bound proteins exhibiting increased abundance after cold acclimation [101]. In earlier studies on AK, a possible link was found between activities of certain AK isozymes and adenylate transport during plant flowering [102]. Upon flower induction, although total AK activity in the leaves and stems remained the same, the intracellular distribution of AK activity changed, with the most prominent being a strong decrease in activity of one of chloroplast AK isozymes. This AK was proposed to functionally interact with the chloroplast adenylate translocator, responding to alterations in energy distribution between chloroplast and cytosol during floral induction [23,102].

9. Role of [Mg^{2+}] in Metabolism and Signaling

There are several aspects to the involvement of Mg in cell energetics: (i) Cellular ATP (and to some extent ADP) is strongly chelated by Mg^{2+}, and the chelated and free nucleotides are frequently key substrates/effectors in metabolism. The same concerns pyrophosphate (PP_i), an alternative energy currency, which is active as an Mg-chelated or Mg-free species; (ii) Binding of Mg^{2+} frequently modulates and stabilizes activities of enzymatic proteins involved in cell energetics processes, but also in DNA replication, transcription and translation and other processes; (iii) The Mg-chelated and free adenylates govern various aspects of cell energetics, such as rates of energy metabolism, translocation of adenylates across membranes, or contribute to allosteric regulation of metabolism; (iv) The substantial changes in intracellular $[Mg^{2+}]$, as in the cytosol of cells under anoxia, may reflect switches in metabolism between MgATP-based and $MgPP_i$-dependent; and (v) Mg^{2+} affects concentrations of Ca^{2+} and other cations, and alleviates the effects of stress by excess $[Na^+]$. Below, we will briefly describe each of these aspects.

9.1. Mg^{2+} and Chelation of Adenylates and PP_i

Changes in subcellular $[Mg^{2+}]$ have significance in establishing the approximate distribution of ATP and ADP among the Mg-free and Mg-complexed forms. For instance, certain kinases, e.g., pyruvate kinase or phosphoglycerate kinase, that use ADP to produce ATP in the process of substrate phosphorylation, react with MgADP rather than ADP as their substrate [53,103]. Under low $[Mg^{2+}]$, these reactions will be limited because of shortages of MgADP and an excess of free ADP, which likely acts as inhibitor. On the other hand, the MgADP complex will be sensitive to changes in concentrations of Mg^{2+} and total ADP, both of which change reciprocally with changes in total ATP. In some instances, MgADP can act as an inhibitor, as is the case for several MgATP-utilizing enzymes in the cytosol which are competitively inhibited by MgADP [63,104]. For these enzymes to operate effectively, it is very important that cytosolic $[Mg^{2+}]$ is maintained at low levels, which implies that MgADP will also be low. Under hypoxia or anoxia conditions, however, cytosolic $[Mg^{2+}]$ markedly increases, which leads to increases in [MgADP], which may affect hexokinase activity, and thus glycolysis [105]. Additionally, changes in cytosolic $[Mg^{2+}]$ may impact protein kinase activities and subsequent signal transduction. Mg^{2+} plays an important role in the ATP binding in the active site of the kinase and facilitates phosphoryl transfer reactions [106].

PP_i, an alternative (to ATP) energy currency, is produced mainly during fatty acid and amino acid activation for the degradation of fatty acids and for protein synthesis, respectively, and during nucleic acids synthesis [107]. Important additional sources of PP_i are various pyrophosphorylases which, in addition to PP_i, produce a variety of nucleotide sugars (e.g., UDP-glucose). In all these reactions, a subsequent hydrolysis of PP_i into two P_i molecules by a pyrophosphatase (PPase) [108,109] or PP_i removal by other PP_i-utilizing enzymes [44] drives the overall metabolism toward the activated substrate formation.

PP_i can be used as an energy source instead of ATP, when the latter supply is low and when cytosolic $[Mg^{2+}]$ increases, as in anoxia/hypoxia. In such cases, PP_i is frequently used as a substrate, as $MgPP_i$, rather than free PP_i (Figure 1B), as in the reaction of PP_i-dependent phosphofructokinase [110]; free PP_i acts as the inhibitor of this enzyme [111]. The vacuolar H^+-PPase uses Mg_2PP_i (Figure 1B) as its substrate [112,113] and is allosterically activated by Mg^{2+} [114]. This means that the PPase needs an increased $[Mg^{2+}]$ for its optimal operation, providing a link to the hypoxic metabolism characterized by Mg^{2+} release upon the decrease in ATP production [50]. Another example of the different requirements for Mg^{2+} is provided for several aminoacyl-tRNA synthetases, key activities producing direct precursors for protein synthesis. In their reverse reaction (pyrophosphorolysis), one group of aminoacyl-tRNA synthetases uses $MgPP_i$ and the other prefers Mg_2PP_i [115], implying that they will be fully active only at specific (and different) Mg^{2+} concentrations. $MgPP_i$ serves also as a substrate for non-proton pumping PPases [116,117] and several other PP_i-utilizing enzymes [42,50,109].

Stability constant for formation of $MgPP_i$ is much lower than that for MgATP (K_{MgPPi} of 1.2 mM^{-1} vs. K_{MgATP} of 73 mM^{-1}) [15]; therefore, this implies that $MgPP_i$-utilizing enzymes will operate effectively only at a relatively high $[Mg^{2+}]$, and even small changes in intracellular $[Mg^{2+}]$ (below ca. 0.7 mM) may have significant effect on the $MgPP_i$ availability (Figure 1B). Excess of PP_i, however, can be lethal, disrupting metabolic pathways. In Arabidopsis plants impaired in cytosolic pyrophosphatase, the accumulated PP_i inhibited UDP-glucose formation by UGPase [118]. It is unknown whether it is free or Mg-chelated PP_i acting as the UGPase inhibitor, but the free PP_i appears to be a better candidate, given the low $[Mg^{2+}]$ in the cytosol. UDP-glucose is a key direct or indirect precursor to myriads of glycosylation reactions, including the formation of sucrose, starch, but also cellulose, hemicellulose, glycoproteins, and many other carbohydrate-containing end-products [119,120]. All these pathways may, thus, be affected by fine changes in cytosolic [Mg^{2+}].

9.2. Mg^{2+} as a Regulator of Enzymatic Activities

Earlier, we have identified several reactions involved in carbohydrate synthesis that require magnesium either via complexation with NTP to form MgNTP, a true substrate, or as an effector of a given enzyme (i.e., stimulating or inhibiting a given activity). These reactions included, among others, several NDP-sugar producing pyrophosphorylases and sucrose synthase (SuSy) (for details see ref [42]). The pyrophosphorylases produce nucleotide sugars, which are substrates for glycosylation reactions, whereas SuSy is involved in the metabolism of sucrose, a soluble sugar [120]. Mg^{2+} activates SuSy toward UDP-Glc production and it inhibits the reverse reaction (sucrose formation) [121]. Stimulation by Mg^{2+} was also found for the activity of phosphorylated (soluble), but not for non-phosphorylated (membrane-bound) SuSy [122], suggesting that Mg^{2+} affects sucrose breakdown (soluble SuSy), but not cellulose synthesis (membrane-bound SuSy).

The apparent control exerted by Mg^{2+} over carbohydrate metabolism strongly suggests a dual role of AK and NDPK in this process. Firstly, AK and NDPK produce nucleoside triphosphates as substrates for the pyrophosphorylase reactions; this happens either directly (production of MgATP by AK) or indirectly (production of MgUTP, MgGTP and MgCTP) by linking AK activity, via NDPK, with kinases of uridylate, guanylate, and cytidylate metabolism [42]. The second role of AK (and possibly of NDPK) is its control of [Mg^{2+}], which acts as a substrate (complexed with NTP) for the pyrophosphorylases and as an effector for both pyrophosphorylases and SuSy [42].

Among many other examples of control exerted by magnesium [17,49–51,55,62], an important case is the functioning of ribozymes [123], in particular in the process of protein synthesis on ribosomes. Most of the studies in this area have been performed in prokaryotes. As we mentioned earlier, aminoacyl-tRNA synthetases belong to two groups with different requirements for Mg^{2+} [115]. In fact, four types of dependencies on Mg^{2+} were observed in these two groups. The class I synthetases require only one Mg^{2+} for the activation reaction (in MgATP), while the class II synthetases require three Mg^{2+} ions (one in MgATP and two in Mg_2PP_i). In class II synthetases, both $MgPP_i$ and Mg_2PP_i participate in the pyrophosphorolysis of the aminoacyl adenylate, but some of them show a better fit if Mg_2PP_i reacts and others when only $MgPP_i$ but not Mg_2PP_i is used in the pyrophosphorolysis. The data for eukaryotic and, in particular, plant enzymes participating in protein synthesis are quite limited; however, the key role of Mg^{2+} has been shown during splicing for the functioning of spliceosome [124].

Aside from the examples presented above, magnesium is also essential for DNA replication and for transcription. Most of the enzymes involved in these processes require Mg either chelated to NTP (which then acts as substrate) or acting as an effector. Crystal structures of DNA polymerases, involved both in replicating DNA and in DNA repair, revealed a crucial role of Mg ions in faithfully positioning a given nucleotide in the active site of the enzyme and promoting phosphoryl transfer [125]. A similar role for Mg was found for RNA polymerase [126]. Biologically active structures of both DNA and RNA are stabilized by Mg [127].

9.3. Mg^{2+} Regulates Energy Metabolism, Adenylate Transport, and Allosteric Regulation

Earlier, we summarized effects of different ratios of adenylate species, both free and Mg-complexed, on metabolism, along with formulas for the calculation of adenylate ratios upon AK equilibrium [62]. Thus, the MgATP/MgADP ratio reflects anabolism-driving potential; ATP_{free}/ADP_{free} adenylate translocation potential; and $MgATP/AMP_{free}$ allosteric regulation driving potential. Numerous enzymes are regulated by the MgATP/MgADP ratio, while adenylates are translocated via membranes as free species, and free AMP and/or free ADP operate in cell metabolism as allosteric effectors [17,55]. The set of free and Mg-bound adenylates plus free magnesium, reflecting the real energy charge of the cell, is established in cell compartments depending on the metabolic fluxes of synthesis and utilization of adenosine phosphates within the pool of total adenylates and magnesium.

Transporters for Mg^{2+} and adenylates in all kinds of membranes (Figures 3 and 4) are most likely involved in Mg^{2+}-signaling simply by regulating the intracellular concentrations of Mg-free and Mg-chelated adenylates, which are equilibrated by AK. The Mg^{2+} transporters are perhaps even more important, because there must be an upper limit to $[Mg^{2+}]$ in a given metabolically active cellular compartment to prevent Mg^{2+} toxic effects. On the other hand, excess of Mg^{2+} may have some beneficial effects under specific stress conditions, e.g., in alleviating the sensitivity of plants to salinity (see below) [117]. Additionally, because adenylates are transported as Mg-free species [100], the $[Mg^{2+}]$ on both sides of a given membrane must have a crucial effect on the rates of the translocation.

It has been proposed that differences in intracellular $[Mg^{2+}]$ between cytosol and mitochondria are the key factor in the regulation of cell respiration [63,67]. It was shown that, in heterotrophic sycamore (*Acer pseudoplatanus* L.) cells, ADP is less complexed with Mg^{2+} in the cytosol than in mitochondrial matrix due to a low $[Mg^{2+}]$ in the cytosol, while ATP is mostly complexed by Mg^{2+} in both compartments. Depletion of Mg^{2+} (after growth on Mg-free media) increases free ADP concentration in the cytosol and matrix, leading to a decrease in coupled respiration and a suppression of cell growth. The $[Mg^{2+}]$ established under the control of AK mediates the ADP/ATP exchange between the cytosol and matrix, MgADP-dependent mitochondrial ATP synthase activity, and cytosolic free ADP homeostasis [67].

Marked changes in $[Mg^{2+}]$ also accompany the so-called induction phase of photosynthesis, reflecting an early response of the photosynthetic apparatus to dark-to-light transitions [55,128]. The buildup of ATP upon illumination causes the depletion of Mg^{2+} to very low values, initially equilibrated by AK to the level of ~0.2 mM in chloroplasts and cytosol. Then, in the course of transition to steady-state photosynthesis, chloroplastic $[Mg^{2+}]$ increases to 1–3 mM upon the involvement of mitochondria in the reoxidation of photosynthetically formed redox equivalents via the malate valve [55]. The estimation of intracellular $[Mg^{2+}]$ during photosynthetic induction and steady-state photosynthesis was possible because of the data on total adenylate contents obtained by rapid fractionation of protoplasts [129–131]. The phenomenon of photosynthetic induction, which is characterized by the delay of photosynthesis upon illumination, can be also partly explained by the depletion of Mg^{2+}, affecting the activity of essential photosynthetic and respiratory enzymes.

AK controls the concentration of AMP, which serves as a cofactor of the mammalian and yeast AMP-activated protein kinase (AMPK), which in turn plays a central role in the regulation of energy metabolism [132]. Plants contain SnRK1 protein, which is an ortholog of AMPK [133]. Although free AMP allosterically activates AMPK, Mg^{2+} may participate in the catalytic mechanism of this enzyme as an indispensable cofactor [134,135]. The energy status of a cell regulates AMPK activity in a complex way, which involves tighter binding of AMP_{free} than of ADP_{free} and of Mg-bound nucleotides [136], while Mg^{2+} likely exerts a regulatory role on the enzyme. Even at high $[Mg^{2+}]$, most AMP exists in a free form (Figure 1A), and thus AMPK can be efficiently activated by AMP upon wide ranges of $[Mg^{2+}]$. The interplay between AMP release, free Mg, and ATP production needs further investigation, in particular for plant SnRK1.

9.4. [Mg²⁺] under Anoxia

Under normal conditions, ATP (and other nucleoside triphosphates) is tightly bound to magnesium, thus contributing to a relatively low $[Mg^{2+}]$ status. Under stress conditions, however, there is frequently an increase in $[Mg^{2+}]$ and other divalent cations, including Ca^{2+}, reflecting lower levels of ATP produced in stressed tissues. This happens especially during anoxia (lack of oxygen), when mitochondrial oxidative phosphorylation is not working, and most energy can be acquired only via glycolysis. Under these conditions, the increased [Mg^{2+}] leads to the activation of Mg^{2+}-requiring enzymes and redirects the energy metabolism from ATP to PP_i-utilization [50].

The decrease in ATP production and the subsequent release of Mg^{2+} under anoxia make PP_i an efficient alternative energy currency. Mg^{2+} can bind to PP_i in two ways—as $MgPP_i$ and Mg_2PP_i (Figure 1B)—and it does so at a higher concentration than with ATP. The ratio between PP_i, $MgPP_i$ and Mg_2PP_i is under pH control [15]. Proton pumping vacuolar PPase uses Mg_2PP_i as a substrate and thus becomes active under oxygen deficiency [112,113]. Other enzymes active under anoxia use Mg-complexed substrates which bind magnesium weakly, such as phosphoenolpyruvate (PEP) or isocitrate [15]—thus, the [Mg^{2+}] parameter is critical for their operation. The importance of PEP turnover under anoxia is determined by the availability of Mg^{2+}, and this metabolite is directly involved in the production of PP_i. Under anoxia, the formation of PP_i by "coupled" reactions of pyruvate phosphate dikinase and pyruvate kinase supports glycolysis under conditions of low [ATP] [50,137]. Another important protein, nitrate reductase, is upregulated under oxygen deficiency, whose activity is controlled by phosphorylation, in a process mediated by Mg^{2+} and 14-3-3 proteins [138].

9.5. Magnesium versus Calcium, Sodium and Aluminum

Changes in [Mg^{2+}] as a feedback of the equilibrium governed by AK (and perhaps also NDPK and other nucleotide kinases) result in corresponding changes in internal [Ca^{2+}] due to the chelation of Ca^{2+} with nucleotides to nearly the same extent as with Mg^{2+}. Magnesium allosterically activates Ca^{2+} binding to calmodulin, with the latter regulating target proteins in response to sub-micromolar changes in [Ca^{2+}] [139]. In turn, changes in [Ca^{2+}] in a given compartment can modulate internal [Mg^{2+}], in a millimolar range [42,140]. Ca^{2+} is chelated by adenine nucleotides to nearly the same extent as Mg^{2+} [141]; therefore, the intracellular [Ca^{2+}] is controlled by the AK equilibrium and that of other nucleoside kinases [142]. This keeps the ratio of [Ca^{2+}]/[Ca_{total}] at the same level as [Mg^{2+}]/[Mg_{total}] despite the fact that the total concentration of Ca^{2+} is a few orders of magnitude lower than that of magnesium [142,143]. The release of Mg^{2+} when ATP level drops corresponds to an increase in internal Ca^{2+} [144].

In the IMS of mitochondria, the increase in [Ca^{2+}] that accompanies Mg^{2+} release subsequently leads to the activation of multiple Ca^{2+}-regulated enzymes. These enzymes include the external NADPH and NADH dehydrogenases of mitochondria, internal NADPH dehydrogenase of mitochondria [145], NAD kinase of the IMS of mitochondria [146], glutamate decarboxylase, cysteine proteases (calpain), Ca/phospholipid-dependent protein kinases, etc. [140]. Mg^{2+} counteracts with Ca^{2+} in the regulation of guard cell opening [147].

High [Mg^{2+}] is known to ease saline (NaCl) stress [148,149]. Interestingly, salinity (NaCl) stress was reported to affect, in a tissue-dependent manner, the ratio of AK/NDPK [150]. This suggested that, under Na^+ excess, different tissues fine-tune their levels of nucleotides to cope with new metabolic requirements. It is possible that internal Mg^{2+} may be involved in these rearrangements, because [Na^+] is known to affect [Mg^{2+}], and vice versa [117]. Na^+ has been reported to displace binding of Mg^{2+} to several enzymes which specifically require Mg for activity [148].

In rice, sorghum, and several other species, even a relatively low cytosolic [Mg^{2+}] can ameliorate toxic effects of aluminum ions (Al^{3+}). For instance, the activity of the plasma membrane MGT1 transporter in rice increases upon Al addition to the roots, to prevent the Al-dependent inhibition of root elongation [151]. Both Al^{3+} and Mg^{2+} ions are believed to compete in binding to various cellular components, including the cell wall and plasma membrane [151]. Mg-dependent processes have also been implicated in an increase in organic acids, e.g., citrate, which is involved in alleviating Al toxicity by exudation, or in controlling cytosolic pH via regulating H^+–ATPase activity [2,68].

9.6. Summary of Mg^{2+} Signaling in Plants

As outlined in Figure 5, ATP is synthesized via the oxidative and photosynthetic phosphorylation in mitochondria and chloroplasts, respectively, which is a consequence of electron transport activity and the generation of membrane potential ($\Delta\mu_H+$) [12,17,62]. The AK then equilibrates adenylates and establishes [Mg^{2+}] in cell compartments. Mg^{2+}, in turn, regulates Mg-dependent enzymes and controls activities of adenylate transporters, which use free adenylates [100]. The Mg-dependent enzymes have a direct impact on various biochemical reactions and physiological processes, including the regulation of transcription and translation [125], photosynthesis and respiration [12,17,50,51,55,62], polysaccharide synthesis [42], and eventually affecting overall growth and development (see ref 49 and references therein). The operation of electron transport chain (ETC), ATP synthases and AK is itself under the feedback control of free Mg^{2+} concentration (shown by dotted arrows in Figure 5).

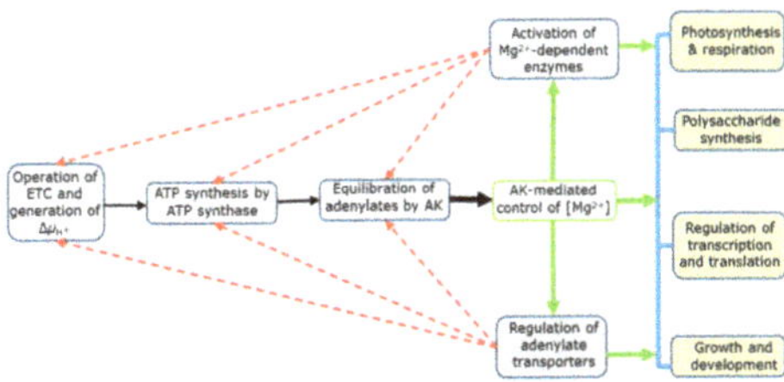

Figure 5. A simplified view of the Mg^{2+} role as enzyme substrate/cofactor and as a signal arising from adenylate pools. Abbreviations: AK, adenylate kinase; $\Delta\mu_H+$, membrane electrochemical potential; ETC, electron transport chain. Red dotted arrows refer to feedback control by Mg^{2+}.

10. Does AK Control Mg^{2+} Signaling in Other Organisms?

Although in this review we focused on plants, we are confident that Mg^{2+} signaling resulting from AK equilibrium and equilibria of related nucleotide-metabolizing enzymes is operating in all types of organisms. AK is an ancient enzyme which is widespread in all three kingdoms of life—archaea, bacteria, and eukarya [152]—and we are not aware of any group of organisms lacking AK activity. Even though the first determination of adenylate-related changes in internal [Mg^{2+}], based on AK K_{app}, was done for blood erythrocytes [53], those studies were not followed up by experimental nor theoretical research focused on AK-control of cellular [Mg^{2+}] in animals. While it is unknown whether AKs in other organisms have as prominent a role in Mg^{2+} signaling as in plant cells, those AKs, by definition, are certainly involved in adenylate equilibrium and, perhaps, in equilibrating other nucleotides. The latter property was shown, for example, for *Escherichia coli* AK, which was found to have a bifunctional role as both AK and NDPK [43]. AKs equilibrate the nucleotide pools, therefore the resulting changes of [Mg^{2+}] can be regarded as an unavoidable consequence of this equilibrium.

More studies, both experimental and theoretical, are required to assess the AK-controlled Mg^{2+} signaling in different organisms, both eukaryotic and prokaryotic. This should also take into account organ- or organism-specific types of metabolism. For instance, the essential differences in plant and animal metabolism are partially grounded in the aspects of nucleotide equilibria, which in many animal cell types are under the buffering control of creatine kinase reaction, which is enzymatically "coupled" with that of AK., e.g., as in a muscle. In the process, creatine kinase converts ADP back to ATP, assuring fast and active energy conversion. This coupling in animal cells pushes ATP/ADP ratios to high values, affects adenylate translocation and keeps Mg^{2+} at low levels [153]. The equilibrium constant of creatine kinase, considering its dependence on pH and other parameters, enables the calculation of free and Mg-bound adenylates, and performing the quantification of organ and tissue bioenergetics [154].

11. Conclusions

The most obvious and basic case of magnesium signaling is the increase in $[Mg^{2+}]$ upon the decrease in energy charge [55]. It results in the increase in $[Mg^{2+}]$ from the sub-millimolar to millimolar values in cell compartments, which leads to the regulation of many Mg-dependent enzymes and affects the operation of adenylate transporters. The release of Mg^{2+} occurs via the action of AK and triggers many processes that are regulated by $[Mg^{2+}]$, including the rate of photosynthesis, respiration, polysaccharide synthesis, stomatal opening, etc. The concentrations of other essential cations, such as Ca^{2+}, Mn^{2+} or K^+ [56], are also dependent on AK equilibrium as well as on equilibria of some other enzymes using nucleotides as substrates. The whole cellular metallome [155] depends on the balance of free and cation-bound nucleotides. Binding constants of metals with adenylates and other compounds depend on changes of $[H^+]$, which is a feedback signal of the equilibrium of pyridine nucleotides [14,156]. The interplay between redox and energy transformations triggers many signaling events (e.g., Mg^{2+}-, sugar- and Ca^{2+}-signaling, among others) [140,157,158] that initiate and regulate growth and development. Nucleotides represent the core of this signaling system, and their equilibria determine its stable and predictable operation that can be computed in the plant metabolomics framework [17]. The role of thermodynamic buffering (and adenylate equilibrium in particular) is also becoming evident in the evolutionary context [49,159].

The unique role of AK is related to its fast equilibration of adenylates and, as it is now apparent, affecting $[Mg^{2+}]$ as a feedback signal. By means of this equilibration, AK monitors and integrates different signals to ensure energy homeostasis in response to a broad range of challenges. It acts as a powerful thermodynamic buffer enzyme [48] that optimizes energy metabolism and maintains the stable and continuous operation of MgATP synthesis and consumption. It is a unique hub regulating [AMP], which itself serves as allosteric effector of essential reactions of cellular metabolism [20,21]. It appears now that AK is also at the heart of Mg^{2+} signaling. The protein that at one time was considered as just a housekeeping enzyme has indeed come a long way.

Author Contributions: Both authors (L.A.K., A.U.I.) have made substantial, direct, and intellectual contributions to the work, and approved it for publication. All authors have read and agreed to the published version of the manuscript.

Funding: The research on the topics related to this manuscript was funded by the strategic fund from Umeå University (to L.A.K.) and by the internal grant of Memorial University of Newfoundland (to A.U.I.).

Institutional Review Board Statement: Not applicable.

Informed Consent Statement: Not applicable.

Data Availability Statement: Not applicable.

Acknowledgments: The authors thank Daria Chrobok for help with the preparation of Figures 3 and 4.

Conflicts of Interest: The authors declare no conflict of interest.

Abbreviations

AEC	Adenylate energy charge
AK	Adenylate kinase
AMPK	AMP-activated protein kinase
CBL	Calcineurin B-like protein
CIPK	CBL-interacting protein kinase
ER	Endoplasmic reticulum
IMS	Intermembrane space
K_{app}	Apparent equilibrium constant
K_{true}	True equilibrium constant

MGT	Magnesium translocator
NDP	Nucleoside diphosphate
NMP	Nucleoside monophosphate
NTP	Nucleoside triphosphate
NDPK	Nucleoside diphosphate kinase
P_i	Inorganic phosphate
PPase	Pyrophosphatase
PP_i	Inorganic pyrophosphate
SuSy	Sucrose synthase
UGPase	UDP-glucose pyrophosphorylase

References

1. Kobayashi, N.I.; Tanoi, K. Critical issues in the study of magnesium transport systems and magnesium deficiency symptoms in plants. *Int. J. Mol. Sci.* **2015**, *16*, 23076–23093. [CrossRef]
2. Hermans, C.; Conn, S.J.; Chen, J.; Xiao, Q.; Verbruggen, N. An update on magnesium homeostasis mechanisms in plants. *Metallomics* **2013**, *5*, 1170–1183. [CrossRef] [PubMed]
3. Dorenstouter, H.; Pieters, G.; Findenegg, G. Distribution of magnesium between chlorophyll and other photosynthetic functions in magnesium deficient "sun"; and "shade"; leaves of poplar. *J. Plant Nutr.* **1985**, *8*, 1089–1101. [CrossRef]
4. White, P.J.; Broadley, M.R. Biofortification of crops with seven mineral elements often lacking in human diets—Iron, zinc, cop-per, calcium, magnesium, selenium and iodine. *New Phytol.* **2009**, *182*, 49–84. [CrossRef] [PubMed]
5. Martin, M.H.; Marschner, H. The mineral nutrition of higher plants. *J. Ecol.* **1988**, *76*, 1250. [CrossRef]
6. Cakmak, I.; Hengeler, C.; Marschner, H. Partitioning of shoot and root dry matter and carbohydrates in bean plants suffering from phosphorus, potassium and magnesium deficiency. *J. Exp. Bot.* **1994**, *45*, 1245–1250. [CrossRef]
7. Hermans, C.; Bourgis, F.; Faucher, M.; Strasser, R.J.; Velrot, S.; Verbruggen, N. Magnesium deficiency in sugar beets alters sugar partitioning and phloem loading in young mature leaves. *Planta* **2004**, *220*, 541–549. [CrossRef]
8. Zhang, B.; Cakmak, I.; Feng, J.; Yu, C.; Chen, X.; Xie, D.; Wu, L.; Song, Z.; Cao, J.; He, Y. Magnesium deficiency reduced the yield and seed germination in wax gourd by affecting the carbohydrate translocation. *Front. Plant Sci.* **2020**, *11*, 797. [CrossRef]
9. Tanoi, K.; Kobayashi, N.I. Leaf senescence by magnesium deficiency. *Plants* **2015**, *4*, 756–772. [CrossRef]
10. Bruggemann, L.I.; Pottosin, I.I.; Schönknecht, G. Cytoplasmic magnesium regulates the fast activating vacuolar cation chan-nel. *J. Exp. Bot.* **1999**, *50*, 1547–1552. [CrossRef]
11. Shaul, O. Magnesium transport and function in plants: The tip of the iceberg. *Biometals* **2002**, *15*, 307–321. [CrossRef] [PubMed]
12. Igamberdiev, A.U.; Kleczkowski, L.A. Optimization of ATP synthase function in mitochondria and chloroplasts via the adenylate kinase equilibrium. *Front. Plant Sci.* **2015**, *6*, 10. [CrossRef] [PubMed]
13. Voon, C.P.; Guan, X.; Sun, Y.; Sahu, A.; Chan, M.N.; Gardeström, P.; Wagner, S.; Fuchs, P.; Nietzel, T.; Versaw, W.K.; et al. ATP compartmentation in plastids and cytosol of *Arabidopsis thaliana*revealed by fluorescent protein sensing. *Proc. Natl. Acad. Sci. USA* **2018**, *115*, E10778–E10787. [CrossRef] [PubMed]
14. Igamberdiev, A.U.; Kleczkowski, L.A. Thermodynamic buffering, stable non-equilibrium and establishment of the computable structure of plant metabolism. *Prog. Biophys. Mol. Biol.* **2019**, *146*, 23–36. [CrossRef] [PubMed]
15. O'Sullivan, W.J.; Smithers, G.W. Stability constants for biologically important metal-ligand complexes. *Methods Enzymol.* **1979**, *63*, 294–336. [CrossRef]
16. Purich, D.L.; Fromm, H.J. Studies on factors influencing enzyme responses to adenylate energy charge. *J. Biol. Chem.* **1972**, *247*, 249–255. [CrossRef]
17. Igamberdiev, A.U.; Kleczkowski, L.A. Membrane potential, adenylate levels and Mg^{2+} are interconnected via adenylate kinase equilibrium in plant cells. *Biochim. Biophys. Acta (BBA) Bioenergy* **2003**, *1607*, 111–119. [CrossRef]
18. Li, F.Y.; Chaigne-Delalande, B.; Kanellopoulou, C. Second messenger role for Mg^{2+} revealed by human T-cell immunodeficien-cy. *Nature* **2011**, *475*, 471–476. [CrossRef]
19. Atkinson, D.E. Energy charge of the adenylate pool as a regulatory parameter. Interaction with feedback modifiers. *Biochemistry* **1968**, *7*, 4030–4034. [CrossRef]
20. Halford, N.G.; Paul, M.J. Carbon metabolite sensing and signalling. *Plant Biotechnol. J.* **2003**, *1*, 381–398. [CrossRef]
21. Dzeja, P.P.; Terzic, A. Adenylate kinase and AMP signaling networks: Metabolic monitoring, signal communication and body energy sensing. *Int. J. Mol. Sci.* **2009**, *10*, 1729–1772. [CrossRef] [PubMed]
22. Schlattner, U.; Wagner, E.; Greppin, H.; Bonzon, M. Chloroplast adenylate kinase from tobacco. Purification and partial characterization. *Phytochemistry* **1996**, *42*, 589–594. [CrossRef]
23. Schlattner, U.; Wagner, E.; Greppin, H.; Bonzon, M. Changes in distribution of chloroplast adenylate kinase isoforms during floral induction. *Physiol. Plant.* **1996**, *96*, 319–323. [CrossRef]
24. Carrari, F.; Coll Garcia, D.; Schauer, N.; Lytovchenko, A.; Palacios-Rojas, N.; Balbo, I.; Rosso, M.; Fernie, A.R. Deficiency of a plastidial adenylate kinase in arabidopsis results in elevated photosynthetic amino acid biosynthesis and enhanced growth. *Plant Physiol.* **2004**, *137*, 70–82. [CrossRef]

25. Lange, P.R.; Geserick, C.; Tischendorf, G.; Zrenner, R. Functions of chloroplastic adenylate kinases in Arabidopsis. *Plant Physiol.* **2007**, *146*, 492–504. [CrossRef]

26. Feng, X.; Yang, R.; Zheng, X.; Zhang, F. Identification of a novel nuclear-localized adenylate kinase 6 from *Arabidopsis thaliana* as an essential stem growth factor. *Plant Physiol. Biochem.* **2012**, *61*, 180–186. [CrossRef]

27. Deppert, W.R.; Wagner, E. Purification of adenylate kinase from green leaves of barley, maize and *Chenopodium rubrum* L. *J. Plant Physiol.* **1995**, *145*, 17–23. [CrossRef]

28. Lee, C.P.; Taylor, N.L.; Millar, A.H. Recent Advances in the Composition and Heterogeneity of the Arabidopsis Mitochondrial Proteome. *Front. Plant Sci.* **2013**, *4*, 4. [CrossRef]

29. Noma, T.; Fujisawa, K.; Yamashiro, Y.; Shinohara, M.; Nakazawa, A.; Gondo, T.; Ishihara, T.; Yoshinobu, K. Structure and expression of human mitochondrial adenylate kinase targeted to the mitochondrial matrix. *Biochem. J.* **2001**, *358*, 225–232. [CrossRef]

30. Salvato, F.; Havelund, J.F.; Chen, M.; Rao, R.S.P.; Rogowska-Wrzesinska, A.; Jensen, O.N.; Gang, D.R.; Thelen, J.J.; Møller, I.M. The potato tuber mitochondrial proteome. *Plant Physiol.* **2013**, *164*, 637–653. [CrossRef]

31. Slovak, R.; Setzer, C.; Roiuk, M.; Bertels, J.; Göschl, C.; Jandrasits, K.; Beemster, G.T.; Busch, W. Ribosome assembly factor Adenylate Kinase 6 maintains cell proliferation and cell size homeostasis during root growth. *New. Phytol.* **2019**, *225*, 2064–2076. [CrossRef] [PubMed]

32. Kleczkowski, L.A.; Randall, D.D. Maize leaf adenylate kinase: Purification and partial characterization. *Plant Physiol.* **1986**, *81*, 1110–1114. [CrossRef] [PubMed]

33. Hamada, M.; Sumida, M.; Kurokawa, Y.; Sunayashiki-Kusuzaki, K.; Okuda, H.; Watanabe, T.; Kuby, S.A. Studies on the adenylate kinase isozymes from the serum and erythrocyte of normal and Duchenne dystrophic patients. Isolation, physicochemical properties, and several comparisons with the Duchenne dystrophic aberrant enzyme. *J. Biol. Chem.* **1985**, *260*, 11595–11602. [CrossRef]

34. Zhang, Y.; Launay, H.; Liu, F.; Lebrun, R.; Gontero, B. Interaction between adenylate kinase 3 and glyceraldehyde-3-phosphate dehydrogenase from *Chlamydomonas reinhardtii*. *FEBS J.* **2018**, *285*, 2495–2503. [CrossRef]

35. Peña, C.; Hurt, E.; Panse, V.G. Eukaryotic ribosome assembly, transport and quality control. *Nat. Struct. Mol. Biol.* **2017**, *24*, 689–699. [CrossRef]

36. Schlattner, U.; Wagner, E.; Greppin, H.; Bonzon, M. Binding of adenylate kinase to RNA. *Biochem. Biophys. Res. Commun.* **1995**, *217*, 509–514. [CrossRef]

37. Raveneau, M.-P.; Benamar, A.; Macherel, D. Water content, adenylate kinase, and mitochondria drive adenylate balance in dehydrating and imbibing seeds. *J. Exp. Bot.* **2017**, *68*, 3501–3512. [CrossRef]

38. Regierer, B.; Fernie, A.R.; Springer, F.; Perez-Melis, A.; Leisse, A.; Koehl, K.; Willmitzer, L.; Geigenberger, P.; Kossmann, J. Starch content and yield increase as a result of altering adenylate pools in transgenic plants. *Nat. Biotechnol.* **2002**, *20*, 1256–1260. [CrossRef]

39. Kleczkowski, L.A. Is leaf ADP-glucose pyrophosphorylase an allosteric enzyme? *Biochim. Biophys. Acta (BBA) Protein Struct. Mol. Enzym.* **2000**, *1476*, 103–108. [CrossRef]

40. Zala, D.; Schlattner, U.; Desvignes, T.; Bobe, J.; Roux, A.; Chavrier, P.; Boissan, M. The advantage of channeling nucleotides for very processive functions. *F1000Research* **2017**, *6*, 724. [CrossRef] [PubMed]

41. Dorion, S.; Rivoal, E. Clues to the functions of plant NDPK isoforms. *Naunyn-Schmiedeberg's Arch. Pharmacol.* **2015**, *388*, 119–132. [CrossRef] [PubMed]

42. Kleczkowski, L.A.; Igamberdiev, A.U. Optimization of nucleotide sugar supply for polysaccharide formation via thermodynamic buffering. *Biochem. J.* **2020**, *477*, 341–356. [CrossRef] [PubMed]

43. Lu, Q.; Inouye, M. Adenylate kinase complements nucleoside diphosphate kinase deficiency in nucleotide metabolism. *Proc. Natl. Acad. Sci. USA* **1996**, *93*, 5720–5725. [CrossRef] [PubMed]

44. Kleczkowski, L. Back to the drawing board: Redefining starch synthesis in cereals. *Trends Plant Sci.* **1996**, *1*, 363–364. [CrossRef]

45. Roberts, J.; Aubert, S.; Gout, E.; Bligny, R.; Douce, R. Cooperation and competition between adenylate kinase, nucleoside diphosphokinase, electron transport, and ATP synthase in plant mitochondria studied by ^{31}P-nuclear magnetic resonance. *Plant Physiol.* **1997**, *113*, 191–199. [CrossRef] [PubMed]

46. Johansson, M.; Hammargren, J.; Uppsäll, E.; MacKenzie, A.; Knorpp, C. The activities of nucleoside diphosphate kinase and adenylate kinase are influenced by their interaction. *Plant Sci.* **2008**, *174*, 192–199. [CrossRef]

47. Valenti, D.; Vacca, R.A.; Romero-Puertas, M.D.C.; De Gara, L.; Marra, E.; Passarella, S. In the early phase of programmed cell death in Tobacco Bright Yellow 2 cells the mitochondrial adenine nucleotide translocator, adenylate kinase and nucleoside diphosphate kinase are impaired in a reactive oxygen species-dependent manner. *Biochim. Biophys. Acta (BBA) Gen. Subj.* **2007**, *1767*, 66–78. [CrossRef]

48. Stucki, J.W. The thermodynamic-buffer enzymes. *Eur. J. Biochem.* **1980**, *109*, 257–267. [CrossRef]

49. Igamberdiev, A.U.; Kleczkowski, L.A. Metabolic systems maintain stable non-equilibrium via thermodynamic buffering. *BioEssays* **2009**, *31*, 1091–1099. [CrossRef]

50. Igamberdiev, A.U.; Kleczkowski, L.A. Magnesium and cell energetics in plants under anoxia. *Biochem. J.* **2011**, *437*, 373–379. [CrossRef]

51. Igamberdiev, A.U.; Kleczkowski, L.A. Optimization of CO_2 fixation in photosynthetic cells via thermodynamic buffering. *Biosystems* **2011**, *103*, 224–229. [CrossRef] [PubMed]
52. Dorion, S.; Clendenning, A.; Rivoal, J. Engineering the expression level of cytosolic nucleoside diphosphate kinase in transgenic Solanum tuberosum roots alters growth, respiration and carbon metabolism. *Plant J.* **2017**, *89*, 914–926. [CrossRef] [PubMed]
53. Rose, I.A. The state of magnesium in cells as estimated from the adenylate kinase equilibrium. *Proc. Natl. Acad. Sci. USA* **1968**, *61*, 1079–1086. [CrossRef] [PubMed]
54. Kleczkowski, L.A.; Randall, D.D. Equilibration of adenylates by maize leaf adenylate kinase—Effects of magnesium on appar-ent and true equilibria. *J. Exp. Bot.* **1991**, *42*, 537–540. [CrossRef]
55. Igamberdiev, A.U.; Kleczkowski, L.A. Implications of adenylate kinase-governed equilibrium of adenylates on contents of free magnesium in plant cells and compartments. *Biochem. J.* **2001**, *360*, 225–231. [CrossRef]
56. Blair, J.M. Magnesium, potassium, and the adenylate kinase equilibrium. Magnesium as a feedback signal from the adenine nucleotide pool. *Eur. J. Biochem.* **1970**, *13*, 384–390. [CrossRef]
57. Kholodenko, N.Y.; Kartashov, I.M.; Makarov, A.D. Some kinetic characteristics of chloroplast adenylate kinase. *Biochemistry (Moscow)* **1983**, *48*, 411–416.
58. Zhou, L.; Lacroute, F.; Thornburg, R. Cloning, expression in *Escherichia coli*, and characterization of *Arabidopsis thaliana* UMP/CMP kinase. *Plant Physiol.* **1998**, *117*, 245–254. [CrossRef]
59. Purich, D.L.; Fromm, H.J. Additional factors influencing enzyme responses to the adenylate energy charge. *J. Biol. Chem.* **1973**, *248*, 461–466. [CrossRef]
60. Pradet, A.; Raymond, P. Adenine nucleotide ratios and adenylate energy charge in energy metabolism. *Annu. Rev. Plant Physiol.* **1983**, *34*, 199–224. [CrossRef]
61. Waters, B.M. Moving magnesium in plant cells. *New Phytol.* **2011**, *190*, 510–513. [CrossRef] [PubMed]
62. Igamberdiev, A.U.; Kleczkowski, L.A. Equilibration of adenylates in the mitochondrial intermembrane space maintains respi-ration and regulates cytosolic metabolism. *J. Exp. Bot.* **2006**, *57*, 2133–2141. [CrossRef] [PubMed]
63. Bligny, R.; Gout, E. Regulation of respiration by cellular key parameters: Energy demand, ADP, and Mg^{2+}. In *Plant Respiration: Metabolic Fluxes and Carbon Balance. Advances in Photosynthesis and Respiration (Including Bioenergy and Related Processes)*; Springer: Cham, The Netherland, 2017; Volume 43, pp. 19–41.
64. Yazaki, Y.; Asukagawa, N.; Ishikawa, Y.; Ohta, E.; Sakata, M. Estimation of cytoplasmic free Mg^{2+} levels and phosphorylation potentials in mung bean root tips by in vivo ^{31}P NMR spectroscopy. *Plant Cell Physiol.* **1988**, *29*, 919–924. [CrossRef]
65. Portis, A.R. Evidence of a low stromal Mg^{2+} concentration in intact chloroplasts in the dark. *Plant Physiol.* **1981**, *67*, 985–989. [CrossRef]
66. Ishijima, S.; Uchibori, A.; Takagi, H.; Maki, R.; Ohnishi, M. Light-induced increase in free Mg^{2+} concentration in spinach chlo-roplasts: Measurement of free Mg^{2+} by using a fluorescent probe and necessity of stromal alkalinization. *Arch. Biochem. Biophys.* **2003**, *412*, 126–132. [CrossRef]
67. Gout, E.; Rébeillé, F.; Douce, R.; Bligny, R. Interplay of Mg^{2+}, ADP, and ATP in the cytosol and mitochondria: Unravelling the role of Mg^{2+} in cell respiration. *Proc. Natl. Acad. Sci. USA* **2014**, *111*, E4560–E4567. [CrossRef]
68. Bose, J.; Babourina, O.; Shabala, S.; Rengel, Z. Low-pH and aluminum resistance in arabidopsis correlates with high cytosolic magnesium content and increased magnesium uptake by plant roots. *Plant Cell Physiol.* **2013**, *54*, 1093–1104. [CrossRef]
69. Conn, S.J.; Conn, V.; Kaiser, B.N.; Leigh, R.A.; Tyerman, S.; Gilliham, M. Magnesium transporters, MGT2/MRS2-1 and MGT3/MRS2-5, are important for magnesium partitioning within Arabidopsis thaliana mesophyll vacuoles. *New Phytol.* **2011**, *190*, 583–594. [CrossRef]
70. Koch, M.; Winkelmann, M.K.; Hasler, M.; Pawelzik, E.; Naumann, M. Root growth in light of changing magnesium distribution and transport between source and sink tissues in potato (*Solanum tuberosum* L.). *Sci. Rep.* **2020**, *10*, 8796. [CrossRef]
71. Romani, A. Cellular magnesium homeostasis. *Arch. Biochem. Biophys.* **2011**, *512*, 1–23. [CrossRef]
72. Vishnu, N.; Jadoon, K.M.; Karsten, F.; Groschner, L.N.; Waldeck-Weiermair, M.; Rost, R.; Hallström, S.; Imamura, H.; Graier, W.F.; Malli, R. ATP increases within the lumen of the endoplasmic reticulum upon intracellular Ca^{2+} release. *Mol. Biol. Cell* **2014**, *25*, 368–379. [CrossRef] [PubMed]
73. DePaoli, M.R.; Hay, J.C.; Graier, W.F.; Malli, R. The enigmatic ATP supply of the endoplasmic reticulum. *Biol. Rev.* **2018**, *94*, 610–628. [CrossRef] [PubMed]
74. Komatsu, H.; Iwasawa, N.; Citterio, D.; Suzuki, Y.; Kubota, T.; Tokuno, K.; Kitamura, Y.; Oka, A.K.; Suzuki, K. Design and Synthesis of highly sensitive and selective fluorescein-derived magnesium fluorescent probes and application to intracellular 3D Mg^{2+} imaging. *J. Am. Chem. Soc.* **2004**, *126*, 16353–16360. [CrossRef] [PubMed]
75. Liu, M.; Yu, X.; Li, M.; Liao, N.; Bi, A.; Jiang, Y.; Liu, S.; Gong, Z.; Wenbin, Z. Fluorescent probes for the detection of magnesium ions (Mg^{2+}): From design to application. *RSC Adv.* **2018**, *8*, 12573–12587. [CrossRef]
76. Gupta, R.; Benovic, J.; Rose, Z. The determination of the free magnesium level in the human red blood cell by ^{31}P NMR. *J. Biol. Chem.* **1978**, *253*, 6172–6176. [CrossRef]
77. Romani, A.; Scarpa, A. Regulation of cell magnesium. *Arch. Biochem. Biophys.* **1992**, *298*, 1–12. [CrossRef]
78. Packer, L.; Douce, R. Plant cell membranes. In *Methods in Enzymology*; Academic Press: London, UK, 1987; Volume 148.

79. Farré, E.M.; Tiessen, A.; Roessner, U.; Geigenberger, P.; Trethewey, R.N.; Willmitzer, L. Analysis of the compartmentation of glycolytic intermediates, nucleotides, sugars, organic acids, amino acids, and sugar alcohols in potato tubers using a nonaqueous fractionation method. *Plant Physiol.* **2001**, *127*, 685–700. [CrossRef] [PubMed]

80. Nieman, R.H.; Clark, R.A.; Pap, D.; Ogata, G.; Maas, E.V. Effects of salt stress on adenine and uridine nucleotide pools, sugar and acid-soluble phosphate in shoots of pepper and safflower. *J. Exp. Bot.* **1988**, *39*, 301–309. [CrossRef]

81. Díaz-Troya, S.; Roldán, M.; Mallén-Ponce, M.J.; Ortega-Martínez, P.; Florencio, F.J. Lethality caused by ADP-glucose accu-mulation is suppressed by salt-induced carbon flux redirection in cyanobacteria. *J. Exp. Bot.* **2020**, *71*, 2005–2017. [CrossRef]

82. Tang, R.J.; Luan, S. Regulation of calcium and magnesium homeostasis in plants: From transporters to signaling network. *Curr. Opin. Plant Biol.* **2017**, *39*, 97–105. [CrossRef]

83. Chen, Z.C.; Peng, W.T.; Li, J.; Liao, H. Functional dissection and transport mechanism of magnesium in plants. *Semin. Cell Dev. Biol.* **2018**, *74*, 142–152. [CrossRef] [PubMed]

84. Li, L.; Tutone, A.F.; Drummond, R.S.; Gardner, R.C.; Luan, S. A novel family of magnesium transport genes in Arabidopsis. *Plant Cell* **2001**, *13*, 2761–2775. [CrossRef] [PubMed]

85. Schock, I.; Gregan, J.; Steinhauser, S.; Schweyen, R.; Brennicke, A.; Knoop, V. A member of a novel *Arabidopsis thaliana* gene family of candidate Mg^{2+} ion transporters complements a yeastmitochondrial group II intron-splicing mutant. *Plant J.* **2000**, *24*, 489–501. [CrossRef] [PubMed]

86. Li, L.-G.; Sokolov, L.N.; Yang, Y.-H.; Li, D.-P.; Ting, J.; Pandy, G.K.; Luan, S. A mitochondrial magnesium transporter functions in arabidopsis pollen development. *Mol. Plant* **2008**, *1*, 675–685. [CrossRef] [PubMed]

87. Saito, T.; Kobayashi, N.I.; Tanoi, K.; Iwata, N.; Suzuki, H.; Iwata, R.; Nakanishi, T.M. Expression and functional analysis of the CorA-MRS2-ALR-type magnesium transporter family in rice. *Plant Cell Physiol.* **2013**, *54*, 1673–1683. [CrossRef]

88. Shabala, S.; Hariadi, Y. Effects of magnesium availability on the activity of plasma membrane ion transporters and light-induced responses from broad bean leaf mesophyll. *Planta* **2005**, *221*, 56–65. [CrossRef]

89. Walker, R.B.; Walker, H.M.; Ashworth, P.R. Calcium-magnesium nutrition with special reference to serpentine soils. *Plant Physiol.* **1955**, *30*, 214–221. [CrossRef]

90. Tang, R.-J.; Zhao, F.-G.; Garcia, V.J.; Kleist, T.J.; Yang, L.; Zhang, H.-X.; Luan, S. Tonoplast CBL–CIPK calcium signaling network regulates magnesium homeostasis in Arabidopsis. *Proc. Natl. Acad. Sci. USA* **2015**, *112*, 3134–3139. [CrossRef]

91. Gao, C.; Zhao, Q.; Jiang, L. Vacuoles protect plants from high magnesium stress. *Proc. Natl. Acad. Sci. USA* **2015**, *112*, 2931–2932. [CrossRef]

92. Ma, X.; Li, Q.-H.; Yu, Y.-N.; Qiao, Y.-M.; Haq, S.U.; Gong, Z.-H. The CBL–CIPK pathway in plant response to stress signals. *Int. J. Mol. Sci.* **2020**, *21*, 5668. [CrossRef]

93. Yan, Y.-W.; Mao, D.-D.; Yang, L.; Qi, J.-L.; Zhang, X.-X.; Tang, Q.-L.; Li, Y.-P.; Tang, R.; Luan, S. Magnesium transporter MGT6 plays an essential role in maintaining magnesium homeostasis and regulating high magnesium tolerance in Arabidopsis. *Front. Plant Sci.* **2018**, *9*, 274. [CrossRef] [PubMed]

94. Mao, D.; Chen, J.; Tian, L.; Liu, Z.; Yang, L.; Tang, R.; Li, J.; Lu, C.; Yang, Y.; Shi, J.; et al. Arabidopsis transporter MGT6 mediates magnesium uptake and is required for growth under magnesium limitation. *Plant Cell* **2014**, *26*, 2234–2248. [CrossRef] [PubMed]

95. Shaul, O.; Hilgemann, D.W.; De-Almeida-Engler, J.; Van Montagu, M.; Inzé, D.; Galili, G. Cloning and characterization of a novel Mg^{2+}/H^+ exchanger. *EMBO J.* **1999**, *18*, 3973–3980. [CrossRef] [PubMed]

96. Gaash, R.; Elazar, M.; Mizrahi, K.; Avramov-Mor, M.; Berezin, I.; Shaul, O. Phylogeny and a structural model of plant MHX transporters. *BMC Plant Biol.* **2013**, *13*, 75. [CrossRef] [PubMed]

97. David-Assael, O.; Mizrachy-Dagri, T.; Berezin, I.; Brook, E.; Shaul, O.; Saul, H. Expression of AtMHX, an Arabidopsis vacuolar metal transporter, is repressed by the 5′ untranslated region of its gene. *J. Exp. Bot.* **2005**, *56*, 1039–1047. [CrossRef]

98. Vreugdenhil, D.; Aarts, M.G.M.; Koorneef, M. Exploring natural genetic variation to improve plant nutrient content. In *Plant Nutritional Genomics*; Broadley, M.R., White, P.J., Eds.; Blackwell: Oxford, UK, 2005; pp. 201–219.

99. Da Fonseca-Pereira, P.; Neri-Silva, R.; Cavalcanti, J.H.F.; Brito, D.S.; Weber, A.P.M.; Araújo, W.L.; Nunes-Nesi, A. Da-ta-mining bioinformatics: Connecting adenylate transport and metabolic responses to stress. *Trends Plant Sci.* **2018**, *23*, 961–974. [CrossRef] [PubMed]

100. Kramer, R. Influence of divalent cations on the reconstituted ADP, ATP exchange. *Biochim. Biophys. Acta (BBA) Bioenergy* **1980**, *592*, 615–620. [CrossRef]

101. Trentmann, O.; Mühlhaus, T.; Zimmer, D.; Sommer, F.; Schroda, M.; Haferkamp, I.; Keller, I.; Pommerrenig, B.; Neuhaus, H.E. Identification of chloroplast envelope proteins with critical importance for cold acclimation. *Plant Physiol.* **2020**, *182*, 1239–1255. [CrossRef]

102. Schlattner, U.; Wagner, E. The adenylate kinase family in plants: Isozyme activity is related to flower induction. *Endocytobiosis Cell Res.* **2001**, *14*, 67–73.

103. Scopes, R.K. Binding of substrates and other anions to yeast phosphoglycerate kinase. *Eur. J. Biochem.* **1978**, *91*, 119–129. [CrossRef]

104. Renz, A.; Stitt, M. Substrate specificity and product inhibition of different forms of fructokinases and hexokinases in devel-oping potato tubers. *Planta* **1993**, *190*, 166–175. [CrossRef]

105. Monasterio, O.; Cárdenas, M.L. Kinetic studies of rat liver hexokinase D (glucokinase) in non-co-operative conditions show an ordered mechanism with MgADP as the last product to be released. *Biochem. J.* **2003**, *371*, 29–38. [CrossRef] [PubMed]

106. Yu, L.; Xu, L.; Xu, M.; Wan, B.; Yu, L.; Huang, Q. Role of Mg^{2+} ions in protein kinase phosphorylation: Insights from molecu-lar dynamics simulations of ATP-kinase complexes. *Mol. Simul.* **2011**, *37*, 1143–1150. [CrossRef]

107. Stitt, M. Pyrophosphate as an energy donor in the cytosol of plant cells: An enigmatic alternative to ATP. *Bot. Acta* **1998**, *111*, 167–175. [CrossRef]

108. Davies, J.M.; Poole, R.J.; Sanders, D. The computed free energy change of hydrolysis of inorganic pyrophosphate and ATP: Apparent significance. For inorganic-pyrophosphate-driven reactions of intermediary metabolism. *Biochim. Biophys. Acta (BBA) Bioenergy* **1993**, *1141*, 29–36. [CrossRef]

109. Gutiérrez-Luna, F.M.; Hernández-Domínguez, E.-E.; Valencia-Turcotte, L.G.; Rodríguez-Sotres, R. Pyrophosphate and pyrophos-phatases in plants, their involvement in stress responses and their possible relationship to secondary metabolism. *Plant Sci.* **2017**, *267*, 11–19. [CrossRef]

110. Horder, M. Complex formation of inorganic pyrophosphate with magnesium –influence of ionic strength, supporting me-dium and temperature. *Biochim. Biophys. Acta* **1974**, *358*, 319–328. [CrossRef]

111. Leigh, R.A.; Pope, A.J.; Jennings, I.R.; Sanders, D. Kinetics of the vacuolar H$^+$-pyrophosphatase: The roles of magnesium, pyrophosphate, and their complexes as substrates, activators, and inhibitors. *Plant Physiol.* **1992**, *100*, 1698–1705. [CrossRef]

112. Maeshima, M. Vacuolar H$^+$-pyrophosphatase. *Biochim. Biophys. Acta* **2000**, *1465*, 37–51. [CrossRef]

113. Nilima, K.; Vinay, S. V-PPase in plants: An overview. *Res. J. Biotechnol.* **2008**, *3*, 57–63.

114. Fraichard, A.; Trossat, C.; Perotti, E.; Pugin, A. Allosteric regulation by Mg^{2+} of the vacuolar H$^+$-PPase from *Acer pseudoplatanus* cells. Ca^{2+}/Mg^{2+} interactions. *Biochimie* **1996**, *78*, 259–266. [CrossRef]

115. Airas, R.K. Differences in the magnesium dependences of the class I and class II aminoacyl-tRNA synthetases from *Escherichia coli*. *Eur. J. Biochem.* **1996**, *240*, 223–231. [CrossRef]

116. Grzechowiak, M.; Ruszkowski, M.; Śliwiak, J.; Szpotkowski, K.; Sikorski, M.; Jaskolski, M. Crystal structures of plant inorganic pyrophosphatase, an enzyme with a moonlighting autoproteolytic activity. *Biochem. J.* **2019**, *476*, 2297–2319. [CrossRef]

117. Pérez-Castiñeira, J.R.; Serrano, A. The H$^+$-Translocating inorganic pyrophosphatase from *Arabidopsis thaliana* is more sensitive to sodium than its Na$^+$-translocating counterpart from *Methanosarcina mazei*. *Front. Plant Sci.* **2020**, *11*, 1240. [CrossRef] [PubMed]

118. Ferjani, A.; Kawade, K.; Asaoka, M.; Oikawa, A.; Okada, T.; Mochizuki, A.; Maeshima, M.; Hirai, M.Y.; Saito, K.; Tsukaya, H. Pyrophosphate inhibits gluconeogenesis by restricting UDP-glucose formation in vivo. *Sci. Rep.* **2018**, *8*, 1–10. [CrossRef] [PubMed]

119. Decker, D.; Öberg, C.; Kleczkowski, L.A. Identification and characterization of inhibitors of UDP-glucose and UDP-sugar pyrophosphorylases for in vivo studies. *Plant J.* **2017**, *90*, 1093–1107. [CrossRef] [PubMed]

120. Decker, D.; Kleczkowski, L.A. UDP-sugar producing pyrophosphorylases—Distinct and essential enzymes with overlap-ping substrate specificities, providing *de novo* precursors for glycosylation reactions. *Front. Plant Sci.* **2019**, *9*, 1822. [CrossRef] [PubMed]

121. Delmer, D.P. The purification and properties of sucrose synthetase from etiolated *Phaseolus aureus* seedlings. *J. Biol. Chem.* **1972**, *247*, 3822–3828. [CrossRef]

122. Takeda, H.; Niikura, M.; Narumi, A.; Aoki, H.; Sasaki, T.; Shimada, H. Phosphorylation of rice sucrose synthase isoforms promotes the activity of sucrose degradation. *Plant Biotechnol.* **2017**, *34*, 107–113. [CrossRef]

123. Saito, H. Outersphere and innersphere coordinated metal ions in an aminoacyl-tRNA synthetase ribozyme. *Nucleic Acids Res.* **2002**, *30*, 5151–5159. [CrossRef]

124. Villa, T.; Pleiss, J.A.; Guthrie, C. Spliceosomal snRNAs: Mg^{2+}-dependent chemistry at the catalytic core? *Cell* **2002**, *109*, 149–152. [CrossRef]

125. Hartwig, A. Role of magnesium in genomic stability. *Mutat. Res. Mol. Mech. Mutagen.* **2001**, *475*, 113–121. [CrossRef]

126. Svetlov, V.; Nudler, E. Basic mechanism of transcription by RNA polymerase II. *Biochim. Biophys. Acta (BBA) Bioenergy* **2013**, *1829*, 20–28. [CrossRef] [PubMed]

127. Misra, V.K.; Draper, D.E. On the role of magnesium ions in RNA stability. *Biopolymers* **1998**, *48*, 113–135. [CrossRef]

128. Igamberdiev, A.U.; Hurry, V.; Krömer, S.; Gardeström, P. The role of mitochondrial electron transport during photosynthetic induction. A study with barley (*Hordeum vulgare*) protoplasts incubated with rotenone and oligomycin. *Physiol. Plant.* **1998**, *104*, 431–439. [CrossRef]

129. Santarius, K.A.; Heber, U. Changes in the intracellular levels of ATP, ADP, AMP and Pi and regulatory function of the adenylate system in leaf cells during photosynthesis. *Biochim. Biophys. Acta (BBA) Biophys. Incl. Photosynth.* **1965**, *102*, 39–54. [CrossRef]

130. Hampp, R.; Goller, M.; Ziegler, H. Adenylate levels, energy charge, and phosphorylation potential during dark-light and light-dark transition in chloroplasts, mitochondria, and cytosol of mesophyll protoplasts from *Avena sativa* L. *Plant Physiol.* **1982**, *69*, 448–455. [CrossRef]

131. Stitt, M.; Lilley, R.M.; Heldt, H.W. Adenine nucleotide levels in the cytosol, chloroplasts, and mitochondria of wheat leaf protoplasts. *Plant Physiol.* **1982**, *70*, 971–977. [CrossRef]

132. Hardie, D.G.; Hawley, S.A. AMP-activated protein kinase: The energy charge hypothesis revisited. *BioEssays* **2001**, *23*, 1112–1119. [CrossRef]

133. Broeckx, T.; Hulsmans, S.; Rolland, F. The plant energy sensor: Evolutionary conservation and divergence of SnRK1 structure, regulation, and function. *J. Exp. Bot.* **2016**, *67*, 6215–6252. [CrossRef]

134. Piattoni, C.V.; Bustos, D.M.; Guerrero, S.A.; Iglesias, A.A. Nonphosphorylating glyceraldehyde-3-phosphate dehydrogenase is phosphorylated in wheat endosperm at serine-404 by an SNF1-related Protein kinase allosterically inhibited by ribose-5-phosphate. *Plant Physiol.* **2011**, *156*, 1337–1350. [CrossRef] [PubMed]

135. Ouyang, Y.; Zhu, L.; Li, Y.; Guo, M.; Liu, Y.; Cheng, J.; Zhao, J.; Wu, Y. Architectural plasticity of AMPK revealed by electron microscopy and X-ray crystallography. *Sci. Rep.* **2016**, *6*, 24191. [CrossRef]

136. Xiao, B.; Sanders, M.J.; Underwood, E.; Heath, R.J.; Mayer, F.V.; Carmena, D.; Jing, C.; Walker, P.A.; Eccleston, J.F.; Haire, L.F.; et al. Structure of mammalian AMPK and its regulation by ADP. *Nat. Cell Biol.* **2011**, *472*, 230–233. [CrossRef]

137. Lasanthi-Kudahettige, R.; Magneschi, L.; Loreti, E.; Gonzali, S.; Licausi, F.; Novi, G.; Beretta, O.; Vitulli, F.; Alpi, A.; Perata, P. Transcript profiling of the anoxic rice coleoptile. *Plant Physiol.* **2007**, *144*, 218–231. [CrossRef]

138. Lambeck, I.C.; Fischer-Schrader, K.; Niks, D.; Roeper, J.; Chi, J.-C.; Hille, R.; Schwarz, G. Molecular Mechanism of 14-3-3 Protein-mediated Inhibition of Plant Nitrate Reductase. *J. Biol. Chem.* **2012**, *287*, 4562–4571. [CrossRef] [PubMed]

139. Gilli, R.; Lafitte, D.; Lopez, C.; Kilhoffer, M.-C.; Makarov, A.; Briand, A.C.; Haiech, J. Thermodynamic analysis of calcium and magnesium binding to calmodulin. *Biochemistry* **1998**, *37*, 5450–5456. [CrossRef] [PubMed]

140. Igamberdiev, A.U.; Hill, R.D. Elevation of cytosolic Ca^{2+} in response to energy deficiency in plants: The general mechanism of adaptation to low oxygen stress. *Biochem. J.* **2018**, *475*, 1411–1425. [CrossRef]

141. Sillen, L.G.; Martell, A.E. Stability constants of metal ion complexes. *J. Chem. Educ.* **1964**, *42*, 521.

142. Vincent, A.; Blair, J. The coupling of the adenylate kinase and creatine kinase equilibria. Calculation of substrate and feedback signal levels in muscle. *FEBS Lett.* **1970**, *7*, 239–244. [CrossRef]

143. Malmendal, A.; Linse, S.; Evenäs, J.; Forsén, S.; Drakenberg, T. Battle for the EF-Hands: Magnesium−Calcium Interference in Calmodulin. *Biochemistry* **1999**, *38*, 11844–11850. [CrossRef]

144. Banti, V.; Giuntoli, B.; Gonzali, S.; Loreti, E.; Magneschi, L.; Novi, G.; Paparelli, E.; Parlanti, S.; Pucciariello, C.; Santaniello, A.; et al. Low Oxygen Response Mechanisms in Green Organisms. *Int. J. Mol. Sci.* **2013**, *14*, 4734–4761. [CrossRef] [PubMed]

145. Møller, I.M.; Rasmusson, A.G. The role of NADP in the mitochondrial matrix. *Trends Plant Sci.* **1998**, *3*, 21–27. [CrossRef]

146. Zielinski, R.E. Calmodulin and calmodulin-binding proteins in plants. *Annu. Rev. Plant Biol.* **1998**, *49*, 697–725. [CrossRef] [PubMed]

147. Lemtiri-Chlieh, F.; Arold, S.T.; Gehring, C.A. Mg^{2+} is a missing link in plant cell Ca^{2+} signalling and homeostasis—A study on vicia faba guard cells. *Int. J. Mol. Sci.* **2020**, *21*, 3771. [CrossRef] [PubMed]

148. Serrano, R.; Mulet, J.M.; Rios, G.; Marquez, J.A.; De Larrinoa, I.I.F.; Leube, M.P.; Mendizabal, I.; Pascual-Ahuir, A.; Proft, M.; Ros, R.; et al. A glimpse of the mechanisms of ion homeostasis during salt stress. *J. Exp. Bot.* **1999**, *50*, 1023–1036. [CrossRef]

149. Chen, Z.C.; Yamaji, N.; Horie, T.; Che, J.; Li, J.; An, G.; Ma, J.F. A magnesium transporter OsMGT1 plays a critical role in salt tolerance in rice. *Plant Physiol.* **2017**, *174*, 1837–1849. [CrossRef]

150. Jacoby, R.P.; Millar, A.H.; Taylor, N.L. Investigating the role of respiration in plant salinity tolerance by analyzing mitochondrial proteomes from wheat and a salinity-tolerant amphiploid (Wheat × *Lophopyrum elongatum*). *J. Proteome Res.* **2013**, *12*, 4807–4829. [CrossRef]

151. Chen, Z.C.; Yamaji, N.; Motoyama, R.; Nagamura, Y.; Ma, J.F. Up-regulation of a magnesium transporter gene *OsMGT1* is required for conferring aluminum tolerance in rice. *Plant Physiol.* **2012**, *159*, 1624–1633. [CrossRef]

152. Moon, S.; Kim, J.; Bae, E. Structural analyses of adenylate kinases from Antarctic and tropical fishes for understanding cold adaptation of enzymes. *Sci. Rep.* **2017**, *7*, 16027. [CrossRef]

153. Barbour, R.L.; Ribaudo, J.; Chan, S.H. Effect of creatine kinase activity on mitochondrial ADP/ATP transport. Evidence for a functional interaction. *J. Biol. Chem.* **1984**, *259*, 8246–8251. [CrossRef]

154. Golding, E.M.; Teague, W.E.; Dobson, G.P. Adjustment of K' to varying pH and pMg for the creatine kinase, adenylate kinase and ATP hydrolysis equilibria permitting quantitative bioenergetic assessment. *J. Exp. Biol.* **1995**, *198*, 1775–1782. [PubMed]

155. Williams, R.J. Chemical advances in evolution by and changes in use of space during time. *J. Theor. Biol.* **2011**, *268*, 146–159. [CrossRef] [PubMed]

156. Sakano, K. Revision of biochemical pH-stat: Involvement of alternative pathway metabolisms. *Plant Cell Physiol.* **1998**, *39*, 467–473. [CrossRef]

157. Kunz, S.; Gardeström, P.; Pesquet, E.; Kleczkowski, L.A. Hexokinase 1 is required for glucose-induced repression of bZIP63, At5g22920, and BT2 in Arabidopsis. *Front. Plant Sci.* **2015**, *6*, 525. [CrossRef] [PubMed]

158. Ciereszko, I. Regulatory roles of sugars in plant growth and development. *Acta Soc. Bot. Pol.* **2018**, *87*, 3583. [CrossRef]

159. Freire, M.Á. Phosphorylation and acylation transfer reactions: Clues to a dual origin of metabolism. *Biosystems* **2020**, *198*, 104260. [CrossRef]

International Journal of **Molecular Sciences**

Review

The Regulation of Nodule Number in Legumes Is a Balance of Three Signal Transduction Pathways

Diptee Chaulagain and Julia Frugoli *

Department of Genetics & Biochemistry, Clemson University, Clemson, SC 29634, USA; dchaula@g.clemson.edu
* Correspondence: jfrugol@clemson.edu; Tel.: +1-864-656-1859

Abstract: Nitrogen is a major determinant of plant growth and productivity and the ability of legumes to form a symbiotic relationship with nitrogen-fixing rhizobia bacteria allows legumes to exploit nitrogen-poor niches in the biosphere. But hosting nitrogen-fixing bacteria comes with a metabolic cost, and the process requires regulation. The symbiosis is regulated through three signal transduction pathways: in response to available nitrogen, at the initiation of contact between the organisms, and during the development of the nodules that will host the rhizobia. Here we provide an overview of our knowledge of how the three signaling pathways operate in space and time, and what we know about the cross-talk between symbiotic signaling for nodule initiation and organogenesis, nitrate dependent signaling, and autoregulation of nodulation. Identification of common components and points of intersection suggest directions for research on the fine-tuning of the plant's response to rhizobia.

Keywords: autoregulation of nodulation; nodulation; nitrogen response in nodulation; *Medicago truncatula*

Citation: Chaulagain, D.; Frugoli, J. The Regulation of Nodule Number in Legumes Is a Balance of Three Signal Transduction Pathways. *Int. J. Mol. Sci.* **2021**, *22*, 1117. https://doi.org/10.3390/ijms22031117

Academic Editor: Jen-Tsung Chen
Received: 31 December 2020
Accepted: 21 January 2021
Published: 23 January 2021

1. Introduction

Nitrogen (N) is a major determinant of plant growth and productivity. In addition, N is required as a constituent of nitric oxide (NO) and polyamines that influence constitutive and induced plant defense [1]. While N is the most abundant gas in the atmosphere, it is unusable as a direct source of plant nutrients because of the inability of plants and most organisms to enzymatically break the triple bond of N_2 and convert it into the main forms that plant roots can take up: NO_3^- and NH_4^+. Thus, N as a plant nutrient must be obtained from decomposition products in the soil or added to soil in plant-absorbable forms.

The largest natural source of N input to the biosphere is biological nitrogen fixation, adding approximately 50–70 Tg of N globally to agricultural systems [2]. Biological nitrogen fixation is the conversion of N_2 to NH_3 catalyzed by nitrogenase enzyme in diazotrophs. These diazotrophs are both free-living and in symbiotic associations between plants and nitrogen-fixing bacteria (legume-rhizobia, *Azolla*-cyanobacteria, nonlegume-*Frankia*). A smaller amount of N input to the biosphere is contributed by nitrates in the rainwater and by organic nitrogen through manure. Non-legume plants take up on average 20–50 g of N per1 Kg of dry biomass produced [3]. In contrast, soybean, a widely cultivated legume for human consumption and animal feed, grown in unfertilized soil contains 55–70% of fixed nitrogen in its aboveground parts during the nodulation period [4]. Thus, symbiotic nitrogen fixation (SNF) is of intense interest as an alternative to chemical fertilizer. Because only a small proportion of commercial legume crop production relies on biological nitrogen fixation, a better understanding of the legume-rhizobia symbiosis could enable the efficient use of the natural process of SNF and reduce dependence on chemical N fertilizer.

SNF is the result of a mutualistic interaction between a compatible plant and diazotrophs in which the plant provides a niche and fixed carbon to bacteria in exchange for fixed nitrogen. The plant family Fabaceae (Legumes) is the third-largest family of flowering plants consisting of ~19,000 known species, 88% of which form nitrogen-fixing root nodules in symbiosis with rhizobia [5]. Legumes are commonly cultivated as food crops, forage, or green manure; *Glycine max* (soybean), *Phaseolus vulgaris* (bean), *Arachis hypogaea* (peanut), *Medicago sativa* (alfalfa) are just a few among many cultivated legumes. Combined with two legumes adopted as model systems, *Medicago truncatula* and *Lotus japonicus*, genetic studies have led to a wealth of information on the signaling involved in establishing and regulating nodule development [6]. This review addresses two systemic pathways, autoregulation of nodulation (AON), which involve control of nodule numbers in response to the establishment of symbiosis, the systemic pathway that controls nodule number in response to available soil nitrogen and addresses what is known about how the two systemic pathways interact with the local pathway for the initiation of nodules. The cross-talk between initiation of nodulation pathways and the inhibitory pathways to prevent excess nodulation occurring at the same time results in a complex array and intertwining of signals that are just beginning to be understood and appreciated.

2. Signaling to Initiate and Form Symbiotic Nodules

Legumes like *M. truncatula*, *M. sativa*, and *Pisum sativum* form indeterminate nodules (contain persistent meristem resulting in cylindrical shaped nodules), whereas legumes like *Lotus japonicus*, *Glycine max*, and *Phaseolus vulgaris* form determinate nodules (spherical nodules lacking persistent meristem) [7]. The indeterminate nodules formed by *M. truncatula* are initiated from the inner cortex next to xylem poles, and at maturity, the meristem continues to produce new cells that are eventually infected, resulting in five developmental zones within a nodule. Named as follows, from the distal to proximal end of a nodule, they are (i) the meristematic zone followed by (ii) the invasion zone-characterized by actively growing infection threads, (iii) the interzone-consisting of differentiating bacteroids, (iv) the N-fixing zone-the site of mature bacteroids in the symbiosomes that fix nitrogen, and (v) the senescence zone- containing old degrading symbiosome [7]. In contrast, determinate nodules are initiated from cell division in the outer cortex; at maturity, nodules contain a relatively homozygous bacteroid population for nitrogen fixation and the nodules senesce within a few weeks of initiation [7]. Both nodule types contain leghemoglobin, providing pink coloration to nodules, which creates a near-anoxic environment in the nodule for nitrogen fixation by the oxygen intolerant nitrogenase enzyme [8]. Despite the differences in nodule type, *M. truncatula* and *L. japonicus* share many known genetic components involved in nodule formation and autoregulation of nodulation and are used as models for their respective nodulation types. The following summary focuses on *M. truncatula* (Mt) nodulation for simplicity, referring to *L. japonicus* (Lj) or other legumes *G. max* (Gm) and *P. vulagaris* (Pv) only where knowledge from the system has been applied to bridge the gap in knowledge in *M. truncatula* or the well-documented difference in between the two systems is considered important for this review.

2.1. Symbiotic Partner Selection

The legume-rhizobia symbiosis for both models relies on chemical communication between the plant and the microbe. The communication starts as plant roots constitutively secrete specific flavonoids, which act as signals to rhizobia, inducing the production of rhizobial nodulation factors (Nod factors) [9,10] (Figure 1). Nod factors are lipochitooligosaccharides decorated by different substituents like methyl, fucosyl, acetyl, etc., giving them unique chemical structures [11,12]. The recognition of specific Nod factors by specific receptors in plants is the major determinant of host-rhizobia specificity [11].

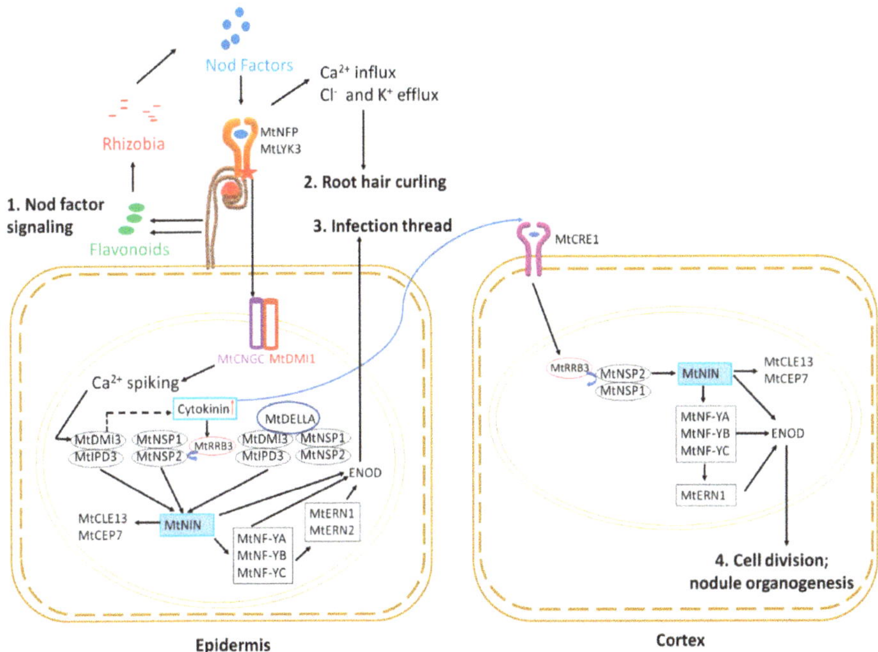

Figure 1. Early symbiotic signaling in epidermis and cortex during nodulation. Nod factor signaling: (1) the process starts with Nod factor signaling initiated by plants by producing flavonoids that attract rhizobia. Rhizobia produce Nod factors that bind to root hair receptors (MtNFP and MtLYK3), initiating the intracellular signaling cascade in the nucleus described in the text. Nuclear signaling leads to altered ion fluxes resulting in root hair curling (2). Simultaneously, Ca2+ spiking activates MtNIN through activation of a series of transcription factors; expression of early nodulin (ENOD) genes, such as *ENOD11* and *ENOD12*, facilitates infection thread progression (3). MtNIN also induces the expression of two small peptides, MtCEP7 and MtCLE13. Cytokinin signaling through MtCRE1 mediates MtNIN, MtNF-Y, and MtERN production in the cortex, which upregulates *ENOD* expression leading to cortical cell division; (4) and nodule organogenesis (see text for details).

2.2. Early Symbiotic Signaling

Nod factors (NFs) are perceived by LysM receptor-like kinases (LYKs) in root epidermal cells. Two forms of LysM receptor-like kinases play a role: MtLYK3/MtLYK4 [13] and MtNFP (NOD FACTOR PERCEPTION) [14,15]. Downstream of NF perception, three members of the MtCNGC15 family (CYCLIC NUCLEOTIDE GATED CHANNEL): CNGC15a, CNGC15b, and CNGC15c, form a complex with the potassium gated ion channel MtDMI1 (DOES NOT MAKE INFECTION 1) in the nuclear envelope to modulate nuclear calcium release [16]. The resulting calcium oscillation initiates a nuclear signaling cascade. The first step is the activation of the nuclear-localized calcium and calmodulin-dependent serine/threonine kinase MtDMI3 [17,18]. The Ca²⁺ influx and Cl⁻ and K⁺ efflux also results in cytoskeletal changes leading to root hair curling and entrapping of bacteria to form nodulation foci (Figure 1) (see Roy et al. [6] for review of genes involved in cytoskeletal changes described below).

For successful nitrogen-fixing nodules, rhizobia must reach the newly divided cortical cells that eventually develop into a nodule (discussed in nodule organogenesis section below). Rhizobia trapped in the root hair curl enter the cell by the degradation of the cell wall and move through the root hair epidermal cell into the cortex by invagination of plasma membrane forming infection thread. The formation of an infection thread requires the expression of early nodulin genes and is facilitated by transcription factor signaling in the epidermal root hair cell nucleus (Figure 1). The MtDMI3 protein, activated by calcium spiking, binds to the transcription factor MtIPD3 (INTERACTING PROTEIN OF DMI3) [19] (CCaMK-CYCLOPS in *Lotus japonicus*) and activates it by phosphorylation [20]. A mutant of *MtDMI3*, *dmi3-1*, results in plants that fail to produce infection threads and cortical cell division, leading to an absence of nodulation phenotype [21]. In contrast, *ipd3-1* mutant plants have delayed, defective infection threads, and nodulation is delayed, eventually forming non-infected nodules [22]. Interestingly, *MtIPD3L* (*IPD3-LIKE*) gene expression under the *MtIPD3* promoter rescued the *ipd3-2* mutant phenotype, showing a functional redundancy between the genes, but *ipd3l* mutants make normal infection threads and wild type nodules, while the *ipd3l ipd3-2* double mutant phenocopies *dmi3-1* [23].

The two GRAS family transcription factors MtNSP1 and MtNSP2 (*NODULATION SIGNALLING PATHWAY*) are also required for nodulation [24,25] and work as homodimers [26]. Null mutants of *MtNSP1* or *MtNSP2* are non-nodulating, whereas a weak allele of *MtNSP2* harboring a non-conservative single amino acid change results in plants with small white nodules [24,25]. MtDELLA, a negative regulator of gibberellic acid signaling, controls rhizobial infection and nodule number potentially by linking the MtDMI3-MtIPD3 and MtNSP2-MtNSP1 complexes, forming a bridge between MtIPD3 and MtNSP2 [27]. Mutants of *MtDELLA1*, *MtDELLA2*, and *MtDELLA3* have reduced nodule numbers and nodule density compared to wild type [28]. Cytokinin also regulates rhizobial infection and nodule primordia formation in multiple ways. The transcription factor MtKNOX3 (KNOTTED-1 LIKE HOMEOBOX3) is upregulated during nodule initiation, and knockdown of *MtKNOX3* causes downregulation of type A cytokinin response genes [29]. Since MtKNOX3 binds to the promoters of the *LONELY GUY* genes *MtLOG1* and *MtLOG2*, as well as *MtIPT3* (*ISOPENTYL TRANSFERASE* 3), *MtKNOX3* may increase cytokinin synthesis in developing nodules [30]. The type-B Response Regulator (MtRRB3), another transcription factor involved in cytokinin signaling, interacts with and trans-activates MtNSP2 and Cell Cycle Switch 52A (MtCCS52A), supported by the observation that *rrb3* mutants form a lower number of infection threads and nodules [31]. The symbiotic pathway can be independently activated by MtDMI3-MtIPD3, MtNSP1-MtNSP2 and MtDMI3-MtIPD3-MtDELLA-MtNSP1-MtNSP2 complexes by upregulating expression of the transcription factor MtNIN (NODULE INCEPTION) [26,27,32].

Genetically positioned downstream of *NIN*, other transcription factors also act in the pathway, including *Nuclear Factor Y* consisting of a heterotrimeric complex of NF-YA, NF-YB, and NF-YC, and *ERN1* (*ETHYLENE RESPONSE FACTOR REQUIRED FOR NODULATION 1*) and *ERN2* (Figure 1 and reviewed by Roy et al. [6]). MtNIN and MtERN1 control the expression of cell wall-associated *MtENOD11* (*EARLY NODULIN 11*) and *MtENOD12*, which are critical for infection thread development [33]. Furthermore, *MtENOD11* transcription is abolished in the *ern1* mutant [34], and upregulation of rhizobia-dependent *MtENOD11* is absent in *nfp*, *dmi1*, *dmi2*, *dmi3*, *nsp1*, and *nsp2* mutants [35], establishing early nodulin gene expression as crucial in symbiosis. Interestingly, as shown in Figure 1, *MtNIN* expression (dependent on cytokinin signaling through the MtCRE1 receptor) is sufficient to induce expression of two small peptides in the root epidermis: MtCEP7-involved in nodulation and MtCLE13-involved in inhibition of nodulation soon after rhizobial inoculation (detected at 4 h post-inoculation) and then later in the nodule primordia (4 days post-inoculation) [36].

Recently, chromatin remodeling has been preliminarily shown to be critical to the development of nodule primordia. In a report on bioRxiv, the Bisseling lab suggests *M. truncatula* histone deacetylases (MtHDTs) are required in nodule primordia, based on conditional RNAi [37]. However, this requirement appears to be because of the effect of reduced MtHDTs on a single gene: MtHDTs positively regulate 3-hydroxy-3-methylglutaryl coenzyme a reductase 1 (*MtHMGR1*) in a cell-autonomous manner [37]. The *MtHMGR* genes encode enzymes that catalyze the rate-limiting step in a pathway that synthesizes precursors to multiple plant hormones, including cytokinin, brassinosteriods, gibberellin, and abscisic acid [38] and *M. truncatula* plants carrying mutations in *HMGR1* do not initiate calcium spiking or form nodules [39].

2.3. Nodule Initiation and Organogenesis

Nodule organogenesis includes both bacterial colonization to form symbiosomes and the autoregulation of nodulation to control nodule number, which occurs simultaneously with the cell division leading to the development of the nodule as a visible root organ [6]. A gain of function mutation in the *L. japonicus HISTIDINE KINASE* gene (*LjLHK1*, a cytokinin receptor corresponding to *M. truncatula MtCRE1*) results in spontaneous nodule formation even in the absence of rhizobia, whereas loss of function mutation in the same gene causes hyperinfection and failure to timely initiate nodule primordia after rhizobia inoculation, indicating cytokinin signaling through LjLHK1 is sufficient for cell division leading to nodule development [40–42]. MtCRE1, the *M. truncatula* equivalent of LjLHK1, functions in the initial cortical cell division and later in the transition between meristematic and differentiation zones of the mature nodule [43]. MtCRE1 signaling also activates the downstream nodulation-related transcription factors MtERN1, MtNSP2, and MtNIN, as well as regulates the expression and accumulation of PINFORMED (MtPIN) auxin efflux carriers [40,43]. Furthermore, MtNIN is also required for the cortical cell division and progression of infection threads in cortical cells, as indicated by excessive nodulation foci and infection threads, but the absence of cell division and nodule primordia in a *MtNIN* promoter mutant lacking cis-regulatory cytokinin responsive elements [44]. Downstream of *MtNIN*, similar signaling events as described above in early symbiotic signaling lead to the activation of *ENOD11* (*EARLY NODULIN 11*) (Figure 1). In addition, the MtNSP1-MtNSP2 complex can bind directly to the *ENOD11* promoter to enhance its expression [25]. Both MtERN1 and MtERN2 function in the epidermis for infection thread development, whereas only MtERN1 is proposed to function in the cortex for nodule organogenesis. The *ern1* mutant displays limited root hair infection and cortical cell division, leading to growth-arrested non-infected nodules [34], whereas *ern2* (a mutant of *MtERN2*, a close sequence homolog of *MtERN1*) forms prematurely senescing nitrogen-fixing nodules that are partially defective in rhizobial colonization [45]. In contrast, *ern1 ern2* double mutant plants are impaired in root hair infection and do not display any symbiotic interactions, leading to the proposal that only MtERN1 functions in nodule organogenesis [45]. In addition to the role of transcription factors, nodule primordia formation in *M. truncatula* requires a local accumulation of auxin at the site of nodule initiation in the inner cortex, generated by inhibition of polar auxin transport (PAT) [46,47]. N signaling was recently linked to root growth in Arabidopsis through phosphorylation/dephosphorylation of PINs (auxin efflux carriers) [48], and this provides a way for PAT to be involved in both initiation and inhibition of nodule formation.

3. Signaling to Inhibit Nodule Formation

While the legume-rhizobia symbiosis involves the plant providing carbon and a niche for bacteria in exchange for nitrogen, the respiratory cost of a nodule associated with nitrogenase activity is 2–3 g carbon per nodule, whereas the total respiratory cost of nitrogen-fixing nodule is 3–5 g of carbon per nodule (or 6–12 g carbon per gram nitrogen) [49]. Increasing the number of nodules does not necessarily increase the total amount of nitrogen fixed, as observed in mutants that make a higher number of nodules (hypernodulating mutants) compared to wild type [50]. Thus, it is ecologically advantageous for legumes to suppress nodulation and SNF in the presence of soil nitrogen, as well as regulate the number of nodules formed according to the plant's nitrogen requirements when growing in soil depleted of nitrogen. Two independent mechanisms exist for regulating nodulation: (i) N dependent control of nodulation, and (ii) autoregulation of nodulation (AON).

3.1. Control of Nodulation Based on Nitrogen Need and Soil Availability

High levels of soil N suppress nodulation and inhibit nitrogen fixation in already formed nodules rapidly after the addition of nitrogen fertilizer [51]. The response to soil nitrate in plants is both local and systemic. Nitrate transporters have an important role in nitrate sensing and nitrogen demand signaling in Arabidopsis (reviewed in [52]) but only two members of the NITRATE TRANSPORTER 1 (NRT1)/PEPTIDE TRANSPORTER (PTR) family have been studied in *M. truncatula*. Of the two, MtNPF6.8 affects nitrate-dependent regulation of primary root growth via abscisic acid signaling, suggesting its role as a nitrate sensor, whereas MtNPF1.7 is essential for nodule formation but does not have a known role in N demand signaling [53]. Externally sourced high N leads to higher shoot concentrations of N, resulting in increased shoot-to-root auxin transport in wild type, which is correlated with the reduced lateral root density. In contrast, nodule density in response to external N is not correlated to root-to-shoot auxin transport [54]. Two recent studies showed that expression of a small peptide of the CLAVATA3 (CLV)/EMBRYO SURROUNDING REGION (ESR)-RELATED (CLE) family in roots, MtCLE35, is induced in the presence of high nitrate and by rhizobia (Figure 2a) [55,56]. The overexpression of *MtCLE35* in roots reduces the nodule number in wild type systemically, depending on the MtSUNN receptor [55–57] demonstrated by the reduction of nodule number in non-transgenic roots of composite plants harboring transgenic roots overexpressing *MtCLE35* [55]. The *MtCLE35* overexpression was unable to reduce nodule number in *rdn1* mutant [56] suggesting MtCLE35 requires a similar post-transcriptional modification as MtCLE12 [56,58].

High N reduces the accumulation of miR2111 in both the shoot and root of the plant, and while *MtCLE35* overexpression reduces the accumulation of miR2111 in both the shoot and root, this accumulation is independent of external N availability, indicating nitrate control of nodulation shares the miR2111-TML components of the AON pathway of repression discussed in the next section [57]. Interestingly, the downregulation of *MtCLE35* using RNAi resulted in significant accumulation of miR2111 in the root but repression of only the *MtTML2* (TOO MUCH LOVE 2) transcript was observed [57]. Since miR2111 can target both *MtTML1* and *MtTML2* [59], the repression of only *MtTML2* combined with a partial but not complete bypass of N inhibition of nodulation by ectopic expression of miR2111, suggests the possibility of an alternate pathway for N inhibition or miR2111 independent differential post-transcriptional regulation of the two TMLs [57]. Thus, MtCLE35 is the systemic nitrate signal inhibiting nodulation in response to nitrate (Figure 2a) and also controls nodule number in AON in the presence of rhizobia (see AON section below and Figure 2c). Locally, in response to high nitrate, the NIN-LIKE PROTEIN (MtNLP1) re-localizes from the cytosol to the nucleus in root cells to inhibit rhizobial infection and nodule formation by physically interacting with MtNIN to suppress *CYTOKININ RESPONSE1 (MtCRE1)* expression, thus inhibiting nodulation [60] (Figure 2a). Additionally, functional *MtNLP1* is required for N-dependent induction of the *MtCLE35* transcript [57], suggesting the transcription factor MtNLP1 controls both local and systemic response to N.

Currently, peptides, miRNAs, and hormones are all known to mediate nitrogen demand signaling under low nitrate conditions by binding to the receptors or by affecting gene expression in root and or shoot [59,61,62] (see Figure 2b). A root generated peptide from the C-TERMINALLY ENCODED PEPTIDE (CEP) family, MtCEP1, is the only well-studied peptide signal of *M. truncatula* nitrogen demand signaling, exhibit negative effects on lateral root formation and positive effect on nodulation under low N conditions [61] (Figure 2b). CEP1 is generated in roots under low soil N availability and binds to the shoot receptor COMPACT ROOT (MtCRA2) to control nodulation from the shoot and lateral root formation from the root [63,64].

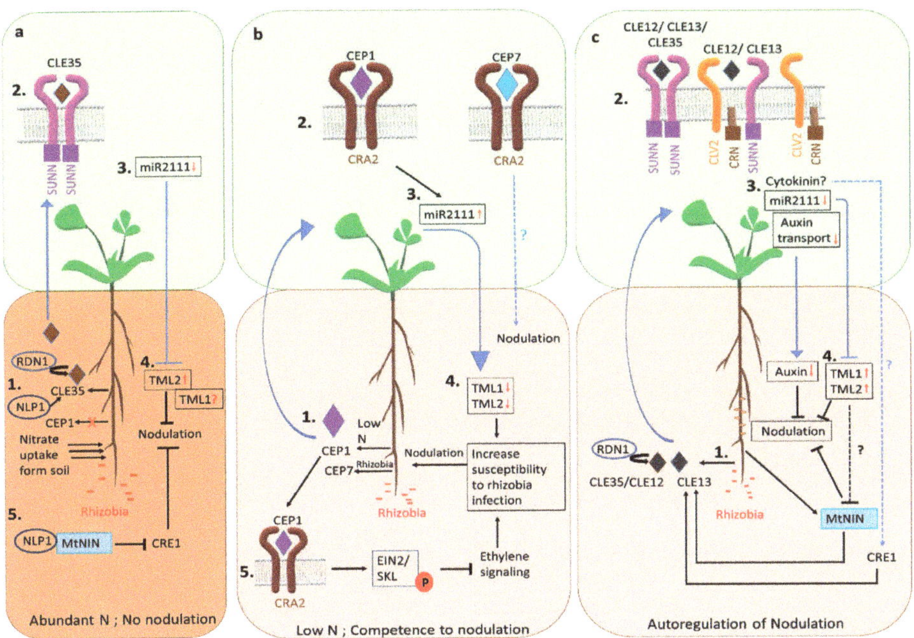

Figure 2. Mechanisms of controlling nodulation in *M. truncatula*. Under abundant soil N availability, plants take up nitrate from soil and do not form symbiotic nodules. (**a**) In response to nitrate (1), the expression of *MtCLE35* is induced in the roots, dependent on the transcription factor MtNLP1. The small peptide MtCLE35 reduces nodule number systemically dependent on MtSUNN (2). This results in reduced expression of miR2111 in the shoot (3), releasing the repression of the *MtTML2* transcript by miR2111 in the root (4). The reason for continued repression of *MtTML1*, which is also the target of miR2111, is not understood; whether MtTML1 is involved in nitrate control of nodulation is not clear (see text). Locally in the roots (5), MtNLP1 binds to MtNIN, inhibiting MtCRE1 expression to inhibit nodulation. (**b**) Under N limited conditions, the small peptide MtCEP1 is generated in the roots (1), which binds to the MtCRA2 receptor in the shoot (2), upregulating the expression of miR2111 in the shoot (3). Evidence suggests miR2111 is transported through the phloem to the roots, increasing the abundance of mature miR2111 in the roots. Mature miR2111 targets and lowers the transcript levels of *MtTML1* and *MtTML2* (4), increasing susceptibility to rhizobia and reducing AON. MtCEP1 activates MtCRA2 in the roots to phosphorylate MtEIN2, preventing its cleavage, thus repressing the ethylene response and promoting susceptibility to rhizobia (5). Another peptide, MtCEP7, generated in response to rhizobia, promotes nodulation dependent on the MtCRA2 receptor. (**c**) autoregulation of nodulation (AON). Rhizobial inoculation leading to nodule initiation results in the generation of MtCLE12 and MtCLE13 signaling peptides in the roots (1), which are transported in the xylem and bind to a shoot receptor complex containing MtSUNN, MtCRA2, and MtCRN (2). Together, the complex causes downregulation of miRNA2111 expression in the shoot (3). The result, perhaps through the transport of cytokinin as well as miRNA2111, is decreased miR2111 abundance in the roots and increased transcript levels of its targets *MtTML1* and *MtTML2* (4), inhibiting further nodulation. Another small peptide MtCLE35, which is also induced by high nitrate, controls the nodule number depending on MtRDN1 and MtSUNN in a similar manner as MtCLE12. MtNIN, involved in nodule organogenesis, also activates *MtCLE13* expression to initiate AON. MtNIN might be under MtTML1 and MtTML2 regulation, maintaining a feedback control between nodule organogenesis and AON. Another potential shoot to root signal, cytokinin, may function through MtCRE1, which is required for *MtCLE13* expression. Dashed lines represent proposed mechanisms; blue lines indicate systemic action and black lines indicate local events. All gene names in the figure are *M. truncatula* gene names, shown without the initial *Mt* for simplicity.

Recent research reveals some of the aspects of root competence for nodulation under low nitrogen conditions. The CEP1-CRA2 dependent enhanced expression of miR2111 in shoots (see AON section below) was shown to target the mRNAs of the genes *TOO MUCH LOVE 1* (*TML1*) and *TML2* in roots, lowering the transcript levels of both genes presumably to maintain susceptibility to symbiotic nodulation [59] (Figure 2b). A separate study showed that CEP1-activated MtCRA2 phosphorylates MtEIN2 (ETHYLENE INSENSITIVE 2), preventing its cleavage and repressing an ethylene response, thus promoting the root susceptibility to rhizobia [62] (Figure 2b). Exogenous ethylene treatment during the first 24 to 48 h of rhizobial inoculation is enough to suppress nodulation in wild type plants [65]. Although MtCRA2 is the common component in the systemic peptide-miRNA pathway and the root-localized ethylene pathway of maintaining root susceptibility to rhizobia, the exact mechanism of how the two pathways coordinate susceptibility is still unknown.

3.2. Autoregulation of Nodulation Signaling

In addition to local signaling, the existence of a long-distance systemic signaling mechanism controlling the nodule number was demonstrated by split-root experiments, in which prior inoculation of one-half of the root system suppressed nodulation in the other half [66]. Such long-distance signaling controlling overall nodule number depending on early nodulation events was termed "autoregulation of nodulation (AON)" [67]. AON is now known to involve root and shoot components and root-to-shoot-to-root signals through combined evidence from genetic and biochemical studies carried out by many researchers (Figure 2c). Split root experiments in *M. truncatula* demonstrated that AON occurs between two and three days after inoculation with rhizobia, and the same level of suppression is maintained for at least 15 days [50].

3.2.1. Components of AON

AON involves receptors, modifying enzymes, and transcription factors that are known to act specifically from root or shoot to control nodule number. In addition to the local signaling, peptides, hormones, and miRNAs are signaling systemically. A defect in a component of AON reduces the suppression of nodulation, resulting in a hypernodulating phenotype [67]. Root to shoot reciprocal grafting experiments using hypernodulating mutants have provided insights into whether the action of an AON gene is root or shoot dependent and split root and grafting experiments (Y grafts) are used to determine the involvement of a root component in generating a root signal or receiving a shoot signal [50].

In *M. truncatula*, three hypernodulating mutants contain lesions in genes encoding components of the AON pathway: *sunn*, *rdn1*, and *crn*. The *sunn* mutants contain lesions in *SUPER NUMERARY NODULES (SUNN)* encoding a CLAVATA1-like leucine-rich receptor kinase [68,69], while *rdn1* mutants lack expression of *ROOT DETERMINED NODULATOR (MtRDN1)* encoding an arabinosyl transferase enzyme [58], and the *crn* mutant contains a *Tnt1* insertion in *CORYNE (MtCRN)* encoding a pseudokinase [70]. In *Lotus japonicus*, in addition to mutants in orthologs of *SUNN* and *MtRDN1* which result in hypernodulation, three cloned hypernodulators are described harboring defects in (i) *CLAVATA2 (LjCLV2)*, a receptor-like protein without a kinase domain [71], (ii) *KLAVIER (LjKLV)*, a receptor-like kinase [72,73] and (iii) *TOO MUCH LOVE (LjTML)*, a nuclear-localized Kelch repeat-containing F-box protein [74,75]. *M. truncatula* contains two sequence homologs of LjTML, MtTML1, and MtTML2; downregulation of either gene using RNAi resulted in a slight increase in nodule number [76]. The location of the effects generating the hypernodulation phenotype for all mutants except *rdn1* and *tml* are shoot-determined (reviewed in [46]). An *M. truncatula* shoot-determined hypernodulating mutant with an unknown causative mutation named *like sunn supernodulator (lss)* may function by epigenetic modification at the *SUNN* locus, as the defect is a lack of *SUNN* expression even though the *SUNN* sequence is a wild type [77]. The plasma-membrane-localized SUNN protein exists as a homomer, and in heteromeric form with MtCLV2 or MtCRN, hence it is likely to function in a receptor complex [70].

3.2.2. Systemic Signals of AON

Two members of the CLE peptide family, MtCLE12 and MtCLE13 (LjCLE-RS1/LjCLE-RS2; GmRIC1/GmRIC2; PvRIC1/PvRIC2) are known root to shoot signals of AON

(Figure 2c) [78–80]. In addition, MtCLE35 was recently identified as a root signal in response to rhizobia as well as high soil nitrate [55,56]. The mature CLEs functional in AON are post-transcriptionally modified 13 amino acid long peptides containing three arabinose residues added to the 7th hydroxyproline amino acid residue [81].

A mobile microRNA, miR2111, which is downregulated in *L. japonicus* shoots after rhizobial inoculation, moves from shoot to root through the phloem and functions as a shoot-to-root signal [82]. The expression of miR2111 in the shoot decreases after inoculation in *M. truncatula*, dependent on the shoot receptor kinase SUNN [59]. Rhizobial inoculation leads to reduced shoot to root auxin transport overlapping with the onset of AON in wild type plants but not in hypernodulating *sunn-1* plants, which have constitutively higher levels of the shoot to root auxin transport [83]. The study also found that the application of an auxin transport inhibitor at the shoot/root junction reduces nodule number in *sunn-1*, indicating that auxin can be a signal inhibiting nodulation. Rhizobial inoculation also results in upregulation of cytokinin biosynthesis in the shoot in a LjHAR1 dependent manner, which can inhibit nodulation; thus, cytokinin has also been proposed as a shoot-to-root signal [84]. Unlike the conserved and well-accepted CLE peptides as a signal for root-to-shoot signaling, agreement on a single shoot-to-root signal is lacking and the possibility of multiple shoot generated signals cannot be ruled out.

3.2.3. Mechanism of AON

In response to rhizobia, expression of the transcription factor MtNIN is induced (see early symbiotic signaling and nodule organogenesis sections), which binds to the promoter of *MtCLE13* activating its transcription [36] (Figure 2c). In addition, CLE12 and CLE13 expression is also dependent on cytokinin through MtCRE1 [36,40,85]. The induction of CLE expression after rhizobial inoculation is the first known step of the AON pathway. Knockdown of *MtCLE12* and *MtCLE13* using RNAi results in a significant increase in nodule number [85]. Overexpression of *MtCLE12* or *MtCLE13* or *MtCLE35* in roots inhibits nodulation in wild type, depending on the SUNN receptor in the shoot, potentially through xylem mediated translocation of peptides [46,69,70,79]. The post-transcriptional modification of CLEs resulting in triarabinosylated peptides is essential for the function in the negative regulatory pathway to control nodules as demonstrated by the absence of a nodule suppression phenotype upon ectopic expression of non-arabinosylated CLEs [81]. The MtRDN1 mediated arabinosylation of MtCLE12, but not MtCLE13, is required for AON [58]. Similarly, MtCLE35 overexpression in *rdn1* mutant plants does not suppress nodulation, indicating a similar requirement of MtRDN1 mediated modification for its function in AON [56]. Whether or how MtCLE13 is modified is still unknown. There is likely a critical spatiotemporal regulation between the CLEs functioning in AON and those integrated with nitrogen demand signaling; further studies in the field are required to improve our current understanding.

The binding of a root-derived signal to a shoot receptor results in the generation of two potential shoot-derived signals mentioned above, cytokinin and miRNA2111 [59,82,84]. Rhizobial inoculation reduces the expression of miR2111 in the shoots (dependent on MtSUNN/LjHAR1), resulting in decreased abundance of the mature miR2111 in the roots, which in turn prevents degradation of *MtTML1* and *MtTML2* transcripts, leading to inhibition of further nodulation [59,82]. However, the mechanism of reduction in miR2111 expression by MtSUNN/LjHAR1 remains unknown. The enhanced cytokinin production in shoots after rhizobia inoculation [84] and the involvement of cytokinin through the cytokinin receptor MtCRE1 in the generation of CLE peptides [36,85] suggest a potential involvement of cytokinin in AON, but the transport of cytokinin to the roots and whether the cytokinin that affects the CLEs is shoot derived has not been established, due to technical difficulties in experimental design. Furthermore, while the altered shoot to root transport and root accumulation of auxin depending on MtSUNN is correlated with AON [54,83] (Figure 2c), the exact mechanism of how auxin controls the number of nodules in AON is unknown.

Interestingly, hypernodulating mutants carrying mutations in *SUNN* and *RDN1* that lack early suppression of nodulation are also tolerant to environmental nitrate [77,86]. However, they do show a suppression effect at 10 days or more post-inoculation, an effect synchronous to the formation of mature nodules, indicating suppression by biologically fixed nitrogen occurs in these mutants [50], but again, the underlying mechanism is unknown.

In addition to effects from auxin and cytokinin, high levels of ethylene suppress nodulation in wild type plants and in all hypernodulating mutants studied to date for this effect of ethylene in nodulation, with the exception of a mutant defective in ethylene signaling, *sickle* (*skl*) in *M. truncatula* [46,65]. The hypernodulation and lack of ethylene sensitivity to nodulation in *skl* are caused by a mutation in *M. truncatula* ortholog of Arabidopsis ethylene signaling protein EIN2 [87]. Ethylene sensitivity in *L. japonicus* requires two copies of *EIN2*, *LjEIN2-1*, and *LjEIN2-2*, for the conserved function [88]. The ethylene-mediated suppression of nodulation in most hypernodulating AON mutants contrasted with the *EIN* mutants suggests an independent role of ethylene and AON in controlling nodulation [65,68]. As opposed to maintaining susceptibility (described in nitrogen dependent control of nodulation, also see Figure 2b), rhizobial inoculation leads to rapid upregulation of ethylene biosynthesis that promotes MtEIN2 cleavage, activating the ethylene pathway to inhibit rhizobia infection [62]. Taken together, the data suggest that ethylene controls the number of nodules through MtEIN2 locally by controlling susceptibility to rhizobial infection independent of the AON pathway.

4. Perspective

In summary, legumes balance the number of nodules formed with the plant's need for N by the integration of the outputs of at least three signaling pathways. Both local and systemic, these pathways include a pathway for nodule initiation, an inhibitory pathway for nodule number (AON), and an inhibitory pathway based on N sufficiency. The more we know, the more questions there are to ask. For example, how do the specific cortical cells that become the nodule enter the cell cycle and divide into nodule primordia, while the adjacent cortical cell does not? Is this a point of regulation, and is it related to chromatin modification? AON occurs within 48 h of inoculation, but does it halt cell division that has already begun, or does it just prevent further initiation? N signaling has been linked to root growth in Arabidopsis through phosphorylation/dephosphorylation of PINs (auxin efflux carriers)-could this explain how N controls nodule formation as well? All of the hormones involved in regulation can be further regulated at the levels of synthesis, transport, and modification/degradation, suggesting areas ripe for future research. Finally, the observation that much of the regulation of nodule development may occur at the level of mRNA stability and translation [89] leaves much more to be discovered. As noted throughout this review, there are interesting hints about how cross-talk occurs between the three pathways, but precisely how the fine-tuning of nodule number is determined remains an open question. Research on the shared components between these pathways, the generation of mutants in multiple individual components of these pathways, and -omics experiments beyond transcriptomes hold the potential to fill in the picture of how nodule number is controlled.

Author Contributions: Writing—original draft preparation, D.C.; writing—review and editing, J.F. and D.C.; funding acquisition, J.F. Both authors have read and agreed to the published version of the manuscript.

Funding: This research was funded by the NATIONAL SCIENCE FOUNDATION, United States, gran numbers IOS 1733470 and 1444461.

Data Availability Statement: This review did not report any new data.

Conflicts of Interest: The authors declare no conflict of interest. The funders had no role in the design of the study; in the collection, analyses, or interpretation of data; in the writing of the manuscript, or in the decision to publish the results.

Abbreviations

AON   autoregulation of nodulation
N   nitrogen
SNF   symbiotic nitrogen fixation

References

1. Mur, L.A.; Simpson, C.; Kumari, A.; Gupta, A.K.; Gupta, K.J. Moving Nitrogen to the Centre of Plant Defence Against Pathogens. *Ann. Bot.* **2017**, *119*, 703–709. [CrossRef]
2. Herridge, D.F.; Peoples, M.B.; Boddey, R.M. Global Inputs of Biological Nitrogen Fixation in Agricultural Systems. *Plant Soil* **2008**, *311*, 1–18. [CrossRef]
3. Xu, G.; Fan, X.; Miller, A.J. Plant Nitrogen Assimilation and use Efficiency. *Ann. Rev. Plant Biol.* **2012**, *63*, 153–182. [CrossRef] [PubMed]
4. Ruschel, A.P.; Vose, P.; Victoria, R.; Salati, E. Comparison of Isotope Techniques and Non-Nodulating Isolines to Study the Effect of Ammonium Fertilization on Dinitrogen Fixation in Soybean, *Glycine max*. *Plant Soil* **1979**, *53*, 513–525. [CrossRef]
5. Graham, P.H.; Vance, C.P. Legumes: Importance and Constraints to Greater Use. *Plant Physiol.* **2003**, *131*, 872–877. [CrossRef] [PubMed]
6. Roy, S.; Liu, W.; Nandety, R.S.; Crook, A.; Mysore, K.S.; Pislariu, C.I.; Frugoli, J.; Dickstein, R.; Udvardi, M.K. Celebrating 20 Years of Genetic Discoveries in Legume Nodulation and Symbiotic Nitrogen Fixation. *Plant Cell* **2020**, *32*, 15–41. [CrossRef]
7. Ferguson, B.J.; Indrasumunar, A.; Hayashi, S.; Lin, M.; Lin, Y.; Reid, D.E.; Gresshoff, P.M. Molecular Analysis of Legume Nodule Development and Autoregulation. *J. Int. Plant Biol.* **2010**, *52*, 61–76. [CrossRef]
8. Garrocho-Villegas, V.; Gopalasubramaniam, S.K.; Arredondo-Peter, R. Plant Hemoglobins: What We Know Six Decades after Their Discovery. *Gene* **2007**, *398*, 78–85. [CrossRef]
9. Redmond, J.W.; Batley, M.; Djordjevic, M.A.; Innes, R.W.; Kuempel, P.L.; Rolfe, B.G. Flavones Induce Expression of Nodulation Genes in Rhizobium. *Nature* **1986**, *323*, 632–635. [CrossRef]
10. Dong, W.; Song, Y. The Significance of Flavonoids in the Process of Biological Nitrogen Fixation. *Int. J. Mol. Sci.* **2020**, *21*, 5926. [CrossRef]
11. Roche, P.; Debellé, F.; Maillet, F.; Lerouge, P.; Faucher, C.; Truchet, G.; Dénarié, J.; Promé, J. Molecular Basis of Symbiotic Host Specificity in Rhizobium Meliloti: nodH and nodPQ Genes Encode the Sulfation of Lipo-Oligosaccharide Signals. *Cell* **1991**, *67*, 1131–1143. [CrossRef]
12. Denarie, J.; Debelle, F.; Prome, J. Rhizobium Lipo-Chitooligosaccharide Nodulation Factors: Signaling Molecules Mediating Recognition and Morphogenesis. *Annu. Rev. Biochem.* **1996**, *65*, 503–535. [CrossRef] [PubMed]
13. Limpens, E.; Franken, C.; Smit, P.; Willemse, J.; Bisseling, T.; Geurts, R. LysM Domain Receptor Kinases Regulating Rhizobial Nod Factor-Induced Infection. *Science* **2003**, *302*, 630–633. [CrossRef] [PubMed]
14. Radutoiu, S.; Madsen, L.H.; Madsen, E.B.; Felle, H.H.; Umehara, Y.; Grønlund, M.; Sato, S.; Nakamura, Y.; Tabata, S.; Sandal, N. Plant Recognition of Symbiotic Bacteria Requires Two LysM Receptor-Like Kinases. *Nature* **2003**, *425*, 585–592. [CrossRef]
15. Amor, B.B.; Shaw, S.L.; Oldroyd, G.E.; Maillet, F.; Penmetsa, R.V.; Cook, D.; Long, S.R.; Dénarié, J.; Gough, C. The NFP Locus of Medicago Truncatula Controls an Early Step of Nod Factor Signal Transduction Upstream of a Rapid Calcium Flux and Root Hair Deformation. *Plant J.* **2003**, *34*, 495–506. [CrossRef] [PubMed]
16. Charpentier, M.; Sun, J.; Vaz Martins, T.; Radhakrishnan, G.V.; Findlay, K.; Soumpourou, E.; Thouin, J.; Very, A.A.; Sanders, D.; Morris, R.J.; et al. Nuclear-Localized Cyclic Nucleotide-Gated Channels Mediate Symbiotic Calcium Oscillations. *Science* **2016**, *352*, 1102–1105. [CrossRef] [PubMed]
17. Levy, J.; Bres, C.; Geurts, R.; Chalhoub, B.; Kulikova, O.; Duc, G.; Journet, E.P.; Ane, J.M.; Lauber, E.; Bisseling, T.; et al. A Putative Ca^{2+} and Calmodulin-Dependent Protein Kinase Required for Bacterial and Fungal Symbioses. *Science* **2004**, *303*, 1361–1364. [CrossRef]
18. Oldroyd, G.E. Speak, Friend, and Enter: Signalling Systems that Promote Beneficial Symbiotic Associations in Plants. *Nat. Rev. Microbiol.* **2013**, *11*, 252. [CrossRef]
19. Messinese, E.; Mun, J.; Yeun, L.H.; Jayaraman, D.; Rougé, P.; Barre, A.; Lougnon, G.; Schornack, S.; Bono, J.; Cook, D.R. A Novel Nuclear Protein Interacts with the Symbiotic DMI3 Calcium-and Calmodulin-Dependent Protein Kinase of *Medicago truncatula*. *Mol. Plant-Microbe Interact.* **2007**, *20*, 912–921. [CrossRef]
20. Yano, K.; Yoshida, S.; Muller, J.; Singh, S.; Banba, M.; Vickers, K.; Markmann, K.; White, C.; Schuller, B.; Sato, S.; et al. CYCLOPS, a Mediator of Symbiotic Intracellular Accommodation. *Proc. Natl. Acad. Sci. USA* **2008**, *105*, 20540–20545. [CrossRef]
21. Catoira, R.; Galera, C.; de Billy, F.; Penmetsa, R.V.; Journet, E.P.; Maillet, F.; Rosenberg, C.; Cook, D.; Gough, C.; Denarie, J. Four Genes of *Medicago truncatula* Controlling Components of a Nod Factor Transduction Pathway. *Plant Cell* **2000**, *12*, 1647–1666. [PubMed]
22. Horváth, B.; Yeun, L.H.; Domonkos, Á.; Halász, G.; Gobbato, E.; Ayaydin, F.; Miró, K.; Hirsch, S.; Sun, J.; Tadege, M. *Medicago truncatula* IPD3 is a Member of the Common Symbiotic Signaling Pathway Required for Rhizobial and Mycorrhizal Symbioses. *Mol. Plant-Microbe Interact.* **2011**, *24*, 1345–1358. [CrossRef] [PubMed]

23. Jin, Y.; Chen, Z.; Yang, J.; Mysore, K.S.; Wen, J.; Huang, J.; Yu, N.; Wang, E. IPD3 and IPD3L Function Redundantly in Rhizobial and Mycorrhizal Symbioses. *Front. Plant Sci.* **2018**, *9*, 267. [CrossRef] [PubMed]
24. Smit, P.; Raedts, J.; Portyanko, V.; Debelle, F.; Gough, C.; Bisseling, T.; Geurts, R. NSP1 of the GRAS Protein Family is Essential for Rhizobial Nod Factor-Induced Transcription. *Science* **2005**, *308*, 1789–1791. [CrossRef] [PubMed]
25. Kalo, P.; Gleason, C.; Edwards, A.; Marsh, J.; Mitra, R.M.; Hirsch, S.; Jakab, J.; Sims, S.; Long, S.R.; Rogers, J.; et al. Nodulation Signaling in Legumes Requires NSP2, a Member of the GRAS Family of Transcriptional Regulators. *Science* **2005**, *308*, 1786–1789. [CrossRef] [PubMed]
26. Hirsch, S.; Kim, J.; Munoz, A.; Heckmann, A.B.; Downie, J.A.; Oldroyd, G.E. GRAS Proteins Form a DNA Binding Complex to Induce Gene Expression during Nodulation Signaling in *Medicago truncatula*. *Plant Cell* **2009**, *21*, 545–557. [CrossRef]
27. Jin, Y.; Liu, H.; Luo, D.; Yu, N.; Dong, W.; Wang, C.; Zhang, X.; Dai, H.; Yang, J.; Wang, E. DELLA Proteins are Common Components of Symbiotic Rhizobial and Mycorrhizal Signalling Pathways. *Nat. Commun.* **2016**, *7*, 1–14.
28. Fonouni-Farde, C.; Tan, S.; Baudin, M.; Brault, M.; Wen, J.; Mysore, K.S.; Niebel, A.; Frugier, F.; Diet, A. DELLA-Mediated Gibberellin Signalling Regulates Nod Factor Signalling and Rhizobial Infection. *Nat. Commun.* **2016**, *7*, 1–13.
29. Azarakhsh, M.; Kirienko, A.; Zhukov, V.; Lebedeva, M.; Dolgikh, E.; Lutova, L. KNOTTED1-LIKE HOMEOBOX 3: A New Regulator of Symbiotic Nodule Development. *J. Exp. Bot.* **2015**, *66*, 7181–7195. [CrossRef]
30. Azarakhsh, M.; Rumyantsev, A.M.; Lebedeva, M.A.; Lutova, L.A. Cytokinin Biosynthesis Genes Expressed during Nodule Organogenesis are Directly Regulated by the KNOX3 Protein in *Medicago truncatula*. *PLoS ONE* **2020**, *15*, e0232352.
31. Tan, S.; Sanchez, M.; Laffont, C.; Boivin, S.; Le Signor, C.; Thompson, R.; Frugier, F.; Brault, M. A Cytokinin Signaling Type-B Response Regulator Transcription Factor Acting in Early Nodulation. *Plant Physiol.* **2020**, *183*, 1319–1330. [CrossRef] [PubMed]
32. Singh, S.; Katzer, K.; Lambert, J.; Cerri, M.; Parniske, M. CYCLOPS, a DNA-Binding Transcriptional Activator, Orchestrates Symbiotic Root Nodule Development. *Cell Host Microbe* **2014**, *15*, 139–152. [CrossRef] [PubMed]
33. Andriankaja, A.; Boisson-Dernier, A.; Frances, L.; Sauviac, L.; Jauneau, A.; Barker, D.G.; de Carvalho-Niebel, F. AP2-ERF Transcription Factors Mediate Nod Factor Dependent Mt ENOD11 Activation in Root Hairs Via a Novel Cis-Regulatory Motif. *Plant Cell* **2007**, *19*, 2866–2885. [CrossRef] [PubMed]
34. Middleton, P.H.; Jakab, J.; Penmetsa, R.V.; Starker, C.G.; Doll, J.; Kalo, P.; Prabhu, R.; Marsh, J.F.; Mitra, R.M.; Kereszt, A.; et al. An ERF Transcription Factor in *Medicago truncatula* that is Essential for Nod Factor Signal Transduction. *Plant Cell* **2007**, *19*, 1221–1234. [CrossRef] [PubMed]
35. Mitra, R.M.; Shaw, S.L.; Long, S.R. Six Nonnodulating Plant Mutants Defective for Nod Factor-Induced Transcriptional Changes Associated with the Legume-Rhizobia Symbiosis. *Proc. Natl. Acad. Sci. USA* **2004**, *101*, 10217–10222. [CrossRef]
36. Laffont, C.; Ivanovici, A.; Gautrat, P.; Brault, M.; Djordjevic, M.A.; Frugier, F. The NIN Transcription Factor Coordinates CEP and CLE Signaling Peptides that Regulate Nodulation Antagonistically. *Nat. Commun.* **2020**, *11*, 1–13. [CrossRef]
37. Li, H.; Schilderink, S.; Cao, Q.; Kulikova, O.; Bisseling, T. Plant-Specific Histone Deacetylases are Essential for Early as Well as Late Stages of Medicago Nodule Development. *bioRxiv* **2020**. [CrossRef]
38. Chappell, J.; Wolf, F.; Proulx, J.; Cuellar, R.; Saunders, C. Is the Reaction Catalyzed by 3-Hydroxy-3-Methylglutaryl Coenzyme A Reductase a Rate-Limiting Step for Isoprenoid Biosynthesis in Plants? *Plant Physiol.* **1995**, *109*, 1337–1343. [CrossRef]
39. Venkateshwaran, M.; Jayaraman, D.; Chabaud, M.; Genre, A.; Balloon, A.J.; Maeda, J.; Forshey, K.; den Os, D.; Kwiecien, N.W.; Coon, J.J.; et al. A Role for the Mevalonate Pathway in Early Plant Symbiotic Signaling. *Proc. Natl. Acad. Sci. USA* **2015**, *112*, 9781–9786. [CrossRef]
40. Gonzalez-Rizzo, S.; Crespi, M.; Frugier, F. The Medicago Truncatula CRE1 Cytokinin Receptor Regulates Lateral Root Development and Early Symbiotic Interaction with Sinorhizobium Meliloti. *Plant Cell* **2006**, *18*, 2680–2693. [CrossRef]
41. Tirichine, L.; Sandal, N.; Madsen, L.H.; Radutoiu, S.; Albrektsen, A.S.; Sato, S.; Asamizu, E.; Tabata, S.; Stougaard, J. A Gain-of-Function Mutation in a Cytokinin Receptor Triggers Spontaneous Root Nodule Organogenesis. *Science* **2007**, *315*, 104–107. [CrossRef] [PubMed]
42. Murray, J.D.; Karas, B.J.; Sato, S.; Tabata, S.; Amyot, L.; Szczyglowski, K. A Cytokinin Perception Mutant Colonized by Rhizobium in the Absence of Nodule Organogenesis. *Science* **2007**, *315*, 101–104. [CrossRef] [PubMed]
43. Plet, J.; Wasson, A.; Ariel, F.; Le Signor, C.; Baker, D.; Mathesius, U.; Crespi, M.; Frugier, F. MtCRE1-dependent Cytokinin Signaling Integrates Bacterial and Plant Cues to Coordinate Symbiotic Nodule Organogenesis in *Medicago truncatula*. *Plant J.* **2011**, *65*, 622–633. [CrossRef] [PubMed]
44. Liu, J.; Rutten, L.; Limpens, E.; van der Molen, T.; van Velzen, R.; Chen, R.; Chen, Y.; Geurts, R.; Kohlen, W.; Kulikova, O.; et al. A Remote Cis-Regulatory Region is Required for NIN Expression in the Pericycle to Initiate Nodule Primordium Formation in *Medicago truncatula*. *Plant Cell* **2019**, *31*, 68–83. [CrossRef]
45. Cerri, M.R.; Frances, L.; Kelner, A.; Fournier, J.; Middleton, P.H.; Auriac, M.C.; Mysore, K.S.; Wen, J.; Erard, M.; Barker, D.G.; et al. The Symbiosis-Related ERN Transcription Factors Act in Concert to Coordinate Rhizobial Host Root Infection. *Plant Physiol.* **2016**, *171*, 1037–1054. [CrossRef]
46. Mortier, V.; Holsters, M.; Goormachtig, S. Never Too Many? How Legumes Control Nodule Numbers. *Plant Cell Environ.* **2012**, *35*, 245–258. [CrossRef]
47. Gamas, P.; Brault, M.; Jardinaud, M.; Frugier, F. Cytokinins in Symbiotic Nodulation: When, Where, what for? *Trends Plant Sci.* **2017**, *22*, 792–802. [CrossRef]

48. Ötvös, K.; Marconi, M.; Vega, A.; O'Brien, J.; Johnson, A.; Abualia, R.; Antonielli, L.; Montesinos, J.C.; Zhang, Y.; Tan, S. Modulation of Plant Root Growth by Nitrogen Source-defined Regulation of Polar Auxin Transport. *EMBO J.* **2020**, e106862.

49. Minchin, F.R.; Witty, J.F. Respiratory/carbon costs of symbiotic nitrogen fixation in legumes. In *Plant Respiration*; Springer: Berlin/Heidelberg, Germany, 2005; pp. 195–205.

50. Kassaw, T.; Bridges, J.W.; Frugoli, J. Multiple Autoregulation of Nodulation (AON) Signals Identified through Split Root Analysis of *Medicago truncatula* Sunn and Rdn1 Mutants. *Plants* **2015**, *4*, 209–224. [CrossRef]

51. Streeter, J.; Wong, P.P. Inhibition of Legume Nodule Formation and N2 Fixation by Nitrate. *Crit. Rev. Plant Sci.* **1988**, *7*, 1–23. [CrossRef]

52. Xuan, W.; Beeckman, T.; Xu, G. Plant Nitrogen Nutrition: Sensing and Signaling. *Curr. Opin. Plant Biol.* **2017**, *39*, 57–65. [CrossRef] [PubMed]

53. Pellizzaro, A.; Alibert, B.; Planchet, E.; Limami, A.M.; Morère-Le Paven, M. Nitrate Transporters: An Overview in Legumes. *Planta* **2017**, *246*, 585–595. [CrossRef] [PubMed]

54. Jin, J.; Watt, M.; Mathesius, U. The Autoregulation Gene SUNN Mediates Changes in Root Organ Formation in Response to Nitrogen through Alteration of Shoot-to-Root Auxin Transport. *Plant Physiol.* **2012**, *159*, 489–500. [CrossRef] [PubMed]

55. Lebedeva, M.; Azarakhsh, M.; Yashenkova, Y.; Lutova, L. Nitrate-Induced CLE Peptide Systemically Inhibits Nodulation in *Medicago truncatula*. *Plants* **2020**, *9*, 1456. [CrossRef] [PubMed]

56. Mens, C.; Hastwell, A.H.; Su, H.; Gresshoff, P.M.; Mathesius, U.; Ferguson, B.J. Characterisation of *Medicago truncatula* CLE34 and CLE35 in Nitrate and Rhizobia Regulation of Nodulation. *New Phytol.* **2020**. [CrossRef] [PubMed]

57. Moreau, C.; Gautrat, P.; Frugier, F. Nitrate-Induced CLE35 Signaling Peptides Inhibit Nodulation through the SUNN Receptor and miR2111 Repression. *Plant Physiol.* **2021**. [CrossRef]

58. Kassaw, T.; Nowak, S.; Schnabel, E.; Frugoli, J. ROOT DETERMINED NODULATION1 is Required for M. truncatula CLE12, but Not CLE13, Peptide Signaling through the SUNN Receptor Kinase. *Plant Physiol.* **2017**, *174*, 2445–2456. [CrossRef]

59. Gautrat, P.; Laffont, C.; Frugier, F. Compact Root Architecture 2 Promotes Root Competence for Nodulation through the miR2111 Systemic Effector. *Cur. Biol.* **2020**, *30*, 1339–1345. [CrossRef]

60. Lin, J.; Li, X.; Luo, Z.; Mysore, K.S.; Wen, J.; Xie, F. NIN Interacts with NLPs to Mediate Nitrate Inhibition of Nodulation in *Medicago truncatula*. *Nature plants* **2018**, *4*, 942–952. [CrossRef]

61. Imin, N.; Mohd-Radzman, N.A.; Ogilvie, H.A.; Djordjevic, M.A. The Peptide-Encoding CEP1 Gene Modulates Lateral Root and Nodule Numbers in *Medicago truncatula*. *J. Exp. Bot.* **2013**, *64*, 5395–5409. [CrossRef]

62. Zhu, F.; Deng, J.; Chen, H.; Liu, P.; Zheng, L.; Ye, Q.; Li, R.; Brault, M.; Wen, J.; Frugier, F.; et al. A CEP Peptide Receptor-Like Kinase Regulates Auxin Biosynthesis and Ethylene Signaling to Coordinate Root Growth and Symbiotic Nodulation in *Medicago truncatula*. *Plant Cell* **2020**, *32*, 2855–2877. [CrossRef] [PubMed]

63. Huault, E.; Laffont, C.; Wen, J.; Mysore, K.S.; Ratet, P.; Duc, G.; Frugier, F. Local and Systemic Regulation of Plant Root System Architecture and Symbiotic Nodulation by a Receptor-Like Kinase. *PLoS Genet.* **2014**, *10*, e1004891. [CrossRef]

64. Mohd-Radzman, N.A.; Laffont, C.; Ivanovici, A.; Patel, N.; Reid, D.; Stougaard, J.; Frugier, F.; Imin, N.; Djordjevic, M.A. Different Pathways Act Downstream of the CEP Peptide Receptor CRA2 to Regulate Lateral Root and Nodule Development. *Plant Physiol.* **2016**, *171*, 2536–2548. [CrossRef] [PubMed]

65. Penmetsa, R.V.; Cook, D.R. A Legume Ethylene-Insensitive Mutant Hyperinfected by its Rhizobial Symbiont. *Science* **1997**, *275*, 527–530. [CrossRef] [PubMed]

66. Kosslak, R.M.; Bohlool, B.B. Suppression of Nodule Development of One Side of a Split-Root System of Soybeans Caused by Prior Inoculation of the Other Side. *Plant Physiol.* **1984**, *75*, 125–130. [CrossRef] [PubMed]

67. Olsson, J.E.; Nakao, P.; Bohlool, B.B.; Gresshoff, P.M. Lack of Systemic Suppression of Nodulation in Split Root Systems of Supernodulating Soybean (*Glycine max* [L.] Merr.) Mutants. *Plant Physiol.* **1989**, *90*, 1347–1352. [CrossRef]

68. Penmetsa, R.V.; Frugoli, J.A.; Smith, L.S.; Long, S.R.; Cook, D.R. Dual Genetic Pathways Controlling Nodule Number in *Medicago truncatula*. *Plant Physiol.* **2003**, *131*, 998–1008. [CrossRef]

69. Schnabel, E.; Journet, E.; de Carvalho-Niebel, F.; Duc, G.; Frugoli, J. The Medicago Truncatula SUNN Gene Encodes a CLV1-Like Leucine-Rich Repeat Receptor Kinase that Regulates Nodule Number and Root Length. *Plant Mol. Biol.* **2005**, *58*, 809–822. [CrossRef]

70. Crook, A.D.; Schnabel, E.L.; Frugoli, J.A. The Systemic Nodule Number Regulation Kinase SUNN in *Medicago truncatula* Interacts with MtCLV2 and MtCRN. *Plant J.* **2016**, *88*, 108–119. [CrossRef]

71. Krusell, L.; Sato, N.; Fukuhara, I.; Koch, B.E.; Grossmann, C.; Okamoto, S.; Oka-Kira, E.; Otsubo, Y.; Aubert, G.; Nakagawa, T. The Clavata2 Genes of Pea and Lotus Japonicus Affect Autoregulation of Nodulation. *Plant J.* **2011**, *65*, 861–871. [CrossRef]

72. Oka-Kira, E.; Tateno, K.; Miura, K.; Haga, T.; Hayashi, M.; Harada, K.; Sato, S.; Tabata, S.; Shikazono, N.; Tanaka, A. Klavier (Klv), a Novel Hypernodulation Mutant of Lotus Japonicus Affected in Vascular Tissue Organization and Floral Induction. *Plant J.* **2005**, *44*, 505–515. [CrossRef] [PubMed]

73. Miyazawa, H.; Oka-Kira, E.; Sato, N.; Takahashi, H.; Wu, G.J.; Sato, S.; Hayashi, M.; Betsuyaku, S.; Nakazono, M.; Tabata, S.; et al. The Receptor-Like Kinase KLAVIER Mediates Systemic Regulation of Nodulation and Non-Symbiotic Shoot Development in *Lotus japonicus*. *Development* **2010**, *137*, 4317–4325. [CrossRef] [PubMed]

74. Magori, S.; Oka-Kira, E.; Shibata, S.; Umehara, Y.; Kouchi, H.; Hase, Y.; Tanaka, A.; Sato, S.; Tabata, S.; Kawaguchi, M. TOO MUCH LOVE, a Root Regulator Associated with the Long-Distance Control of Nodulation in *Lotus japonicus*. *Mol. Plant-Microbe Interact.* **2009**, *22*, 259–268. [CrossRef] [PubMed]

75. Takahara, M.; Magori, S.; Soyano, T.; Okamoto, S.; Yoshida, C.; Yano, K.; Sato, S.; Tabata, S.; Yamaguchi, K.; Shigenobu, S. TOO MUCH LOVE, a Novel Kelch Repeat-Containing F-Box Protein, Functions in the Long-Distance Regulation of the legume–Rhizobium Symbiosis. *Plant Cell Physiol.* **2013**, *54*, 433–447.

76. Gautrat, P.; Mortier, V.; Laffont, C.; De Keyser, A.; Fromentin, J.; Frugier, F.; Goormachtig, S. Unraveling New Molecular Players Involved in the Autoregulation of Nodulation in *Medicago truncatula*. *J. Exp. Bot.* **2019**, *70*, 1407–1417.

77. Schnabel, E.; Mukherjee, A.; Smith, L.; Kassaw, T.; Long, S.; Frugoli, J. The Lss Supernodulation Mutant of *Medicago truncatula* Reduces Expression of the SUNN Gene. *Plant Physiol.* **2010**, *154*, 1390–1402. [CrossRef]

78. Okamoto, S.; Ohnishi, E.; Sato, S.; Takahashi, H.; Nakazono, M.; Tabata, S.; Kawaguchi, M. Nod Factor/Nitrate-Induced CLE Genes that Drive HAR1-Mediated Systemic Regulation of Nodulation. *Plant Cell Physiol.* **2008**, *50*, 67–77. [CrossRef]

79. Mortier, V.; Den Herder, G.; Whitford, R.; Van de Velde, W.; Rombauts, S.; D'Haeseleer, K.; Holsters, M.; Goormachtig, S. CLE Peptides Control *Medicago truncatula* Nodulation Locally and Systemically. *Plant Physiol.* **2010**, *153*, 222–237. [CrossRef]

80. Ferguson, B.J.; Li, D.; Hastwell, A.H.; Reid, D.E.; Li, Y.; Jackson, S.A.; Gresshoff, P.M. The Soybean (*Glycine max*) Nodulation-suppressive CLE Peptide, Gm RIC 1, Functions Interspecifically in Common White Bean (*Phaseolus vulgaris*), but Not in a Supernodulating Line Mutated in the Receptor Pv NARK. *Plant Biotechnol. J.* **2014**, *12*, 1085–1097.

81. Okamoto, S.; Shinohara, H.; Mori, T.; Matsubayashi, Y.; Kawaguchi, M. Root-Derived CLE Glycopeptides Control Nodulation by Direct Binding to HAR1 Receptor Kinase. *Nat. Commun.* **2013**, *4*, 2191. [CrossRef]

82. Tsikou, D.; Yan, Z.; Holt, D.B.; Abel, N.B.; Reid, D.E.; Madsen, L.H.; Bhasin, H.; Sexauer, M.; Stougaard, J.; Markmann, K. Systemic Control of Legume Susceptibility to Rhizobial Infection by a Mobile microRNA. *Science* **2018**, *362*, 233–236. [CrossRef] [PubMed]

83. van Noorden, G.E.; Ross, J.J.; Reid, J.B.; Rolfe, B.G.; Mathesius, U. Defective Long-Distance Auxin Transport Regulation in the *Medicago truncatula* Super Numeric Nodules Mutant. *Plant Physiol.* **2006**, *140*, 1494–1506. [CrossRef] [PubMed]

84. Sasaki, T.; Suzaki, T.; Soyano, T.; Kojima, M.; Sakakibara, H.; Kawaguchi, M. Shoot-Derived Cytokinins Systemically Regulate Root Nodulation. *Nat. Commun.* **2014**, *5*, 1–9. [CrossRef] [PubMed]

85. Mortier, V.; De Wever, E.; Vuylsteke, M.; Holsters, M.; Goormachtig, S. Nodule Numbers are Governed by Interaction between CLE Peptides and Cytokinin Signaling. *Plant J.* **2012**, *70*, 367–376. [CrossRef] [PubMed]

86. Schnabel, E.L.; Kassaw, T.K.; Smith, L.S.; Marsh, J.F.; Oldroyd, G.E.; Long, S.R.; Frugoli, J.A. The ROOT DETERMINED NODULA-TION1 Gene Regulates Nodule Number in Roots of Medicago Truncatula and Defines a Highly Conserved, Uncharacterized Plant Gene Family. *Plant Physiol.* **2011**, *157*, 328–340. [CrossRef] [PubMed]

87. Varma Penmetsa, R.; Uribe, P.; Anderson, J.; Lichtenzveig, J.; Gish, J.; Nam, Y.W.; Engstrom, E.; Xu, K.; Sckisel, G.; Pereira, M. The *Medicago truncatula* Ortholog of Arabidopsis EIN2, Sickle, is a Negative Regulator of Symbiotic and Pathogenic Microbial Associations. *Plant J.* **2008**, *55*, 580–595. [CrossRef]

88. Miyata, K.; Kawaguchi, M.; Nakagawa, T. Two Distinct EIN2 Genes Cooperatively Regulate Ethylene Signaling in *Lotus japonicus*. *Plant Cell Physiol.* **2013**, *54*, 1469–1477. [CrossRef]

89. Zanetti, M.E.; Blanco, F.; Reynoso, M.; Crespi, M. To Keep Or Not to Keep: mRNA Stability and Translatability in Root Nodule Symbiosis. *Curr. Opin. Plant Biol.* **2020**, *56*, 109–117. [CrossRef]

International Journal of
Molecular Sciences

MDPI

Article

From Spaceflight to Mars *g*-Levels: Adaptive Response of *A. Thaliana* Seedlings in a Reduced Gravity Environment Is Enhanced by Red-Light Photostimulation

Alicia Villacampa [1], Malgorzata Ciska [1], Aránzazu Manzano [1], Joshua P. Vandenbrink [2], John Z. Kiss [3], Raúl Herranz [1,*] and F. Javier Medina [1,*]

[1] Centro de Investigaciones Biológicas Margarita Salas (CSIC), Ramiro de Maeztu 9, 28040 Madrid, Spain; avillacampa@cib.csic.es (A.V.); mciska@cib.csic.es (M.C.); aranzazu@cib.csic.es (A.M.)
[2] School of Biological Sciences, Louisiana Tech University, Ruston, LA 71272, USA; jpvdb@latech.edu
[3] Department of Biology, University of North Carolina-Greensboro, Greensboro, NC 27402, USA; jzkiss@uncg.edu
* Correspondence: rherranz@cib.csic.es (R.H.); fjmedina@cib.csic.es (F.J.M.)

Abstract: The response of plants to the spaceflight environment and microgravity is still not well understood, although research has increased in this area. Even less is known about plants' response to partial or reduced gravity levels. In the absence of the directional cues provided by the gravity vector, the plant is especially perceptive to other cues such as light. Here, we investigate the response of *Arabidopsis thaliana* 6-day-old seedlings to microgravity and the Mars partial gravity level during spaceflight, as well as the effects of red-light photostimulation by determining meristematic cell growth and proliferation. These experiments involve microscopic techniques together with transcriptomic studies. We demonstrate that microgravity and partial gravity trigger differential responses. The microgravity environment activates hormonal routes responsible for proliferation/growth and upregulates plastid/mitochondrial-encoded transcripts, even in the dark. In contrast, the Mars gravity level inhibits these routes and activates responses to stress factors to restore cell growth parameters only when red photostimulation is provided. This response is accompanied by upregulation of numerous transcription factors such as the environmental acclimation-related WRKY-domain family. In the long term, these discoveries can be applied in the design of bioregenerative life support systems and space farming.

Keywords: microgravity; partial gravity; transcription factors; gene expression; root meristem

Citation: Villacampa, A.; Ciska, M.; Manzano, A.; Vandenbrink, J.P.; Kiss, J.Z.; Herranz, R.; Medina, F.J. From Spaceflight to Mars *g*-Levels: Adaptive Response of *A. Thaliana* Seedlings in a Reduced Gravity Environment Is Enhanced by Red-Light Photostimulation. *Int. J. Mol. Sci.* **2021**, *22*, 899. https://doi.org/10.3390/ijms22020899

Received: 15 December 2020
Accepted: 14 January 2021
Published: 18 January 2021

Publisher's Note: MDPI stays neutral with regard to jurisdictional claims in published maps and institutional affiliations.

1. Introduction

The achievement of plant cultivation in space, also called "space farming," is an important step in the development of bioregenerative life support systems to enable long-term space exploration, since plants are fundamental elements for oxygen and nutrient supplies as well as waste recycling [1]. With this objective, it is important to study the response of plants to the space environment. Plants have been successfully grown in space on numerous occasions [2], even though major physiological changes, such as the alteration of cell proliferation rate and ribosome biogenesis, have been reported [3]. Most major physiological changes are regulated and tuned by phytohormones and transcription factors (TFs). These latter function as molecular switches activating or repressing the expression of genes or sets of genes in response to different stimuli, e.g., changes in the environmental conditions. Some changes in the phytohormone levels have been previously reported in experiments performed in real and simulated microgravity, such as a different distribution of cytokinin in real microgravity [4] and auxin accumulation in simulated microgravity [5]. However, more attention has been given to the changes in plant physiology (e.g., response to hypoxia, cell wall modifications, accelerated cell cycle) rather than to the hormone regulatory pathways and TFs.

Space exploration involves the exposure of plants to microgravity conditions, as they exist on spacecraft and stations orbiting the Earth, such as the International Space Station (ISS). Although microgravity effects have been extensively studied in living organisms, they are difficult to overcome since plants, like any other terrestrial organisms, have evolved in a constant gravity vector. Plants orient their growth according to the gravity vector (gravitropism), with positive root gravitropism and negative shoot gravitropism. Nevertheless, in microgravity, the cue for this tropism (i.e., the gravity vector) is not present.

Other tropisms are also involved in directing plant growth. For instance, using light as the tropistic cue, phototropism drives plant growth orientation with a negative root phototropism and positive shoot phototropism [6]. The interaction among gravitropism, phototropism and other tropisms, such as hydrotropism [7] and thigmotropism [8], produces the overall direction of plant growth [9], which is constantly adapted to the changing environmental conditions. These well-established positive and negative tropisms on Earth must be reevaluated in space since, in the absence of the gravity vector, new phototropic responses can be observed that were masked by gravitropism in the Earth. In fact, root positive phototropic response to red light [10] and blue light [11] have been reported in spaceflight studies. Different wavelengths of light are known to promote different responses on plant growth and development [12,13]. Thus, specific light conditions could be applied to overcome some of the deleterious effects on plant growth and development induced by microgravity. For example, red light is known to stimulate cell proliferation and promote ribosome biogenesis [14], both processes affected by spaceflight. In fact, red light has already been used in a similar experiment in simulated microgravity and was applied to *Arabidopsis thaliana* seedlings as a part of the Seedling Growth (SG) series (SG1 and SG2 experiments on the ISS) [15]. The results obtained were positive, in the sense of compensating at least a part of the alterations induced by microgravity.

Furthermore, the influence of partial or reduced gravity levels on the plant physiology should be investigated to enable human settlements on nearby planets [16]. In recent years, special attention has been given to Mars. Little is known so far on the plant response to partial gravity levels, which is important considering space agencies' plans to travel back to the Moon (Deep Space Gateway, DSG) in 2024 [17] and to Mars in the near future. With the purpose of studying how partial gravity levels can affect plant development, some studies have used analogs, such as random positioning machines (RPMs), to reproduce Moon or Mars gravity levels and study their effect on Earth [18]. Simultaneously, the European Modular Cultivation System (EMCS, [19]), which was installed in the ISS from 2008 to 2018, provided the ability to apply different g-forces in space by means of a built-in centrifuge. The SG experiment was executed in this hardware to test the contribution of red and blue light stimulation interaction with the reduced gravity stimuli [20]. Firstly, we used the EMCS to investigate transcriptomic changes in *A. thaliana* seedlings exposed to different g-levels for the last two days with blue light stimulation in the SG series (SG1 and SG2). We applied different gravity levels (microgravity, 0.1g; Moon; Mars; near earth g-level; 1g) to blue-light stimulated wild-type (WT) Landsberg ecotype *A. thaliana* seedlings and demonstrated a replacement of gravitropism by blue-light-based phototropism signaling at microgravity level [21], but a striking stress response was found at 0.1g. We also determined different components of the transcriptional response to the lack of gravity as the g-gradient is progressively reduced [22].

The use of transcriptomic techniques has provided vast data on gene expression in plants grown in space. *A. thaliana* is so far the most widely studied plant in space biology using omics techniques [21–31] and microscopic methods [4,32], although a few crop species have been recently incorporated to space studies [32–36]. Scarce material, high cost, and extensive logistics are highly limiting factors for space experiments, making the investigation of plant response to the space environment challenging, reinforcing the requirement of better controls and complementary research in ground simulation and reference facilities [37]. Moreover, there is a growing awareness in the space biology community to define and use the same criteria when describing the spaceflight experiment metadata so the cross-comparisons between the spaceflight experiments can be performed more rigorously [38,39].

Here, we combined morphological and molecular approaches to describe the changes in 6-day-old *A. thaliana* seedlings (Col-0) grown in the SG experiments (SG2 and SG3) in three *g*-levels (microgravity, Mars gravity level and 1*g* ground reference run (GRR)) under red light photostimulation, and a control in darkness for the last two days of the experiment. In addition, we compared our results to other transcriptomic data obtained from *A. thaliana* seedlings grown during spaceflight and available in the GeneLab database ([40]; https://genelab.nasa.gov/) for further validation of our results. In the long term, our studies will pave the way to understand the molecular mechanisms to improve the cultivation conditions of plants on other planets.

2. Results

2.1. Anatomic Changes in Microgravity and Partial Gravity in Different Light Conditions

We analyzed the morphology of *A. thaliana* seedlings grown on the EMCS in the ISS. In our experimental conditions, the seeds germinated in altered gravity levels giving us a chance to observe how the plant deals with the new environment, meaning how they acclimate. The germination rate during the space experiment was similar to the one in the GRR (ISS samples 96.7%; GRR 93.93%) suggesting the plant activates these acclimation mechanisms already during germination.

We investigated the influence of different gravity levels and light conditions on meristem organization and size expressed in the length of the meristem (the distance from quiescent center to the first elongated cell in the epidermis) and in the number of meristematic cells in the epidermis (Figure 1A). The typical organization of the meristem with easily distinguished quiescent center and three layers of meristematic cells (epidermis, endodermis and cortex) was observed in all the conditions. These results suggest that altered gravity levels, and in general spaceflight conditions, do not disturb the well-conserved organization of the meristem in *A. thaliana*. The root cap columella also displayed its typical organization with the first meristematic layer followed by three to four layers of gravity-perceiving statocytes [41]. No difference in the number of layers of columella cells was observed among the conditions. In respect to the meristem length, in the seedlings grown in the dark, no significant changes were observed at any *g*-level. However, in the seedlings photostimulated with red light, a gradual increase in the meristem size was observed with the decrease of *g*-level (1*g* GRR < Mars < μ*g*), although the difference between Mars and μ*g* was not significant (Figure 1B). In addition, the length of the meristem and the number of meristematic cells per meristematic layer were increased in the photostimulated seedlings in comparison to the seedlings grown at the same *g*-level in darkness, although the difference was only statistically significant at the Mars gravity level (Figure 1B). These observations were similar to those published in previous reports showing that red light stimulated proliferation [14,15].

Next, we estimated the nucleolar activity by measuring the area of immunofluorescent staining using an antibody against the nucleolar protein fibrillarin, in different *g*-levels and light conditions (Figure 1C,D). Fibrillarin is a well-known and abundant nucleolar protein involved in pre-rRNA processing regulation [42], and it can be used as a nucleolar marker. Under standard conditions of growth, the size of the nucleolus in the meristematic cells is directly related to its activity, determined by the production of the ribosomal units [43]. Therefore, the size and the structural features of the nucleolus are a reliable marker of the rate of ribosome biogenesis [44], which is determined by the demand in protein synthesis, meaning the higher the nucleolar size the higher protein production. A reduction in the size of the nucleolus was observed in meristems of the seedlings grown at Mars *g*-level without photostimulation. This reduction was significant in comparison to the nucleolus in red-light photostimulated seedlings at the same gravity level, and in comparison to the seedlings grown in 1*g* GRR and in microgravity without photostimulation. This indicates that the protein biosynthesis was also reduced in this condition. Red light seems to have a positive effect on the nucleolar activity at the Mars *g*-level, since meristematic nucleoli in red photostimulated seedlings at this gravity level display similar size to the GRR seedlings. Surprisingly, nucleoli in seedlings grown in microgravity in both light conditions also had similar size as nucleoli in GRR seedlings (Figure 1D).

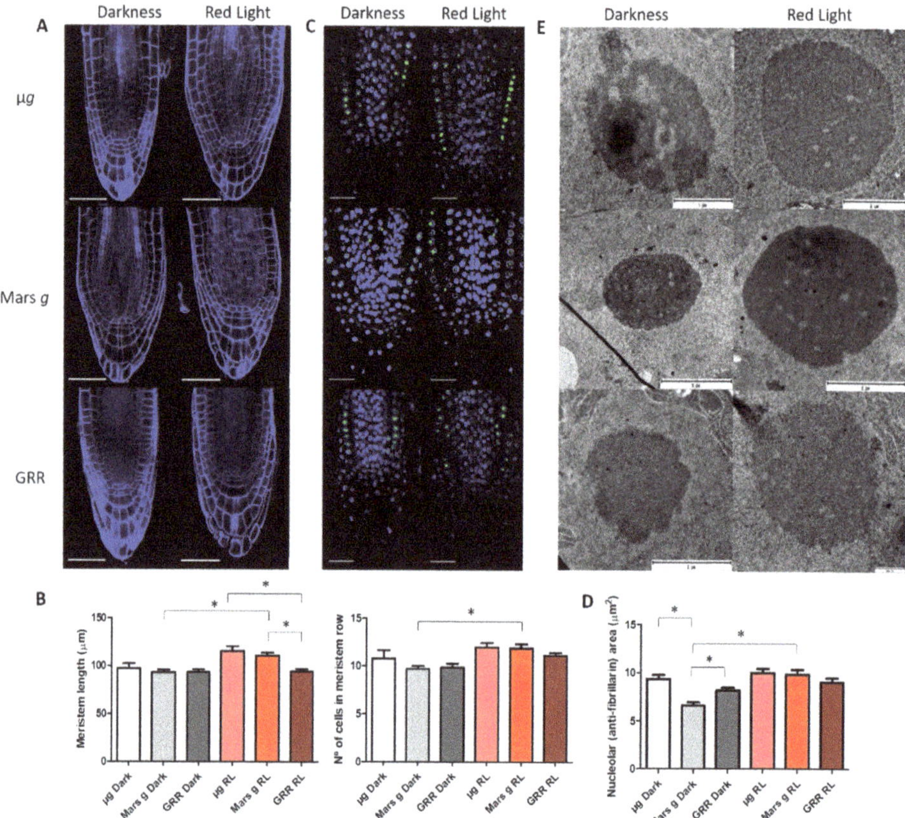

Figure 1. Meristem and nucleolus effects of microgravity and Mars gravity. (**A**) Confocal microscope images of cell-wall-stained root meristems of the different gravity and photostimulation conditions. Scale bar represents 50 µm. (**B**) Quantification of meristem length and number of cells per meristematic layer. Meristem length statistical analysis was made with ANOVA with Scheffe test post-hoc. Number of cells in the meristem row analysis was ANOVA with T3-Dunnett post-hoc (data not homoscedastic). In both cases the *p*-value for significance is 0.05. (**C**) Confocal images of immunostained root meristems; green: anti-fibrillarin, blue: DAPI. Scale bar represents 25 µm. (**D**) Nucleolar area (anti-fibrillarin immunostaining quantification) in µm². Statistical analysis was made using a non-parametric test with corrected *p*-value of 0.0033 (data without normal distribution. Bars represent mean + SED. * indicates significant statistical differences. (**E**) Electron microscope images of meristematic cell nucleoli. The difference in nucleolar size among different conditions is clearly observed. Furthermore, typical structural models corresponding to inactive nucleoli are seen in darkness conditions, especially in microgravity.

To investigate in more detail the changes that nucleolus undergoes in different light and gravity conditions, we analyzed nucleolar ultrastructure using TEM. In most conditions, nucleoli had a regular round shape and a typical structure, where three components could be distinguished: granular component (GC), dense fibrillar component (DFC) and fibrillar centers (FC) (Figure 1E). Despite the fact that we have not observed in microgravity conditions without photostimulation the reduction in nucleolar size that was observed before in etiolated plants in the ROOT experiment [3], we could confirm in TEM images that nucleoli presented features typical for low-active nucleoli (low content of GC and heterogeneous FCs with condensed chromatin inside) [45]. However, in nucleoli of red-light photostimulated seedlings exposed to microgravity and partial gravity, features of active nucleoli, such as abundant GC intermingled with DFC and small FCs, were observed. No condensed intranucleolar chromatin was observed in these samples. This observation agrees with previous reports from our ROOT spaceflight experiment, where the combined effect of etiolation and microgravity caused a significant reduction in nucleolar activity [3]. Since, in the SG2 and SG3 experiments, the seedlings germinated and grew for four days with a photoperiod, the inhibitory effect of microgravity and darkness (last two days of culture) on nucleolar activity and nucleolar size could be diminished with respect to the experiments using etiolated seedlings.

2.2. Global Transcriptomics

Principal component analysis (PCA) of the replicates is shown in Figure 2A. The clustering of the replicates is consistent. There is a very clear separation among the three *g*-levels in the samples exposed to red-light photostimulation (µgrl, Marsrl and grrrl) but it is not well distinguished between the Marsd and grrd samples.

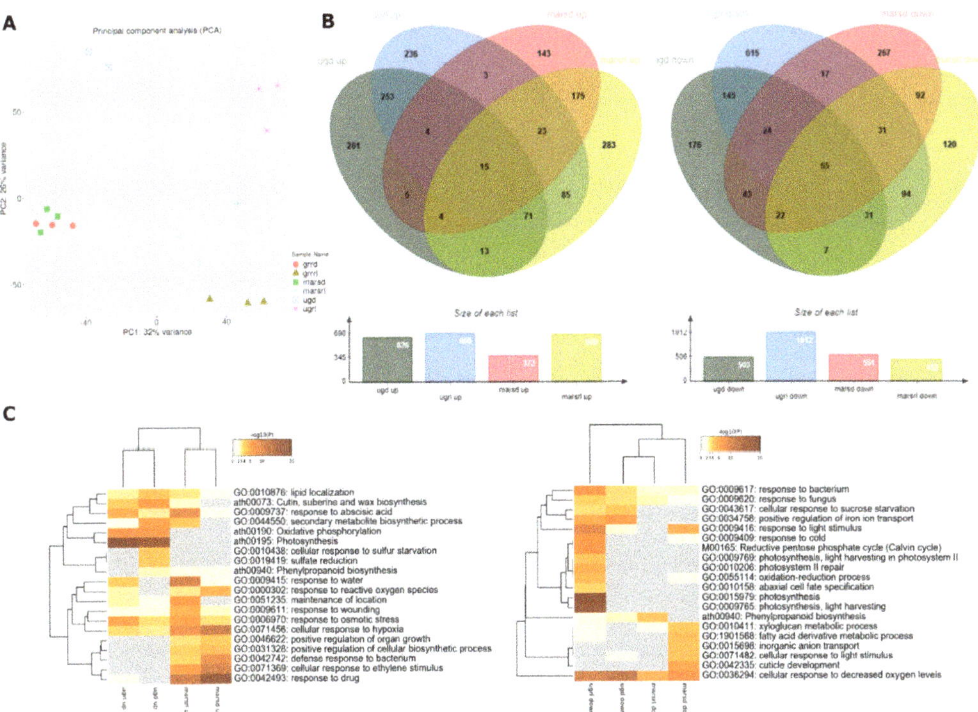

Figure 2. Global transcriptomic effects of microgravity and Mars gravity. (**A**) Principal component analysis (PCA) where all replicates included in RNASeq analysis are represented. (**B**) Venn diagrams with differentially expressed genes (DEGs) with q-value < 0.05 and Log$_2$FC > 1.5 (and < −1.5), separated in upregulated (**left**) and downregulated (**right**) genes. (**C**) Metascape Gene Ontology Heatmaps of top 20 enriched clusters for upregulated (**left**) and downregulated (**right**) DEGs.

To elaborate on this, we applied two different approaches to perform the comparisons in the transcriptomics data. First, to investigate the effect of different gravity levels on *A. thaliana* seedlings, we compared the transcriptomes of samples grown in microgravity or

Mars gravity level to the same light condition in 1-*g* GRR transcriptome as the reference; µgd-grrd, Marsd-grrd, µgrl-grrrl; Marsrl-grrrl. Next, to dissect the effect of red light at different *g*-levels, we compared the transcriptomes of seedlings grown in the same *g*-levels but in different light conditions; µgrl-µgd; Marsrl-Marsd, grrrl-grrd. In both cases, we used a q-value < 0.05 and a threshold fold change of $Log_2FC \pm 1.5$.

In the gravity level comparisons, more than 600 differentially expressed genes (DEGs) were upregulated in µgd, µgrl, and Marsrl, whereas in Marsd only half of this number (372) (Figure 2B). These results show that transcriptomes of the seedlings photostimulated by red light presented a similar number of upregulated DEGs at both gravity levels, but in seedlings grown in dark, twice as many DEGs were upregulated in microgravity than in Mars gravity. Approximately half of the upregulated DEGs are common for each gravity level independently of being exposed to darkness or photostimulation (253 and 175 in microgravity and Mars, respectively). Gene Ontology (GO) analysis of biological process categories of those sub-lists are shown in Figure S1. The low number of upregulated genes in the Marsd sample is in accordance with a small size of meristematic nucleoli in this condition, suggesting a reduced rate of protein biosynthesis, meaning that the transcriptome status is reflected at the proteome level. This effect could be a result of the upregulation of Ovate Family Protein 10 (OFP10), a transcription repressor [46] upregulated only in Marsd condition (Log_2FC 2; all Log_2FC values given in the text are statistically significant; q-value < 0.05).

Only 15 DEGs are upregulated in all four comparisons. Around 500 genes were downregulated in µgd, Marsd, and Marsrl, whereas in µgrl, there were twice as many (1012 genes). In this case, we observed twice the number of downregulated genes in the red-light photostimulated seedlings in microgravity in comparison to Mars *g*-level. This number is particularly high (615 DEGs) in µgrl only, and specifically enriched in photosynthesis function (Figure S1). A total of 55 DEGs are downregulated in the four conditions. There are also nearly a hundred up- and down-regulated genes in the red-light conditions (µgrl and Marsrl), which include abiotic stress responses in the upregulated DEG and metabolic biosynthetic pathways in the downregulated genes (Figure S1). In summary, seedlings grown at Mars *g*-level in darkness seem to be the least altered samples (in agreement with the PCA), while the ones grown in microgravity and photostimulated with red light show a high number of DEGs. Given the small variations observed in the plant anatomy, as reported in the preceding section, this dysregulation does not necessarily mean an adverse effect on the plant. Most likely, the changes in transcript levels also involve genes associated with the acclimation that the seedlings experience from the germination and during six-day exposure to altered gravity level. The red-light photostimulation comparison is discussed below.

In the Gene Ontology (GO) analysis, we used the total number of upregulated or downregulated DEGs to look for common and specific altered molecular functions in each condition. We observed that in the upregulated genes there was a clear clustering of common categories by *g*-levels rather than by light conditions (Figure 2C, left). Among the categories upregulated in all conditions, response to osmotic stress, wounding and cellular response to hypoxia could be identified. Photosynthesis categories are highly upregulated in microgravity, but not at Mars *g*-level. It is surprising that this category is equally upregulated in both light conditions, the red-light photostimulated sample and seedlings grown in darkness for the last two days. Another strongly upregulated category in microgravity was oxidative phosphorylation. On the other hand, categories specific for Mars conditions (both light treatments) include response to ethylene, drug, defense response and positive regulation of biosynthetic processes and organ growth.

Among the downregulated DEGs (Figure 2C, right), the common functions are related to hypoxia, while the rest of the downregulated functions seem to be specific for each *g*-level and light condition. In addition, in the µgrl sample, photosynthesis and light harvesting category was strongly downregulated. This result is in agreement with the previous results obtained in the SG1–2 experiments in the Ler ecotype, using blue-light stimulated seedlings [21], confirming the role of light and phototropism as an alternative cue for plant development in the total absence of gravity. To determine if the same genes are being downregulated, we compared the two datasets and found 16 common genes. Even if the overlap of downregulated genes was not striking, there was an enrichment of photosynthesis function in those 16 genes specifically related to light harvesting in photosystem I and II, and five downregulated light-harvesting chlorophyll *a*/*b* binding (LHCB) proteins (Figure S2). These results suggest that the same function is affected by microgravity in both spaceflight experiments (i.e., one with red-light stimulation and the other with blue light).

Extended heatmaps with top 100 enriched clusters are shown in Figure S1. Common GO categories are either upregulated in both gravity conditions, such as the response to water, response to reactive oxygen species, response to osmotic stress, response to wounding and cellular response to hypoxia, or downregulated, such as cellular response to decreased oxygen levels. However, only a few or none of the dysregulated transcripts are common for both conditions in each category (Figure S3).

2.3. Dysregulation of Transcriptional Factors (TFs) and Hormonal Pathways in Microgravity and Partial Gravity

Many of the functional categories that are dysregulated in both microgravity and Mars gravity conditions such as osmotic and biotic stresses or response to hypoxia are regulated by both phytohormones and families of transcription factors [47–50]. Among the most over-represented TF families in the DEGs, we encountered the WRKY domain family, which forms one of the largest TF families in flowering plants, as well as other large families of TFs such as ethylene responsive factor (ERF), ATAF1/2 CUC2 (cup-shaped cotyledon) (NAC) and myeloblastosis (MYB). We tested by Chi-squared analyses whether a g iven TF family is overrepresented in each of the conditions and discovered that WRKY and NAC TFs are overrepresented in Mars *g*-level (both light conditions), and MYB TFs are overrepresented in Mars *g*-level (both light conditions) and in microgravity red-light photostimulated samples. In contrast, ERFs are overrepresented in all four conditions. Since WRKY TFs have an important role in plant acclimation and are "multifunctional switches," we focused on this family of TFs (reviewed in [47]). Data on WRKYs upregulated in the Mars conditions (both lights) are presented in Table 1. Although in microgravity WRKY were not overrepresented, one family member, *At*WRKY63, was significantly upregulated in both light conditions (µgd-grrd Log$_2$FC 2.56 and µgrl-grrrl Log$_2$FC 2.79).

We evaluated the influence of microgravity and Mars *g*-level on hormonal pathways using Kyoto Encyclopedia of Genes and Genomes (KEGG) pathway analyses. Few important hormonal pathways were significantly affected by microgravity and partial gravity, as seen in Figure 3 for the "plant hormone signal transduction pathway" (ath04075), including the expression level and significance for the four light/gravity conditions under study (µgd-grrd, µgrl-grrrl, Marsd-grrd, Marsrl-grrrl). Each signaling step includes the information of transcription levels of one or more genes involved in signal transduction (full list of genes is available at KEGG database under the ath04075 pathway identifier). According to KEGG Pathway analysis, the auxin pathway was activated at different steps in microgravity conditions (GRETCHEN HAGEN 3 (GH3) step in µgd and small auxin upregulated RNA (SAUR) in µgrl and repressed in Marsrl condition (GH3 step), suggesting cell enlargement and plant growth are promoted in microgravity but not at Mars *g*-level. *GH3* genes, downregulated in µgd (*DFL1*) and upregulated in Marsrl (*GH3.3* and *AT1G48660*) encode auxin-amido synthetases and promote the inactivation of indole acetic acid (IAA) [51]. Eleven SAUR genes which regulate auxin-mediated growth are upregulated in µgrl sample.

Table 1. Upregulated WRKY-domain transcription factors (TFs) in Mars gravity level. Statistically significant (q-value < 0.05) WRKY TF in marsd-grrd and marsrl-grrrl comparisons. Log_2FC for each TF in each comparison is shown. Reference list from this table is provided in the Supplementary Material.

Name	Group	Functions	Marsd-grrd	Marsrl-grrrl
*At*WRKY38	III	- negative roles in plant defense [1] - involved in SA signaling pathway [1] - JA-signaling repressor [2]	2.34	2.19
*At*WRKY40	II-a	- role in response to salinity/osmotic [3] - response to touch [4] - defense response	2.13	1.82
*At*WRKY45	I	- flooding stress [5] - phosphate ion transport [6]	1.19	0.81
*At*WRKY46	III	- regulates development, stress and hormonal response by facilitating growth of lateral roots in osmotic/salt stress through ABA signaling and auxin homeostasis [7] - role in the immune process; induced by *P. syringae* or SA [8] - cellular response to hypoxia [9]	3.73	3.61
*At*WRKY51	III	- mediates SA- and low oleic acid-dependent repression of JA signaling involved in plant defense [10]	1.85	2.14
*At*WRKY53	III	- positive effect on plant senescence [11] - role in the immune process - may play a role in SA signaling pathway [8]	2.35	1.58
*At*WRKY54	III	- negative regulation of senescence [12] - defense response - regulation of brassinosteroid, JA, SA and ethylene pathways (arabidopsis.org) - osmotic stress	2.82	2.00
*At*WRKY59	II-c	- regulation of transcription - induced by *P. syringae* or SA, [1]	4.34	3.02
*At*WRKY62	III	- negative role in plant defense.	2.77	2.48
*At*WRKY66	III	- regulation of transcription	1.95	3.56
*At*WRKY75	II-c	- (with AtWRKY44) development of the root hairs [13] - has a positive effect on leaf senescence [14] - participate in the regulation of phosphorus deficiency signaling [13]	1.78	1.61
*At*WRKY33	I_C	- binds to SIB1, JAZ1 and JAZ5 affecting JA-mediated defense signal pathway [15] - involved in abiotic stress response, in particular salt/osmotic stress (arabidopsis.org)	1.84	1.45
*At*WRKY70	III	- negative regulation of senescence [12] - role in the immune process may play a role in SA signaling pathway [8]	2.28	1.49

The cytokinin pathway was significantly activated ($Log_2FC > 1.5$) in microgravity through histidine phosphotransfer proteins (AHPs), which function as positive regulators of cytokinin signaling [52]. From the six members of this gene family expressed in *A. thaliana*, two were upregulated in microgravity: *AHP3* and *AHP4*. In Marsd conditions, this route was significantly inhibited at the CRE1 step (histidine kinase 2, *HK2*). The gibberellin pathway was significantly repressed at the TF step in microgravity (*PIF4, PIL6*).

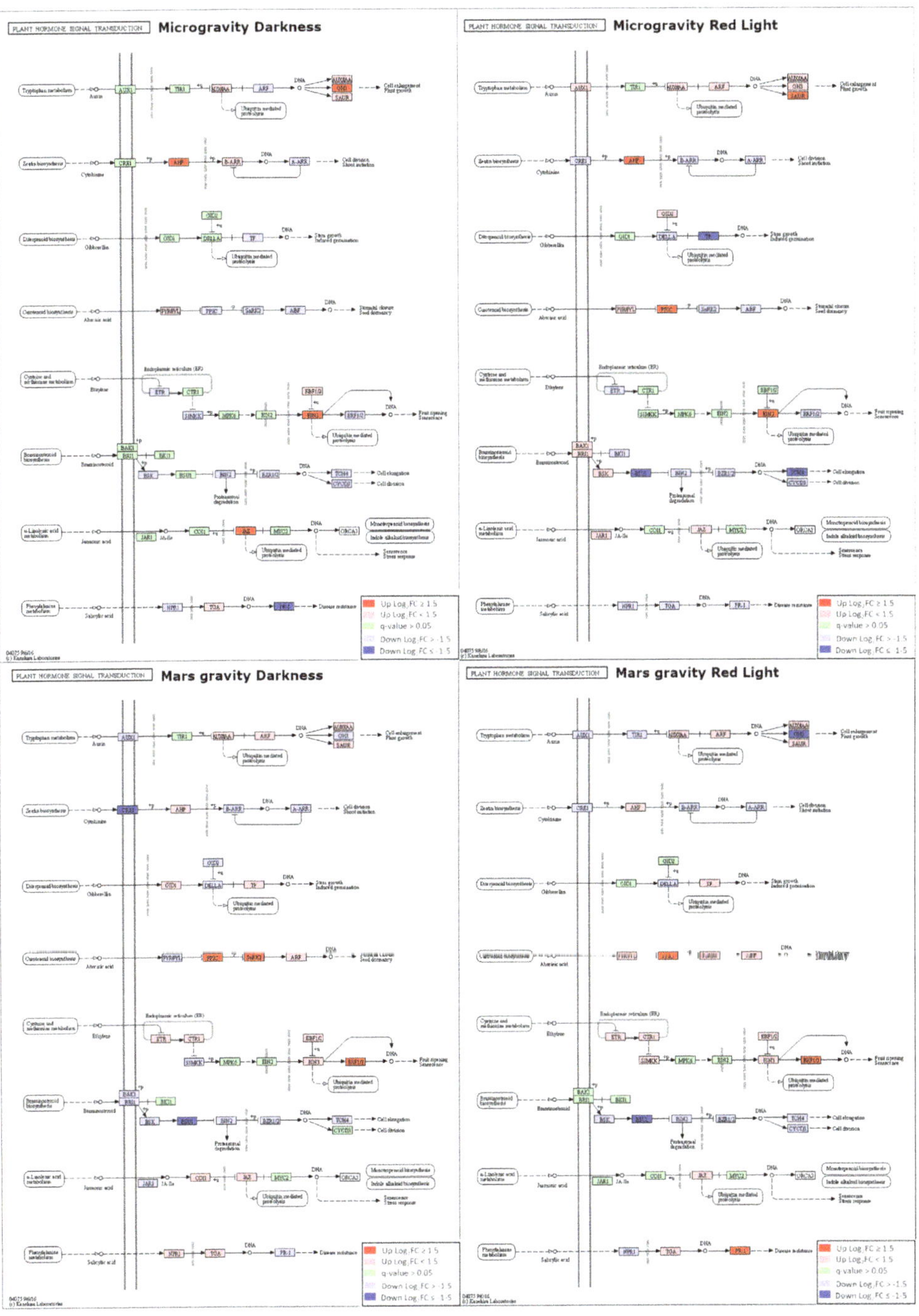

Figure 3. Hormone signaling changes in microgravity and Mars gravity. Plant hormone signal transduction (ath04075) Kyoto Encyclopedia of Genes and Genomes (KEGG) representation with color-coded changes in each experimental condition: microgravity darkness (µgd-grrd), microgravity red light (µgrl-grrrl), Mars gravity darkness (marsd-grrd) and Mars gravity red light (marsrl-grrrl).

At least one step of the abscisic acid (ABA) pathway is clearly activated in all conditions except for μgd (Figure 3). Among the upregulated genes that contribute to this activation are highly ABA-induced *PP2C* gene 2 (*HAI2*) and *HAI3* proteins and, in the case of the Marsrl samples, also SNF1-related protein kinase 2.9 (*SNRK2.9*) and *SNRK2.5*. The ethylene pathway was activated in all conditions; at EIN3 step in microgravity (*AT5G65100*) and ERF1/2 at Mars *g*-level (*ERF1, ERF2*), which is in agreement with the GO analysis and the overrepresentation of ERFs in the upregulated genes in all conditions.

The brassinosteroid pathway is significantly inhibited through BRI1 suppressor 1 (BSU1) protein in μgrl and Mars samples. In addition, *TOUCH 4 (TCH4)*, which is involved in the response to mechanical stimulus, cold, hypoxia and regulates cell elongation (arabidopsis.org), is downregulated in all light and gravity conditions (although with threshold Log$_2$FC > 1.5 only in μgrl sample). Both microgravity and partial gravity influenced the transcription of jasmonate-ZIM domain (JAZ) proteins, which are transcriptional repressors in jasmonic acid (JA) signaling. *JAZ1* was upregulated in Mars gravity and *JAZ4* in microgravity. Additionally, *ABA-inducible BHLH-type transcription factor (AIB)* that interconnects JA-ABA pathways was also downregulated in microgravity conditions. The salicylic acid pathway was repressed in microgravity (*PR1-like, AT1G50060*) and activated in the Marsrl sample (*AT4G33720*) at the PR-1 step.

Similar analysis with light comparisons (μgrl-μgd, Marsrl-Marsd, grrrl-grrd) showed that red light activates the pathways regulating proliferation and growth: cytokinin pathway in microgravity and GRR samples and auxin pathway at Mars *g*-level (GH3 step) but does not alter significantly other hormonal pathways (Figure S4). These results suggest that the gravity level has more impact on hormonal pathways regulation than light conditions.

In summary, an activation of proliferation-promoting pathways (cytokinin and auxin) is evident in microgravity but not at Mars *g*-level. Further activation of the cytokinin pathway was observed with red-light photostimulation (μgrl-μgd comparison). At Mars *g*-level, red light reverses partial inhibition of auxin pathway (GH3 step) which is in agreement with its proliferation-activating effect. On the other hand, stress-related pathways, in particular ABA, ethylene and salicylic acid (SA) seem to be more activated at Mars *g*-level, which could suggest that in partial gravity, the plant perceives the stress signal and responds with activating acclimation mechanisms.

2.4. Plastid and Mitochondrial Genome Expression

We compared the distribution in the genome of DEGs in each condition using ShinyGO analyses [53] and found that in microgravity, the expression of genes encoded in the chloroplastic and mitochondrial genomes were over-represented. This enrichment in plastid and mitochondrial gene expression was specific to microgravity (Figure 4A) and not present in Mars *g*-level. Furthermore, we compared these results with the WT (Ler ecotype) blue-light dataset from the SG experiments (GLDS 251) containing transcriptomic data from seedlings exposed to the following partial gravity levels: microgravity, low gravity level (0.09 ± 0.02 g), Moon gravity (0.18 ± 0.04 g) and Mars gravity (0.36 ± 0.02 g) levels, and found a similar enrichment in chloroplastic and mitochondrial gene expression exclusively in microgravity, but not in any partial gravity including low gravity level (Figure 4B). This enrichment was detected in the upregulated DEGs, specifically 70 chloroplast-encoded genes were upregulated in microgravity dark and 83 in microgravity red-light conditions in our dataset (45 in the blue-light exposed samples from GLDS-251 dataset), and 10 mitochondrial genes were upregulated in microgravity dark and 30 in microgravity red light (5 in the blue dataset). The upregulated chloroplast-encoded transcripts involved multiple subunits of photosystem I and II and NAD(P)H dehydrogenase complex and electron transporters PETA (photosynthetic electron transfer A), PETB and PETD. The upregulated mitochondrion-encoded transcripts involved ribosomal proteins L16 and S3R, cytochrome oxidase 1 and 2 (COX1, COX2), NADH dehydrogenase subunits: 4, 5A and 5C and ATP-binding cassette I2 (ABCI2), a cytochrome C biogenesis protein.

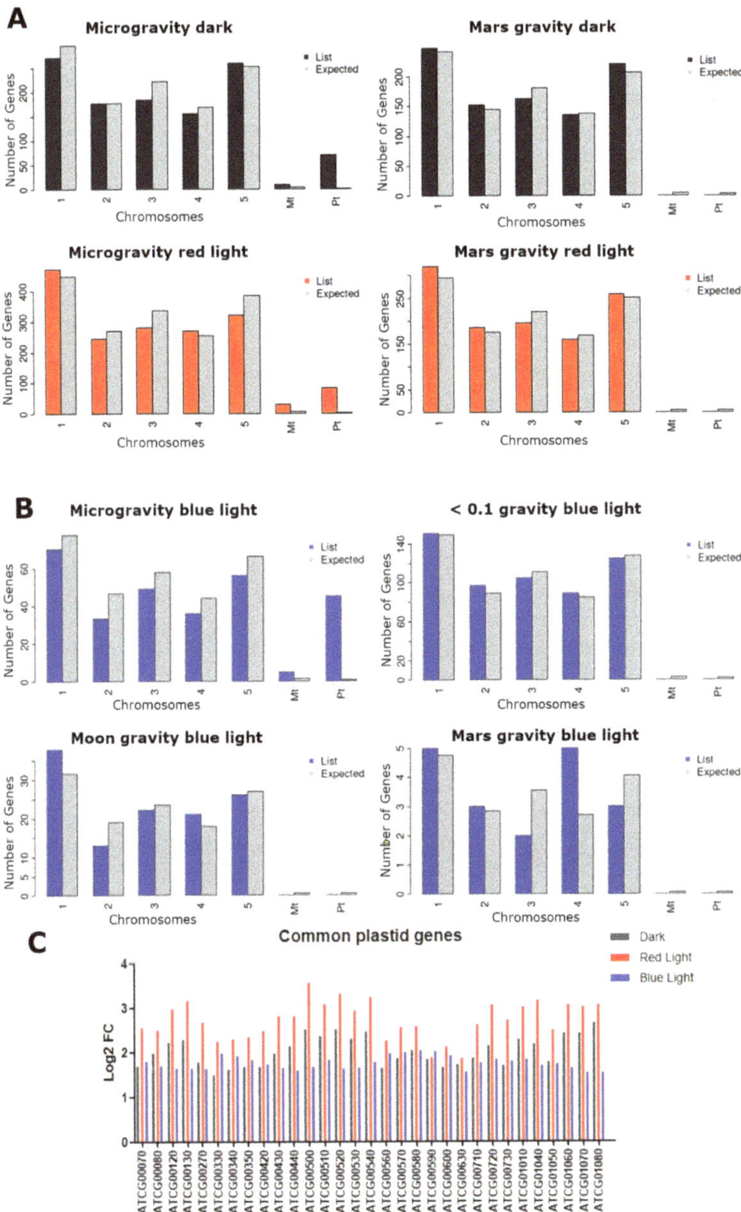

Figure 4. Plastid and mitochondrial genome expression in microgravity. (**A**) Distribution of DEGs across chromosomes in different comparisons: μgd-grrd, marsd-grrd, μgrl-grrrl and marsrl-grrrl. (**B**) Distribution of DEGs in GLDS-251 (NASA GeneLab Accession No.) blue-light stimulated seedlings in different gravity levels: microgravity, <0.1g, Moon gravity and Mars gravity. Bars (**A**,**B**) represent number of genes in query list (dark grey, red or blue) and expected number of genes (light grey). The distribution of the query genes is statistically significant (Chi-squared test) in microgravity dark (p-value: $1.1 \cdots 10^{-209}$), microgravity red light (p-value: $1.5 \cdots 10^{-207}$), mars red light (p-value: 0.014), microgravity blue light (p-value: 0). (**C**) Log$_2$FC of the common genes of the three microgravity comparisons: μgd-grrd (grey), μgrl-grrrl (red) and GLDS251 μg-1g control with blue light stimulation (blue).

We also compared our results with datasets of two additional spaceflight experiments in which enrichment of organelle-encoded genes was reported in the transcriptomic data available in GeneLab (GLDS-38 and GLDS-44) [29,30]. These experiments were performed using the Biological Research in Canisters (BRIC) hardware without any lighting, so etiolated *A. thaliana* WT Col-0 seedlings grown in microgravity were used. Although not completely, there is a significant overlap in some of these genes in two or more experiment datasets (Table S3), even considering that it is not uncommon to see differences between spaceflight experiments [54]. The fact that this phenomenon is observed in experiments where different *A. thaliana* WT lines were used and different hardware and environmental conditions were applied, strongly suggests that it is one of the major microgravity effects. These results are consistent with studies of impaired mitochondrial function from drosophila [55] to humans [56].

This plastid and mitochondrial genome expressions were observed in the three different light conditions (blue light, red light, darkness). However, red light leads to generally higher upregulation levels and a greater number of genes were upregulated in our dataset of the red-light photostimulated seedlings compared to plants grown in darkness, or in the Ler dataset of blue-light photostimulated seedlings, as shown by the fold change of the common upregulated genes (Figure 4C). The level of change of all plastid and mitochondrial genes in the three light conditions is shown in Figure S5.

2.5. Dissecting the Contribution of the Red-Light Photostimulation to the Response to Each g-Level

To dissect the transcriptomic changes provoked by red-light photostimulation at each *g*-level, we performed an additional set of transcriptomic comparisons of the photostimulated seedlings versus the seedlings grown in darkness at 1*g*, Mars and microgravity levels (grrrl-grrd; Marsrl-Marsd; μgrl-μgd) (Figure 5). The first observation is the fact that the number of upregulated genes was much higher than the number of downregulated at all gravity levels. As expected, we have observed upregulation of genes related to photosynthesis, light harvesting, pigment biosynthesis and response to light stimulus in all gravity conditions including GRR. Other categories upregulated in all gravity conditions included transcripts involved in response to karrikin (which help stimulate seed germination and plant development), phenylpropanoid metabolic processes, secondary metabolic processes and, in microgravity (to a lesser degree in GRR), the reductive pentose phosphate cycle (Calvin cycle). Surprisingly among downregulated transcript categories, we found the ones related to red to far-red signaling pathway and the response to far red light, which indicates that feedback regulatory mechanisms were triggered to tune down the seedling response. Metascape analysis of upregulated transcript heatmap profiles clusters Mars comparison (Marsrl-Marsd) together with GRR comparison (grrrl-grrd), which suggests a more similar plant response in these conditions. In contrast, in downregulated transcript profiles, Mars comparison clusters together with microgravity comparison (μgrrl-μgd). These observations are also reflected in the higher number of upregulated genes common in Mars and GRR than the ones common in Mars and microgravity and a higher number of common genes downregulated between Mars and microgravity than those common for Mars and GRR, or microgravity and GRR (Figure 5). Taken together, these results could suggest that red-light photostimulation has a positive effect on seedlings in Mars gravity level, which is similar as in Earth conditions. Extended heatmaps are shown in Figure S6.

The processes that are the most affected in microgravity by red light include cell wall organization and biogenesis, cell cycle, microtubule-based processes, DNA replication initiation and trichoblast differentiation (Figure 5). This effect is less pronounced in the other gravity conditions, which suggests that red-light photostimulation has especially positive influence on seedlings in microgravity.

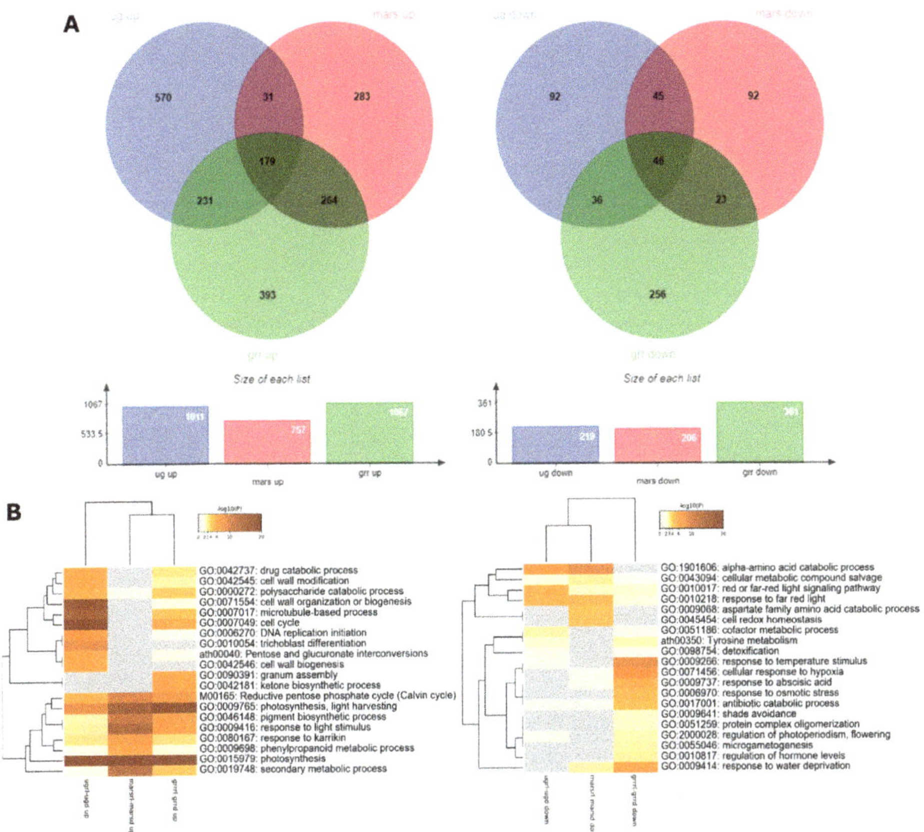

Figure 5. Red-Light induced changes in gene expression across gravity conditions. (**A**) Venn diagrams representing DEGs (with q-value < 0.05 and $Log_2FC > 1.5$ (or <-1.5). Left: upregulated genes. Right: downregulated genes. (**B**) Metascape Gene Ontology (GO) heatmaps of top 20 enriched clusters. Left: upregulated genes. Right: downregulated genes.

On the other hand, red light has an inhibitory effect on proteolysis in microgravity and Mars conditions, as seen by downregulation of the genes related to amino acid catabolic processes. This observation is in line with the positive effect of red light on biosynthesis and proliferation [14] and is reflected in the increased size of the nucleolus (and meristem) in the red-light photostimulated samples.

Among the most important transcripts in the cell cycle category upregulated in microgravity conditions were PROLIFERA (*PRL*; Log_2FC µgrl-µgd 2.16) and multiple cyclins: A2, A3, B1 (B1:1, :3, :4 (cyclin2)), B2(:1, :2, :3), D4 (:1; :2), P4(:1; :2, :3) as well as proliferation markers *AURORA1* and *AURORA2*. Upregulated genes encoding microtubule-related proteins included many subunits of the mitotic spindle, but also some important cytoskeleton-related proteins such as tubulin beta-1 chain (*TUB1*; Log_2FC µgrl-µgd 1.84) and SPIRAL1-LIKE4 (*SP1L4*; Log_2FC µgrl-µgd 1.69). *TUB1* encodes beta tubulin regulated by phytochrome A (phyA)-mediated far-red light high-irradiance and the phytochrome B (phyB)-mediated red-light high-irradiance responses [57]. SP1L4 regulates cortical microtubule organization essential for anisotropic cell growth [58].

In terms of cell wall enzymes and proteins, among the upregulated transcripts in microgravity light comparison were present numerous enzymes from the group of hydrolases (*Xyloglucan endotransglucosylase/hydrolase 12, XTH12; XTH13; XTH14; XTH20; XTH26;*

glycosyl hydrolase 9B13, GH9B13) and transferases (*galacturonosyltransferase 12, GAUT12; rhamnogalacturonan xylosyltransferase 1, RGXT1; glucuronoxylan methyltransferase 2, GXM2*) as well as *pectin methylesterase 46* (*PME46*), *pectin acetylesterase 10* (*PAE19*) and *polygalacturonase involved in expansion 1* (*PGX1*). In addition, numerous extensins (*LRX1, EXT2, EXT7, EXT8, EXT9, EXT10, EXT12, EXT13, EXT15, EXT16* and *EXT17*) and two fasciclin-like arabinogalactans (*FLA6, FLA7*) were present. These results suggest the cell wall undergoes profound modifications in this condition that might be related to increased cell growth and expansion [54,59].

3. Discussion

3.1. Red-Light Photostimulation has a Positive Effect on Cell Proliferation in Both Microgravity and Mars Gravity Conditions

Our morphometric studies indicate that features like meristem and nucleolus size are more robust in the red-light photostimulated seedlings. In the case of the nucleolus, although the difference in size between dark-grown and red-light photostimulated seedlings is not significant in microgravity, the features of nucleolar ultrastructure clearly indicate that red-light photostimulation increases nucleolar activity. Increased root meristem size (expressed as the length and the number of meristematic cells), nucleolar size and changes in nucleolar ultrastructure confirm positive effect of red light at Mars *g*-level, suggesting that the rates of both the meristematic cell proliferation and protein production were increased. On the other hand, the transcriptomic data from different light condition comparisons (grrrl-grrd; Marsrl- Marsd; μgrl-μgd) confirmed that in microgravity, cell cycle and proliferation related genes are upregulated and the hormonal routes promoting proliferation activated (Figure 5B and Figure S3).

Red and far-red light are perceived by photoreceptors termed phytochromes, which are expressed in different zones of the root (reviewed in [60]). PhyA and phyB, expressed mainly in the root tip, are involved in both red-light-induced positive root phototropism and gravitropism [61,62]. Red-light photostimulation of seedlings in GRR has a positive effect on the overall physiology in comparison to the seedlings kept in darkness for the last two days of cultivation [37]. In addition, red light is known to stimulate cell proliferation and ribosome biogenesis [14], which are observed in our results, particularly when the gravity vector cannot completely guide the plant development. In addition, red light also restored the meristematic competence balance, which was extensively described to be affected in early plant development in our previous ROOT experiment in the ISS in etiolated *A. thaliana* seedlings [3].

Thus, we conclude that red-light photostimulation could help plants to overcome some of the deleterious effects of the spaceflight environment. Similar effects were seen previously in our experiments with blue-light photostimulation [22].

3.2. Microgravity has a Deleterious Effect on Plant Physiology: Elevated Plastid and Mitochondrial Genome Expression is Observed in Microgravity, but Not in Partial Gravity

Dysregulation of the genes involved in photosynthesis was specific for microgravity condition. Photosynthesis-related genes were upregulated in both light conditions, which was related to the increased plastid genome expression (to be discussed further). On the other hand, in the μgrl sample, genes involved in photosynthesis were also downregulated. This is in agreement with the results of the previous SG experiment [21], although only 16 genes were common for both datasets obtained from different light conditions. Reduction in photosynthesis activity, and specifically in photosystem I complex, was observed in previous studies of *Brassica rapa* plants grown in space [63], and in *Oryza sativa* plants grown in simulated microgravity [64]. Furthermore, structural changes in chloroplasts, such as alterations of thylakoid membranes in seedlings grown in real [65] and simulated microgravity [66] were reported, as well as a reduction of chloroplast size in simulated microgravity [67]. Downregulation of LHDB proteins, which was observed in microgravity in red- (our dataset) and blue-light stimulated seedlings (GLDS-251) (Figure S2), is known to affect stomatal closure in effect reducing photosynthetic activity. This downregulation also decreases plant tolerance to drought stress [68,69].

Dark-grown samples showed a general decrease in photosynthetic activity as demonstrated by the transcriptomic analysis of 1*g* GRR samples [37]. It is therefore not surprising that we have not observed this category in dysregulated genes in μgd-grrd comparison.

Additionally, few ERFs important for development of tolerance to a number of abiotic stressors were downregulated in microgravity: *Translucent Green* (*TG*), downregulated in microgravity and darkness, is involved in drought tolerance [70], *C-repeat/DRE binding factor 1* (*CBF1*) and *CBF2*, downregulated in microgravity in both light conditions, are involved in tolerance to freezing [71,72], *AtERF72*, also downregulated in both light conditions, is involved in the tolerance to peroxide (H_2O_2) and heat stress [73]. In summary, the downregulation of these ERFs in microgravity may have a negative effect on *A. thaliana* tolerance to adverse conditions.

Two organelles seem to be particularly affected in microgravity during space flight. Plastids contain 3000–4000 proteins and most of them are encoded in the nucleus [74]. Around 90 to 100 are encoded in the chloroplast genome [75]. The majority of these genes are upregulated in microgravity (70 in μgd and 83 μgrl). Mitochondria contain around 3000 proteins, from which 57 were identified in the mitochondrial genome [10,76]. In our dataset, we identified 10 of them in μgd, peaking to 30 out of 57 in μgrl and 10 in microgravity and darkness. Various factors influence plastid gene expression, such as light, temperature, plastid development or circadian clock [74]. Abscisic acid (ABA) represses transcription of chloroplast genes [77]. The ABA pathway was activated in the Mars samples, which could explain why this phenomenon is not observed in this condition. Upregulation of plastid-encoded genes was reported before in spaceflight experiments [29].

Mitochondria and chloroplasts are tightly involved in cellular metabolism and are thought to be initial sensors for cellular dysfunction caused by external stress. Research to date suggests that factors that participate in signaling between these organelles and the nucleus (anterograde communication: communication from nucleus to organelle; retrograde communication: from organelle to nucleus) also participate in the recognition of the stress level. The decision is whether to adjust the metabolism, or to execute programmed cell death (PCD) [78]. The key signaling molecules in mitochondrial dysfunction are ANAC017, ANAC013 and Alternative Oxidase 1a (AOX1a) [79]. *ANAC017* is only slightly downregulated in microgravity and *AOX1a* and *ANAC13* is only upregulated in Marsrl condition (around 1-fold). In addition, there is a set of mitochondrial proteins which are consistently upregulated in stress conditions when the dysfunction of mitochondria takes place [80–82], but only one of these is upregulated in the μgrl sample (*AT3G50930*) and three downregulated (μgd, μgrl: *AT1G20350, AT1G21400*; μgd: *AT4G15690*).

Based on our results, it would seem that there is a dysregulation of the chloroplast and mitochondrial genome expression, and this dysregulation is not perceived and corrected in the typical retrograde communication in response to organelle dysfunction. Moreover, this organelle dysfunction is not present in partial gravity level, probably because the retrograde communication is not disturbed. Supporting the last assumption is the fact that Sigma factor binding protein (SIB1), which binds sig1R factor (nuclear encoded factor that regulates the chloroplast genome expression) and has an important role in the retrograde communication, is downregulated in μgd, but upregulated in Mars conditions. *At*WRKY40 was shown to be a repressor of retrograde-mediated expression while *At*WRKY63 has an opposite activating effect [79]. The antagonistic functioning of WRKY40 and WRKY63 could also play a role in this dysregulation. At Mars *g*-level, *AtWRKY40* is upregulated, subduing the expression of stress-responsive genes (genes responding to mitochondrial and chloroplast dysfunction). On the other hand, upregulation of *AtWRKY63* could play a role in the dysfunction of both organelles. It has been reported that disturbed mitochondrial retrograde signaling leads to increased sensitivity of plants to stress conditions [83]. Mitochondrial retrograde signaling is involved in acclimation to flooding, and *At*WRKY40 is involved in promoting this acclimation together with *At*WRKY45 [83], which is also upregulated at Mars *g*-level and downregulated in microgravity.

Mitochondria and WRKY40 also participate in response to touch and wounding. This response involves other signaling factors such as OM66 and mitochondrial dicarboxylate carriers DIC2, DIC1 [84,85]. The *OM66, AtWRKY40, DIC1* and *DIC2* are upregulated at the Mars *g*-level (both light conditions). It is possible that at the Mars *g*-level, the seedling is activating the response to touch in the search for the direction stimuli, whereas in microgravity this route is not activated (as *OM66, AtWRKY40, DIC1* and *DIC2* are not upregulated).

However, the mechanism involved in the perception of microgravity and the mechanism that causes the dysregulation of the organellar genome transcription remains unresolved. Nevertheless, our results show that by the combination of light with applying even low *g*-level, this stress response observed at the Mars *g*-level (and even more intense at the Moon *g*-level [18,22]) can be corrected.

3.3. Seedlings Grown at Mars g-Level Activate Stress Responses Involving WRKY TFs Possibly Leading to Acclimation

The acceleration similar to the Mars *g*-level in the EMCS centrifuge in orbit is enough to provide gravitropism cue to plants on the ISS as seen in Figure 6 and previously reported [15]. Consistently, the transcriptome changes observed at Mars *g*-level are very different from those found in *A. thaliana* exposed to microgravity, which is especially clear in the seedlings grown without photostimulation. DEGs involved in multiple stress responses, like hypoxia (decreased oxygen levels), drought, reactive oxygen species, osmotic and biotic stresses are altered in both gravity levels. These observations are in accordance with previous experiments in the microgravity and partial gravity conditions [22,25,54]. However, when we compare the number of common DEGs in microgravity and Mars conditions (Figure 2), we observe only 15 upregulated genes and 55 downregulated genes. In fact, even though common GO categories are upregulated and downregulated for both conditions, only a few or none of the specific DEGs are common for both gravity levels (Figure S2). Moreover, photosynthesis, which is highly affected by microgravity [21], does not seem to be disturbed at Mars *g*-level. It is evident that both conditions induce different responses in seedlings and, therefore, the strategies to grow plants, either during spaceflight or on a planet with reduced gravity, but enough in magnitude to trigger a full gravitropic response, should also be different.

GO categories common for all the samples are most likely related to the spaceflight conditions; however, the plant responds to these environmental factors differently in microgravity and Mars *g*-level, as suggested by the low number of common DEGs. The bioavailability of oxygen in the spaceflight environment is reduced and very dependent on the hardware used for the experiment in orbit, provoking the plant response to hypoxia [54,86]. This stress is closely related to waterlogging response (or water stress), also present in the upregulated group in the GO analysis; in fact, when a plant is submerged under water the availability of oxygen is reduced and response to hypoxia activated [87]. Morphological changes typical for plants in response to flooding, such as the appearance of adventitious roots in *A. thaliana* ("roots on the stem"; [87]), were observed in BRIC-16-Cyt experiment [88]. In addition, Stout et al. [89] demonstrated increased activity of fermentative enzymes in the roots of *B. rapa* grown in the space environment, which indicates root zone hypoxia. It is possible that the plant activates these known mechanisms to enhance oxygen intake. The response to the osmotic stress is also a category frequently dysregulated in space experiments [29]. The nature of this response is not well understood but it was suggested that plants could activate in microgravity the response to osmotic stress due to the absence of structural guide, compensating it with the stabilization of microtubules [29,90]. Nevertheless, the response to this stressor is also present in partial gravity where the seedling seems to perceive gravitational cues (Figure 6). Our results suggest that the hormonal and transcriptional routes involved in response to osmotic stress, such as the ABA pathway and ERF TFs upregulation, are activated. ERFs are characterized by the presence of ERF DNA binding domain [91] and fulfill a wide range of functions in response to multiple stresses. Differential expression of ERFs was reported in adverse conditions such as waterlogging and hypoxia [92,93].

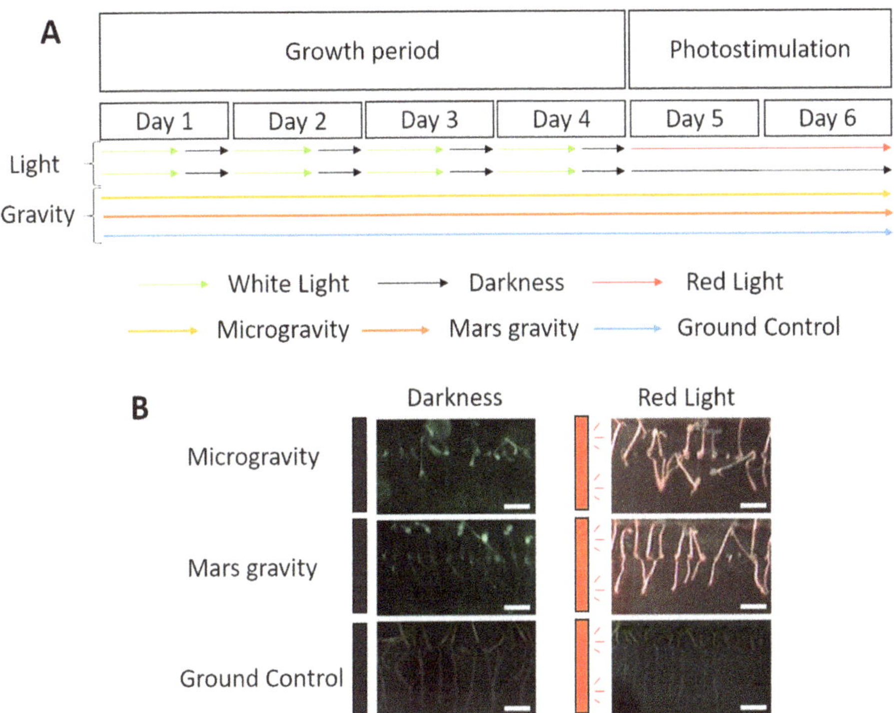

Figure 6. Experiment design. (**A**) Experimental timeline. Seedlings grown during six days in a long day photoperiod regime followed by two days of red-light photostimulation or darkness. (**B**) Images of seedlings at the end of the experiment. Grey and red rectangles at the left of the images represent light-emitting diode (LED) light position for photostimulation. Scale bar represents 3 mm.

At the Mars g-level, we observed a clear enrichment in WRKY TFs. They are involved in a wide variety of functions from abiotic and biotic stress response to developmental and multiple physiological processes on Earth [83,94–97]; reviewed in [47]. They also participate in hormonal response for example in JA/SA hormonal signaling. The WRKY family is defined by the presence of at least one WRKY DNA binding domain (DBD). They interact with W-box (with TTGACC/T motif) and clustered W-boxes located in the target genes that are activated or repressed under a specific condition. Most WRKY TFs are multifunctional meaning they play a role in a number of responses (see Table 1) thanks to multiple functional domains they contain (zinc-finger motifs, leucine zippers, kinase domain, (CaM)-binding domain etc.) [47,98,99]. For example, *AtWRKY75* which is upregulated at the Mars gravity level, plays a role in the immune process, response to osmotic stress, regulation of phosphorus deficiency signaling, development of the root hairs and has a positive effect on leaf senescence [100].

The multiple functional domains that each WRKY contains enable them to form complexes with numerous proteins and fulfil a wide range of functions [47]. This multi-functionality makes WRKY TFs a perfect target for genetic manipulation to create more resistant breeds that can be used in future space experiments. By modifying just one WRKY, a resistance to a set of abiotic and biotic stresses and developmental traits can be achieved. Since TFs such as WRKY are key players in molecular breeding of crops due to their important role in the process of crop domestication [101,102], they have received a lot of attention in recent years. Genetic modification to obtain a specific positive trait was successfully used before, for example, to produce drought-resistant rice overexpressing *Os*WRKY30/70 [103,104]. Similar strategies could be applied also to develop cultivars for "space farming."

The upregulation of WRKY and other important TFs families, such as NACs, in Mars samples suggests that seedlings grown in partial gravity level activate multiple routes to cope with stress associated with space environment, and they can acclimate by modulating genome expression. On the other hand, among the genes upregulated in microgravity, a lower number of TFs can be found. In fact, as mentioned before, a number of ERFs which participate in development of tolerance to various adverse factors are downregulated in this condition.

Hormonal pathways promoting growth and proliferation are activated in microgravity and hormonal pathways promoting stress response are activated at Mars *g*-level

A summary of anatomical changes and changes in hormonal pathways is presented in Tables 2 and 3. Auxin and cytokinin pathways, two hormones regulating cell growth and proliferation, were disrupted differently in microgravity and Mars *g*-level. Our results suggest that both pathways were activated in the microgravity conditions, and few elements of these routes were repressed at Mars *g*-level. Dysregulation of auxin pathway in space-grown seedlings was reported in previous studies [23,29].

Table 2. Summary of anatomic changes in the root compared to the corresponding ground control. Arrow pointing up: increase. Arrow pointing down: decrease. Statistically significant changes are highlighted in red.

		Meristem Length	Meristem No. of Cells	Fibrillarin Area	Nucleolar Ultrastructure
Microgravity	Darkness	↑	↑	↑	↓
	Red Light	↑	↑	↑	↑
Mars gravity	Darkness	-	-	↓	↑
	Red Light	↑	↑	↑	↑

Table 3. Summary of phytohormone signaling changes in gene expression of each condition compared to the ground control. Arrow pointing up: upregulation. Arrow pointing down: downregulation. Red indicates stronger changes according to fold change (at least one step $Log_2 FC > 1.5$).

		Auxin	CK	Brassino-steroids	ABA	Ethylene	JA	SA
Microgravity	Darkness	↑	↑	↓	↓	↑	↑	↓
	Red Light	↑	↑	↓	↑	↑	↑	↓
Mars gravity	Darkness	↓	↓	↓	↑	↑	↑	↑
	Red Light	↓	↓	↓	↑	↑	-	↑

The SA pathway was activated in red light at Mars *g* and inhibited in the microgravity darkness conditions, and the JA pathway was inhibited through JAZ proteins in all samples. Both SA and JA play major roles in the defense response to pathogens. The role of SA is to activate the resistance against biotrophic pathogens, whereas JA is involved in the activation of defense mechanisms [105]. Specifically, the principal function of JA is the promotion of the resistance to plant pathogens by production of defense compounds and, at the same time, it inhibits plant growth. Both, JA and SA, also participate in the response to abiotic stressors and development of tolerance. JA is involved in response to cold, drought, salinity and light (reviewed in [106]). SA was reported to be involved in response to environmental stressors such as high and low temperature, drought, salinity and UV-B radiation (reviewed in [107]). SA-mediated mechanisms together with reactive oxygen species (ROS) and glutathione (GSH) regulate the transcription of different sets of defense genes in a spatio-temporal manner [108]. On the other hand, JA regulatory pathway act through the crosstalk with other phytohormone pathways including ABA, SA and ethylene [106] and is fine-tuned by numerous JA compounds and their different modes of action [109]. Hormones that promote plant growth, such as auxin, gibberellins and cytokinins, repress JA/SA mediated defense-response to prioritize the growth. On the other hand, activation of SA and JA routes can suppress these growth-promoting pathways to activate the defense [110].

JAZ proteins, which were upregulated in both gravity conditions, are transcriptional repressors in JA signaling [111], and by tuning down the response to JA, they enable the recovery of the organ growth [112]. They act on diverse TF families including bHLH, MYB or WRKY [113]. Furthermore, they interact with DELLA proteins to regulate JA and Gibberellin acid (GA) signaling [114], which leads to regulation of plant growth and plant defense response upon environmental conditions. JAZ1 was upregulated at Mars gravity and JAZ4 in microgravity. Apart from the leading role in plant resistance and defense, JAZ proteins are also implicated in the response to abiotic stresses. JAZ1 confers tolerance to alkaline stress [115] and JAZ4, which is not induced as other JAZ proteins by insects or wounding [116], was shown to be involved in control of leaf senescence [117], freezing tolerance [118], growth and development [113]. AIB, which interconnects JA-ABA pathways, was downregulated in microgravity conditions. AIB interacts with JAZ proteins to negatively regulate jasmonate responses. It is induced by ABA and participates in developing a drought tolerance [49]. The ABA pathway is strongly activated at Mars g-level. It is the most important regulator of the response to drought and osmotic stress in plants and a positive regulator of root hydrotropism [119]. ABA, together with MAP-kinase (MAPK) perception and signaling pathways, are involved in all abiotic stresses which cause the decrease of turgor pressure and water loss [48].

The ethylene pathway, together with ERFs, are involved in a wide range of stress responses. Although it has been suggested that the detection of an ethylene response in spaceflight experiments was an effect of the ethylene accumulated from the previous experiments, any ethylene accumulation during this experiment would have been erased from the hardware by a flushing procedure performed before the initiation of the experiment as well as by constant air flow by a connected gas removal module [20,21,120].

In summary, the dysregulation of different hormonal routes in our samples suggests that, in microgravity, the seedlings do not address the external stress and take the option of growing by activating growth and proliferation promoting auxin and cytokinin routes. On the other hand, in the Mars g samples, the growth-promoting routes are inhibited and ABA, ethylene and salicylic acid routes, known for their crucial role in response to osmotic, drought and biotic stresses, are activated.

4. Materials and Methods

4.1. Spaceflight Experiment and Procedures

The SG experiments were a series of spaceflight experiments aimed at investigating the response of young seedlings of *A. thaliana* to the joint stimuli of different levels of gravity and light. Here, we present results corresponding to part 2 and part 3 of the SG series (SG2 and SG3 experiments).

Experimental containers (ECs), each containing 5 culture chambers (cassettes) with 28 seeds attached to gridded nitrocellulose membrane with guar gum (as described in [37,121]) were used. SG2 was sent to the International Space Station (ISS) during the SpaceX CRS-4 (September 2014) and returned on CRS-5 (February 2015), and SG3 was sent to the ISS during the SpaceX CRS-11 campaign (June 2017) and returned a month later, on the same mission.

The experiment did not start until the ECs were loaded into the EMCS and the cassettes were hydrated [20]. Experimental conditions such as hydration, environmental humidity (>80%), gas exchange (O_2 levels kept at 10%; CO_2 at 0.45%) and temperature ($22.5\,^{\circ}\text{C} \pm 2\,^{\circ}\text{C}$), were monitored and controlled remotely from the Norwegian User Support and Operation Centre (N-USOC, Trondheim, Norway). The GRR was performed at this site in the identical hardware using the same experimental conditions a few months later. During the experiment, seedlings were grown for four days in a long day photoperiod (16 h white light, 30–40 μmol/m^2s and 8 h darkness) at two nominal *g*-levels on board the ISS (microgravity and Mars gravity level nominally 0.3 *g* (0.34 ± 0.05 *g*), as provided by the EMCS centrifuge) and the Ground Reference Run 1*g* control. In the last two days of the timeline of the experiment, a change in the light conditions was introduced, half of the material was photostimulated with red light-emitting diodes (LEDs) on one lateral of the ECs, and the rest of the material was grown in darkness.

The timeline of the experiment is shown in Figure 6. The three gravity levels were constant for each EC throughout the duration of the experiment. Before the experiment began, flushing of the EMCS was performed to erase any possible traces of ethylene and other gases from previous experiments. When the six days of seedling growth were completed, the ISS astronauts then removed each EC from the EMCS and froze the samples at $-80\,^{\circ}\text{C}$ in orbit (in the MELFI) or used the FixBox to fix the samples with aldehydes for morphological studies (as described in [37]). Germination rate was calculated as the percentage of germinated seeds.

4.2. Confocal Microscopy

The details of the spaceflight device (termed the FixBox) used for fixation of the samples in 5% (w/v) formaldehyde and the procedure are described in [122]. Briefly, seedlings were fixed in 5% (w/v) FA for 3 h at room temperature (RT) and then kept at 4 °C until return to Earth (μg and Mars *g*-level) or directly processed in the GRR. Fixative was rinsed three times in PBS and then seedlings were digested with digestion solution, containing: 2% (w/v) cellulase, 1% (w/v) pectinase, 0.05% (w/v) macerozyme, 0.4% (w/v) mannitol, 10% (v/v) glycerol and 0.2% (v/v) Triton x-100 in PBS for immunofluorescence with an anti-fibrillarin antibody (Abcam, Cambridge, UK, ab4566 [38F3]). For cell wall staining, the tissues were digested with enzymes not containing cellulose and with 0.5% macerozyme (w/v) and stained with SCRI 2200 a cellulose specific stain (Renaissance Chemicals, North Duffield, UK) [123,124]. Fibrillarin area and meristem size were measured with ImageJ v1.53c. Statistical analyses of the measurements were made using SPSS v25 software.

4.3. Electron Microscopy

The FixBox [122] was also used for fixation of the samples with 4.5% (v/v) glutaraldehyde (Sigma-Aldrich, St. Louis, MO, USA, #G5882) and 1.5% (w/v) formaldehyde (Electron Microscopy Sciences, Hatfield, PA, USA, #15710) in PBS for electron microscopy analysis. Following aldehyde fixation, samples were post-fixed in 1% (w/v) osmium tetroxide in PBS for 1 h and dehydrated in ethanol. Root tips were embedded in epoxy resin and then sectioned. Ultrathin sections were mounted on nickel grids coated with a 0.5% (w/v) Formvar film (Sigma-Aldrich, St. Louis, MO, USA, #09823) and stained with 5% (w/v) uranyl acetate (Electron Microscopy Sciences, Hatfield, PA, USA, #22400) and 0.3% (w/v) lead citrate (Electron Microscopy Sciences, Hatfield, PA, USA, #17800). Next, the samples were examined in a JEOL 1230 transmission electron microscope (TEM) at 80 kV.

4.4. RNA Extraction and Sequencing

Details of RNA extraction and sequencing are described in [37]. Briefly, the RNA extraction kit MACHEREY-NAGEL (Macherey-Nagel (MN), Düren, Germany, #740949) was used to extract RNA for pools of 8–10 seedlings. The RNA extraction kit includes DNAse treatment for 15 min. RNA quality was analyzed with the Bioanalyzer 2100 expert Plant RNA nano with Agilent RNA 6000 Nano Kit (Agilent Technologies, Santa Clara,

CA, USA, #5067-1511). Sequencing was performed on the Illumina HiSeq2500 sequencer (Center for Genomic Regulation, Barcelona, Spain) with stranded RNA read type and 50 bp read length. Seventeen total RNA samples were used to generate sequencing libraries using the Illumina TruSeq RNA Library Preparation Kit (Illumina, San Diego, CA, USA, #RS-122-2001). Samples were individually indexed. The samples then were combined at equimolar proportions into two pools. Each pool was loaded onto two lanes of a flow cell. Sequencing was performed until the 25 million reads per sample objective were reached (27.5 $\pm$ 1 millions of sequences obtained). Results from these studies have been deposited as the GLDS-314 and are available at NASA's GENELAB repository (DOI:10.26030/z5yf-jx91, https://genelab-data.ndc.nasa.gov/genelab/accession/GLDS-314, [40]).

4.5. Functional Analysis

Differential expression analysis was done using Deseq2 [125]. PCA was made using iDEP.91 [126]. Gene Ontology analysis of the DEGs was done using ShinyGO [53] with default settings and Metascape [127] with custom analysis adding molecular function and cellular component. ShinyGO was also used to analyze the distribution of query genes across the genome. Venn diagrams were made using jvenn [128]. String v11.0 was used for protein–protein interaction analysis [129]. Functional analysis of the dysregulated genes involved in phytohormone signaling pathway was performed using KEGG Pathway, the reference database for pathway mapping in KEGG MAPPER [130] available at Kyoto Encyclopedia of Genes and Genomes (KEGG). For the analysis of enrichment in a comparison of a specific gene family, a Chi-squared test was used using GraphPad software v5 (San Diego, CA, USA). The list of the identifiers and Log_2FC values of the genes dysregulated in microgravity (Table S1) and Mars *g*-level (Table S2) are given in the Supplementary Material.

5. Conclusions

We conclude that the response to reduced gravity does not show a gradual decrease in the intensity of the effects observed at microgravity, but clearly differentiated effects on plant growth and physiology are detected, as shown by anatomical and transcriptomic changes. In microgravity, *A. thaliana* accelerates cell proliferation and growth in the root meristem, even though some of the cellular processes, such as retrograde and anterograde communication, appear to be disturbed. This strategy could be activated by applying alternative directional cues, such as light (in particular, red light), and it could lead to adverse effects on the long-term plant development, considering the high energetic cost that it entails. On the other hand, at the Mars gravity level, the seedling perceives external stress and activates responses cooperating with the acclimation of the plant to the environmental conditions, such as upregulation of WRKY TFs. Red light increases cell proliferation at all gravity levels, as shown by microscopic and transcriptional analyses and it is particularly required to prime the adaptive stress response to the Mars *g*-level. In long-term applications, the combination of partial gravity level and red-light photostimulation could be used in space farming to avoid dysregulation of those pathways appearing affected in microgravity and to promote robust seedling growth.

Supplementary Materials: This article contains supplementary materials. Supplementary materials can be found at https://www.mdpi.com/1422-0067/22/2/899/s1. Figure S1. Extended Gene Ontology clusters affected by microgravity or Mars gravity, Figure S2. Common microgravity downregulated photosynthesis genes in GLDS-314 (µgrlgrrrl comparison, red photostimulated) and GLDS-251 (blue photostimulated), Figure S3. Venn diagrams of DEGs in common GO categories to the four comparisons, Figure S4. Hormone signal transduction changes in red light compared to darkness, Figure S5. Plastid and mitochondrial genome expression, Figure S6. Extended heatmaps of Gene Ontology analysis of the effect of red-light photostimulation across gravity levels, Table S1. List of the genes and the Log_2FC values of the dysregulated in microgravity, Table S2. List of the genes and the Log_2FC values of the dysregulated in Mars gravity, Table S3. Upregulated plastid and mitochondrial genes. List of references from Table 1.

Author Contributions: Conceptualization, R.H. and F.J.M.; methodology and investigation, A.V., M.C., A.M. and J.P.V.; writing—original draft, M.C., A.V. and R.H.; writing—review and editing, R.H., F.J.M. and J.Z.K.; supervision and funding acquisition, F.J.M., R.H. and J.Z.K. All authors have read and agreed to the published version of the manuscript.

Funding: This research was funded by the Agencia Estatal de Investigación of the Spanish Ministry of Science an Innovation, Grants #ESP2015-64323-R and #RTI2018-099309-B-I00 (co-funded by EU-ERDF) to F.J.M., by pre-doctoral fellowships to A.M. and A.V. from the Spanish National Program for Young Researchers Training (MINECO, Ref. #BES-2013-063933, #BES-2016-077976) and the Seedling Growth Project for ISS experimentation #LSRA2009-0932/ 1177, a shared project of ESA-ELIPS Program and NASA, to F.J.M. and J.Z.K. In addition, J.Z.K. is funded by Grants NNX12A065G and 80NSSC17K0546. This research is related to the Space Omics TT funded by the ESA contract 4000131202/20/NL/PG to R.H.

Data Availability Statement: The original sequencing data described in this study have been deposited at NASA's GENELAB repository as the GLDS-314 dataset (DOI: 10.26030/z5yf-jx91, https://genelab-data.ndc.nasa.gov/genelab/accession/GLDS-314, [40]).

Acknowledgments: We want to acknowledge the collaboration and support of many people who have contributed to the success of the experiments of the 'Seedling Growth' project, which are reported in this paper. This includes payload developers (Airbus), ESA and NASA managers, scientific collaborators (Eugenie Carnero-Diaz, Richard E. Edelmann, Julio Saez-Vasquez, Katherine D.L. Millar and Miguel Angel Valbuena) and the astronauts that performed the experiments on board the International Space Station, as well as the EMCS engineers and technicians (N-USOC) that performed the ground control operations.

Conflicts of Interest: The authors declare no conflict of interest. The funders had no role in the design of the study; in the collection, analyses, or interpretation of data; in the writing of the manuscript, or in the decision to publish the results.

Abbreviations

GRR	Ground Reference Run
EMCS	European Modular Cultivation System
ISS	International Space Station
SG	Seedling Growth experiment
DAPI	2-(4-amidinophenyl)-1H -indole-6-carboxamidine
GO	Gene Ontology
IAA	Indole acetic acid
JA	Jasmonic acid
SA	Salicylic acid
N-USOC	Norwegian User Support and Operation Center
PBS	Phosphate-Buffered Saline
RT	Room temperature
SED	Standard Error of the Difference
TEM	Transmission Electron Microscope
GC	Granular Component of the nucleolus
DFC	Dense Fibrillar Component of the nucleolus
FC	Fibrillar Centers of the nucleolus
µgd-grrd	comparison of transcriptome of seedlings grown in microgravity without photostimulation to the corresponding GRR
µgrl-grrrl	comparison of the transcriptome of seedlings photostimulated with red light for the last two days grown in microgravity to the corresponding GRR
Marsd-grrd	comparison of transcriptome of seedlings grown at Mars gravity level without photostimulation to the corresponding GRR

Marsrl-grrrl	comparison of the transcriptome of seedlings photostimulated with red light for the last two days grown at Mars gravity level to the corresponding GRR
μgrl-μgd, Marsrl-Marsd grrrl-grrd	transcriptomic data comparisons between two light conditions (red photostimulation and darkness) at each gravity level (μg, Mars *g*-level and 1*g* GRR)

References

1. Fu, Y.; Li, L.; Xie, B.; Dong, C.; Wang, M.; Jia, B.; Shao, L.; Dong, Y.; Deng, S.; Liu, H.; et al. How to establish a bioregenerative life support system for long-term crewed missions to the moon or mars. *Astrobiology* **2016**, *16*, 925–936. [CrossRef] [PubMed]
2. Vandenbrink, J.P.; Kiss, J.Z. Space, the final frontier: A critical review of recent experiments performed in microgravity. *Plant Sci.* **2016**, *243*, 115–119. [CrossRef] [PubMed]
3. Matía, I.; González-Camacho, F.; Herranz, R.; Kiss, J.Z.; Gasset, G.; van Loon, J.J.W.A.; Marco, R.; Medina, F.J. Plant cell proliferation and growth are altered by microgravity conditions in spaceflight. *J. Plant Physiol.* **2010**, *167*, 184–193. [CrossRef]
4. Ferl, R.J.; Paul, A.-L. The effect of spaceflight on the gravity-sensing auxin gradient of roots: GFP reporter gene microscopy on orbit. *NPJ Microgravity* **2016**, *2*, 15–23. [CrossRef]
5. Manzano, A.I.; Larkin, O.J.; Dijkstra, C.E.; Anthony, P.; Davey, M.R.; Eaves, L.; Hill, R.J.A.; Herranz, R.; Medina, F.J. Meristematic cell proliferation and ribosome biogenesis are decoupled in diamagnetically levitated *Arabidopsis* seedlings. *BMC Plant Biol.* **2013**, *13*, 124. [CrossRef]
6. Wyatt, S.E.; Kiss, J.Z. Plant tropisms: From Darwin to the International Space Station. *Am. J. Bot.* **2013**, *100*, 1–3. [CrossRef]
7. Kiss, J.Z. Where's the water? Hydrotropism in plants. *Proc. Natl. Acad. Sci. USA* **2007**, *104*, 4247–4248. [CrossRef]
8. Braam, J. In touch: Plant responses to mechanical stimuli. *New Phytol.* **2005**, *165*, 373–389. [CrossRef] [PubMed]
9. Muthert, L.W.F.; Izzo, L.G.; van Zanten, M.; Aronne, G. Root tropisms: Investigations on Earth and in space to unravel plant growth direction. *Front. Plant Sci.* **2020**, *10*, 1–22. [CrossRef] [PubMed]
10. Millar, K.D.L.; Kumar, P.; Correll, M.J.; Mullen, J.L.; Hangarter, R.P.; Edelmann, R.E.; Kiss, J.Z. A novel phototropic response to red light is revealed in microgravity. *New Phytol.* **2010**, *186*, 648–656. [CrossRef] [PubMed]
11. Vandenbrink, J.P.; Herranz, R.; Medina, F.J.; Edelmann, R.E.; Kiss, J.Z. A novel blue-light phototropic response is revealed in roots of *Arabidopsis thaliana* in microgravity. *Planta* **2016**, *244*, 1201–1215. [CrossRef] [PubMed]
12. Sullivan, J.A.; Deng, X.W. From seed to seed: The role of photoreceptors in *Arabidopsis* development. *Dev. Biol.* **2003**, *260*, 289–297. [CrossRef]
13. Yang, C. The effects of red, blue, and white light-emitting diodes on the growth, development, and edible quality of hydroponically grown lettuce (*Lactuca sativa* L. var. *capitata*). *Sci. Hortic.* **2013**, *150*, 86–91.
14. Reichler, S.A.; Balk, J.; Brown, M.E.; Woodruff, K.; Clark, G.B.; Roux, S.J. Light differentially regulates cell division and the mRNA abundance of pea nucleolin during de-etiolation. *Plant Physiol.* **2001**, *125*, 339–350. [CrossRef] [PubMed]
15. Valbuena, M.A.; Manzano, A.; Vandenbrink, J.P.; Pereda-Loth, V.; Carnero-Diaz, E.; Edelmann, R.E.; Kiss, J.Z.; Herranz, R.; Medina, F.J. The combined effects of real or simulated microgravity and red-light photoactivation on plant root meristematic cells. *Planta* **2018**, *248*, 691–704. [CrossRef]
16. Kiss, J.Z. Plant biology in reduced gravity on the Moon and Mars. *Plant Biol.* **2014**, *16*, 12–17. [CrossRef]
17. Crusan, J.; Bleacher, J.; Caram, J.; Craig, D.; Goodliff, K.; Herrmann, N.; Mahoney, E.; Smith, M. NASA's Gateway: An update on progress and plans for extending human presence to cislunar space. In Proceedings of the IEEE Conference on Aerospace, Big Sky, MT, USA, 2–9 March 2019.
18. Manzano, A.; Herranz, R.; den Toom, L.A.; te Slaa, S.; Borst, G.; Visser, M.; Medina, F.J.; van Loon, J.J.W.A. Novel, Moon and Mars, partial gravity simulation paradigms and their effects on the balance between cell growth and cell proliferation during early plant development. *NPJ Microgravity* **2018**, *4*, 1–11. [CrossRef]
19. Brinckmann, E. ESA hardware for plant research on the International Space Station. *Adv. Space Res.* **2005**, *36*, 1162–1166. [CrossRef]
20. Kiss, J.Z.; Aanes, G.; Schiefloe, M.; Coelho, L.H.F.; Millar, K.D.L.; Edelmann, R.E. Changes in operational procedures to improve spaceflight experiments in plant biology in the European Modular Cultivation System. *Adv. Space Res.* **2014**, *53*, 818–827. [CrossRef]
21. Vandenbrink, J.P.; Herranz, R.; Poehlman, W.L.; Feltus, F.A.; Villacampa, A.; Ciska, M.; Medina, F.J.; Kiss, J.Z. RNA-seq analyses of *Arabidopsis thaliana* seedlings after exposure to blue-light phototropic stimuli in microgravity. *Am. J. Bot.* **2019**, *106*, 1466–1476. [CrossRef]
22. Herranz, R.; Vandenbrink, J.P.; Villacampa, A.; Manzano, A.; Poehlman, W.L.; Feltus, F.A.; Kiss, J.Z.; Medina, F.J. RNAseq analysis of the response of *Arabidopsis thaliana* to fractional gravity under blue-light stimulation during spaceflight. *Front. Plant Sci.* **2019**, *10*, 1–11. [CrossRef] [PubMed]
23. Paul, A.-L.; Zupanska, A.K.; Schultz, E.R.; Ferl, R.J. Organ-specific remodeling of the *Arabidopsis* transcriptome in response to spaceflight. *BMC Plant Biol.* **2013**, *13*, 112. [CrossRef] [PubMed]
24. Basu, P.; Kruse, C.P.S.; Luesse, D.R.; Wyatt, S.E. Growth in spaceflight hardware results in alterations to the transcriptome and proteome. *Life Sci. Sp. Res.* **2017**, *15*, 88–96. [CrossRef] [PubMed]

25. Choi, W.-G.; Barker, R.J.; Kim, S.-H.; Swanson, S.J.; Gilroy, S. Variation in the transcriptome of different ecotypes of *Arabidopsis thaliana* reveals signatures of oxidative stress in plant responses to spaceflight. *Am. J. Bot.* **2019**, *106*, 123–136. [CrossRef] [PubMed]
26. Correll, M.J.; Pyle, T.P.; Millar, K.D.L.; Sun, Y.; Yao, J.; Edelmann, R.E.; Kiss, J.Z. Transcriptome analyses of *Arabidopsis thaliana* seedlings grown in space: Implications for gravity-responsive genes. *Planta* **2013**, *238*, 519–533. [CrossRef]
27. Fengler, S.; Spirer, I.; Neef, M.; Ecke, M.; Nieselt, K.; Hampp, R. A whole-genome microarray study of *Arabidopsis thaliana* semisolid callus cultures exposed to microgravity and nonmicrogravity related spaceflight conditions for 5 days on board of Shenzhou 8. *Biomed. Res. Int.* **2015**, *2015*. [CrossRef]
28. Hausmann, N.; Fengler, S.; Hennig, A.; Franz-Wachtel, M.; Hampp, R.; Neef, M. Cytosolic calcium, hydrogen peroxide and related gene expression and protein modulation in *Arabidopsis thaliana* cell cultures respond immediately to altered gravitation: Parabolic flight data. *Plant Biol.* **2014**, *16*, 120–128. [CrossRef]
29. Kruse, C.P.S.; Meyers, A.D.; Basu, P.; Hutchinson, S.; Luesse, D.R.; Wyatt, S.E. Spaceflight induces novel regulatory responses in *Arabidopsis* seedling as revealed by combined proteomic and transcriptomic analyses. *BMC Plant Biol.* **2020**, *20*, 1–16. [CrossRef]
30. Kwon, T.; Sparks, J.A.; Nakashima, J.; Allen, S.N.; Tang, Y.; Blancaflor, E.B. Transcriptional response of *Arabidopsis* seedlings during spaceflight reveals peroxidase and cell wall remodeling genes associated with root hair development. *Am. J. Bot.* **2015**, *102*, 21–35. [CrossRef]
31. Paul, A.L.; Popp, M.P.; Gurley, W.B.; Guy, C.; Norwood, K.L.; Ferl, R.J. *Arabidopsis* gene expression patterns are altered during spaceflight. *Adv. Space Res.* **2005**, *36*, 1175–1181. [CrossRef]
32. Yamazaki, C.; Fujii, N.; Miyazawa, Y.; Kamada, M.; Kasahara, H.; Osada, I.; Shimazu, T.; Fusejima, Y.; Higashibata, A.; Yamazaki, T.; et al. The gravity-induced re-localization of auxin efflux carrier CsPIN1 in cucumber seedlings: Spaceflight experiments for immunohistochemical microscopy. *NPJ Microgravity* **2016**, *2*, 1–7. [CrossRef] [PubMed]
33. Chandler, J.O.; Haas, F.B.; Khan, S.; Bowden, L.; Ignatz, M.; Enfissi, E.M.A.; Gawthorp, F.; Griffiths, A.; Fraser, P.D.; Rensing, S.A.; et al. Rocket science: The effect of spaceflight on germination physiology, ageing, and transcriptome of *Eruca sativa* seeds. *Life* **2020**, *10*, 49. [CrossRef] [PubMed]
34. Hoson, T.; Soga, K.; Wakabayashi, K.; Kamisaka, S.; Tanimoto, E. Growth and cell wall changes in rice roots during spaceflight. *Plant Soil* **2003**, *255*, 19–26. [CrossRef] [PubMed]
35. Jin, J.; Chen, H.; Cai, W. Transcriptome analysis of *Oryza sativa* calli under microgravity. *Microgravity Sci. Technol.* **2015**, *27*, 437–453. [CrossRef]
36. Sugimoto, M.; Oono, Y.; Gusev, O.; Matsumoto, T.; Yazawa, T.; Levinskikh, M.A.; Sychev, V.N.; Bingham, G.E.; Wheeler, R.; Hummerick, M. Genome-wide expression analysis of reactive oxygen species gene network in *Mizuna* plants grown in long-term spaceflight. *BMC Plant Biol.* **2014**, *14*, 1–11. [CrossRef]
37. Manzano, A.; Villacampa, A.; Sáez-Vásquez, J.; Kiss, J.Z.; Medina, F.J.; Herranz, R. The importance of Earth reference controls in spaceflight—Omics research: Characterization of nucleolin mutants from the seedling growth experiments. *iScience* **2020**, *23*, 101686. [CrossRef]
38. Rutter, L.; Barker, R.; Bezdan, D.; Cope, H.; Costes, S.V.; Degoricija, L.; Fisch, K.M.; Gabitto, M.I.; Gebre, S.; Giacomello, S.; et al. A new era for space life science: International standards for space omics processing. *Patterns* **2020**, *1*, 100148. [CrossRef]
39. Madrigal, P.; Gabel, A.; Villacampa, A.; Manzano, A.; Deane, C.S.; Bezdan, D.; Carnero-Diaz, E.; Medina, F.J.; Hardiman, G.; Grosse, I.; et al. Revamping space-omics in Europe. *Cell Syst.* **2020**, *11*, 555–556. [CrossRef]
40. Ray, S.; Gebre, S.; Fogle, H.; Berrios, D.C.; Tran, P.B.; Galazka, J.M.; Costes, S.V. GeneLab: Omics database for spaceflight experiments. *Bioinformatics* **2019**, *35*, 1753–1759. [CrossRef]
41. Kiss, J.Z. Mechanisms of the early phases of plant gravitropism. *CRC Crit. Rev. Plant Sci.* **2000**, *19*, 551–573. [CrossRef]
42. Barneche, F.; Steinmetz, F.; Echeverria, M. Fibrillarin genes encode both a conserved nucleolar protein and a novel snoRNA involved in rRNA methylation in *Arabidopsis thaliana*. *J. Biol. Chem.* **2000**. [CrossRef] [PubMed]
43. Guerrero, F.; De la Torre, C.; García-Herdugo, G. Control of nucleolar growth during interphase in higher plant meristem cells. *Protoplasma* **1989**, *152*, 96–100. [CrossRef]
44. Sáez-Vásquez, J.; Medina, F.J. The plant nucleolus. *Adv. Bot. Res.* **2008**, *47*, 1–46.
45. Manzano, A.I.; Herranz, R.; Manzano, A.; van Loon, J.J.W.A.; Medina, F.J. Early effects of altered gravity environments on plant cell growth and cell proliferation: Characterization of morphofunctional nucleolar types in an *Arabidopsis* cell culture system. *Front. Astron. Sp. Sci.* **2016**, *3*, 2. [CrossRef]
46. Wang, S.; Chang, Y.; Guo, J.; Zeng, Q.; Ellis, B.E.; Chen, J. *Arabidopsis* ovate family proteins, a novel transcriptional repressor family, control multiple aspects of plant growth and development. *PLoS ONE* **2011**, *6*, e23896. [CrossRef] [PubMed]
47. Phukan, U.J.; Jeena, G.S.; Shukla, R.K. WRKY transcription factors: Molecular regulation and stress responses in plants. *Front. Plant Sci.* **2016**, *7*, 1–14. [CrossRef]
48. Danquah, A.; de Zelicourt, A.; Colcombet, J.; Hirt, H. The role of ABA and MAPK signaling pathways in plant abiotic stress responses. *Biotechnol. Adv.* **2014**, *32*, 40–52. [CrossRef]
49. Li, H.; Sun, J.; Xu, Y.; Jiang, H.; Wu, X.; Li, C. The bHLH-type transcription factor AtAIB positively regulates ABA response in *Arabidopsis*. *Plant Mol. Biol.* **2007**, *65*, 655–665. [CrossRef]
50. Burke, R.; Schwarze, J.; Sherwood, O.L.; Jnaid, Y.; Mccabe, P.F.; Kacprzyk, J. Stressed to death: The role of transcription factors in plant programmed cell death induced by abiotic and biotic stimuli. *Front. Plant Sci.* **2020**, *11*, 1–12. [CrossRef]

51. Aoi, Y.; Tanaka, K.; Cook, S.D.; Hayashi, K.-I.; Kasahara, H. GH3 auxin-amido synthetases alter the ratio of indole-3-acetic acid and phenylacetic acid in *Arabidopsis*. *Plant Cell. Physiol.* **2020**, *61*, 596–605. [CrossRef]

52. Hutchison, C.E.; Li, J.; Argueso, C.; Gonzalez, M.; Lee, E.; Lewis, M.W.; Maxwell, B.B.; Perdue, T.D.; Schaller, G.E.; Alonso, J.M.; et al. The *Arabidopsis* histidine phosphotransfer proteins are redundant positive regulators of cytokinin signaling. *Plant Cell* **2006**, *18*, 3073–3087. [CrossRef] [PubMed]

53. Ge, S.X.; Jung, D.; Yao, R. ShinyGO: A graphical gene-set enrichment tool for animals and plants. *Bioinformatics* **2020**, *36*, 2628–2629. [CrossRef] [PubMed]

54. Johnson, C.M.; Subramanian, A.; Pattathil, S.; Correll, M.J.; Kiss, J.Z. Comparative transcriptomics indicate changes in cell wall organization and stress response in seedlings during spaceflight. *Am. J. Bot.* **2017**, *104*, 1219–1231. [CrossRef]

55. Herranz, R.; Benguría, A.; Laván, D.A.; López-Vidriero, I.; Gasset, G.; Medina, F.J.; van Loon, J.J.W.A.; Marco, R. Spaceflight-related suboptimal conditions can accentuate the altered gravity response of *Drosophila* transcriptome. *Mol. Ecol.* **2010**, *19*, 4255–4264. [CrossRef]

56. Silveira, W.A.; Fazelinia, H.; Rosenthal, S.B.; Mason, C.E.; Costes, S.V. Comprehensive multi-omics analysis reveals mitochondrial stress as a central biological hub for spaceflight impact. *Cell* **2020**, *183*, 1185–1201.e20. [CrossRef]

57. Leu, W.M.; Cao, X.L.; Wilson, T.J.; Snustad, D.P.; Chua, N.H. Phytochrome A and phytochrome B mediate the hypocotyl-specific downregulation of TUB1 by light in *Arabidopsis*. *Plant Cell* **1995**, *7*, 2187–2196. [PubMed]

58. Nakajima, K.; Kawamura, T.; Hashimoto, T. Role of the *SPIRAL$_1$* gene family in anisotropic growth of *Arabidopsis thaliana*. *Plant Cell Physiol.* **2006**, *47*, 513–522. [CrossRef]

59. Wakabayashi, K.; Soga, K.; Hoson, T.; Kotake, T.; Yamazaki, T.; Ishioka, N.; Shimazu, T.; Kamada, M. Microgravity affects the level of matrix polysaccharide 1,3:1,4-β-glucans in cell walls of rice shoots by increasing the expression level of a gene involved in their breakdown. *Astrobiology* **2020**, *20*, 820–829. [CrossRef]

60. Mo, M.; Yokawa, K.; Wan, Y.; Baluška, F. How and why do root apices sense light under the soil surface? *Front. Plant Sci.* **2015**, *6*, 1–8. [CrossRef]

61. Kiss, J.Z.; Mullen, J.L.; Correll, M.J.; Hangarter, R.P. Phytochromes A and B mediate red-light-induced positive phototropism in roots. *Plant Physiol.* **2003**, *131*, 1411–1417. [CrossRef]

62. Correll, M.J.; Edelmann, R.E.; Hangarter, R.P.; Mullen, J.L.; Kiss, J.Z. Ground-based studies of tropisms in hardware developed for the European Modular Cultivation System (EMCS). *Adv. Space Res.* **2005**, *36*, 1203–1210. [CrossRef]

63. Jiao, S.; Hilaire, E.; Paulsen, A.Q.; Guikema, J.A. *Brassica rapa* plants adapted to microgravity with reduced photosystem I and its photochemical activity. *Physiol. Plant* **2004**, 281–290. [CrossRef]

64. Chen, B.; Zhang, A.; Lu, Q. Characterization of photosystem I in rice (*Oryza sativa* L.) seedlings upon exposure to random positioning machine. *Photosynth. Res.* **2013**, *116*, 93–105. [CrossRef] [PubMed]

65. Stutte, G.W.; Monje, O.; Hatfield, R.D.; Paul, A.L.; Ferl, R.J.; Simone, C.G. Microgravity effects on leaf morphology, cell structure, carbon metabolism and mRNA expression of dwarf wheat. *Planta* **2006**, *224*, 1038–1049. [CrossRef] [PubMed]

66. Mikhaylenko, N.F.; Sytnik, S.K.; Zolotareva, E.K. Effects of slow clinorotation on lipid contents and proton permeability of thylakoid membranes of pea chloroplasts. *Adv. Space Res.* **2001**, *27*, 1007–1010. [CrossRef]

67. Adamchuk, N.I. Ultrastructural and functional changes of photosynthetic apparatus of *Arabidopsis thaliana* (L.) Heynh induced by clinorotation. *Adv. Space Res.* **1998**, *177*, 1131–1134. [CrossRef]

68. Staneloni, R.J.; Rodriguez-Batiller, M.J.; Casal, J.J. Abscisic acid, high-light, and oxidative stress down-regulate a photosynthetic gene via a promoter motif not involved in phytochrome-mediated transcriptional regulation. *Mol. Plant* **2008**, *1*, 75–83. [CrossRef]

69. Xu, Y.; Liu, R.; Yan, L.; Liu, Z.; Jiang, S.; Shen, Y. Light harvesting chlorophyll a/b-binding proteins are required for stomatal response to abscisic acid in *Arabidopsis*. *J. Exp. Bot.* **2012**, *63*, 1095–1106. [CrossRef] [PubMed]

70. Zhu, D.; Wu, Z.; Cao, G.; Li, J.; Wei, J.; Tsuge, T.; Hongya, G.; Takashi, A.; Li-Jia, Q. Translucent green, an ERF family transcription factor, controls water balance. *Mol. Plant* **2014**, *7*, 601–615. [CrossRef]

71. Gilmour, S.J.; Fowler, S.G.; Thomashow, M.F. *Arabidopsis* transcriptional activators CBF1, CBF2, and CBF3 have matching functional activities. *Plant Mol. Biol.* **2004**, *54*, 767–781. [CrossRef]

72. Jaglo-Ottosen, K.R.; Gilmour, S.J.; Zarka, D.G.; Schabenberger, O.; Thomashow, M.F. *Arabidopsis* CBF1 overexpression induces COR genes and enhances freezing tolerance. *Science* **1998**, *280*, 104–106. [CrossRef] [PubMed]

73. Ogawa, T.; Pan, L.; Kawai-Yamada, M.; Yu, L.H.; Yamamura, S.; Koyama, T.; Kitajima, S.; Ohme-Takagi, M.; Sato, F.; Uchimiya, H. Functional analysis of *Arabidopsis* ethylene-responsive element binding protein conferring resistance to Bax and abiotic stress-induced plant cell death. *Plant Physiol.* **2005**, *138*, 1436–1445. [CrossRef] [PubMed]

74. Leister, D.; Wang, L.; Kleine, T. Organellar gene expression and acclimation of plants to environmental stress. *Front. Plant Sci.* **2017**, *8*. [CrossRef] [PubMed]

75. Wicke, S.; Schneeweiss, G.M.; dePamphilis, C.W.; Müller, K.F.; Quandt, D. The evolution of the plastid chromosome in land plants: Gene content, gene order, gene function. *Plant Mol. Biol.* **2011**, *76*, 273–297. [CrossRef] [PubMed]

76. Unseld, M.; Marienfeld, J.R.; Brandt, P.; Brennicke, A. The mitochondrial genome of *Arabidopsis* thaliana contains 57 genes in 366,924 nucleotides. *Nat. Genet.* **1997**, *15*, 1–5. [CrossRef]

77. Yamburenko, M.V.; Zubo, Y.O.; Vanková, R.; Kusnetsov, V.V.; Kulaeva, O.N.; Börner, T. Abscisic acid represses the transcription of chloroplast genes. *J. Exp. Bot.* **2013**, *64*, 4491–4502. [CrossRef]

78. Vanlerberghe, G.C. Alternative oxidase: A mitochondrial respiratory pathway to maintain metabolic and signaling homeostasis during abiotic and biotic stress in plants. *Int. J. Mol. Sci.* **2013**, *14*, 6805–6847. [CrossRef]

79. Van Aken, O.; Zhang, B.; Law, S.; Narsai, R.; Whelan, J. AtWRKY40 and AtWRKY63 modulate the expression of stress-responsive nuclear genes encoding mitochondrial and chloroplast proteins. *Plant Physiol.* **2013**, *162*, 254–271. [CrossRef]

80. Van Aken, O.; Zhang, B.; Carrie, C.; Uggalla, V.; Paynter, E.; Giraud, E.; Whelan, J. Defining the mitochondrial stress response in *Arabidopsis thaliana*. *Mol. Plant* **2009**, *2*, 1310–1324. [CrossRef]

81. Schwarzländer, M.; König, A.C.; Sweetlove, L.J.; Finkemeier, I. The impact of impaired mitochondrial function on retrograde signalling: A meta-analysis of transcriptomic responses. *J. Exp. Bot.* **2012**, *63*, 1735–1750. [CrossRef]

82. Wang, Y.; Berkowitz, O.; Selinski, J.; Xu, Y.; Hartmann, A.; Whelan, J. Stress responsive mitochondrial proteins in *Arabidopsis thaliana*. *Free Radic. Biol. Med.* **2018**, *122*, 28–39. [CrossRef] [PubMed]

83. Meng, X.; Li, L.; Narsai, R.; De Clercq, I.; Whelan, J.; Berkowitz, O. Mitochondrial signalling is critical for acclimation and adaptation to flooding in *Arabidopsis thaliana*. *Plant J.* **2020**, *103*, 227–247. [CrossRef] [PubMed]

84. Van Aken, O.; De Clercq, I.; Ivanova, A.; Law, S.R.; van Breusegem, F.; Millar, A.H.; Whelan, L. Mitochondrial and chloroplast stress responses are modulated in distinct touch and chemical inhibition phases. *Plant Physiol.* **2016**, *171*, 2150–2165. [CrossRef] [PubMed]

85. Lee, D.; Polisensky, D.H.; Braam, J. Genome-wide identification of touch- and darkness-regulated *Arabidopsis* genes: A focus on calmodulin-like and XTH genes. *New Phytol.* **2005**, *165*, 429–444. [CrossRef] [PubMed]

86. Liao, J.; Liu, G.; Monje, O.; Stutte, G.W.; Porterfield, D.M. Induction of hypoxic root metabolism results from physical limitations in O_2 bioavailability in microgravity. *Adv. Space Res.* **2004**, *34*, 1579–1584. [CrossRef]

87. Pedersen, O.; Sauter, M.; Colmer, T.D.; Nakazono, M. Regulation of root adaptive anatomical andmorphological traits during low soil oxygen. *New Phytol.* **2020**. [CrossRef]

88. Millar, K.D.L.; Johnson, C.M.; Edelmann, R.E.; Kiss, J.Z. An endogenous growth pattern of roots is revealed in seedlings grown in microgravity. *Astrobiology* **2011**, *11*, 787–797. [CrossRef]

89. Stout, S.C.; Porterfield, D.M.; Briarty, L.G.; Kuang, A.; Musgrave, M.E. Evidence of root zone hypoxia in *Brassica rapa* L. grown in microgravity. *Int. J. Plant Sci.* **2001**, *162*, 249–255. [CrossRef]

90. Ban, Y.; Kobayashi, Y.; Hara, T.; Hamada, T.; Hashimoto, T.; Takeda, S.; Hattori, T. α-Tubulin is rapidly phosphorylated in response to hyperosmotic stress in rice and *Arabidopsis*. *Plant Cell Physiol.* **2013**, *54*, 848–858. [CrossRef]

91. Nakano, T.; Suzuki, K.; Fujimura, T.; Shinshi, H. Genome-wide analysis of the ERF gene family in *Arabidopsis* and rice. *Plant Physiol.* **2006**, *140*, 411–432. [CrossRef]

92. Rajhi, I.; Yamauchi, T.; Takahashi, H.; Nishiuchi, S.; Shiono, K. Identification of genes expressed in maize root cortical cells during lysigenous aerenchyma formation using laser microdissection and microarray analyses. *New Phytol.* **2011**, 351–368. [CrossRef] [PubMed]

93. Safavi-Rizi, V.; Herde, M.; Stöhr, C. RNA-Seq reveals novel genes and pathways associated with hypoxia duration and tolerance in tomato root. *Sci. Rep.* **2020**, *10*, 1–17.

94. Ding, Z.J.; Yan, J.Y.; Li, C.X.; Li, G.X.; Wu, Y.R.; Zheng, S.J. Transcription factor WRKY46 modulates the development of *Arabidopsis* lateral roots in osmotic/salt stress conditions via regulation of ABA signaling and auxin homeostasis. *Plant J.* **2015**, *84*, 56–69. [CrossRef] [PubMed]

95. Zhang, Y.; Yu, H.; Yang, X.; Li, Q.; Ling, J.; Wang, H.; Gu, X.; Huang, S.; Jiang, W. CsWRKY46, a WRKY transcription factor from cucumber, confers cold resistance in transgenic-plant by regulating a set of cold-stress responsive genes in an ABA-dependent manner. *Plant Physiol. Biochem.* **2016**, *108*, 478–487. [CrossRef] [PubMed]

96. Kim, K.C.; Lai, Z.; Fan, B.; Chen, Z. *Arabidopsis* WRKY38 and WRKY62 transcription factors interact with histone deacetylase 19 in basal defense. *Plant Cell* **2008**, *20*, 2357–2371. [CrossRef]

97. Finatto, T.; Viana, V.E.; Woyann, L.G.; Busanello, C.; da Maia, L.C.; de Oliveira, A.C. Can WRKY transcription factors help plants to overcome environmental challenges? *Genet. Mol. Biol.* **2018**, *41*, 533–544. [PubMed]

98. Chen, L.; Song, Y.; Li, S.; Zhang, L.; Zou, C.; Yu, D. The role of WRKY transcription factors in plant abiotic stresses. *Biochimica Biophysica Acta (BBA)-Gene Regul. Mech.* **2012**, *1819*, 120–128. [CrossRef]

99. Park, C.Y.; Lee, J.H.; Yoo, J.H.; Moon, B.C.; Choi, M.S.; Kang, Y.H.; Lee, S.M.; Kim, H.S.; Kang, K.Y.; Chung, W.S.; et al. WRKY group IId transcription factors interact with calmodulin. *FEBS Lett.* **2005**, *579*, 1545–1550. [CrossRef]

100. Li, Z.; Peng, J.; Wen, X.; Guo, H. Gene network analysis and functional studies of senescence-associated genes reveal novel regulators of *Arabidopsis* leaf senescence. *J. Integr. Plant Biol.* **2012**, *54*, 526–539. [CrossRef]

101. Century, K.; Reuber, T.L.; Ratcliffe, O.J. Regulating the regulators: The future prospects for transcription-factor-based agricultural biotechnology products. *Plant Physiol.* **2008**, *147*, 20–29. [CrossRef]

102. Doebley, J.F.; Gaut, B.S.; Smith, B.D. The molecular genetics of crop domestication. *Cell* **2006**, *127*, 1309–1321. [CrossRef] [PubMed]

103. Raineri, J.; Wang, S.; Peleg, Z.; Blumwald, E.; Chan, R.L. The rice transcription factor OsWRKY47 is a positive regulator of the response to water deficit stress. *Plant Mol. Biol.* **2015**, *88*, 401–413. [CrossRef] [PubMed]

104. Shen, H.; Liu, C.; Zhang, Y.; Meng, X.; Zhou, X.; Chu, C.; Wang, X. OsWRKY30 is activated by MAP kinases to confer drought tolerance in rice. *Plant Mol. Biol.* **2012**, *80*, 241–253. [CrossRef] [PubMed]

105. Caarls, L.; Pieterse, C.M.J.; Van Wees, S.C.M. How salicylic acid takes transcriptional control over jasmonic acid signaling. *Front. Plant Sci.* **2015**, *6*, 1–11. [CrossRef]

106. Wang, J.; Song, L.; Gong, X.; Xu, J.; Li, M. Functions of jasmonic acid in plant regulation and response to abiotic stress. *Int. J. Mol. Sci.* **2020**, *21*, 1446. [CrossRef]

107. Khan, M.I.R.; Fatma, M.; Per, T.S.; Anjum, N.A.; Khan, N.A. Salicylic acid-induced abiotic stress tolerance and underlying mechanisms in plants. *Front. Plant Sci.* **2015**, *6*, 1–17. [CrossRef]

108. Herrera-Vásquez, A.; Salinas, P.; Holuigue, L. Salicylic acid and reactive oxygen species interplay in the transcriptional control of defense genes expression. *Front. Plant Sci.* **2015**, *6*, 1–9. [CrossRef]

109. Wasternack, C.; Strnad, M. Jasmonate signaling in plant stress responses and development—Active and inactive compounds. *New Biotechnol.* **2016**, *33*, 604–613. [CrossRef]

110. Pieterse, C.M.J.; Van Der Does, D.; Zamioudis, C.; Leon-Reyes, A.; Van Wees, S.C.M. Hormonal modulation of plant immunity. *Annu. Rev. Cell Dev. Biol.* **2012**, *28*, 489–521. [CrossRef]

111. Kazan, K.; Manners, J.M. JAZ repressors and the orchestration of phytohormone crosstalk. *Trends Plant Sci.* **2012**, *17*, 22–31. [CrossRef]

112. Guo, Q.; Yoshida, Y.; Major, I.T.; Wang, K.; Sugimoto, K.; Kapali, G.; Havko, N.E.; Benning, C.; Howe, G.A. JAZ repressors of metabolic defense promote growth and reproductive fitness in *Arabidopsis*. *Proc. Natl. Acad. Sci. USA* **2018**, *115*, E10768–E10777. [CrossRef] [PubMed]

113. Oblessuc, P.R.; Obulareddy, N.; DeMott, L.; Matiolli, C.C.; Thompson, B.K.; Melotto, M. JAZ4 is involved in plant defense, growth, and development in *Arabidopsis*. *Plant J.* **2020**, *101*, 371–383. [CrossRef] [PubMed]

114. Qi, T.; Huang, H.; Wu, D.; Yan, J.; Qi, Y.; Song, S.; Xie, D. *Arabidopsis* DELLA and JAZ proteins bind the WD-Repeat/bHLH/MYB complex to modulate gibberellin and jasmonate signaling synergy. *Plant Cell* **2014**, *26*, 1118–1133. [CrossRef] [PubMed]

115. Zhu, D.; Li, R.; Liu, X.; Sun, M.; Wu, J.; Zhang, N.; Zhu, Y. The positive regulatory roles of the TIFY10 proteins in plant responses to alkaline stress. *PLoS ONE* **2014**, *9*, e111984. [CrossRef] [PubMed]

116. Hoo, S.C.; Koo, A.J.K.; Gao, X.; Jayanty, S.; Thines, B.; Jones, A.D.; Howe, G.A. Regulation and function of Arabidopsis JASMONATE ZIM-domain genes in response to wounding and herbivory. *Plant Physiol.* **2008**, *146*, 952–964.

117. Jiang, Y.; Liang, G.; Yang, S.; Yu, D. *Arabidopsis* WRKY57 functions as a node of convergence for jasmonic acid- and auxin-mediated signaling in jasmonic acid-induced leaf senescence. *Plant Cell* **2014**, *26*, 230–245. [CrossRef]

118. Hu, Y.; Jiang, L.; Wang, F.; Yu, D. Jasmonate regulates the INDUCER OF CBF expression-C-repeat binding factor/dre binding factor1 cascade and freezing tolerance in *Arabidopsis*. *Plant Cell* **2013**, *25*, 2907–2924. [CrossRef]

119. Takahashi, N.; Goto, N.; Okada, K.; Takahashi, H. Hydrotropism in abscisic acid, wavy, and gravitropic mutants of *Arabidopsis thaliana*. *Planta* **2002**, *216*, 203–211. [CrossRef]

120. Kiss, J.Z. Conducting plant experiments in space. In *Plant Gravitropism: Methods and Protocols*; Blancaflor, E.B., Ed.; Springer: New York, NY, USA, 2015; pp. 255–283.

121. Vandenbrink, J.P.; Kiss, J.Z. Preparation of a spaceflight experiment to study tropisms in Arabidopsis seedlings on the International Space Station. *Methods Mol. Biol.* **2019**, *1924*, 207–214.

122. Manzano, A.; Creus, E.; Tomás, A.; Valbuena, M.A.; Villacampa, A.; Ciska, M.; Edelmann, R.E.; Kiss, J.Z.; Medina, F.J.; Herranz, R. The fixbox: Hardware to provide on-orbit fixation capabilities to the EMCS on the ISS. *Microgravity Sci. Technol.* **2020**, *32*, 1105–1120. [CrossRef]

123. Musielak, T.J.; Schenkel, L.; Kolb, M.; Henschen, A.; Bayer, M. A simple and versatile cell wall staining protocol to study plant reproduction. *Plant Reprod.* **2015**, *28*, 161–169. [CrossRef] [PubMed]

124. Robert, H.S.; Grunewald, W.; Sauer, M.; Cannoot, B.; Soriano, M.; Swarup, R.; Weijers, D.; Bennett, M.; Boutilier, K.; Friml, J. Plant embryogenesis requires aux/lax-mediated auxin influx. *Development* **2015**, *142*, 702–711. [CrossRef] [PubMed]

125. Love, M.I.; Huber, W.; Anders, S. Moderated estimation of fold change and dispersion for RNA-seq data with DESeq2. *Genome Biol.* **2014**, *15*, 1–21. [CrossRef] [PubMed]

126. Ge, S.X.; Son, E.W.; Yao, R. iDEP: An integrated web application for differential expression and pathway analysis of RNA-Seq data. *BMC Bioinform.* **2018**, *19*, 1–24. [CrossRef]

127. Zhou, Y.; Zhou, B.; Pache, L.; Chang, M.; Khodabakhshi, A.H.; Tanaseichuk, O.; Benner, C.; Chanda, S.K. Metascape provides a biologist-oriented resource for the analysis of systems-level datasets. *Nat. Commun.* **2019**, *10*. [CrossRef]

128. Bardou, P.; Mariette, J.; Escudié, F.; Djemiel, C.; Klopp, C. Jvenn: An interactive Venn diagram viewer. *BMC Bioinform.* **2014**, *15*, 1–7. [CrossRef]

129. Szklarczyk, D.; Gable, A.L.; Lyon, D.; Junge, A.; Wyder, S.; Huerta-Cepas, J.; Simonovic, M.; Doncheva, N.T.; Morris, J.H.; Bork, P.; et al. STRING v11: Protein-protein association networks with increased coverage, supporting functional discovery in genome-wide experimental datasets. *Nucleic Acids Res.* **2019**, *47*, D607–D613. [CrossRef]

130. Kanehisa, M.; Sato, Y. KEGG mapper for inferring cellular functions from protein sequences. *Protein Sci.* **2020**, *28*, 28–35. [CrossRef]

International Journal of
Molecular Sciences

Review

The Complex Story of Plant Cyclic Nucleotide-Gated Channels

Edwin Jarratt-Barnham, Limin Wang, Youzheng Ning and Julia M. Davies *

Department of Plant Sciences, University of Cambridge, Cambridge CB2 3EA, UK; ecj39@cam.ac.uk (E.J.-B.);
lw577@cam.ac.uk (L.W.); yn283@cam.ac.uk (Y.N.)
* Correspondence: jmd32@cam.ac.uk; Tel.: +44-1223-333-939

Abstract: Plant cyclic nucleotide-gated channels (CNGCs) are tetrameric cation channels which may be activated by the cyclic nucleotides (cNMPs) adenosine $3',5'$-cyclic monophosphate (cAMP) and guanosine $3',5'$-cyclic monophosphate (cGMP). The genome of *Arabidopsis thaliana* encodes 20 CNGC subunits associated with aspects of development, stress response and immunity. Recently, it has been demonstrated that CNGC subunits form heterotetrameric complexes which behave differently from the homotetramers produced by their constituent subunits. These findings have widespread implications for future signalling research and may help explain how specificity can be achieved by CNGCs that are known to act in disparate pathways. Regulation of complex formation may involve cyclic nucleotide-gated channel-like proteins.

Keywords: calcium signalling; CaM; calmodulin; cAMP; cGMP; CNGC; cyclic nucleotide-gated channel; CNGCL; cyclic nucleotide-gated channel like

check for
updates

Citation: Jarratt-Barnham, E.; Wang, L.; Ning, Y.; Davies, J.M. The Complex Story of Plant Cyclic Nucleotide-Gated Channels. *Int. J. Mol. Sci.* **2021**, *22*, 874. https://doi.org/10.3390/ijms22020874

Received: 11 December 2020
Accepted: 15 January 2021
Published: 16 January 2021

Publisher's Note: MDPI stays neutral with regard to jurisdictional claims in published maps and institutional affiliations.

1. Introduction

Plant cyclic nucleotide-gated channels (CNGCs) are held to be tetrameric cation channels formed by four subunits which may be activated by the cyclic nucleotide monophosphates (cNMPs) adenosine $3',5'$-cyclic monophosphate (cAMP) and guanosine $3',5'$-cyclic monophosphate (cGMP) [1–10]. The ability of cNMPs to act as signalling molecules in plants has been questioned. However, enhanced detection methods are revealing stimulus-induced increases and the ability to lower cAMP in cellular compartments using a "cAMP sponge" is now allowing the consequences of depletion to be investigated [11]. CNGCs are integral not only to plant nutrition, but also to calcium (Ca^{2+}) signalling in development, abiotic stress and immunity [9,10,12–15]. Most research into CNGC contribution to signalling has focused on the 20 *cngc* loss-of-function mutants in *Arabidopsis thaliana* (summarised in Table 1) and on the functional characteristics of homotetrameric CNGCs in heterologous expression systems such as *Escherichia coli*, yeast, *Xenopus* oocytes and HEK293 cells. The 20 *AtCNGC* gene sequences [16] have been used to predict that *Glycine max* has 39 [17], *Hordeum vulgare* has 9 [7], *Nicotiana tabacum* has 35 [18], *Oryza sativa* has 16 [19], *Triticum aestivum* has 47 [20], *Zea mays* has 12 [21], *Brassica oleracea* has 26 [22] and *Brassica rapa* has 30 *CNGCs* [23]. Advances made with *Arabidopsis* may well have implications for crop species.

Table 1. *AtCNGCs* are involved in diverse signalling pathways ranging from development to stress responses. For each CNGC in *Arabidopsis thaliana*, the reported physiological or developmental roles are presented, based on phenotyping loss-of-function mutants. Where the roles of two or more CNGCs overlap, it may be postulated that a CNGC complex might form between these subunits if they co-localise. Currently, complex formation has only been investigated in a few of these cases.

Gene	Proposed Physiological or Developmental Process	References
AtCNGC1	Negative regulation of Pb^{2+} tolerance; primary root growth; gravitropism	[24–26]

125

Table 1. *Cont.*

Gene	Proposed Physiological or Developmental Process	References
AtCNGC2	Pathogen defence; programmed cell death; nitric oxide generation; suppression of leaf senescence; flowering time; thermotolerance (heat and chill); Ca^{2+} transport in leaves and Ca^{2+} sensitivity; jasmonic acid-induced Ca^{2+} entry	[27–39]
AtCNGC3	Germination; salt tolerance; Na^+ and K^+ uptake	[40]
AtCNGC4	Pathogen defence; programmed cell death; flowering time; thermotolerance (heat and chill); Ca^{2+} tolerance	[27–29,32,38,41]
AtCNGC5	cGMP-activated Ca^{2+} entry in guard cells; salt tolerance; root hair growth; auxin signalling	[4,42,43]
AtCNGC6	cGMP-activated Ca^{2+} entry in guard cells; thermotolerance (heat); root hair growth; auxin signalling	[4,43–45]
AtCNGC7	Pollen tube growth	[46,47]
AtCNGC8	Pollen tube growth	[46,47]
AtCNGC9	Root hair growth; auxin signalling	[43,44]
AtCNGC10	Negative regulation of salt tolerance; K^+, Na^+ and Pb^{2+} uptake; K^+ homeostasis; negative regulation of Pb^{2+} tolerance; regulation of starch granule size; gravitropism; flowering time; hypocotyl elongation	[25,48,49]
AtCNGC11	Pathogen defence; programmed cell death; Pb^{2+} and Cd^{2+} uptake; Pb^{2+} tolerance; negative regulation of Cd^{2+} tolerance	[15,25,50–53]
AtCNGC12	Pathogen defence; programmed cell death	[15,50–53]
AtCNGC13	Pb^{2+} uptake; negative regulation of Pb^{2+} tolerance	[25]
AtCNGC14	Root hair growth; gravitropism; auxin signalling	[43,44,54–57]
AtCNGC15	Pb^{2+} and Cd^{2+} uptake; Pb^{2+} tolerance; root development	[25,58]
AtCNGC16	Heat and drought tolerance in pollen; negative regulation of Cd^{2+} tolerance	[25,59]
AtCNGC17	Growth regulation; salt tolerance	[42,60]
AtCNGC18	Pollen tube growth and guidance	[6,46,61–63]
AtCNGC19	Response to salt; Pb^{2+} and Cd^{2+} uptake; negative regulation of Pb^{2+} tolerance; herbivory response; pathogen defence; endophyte response; regulating cell death	[25,64–67]
AtCNGC20	Response to salt; pathogen defence; regulating cell death	[65,67]

There is increasing evidence to suggest that CNGCs form heterotetrameric complexes which may have unique functional characteristics, compared to homotetrameric channels [9,10,27–29,44,46]. These may help facilitate the generation of stimulus-specific Ca^{2+} signatures (as monophasic, biphasic or oscillatory increases in this second messenger in a given cellular compartment) that could be decoded by specific complements of Ca^{2+}-binding proteins to cause a stimulus-specific response [9,10]. These discoveries, in combination with recent advances in the model of CNGC structure, have major implications for our understanding of CNGC function and generate new areas for future research. Here, precedents for diverse channel subunit interactions are reviewed, with consideration of in vivo factors that may determine the subunits of CNGC complexes. Additionally, regulatory diversity of AtCNGC subunits is reviewed as a critical determinant of heterotetrameric channel function, with the possibility that CNGC-like proteins (if present) may restrict complex formation.

2. Heteromeric Channel Complexes Also Occur in Plants

It has been known for some time that animal cyclic nucleotide-gated (CNG) channels form heteromeric complexes and that the combinations of these subunits define the functional characteristics of the channel [68–71]. In mammalian retinal phototransduction, CNG channels in rod cells are formed by three subunits of CNGA1 and one CNGB1 subunit, with the C-terminal leucine zipper region of the CNGA1 subunits interacting to set the stoichiometry [70,71]. In contrast, in cone cells, the channel is formed by three subunits of CNGA3 and one CNGB3 subunit. It is held that the CNGB subunits "fine tune" channel behaviour by regulating opening/closing kinetics, affinity for cyclic nucleotides and ability to be regulated by Ca^{2+} [70]. Precedents for heterotetrameric channel assembly in plants have come from members of the Shaker voltage-dependent K^+ channel family, underpinning K^+ uptake and distribution. Interactions amongst Shaker channel subunits are dependent on specific regions of the cytosolic C-terminal domain (CT) and specific subunits may have inhibitory effects on the overall channel complex [72–76]. A breakthrough study on *Medicago truncatula* nodulation nuclear signalling revealed an interaction between the K^+ channel DMI1 (Does not Make Infections 1) and MtCNGC15s, probably to enable voltage change mediated by DMI1 to promote MtCNGC15s opening and Ca^{2+} flux [77]. This opens up the possibility of interaction between entirely different channel families. Although there are no reports of interaction between K^+ channels and CNGCs in *Arabidopsis* or other plants, there is now a good body of evidence that *Arabidopsis* CNGC subunits (AtCNGCs) also form heterotetrameric channels within the family. CNGC–CNGC interactions have also been proposed in *Zea mays* and the moss *Physcomitrella patens* [21,28].

Bifluorescence complementation (BiFC) in *Nicotiana benthamiana* has shown interactions occur in planta between AtCNGC2 and AtCNGC4 [27], between AtCNGC8 and AtCNGC18 [46], between AtCNGC7 and AtCNGC18 [46], and between AtCNGC19 and AtCNGC20 [67]. BiFC analyses also suggest that AtCNGC6, AtCNGC9 and AtCNGC14 interact with each other (AtCNGC6/9, AtCNGC6/14, AtCNGC9/14) [44]. To demonstrate that BiFC signals are produced by heterotetrameric complexes, instead of clustered homotetramers, Pan et al. analysed single-molecule fluorescence to determine that co-expression of *AtCNGC7* or *AtCNGC8* with *AtCNGC18* resulted in heterotetrameric AtCNGC7/18 or AtCNGC8/18 complexes with a 2:2 stoichiometry [46]. This was an important breakthrough in establishing tetramer formation. Prior to this, the need for four CNGC subunits to combine to make a channel had been assumed by analogy with animal and bacterial channels, then supported by molecular modelling [78] (see Section 4 on channel structure). The range of subunit interactions found to date shows that they are not limited to co-members of the five phylogenetic groups within the family (I, II, III, IV-A and IV-B; Figure 1 [16–19,21–23,79]) but can occur across groups. Therefore, CNGC–CNGC interactions are isoform specific but are not restricted to closely related CNGC isoforms. In contrast, AtCNGC16 and AtCNGC18 interactions have not been observed [46], demonstrating that members of the same group (in this case group III) may not work together and that phylogeny alone may not be a useful tool in predicting interactions. Rather, a consideration of co-localisation is needed.

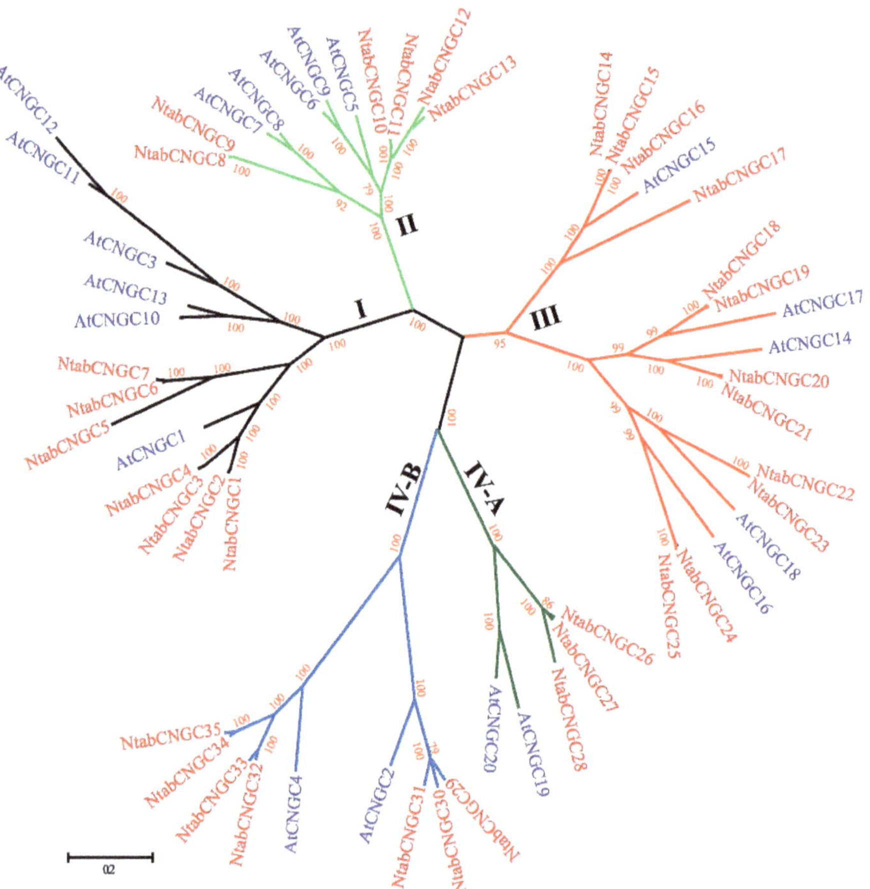

Figure 1. The phylogenetic relationship between CNGCs of *Arabidopsis thaliana* and *Nicotiana tabacum*. Multiple sequence alignment was carried out with the MUSCLE program and MEGA 6.0 used to generate the tree using the Jones–Taylor–Thornton (JTT) model. Bootstrap values from 1000 repeats are shown. Figure reproduced from Nawaz et al. 2019 [18], with permission of the Publisher.

3. Complex Formation Would Depend on the Co-Localisation and Relative Abundance of CNGC Isoforms

Whether a CNGC is present in the plasma membrane or an endomembrane (as a homotetramer or in a heterotetrameric complex) has consequences for the generation of a signal-specific $[Ca^{2+}]_{cyt}$ signature. It is envisaged that CNGC plasma membrane localisation would play a part in signature initiation driven by receptors in that membrane whilst CNGCs in the tonoplast, for example, would be downstream and acting to amplify the $[Ca^{2+}]_{cyt}$ signal. Most CNGCs are believed to localise to the plasma membrane, and this has been reported for many AtCNGCs, including AtCNGC2 [33], AtCNGC3 [40], AtCNGC6 [44], AtCNGC7 [46], AtCNGC8 [46], AtCNGC9 [44], AtCNGC10 [48,80], AtCNGC11 [51], AtCNGC12 [51,81], AtCNGC14 [44], AtCNGC17 [60], AtCNGC18 [46,62,63], AtCNGC19 [66] and AtCNGC20 [82]. AtCNGC5 is reported to localise in microdomains at the periphery of *N. benthamiana* protoplasts when expressed heterologously [4]. It must be noted, however, that in contradiction of the findings by Meena et al. [66] and Fischer et al. [82], Yuen and Christopher reported that AtCNGC19 and AtCNGC20 localised to the vacuolar membrane, not the plasma membrane [83]. Similarly, Chang et al. found

AtCNGC7, AtCNGC8 and AtCNGC16 to localise to endomembrane in pollen tubes, as opposed to the plasma membrane [61]. In silico predictions suggest that 11 of 12 *Zea mays* CNGCs could localise to the plasma membrane [21]. The importance of localisation is exemplified by the positioning of nuclear CNGCs to enable nuclear Ca^{2+} "spiking" in symbiosis signalling [77]. MtCNGC15a, MtCNGC15b and MtCNGC15c of *M. truncatula* localise to the nuclear envelope, with 14 of 21 MtCNGCs predicted to contain nuclear localisation sequences [77]. Recently, AtCNGC15 was also found to localise to the nuclear membrane and to be involved in root development [58]. From these results, it is clear that many CNGCs could co-localise, which would facilitate extensive CNGC–CNGC subunit interactions, including the formation of heterotetrameric complexes.

These interactions will necessarily be dependent on the co-expression of *CNGCs* within individual cells. Transcriptomic data indicate that many *CNGCs* (23 of 35 in *N. tabacum* [18]) are expressed throughout the plant. Expression levels, however, vary greatly, suggesting that many CNGCs have tissue-specific or cell-specific functions [4,18,21–23,44,47]. Expression of an individual CNGC can vary with development and growth conditions. For example, leaf expression of *AtCNGC3* increases with leaf age and lessens when the plant is grown in soil rather than on an agar plate [40]. *CNGC* promotors have been found to contain elements associated with responses to abscisic acid, auxin, ethylene, gibberellin, salicylic acid and methyl-jasmonate [17–21], and *CNGC* transcript levels are highly sensitive to abiotic and biotic stressors [17–20,22,23,79]. Notably, it has also been found that *Ziziphus jujube CNGC2* expression is rapidly induced by application of cAMP to callus [79], potentially providing a mechanism for priming signalling pathways involving ZjCNGC2. In silico analysis has also predicted that *NtabCNGC* expression may be regulated extensively by micro-RNAs and *cis*-acting regulatory elements [18]. Consequently, the abundance of different CNGC isoforms in each cell type is likely to be variable, and dependent on environmental conditions, which may result in the formation of different CNGC heterotetramers in different cell types.

A further consideration is that studies typically measure transcript levels from entire organs and may lack sufficient spatial resolution to detect low, cell-specific expression [17–19,22,23,44]. *AtCNGC5*, *AtCNGC6*, *AtCNGC9* and *AtCNGC14* have been implicated in root hair growth and their transcripts are abundant in roots [43,44,54–57]. However, *CNGC*promotor::GUS fusions of *AtCNGC6*, *AtCNGC9* and *AtCNGC14* also suggest that these *AtCNGCs* are expressed throughout the plant [44]. *AtCNGC6* expression is predicted in root, shoot, leaf and guard cells [4,44,45]. *AtCNGC9* expression is predicted in root hairs and guard cells but not leaves or shoots, and *AtCNGC14* expression is predicted in roots, shoots and the leaf [44]. As stated earlier, BiFC suggests that AtCNGC6, AtCNGC9 and AtCNGC14 may form a range of heterotetrameric complexes with each other [44]. From these data, it may be predicted that different complexes form in different cell types depending on the expression patterns of the interacting partners. Consequently, the signalling function of these AtCNGCs may be cell type specific. Notably, *AtCNGC6* has been implicated not only in root hair growth polarity but also in thermotolerance and cGMP-induced Ca^{2+} influx in guard cells (in which it could partner AtCNGC5) [4,28,44]. Additionally, whilst *AtCNGC9* has so far only been implicated in root hair growth polarity, its expression in guard cells would suggest a specific activity there as part of a complex with AtCNGC6. Both guard cell and mesophyll plasma membrane contain Ca^{2+} channels that are activated by cyclic nucleotides [84]. It is, therefore, possible that the formation of different CNGC complexes with unique functional characteristics in different cell types helps permit stimulus-specific signalling in the diverse pathways that AtCNGC6 and other AtCNGCs work in (Table 1).

The relative abundance of CNGC subunits is also likely to be a key determinant of CNGC complex formation and function. Yoshioka et al. discovered that the phenotypes associated with the *cpr22* mutant (a gene fusion between *AtCNGC11* and *AtCNGC12*) could be supressed by overexpression of *AtCNGC12*, and it was proposed that this was due to disruption of heterotetrameric complex formation [53]. Notably, *AtCNGC6* transcripts are >4.5 times more abundant than *AtCNGC9* transcripts in guard cells [4]. If these two CNGCs were to form heterotetrameric complexes (as suggested) and transcript abundance were proportional to protein abundance, then excess AtCNGC6 subunits must either form homotetramers, be degraded, sequestered in the membrane, or interact with additional AtC-NGC subunits. Following a study of *AtCNGC2* homologues in *P. patens* by Finka et al., such an interaction between AtCNGC2 and AtCNGC6 has been proposed for heat signalling [28]. In pollen tube growth, the ratio of *AtCNGC18* to *AtCNGC8* expression is expected to determine the proportion of AtCNGC8/18 complexes relative to AtCNGC8 or AtCNGC18 homotetramers, leading to changes in cell permeability [46]. When expressed singly in *Xenopus* oocytes, AtCNCG18 forms a constitutively voltage-dependent, Ca^{2+}-permeable channel but AtCNGC8 is electrically silent (as is AtCNGC7). Equimolar co-expression of *AtCNGC18* and *AtCNGC8* (presumably with heterotetramer formation) resulted in greatly reduced Ca^{2+} influx compared to *AtCNGC18* expression alone [46]. Pan et al. proposed that, by recruiting AtCNGC18 into heterotetrameric AtCNGC8/18 complexes, AtCNGC8 represses AtCNGC18 activity [46]. Consequently, whilst the overexpression of *AtCNGC18* in *A. thaliana* disrupts Ca^{2+}-dependent pollen tube growth, potentially by forming deregulated AtCNGC18 homotetramers, this phenotype can be rescued by overexpressing *AtCNGC8*, presumably by recruiting AtCNGC18 subunits from homotetramers to generate heterotetramers [46]. However, it should be noted that Gao et al. reported that expression of *AtCNGC8* or *AtCNGC7* in HEK293T cells produced constitutively voltage-dependent, Ca^{2+}-permeable channels which were activated further by addition of 8Br-cNMPs [5]. This contradiction may be caused by the different bathing solutions used in each study or the use of different heterologous expression systems. Examples of key findings from transport studies on heterologously expressed CNGC genes are shown in Table 2. This further shows that the choice of expression system may have an effect on the outcome. For example, K^+ selectivity over Na^+ of AtCNGC2 was greater in HEK293 cells than in *Xenopus* oocytes [2]. Nevertheless, AtCNGC7/18, and AtCNGC8/18 complexes appear to form spontaneously in a heterologous expression system, yielding different transport characteristics to the homomeric forms [46]. If single-molecule fluorescence were to confirm the existence of an AtCNGC2/4 complex, as well as complexes amongst the AtCNGC6, AtCNGC9 and AtCNGC14 triad and AtCNGC19/20 couple as predicted by BiFC [27,44,67], then it is likely that CNGC complexes could be widespread if co-localisation permits.

Table 2. Examples of transport characteristics and response to cNMPs of CNGCs when either heterologously expressed or present in native membrane. Most studies have focused on the CNGCs of *Arabidopsis thaliana* (AtCNGC). However, there are also studies reporting the activity of CNGCs in *Physcomitrella patens* (PpCNGC), *Hordeum vulgare* (HvCNGC), *Oryza sativa* (OsCNGC) and *Medicago truncatula* (MtCNGC) as detailed below. "Whole cell" refers to the recording configuration in which channel activity is captured from effectively the entire plasma membrane. "Inside-out patch" refers to the configuration in which the cytosolic face of a membrane patch (held in the electrode tip) faces the bathing medium. "Cell attached" refers to the configuration in which the membrane patch (held in the electrode tip) remains undetached from the remaining membrane.

CNGCs	System	Tested Cations	Tested with cNMPs?	Results	References
AtCNGC1	HEK293—whole cell	Tested K^+ and Na^+ conductance	Yes	Application of 100 µM db-cAMP stimulated AtCNGC1 K^+ and Na^+ conductance. No K^+ or Na^+ conductance was observed in the absence of db-cAMP.	[85]

Table 2. *Cont.*

CNGCs	System	Tested Cations	Tested with cNMPs?	Results	References
AtCNGC1	Yeast	Tested Ca^{2+} uptake	No	In the presence of yeast pheromone α factor, AtCNGC1 in a Ca^{2+} uptake-deficient yeast mutant increased colony growth, indirectly demonstrating Ca^{2+} conduction.	[24]
AtCNGC1, AtCNGC2, AtCNGC4	Yeast	Tested K^+ uptake	Yes	Addition of 100 μM db-cAMP stimulated growth of a K^+ uptake-deficient yeast mutant expressing *AtCNGC1*, *AtCNGC2*, and *AtCNGC4*.	[86]
AtCNGC1, AtCNGC2, AtCNGC4	Yeast	Tested K^+ and Ca^{2+} uptake	Yes	*AtCNGC1M2* (deletion in C-terminal domain) expression in a Ca^{2+} uptake yeast mutant resulted in growth, indicating Ca^{2+} permeability of AtCNGC1. Expression of *AtCNGC2* and *AtCNGC4* enhanced growth of a K^+ uptake-deficient yeast mutant. Application of 100 μM db-cAMP increased growth of yeast mutant transformed with *AtCNGC1M2*.	[87]
AtCNGC2	Yeast	Tested K^+ uptake	Yes	In the presence of 10 μM db-cAMP or db-cGMP, transfection with *AtCNGC2* enhanced growth of a K^+ uptake-deficient yeast mutant.	[1]
AtCNGC2	*Xenopus* oocytes—whole cell	Tested K^+ conductance	Yes	Application of 10 μM db-cAMP stimulated AtCNGC2 K^+ conductance. No K^+ conductance was observed in the absence of db-cNMPs.	[1]
AtCNGC2	*Xenopus* oocytes—whole cell	Tested K^+, Na^+, Li^+, Cs^+ and Rb^+ conductance	Yes	In the presence of 100 μM db-cAMP, AtCNGC2 conducted K^+, Li^+, Cs^+ and Rb^+. Na^+ conductance was significantly less. No data were reported concerning conductance in the absence of db-cAMP.	[2]
AtCNGC2	*Xenopus* oocytes—inside-out patch	Tested K^+ conductance	Yes	Application of 100 μM cAMP stimulated AtCNGC2 K^+ conductance. No K^+ conductance was observed in the absence of db-cAMP.	[2]

Table 2. *Cont.*

CNGCs	System	Tested Cations	Tested with cNMPs?	Results	References
AtCNGC2	*Xenopus* oocytes—inside-out patch	Tested K^+ and Na^+ conductance	Yes	Application of 100 μM cAMP stimulated AtCNGC2 K^+ conductance, but not Na^+ conductance. No K^+ or Na^+ conductance was observed in the absence of cAMP. Mutation of N416 and D417 in the pore resulted in Na^+ conductance similar to K^+ conductance.	[85]
AtCNGC2	HEK293—whole cell and inside-out patch	Tested K^+ and Na^+ conductance	Yes	Application of 100 μM db-cAMP stimulated AtCNGC2 K^+ conductance, but not Na^+ conductance. No K^+ or Na^+ conductance was observed in the absence of db-cAMP. Mutation of N416 and D417 in the pore region resulted in Na^+ conductance similar to K^+ conductance.	[2,85]
AtCNGC2	Guard cell protoplasts—whole cell	Tested Ba^{2+} conductance (as a proxy for Ca^{2+})	Yes	Application of 1 mM db-cAMP stimulated AtCNGC2-dependent Ca^{2+} conductance.	[30]
AtCNGC2	HEK293T—whole cell	Tested Ca^{2+} conductance	Yes	Application of 200 μM 8Br-cAMP stimulated AtCNGC2 Ca^{2+} conductance. No data were reported concerning Ca^{2+} conductance in the absence of db-cAMP.	[38]
AtCNGC2, AtCNGC4	Mesophyll cell protoplasts—whole cell	Tested Ba^{2+} conductance (as a proxy for Ca^{2+})	No	Wild-type mesophyll cell protoplasts conducted Ca^{2+} in response to H_2O_2 or flg22. Ca^{2+} conductance was lost in *Atcngc2* or *Atcngc4* loss-of function mutants, as well as the *Atcngc2 Atcngc4* double mutant.	[29]
AtCNGC2, AtCNGC4	*Xenopus* oocytes—whole cell	Tested Ca^{2+}, Mg^{2+}, Ba^{2+}, Sr^{2+}, K^+ and Na^+ conductance	No	Independently, AtCNGC2 or AtCNGC4 did not conduct Ca^{2+} in the absence of cNMPs. Co-expression of *AtCNGC2* and *AtCNGC4* produced Ca^{2+}, Sr^{2+}, Ba^{2+} and K^+-permeable (Na^+ and Mg^{2+}-impermeable) channels in the absence of cNMPs.	[29]
AtCNGC3	Yeast	Tested Na^+ and K^+ uptake	No	Yeast expressing *CNGC3* accumulated more Na^+ and K^+, suggesting a pathway for Na^+ and K^+ transport.	[40]

Table 2. *Cont.*

CNGCs	System	Tested Cations	Tested with cNMPs?	Results	References
AtCNCG4	*Xenopus* oocytes—inside-out patch	Tested K^+, Na^+ and Cs^+ conductance	Yes	Application of 500 µM cAMP or cGMP stimulated AtCNGC4 K^+, Na^+ and Cs^+ conductance. Compared to K^+, outward conductance of Cs^+ was significantly lower. No conduction of K^+, Na^+ or Cs^+ was observed in the absence of cNMPs.	[41]
AtCNGC5, AtCNGC6	Guard cell protoplasts—whole cell	Tested Mg^{2+}, Ba^{2+} and Ca^{2+} conductance	Yes	Application of 500 µM 8Br-cGMP stimulated Mg^{2+}, Ca^{2+} and Ba^{2+} conductance. Mg^{2+} conductance was lost in *Atcngc5 Atcngc6* double mutants. AtCNGC1, AtCNGC2 and AtCNGC20 did not appear to contribute to these guard cell 8Br-cGMP-activated currents.	[4]
AtCNGC5, AtCNGC6	HEK293T—whole cell	Tested Ca^{2+} and Na^+ conductance	No	HEK293 cells expressing *CNGC5* or *CNGC6* displayed inward currents carried by Ca^{2+}, not Na^+.	[43]
AtCNGC6	Root protoplasts—whole cell	Tested Ca^{2+} conductance	Yes	Application of 50 µM db-cAMP stimulated AtCNGC6-dependent Ca^{2+} conductance, application of a phosphodiesterase inhibitor also stimulated Ca^{2+} conductance.	[45]
AtCNGC7, AtCNGC8	*Xenopus* oocytes—whole cell	Tested Ca^{2+} conductance	No	AtCNGC7 or AtCNGC8 Ca^{2+} conductivity was not observed in the absence of cNMPs.	[46]
AtCNGC7, AtCNGC8, AtCNGC9, AtCNGC10, AtCNGC16 and AtCNGC18	HEK293T—whole cell	Tested Ca^{2+} and K^+ conductance	Yes	Application of 100 µM 8Br-cAMP or 100 µM 8Br-cGMP stimulated AtCNGC7, AtCNGC8, AtCNGC9, AtCNGC10, AtCNGC16 and AtCNGC18 Ca^{2+} conductance. Ca^{2+} conductance did not require 8Br-cNMP application. No significant K^+ conductance reported.	[6]
AtCNGC10	*E. coli* and yeast	Tested K^+ uptake	Yes	AtCNGC10 complemented *E. coli* and yeast K^+ uptake mutants. In *E. coli*, co-expression of *AtCNGC10* and *CaM* inhibited cell growth, but cGMP overcame this.	[88]

Table 2. *Cont.*

CNGCs	System	Tested Cations	Tested with cNMPs?	Results	References
AtCNGC10	HEK293— whole cell	Tested K^+ conductance	Yes	In the presence of 100 µM db-cGMP, AtCNGC10 conducted K^+. No data were reported concerning conductance in the absence of db-cAMP.	[80]
AtCNGC10	Yeast	Tested K^+ and Na^+ uptake	No	*AtCNGC10*-transformed yeast accumulated more Na^+ in the presence of 20 mM NaCl. Expression rescued growth of K^+ uptake-deficient yeast.	[49]
AtCNGC11, AtCNGC12	Yeast	Tested K^+ uptake	Yes	Growth of K^+ uptake-deficient yeast was complemented by AtCNGC11, AtCNGC12 or the chimeric AtCNGC11/12. Growth was enhanced by 100 µM db-cAMP but not db-cGMP.	[53]
AtCNGC11, AtCNGC12	Yeast	Tested Ca^{2+} uptake	No	Expression of *AtCNGC11*, *AtCNGC12* or *AtCNGC11/12* complemented growth of Ca^{2+} uptake-deficient yeast.	[51]
AtCNGC11, AtCNGC12	Yeast	Tested K^+ uptake	No	AtCNGC11/12 or AtCNGC12 restored growth of K^+ uptake-deficient yeast.	[81]
AtCNGC11, AtCNGC12	*Xenopus* oocytes— whole cell	Tested Ca^{2+} conductance	Yes	Expression of *AtCNGC12* caused a Ca^{2+} conductance that was not enhanced by cNMPs. *AtCNGC11* expression did not cause a Ca^{2+} conductance, even with cNMPs. Co-expression did not affect the AtCNGC12-dependent conductance.	[8]
AtCNGC14	*Xenopus* oocytes— whole cell	Tested Ca^{2+} conductance	No	AtCNGC14 Ca^{2+} conductivity was observed in the absence of cNMPs. It was not tested whether application of cNMPs would increase Ca^{2+} conductance.	[57]
AtCNGC14	*Xenopus* oocytes— whole cell	Tested Ca^{2+} conductance	No	AtCNGC14 inward Ca^{2+} currents were observed in the absence of cNMPs.	[56]
AtCNGC18	*E. coli*	Tested Ca^{2+} uptake	No	Expression of *AtCNGC18* in *E. coli* caused Ca^{2+} accumulation.	[62]
AtCNGC18	HEK293T— whole cell	Tested Ca^{2+} conductance	Yes	Application of 100 µM 8Br-cAMP or 100 µM 8Br-cGMP stimulated greater AtCNGC18 Ca^{2+} conductance, but not 20 µM 8Br-cNMP. Ca^{2+} conductance did not require 8Br-cNMP application.	[5]

Table 2. *Cont.*

CNGCs	System	Tested Cations	Tested with cNMPs?	Results	References
AtCNGC18	*Xenopus* oocytes—whole cell	Tested Ca^{2+} conductance	Yes	In the presence of 100 µM db-cAMP, AtCNGC18 conducted Ca^{2+}. No data were reported concerning conductance in the absence of db-cAMP.	[63]
AtCNGC18	Pollen tube protoplasts—whole cell	Tested Ca^{2+} conductance	Yes	Application of 100 µM 8Br-cGMP stimulated AtCNGC18-dependent Ca^{2+} conductance. Ca^{2+} conductance was not apparent in the absence of 8Br-cGMP.	[6]
AtCNGC18	*Xenopus* oocytes—whole cell	Tested Ca^{2+} conductance	No	AtCNGC18 Ca^{2+} conductivity was observed in the absence of cNMPs. Co-expression of AtCNGC18 with AtCNGC7 or AtCNGC8 eliminated Ca^{2+} conductivity.	[46]
AtCNGC19	*Xenopus* oocytes—whole cell	Tested Ca^{2+}, Na^+ and K^+ conductance	Yes	In the presence of 300 µM db-cAMP, AtCNGC19 elicited Ca^{2+} inward currents but not K^+ and Na^+ currents.	[66]
AtCNGC19, AtCNGC20	*Xenopus* oocytes—whole cell	Tested Ca^{2+} conductance	No	AtCNGC19 and AtCNGC20 conductivity was observed in the absence of cNMPs. It was not tested whether application of cNMPs would increase Ca^{2+} conductance. Co-expression of *AtCNGC19* and *AtCNGC20* increased conductance compared to independently expressed *AtCNGC19* and *AtCNGC20*.	[67]
PpCNGCb	Moss protoplasts cell attached	Tested Ba^{2+} conductance (as a proxy for Ca^{2+})	No	Ba^{2+} conductivity did not require application of cNMPs, but it is possible that there were endogenous cNMPs. Ba^{2+} conductivity was altered in *Ppcngcb* mutants.	[28]
HvCNGC2-3	*Xenopus* oocytes—whole cell	Tested K^+ and Na^+ conductance	Yes	Application of 10 µM 8Br-cGMP stimulated HvCNGC2-3 Na^+ and K^+ conductivity only in the co-presence of both ions. No Na^+ or K^+ conductance was observed in the absence of cGMP, or 10 µM 8Br-cAMP. Ca^{2+} conductivity was not observed.	[7]

Table 2. *Cont.*

CNGCs	System	Tested Cations	Tested with cNMPs?	Results	References
OsCNGC9	HEK293—whole cell	Tested Ca^{2+} and K^+ conductance	No	OsCNGC9 Ca^{2+} conductivity was observed in the absence of cNMPs. Comparatively little K^+ conductivity was observed. It was not tested whether application of cNMPs would increase Ca^{2+} or K^+ conductance.	[89]
OsCNGC13	HEK293—whole cell	Tested Ca^{2+} and K^+ conductance	No	OsCNGC13 mediated Ca^{2+} inward currents but not K^+ currents. It was not tested whether application of cNMPs would increase Ca^{2+} conductance.	[90]
MtCNGC15	*Xenopus* oocytes—whole cell	Tested Ba^{2+} and Ca^{2+} conductance	No	MtCNGC15 Ca^{2+} conductivity did not require application of cNMPs. It was not tested whether application of cNMPs would increase Ca^{2+} conductance.	[77]

4. CNGCs Are Extensively Regulated—Formation of CNGC Complexes Generates Further Regulatory and Functional Complexity

Here we summarise the current understanding of CNGC structure and regulation, discussing how the formation of CNGC complexes may further affect CNGC regulation and function. The breakthrough studies on plant CNGCs used primary structures of potassium channels and animal CNG subunits to assign domains [91–93]. The overall model CNGC subunit has six transmembrane domains (S1–S6; Figure 2) with a pore region (P loop) between S5 and S6 that permits ion transport [13]. Animal and bacterial cation channel subunits that contain a single P loop form tetramers; this includes animal CNGs, with clear evidence from cryo-electron microscopy showing tetramer formation in a lipid environment [94]. Triplet amino acid residues in the P loop that could act as selectivity filters (AGN, AND, GNL, GQG, GQN, GQS) vary between the *Arabidopsis* CNGCs [95], with AND or GQs thought to confer some level of Ca^{2+} selectivity [4–6,30,35,45,78]. Recent analysis of AtCNGCs has revealed the presence of a diacidic motif for Mg^{2+} binding close to the pore region in the cytosolic CT in all but AtCNGC2. It has been proposed that this could account for channel blocking by cytosolic Mg^{2+}; the consequences for signalling and nutrition have yet to be explored [96]. The CT contains a cyclic nucleotide-binding domain (CNBD) which is believed to be formed of four α-helices (αA, αP, αB, αC) and eight β-sheets (β1–β8). Overlapping with the C-terminal side of the CNBD is a calmodulin (CaM)-binding domain (CaMBD) [97,98]. A CaM-binding IQ (isoleucine-glutamine) motif is also present [18,21,23,82] and the CT can contain multiple phosphorylation sites [29,67,89]. The CT of AtCNGC8 appears necessary and sufficient for interactions between AtCNGC8 and AtCNGC18 subunits in *Xenopus* oocytes [46]. There is also a short, cytosolic N-terminal domain (NT) that is predicted to harbour CNGC–CNGC interaction domains [99] and may contain phosphorylation sites [29,67]. The NT of AtCNGC12 contains a CaMBD [50]. The AtCNGC19 and AtCNGC20 NTs are predicted to have cysteine residues that could form an Fe/Cu-binding site to act as a Fenton catalyst in the production of hydroxyl radicals regulate plant Ca^{2+} channel activity in growth and stress responses [99,100].

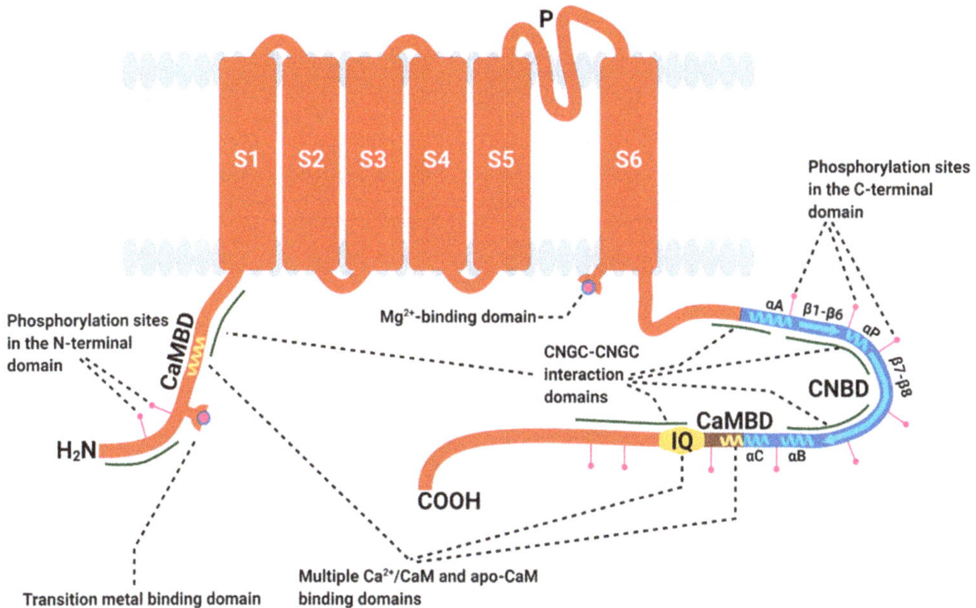

Figure 2. Cyclic nucleotide-gated channel (CNGC) subunit structure. Plant CNGCs consist of six transmembrane domains (S1–S6) with a pore region between S5 and S6. Both N- and C-terminal domains are cytosolic. In the C-terminal domain is a cyclic nucleotide-binding domain (CNBD), which is believed to be formed of four α-helices (αA, αP, αB, αC) and eight β-sheets (β1–β8). Overlapping with the C-terminal side of the CNBD is a calmodulin (CaM)-binding domain (CaMBD). CNGCs also contain a CaM-binding IQ (isoleucine-glutamine) motif. From the data presented by Pan et al. 2019 [46] and Chiasson et al. 2017 [99], the N- and C-terminal regions are predicted to contain CNGC–CNGC interaction domains. Lemtiri-Chlieh et al. 2020 [96] predict the presence of a Mg²⁺-binding domain downstream of the CNGC pore. Demidchik et al. 2014 [99] predict a transition metal-binding domain in the N-terminus of AtCNGC19 and AtCNGC20. CNGC phosphorylation is likely for AtCNGC4 [29], AtCNGC18 [63], AtCNGC19 [67], AtCNGC20 [67] and OsCNGC9 [89]. An N-terminal CaMBD has been identified in AtCNGC12 [50]. Not all plant CNGCs contain all the structures displayed on this image. Structure was adapted from Chin et al. 2009 [12] and Kaplan et al. 2007 [101]. Figure created with BioRender.com.

4.1. CNGCs Are Regulated by cNMPs That May Be Generated by Soluble or Membrane Proteins

cAMP and cGMP are secondary messengers which are synthesised by adenylyl cyclases (ACs) and guanylyl cyclases (GCs), respectively. cNMP gating of CNGCs is well documented and is summarised in Table 2, but historically the physiological importance of cNMPs has been controversial [11,102–104]. cNMP levels in plants are significantly lower than in animals and, until the development of more sensitive assays, it was doubted whether cNMPs were present at all [102]. To date, cAMP has been implicated in seed germination and cell cycle progression, pollen tube growth and orientation, stomatal kinetics, photosynthesis and photorespiration, abiotic stress responses (heat and chill stress, salinity, drought, aluminium, nutrient deficiency), wounding and immunity [11]. A range of soluble and membrane proteins has now been identified with potential AC or GC activity, in their cytosolic domains for the membrane proteins. For ACs, these include the K⁺ uptake transporters AtKUP5 and AtKUP7 (K⁺ uptake permease) [105,106], and AtLRRAC1 (leucine-rich repeat adenylyl cyclase1) [107]. For GCs, AtGC1 (guanylyl cyclase1) [108], AtNOGC1 (nitric oxide-dependent guanylate cyclase1) [109], AtPSKR1 (phytosulfokine receptor1) [110], AtPepR1 (plant elicitor peptide receptor1) [3], AtBRI1 (brassinosteroid insenstive1) [111] and AtWAKL10 (wall-associated kinase (WAK)-like10) [112] have all been identified and studied in vitro. Two homologues of AtPepR1 have now been identified in tomato (*Solanum lycopersicum* L.; SlGC17, SlGC18) and have been reported to have GC activity that is required for $[Ca^{2+}]_{cyt}$ increase in response to flg22, chitin and AtPep1 [113]. Analysis of recombinant protein activity suggests that AC and GC activity in plants is typically much lower than in animal counterparts, with pmol or fmol cNMP μg^{-1} protein min^{-1} values reported [3,107–109,111]. However, the membrane-bound AtWAKL10 and AtPSKR1 have a V_{max} of approximately 2 μmol mg^{-1} min^{-1} [110,112]. It is possible that, in planta, CNGC cNMP sensitivity is increased by CaM, phosphorylation, or formation of heterotetrameric complexes, and so lower cNMP concentrations are required than those used in heterologous expression systems or native membranes in transport studies.

It may be that the low activity of plant ACs and GCs is central to signal specificity. If the domains were in close proximity to specific CNGCs, generating cNMP concentrations sufficient to activate those channels [3], then crosstalk between separate CNGC-dependent signalling pathways could be eliminated—only CNGCs co-localising with the AC or GC would be activated. How close is close enough? At the plasma membrane, both AtCNGC17 and AtPSKR1 have been found to interact with AtBAK1 (BRI-associated receptor kinase1) and although AtCNGC17 does not interact with the PSKR1 receptor that generates cGMP, this channel is essential for phytosulfokine/PSKR1-dependent protoplast expansion involving the H$^+$-ATPases AtAHA1 and AtAHA2 and is thought to form part of this multi-protein complex [60]. Increased $[Ca^{2+}]_{cyt}$ promotes PSKR1's GC activity but inhibits its kinase activity [114], raising the possibility that AtCNGC17-mediated $[Ca^{2+}]_{cyt}$ elevation not only generates a positive feedback loop for the cGMP pathway but could also curtail any phosphorylation-dependent pathway to ensure signalling specificity. A positive feedback loop may also explain the jasmonic acid-induced rise in cAMP in leaf epidermal cells that requires AtCNGC2 [35]. It is possible that AtCNGC2-mediated Ca^{2+} influx activates ACs either directly or via intermediates such as calcium-dependent protein kinases and CaMs.

The recent finding that *Arabidopsis* root hair K$^+$ influx precedes increased growth rate and can cause $[Ca^{2+}]_{cyt}$ increase [115] begs the question of whether AC activity of AtKUP5 and AtKUP7 is involved. Both these K$^+$ transporters are expressed in root hairs [116], and AtKUP7 is in the plasma membrane [117]. When expressed in yeast, AtKUP5-mediated K$^+$ influx causes cAMP accumulation [106]. This suggests a model in which KUP-mediated K$^+$ influx to the root hair causes cAMP increase to activate the CNGCs (AtCNGC5,6,9,14) implicated in Ca^{2+} influx and polar growth [43,44,55–57]. As AtCNGC5, AtCNGC6 and AtCNGC9 appear to transport Ca^{2+} rather than monovalent cations [4,6,43] (Table 2), it seems likely these subunits are relevant to Ca^{2+} signalling. The spatial localisation of the KUPs relative to the CNGCs is worthy of attention. Activation of CNGCs with strong K$^+$ permeation could conceivably contribute to K$^+$ uptake in root hairs and other cells, indeed AtCNGC3 and AtCNGC10 are held to be important for root K$^+$ acquisition [48,101].

Salt stress causes cGMP accumulation within seconds in *Arabidopsis* seedlings [118] and CNGCs have been proposed to be part of the salt-induced $[Ca^{2+}]_{cyt}$ signalling response [119]. In *Arabidopsis*, cNMPs can restrict Na$^+$ influx rather than promote it, ostensibly by reducing the open probability of root plasma membrane Na$^+$-permeable channels (an effect observed in approximately half of the patch clamp trials) [120]. This would imply negative regulation by cNMPs of a putative CNGC channel. Notably, AtCNGC3 and AtCNGC10 appear to contribute to Na$^+$ uptake [40,49]. Patch clamp electrophysiological analysis of *Arabidopsis* root epidermal plasma membrane has also revealed a Na$^+$ influx channel that could not discriminate against K$^+$ [121] (a "non-selective" cation channel [122]) and this was proposed to be Ca^{2+} permeable in a later study [123]. It is not known whether the channel is regulated by cNMPs and could account for the negative effects of cNMPs on Na$^+$ influx reported by Maathuis and Sanders, 2001 [120]. The roles of cNMPs and CNGCs in salt stress urgently require further elucidation. Understanding which CNGC subunits and potential heteromeric complexes are expressed in different root cells (which vary in their salt-induced $[Ca^{2+}]_{cyt}$ response [124]), what their functional permeability is to Na$^+$ and Ca^{2+} and how they are regulated by cNMPs is likely to be of great importance.

4.2. CNGCs Are Positively and Negatively Regulated by CaM, Potentially Affording Ca^{2+} Sensing and Feedback

Calmodulins are Ca^{2+}-binding proteins with a major role in Ca^{2+} signal transduction in plants [10,29,46]. In former models of CNGC activity, CaM was believed to have an exclusively inhibitory effect where the Ca^{2+}-bound form of CaM (Ca^{2+}/CaM) inhibited CNGCs by perturbing cNMP gating, competing for a binding site internal to the CNBD [98]. Subsequently, multiple CaMBDs have been identified, with structural divergence amongst CNGC isoforms. For example, AtCNGC12 contains an N-terminal CaMBD which interacts with Ca^{2+}/CaM and could result in channel closure [50]. In addition, an IQ domain has been identified which is C-terminal to the CNBD and conserved in the majority of plant CNGCs [18,21,23,85]. Yeast 2-hybrid assays suggest that interactions between CNGC CTs and CaM isoforms are specific, with the IQ domain contributing to many of these interactions [33]. Indeed, CaM isoform-specific effects are now being documented [10], for example root hair AtCNGC14 is negatively regulated by AtCaM7 binding to its CT but not by AtCaM2 [56]. Additionally, AtCNGC6 is inhibited by AtCaM2,3,5,7 at the IQ domain in heat shock signalling but not by AtCaM1,4 or 6 [125].

It has been proposed that apo-CaM (CaM without Ca^{2+} ligands) constitutively binds to the IQ domain in a Ca^{2+}-independent manner to act as a Ca^{2+} sensor [10,33,50]. The model arising from studies on AtCNGC12 has the channel's opening causing local Ca^{2+} elevation, hence permitting Ca^{2+} binding to apo-CaM [50]. That initial channel opening could be triggered by membrane hyperpolarisation because when expressed in *Xenopus*, AtCNGC12 presents as a hyperpolarisation-activated Ca^{2+} channel that does not require cNMPs [8]. Ca^{2+}-CaM interaction at the IQ domains of adjacent subunits and Ca^{2+}-CaM recruitment to CaMBDs could modulate channel activity [50]. Evidence from *Xenopus* expression points to AtCaM1 as an activating ligand [8]. As Ca^{2+} increases, Ca^{2+}-CaM binding to the NT CaMBD effects channel closure [10,50]. Much, therefore, depends on which CaM isoforms are locally available and their affinities for Ca^{2+} and the CaMBDs. A further model built on AtCNGC8/18 activity in *Xenopus* coupled with analysis of CT binding has apo-CaM2 binding to the IQ domains to counter the inhibitory effect of AtCNGC8 and so promote channel opening. As Ca^{2+} increases as a consequence, Ca^{2+}-CaM2 forms but then dissociates to promote channel closure [46]. Expressing *AtCNGC8* and *AtCNGC18* with *AtCaM2* in HEK293 cells leads to [Ca^{2+}]$_{cyt}$ oscillations [46], which has implications for pollen tube apical [Ca^{2+}]$_{cyt}$ oscillations during growth. It remains to be seen how cNMPs fit into this regulatory complex, given activation of AtCNGC18 by cGMP in native pollen plasma membrane [6] and by cNMPs in heterologous expression [5,63].

CaM regulation of CNGCs may be important in immune signalling. BIK1 is a receptor-like cytoplasmic kinase which acts downstream of FLS2 (Flagellin Sensitive2) [126], an LRR receptor-like kinase which binds to the bacterial flg22 peptide and is required for [Ca^{2+}]$_{cyt}$ elevation [127]. Although at the whole-plant level (which may lack sufficient resolution) AtCNGC2 was reported to have no involvement in flg22-induced [Ca^{2+}]$_{cyt}$ increase [128], a genetic analysis of *AtCNGC2* and *AtCNGC4* concluded that both genes act in the flg22 pathway [27]. At the leaf disc level, use of loss-of-function mutants indicated that both AtCNGC2 and AtCNGC4 are involved in flg22-induced [Ca^{2+}]$_{cyt}$ increase, given a permissive apoplastic Ca^{2+} level [29]. Similarly, patch clamping of mesophyll protoplasts showed that both were needed for flg22-induced plasma membrane Ca^{2+} influx currents [29]. It should be noted, however, that flg22-induced plasma membrane depolarisation of individual mesophyll cells (which involves Ca^{2+} influx) was found to be independent of *AtCNGC2* [129]. Following the results of heterologous co-expression in *Xenopus* oocytes [29], it is likely the subunits form a AtCNGC2/4 complex. Although single expression of either *AtCNGC2* or *AtCNGC4* in *Xenopus* oocytes failed to cause channel activity, their co-expression produced a hyperpolarisation-activated Ca^{2+}-permeable channel that did not require cNMPs [29]. This channel activity could be supressed by the co-expression of *AtCAM7* and this suppression could be overcome by the additional expression of *AtBIK1* [29]. It was subsequently found that application of flg22 induces AtBIK1-mediated phosphorylation of the AtCNGC4-

CT in planta, which is believed to overcome AtCaM7-mediated repression [9,29]. More recently, a split luciferase complementation assay using nano-luciferase suggests that the CT and NT of homomeric AtCNGC2 and AtCNGC4 may move apart when challenged with flg22 in planta [130]. It may be postulated that this change is linked to the disassembly of homomeric complexes and the formation of heterotetrameric complexes. Facultative complex formation may contribute to the ability for CNGC subunits to carry out multiple signalling roles.

Again, it remains to be seen what role, if any, cNMPs play in this pathway. Electrophysiological analyses support cNMP activation of AtCNGC2 either in native membrane or when heterologously expressed (Table 2; [30,38]) and with an apparent ability to discriminate between cAMP and cGMP in guard cells [4,30]. As flg22 can induce guard cell $[Ca^{2+}]_{cyt}$ oscillations [131] and as CNGCs could be involved in $[Ca^{2+}]_{cyt}$ oscillations [46], further consideration of guard cell AtCNGC2 in this immune pathway is warranted. The recent discovery that BIK1 phosphorylates the guard cell plasma membrane Ca^{2+} channel AtOSCA1.3 (hyperOsmolality-induced $[Ca^{2+}]$i increase 1.3) as part of the stomatal flg22 response [132] still leaves room for other Ca^{2+} influx pathways. Moving away from *Arabidopsis*, SlCNGC1 and SlCNGC14 are required for the tomato flg22-induced $[Ca^{2+}]_{cyt}$ increase but it is not yet clear whether these subunits can form a complex [133]. Overall, the study of CNGC regulation by CaM is complicated by the abundance of CaM and CNGC isoforms and the multitude of CaMBDs. It is likely that this complexity contributes to the specificity of signal transduction by CNGCs.

4.3. CNGCs Are Regulated by Phosphorylation

CNGC phosphorylation is emerging as an important regulator of activity. AtCNGC4 contains nine phosphorylation sites within and around the CT CNBD and, as described in Section 4.2, phosphorylation by BIK1 relieves CaM7-mediated inhibition of the putative AtCNGC2/4 complex in flg22 signalling [29]. Similarly, the rice receptor-like cytoplasmic kinase OsRLCK185 (receptor-like cytoplasmic kinase 185) is responsible for activation of OsCNGC9 by phosphorylation, triggering defence responses [89]. *AtCNGC19* and *AtCNGC20* have also been found to play a role in plant defence downstream of BAK1/SERK4 (somatic embryogenesis receptor kinase 4). However, in this case, the authors proposed that the signalling cascade progressed through BAK1-mediated phosphorylation of the AtCNGC20-CT, leading to proteasome-dependent degradation, as opposed to phosphorylation-mediated channel activation [67]. Mutation of $Thr^{560}/Ser^{617}/Ser^{618}/Thr^{619}$ in the AtCNGC20-CT reduced BAK1-mediated phosphorylation, and additional phosphorylation sites were predicted in the C- and N-terminals [67]. It was not reported, however, whether AtCNGC20-CT phosphorylation might also be an activating signal, which may subsequently be followed by signal termination via protein degradation. It is possible, therefore, that AtCNGC20-CT phosphorylation may overcome CaM-mediated inhibition, as found with AtCNGC2/4 in the flg22 pathway. It is also possible that AtCNGC2/4 phosphorylation may promote proteasome-mediated degradation and, as such, CNGC phosphorylation may have dual function in planta.

The role of CNGC phosphorylation is not restricted to defence signalling. Calcium-dependent protein kinase 32 (CPK32) appears to interact with AtCNGC18 in planta, increases the conductance of AtCNGC18 homotetramers when co-expressed in *Xenopus* oocytes and, following overexpression in pollen tubes, leads to increased apical $[Ca^{2+}]_{cyt}$ [63]. Following the identification of AtCNGC8/18 heterotetramers, it would be interesting to test how CPK32 affects AtCNGC8/18 activity. In silico analysis also predicts numerous phosphorylation sites in CNGCs from *N. tabacum* [18] and *Brassica oleracea* [22], suggesting that kinase/phosphatase regulation of CNGC activity is widespread.

4.4. CNGC Complexes Generate Further Functional and Regulatory Complexity

As discussed, the functional characteristics of heterologously expressed AtCNGC8/18 and the putative AtCNGC2/4 complex can be different from the homotetrameric channels of their constituent subunits, including changes to cNMP gating [29,46]. It also appears that CNGC complexes display changes in ion selectivity. AtCNGC2 can conduct K^+, Cs^+, Rb^+ and Li^+ but has a very low permeability to Na^+ which correlates with a change in pore region amino acid sequence, from GQN to AND [85]. It would appear, therefore, that AtCNGC2 has a role in mineral nutrition where Na^+ permeability is particularly deleterious. AtCNGC2 has been implicated in uptake of Ca^{2+} into leaves [38] and it may be that Na^+ exclusion is important in this function. Unlike AtCNGC2, AtCNGC4 appears to be permeable to both K^+ and Na^+ [41]. However, co-expression of *AtCNGC2* and *AtCNGC4* forms channels which are permeable to K^+ but impermeable to Na^+ [29]. This supports the hypothesis that AtCNGC2 and AtCNGC4 do form complexes and suggests that these complexes have unique functional characteristics, including permeability. It is important for future studies of CNGC permeability, therefore, to consider whether the CNGCs being tested exist as complexes in planta. For example, Zhang et al. reported that expression of *AtCNGC14* in *Xenopus* oocytes produced channels which were permeable to Mg^{2+} but effectively impermeable to K^+, Na^+ and Ba^{2+}, under the conditions tested [57]. It should be investigated whether AtCNGC complexes containing AtCNGC14 also display selectivity against these ions.

CNGC complexes will also make allosteric regulation of CNGC activity more intricate. For example, since CNGC complexes will be composed of multiple CNGC isoforms, with non-identical CTs and NTs (Figure 3A), the CNGC complex may interact with new combinations of allosteric regulators, including different CaM isoforms or different protein kinases. Each combination is likely to be unique to each CNGC complex and could result in unique feedback loops. Consequently, CNGC complexes may produce characteristic Ca^{2+} signatures which would enable CNGC complexes, even those that share a CNGC subunit, to participate in different signalling pathways. In addition, it is possible that the selectivity of some CNGCs, such as HvCNGC2-3 [7], for one of cAMP or cGMP may lead to specificity in heterotetrameric complexes.

An additional consideration is how CNGC subunits may compete for interactions with other CNGC isoforms. For example, it is apparent that AtCNGC7 and AtCNGC8 preferentially interact with AtCNGC18 in *Xenopus* to form AtCNGC7/18 or AtCNGC8/18 heterotetrameric complexes, instead of forming homotetrameric complexes [46]. It is also likely that AtCNGC2 and AtCNGC4 also preferentially interact to form AtCNGC2/4 heterotetramers, as opposed to homotetramers. Therefore, in a situation where two or more possible CNGC complexes may be formed, it is likely that particular CNGC complexes will form preferentially over others. Understanding these interaction dynamics may be key to understanding CNGC activity. The *brush* mutation in *Lotus japonicus* is an exemplar of how small changes in CNGC structure can significantly alter CNGC complex function in planta [99]. *BRUSH* is an *LjCNGC* homologous to AtCNGC19 and AtCNGC20 [99]. The *brush* mutation is found within the CNGC N-terminus and leads to a quantitative gain-of-function phenotype associated with the constitutive, voltage-dependent Ca^{2+} permeability of the *brush* homotetramer [99]. Competition between alternative CNGC subunits is believed to restrict formation of this homotetramer except in those plants strongly expressing *brush* [99].

There also remain a number of avenues which have remained unexplored in the study of plant CNGC complexes. For example, it remains unknown whether plant CNGCs form complexes with three or four different subunits (Figure 3B). Furthermore, whilst it has been assumed in models of CNGC complexes that the stoichiometry of CNGC subunits is 2:2 [10,67], there is only evidence supporting that assumption in the case of AtCNGC8/18 and AtCNGC7/18 [46]. It is also possible that CNGC function may be altered by the order in which CNGC subunits are ordered around the channel pore. In their analysis of animal CNG channels, Liu et al. discovered that the order of CNG subunits could alter channel conductance by up to 50% [69]. Two CNGC complexes, therefore, whilst having the same stoichiometry of CNGC subunits, may display different functional characteristics. Perhaps the formation of CNGC complexes, and the order of CNGC subunits, in planta is influenced by allosteric regulators such as CaM, which may promote stronger interactions between different CNGC subunits, and help dictate their order. Alternatively, it is possible that where the order of CNGC subunits differs, the conductance, ion selectivity and interactions with allosteric regulators are all altered, leading to divergent functional outcomes between otherwise similar CNGC complexes.

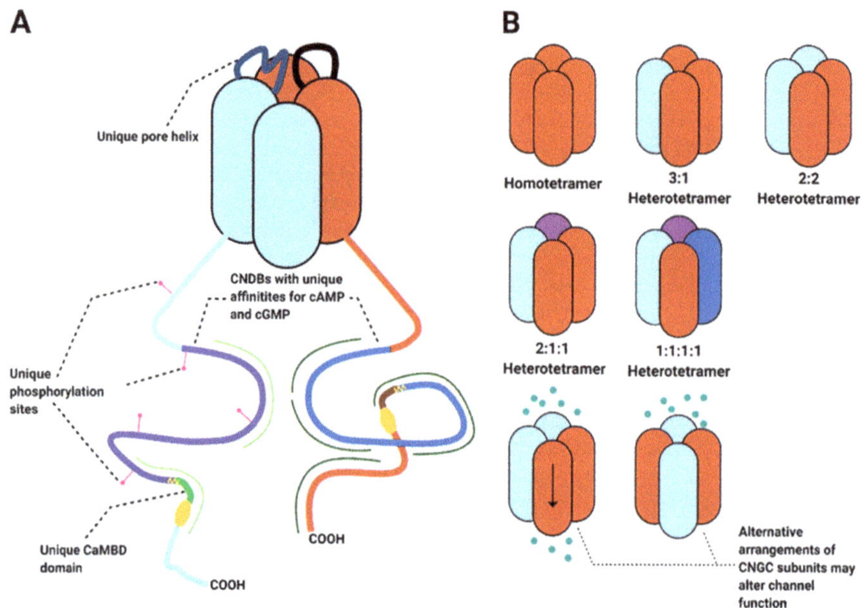

Figure 3. (**A**) Heterotetrameric complexes possess non-identical C-terminal domains. Differences in the cyclic nucleotide binding domain may determine specificity for cGMP or cAMP. Different CaM isoforms show preferential binding with different CNGC C-terminal domains [8,33], so heterotetrameric complexes may show combinatorial control by different CaM isoforms. Different CNGC C-terminal domains will be phosphorylation targets of different protein kinases, providing a mechanism for signal integration. The combination of different pore helixes may lead to changes in CNGC ion selectivity. For simplicity not all pore helices or cytosolic domains are shown. (**B**) Multiple different arrangements of CNGC subunits could be possible. CNGC activity may be regulated by both the different isoforms involved and their arrangement. Figure created with BioRender.com.

In silico analysis of CNGC sequences have identified several uncharacterised motifs which may further enhance CNGC regulation and function. Nawaz et al. identified three uncharacterised NtabCNGC motifs that are approximately 50 amino acids long [18], as well as a raffinose synthase motif in all 35 NtabCNGCs [18]. Furthermore, three uncharacterised motifs in *Brassica rapa* CNGCs are found in a number of closely related BraCNGCs, suggesting that these BraCNGCs have additional functionality [23]. It is possible that some of these uncharacterised motifs contribute to CNGC subunit interactions and, if heterotetrameric complexes were confirmed to be widespread amongst CNGCs, these complexes would significantly increase the complexity of CNGC regulation.

5. Could CNGCLs Modulate Complex Formation?

Angiosperm evolution has seen the loss of several types of Ca^{2+} channel that are still found in animals and an apparent overall reduction in diversity of Ca^{2+} influx mechanisms compared to animals [134]. This suggests a greater reliance on the channels that were retained over evolution such as CNGCs. By comparison, many plant species harbour more cyclic nucleotide-gated channel subunits than animals. Vertebrates (including mammals) and invertebrates have only six CNGs [94,98,135]. Subunits of the hyperpolarisation activated cyclic nucleotide-gated (HCN) cation channels (that operate in cardiac cells) are also present in low numbers (three in invertebrates, four in mammals and four to six in other vertebrates) [136]. An ability to form diverse CNGC complexes from a greater number of subunits could compensate for the reduction in diversity of Ca^{2+} influx mechanism evident in plant genomes and fit each cell to respond appropriately to the diverse and coincident stimuli experienced during their sessile lives. Additionally, a range of homomeric or heteromeric CNGC channels could permit function beyond Ca^{2+} signalling and help explain the role of CNGCs in mineral nutrition. Truncated, CNGC-like (CNGCL) proteins may also provide a further layer of regulation by modulating CNGC complex formation (Figure 4). Pan et al. demonstrated that the AtCNGC8 CT inhibits AtCNGC18 activity [46]. This is likely due to the formation of AtCNGC8/18 heterotetramer-like interactions which prevent formation of the AtCNGC18 homotetramer. In principle, therefore, any CNGC CT could disrupt CNGC–CNGC interactions. Likewise, since the data presented by Chiasson et al. suggest that the CNGC NT also contains CNGC–CNGC interaction domains [137], it is possible that any CNGC NT could also disrupt CNGC–CNGC interactions. Genome-wide analysis of *CNGC* sequences has identified a number of truncated *CNGC* genes in *B. rapa*, *B. oleracea* and *N. tabacum* which were not analysed further in the original studies since they lack key CNGC domains [18,21,22].

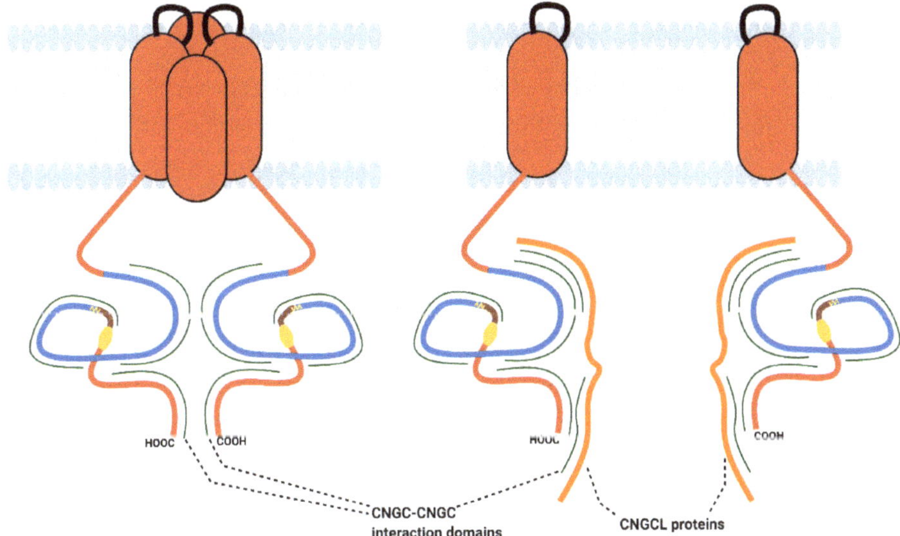

Figure 4. CNGCL proteins may disrupt the interactions between CNGC subunits and destabilise CNGC tetramers. For simplicity, pore loops and CT are shown in only two subunits. Figure created with BioRender.com.

We examined some of these CNGCL sequences in silico to determine whether they may have the potential to disrupt CNGC–CNGC interactions. Six CNGCL genes (*Bra024083*, *Bo3g005110*, *Bo8g027170*, *Bo5g104990*, *Bo3g052670* and *Bo6g074480*) were identified from genome-wide analyses of *B. rapa* and *B. oleracea* and located in the EnsemblPlants database (https://plants.ensembl.org/index.html) [22,23]. For the five *B. oleracea* genes, the protein sequences were extracted from their UniProtKB identifiers. For *Bra024083*, the annotation in EnsemblPlants predicts a 78 amino acid sequence, whereas the NCBI reference sequence for *Bra024083*, XP_009137913.1, predicts a 100 amino acid protein. Both sequences were used in the subsequent analysis. The seven protein sequences were used as queries to search for homologous sequences in the genomes of *B. rapa* and *B. oleracea* using the NCBI BLAST protein program with default parameters (https://blast.ncbi.nlm.nih.gov/Blast.

cgi?PAGE=Proteins). Following this, protein sequences of CNGCLs and the most similar CNGC identified in the BLAST search were submitted to pairwise local sequence using the EMBOSS Water program with default parameters (https://www.ebi.ac.uk/Tools/psa/emboss_water/). These alignments are presented in Figure A1 in Appendix A.

Bra024083 (NCBI reference sequence XP_009137913.1) was initially identified for its homology with *AtCNGC17* [23]. In our analysis, the 78 amino acid prediction aligns to a region in the CT of BraCNGC17 (NCBI reference sequence XP_009127879.2, positions 640–728) with 76.4% identity. The 100 amino acid prediction aligns to an overlapping region of BraCNGC17 (positions 629–728) with 75% identity. In *B. oleracea*, *Bo3g005110* (UniProtKB_A0A0D3DKM8) and *Bo8g027170* (UniProtKB_A0A0D3DKM8) are predicted to encode identical 99 amino acid peptides which align to a 70 amino acid stretch in the CT of BoCNGC7 (NCBI reference sequence XP_013585954.1, positions 579–648) with 80.0% identity. Similarly, *Bo5g104990* (UniProtKB_A0A0D3CHU3) is predicted to encode a 113 amino acid peptide which aligns to a 76 amino acid stretch in the CT of BoCNGC7 (positions 572–648) with 83.1% identity. *Bo3g052670* (UniProtKB_A0A0D3B8V0) is predicted to encode a 412 amino acid peptide which aligns to the NT sequence of BoCNGC12 (NCBI reference sequence XP_013631017.1, positions 1–425) with 68.2% identity and *Bo6g074480* is predicted to encode a 480 amino acid peptide which aligns with BoCNGC12 (positions 103–648) with 60.9% identity. In silico observations may be misleading but these putative CNGCL proteins may warrant further attention to determine whether they are functional in planta and interact with CNGCs.

6. Conclusions

Research into plant CNGCs has historically focused on the role of individual *CNGC* genes. There is increasing evidence, however, to suggest that plant CNGCs function as heterotetrameric complexes.

To understand the role of CNGCs (whether in Ca^{2+} signalling or nutrition), it is necessary to determine which subunits interact and determine whether they form heterotetrameric complexes. If CNGC complexes were common, it will be important to determine which CNGC interactions occur preferentially and to study *CNGC* expression patterns to help predict the composition of CNGC complexes in different cell types. Consequently, a systematic study of CNGC interactions through BiFC and single-molecule fluorescence is needed to understand which complexes may be present in planta. In common with animal studies, cryo-electron microscopy should be able to resolve tetrameric structures. The advent of fluorophore-labelled cyclic nucleotides is now enabling the effect of cNMP binding on channel kinetics to be elucidated for animal homomeric and heterotetrameric CNGs [138] and could be applied to heterologously expressed plant CNGCs to further understand differences between complexes. It will subsequently be important to test the activities of these CNGCs in physiologically relevant conditions and determine how their behaviour is different from homotetrameric channels.

Recent studies, therefore, have significant implications for the future of CNGC research and may herald a major shift in our understanding of CNGC function. The role of CNGCs in plants may, truly, be complex.

Funding: This research was funded by the University of Cambridge Brookes and Isaac Newton Trusts and the University of Cambridge Commonwealth, European and International Trusts.

Institutional Review Board Statement: Not applicable.

Informed Consent Statement: Not applicable.

Data Availability Statement: Data can be obtained by contacting the corresponding author.

Conflicts of Interest: The authors declare no conflict of interest.

Abbreviations

AC	Adenylyl cyclase
AGN	Alanine-glycine-asparagine
AHA	Arabidopsis H$^+$-ATPase
AND	Alanine-asparagine-aspartic acid
Apo-CaM	Calmodulin without Ca^{2+} ligands
At	Arabidopsis thaliana
BAK1	BRI-associated receptor kinase1
BiFC	Bifluorescence complementation
Bo	Brassica oleracea
Bra	Brassica rapa
BRI1	Brassinosteroid insensitive1
Ca^{2+}/CaM	Calmodulin with Ca^{2+} ligands
CaM	Calmodulin
CaMBD	Calmodulin-binding domain
cAMP	Cyclic adenosine monophosphate
cGMP	Cyclic guanosine monophosphate
CNDB	Cyclic nucleotide-binding domain
CNGC	Cyclic nucleotide-gated channel
CNGCL	A gene or protein which contains some, but not all the domains associated with cyclic nucleotide-gated channels
cNMP	Cyclic nucleotide monophosphate
CPK32	Calcium-dependent protein kinase 32
CT	Carboxy-terminal domain
db-cAMP	Dibutyryl-cyclic adenosine monophosphate
db-cGMP	Dibutyryl-cyclic guanosine monophosphate
DMI1	Does not make infections 1
flg22	Flagellin 22 peptide
FLS2	Flagellin Sensitive 2
GC	Guanylyl cyclase
GNL	Glycine-asparagine-leucine
GQG	Glycine-glutamine-glycine
GQN	Glycine-glutamine-asparagine
GQS	Glycine-glutamine-serine
GUS	β-glucuronidase
HEK293	Human embryonic kidney cell line 293
Hv	Hordeum vulgare
IQ	Isoleucine-glutamine
KUP	K$^+$ uptake permease
Lj	Lotus japonicus
LRRAC1	Leucine-rich repeat adenylyl cyclase1
Mt	Medicago truncatula
Ntab	Nicotiana tabacum
NOGC1	Nitric oxide-dependent guanylate cyclase1
NT	N-terminal domain
Os	Oryza sativa
OSCA1.3	hyperOsmolality-induced [Ca^{2+}]i increase 1.3
PepR1	Plant elicitor peptide recptor1
Pp	Physcomitrella patens
PSKR1	Phytosulfokine receptor1
RLCK185	Receptor-like cytoplasmic kinase 185
SERK4	Somatic embryogenesis receptor kinase 4
Sl	Solanum lycopersicum
WAKL10	Wall-associated kinase (WAK)-like 1
Zj	Ziziphus jujube

Appendix A

```
Bra024083 (EnsemblPlants) (78 amino acids long) - 76.4% identity

Bra024083_(EnsemblPlants)             1 MENHLTAVESKQS---DEEKEVEVGG--EEGEECVDSSPKTKMNTGVMVLASRFAANTRRGIAAQR------PRFKKPEDPDFSAEPDD 78
                                        |||||||||||    |||:||.|..  ||-|.||.|.||||||||.|||||||||||||||:|||    ||||||:||||||||
BraCNGC17_XP_009127879.2            640 MENHLTAVESKQSDEDDEEEEVVVRKVVEEEEEGVGSSPKTKMNIGVMVLASRFAANTRRGVAAQRVKDVEMPRFKKPEEPDFSAEPDD 728

Bra024083 (NCBI reference sequence XP 009137913.1) (100 amino acids long) - 75.0% identity

Bra024083_XP_009137913.1             12 QAAWLRNNRSAMENHLTAVESKQS---DEEKEVEVGG--EEGEECVDSSPKTKMNTGVMVLASRFAANTRRGIAAQR------PRFKKPEDPDFSAEPDD 100
                                        ||||.|..|.||||||||||||||   |||:||.|..  ||-|.||.|.||||||||.|||||||||||||||:||||    ||||||:||||||||
BraCNGC17_XP_009127879.2            629 QAAWRRYKRRAMENHLTAVESKQSDEDDEEEEVVVRKVVEEEEEGVGSSPKTKMNIGVMVLASRFAANTRRGVAAQRVKDVEMPRFKKPEEPDFSAEPDD 728

Bo3g005110/Bo8g027170 (99 amino acids long) - 80.0% identity

Bo3g005110/Bo8g027170_UniProtKB_A0A0D3DKM8  28 LETQSRLKVTFSTRTVKALAEVEAFALEAEELKFVASQFRRLHSRQVEQTFRLYSQQWRTWASSFIEAAW 97
                                               |:.::...:..||||||.||||||||||||||||||||||:||||.|||||||||||||:|||
BoCNGC7_XP_013585954.1                     579 LDPKAGSNLPSSTRTVKALTEVEAFALEAEELKFVASQFRRLHSRQVQQTFRFYSQQWRTWASSFIQAAW 648

Bo5g104990 (113 amino acids long) - 83.1% identity

Bo5g104990_UniProtKB_A0A0D3CHU3      35 EEEETKAYTRKAGSNLPFSTRTIKALAEVEAFALEVEELKFVASQFRRLHSRQVEQTFRLYSQQWRTWASSFIEAAW 111
                                        ||..|.|...||||||.||||:|||.|||||||.|||||||||||||||||||:||||.|||||||||||||:|||
BoCNGC7_XP_013585954.1              572 EELLTWALDPKAGSNLPSSTRTVKALTEVEAFALEAEELKFVASQFRRLHSRQVQQTFRFYSQQWRTWASSFIQAAW 648

Bo3g052670 (412 amino acids long) - 68.2% identity

Bo3g052670_UniProtKB_A0A0D3B8V0       1 MERILNKQAVINRVKSLQ---------MNTLRGNWRKTILIVCVVALGIDPLYLFVPVVDSPKFCFTFDKKLATGVSVLRTFIDVFYVVHIILNFMR--- 88
                                        |||....|:|...:||::          :|||. ||||.||:||||||:|||:||:||:|||.|||||||||..||.:|||||.|||:|||.||:.
BoCNGC12_Isoform_X1_XP_013631017.1    1 MERASTMQSVHENIKSVRGQLKKVYKTLNTLE-NWRKAILLVCVVALGVDPLFLFIPVIDSPNFCFTFDKKLAAVVSAIRTFIDTFYVIHIIFNFITEFI 99

Bo3g052670_UniProtKB_A0A0D3B8V0      89 ------YSNGELNLHAKPKRETYYFISYTIVDILSVLPMPQVLVLTLMRRSDSLVSREILKWIVLSQYIPRILRIYPLYKEVTKASGTVAETKWIGAAFN 182
                                          ..|||.|.:...:|..:  |||.|:|..:|..:  ||||||.||:|||||||||.|:|||||||:|||.||||:||||||||||||||||
BoCNGC12_Isoform_X1_XP_013631017.1  100 APRSQVSLRGELIVHSKATRKRLFFFHF-IVDICSVIPIPQVVVLILIHRSDSLVSQAILKWIILTQYVPRIIRIYPLLKEVTRASGTIAETKWVGAAFN 198

Bo3g052670_UniProtKB_A0A0D3B8V0     183 LFLYMLHSHVFGAFWYVSSVEKINKCWRLACPHLPGCSLKFQYCGREGGNNIPYGLNTSCPLIEPDQITNSSVFNFGMYIDALKSGIVEVKPDDFPRKFF 282
                                        ..:|:|.|:..:.||.: ||||||||||.||:|||:|.||:...:|:...|..| |||.||||:|:||.||:.||||||||||.|||||:|.||||||
BoCNGC12_Isoform_X1_XP_013631017.1  199 LFLYMLHSHVFGAFWYVSSVEKKNKCWRLECAKIFGCNLRYQYCARGRQNNGRY-LNTTCPLIDPDQIIGSTVFNFGMYTDALRSGIVESKPRDFPRKFF 297

Bo3g052670_UniProtKB_A0A0D3B8V0     283 YCFWWGLRNISALGQNLRTSNSVGDIVFALIICVSGLLLFAVLIGNIQKYLQSNTTRIDEMEERKRDTEIWMSSKSLPENLKDMIRRHKEEKWQKTRGIE 382
                                        |||||||||||||||||:||||||||||||||||||||||||||||||||.|.|:|||.||||:|||||:||:.|||:.||:|:.:||::.|:|||||
BoCNGC12_Isoform_X1_XP_013631017.1  298 YCFWWGLRNISALGQNLKTSNSVGDIVFALIICVSGLLLFAVLIGNIQKYLQSTTIRLDEMEEKKRDTEKWMSNRMLPEYLKERIRRYENYKWRKTRGIE 397

Bo3g052670_UniProtKB_A0A0D3B8V0     383 EEAFLQSLPDHIRL----HIERNSLNSV      406
                                        |||.|.|||..:||    |:....||||
BoCNGC12_Isoform_X1_XP_013631017.1  398 EEALLHSLPKDLRLETKRHLYLTLLNSV      425

Bo6g074480 (480 amino acids long) - 60.9% identity

Bo6g074480_UniProtKB_A0A0D3CUI0       9 SSVIKRRELVVHSKAIRKRLFFFYFSVDIVSVLPIPQVVVLTLVSRKQMTSVVSK----------------EIYPIFKEVTRASGTVS------------ 80
                                        |.|..|.||:|||||.||||||||:|.|||.||:||||||||.|:.|..  |:||:         .|||:.||||||||||||::
BoCNGC12_Isoform_X1_XP_013631017.1  103 SQVSLRGELIVHSKATRKRLFFFHFIVDICSVIPIPQVVVLILIHRSD--SLVSQAILKWIILTQYVPRIIRIYPLLKEVTRASGTIAETKWVGAAFNLF 200

Bo6g074480_UniProtKB_A0A0D3CUI0      81 -------VIGAFWYLSAIEKKDTCWHEACAKISGCNLTNYLCARGGTGGDNSRFLNTSCPLIDPEQIINSTVFNFGMYIDALKSGVVES--RNFPRKLLY 171
                                        |.|||||:|::||:.||...||||.||||.....|||.  :|.|:||||||:|||:|||||||||.||:||:|||:|||.||:||:||| |:||||..|
BoCNGC12_Isoform_X1_XP_013631017.1  201 LYMLHSHVFGAFWYVSSVEKKNKCWRLECAKIFGCNLRYQYCARGRQ--NNGRYLNTTCPLIDPDQIIGSTVFNFGMYTDALRSGIVESKPRDFPRKFFY 298

Bo6g074480_UniProtKB_A0A0D3CUI0     172 CFWWGLRNL----------SNSEGEILFAIIICVSGVLLFAVLIGNVQKYLQSTTIRVDEWEEKKRDTEKWMSYRKLPENIKERIRKYEDYKWQETRSTEE 262
                                        |||||||:         |||.|:|:||:|||||||:||||||||||:|||||||||:||.||||||||||||:||.||||||||||||:|||:||::||..||
BoCNGC12_Isoform_X1_XP_013631017.1  299 CFWWGLRNISALGQNLKTSNSVGDIVFALIICVSGLLLFAVLIGNIQKYLQSTTIRLDEMEEKKRDTEKWMSNRMLPEYLKERIRRYENYKWRKTRGIEE 398

Bo6g074480_UniProtKB_A0A0D3CUI0     263 EALLRSLPKDIRLETKHHFYMKLLKRVPWLSFMDDGWLLEALCDRVKPVFYSENSYIVRK-----------------------------------GDICG 327
                                        ||||.||||.|.|:.||.|.|:.||.||||:.|||.||||||||||.||||.|||||:::                                   |||||
BoCNGC12_Isoform_X1_XP_013631017.1  399 EALLHSLPKDLRLETKRHLYLTLLNSVPWLNMMDDSWLLEALCDRVKSVFYSANSYIVKEGDPVAEMLIITKGSLKSMIGFSDITGYYDSSYLQAGDICG 498

Bo6g074480_UniProtKB_A0A0D3CUI0     328 DLLFWVLDPHPSSRVPSSVRTVITVTDVEGFILLPDDVKFVASHLNRLHSVKLKHMFRYYSMSWMSWGACYIQAAWRAHCRRKASKTLRAKKDKQQIQQD 427
                                        |||||||||.||.:|:|.|:|:|:||||||||||.|||||||||.||.||||||.||.||.:|:.|..||||||||.|:||.|:.|.||:||.|.|:|.
BoCNGC12_Isoform_X1_XP_013631017.1  499 DLLFWVLDPHSSSSLPTSDRSVLTLTDVEGFILLHDDLKFVASHFNRSHSSRLRHMFRFYSAHWRLWAACFIQAAWREHCKRKLSRILHAKRDYNHIPQG 598

Bo6g074480_UniProtKB_A0A0D3CUI0     428 VQLNLGATLYVSRFVSKALRNRQDDSAECSSFSQMLPLVPHKPADPEFSK 477
                                        .||||||.|||||||||||||||.::|.||....|||.:|||||||||||
BoCNGC12_Isoform_X1_XP_013631017.1  599 PQLNLGAALYVSRFVSKALRNRQKNAANCSISPHMLPPIPHKPADPEFSK 648
```

Figure A1. Local sequence alignments of CNGCL protein sequences against CNGC protein sequences. In each case the CNGCL sequence is displayed above the CNGC sequence. In each subheading the full length of the CNGCL sequence is given as well as the percentage identity with the CNGC sequence shown.

References

1. Leng, Q.; Mercier, R.W.; Yao, W.; Berkowitz, G.A. Cloning and first functional characterization of a plant cyclic nucleotide-gated cation channel. *Plant Physiol.* **1999**, *121*, 753–761. [CrossRef] [PubMed]
2. Leng, Q.; Mercier, R.W.; Hua, B.-G.; Fromm, H.; Berkowitz, G.A. Electrophysiological analysis of cloned cyclic nucleotide-gated ion channels. *Plant Physiol.* **2002**, *128*, 400–410. [CrossRef] [PubMed]
3. Qi, Z.; Verma, R.; Gehring, C.; Yamaguchi, Y.; Zhao, Y.; Ryan, C.A.; Berkowitz, G.A. Ca^{2+} signaling by plant *Arabidopsis thaliana* Pep peptides depends on AtPepR1, a receptor with guanylyl cyclase activity, and cGMP-activated Ca^{2+} channels. *Proc. Natl. Acad. Sci. USA* **2010**, *107*, 21193–21198. [CrossRef] [PubMed]
4. Wang, Y.-F.; Munemasa, S.; Nishimura, N.; Ren, H.-M.; Robert, N.; Han, M.; Puzõrjova, I.; Kollist, H.; Lee, S.; Mori, I.; et al. Identification of cyclic GMP-activated nonselective Ca^{2+}-permeable cation channels and associated *CNGC5* and *CNGC6* genes in Arabidopsis guard cells. *Plant Physiol.* **2013**, *163*, 578–590. [CrossRef]

5. Gao, Q.-F.; Fei, C.-F.; Dong, J.-Y.; Gu, L.-L.; Wang, Y.-F. *Arabidopsis* CNGC18 is a Ca^{2+}-permeable channel. *Mol. Plant* **2014**, *7*, 739–743. [CrossRef] [PubMed]

6. Gao, Q.-F.; Gu, L.-L.; Wang, H.-Q.; Fei, C.-F.; Fang, X.; Hussain, J.; Sun, S.-J.; Dong, J.-Y.; Liu, H.; Wang, Y.-F. Cyclic nucleotide-gated channel 18 is an essential Ca^{2+} channel in pollen tube tips for pollen tube guidance to ovules in *Arabidopsis*. *Proc. Natl. Acad. Sci. USA* **2016**, *113*, 3096–3101. [CrossRef]

7. Mori, I.C.; Nobukiyo, Y.; Nakahara, Y.; Shibasaka, M.; Furuichi, T.; Katsuhara, M. A cyclic nucleotide-gated channel, HvCNGC2-3, is activated by the co-presence of Na$^+$ and K$^+$ and permeable to Na$^+$ and K$^+$ non-selectively. *Plants* **2018**, *7*. [CrossRef]

8. Zhang, Z.; Hou, C.; Tian, W.; Li, L.; Zhu, H. Electrophysiological studies revealed CaM1-mediated regulation of the *Arabidopsis* calcium channel CNGC12. *Front. Plant Sci.* **2019**, *10*. [CrossRef]

9. Tian, W.; Wang, C.; Gao, Q.; Li, L.; Luan, S. Calcium spikes, waves and oscillations in plant development and biotic interactions. *Nat. Plants* **2020**, *6*, 750–759. [CrossRef]

10. Dietrich, P.; Moeder, W.; Yoshioka, K. Plant cyclic nucleotide-gated channels: New insights on their functions and regulation. *Plant Physiol.* **2020**, *184*, 27–38. [CrossRef]

11. Blanco, E.; Fortunato, S.; Viggiano, L.; de Pinto, M.C. Cyclic AMP: A polyhedral signalling molecule in plants. *Int. J. Mol. Sci.* **2020**, *21*. [CrossRef] [PubMed]

12. Chin, K.; Moeder, W.; Yoshioka, K. Biological roles of cyclic-nucleotide-gated ion channels in plants: What we know and don't know about this 20 member ion channel family. *Botany* **2009**, *87*, 668–677. [CrossRef]

13. Duszyn, M.; Świeżawska, B.; Szmidt-Jaworska, A.; Jaworski, K. Cyclic nucleotide gated channels (CNGCs) in plant signalling—Current knowledge and perspectives. *Plant Physiol.* **2019**, *241*. [CrossRef] [PubMed]

14. Jha, S.K.; Sharma, M.; Pandey, G.K. Role of cyclic nucleotide gated channels in stress management in plants. *Curr. Genom.* **2016**, *17*, 315–329. [CrossRef] [PubMed]

15. Moeder, W.; Urquhart, W.; Ung, H.; Yoshioka, K. The role of cyclic nucleotide-gated ion channels in plant immunity. *Mol. Plant* **2011**, *4*, 442–452. [CrossRef] [PubMed]

16. Mäser, P.; Thomine, S.; Schroeder, J.I.; Ward, J.M.; Hirschi, K.; Sze, H.; Talke, I.N.; Amtmann, A.; Maathuis, F.J.; Sanders, D.; et al. Phylogenetic relationships within cation transporter families of Arabidopsis. *Plant Physiol.* **2001**, *126*, 1646–1667. [CrossRef]

17. Zeng, H.; Zhao, B.; Wu, H.; Zhu, Y.; Chen, H. Comprehensive in silico characterization and expression profiling of nine gene families associated with calcium transport in soybean. *Agronomy* **2020**, *10*. [CrossRef]

18. Nawaz, Z.; Kakar, K.U.; Ullah, R.; Yu, S.; Zhang, J.; Shu, Q.-Y.; Ren, X. Genome-wide identification, evolution and expression analysis of cyclic nucleotide-gated channels in tobacco (*Nicotiana tabacum* L.). *Genomics* **2019**, *111*, 142–158. [CrossRef]

19. Nawaz, Z.; Kakar, K.U.; Saand, M.A.; Shu, Q.-Y. Cyclic nucleotide-gated ion channel gene family in rice, identification, characterization and experimental analysis of expression response to plant hormones, biotic and abiotic stresses. *BMC Genom.* **2014**, *15*, 1–18. [CrossRef]

20. Guo, J.; Islam, M.A.; Lin, H.; Ji, C.; Duan, Y.; Liu, P.; Zeng, Q.; Day, B.; Kang, Z.; Guo, J. Genome-wide identification of cyclic nucleotide-gated ion channel gene family in wheat and functional analyses of *TaCNGC14* and *TaCNGC16*. *Front. Plant Sci.* **2018**, *9*, 18. [CrossRef]

21. Hao, L.; Qiao, X. Genome-wide identification and analysis of the *CNGC* gene family in maize. *PeerJ* **2018**, *6*, e5816. [CrossRef] [PubMed]

22. Kakar, K.U.; Nawaz, Z.; Kakar, K.; Ali, E.; Almoneafy, A.A.; Ullah, R.; Ren, X.; Shu, Q.-Y. Comprehensive genomic analysis of the *CNGC* gene family in *Brassica oleracea*: Novel insights into synteny, structures, and transcript profiles. *BMC Genom.* **2017**, *18*, 1–18. [CrossRef] [PubMed]

23. Li, Q.; Yang, S.; Ren, J.; Ye, X.; jiang, X.; Liu, Z. Genome-wide identification and functional analysis of the cyclic nucleotide-gated channel gene family in Chinese cabbage. *3 Biotech* **2019**, *9*, 1–14. [CrossRef] [PubMed]

24. Ma, W.; Ali, R.; Berkowitz, G.A. Characterization of plant phenotypes associated with loss-of-function of AtCNGC1, a plant cyclic nucleotide gated cation channel. *Plant Physiol. Biochem.* **2006**, *44*, 494–505. [CrossRef] [PubMed]

25. Moon, J.Y.; Belloeil, C.; Ianna, M.L.; Shin, R. *Arabidopsis* CNGC family members contribute to heavy metal ion uptake in plants. *Int. J. Mol. Sci.* **2019**, *20*, 413. [CrossRef] [PubMed]

26. Sunkar, R.; Kaplan, B.; Bouché, N.; Arazi, T.; Dolev, D.; Talke, I.N.; Maathuis, F.J.M.; Sanders, D.; Bouchez, D.; Fromm, H. Expression of a truncated tobacco *NtCBP4* channel in transgenic plants and disruption of the homologous *Arabidopsis CNGC1* gene confer Pb^{2+} tolerance. *Plant J.* **2000**, *24*, 533–542. [CrossRef] [PubMed]

27. Chin, K.; DeFalco, T.A.; Moeder, W.; Yoshioka, K. The Arabidopsis cyclic nucleotide-gated ion channels AtCNGC2 and AtCNGC4 work in the same signaling pathway to regulate pathogen defense and floral transition. *Plant Physiol.* **2013**, *163*, 611–624. [CrossRef]

28. Finka, A.; Cuendet, A.F.H.; Maathuis, F.J.M.; Saidi, Y.; Goloubinoff, P. Plasma membrane cyclic nucleotide gated calcium channels control land plant thermal sensing and acquired thermotolerance. *Plant Cell* **2012**, *24*, 3333–3348. [CrossRef]

29. Tian, W.; Hou, C.; Ren, Z.; Wang, C.; Zhao, F.; Dahlbeck, D.; Hu, S.; Zhang, L.; Niu, Q.; Li, L.; et al. A calmodulin-gated calcium channel links pathogen patterns to plant immunity. *Nature* **2019**, *572*, 131–135. [CrossRef]

30. Ali, R.; Ma, W.; Lemtiri-Chlieh, F.; Tsaltas, D.; Leng, Q.; von Bodman, S.; Berkowitz, G.A. Death don't have no mercy and neither does calcium: *Arabidopsis* CYCLIC NUCLEOTIDE GATED CHANNEL2 and innate immunity. *Plant Cell* **2007**, *19*, 1081–1095. [CrossRef]

31. Clough, S.J.; Fengler, K.A.; Yu, I.; Lippok, B.; Smith, R.K.; Bent, A.F. The Arabidopsis *dnd1* "defense, no death" gene encodes a mutated cyclic nucleotide-gated ion channel. *Proc. Natl. Acad. Sci. USA* **2000**, *97*, 9323–9328. [CrossRef] [PubMed]

32. Cui, Y.; Lu, S.; Li, Z.; Cheng, J.; Hu, P.; Zhu, T.; Wang, X.; Jin, M.; Wang, X.; Li, L.; et al. CYCLIC NUCLEOTIDE-GATED ION CHANNELs 14 and 16 promote tolerance to heat and chilling in rice. *Plant Physiol.* **2020**, *183*, 1794–1808. [CrossRef] [PubMed]

33. Fischer, C.; DeFalco, T.A.; Karia, P.; Snedden, W.A.; Moeder, W.; Yoshioka, K.; Dietrich, P. Calmodulin as a Ca^{2+}-sensing subunit of Arabidopsis cyclic nucleotide-gated channel complexes. *Plant Cell Physiol.* **2017**, *58*, 1208–1221. [CrossRef] [PubMed]

34. Genger, R.K.; Jurkowski, G.I.; McDowell, J.M.; Lu, H.; Jung, H.W.; Greenberg, J.T.; Bent, A.F. Signaling pathways that regulate the enhanced disease resistance of *Arabidopsis "Defense, No Death"* mutants. *MPMI* **2008**, *21*, 1285–1296. [CrossRef]

35. Lu, M.; Zhang, Y.; Tang, S.; Pan, J.; Yu, Y.; Han, J.; Li, Y.; Du, X.; Nan, Z.; Sun, Q. AtCNGC2 is involved in jasmonic acid-induced calcium mobilization. *J. Exp. Bot.* **2016**, *67*, 809–819. [CrossRef]

36. Ma, W.; Qi, Z.; Smigel, A.; Walker, R.K.; Verma, R.; Berkowitz, G.A. Ca^{2+}, cAMP, and transduction of non-self perception during plant immune responses. *Proc. Natl. Acad. Sci. USA* **2009**, *106*, 20995–21000. [CrossRef]

37. Ma, W.; Smigel, A.; Walker, R.K.; Moeder, W.; Yoshioka, K.; Berkowitz, G.A. Leaf senescence signaling: The Ca^{2+}-conducting Arabidopsis cyclic nucleotide gated channel2 acts through nitric oxide to repress senescence programming. *Plant Physiol.* **2010**, *154*, 733–743. [CrossRef]

38. Wang, Y.; Kang, Y.; Ma, C.; Miao, R.; Wu, C.; Long, Y.; Ge, T.; Wu, Z.; Hou, X.; Zhang, J.; et al. CNGC2 is a Ca^{2+} influx channel that prevents accumulation of apoplastic Ca^{2+} in the leaf. *Plant Physiol.* **2017**, *173*, 1342–1354. [CrossRef]

39. Yu, I.; Parker, J.; Bent, A.F. Gene-for-gene disease resistance without the hypersensitive response in *Arabidopsis dnd1* mutant. *Proc. Natl. Acad. Sci. USA* **1998**, *95*, 7819–7824. [CrossRef]

40. Gobert, A.; Park, G.; Amtmann, A.; Sanders, D.; Maathuis, F.J.M. *Arabidopsis thaliana* Cyclic Nucleotide Gated Channel 3 forms a non-selective ion transporter involved in germination and cation transport. *J. Exp. Bot.* **2006**, *57*, 791–800. [CrossRef]

41. Balagué, C.; Lin, B.; Alcon, C.; Flottes, G.; Malmström, S.; Köhler, C.; Neuhaus, G.; Pelletier, G.; Gaymard, F.; Roby, D. HLM1, an essential signaling component in the hypersensitive response, is a member of the cyclic nucleotide–gated channel ion channel family. *Plant Cell* **2003**, *15*, 365–379. [CrossRef] [PubMed]

42. Massange-Sánchez, J.A.; Palmeros-Suárez, P.A.; Espitia-Rangel, E.; Rodríguez-Arévalo, I.; Sánchez-Segura, L.; Martínez-Gallardo, N.A.; Alatorre-Cobos, F.; Tiessen, A.; Délano-Frier, J.P. Overexpression of Grain Amaranth (*Amaranthus hypochondriacus*) AhERF or AhDOF transcription factors in *Arabidopsis thaliana* increases water deficit- and salt-stress tolerance, respectively, via contrasting stress-amelioration mechanisms. *PLoS ONE* **2016**, *11*. [CrossRef] [PubMed]

43. Tan, Y.-Q.; Yang, Y.; Zhang, A.; Fei, C.-F.; Gu, L.-L.; Sun, S.-J.; Xu, W.; Wang, L.; Liu, H.; Wang, Y.-F. Three CNGC family members, CNGC5, CNGC6, and CNGC9, are required for constitutive growth of *Arabidopsis* root hairs as Ca^{2+}-permeable channels. *Plant Commun.* **2020**, *1*, 100001. [CrossRef] [PubMed]

44. Brost, C.; Studtrucker, T.; Reimann, R.; Denninger, P.; Czekalla, J.; Krebs, M.; Fabry, B.; Schumacher, K.; Grossmann, G.; Dietrich, P. Multiple cyclic nucleotide-gated channels coordinate calcium oscillations and polar growth of root hairs. *Plant J.* **2019**, *99*, 910–923. [CrossRef] [PubMed]

45. Gao, F.; Han, X.; Wu, J.; Zheng, S.; Shang, Z.; Sun, D.; Zhou, R.; Li, B. A heat-activated calcium-permeable channel—Arabidopsis cyclic nucleotide-gated ion channel 6—Is involved in heat shock responses. *Plant J.* **2012**, *70*, 1056–1069. [CrossRef] [PubMed]

46. Pan, Y.; Chai, X.; Gao, Q.; Zhou, L.; Zhang, S.; Li, L.; Luan, S. Dynamic interactions of plant CNGC subunits and calmodulins drive oscillatory Ca^{2+} channel activities. *Dev. Cell* **2019**, *48*, 710–725. [CrossRef]

47. Tunc-Ozdemir, M.; Rato, C.; Brown, E.; Rogers, S.; Mooneyham, A.; Frietsch, S.; Myers, C.T.; Poulsen, L.R.; Malhó, R.; Harper, J.F. Cyclic nucleotide gated channels 7 and 8 are essential for male reproductive fertility. *PLoS ONE* **2013**, *8*, e55277. [CrossRef]

48. Borsics, T.; Webb, D.; Andeme-Ondzighi, C.; Staehelin, L.A.; Christopher, D.A. The cyclic nucleotide-gated calmodulin-binding channel AtCNGC10 localizes to the plasma membrane and influences numerous growth responses and starch accumulation in *Arabidopsis thaliana*. *Planta* **2007**, *225*, 563–573. [CrossRef]

49. Jin, Y.; Jing, W.; Zhang, Q.; Zhang, W. Cyclic nucleotide gated channel 10 negatively regulates salt tolerance by mediating Na^+ transport in *Arabidopsis*. *J. Plant Res.* **2015**, *128*, 211–220. [CrossRef]

50. DeFalco, T.A.; Marshall, C.B.; Munro, K.; Kang, H.-G.; Moeder, W.; Ikura, M.; Snedden, W.A.; Yoshioka, K. Multiple calmodulin-binding sites positively and negatively regulate Arabidopsis CYCLIC NUCLEOTIDE-GATED CHANNEL12. *Plant Cell* **2016**, *28*, 1738–1751. [CrossRef]

51. Urquhart, W.; Gunawardena, A.H.L.A.N.; Moeder, W.; Ali, R.; Berkowitz, G.A.; Yoshioka, K. The chimeric cyclic nucleotide-gated ion channel ATCNGC11/12 constitutively induces programmed cell death in a Ca^{2+} dependent manner. *Plant Mol. Biol.* **2007**, *65*, 747–761. [CrossRef] [PubMed]

52. Yoshioka, K.; Kachroo, P.; Tsui, F.; Sharma, S.B.; Shah, J.; Klessig, D.F. Environmentally sensitive, SA-dependent defense responses in the *cpr22* mutant of Arabidopsis. *Plant J.* **2001**, *26*, 447–459. [CrossRef] [PubMed]

53. Yoshioka, K.; Moeder, W.; Kang, H.-G.; Kachroo, P.; Masmoudi, K.; Berkowitz, G.; Klessig, D.F. The chimeric *Arabidopsis* CYCLIC NUCLEOTIDE-GATED ION CHANNEL11/12 activates multiple pathogen resistance responses. *Plant Cell* **2006**, *18*, 747–763. [CrossRef] [PubMed]

54. Dindas, J.; Scherzer, S.; Roelfsema, M.R.G.; von Meyer, K.; Müller, H.M.; Al-Rasheid, K.A.S.; Palme, K.; Dietrich, P.; Becker, D.; Bennett, M.J.; et al. AUX1-mediated root hair auxin influx governs $SCF^{TIR1/AFB}$-type Ca^{2+} signaling. *Nat. Commun.* **2018**, *9*, 1174. [CrossRef]

55. Shih, H.-W.; DePew, C.L.; Miller, N.D.; Monshausen, G.B. The cyclic nucleotide-gated channel CNGC14 regulates root gravitropism in *Arabidopsis thaliana*. *Curr. Biol.* **2015**, *25*, 3119–3125. [CrossRef] [PubMed]

56. Zeb, Q.; Wang, X.; Hou, C.; Zhang, X.; Dong, M.; Zhang, S.; Zhang, Q.; Ren, Z.; Tian, W.; Zhu, H.; et al. The interaction of CaM7 and CNGC14 regulates root hair growth in Arabidopsis. *J. Integr. Plant Biol.* **2020**, *62*, 887–896. [CrossRef]

57. Zhang, S.; Pan, Y.; Tian, W.; Dong, M.; Zhu, H.; Luan, S.; Li, L. *Arabidopsis* CNGC14 mediates calcium influx required for tip growth in root hairs. *Mol. Plant* **2017**, *10*, 1004–1006. [CrossRef]

58. Leitão, N.; Dangeville, P.; Carter, R.; Charpentier, M. Nuclear calcium signatures are associated with root development. *Nat. Commun.* **2019**, *10*, 1–9. [CrossRef]

59. Tunc-Ozdemir, M.; Tang, C.; Ishka, M.R.; Brown, E.; Groves, N.R.; Myers, C.T.; Rato, C.; Poulsen, L.R.; McDowell, S.; Miller, G.; et al. A cyclic nucleotide-gated channel (CNGC16) in pollen is critical for stress tolerance in pollen reproductive development. *Plant Physiol.* **2013**, *161*, 1010–1020. [CrossRef] [PubMed]

60. Ladwig, F.; Dahlke, R.I.; Stührwohldt, N.; Hartmann, J.; Harter, K.; Sauter, M. Phytosulfokine regulates growth in Arabidopsis through a response module at the plasma membrane that includes CYCLIC NUCLEOTIDE-GATED CHANNEL17, H$^+$-ATPase, and BAK1. *Plant Cell* **2015**, *27*, 1718–1729. [CrossRef]

61. Chang, F.; Yan, A.; Zhao, L.-N.; Wu, W.-H.; Yang, Z. A putative calcium-permeable cyclic nucleotide-gated channel, CNGC18, regulates polarized pollen tube growth. *J. Integr. Plant Biol.* **2007**, *49*, 1261–1270. [CrossRef]

62. Frietsch, S.; Wang, Y.-F.; Sladek, C.; Poulsen, L.R.; Romanowsky, S.M.; Schroeder, J.I.; Harper, J.F. A cyclic nucleotide-gated channel is essential for polarized tip growth of pollen. *Proc. Natl. Acad. Sci. USA* **2007**, *104*, 14531–14536. [CrossRef] [PubMed]

63. Zhou, L.; Lan, W.; Jiang, Y.; Fang, W.; Luan, S. A calcium-dependent protein kinase interacts with and activates a calcium channel to regulate pollen tube growth. *Mol. Plant* **2014**, *7*, 369–376. [CrossRef] [PubMed]

64. Jogawat, A.; Meena, M.K.; Kundu, A.; Varma, M.; Vadassery, J. Calcium channel CNGC19 mediates basal defense signaling to regulate colonization by *Piriformospora indica* in Arabidopsis roots. *J. Exp. Bot.* **2020**, *71*, 2752–2768. [CrossRef] [PubMed]

65. Kugler, A.; Köhler, B.; Palme, K.; Wolff, P.; Dietrich, P. Salt-dependent regulation of a CNG channel subfamily in Arabidopsis. *BMC Plant Biol.* **2009**, *9*, 140. [CrossRef] [PubMed]

66. Meena, M.K.; Prajapati, R.; Krishna, D.; Divakaran, K.; Pandey, Y.; Reichelt, M.; Mathew, M.K.; Boland, W.; Mithöfer, A.; Vadassery, J. The Ca^{2+} channel CNGC19 regulates Arabidopsis defense against spodoptera herbivory. *Plant Cell* **2019**, *31*, 1539–1562. [CrossRef] [PubMed]

67. Yu, X.; Xu, G.; Li, B.; de Souza Vespoli, L.; Liu, H.; Moeder, W.; Chen, S.; de Oliveira, M.V.V.; Ariádina de Souza, S.; Shao, W.; et al. The receptor kinases BAK1/SERK4 regulate Ca^{2+} channel-mediated cellular homeostasis for cell death containment. *Curr. Biol.* **2019**, *29*, 3778–3790. [CrossRef]

68. Kaupp, U.B.; Seifert, R. Cyclic nucleotide-gated ion channels. *Physiol. Rev.* **2002**, *82*, 769–824. [CrossRef]

69. Liu, D.T.; Tibbs, G.R.; Siegelbaum, S.A. Subunit stoichiometry of cyclic nucleotide-gated channels and effects of subunit order on channel function. *Neuron* **1996**, *16*, 983–990. [CrossRef]

70. Michalakis, S.; Becirovic, E.; Biel, M. Retinal cyclic nucleotide-gated channels: From pathophysiology to therapy. *Int. J. Mol. Sci.* **2018**, *19*, 749. [CrossRef]

71. Shuart, N.G.; Haitin, Y.; Camp, S.S.; Black, K.D.; Zagotta, W.N. Molecular mechanism for 3:1 subunit stoichiometry of rod cyclic nucleotide-gated ion channels. *Nat. Commun.* **2011**, *2*, 457. [CrossRef]

72. Dreyer, I.; Porée, F.; Schneider, A.; Mittelstädt, J.; Bertl, A.; Sentenac, H.; Thibaud, J.-B.; Mueller-Roeber, B. Assembly of plant *Shaker*-like K$_{out}$ channels requires two distinct sites of the channel α-subunit. *Biophys. J.* **2004**, *87*, 858–872. [CrossRef] [PubMed]

73. Lebaudy, A.; Pascaud, F.; Véry, A. A.; Alcon, C.; Dreyer, I.; Thibaud, J.-B.; Lacombe, B. Preferential KAT1-KAT2 heteromerization determines inward K$^+$ current properties in *Arabidopsis* guard cells. *J. Biol. Chem.* **2010**, *285*, 6265–6274. [CrossRef] [PubMed]

74. Naso, A.; Dreyer, I.; Pedemonte, L.; Testa, I.; Gomez-Porras, J.L.; Usai, C.; Mueller-Rueber, B.; Diaspro, A.; Gambale, F.; Picco, C. The role of the C-terminus for functional heteromerization of the plant channel KDC1. *Biophys. J.* **2009**, *96*, 4063–4074. [CrossRef] [PubMed]

75. Nieves-Cordones, M.; Chavanieu, A.; Jeanguenin, L.; Alcon, C.; Szponarski, W.; Estaran, S.; Chérel, I.; Zimmermann, S.; Sentenac, H.; Gaillard, I. Distinct amino acids in the C-linker domain of the Arabidopsis K$^+$ channel KAT2 determine its subcellular localization and activity at the plasma membrane. *Plant Physiol.* **2014**, *164*, 1415–1429. [CrossRef] [PubMed]

76. Wang, X.-P.; Chen, L.-M.; Liu, W.-X.; Shen, L.-K.; Wang, F.-L.; Zhou, Y.; Zhang, Z.; Wu, W.-H.; Wang, Y. AtKC1 and CIPK23 synergistically modulate AKT1-mediated low-potassium stress responses in Arabidopsis. *Plant Physiol.* **2016**, *170*, 2264–2277. [CrossRef]

77. Charpentier, M.; Sun, J.; Martins, T.V.; Radhakrishnan, G.V.; Findlay, K.; Soumpourou, E.; Thouin, J.; Véry, A.-A.; Sanders, D.; Morris, R.J.; et al. Nuclear-localized cyclic nucleotide–gated channels mediate symbiotic calcium oscillations. *Science* **2016**, *352*, 1102–1105. [CrossRef] [PubMed]

78. Zelman, A.K.; Dawe, A.; Gehring, C.; Berkowitz, G.A. Evolutionary and structural perspectives of plant cyclic nucleotide-gated cation channels. *Front. Plant Sci.* **2012**, *3*, 95. [CrossRef]

79. Wang, L.; Li, M.; Liu, Z.; Dai, L.; Zhang, M.; Wang, L.; Zhao, J.; Liu, M. Genome-wide identification of *CNGC* genes in Chinese jujube (*Ziziphus jujuba* Mill.) and *ZjCNGC2* mediated signalling cascades in response to cold stress. *BMC Genomics* **2020**, *21*, 191–216. [CrossRef]

80. Christopher, D.A.; Borsics, T.; Yuen, C.Y.; Ullmer, W.; Andème-Ondzighi, C.; Andres, M.A.; Kang, B.-H.; Staehelin, L.A. The cyclic nucleotide gated cation channel AtCNGC10 traffics from the ER via Golgi vesicles to the plasma membrane of Arabidopsis root and leaf cells. *BMC Plant Biol.* **2007**, *7*, 48. [CrossRef]

81. Baxter, J.; Moeder, W.; Urquhart, W.; Shahinas, D.; Chin, K.; Christendat, D.; Kang, H.-G.; Angelova, M.; Kato, N.; Yoshioka, K. Identification of a functionally essential amino acid for Arabidopsis cyclic nucleotide gated ion channels using the chimeric *AtCNGC11/12* gene. *Plant J.* **2008**, *56*, 457–469. [CrossRef] [PubMed]

82. Fischer, C.; Kugler, A.; Hoth, S.; Dietrich, P. An IQ domain mediates the interaction with calmodulin in a plant cyclic nucleotide-gated channel. *Plant Cell Physiol.* **2013**, *54*, 573–584. [CrossRef] [PubMed]

83. Yuen, C.C.Y.; Christopher, D.A. The group IV-A cyclic nucleotide-gated channels, CNGC19 and CNGC20, localize to the vacuole membrane in *Arabidopsis thaliana*. *AoB PLANTS* **2013**, *5*. [CrossRef]

84. Lemtiri-Chlieh, F.; Berkowitz, G.A. Cyclic adenosine monophosphate regulates calcium channels in the plasma membrane of *Arabidopsis* leaf guard and mesophyll cells. *J. Biol. Chem.* **2004**, *279*, 35306–35312. [CrossRef]

85. Hua, B.-G.; Mercier, R.W.; Leng, Q.; Berkowitz, G.A. Plants do it differently. A new basis for potassium/sodium selectivity in the pore of an ion channel. *Plant Physiol.* **2003**, *132*, 1353–1361. [CrossRef]

86. Mercier, R.W.; Rabinowitz, N.M.; Ali, R.; Gaxiola, R.A.; Berkowitz, G.A. Yeast hygromycin sensitivity as a functional assay of cyclic nucleotide gated cation channels. *Plant Physiol. Biochem.* **2004**, *42*, 529–536. [CrossRef]

87. Ali, R.; Zielinski, R.E.; Berkowitz, G.A. Expression of plant cyclic nucleotide-gated cation channels in yeast. *J. Exp. Bot.* **2006**, *57*, 125–138. [CrossRef]

88. Li, X.; Borsics, T.; Harrington, H.M.; Christopher, D.A. *Arabidopsis* AtCNGC10 rescues potassium channel mutants of *E. coli*, yeast and *Arabidopsis* and is regulated by calcium/calmodulin and cyclic GMP in *E. coli*. *Funct. Plant Biol.* **2005**, *32*, 643–653. [CrossRef]

89. Wang, J.; Liu, X.; Zhang, A.; Ren, Y.; Wu, F.; Wang, G.; Xu, Y.; Lei, C.; Zhu, S.; Pan, T.; et al. A cyclic nucleotide-gated channel mediates cytoplasmic calcium elevation and disease resistance in rice. *Cell Res.* **2019**, *29*, 820–831. [CrossRef]

90. Xu, Y.; Yang, J.; Wang, Y.; Wang, J.; Yu, Y.; Long, Y.; Wang, Y.; Zhang, H.; Ren, Y.; Chen, J.; et al. OsCNGC13 promotes seed-setting rate by facilitating pollen tube growth in stylar tissues. *PLoS Genet.* **2017**, *13*, e1006906. [CrossRef]

91. Schuurink, R.C.; Shartzer, S.F.; Fath, A.; Jones, R.L. Characterization of a calmodulin-binding transporter from the plasma membrane of barley aleurone. *Proc. Natl. Acad. Sci. USA* **1998**, *95*, 1944–1949. [CrossRef] [PubMed]

92. Köhler, C.; Merkle, T.; Neuhaus, G. Characterisation of a novel gene family of putative cyclic nucleotide- and calmodulin-regulated ion channels in *Arabidopsis thaliana*. *Plant J.* **1999**, *18*, 97–104. [CrossRef] [PubMed]

93. Arazi, T.; Sunkar, R.; Kaplan, B.; Fromm, H. A tobacco plasma membrane calmodulin-binding transporter confers Ni^{2+} tolerance and Pb^{2+} hypersensitivity in transgenic plants. *Plant J.* **1999**, *20*, 171–182. [CrossRef] [PubMed]

94. Zheng, X.; Fu, Z.; Su, D.; Zhang, Y.; Li, M.; Pan, Y.; Li, H.; Li, S.; Grassucci, R.A.; Ren, Z.; et al. Mechanism of ligand activation of a eukaryotic cyclic nucleotide-gated channel. *Nat. Struct. Biol.* **2020**, *27*, 625–634. [CrossRef]

95. Jammes, F.; Hu, H.-C.; Villiers, F.; Bouten, R.; Kwak, J.M. Calcium-permeable channels in plant cells. *FEBS J.* **2011**, *278*, 4262–4276. [CrossRef]

96. Lemtiri-Chlieh, F.; Arold, S.T.; Gehring, C. Mg^{2+} is a missing link in plant cell Ca^{2+} signalling and homeostasis—a study on *Vicia faba* guard cells. *Int. J. Mol. Sci.* **2020**, *21*, 3771. [CrossRef]

97. Hua, B.-G.; Mercier, R.W.; Zielinski, R.E.; Berkowitz, G.A. Functional interaction of calmodulin with a plant cyclic nucleotide gated cation channel. *Plant Physiol. Biochem.* **2003**, *41*, 945–954. [CrossRef]

98. Zagotta, W.N.; Siegelbaum, S.A. Structure and function of cyclic nucleotide-gated channels. *Annu. Rev. Neurosci.* **1996**, *19*, 235–263. [CrossRef]

99. Demidchik, V.; Straltsova, D.; Medvedev, S.S.; Pozhvanov, G.A.; Sokolik, A.; Yurin, V. Stress-induced electrolyte leakage: The role of K^+-permeable channels and involvement in programmed cell death and metabolic adjustment. *J. Exp. Bot.* **2014**, *65*, 1259–1270. [CrossRef]

100. Demidchik, V.; Shabala, S.N.; Coutts, K.B.; Tester, M.A.; Davies, J.M. Free oxygen radicals regulate plasma membrane Ca^{2+}- and K^+-permeable channels in plant root cells. *J. Cell Sci.* **2003**, *116*, 81–88. [CrossRef]

101. Kaplan, B.; Sherman, T.; Fromm, H. Cyclic nucleotide-gated channels in plants. *FEBS Lett.* **2007**, *581*, 2237–2246. [CrossRef] [PubMed]

102. Isner, J.-C.; Maathuis, F.J.M. cGMP signalling in plants: From enigma to main stream. *Funct. Plant Biol.* **2018**, *45*, 93–101. [CrossRef] [PubMed]

103. Świeżawska, B.; Duszyn, M.; Jaworski, K.; Szmidt-Jaworska, A. Downstream targets of cyclic nucleotides in plants. *Front. Plant Sci.* **2018**, *9*, 1428. [CrossRef] [PubMed]

104. Gross, I.; Durner, J. In search of enzymes with a role in 3′, 5′-cyclic guanosine monophosphate metabolism in plants. *Front. Plant Sci.* **2016**, *7*, 576. [CrossRef] [PubMed]

105. Al-Younis, I.; Wong, A.; Gehring, C. The *Arabidopsis thaliana* K^+-uptake permease 7 (AtKUP7) contains a functional cytosolic adenylate cyclase catalytic centre. *FEBS Lett.* **2015**, *589*, 3848–3852. [CrossRef]

106. Al-Younis, I.; Wong, A.; Lemtiri-Chlieh, F.; Schmöckel, S.; Tester, M.; Gehring, C.; Donaldson, L. The *Arabidopsis thaliana* K^+-Uptake Permease 5 (AtKUP5) contains a functional cytosolic adenylate cyclase essential for K^+ transport. *Front. Plant Sci.* **2018**, *9*, 1645. [CrossRef]

107. Bianchet, C.; Wong, A.; Quaglia, M.; Alqurashi, M.; Gehring, C.; Ntoukakis, V.; Pasqualini, S. An *Arabidopsis thaliana* leucine-rich repeat protein harbors an adenylyl cyclase catalytic center and affects responses to pathogens. *J. Plant Physiol.* **2019**, *232*, 12–22. [CrossRef]

108. Ludidi, N.; Gehring, C. Identification of a novel protein with guanylyl cyclase activity in *Arabidopsis thaliana*. *J. Biol. Chem.* **2003**, *278*, 6490–6494. [CrossRef]

109. Mulaudzi, T.; Ludidi, N.; Ruzvidzo, O.; Morse, M.; Hendricks, N.; Iwuoha, E.; Gehring, C. Identification of a novel *Arabidopsis thaliana* nitric oxide-binding molecule with guanylate cyclase activity in vitro. *FEBS Lett.* **2011**, *585*, 2693–2697. [CrossRef]

110. Kwezi, L.; Ruzvidzo, O.; Wheeler, J.I.; Govender, K.; Iacuone, S.; Thompson, P.E.; Gehring, C.; Irving, H.R. The phytosulfokine (PSK) receptor is capable of guanylate cyclase activity and enabling cyclic GMP-dependent signaling in plants. *J. Biol. Chem.* **2011**, *286*, 22580–22588. [CrossRef]

111. Kwezi, L.; Meier, S.; Mungur, L.; Ruzvidzo, O.; Irving, H.; Gehring, C. The *Arabidopsis thaliana* Brassinosteroid Receptor (AtBRI1) contains a domain that functions as a guanylyl cyclase *in vitro*. *PLoS ONE* **2007**, *2*, e449. [CrossRef] [PubMed]

112. Meier, S.; Ruzvidzo, O.; Morse, M.; Donaldson, L.; Kwezi, L.; Gehring, C. The Arabidopsis *Wall Associated Kinase-Like 10* gene encodes a functional guanylyl cyclase and is co-expressed with pathogen defense related genes. *PLoS ONE* **2010**, *5*, e8904. [CrossRef] [PubMed]

113. Rahman, H.; Wang, X.-Y.; Xu, Y.-P.; He, Y.-H.; Cai, X.-Z. Characterization of tomato protein kinases embedding guanylate cyclase catalytic center motif. *Sci. Rep.* **2020**, *10*, 1–16. [CrossRef] [PubMed]

114. Muleya, V.; Wheeler, J.I.; Ruzvidzo, O.; Freihat, L.; Manallack, D.T.; Gehring, C.; Irving, H.R. Calcium is the switch in the moonlighting dual function of the ligand-activated receptor kinase phytosulfokine receptor 1. *Cell Commun. Signal.* **2014**, *12*, 60. [CrossRef] [PubMed]

115. Sun, X.; Qiu, Y.; Peng, Y.; Ning, J.; Song, G.; Yang, Y.; Deng, M.; Men, Y.; Zhao, X.; Wang, Y.; et al. Close temporal relationship between oscillating cytosolic K^+ and growth in root hairs of *Arabidopsis*. *Int. J. Mol. Sci.* **2020**, *21*, 6184. [CrossRef]

116. Ahn, S.J.; Shin, R.; Schachtman, D.P. Expression of *KT/KUP* genes in Arabidopsis and the role of root hairs in K^+ uptake. *Plant Physiol.* **2004**, *134*, 1135–1145. [CrossRef]

117. Han, M.; Wu, W.; Wu, W.-H.; Wang, Y. Potassium transporter KUP7 is involved in K^+ acquisition and translocation in *Arabidopsis* root under K^+-limited conditions. *Mol. Plant* **2016**, *9*, 437–446. [CrossRef]

118. Donaldson, L.; Ludidi, N.; Knight, M.R.; Gehring, C.; Denby, K. Salt and osmotic stress cause rapid increases in *Arabidopsis thaliana* cGMP levels. *FEBS Lett.* **2004**, *569*, 317–320. [CrossRef]

119. Shabala, S.; Wu, H.; Bose, J. Salt stress sensing and early signalling events in plant roots: Current knowledge and hypothesis. *Plant Sci.* **2015**, *241*, 109–119. [CrossRef]

120. Maathuis, F.J.M.; Sanders, D. Sodium uptake in Arabidopsis roots is regulated by cyclic nucleotides. *Plant Physiol.* **2001**, *127*, 1617–1625. [CrossRef]

121. Demidchik, V.; Tester, M. Sodium fluxes through nonselective cation channels in the plasma membrane of protoplasts from Arabidopsis roots. *Plant Physiol.* **2002**, *128*, 379–387. [CrossRef] [PubMed]

122. Demidchik, V.; Shabala, S.; Isayenkov, S.; Cuin, T.A.; Pottosin, I. Calcium transport across plant membranes: Mechanisms and functions. *New Phytol.* **2018**, *220*, 49–69. [CrossRef] [PubMed]

123. Demidchik, V.; Bowen, H.C.; Maathuis, F.J.M.; Shabala, S.N.; Tester, M.A.; White, P.J.; Davies, J.M. *Arabidopsis thaliana* root non-selective cation channels mediate calcium uptake and are involved in growth. *Plant J.* **2002**, *32*, 799–808. [CrossRef] [PubMed]

124. Kiegle, E.; Moore, C.A.; Haseloff, J.; Tester, M.A.; Knight, M.R. Cell-type-specific calcium responses to drought, salt and cold in the *Arabidopsis* root. *Plant J.* **2000**, *23*, 267–278. [CrossRef] [PubMed]

125. Niu, W.-T.; Han, X.-W.; Wei, S.-S.; Shang, Z.-L.; Wang, J.; Yang, D.-W.; Fan, X.; Gao, F.; Zheng, S.-Z.; Bai, J.-T.; et al. Arabidopsis cyclic nucleotide-gated channel 6 is negatively modulated by multiple calmodulin isoforms during heat shock. *J. Exp. Bot.* **2020**, *71*, 90–104. [CrossRef] [PubMed]

126. Lu, D.; Wu, S.; Gao, X.; Zhang, Y.; Shan, L.; He, P. A receptor-like cytoplasmic kinase, BIK1, associates with a flagellin receptor complex to initiate plant innate immunity. *Proc. Natl. Acad. Sci. USA* **2010**, *107*, 496–501. [CrossRef] [PubMed]

127. Gómez-Gómez, L.; Boller, T. FLS2: An LRR receptor–like kinase involved in the perception of the bacterial elicitor flagellin in *Arabidopsis*. *Mol. Cell* **2000**, *5*, 1003–1011. [CrossRef]

128. Ma, Y.; Walker, R.K.; Zhao, Y.; Berkowitz, G.A. Linking ligand perception by PEPR pattern recognition receptors to cytosolic Ca^{2+} elevation and downstream immune signaling in plants. *Proc. Natl. Acad. Sci. USA* **2012**, *109*, 19852–19857. [CrossRef]

129. Jeworutzki, E.; Roelfsema, M.R.G.; Anschütz, U.; Krol, E.; Elzenga, J.T.M.; Felix, G.; Boller, T.; Hedrich, R.; Becker, D. Early signaling through the Arabidopsis pattern recognition receptors FLS2 and EFR involves Ca^{2+}-associated opening of plasma membrane anion channels. *Plant J.* **2010**, *62*, 367–378. [CrossRef]

130. Wang, F.-Z.; Zhang, N.; Guo, Y.-J.; Gong, B.-Q.; Li, J.-F. Split nano luciferase complementation for probing protein-protein interactions in plant cells. *J. Integr. Plant Biol.* **2020**, *62*, 1065–1079. [CrossRef]

131. Thor, K.; Peiter, E. Cytosolic calcium signals elicited by the pathogen-associated molecular pattern flg22 in stomatal guard cells are of an oscillatory nature. *New Phytol.* **2014**, *204*, 873–881. [CrossRef] [PubMed]

132. Thor, K.; Jiang, S.; Michard, E.; George, J.; Scherzer, S.; Huang, S.; Dindas, J.; Derbyshire, P.; Leitão, N.; DeFalco, T.A.; et al. The calcium-permeable channel OSCA1.3 regulates plant stomatal immunity. *Nature* **2020**, *585*, 569–573. [CrossRef] [PubMed]

133. Zhang, X.-R.; Xu, Y.-P.; Cai, X.-Z. *SlCNGC1* and *SlCNGC14* suppress *Xanthomonas oryzae* pv. *oryzicola*-induced hypersensitive response and non-host resistance in tomato. *Front. Plant Sci.* **2018**, *9*, 9. [CrossRef]
134. Edel, K.H.; Marchadier, E.; Brownlee, C.; Kudla, J.; Hetherington, A.M. The evolution of calcium-based signalling in plants. *Curr. Biol.* **2017**, *27*, 667–679. [CrossRef] [PubMed]
135. O'Halloran, D.M.; Altshuler-Keylin, S.; Zhang, X.-D.; He, C.; Morales-Phan, C.; Yu, Y.; Kaye, J.A.; Brueggemann, C.; Chen, T.-Y.; L'Etoile, N.D. Contribution of the cyclic nucleotide gated channel subunit, CNG-3, to olfactory plasticity in *Caenorhabditis elegans*. *Sci. Rep.* **2017**, *7*, 169. [CrossRef]
136. Wilson, C.M.; Stecyk, J.A.W.; Couturier, C.S.; Nilsson, G.E.; Farrell, A.P. Phylogeny and effects of anoxia on hyperpolarization-activated cyclic nucleotide-gated channel gene expression in the heart of a primitive chordate, the Pacific hagfish (*Eptatretus stoutii*). *J. Exp. Biol.* **2013**, *216*, 4462–4472. [CrossRef]
137. Chiasson, D.M.; Haage, K.; Sollweck, K.; Brachmann, A.; Dietrich, P.; Parniske, M. A quantitative hypermorphic *CNGC* allele confers ectopic calcium flux and impairs cellular development. *eLife* **2017**, *6*. [CrossRef]
138. Lelle, M.; Otte, M.; Bonus, M.; Gohlke, H.; Benndorf, K. Fluorophore-labeled cyclic nucleotides as potent agonists of cyclic nucleotide-regulated ion channels. *ChemBioChem* **2020**, *21*, 2311–2320. [CrossRef]

International Journal of
Molecular Sciences

Article

Molecular Analysis of *14-3-3* Genes in *Citrus sinensis* and Their Responses to Different Stresses

Shiheng Lyu [1,2], Guixin Chen [1], Dongming Pan [1], Jianjun Chen [2,*] and Wenqin She [1,*]

[1] College of Horticulture, Fujian Agriculture and Forestry University, Fuzhou 350002, Fujian, China; 2140305002@fafu.edu.cn (S.L.); 1170371005@fafu.edu.cn (G.C.); pdm666@fafu.edu.cn (D.P.)

[2] Mid-Florida Research and Education Center, Department of Environmental Horticulture, Institute of Food and Agricultural Sciences, University of Florida, Apopka, FL 32703, USA

[*] Correspondence: jjchen@ufl.edu (J.C.); wenqinshe@fafu.edu.cn (W.S.)

Abstract: 14-3-3 proteins (14-3-3s) are among the most important phosphorylated molecules playing crucial roles in regulating plant development and defense responses to environmental constraints. No report thus far has documented the gene family of *14-3-3s* in *Citrus sinensis* and their roles in response to stresses. In this study, nine *14-3-3* genes, designated as *CitGF14s* (*CitGF14a* through *CitGF14i*) were identified from the latest *C. sinensis* genome. Phylogenetic analysis classified them into ε-like and non-ε groups, which were supported by gene structure analysis. The nine *CitGF14s* were located on five chromosomes, and none had duplication. Publicly available RNA-Seq raw data and microarray databases were mined for 14-3-3 expression profiles in different organs of citrus and in response to biotic and abiotic stresses. RT-qPCR was used for further examining spatial expression patterns of *CitGF14s* in citrus and their temporal expressions in one-year-old *C. sinensis* "Xuegan" plants after being exposed to different biotic and abiotic stresses. The nine *CitGF14s* were expressed in eight different organs with some isoforms displayed tissue-specific expression patterns. Six of the *CitGF14s* positively responded to citrus canker infection (*Xanthomonas axonopodis* pv. *citri*). The *CitGF14s* showed expressional divergence after phytohormone application and abiotic stress treatments, suggesting that 14-3-3 proteins are ubiquitous regulators in *C. sinensis*. Using the yeast two-hybrid assay, CitGF14a, b, c, d, g, and h were found to interact with CitGF14i proteins to form a heterodimer, while CitGF14i interacted with itself to form a homodimer. Further analysis of *CitGF14s* co-expression and potential interactors established a 14-3-3s protein interaction network. The established network identified *14-3-3* genes and several candidate clients which may play an important role in developmental regulation and stress responses in this important fruit crop. This is the first study of 14-3-3s in citrus, and the established network may help further investigation of the roles of 14-3-3s in response to abiotic and biotic constraints.

Keywords: abiotic stress; CitGF14s; citrus canker; *Citrus sinensis*; sweet orange; 14-3-3s

Citation: Lyu, S.; Chen, G.; Pan, D.; Chen, J.; She, W. Molecular Analysis of *14-3-3* Genes in *Citrus sinensis* and Their Responses to Different Stresses. *Int. J. Mol. Sci.* **2021**, *22*, 568. https://doi.org/10.3390/ijms22020568

Received: 15 December 2020
Accepted: 5 January 2021
Published: 8 January 2021

1. Introduction

Plants are constantly exposed to different abiotic and biotic stresses, including drought, extreme temperatures, high salinity, and various pathogens. Due to their sessile nature, plants have evolved a series of mechanisms to cope with the environmental challenges. Plant 14-3-3s, encoded by genes called general regulatory factors [1,2], regulate critical biochemical processes and sophisticated signaling networks in plants though protein–protein interactions by binding to phosphorylated protein clients [3]. 14-3-3 proteins were originally isolated from mammalian brain tissue and were named according to their elution and migration pattern on DEAD-cellulose chromatography and starch-gel electrophoresis [4]. The 14-3-3s are highly conserved proteins and exist in all eukaryotes with multiple isoforms. Yeast has two genes encoding 14-3-3s [5], animals typically have seven [6], and plants have more *14-3-3* genes: 13 in *Arabidopsis* [6], 17 in tobacco (*Nicotiana tabacum*) [7], 12 in tomato (*Solanum lycopersicum*) [8], and eight in rice (*Oryza sativa*) [9].

The protein sequence of 14-3-3s can be divided into three regions: A variable N-terminus, a conserved core region, and a variable C-terminus. Based on the gene structure, plant 14-3-3 proteins are divided into two distinct groups, namely epsilon (ε) and non-epsilon, and the latter is plant-specific [2,10,11]. Crystal structure studies show that 14-3-3 dimers consist of a typical clamp shape structure containing alpha helical amphipathic grooves formed by a monomer [12,13]. The monomer can interact with phosphorylated proteins; thus, the groove is the main target binding site [14]. Large-scale interactomics and mass-spectrometry-based studies have identified more than 300 potential 14-3-3 targets in plant [3]. There are about a dozen 14-3-3 proteins forming homodimers and heterodimers that function to reverse phosphorylation of proteins in plants [15,16]. Three canonical phosphorylation-dependent 14-3-3 binding motifs can be recognized by all isoforms: RSX-pSXP (mode-I), RXXXpSXP (mode-II), and pS/pTX1–2-COOH (mode-III) (where R, S, and P represents arginine, serine, and proline, X is any amino acid, pS is phosphoserine) [12,17,18]. Phosphorylation is an essential posttranslational modification, which is fast and reversible and affects thousands of proteins and regulates a plethora of different processes in plants [19]. Protein phosphorylation occurs mainly on serine (pS), threonine (pT), and tyrosine (pY) residues. The 14-3-3 proteins can also bind non-phosphorylated targets, such as WLDLE [20] and GHSL [20,21].

A growing body of evidence indicates that the 14-3-3s can regulate plant responses to abiotic and biotic stresses [16,19,22–24]. 14-3-3s play important roles in plant tolerance to salinity and drought. *Arabidopsis* 14-3-3s κ and λ have been reported to inhibit the SOS (salt overly sensitive) pathway by repressing SOS2 kinase activity in the absence of salt stress [25]. Rice 14-3-3 family genes were named as *GF14a* through *GF14h*, and four members (*GF14b, GF14c, GF14e,* and *GF14f*) were all induced by PEG6000 (drought-mimic) treatments [26]. The 14-3-3s have been reported to regulate plant cold tolerance. RARE COLD INDUCIBLE 1A (RCI1A) and RCI1B were the first two 14-3-3 proteins that were demonstrated to be induced by cold stress in *Arabidopsis* [27]. The kinetics of RCI1A and RCI1B mRNA accumulation induced by cold stress is correlated with the increased freezing tolerance that occurs during the cold acclimation process in *Arabidopsis*, implying that these genes play pivotal roles in this adaptive process [27,28]. The 14-3-3 proteins are also involved in regulation of nutrient stress, such as low phosphorus stress [29], iron deficiency [30], wounding [28], and ABA signal [31]. 14-3-3 proteins respond to pathogen infection by changing transcript levels or, in some instances, protein abundance or properties, or both [19]. Tomato *TFT1, TFT4,* and *TFT6* genes were upregulated in the Cf-9-mediated hypersensitive response (HR) [32]. Similarly, *14-3-3* genes are expressed during a race-specific HR of soybean inoculated with *Pseudomonas syringae* [33] and upon a resistant reaction to the soybean cyst nematode [34]. Tobacco 14-3-3 isoform h is induced after inoculation with tobacco mosaic virus (TMV) [7]. A *Gossypium hirsutum 14-3-3* is rapidly expressed in response to *Verticillium dahliae* [35] in a cultivar with enhanced wilt resistance, which may suggest a specific role for *14-3-3* in resistance to the pathogen.

Citrus fruits are among the highest value fruit crops in terms of nutritional components and international trade. Citrus crop production constantly encounters both abiotic and biotic stresses, such as drought, salinity, cold, and pathogens, which have significantly affected citrus production worldwide. A better understanding of citrus responses to these constraints will improve breeding strategies and production practices for increased resistance or tolerance to stresses. Plant *14-3-3s* as general regulatory factors may play important roles in citrus responses to these stresses. The genome-wide analysis of *14-3-3* family genes has been identified from various plants, including *Arabidopsis* [6], soybean (*Glycine max*) [36], common bean (*Phaseolus vulgaris*) [37], rice [26], black cottonwood (*Populus trichocarpa*) [38], and foxtail millet (*Setaria italica*) [39]. Up to now, there have been no reports on *14-3-3* family genes in citrus.

In this study, we report a comprehensive genomic identification and phylogenetic analysis of nine members of the *14-3-3* gene family in sweet orange (*Citrus sinensis*) and document their expression profiles in different organs and their responses to abiotic and biotic stresses as well as hormone signal. Our results for the first time provide fundamental information about *14-3-3* genes and their responses to stresses in this citrus species.

2. Results

2.1. 14-3-3 Identification, Phylogenetic Analysis, and Function Prediction

A total of 13 putative *14-3-3* genes were identified from the whole genome of *C. sinensis* using the *14-3-3* genes from *Arabidopsis*, soybean, and black cottonwood as queries. After removing incomplete and redundant sequences, nine *14-3-3* genes were confirmed. They were designated as *CitGF14* (*CitGF14a* through *CitGF14i*) (Table 1). Their open reading frames ranged from 741 to 798 bp encoding 247 to 266 amino acids with putative MW varying from 27.9 to 30.2 kDa. The theoretical isoelectric points ranged from 4.69 to 5.14. Corresponding proteins were predicted to localize in the cytoplasm (cyto), chloroplast (chlo), nucleus plasma (nucl plas), and plasma membrane (plas) depending on individual proteins (Table 1)

Table 1. Identified nine *CitGF14* (*14-3-3*) genes from the whole genome of *Citrus sinensis*.

Name	Gene ID	*Arabidopsis* Orthologue	Chr. No.	Chr. Location	ORF (bp)	Length (aa)	PI	MW (kDa)	Subcellular Localization
CitGF14a	Cs2g04850.1	AT2G42590.1	Chr 2	2,529,656–2,533,444	774	258	4.69	29.433	Cyto
CitGF14b	Cs2g15550.4	AT2G42590.2	Chr 2	12,364,752–12,368,242	798	266	4.72	29.951	Chlo
CitGF14c	Cs3g18200.1	AT1G34760.1	Chr 3	21,756,849–21,761,115	759	253	4.92	28.861	Cyto
CitGF14d	Cs7g11330.1	AT1G26480.1	Chr 7	7,462,893–7,465,189	795	265	5.14	30.227	Cyto
CitGF14e	Cs1g20220.2	AT5G65430.2	Chr 1	23,332,994–23,337,303	741	247	4.83	27.946	Nucl_plas
CitGF14f	Cs3g17470.1	AT5G65430.1	Chr 3	21,178,807–21,181,787	756	252	4.76	28.536	Nucl_plas
CitGF14g	Cs3g17990.1	AT1G78300.1	Chr 3	21,561,464–21,564,313	795	265	4.69	29.742	Nucl_plas
CitGF14h	Cs6g18830.1	AT5G38480.1	Chr 6	18,853,688–18,857,047	789	263	4.75	29.740	Nucl plas
CitGF14i	Or1.1t01991.1	AT1G78300.1	chrUn	31,516,661–31,519,401	783	261	4.84	29.442	Plas

The evolutionary relationships of 44 *14-3-3s* from *Arabidopsis*, rice, black cottonwood, and *C. sinensis* were phylogenetically analyzed. Eighteen of them were clustered into ε-like groups, and 26 were clustered into non-ε groups (Figure 1). Four *CitGF14s* (*CitGF14a, b, c,* and *d*) were grouped into ε-like isoforms, and the remaining five *CitGF14s* were grouped into non-ε isoforms. None of the *CitGF14s* were duplicated.

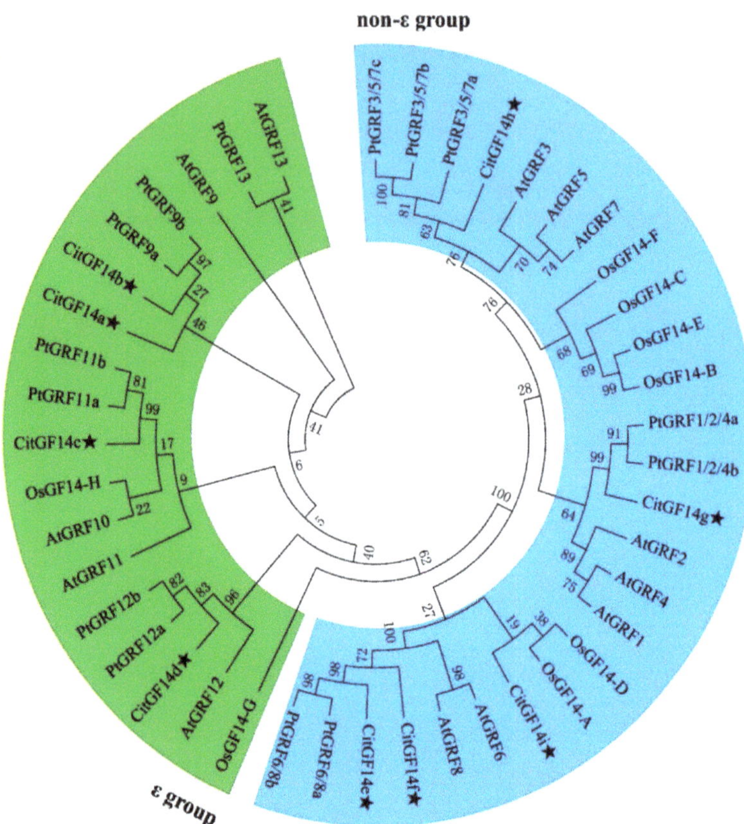

Figure 1. Phylogenetic trees of *14-3-3* family genes in *Citrus sinensis* (*CitGF14s*, indicated by the star symbol), *Arabidopsis* (*AtGRF*), *Populus* trichocarpa (*PtGRF*), and *Oryza sativa* (*OsGRF*) constructed using the MEGA6.0 program with neighbor-joining method within 1000 bootstrap replicates. The green and blue shade separates the ε and non-ε groups.

2.2. Localization in Chromosomes and Gene Structure

Eight *14-3-3* genes (*CitGF14a* through *CitGF14h*) were located on five chromosomes (Chr 1, 2, 3, 6, and 7) of *C. sinensis* (Figure 2), yet *CitGF14i* could not be mapped on a chromosome and remained as unanchored scaffolds. Both *CitGF14a* and *CitGF14b* were situated on chromosome 2; *CitGF14e*, *CitGF14h*, and *CitGF14d* were located on chromosomes 1, 6, and 7, respectively; while *CitGF14c*, *CitGF14f*, and *CitGF14g* were linked on chromosome 3.

Gene structure analysis showed that *CitGR14s* contained 3 to 6 exons, interspersed by highly distinct introns (Figure 3). Four ε-like *CitGF14s* showed six conserved exons interrupted by intron and UTR in different lengths. Non-ε *CitGF14* genes were also interrupted by introns. *CitGF14e*, *CitGF14g*, and *CitGF14i* carried three introns, *CitGF14f* and *CitGF14h* possessed four introns, the remaining had six introns.

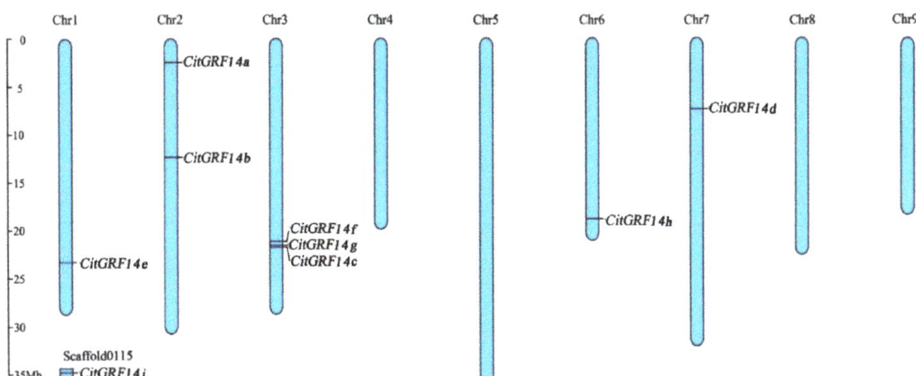

Figure 2. Genomic distribution of *14-3-3* (*CitGF14s*) genes across nine *Citrus sinensis* chromosomes. The chromosome number is indicated at the top of each chromosome. The scale is in megabases (Mb). Chromosomal locations of *CitGF14s* were indicated based on the physical position of each gene.

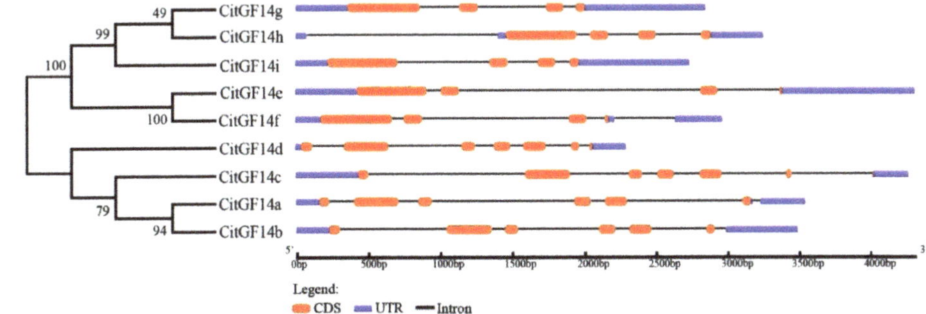

Figure 3. Structure analysis of *CitGF14s* genes in *Citrus sinensis*. The graphic representation is displayed using GSDS (http://gsds.cbi.pku.edu.cn/). The unrooted phylogenetic tree was constructed using the amino sequences of *CitGF14s* genes by the Neighbor-Joining method with 1000 bootstrap replicates. Exons consisted of CDS shown as orange boxes, introns are shown as thin lines, and UTRs are shown as blue boxes.

2.3. CitGF14s Sequence Alignment

Amino acid sequence alignment showed that the deduced *CitGF14s* from *C. sinensis* were highly conserved with the exception of the N-terminal and C-terminal regions. Nine α-helices were shown in green rectangular boxes (Additional file 1: Figure S1A). Two conserved signature motifs RNL(L/V)SV(G/A)YKNV and YKDSTLIMQLLRDNLTLWTS were found in α-helix 3 and α-helix 9, respectively.

Protein CitGF14b, CitGF14d, CitGF14e, CitGF14f, CitGF14g, CitGF14h, and CitGF14i had rather similar three-dimensional structures based on the Swiss-model prediction (Additional file 1: Figure S1B). All these similar proteins contained two potassium channel KAT1 ligands except for CitGF14b. The remaining two proteins (CitGF14a and CitGF14c) had relatively simple structures with no ligands.

2.4. Cis-Regulatory Elements

Different cis-acting elements related to plant growth, development, and stress responses were identified in *CitGR14s* (Additional file 2: Table S1). The circadian element was presented in all *CitGF14* promoters except for *CitGF14d*. The Skn-1 motif required for endosperm expression was highly conserved in six *CitGF14* promoters. The ABA responsive element ABRE was presented in *CitGF14b*, *CitGF14c*, *CitGF14e*, and *CitGF14g*. Box-W1, a fungal elicitor responsive element, was a highly conserved stress-related element which was found in five of nine *CitGF14s*. The upstream flanking regions of *CitGF14g* contained 11 stress-responsive promoters, including ABRE, ARE, CGTCA-motif, ERE, GC-motif, HSE, MBS, P-box, TC-rich repeats, TCA-element, TGA-element, and GACG-motif. Moreover, other cis-acting elements associated with biotic and abiotic stress responses, such as WUN-motif, GARE-motif, AuxRR-core, and SARE were also identified.

2.5. Tissue-Specific Expression Patterns of CitGF14s

The expression of *CitGF14s* in callus, leaf, flower, and fruit, which were mined from the RNA-Seq raw data of *C. sinensis* genome database (http://citrus.hzau.edu.cn/orange/) is presented in Figure 4A. *CitGF14a* and *CitGF14i* were strongly expressed in flowers and leaves. The expression of *CitGF14e* in flower was higher than in callus and then further increased in leaves and fruit. *CitGF14b* expression was induced in callus and flower, slightly decreased in leaves, and then increased in fruit. *CitGF14h* was highly induced and constantly expressed in all four tissues or organs. On the other hand, *CitGF14d* was down regulated in callus, leaves, and fruit. Other *CitGF14s* were either slightly or moderately induced depending on tissue or organ.

The spatial expressions of *CitGF14s* were further analyzed by RT-qPCR in *C. sinensis* "Xuegan" when plants were not exposed to any stresses (Figure 4B). *CitGF14s* were not highly induced in roots and stems except for *CitGF14e* and *CitGF14h* in stems that had over a 1.5-fold increase. The expression of *CitGF14a, c, d, h*, and *i* in shoots and *CitGF14c, e,* and *h* in leaves was highly induced, which was largely similar to those mentioned in above RNA-Seq data with the exception of *CitGF14d* that was primarily down regulated in callus, leaves, and fruit based on the RNA-Seq data (Figure 4A), but it was highly upregulated in flower as well as shoots in the RT-qPCR analysis. Furthermore, the expression of all *CitGF14s* was low in peel, juice, and seeds.

2.6. Responses to Infection of Citrus Canker and Citrus Greening Pathogens

Affymetrix microarrays data were mined in this study for potential roles of *CitGF14* genes in response to citrus canker. Results showed that all citrus probe sets contained less than 10,000 genes, indicating the not all citrus *CitGF14s* had been covered by the microarray data. For example, the probe signal for *CitGF14d* was not detected in the microarray. Based on eight *CitGF14s* from the microarray data, their responses to citrus canker infection are presented in Figure 5A (first four columns). *CitGF14g* was induced by *Xanthomonas axonopodis* pv. *citri* (Xaa) 6 to 48 h after infection and by *Xanthomonas axonopodis* pv. *Aurantifolii* (Xac) 48 h after infection (Figure 5A). The inoculation of Xaa and Xac respectively induced *CitGF14i* expression only 48 h after infection. *CitGF14h* was slightly induced by Xaa and Xac 48 h after infection. The other *CitGF14s* did not respond to the infection of the two pathogens.

The infection of Xac to *C. sinensis* "Xuegan" caused downregulation of *CitGF14a* in 2 to 6 h and then variable expression thereafter till 192 h (Figure 5B). *CitGF14b* was induced 6 h after inoculation, and its expression was then reduced from 12 to 96 h but highly increased at 192 h. *CitGF14d* was highly induced from 48 h to 192 h. *CitGF14g* responded quickly, 2 h after the inoculation, and reached the highest expression level from 48 h to 96 h. There was a downregulation of *CitGF14i* initially, its expression increased from 6 h to 24 h, and attained the highest expression level from 48 h to 192 h. The active responses of *CitGF14g* and *CitGF14i* largely concurred with the above microarray results. The other *CitGF14s* showed varied levels of down or upregulation over the 192-h evaluation period.

Microarray data were also explored for potential roles of *CitGF14s* in response to the infection of citrus greening: *Candidatus* Liberibacter asiaticus (Ca. Las) (Figure 5A, five columns from the right). Compared to healthy organs, there was slight increase in the expressions of *CitGF14c* and *CitGF14f* in leaves and *CitGF14e* in peel. The other genes showed little response to the infection except for *CitGF14a* that was down regulated in leaves and stem and *CitGF14g* and *CitGF14i* that were down regulated in peel.

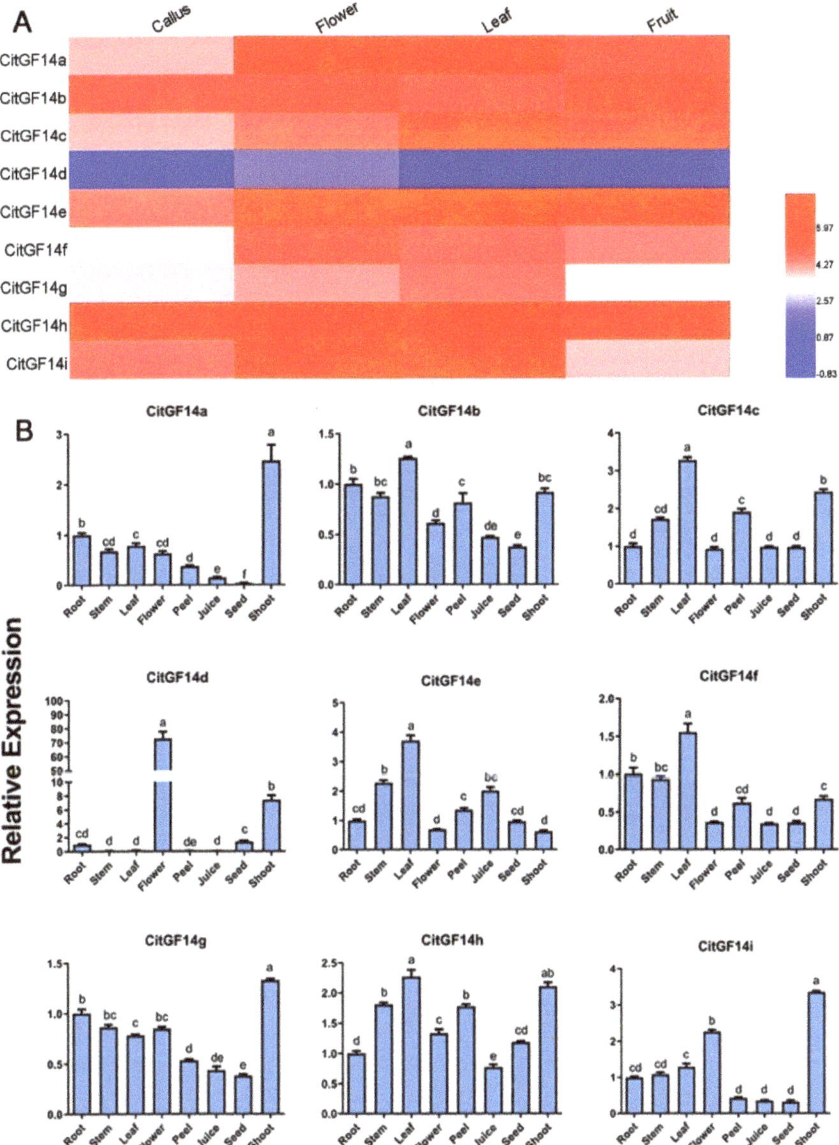

Figure 4. Expression analysis of *CitGF14s* in different tissue and organs of *Citrus sinensis* plants. (**A**) Expression profiles of *CitGF14s* derived from RNA-Seq data where RPKM expression values were log-transformed for normalization. Log2 based RPKM values were used for creating the heat map with clustering by HemI. The scale represents the relative intensity of RPKM values. (**B**) RT-qPCR analysis of the expression patterns of *CitGF14s* in different organs of *C. sinensis*. The relative expression was normalized using the ACTIN and GAPDH genes as references. Each bar represented the mean of four biological replications with standard error. Different letters on the top of bars indicate significant differences analyzed by Tukey's HSD test at $p < 0.05$ level.

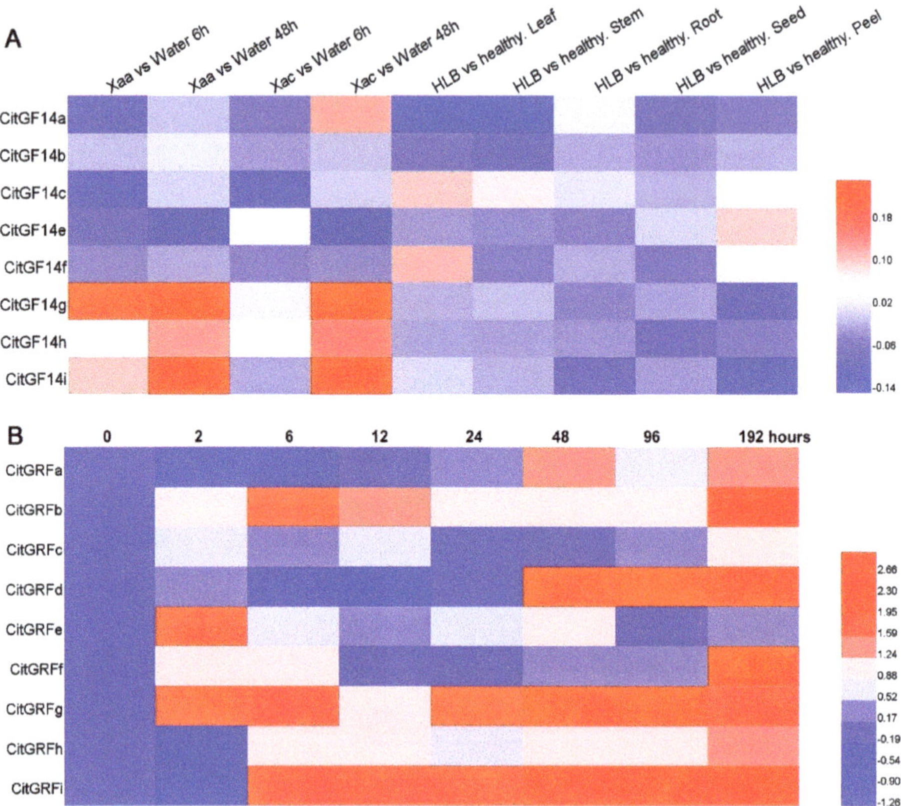

Figure 5. Expression profiles of *CitGF14s* in response to different stresses. (**A**) The microarray data was downloaded from NCBI database. Heatmap shows hierarchical clustering of *CitGF14* expression after plants were inoculated with citrus canker pathogens: *Xanthomonas axonopodis* pv. *citri* (Xaa) and *Xanthomonas axonopodis* pv. *Aurantifolii* (Xac) and citrus greening pathogen: *Candidatus* Liberibacter asiaticus (*Ca.* Las). Gene expression values were calculated based on the ratios between the infection and the mock (control). Heatmap was generated based on log2 (infection expression/mock expression) values by using HemI. The color scale represents the relative intensity level of transcript abundance. (**B**) RT-qPCR analysis of *CitGF14s* expression after *C. sinensis* "Xuegan" was inoculated with Xac. The relative expression was normalized using the ACTIN and GAPDH genes as references using $2^{-\Delta\Delta Ct}$ method. The values were based on the means of four biological replications.

2.7. Responses to Plant Hormone Treatments

Foliar application of jasmonate (JA) on "Xuegan" caused more downregulation of *CitGF14* genes than upregulation (Figure 6A). The upregulation only occurred with *CitGF14a*, *c*, and *f* 12 h after JA application and *CitGF14h* from 6 h to 12 h. Genes with the most pronounced down regulation were *CitGF14a* at 2 h and 48 h; *CitGF14d* from 6 h to 24 h; *CitGF14e* from 6 h to 12 h; and *CitGF14g* and *i* at 24 h. *CitGF14b* and *CitGF14d* showed noticeable responses to ABA (Figure 6A). The former was primarily down regulated from 2 h to 48 h after application, whereas the latter was completely upregulated from 2 h to 48 h after application. ABA application also induced the expression of all the other *CitGF14s* at different time periods except for *CitGF14i* that was largely down regulated.

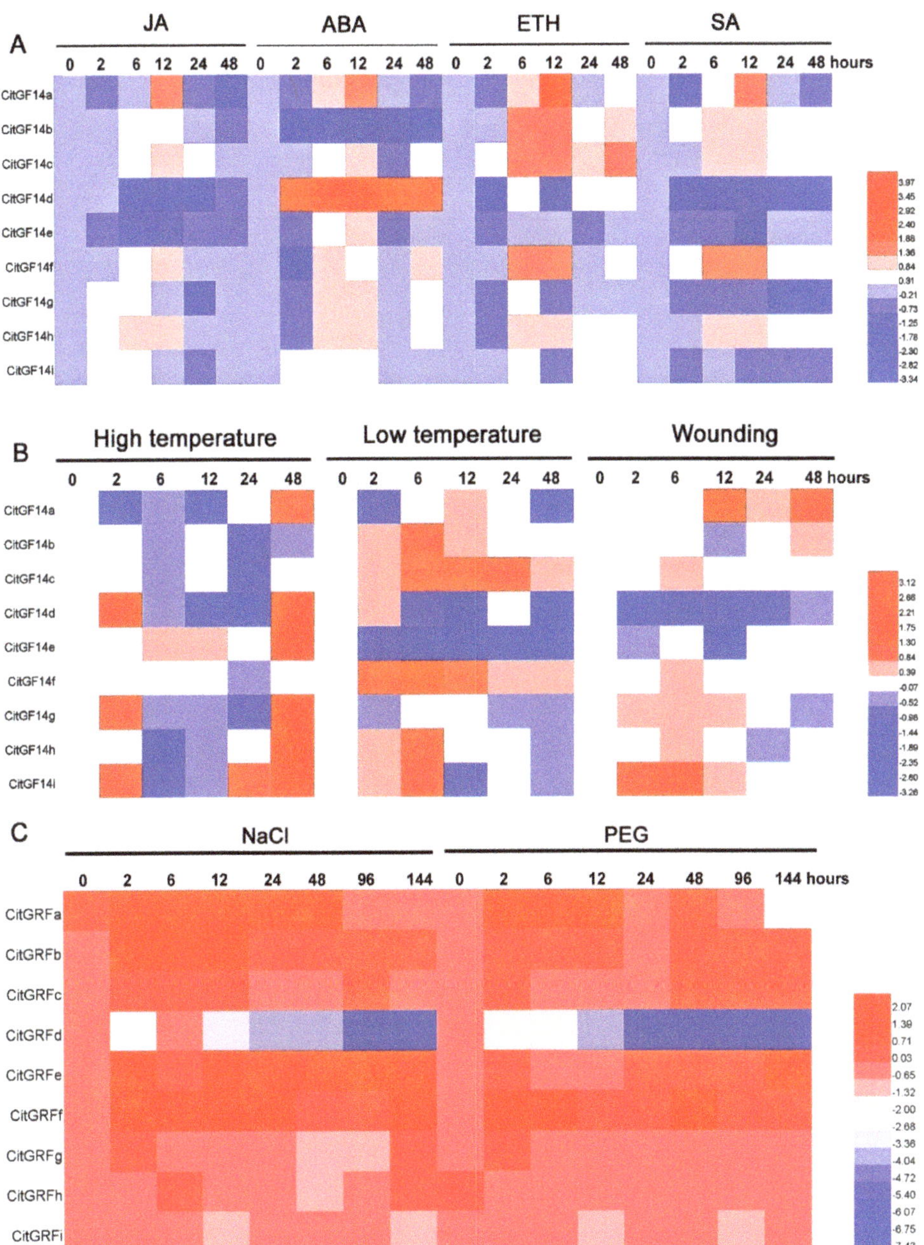

Figure 6. RT-qPCR analysis of *CitGF14* expressions after *C. sinensis* "Xuegan" plants were sprayed with jasmonate (JA), abscisic acid (ABA), ethephon (ETH), and salicylic acid (SA) (**A**), exposed to low (4 °C) and high (42 °C) temperatures as well as wounding (**B**), and treated with NaCl (200 mM) and polyethylene glycol (PEG) (20% PEG6000), a simulated drought stress (**C**). The relative expression was normalized using the ACTIN and F-box gene as references by $2^{-\Delta\Delta Ct}$ method. Heatmaps were generated based on log2 (treatment expression/control expression) with HemI. The values were based on the means of four biological replications.

The application of ethephon (ETH) induced variable expressions of *CitGF14a, b, c, f,* and *h* (Figure 6A). These genes were upregulated from 6 h to 12 h after application with exception of *CitGF14b* and *c* whose expression reduced at 24 h but increased at 48 h. The other genes were primarily down regulated after ETH application. Foliar spraying of salicylic acid (SA) also induced variable expressions of *CitGF14a, b, c, f,* and *h* (Figure 6A). *CitGF14a* was upregulated after 12 h of application, and *CitGF14b, c, f,* and *h* were induced from 6 h to 12 h. The other *CitGF14s* were downregulated or remained unchanged.

2.8. Responses to Low and High Temperatures and Wounding

The exposure of *C. sinensis* "Xuegan" plants to 42 °C led to the upregulations of five *CitGF14* genes at different times during the treatment (Figure 6B). *CitGF14a, e,* and *h* were either uninduced or downregulated from 0 h to 24 h but highly induced at 48 h. On the other hand, the expression levels of *CitGF14d* and *g* were higher at 2 h, decreased thereafter, and then highly increased at 48 h. *CitGF14i* had similar expression pattern as *CitGF14d* and *g*, but its increase started 24 h and sustained to 48 h. The other *CitGF14s* were either downregulated or showed little change.

Chilling treatment of "Xuegan" resulted in *CitGF14* responses opposite to the high temperature treatment. *CitGF14b, c,* and *f*, which were not upregulated at high temperature were strongly induced (Figure 6B). *CitGF14b* was induced at 2 h, and its expression reached the highest at 6 h, and then decreased at 12 h. *CitGF14c* was induced at 2 h, attained the highest from 6 h to 24, but decreased at 48 h. The expression of *CitGF14f* was high at 2 h and peaked at 6 h, and then gradually reduced. Additionally, *CitGF14h* and *CitGF14i* were also highly induced at 6 h. The expression of the other *CitGF14s* were variable, and largely downregulated.

Wounding of "Xuegan" plants induced the upregulation of *CitGF14a* at 12 to 48 h and *CitGF14i* from 2 h to 6 h (Figure 6B). Wounding treatment also moderately induced the expression of *CitGF14b* at 48 h and *CitGF14c, f,* and *h* at 6 h as well as *CitGF14g* from 2 h to 12 h, and *CitGF14i* at 12 h. The other *CitGF14s* either remained unchanged or downregulated.

2.9. Responses to Salinity and Drought Stresses

All *CitGF14* genes responded to the salt treatment (Figure 6C). *CitGF14a* was highly induced from 0 h to 48 h. The expressions of *CitGF14b, e,* and *f* were higher starting from 2 h to 144 h. The other *CitGF14s* were upregulated varying from moderate to high except for *CitGF14d* that fluctuated up and down and then remained downregulation from 24 h to 144 h.

The simulated drought stress by polyethylene glycol (PEG) treatment showed a rather similar expression pattern of *CitGF14s* as NaCl treatment (Figure 6C). All genes were upregulated except for *CitGF14d* that was initially upregulated, then slightly decreased, and finally downregulated from 24 h to 144 h. *CitGF14a* and *b* were highly induced from 2 h to 12 h, and both became moderately induced in 24 h. The expression of *CitGF14a* was higher again at 48 h and then decreased thereafter, but *CitGF14b* maintained a high expression level from 48 h to 144 h. *CitGF14c* and *e* were highly upregulated at 2 h, remained moderate in expression from 6 h to 24 h and 6 h to 12 h, respectively, and finally highly induced thereafter. *CitGF14f* was consistently highly expressed from 2 h to 144 h.

2.10. Interactions among CitGFs

To determine whether 14-3-3 proteins could interact with each other or with other proteins, we systematically assessed the interactions among all the 14-3-3 proteins using yeast two hybrid (Y2H) assay. GAL4 DNA-binding domain (BD) was fused to prey protein (CitGF14a through CitGF14i), GAL4 transcriptional activating domain (AD) was fused to bait protein (CitGF14a through CitGF14i and pGADT7 empty vector), and they were co-transformed into yeast cells. As shown in Figure 7, CitGF14a, b, c, d, g, and h interacted with CitGF14i proteins to form a heterodimer while CitGF14i interacted with itself to form a homodimer in yeast. Interactions were also found in other 14-3-3 protein pairs, CitGF14g interacted with b and c as prey or bait. CitGF14d showed strong interaction with a and c as prey, while when CitGF14*d* was bait, the interaction became weak. CitGF14h interacted with CitGF14c as bait protein, but CitGF14*h* weakly interacted with all target and empty vector. Thus, CitGF14h showed auto-activation as prey vector.

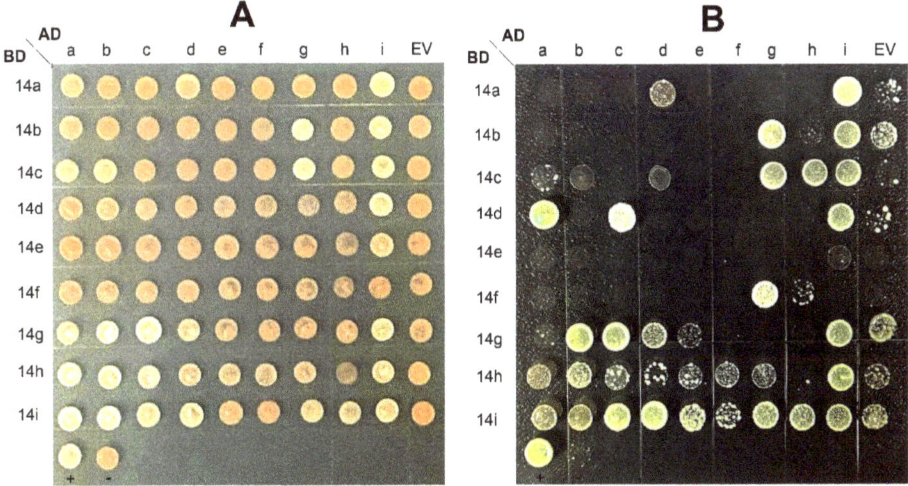

Figure 7. Yeast two-hybrid (Y2H) assay to test interactions among nine CitGF14 (14-3-3) proteins of *Citrus sinensis*. The coding sequences of the *CitGF14s* genes were cloned into the Y2H vectors pGADT7 (AD) and pGBKT7 (BD), and introduced into yeast cells Y2H gold. Transformants were assayed for growth on SD/-Trp-Leu (**A**) and SD/-Trp-Leu-His-Ade-x-α-gal (**B**) nutritional selection media. EV was empty vector pGADT7 used as a negative control.

3. Discussion

Plant 14-3-3 proteins play important roles in regulation of plant responses to abiotic and biotic stresses. Different number of 14-3-3 proteins have been identified and analyzed in a number of land plant species, but such information is not available in citrus. The present study for the first time documented 14-3-3 proteins in *C. sinensis* and evaluated *14-3-3* genes in response to various stresses.

3.1. Nine 14-3-3 Genes and Their Characteristics in C. sinensis

A total of nine *CitGF14* genes were identified in *C. sinensis* by a genome-wide search based on conserved domains and sequence similarities from known *14-3-3s*. They were divided into ε-like and non-ε groups (Figure 1), and eight of the nine were located on five chromosomes (Figure 2). Phylogenetic results were consistent with the clustering of 14-3-3 proteins in *Arabidopsis* [6], rice [40], and wheat [41]. The ε group, also known as "living fossil" 14-3-3 isoforms, is considered essential to eukaryotic biology, while the proteins in the non-ε group generally play organism-specific regulatory roles [42]. Gene structure analysis also supported this classification (Figure 3). Each of ε group genes (*CitGF14a* through *CitGF14d*) has six exons, while each of non-ε group genes (*CitGF14e* through *CitGF14i*) has three exons.

The relatively small number of *CitGF14* genes along with no duplicated ones were not surprising because there was no recent whole-genome duplication in *C. sinensis* evolution except γ event which was shared by all core eudicots [43]. Phylogentic analysis suggest that *C. sinensis* 14-3-3s are evolutionarily close to *Populus* since they belonged to the same clade and they are all ancient. Amino acid sequence alignment indicated that all *C. sinensis* 14-3-3s showed a high level of amino-acid similarity except for the N-terminal and C-terminal regions (Additional file 1: Figure S1), which are similar to all other 14-3-3s [44]. These highly conserved 30-kDa acidic proteins are, each composed of approximately 250 amino acids, nine α-helices, and two conserved signature motifs RNL(L/V)SV(G/A)YKNV and YKDSTLIMQLLRDNLTLWTS [14,44].

The online tool predicted that *C. sinensis* 14-3-3 isoforms were localized in cytoplasm, chloroplast, nucleus plasma, and plasma membrane (Table 1), suggesting distinct and differential patterns of subcellular distribution. These isoforms exhibited a high cell-type specificity. The specificity of cellular and subcellular localization may contribute to their diverse interactions with targets as well as differential functions in cellular activities. *Arabidopsis* 14-3-3/GFP fusions experiment indicated that 14-3-3 localization is both isoform specific and highly dependent upon interaction with cellular clients [45]. Additionally, this interaction can alter the subcellular localization of target proteins [46]. In soybean, SGF14 proteins can regulate the nuclear-cytoplasmic movement of GmMYB176, which can alter the expression of *CHS8*, a gene in isoflavonoid biosynthesis [47].

3.2. CitGF14s Were Differentially Expressed in Various Organs of Citrus

The tissue-specific pattern of gene expression can provide important clues about gene function [48]. The expression of *CitGF14s* in different organs (Figure 4) may suggest that they are involved in various aspects of physiological and developmental processes. *CitGF14a, e, h,* and *i* were highly upregulated in different organs based on RNA-Seq raw data (Figure 4A). RT-qPCR analysis also showed the highly upregulation of *CitGF14a, c, d, h,* and *i* in shoots and *CitGF14c, e,* and *h* in leaves, which were principally similar to those from the RNA-Seq analysis. A discrepancy occurred in *CitGF14d* between RNA-Seq (Figure 4A) and RT-qPCR (Figure 4B) data. RNA-Seq data showed its down regulation in almost all tested tissue and organs, but it was dramatically unregulated in flower and also in shoot in RT-qPCR analysis. Such a disagreement could be attributed to the differences in cultivars and plant growth conditions, which needs further investigation. Nevertheless, the high level of *CitGF14d* transcript in flowers and shoot in contrast to the minimal levels in the other organs (Figure 4B) may suggest the specificity of *CitGF14d* in flower and leaf development. A similar expression pattern also occurred in another woody plant mulberry tree [49] where some *MaGF14s* were specifically expressed in certain organs. Additionally, soybean *14-3-3* isoforms also showed ubiquitous expression in all tissues, and different expression of *SGF14* genes in embryos during seed development indicated that *14-3-3s* may be involved in soybean seed development [36]. In cotton, Northern blotting and RT-qPCR analysis showed that *Gh14-3-3* genes were developmentally regulated in fiber development [50,51]. In an early report, a high level of *Arabidopsis 14-3-3ω* mRNA occurred in flowers. On the other hand, *14-3-3κ* and *14-3-3λ* expression did not show much difference across tissues [52]. The expression of different *CitGF14s* in different organs or in the same organs of citrus may indicate the versality of *CitGF14* involvement in citrus growth and development, which deserve further investigation.

3.3. CitGF14s Were Induced by Citrus Canker and Greening Infections

Citrus canker and citrus greening are two notorious bacterial pathogens and have significantly affected citrus production worldwide. In this study, both microarray and RT-qPCR data clearly showed that *CitGF14s* are involved in citrus responses to canker (Figure 5). Microarray data indicated that both *CitGF14g* and *CitGF14i* were highly upregulated upon the infection of both Xaa and Xac (Figure 5A). RT-qPCR analysis further confirmed the upregulation of both *CitGF14g* and *CitGF14i* (Figure 5B) and found that *CitGF14b, d,* and *f* were also involved in the later period of Xac infection. These results indicate that 14-3-3 proteins may involve in the regulation of citrus canker resistance as well as functional redundancy in stress tolerance.

The responses of *CitGF14s* to citrus greening pathogen were not pronounced as those to citrus canker based on the microarray results (Figure 5A). There was a slight increase in the expressions of *CitGF14c* and *f* in leaves and *CitGF14e* in peel, whereas *CitGF14a, g,* and *i* were downregulated in leaves, stems or fruit peel. How such up or down regulations affects the pathogen development is unknown. Considering the severity of citrus greening in the citrus industry and the role of 14-3-3 in regulation of plant responses to biotic stresses, further research on *CitGF14s* in regulation of citrus response to greening is warranted.

Pathogen infection triggered differential expressions of *14-3-3* genes has been documented in other plants. In tomato, at least 10 *14-3-3* genes are differentially expressed in response to fungal toxin fusicoccin [32]. The tomato 14-3-3 protein TFT7 can bind with MAPKKKα and SIMKK2, resulting in programmed cell death associated with immunity [53]. TFT4 is another member of tomato *14-3-3* proteins that can bind with the effector XopQ from *Xanthomonas euvesicatoria* (Xcv) to suppress the effector-triggered immunity [54].

3.4. CitGF14s Differentially Responded to Abiotic Stresses

Phytohormones play a central role in plant defense responses to environmental stress. RT-qPCR analysis showed that all *CitGF14* genes responded to JA, ABA, ETH, and SA applications by either down or up regulation at different times (Figure 6A). Such variable responses may suggest that 14-3-3s could function as a multiple regulator in plant hormone signaling. Four rice *GF14* genes were induced by benzothiadiazole, JA, ETH, and H_2O_2 during pathogen attach [26]. Quantification of the 20R/16R promoter-driven GUS expression in different transgenic potato plants revealed that *14-3-3* isoforms can be induced by various stimuli, such as ABA, SA, NaCl, and metal ions [55]. The *Arabidopsis* 14-3-3 λ isoform was reported to specifically bind the C-terminal domain of RPW8.2, resulting in the enhanced resistance to powdery mildew fungus via the SA signaling pathway [56]. In the present study, *CitGF14b* was highly downregulated, but *CitGF14d* was strongly upregulated by ABA treatment. This may suggest that citrus *14-3-3s* could link to ABA in mediation of different stress responses. A study with barley 14-3-3 proteins as baits in yeast two-hybrid (Y2H) library resulted in the identification of 132 new molecular targets, including three ABA signal transduction related proteins (AREB/ABF/ABI5-like proteins) [57].

High and low temperatures as well as wounding induced variable expressions of all nine *CitGF14s* (Figure 6B). *CitGF14a, d, e,* and *g* that were upregulated at high temperature became down regulated in low temperature treatment. The reverse is true for *CitGF14b, c,* and *f*. Only *CitGF14h* and *i* had both up and down regulations in high and low temperature treatments. These results may indicate that different isoforms of *CitGF14s* were involved in response to two opposite temperature stresses. A study with *Arabidopsis* showed that overexpression of *14-3-3ε* was ineffective in cold tolerance, but overexpression of both ω and ε produced more cold tolerant plants [58]. Wounding also induced all *CitGF14s* gene expression including the upregulation of *CitGF14a* and *i* but downregulation of *CitGF14d* at variable times. These results concurred with a report by Lapointe et al. [59] that *14-3-3* mRNA was upregulated in poplar plants after wounding treatment.

CitGF14s exhibited similar expression patterns in response to both NaCl and PEG-simulated drought stresses (Figure 6C). All *CitGF14s* genes were highly upregulated except *CitGF14d* that was downregulated, suggesting that all *CitGF14s* participated in citrus responses to NaCl and drought. These results are consistent with 14-3-3 responses to salinity and drought stresses in other plants [3,22,60]. An *Arabidopsis GF14λ* was introduced into cotton plants, resulting in improved drought tolerance with a "stay-green" phenotype. The stomata of the transgenic plants might be regulated by *GF14λ* through some transporters, such as H+-ATPase whose activities are controlled by their interaction with 14-3-3 proteins. In the present study, we noticed that *CitGF14d*, which was highly induced by ABA signal (Figure 6A), were downregulated under salinity and drought stresses (Figure 6C). This result may indicate that *CitGF14d* acted as a negative regulator. The overexpression of *GsGF14o* from *Glycine soja* in *Arabidopsis* resulted in down-regulation of a drought-responsive marker gene, the transgenic line showed reduction of stomatal development under drought treatment. Thus, the *Glycine soja* 14-3-3 gene *GsGF14o* functioned as a negative regulator of drought tolerance [61].

3.5. Prediction of Interactions among CitGF14 Proteins

Different genes with similar expression patterns, commonly known as co-expressed genes were believed to be functionally related [62]. To improve our understanding of *CitGF14s'* functions in *C. sinensis, CitGF14s* were integrated into the Network Inference for Citrus Co-Expression (NICCE) (http://citrus.adelaide.edu.au/nicce/home.aspx). A detailed *14-3-3* gene information and tabs were presented in Additional file 3: Table S2, which contained top 100 expressed genes in *C. sinensis* based on HRR (highest reciprocal ranks). HRR highlighted in black (bold), black, and grey colors signified statistical significance of HRR at $p < 0.01$, $p < 0.05$, and $p > 0.05$ levels, respectively. As described previously, the microarray data did not cover all the *CitGF14* genes, we only obtained co-expressed genes from six member of *CitGF14* genes in *C. sinensis* datasets. We found a number of genes co-expressed with three citrus canker responsible *CitGF14s* (*CitGF14g, CitGF14h,* and *CitGF14i*). They are leucine-rich repeat family proteins, zinc finger (C3HC4-type RING finger) family proteins, and zinc finger (DHHC type) family proteins, WD-40 repeat family proteins, lesion inducing protein-related, and peroxidase 63. These proteins are involved in a variety of functions ranging from signal transduction and transcription regulation to defense responses. For example, the leucine-rich repeat family proteins are associated with innate immunity in plants, serving as the first line of defense against pathogens. Our results may indicate that *CitGF14g, CitGF14h,* and *CitGF14i* are involved in regulation of defense response to the infection of citrus canker pathogens.

The interaction among 14-3-3 isoforms or individual isoforms with other proteins is important for understanding the biological functions of 14-3-3s. Y2H assay is a powerful tool widely used for identifying novel protein–protein interaction [63]. In the present study, CitGF14i was found to be able to interact with CitGF14a, b, c, d, g, or h to form a heterodimer and interact with itself to form a homodimer in yeast (Figure 7). Interactions were also found in other 14-3-3 protein pairs, including CitGF14g with b or c and CitGF14d with a or c. Among the isoforms, CitGF14i appears to be the most active and important one due to its interaction with six isoforms and also with itself. It is known that 14-3-3 isoforms have different affinities to individual targets; thereby, there is a possibility that regulation of specific processes could be accomplished by single 14-3-3 isoforms [16]. In this study, *CitGF14i* was highly up-regulated in flower and leaves or shoots in general (Figure 4), strongly induced by the infection of Xac (Figure 5B), largely down-regulated in response to the application of growth regulators, and variably expressed in abiotic treatments (Figure 6). Intriguingly, *CitGF14i* is the only one localized in plasma membrane. Whether its subcellular location contributes to such active interactions is unknown. Although the function of all *CitGF14s* deserve further investigation, specific attention should be given to *CitGF14i* for its interactions with other proteins. Y2H screens complemented with 14-3-3 protein affinity purification and tandem mass spectrometry are another powerful tool for identifying protein interactions. This method identified five 14-3-3 isoforms in 7-day-old barley. Some of proteins were identified as 14-3-3 targets in both Y2H and affinity purification including 14-3-3 proteins themselves [64]. In order to uncover the 14-3-3 signaling pathway in healthy and disease, a high-throughput data in VisANT graphs (http://visant.bu.edu) was used to graph and validate 14-3-3 protein interactions [65].

Two public citrus databases provided a genome-wide approach to predict 14-3-3 protein–protein interactions (PPI) or gene co-expression. Thus, a 14-3-3s protein interaction network was generated by CitrusNet (Figure 8). Among them, CitGF14a, e, and i play much greater roles than CitGF14g and h, which are more important than the remaining four CitGF14s. CitGF14d, CitGF14g, and CitGF14i contribute to their resistance to Xac infection, and CitGF14d could be a negative regulator in response to drought stress. With more than 150 nodes in the network, most nodes had different degrees of connection. Detailed connection among each CitGF14 and target was provided in Additional file 4: Table S3. More than 70 citrus 14-3-3 targets were identified, which participate in many molecular processes, including those involved in development, hormone, redox, signaling, stress, and transport. Polyubiquitin-A, clathrin heavy chain 1, heat shock protein 81-3, and heat shock protein 83 were protein nodes with the highest degree in CitrusNet [66], and these proteins also were found in 14-3-3 PPI network. Furthermore, casein kinase I, vesicle-fusing ATPase, serine/threonine-protein kinase TOR (TARGET OF RAPAMYCIN), histone deacetylase 6, leucine rich repeat-type serine/threonine receptor-like kinase were predicted to interact with several members of CitGF14s. TOR kinase is a client of CitGF14s. In *Arabidopsis*, TOR kinase was important in controlling plant growth, responding to environmental cues, and regulating cell processes [67]. In genome-wide PPI network, TOR kinase was a central part of citrus hormone cross-talk, which potentially interacted with proteins related to hormone signaling and hormone receptors [66]. Therefore, all these proteins play critical roles in CitGF14s regulating of various aspects of cellular processes. Furthermore, different members of CitGF14 interactions were found in both co-expression network and PPI network. Protein BRASSINOSTEROID INSENSITIVE 1 and leucine rich repeat-type serine/threonine receptor-like kinase appeared in PPI network, and they were also predicted as *14-3-3s* co-expression genes.

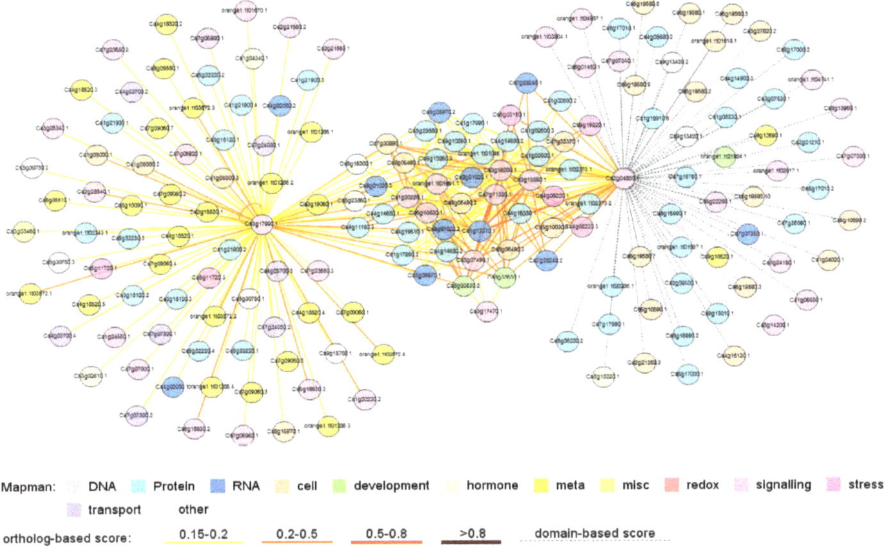

Figure 8. Protein–protein interaction network of CitGF14s and target proteins. The colored nodes represent proteins of miscellaneous functions in this network (see picture detail). Solid line or dotted line represents the predicted interaction based on ortholog or domain.

4. Materials and Methods

4.1. Identification of 14-3-3 Genes in Citrus sinensis

The *14-3-3* genes from *Arabidopsis* [6], soybean [36], and *Populus* [38] were employed as queries to perform a local BLASTP against the database: Orange Genome Annotation Project (http://citrus.hzau.edu.cn/orange/), which were retrieved from previous studies and database respectively [40]. Proteins with e-value belonging to a significant match (e-value $< 10^{-5}$) in the blast analysis were considered as potential 14-3-3 members. The resulting protein sequences were examined with Pfam (http://pfam.sanger.ac.uk/search) and SMART (http://smart.embl-heidelberg.de/) to ensure the presence of 14-3-3-specific domains.

4.2. Chromosome Location of 14-3-3 Genes and Their Protein Properties and Sequence Analyses

The chromosome locations of *14-3-3s* genes were searched in the database of Orange Genome Annotation Project (http://citrus.hzau.edu.cn/cgi-bin/gb2/gbrowse/orange/) [43]. The molecular weight (MW) and isoelectric point (PI) of non-redundant genes were calculated by the online tool ExPASy (http://www.expasy.org/tools/). Subcellular localization was performed using WoLF PSORT at the website http://www.genscript.com/wolf-psort.html. All 14-3-3 amino acid sequences were used for identifying three-dimensional structure of 14-3-3 proteins using the SWISS-MODEL (https://swissmodel.expasy.org). To identify cis-regulatory elements, the 1.5 kb upstream regions to the translation start codon were selected from the Orange Genome Annotation Project and analyzed with PlantCARE (http://bioinformatics.psb.ugent.be/webtools/plantcare/html/) databases. The Gene Structure Display Server (http://gsds.cbi.pku.edu.cn) [68] was used to display gene structure models.

4.3. Sequence Alignment and Phylogenetic Analysis of 14-3-3 Proteins

The amino acid sequences of *14-3-3* genes from different plant species were aligned using the software Clustal_X (version 1.83) with default parameters [69]. The unrooted phylogenetic trees were constructed based on alignments using MEGA 4.0 with the neighbor-joining method [70]. The bootstrap test was carried out with 1000 replicates.

4.4. RNA-Seq and Microarray Data Analysis

After the identification and characterization of *14-3-3* genes in citrus, publicly available RNA-Seq raw data and microarray database were mined for *14-3-3* expression profiles in different organs of citrus and in response to abiotic and biotic stresses.

The expression of *14-3-3s* in callus, leaf, flower, and fruit were analyzed using RNA-Seq raw data available in *C. sinensis* genome database (http://citrus.hzau.edu.cn/orange/). The expression level of *14-3-3* genes was calculated as Log2 based RPKM. The expression data were hierarchically clustered with average linkage and displayed in HemI [71].

Affymetrix microarray data obtained from the NCBI Gene Expression Omnibus (GEO) database under the series accession number GSE33003, GSE33004, and GSE10798 were mined for citrus *14-3-3* response to biotic stresses. In experiments with Ca. Las infection (GSE33003 and GSE33004), young, healthy Valencia sweet orange plants were graft-inoculated with budwood from Ca. Las-infected citrus plants. The leaf, stem, and root samples were collected from three symptomatic and three healthy control trees for RNA extraction and analyzed using microarrays [72,73]. For citrus canker (Xac) or (Xaa) infection (GSE10798), adult leaves of sweet orange were infiltrated with the bacterial suspensions or water (mock control). Leaves samples were collected after bacterial infiltration for RNA extraction and hybridization on Affymetrix microarrays [74]. The microarray CEL files were normalized using Robust Multi-array Average (RMA) in R/Bioconductor (ver 2.15), and normalized data was used for identifying differential expressed genes [75,76]. The heatmap for the *14-3-3* transcripts with their expression values were performed by HemI [66].

4.5. Plant Materials and Treatments

To confirm the expression of *14-3-3* genes in different organs of *C. sinensis*, root, stem, leaf, and shoot samples were taken from one-year old seedling of *C. sinensis*. Flower, fruit peel, juice, and seed samples were collected from eight-year-old adult trees. Four biological samples per organ were frozen in liquid N and stored in −80 °C.

Further analysis of *14-3-3* genes in response to various stresses were conducted using *C. sinensis* "Xuegan". Seeds of "Xuegan" were germinated in a plant growth chamber at 25 °C under a photoperiod of 16-h light/8-h dark with a relative humidity of 70%. The light was provided by fluorescent white-light tubes. At the two-leaf stage, seedlings were transplanted singly into pots (15 × 15 cm) filled with Fafard Professional Potting Mix (Sun Gro Horticulture, Agawam, MA, USA). The following treatments were applied to the plants, each experiment was arranged as a complete randomized design with four replicate per treatment.

Infection with citrus canker: The bacterial strain Xac 29-1 was cultured in NA nutrient broth with 1.5% agar at 28 °C for 36 h. Cultured bacterial cells were washed twice with sterile water and then resuspended in sterile water to a final concentration OD600 = 0.3. The diluted cells were infiltrated into leaves with a needleless syringe, sterile water was injected into leaves as the control treatment. Leaf samples were collected after 0, 2, 6, 12, 48, 96, and 192 h of inoculation, respectively.

Hormone treatments: Seedlings were sprayed with 1 mM SA, 100 µM JA, 100 µM ABA, and 100 µM ETH solutions, respectively. Leaf samples were collected after 0, 2, 6, 12, 24, and 48 h of the application, respectively. Leaf samples collected from seedlings sprayed with distilled water at the corresponding time were considered the control treatment.

Low or high temperature and wounding treatments: Seedlings were incubated in a growth chamber with temperature of 4 °C or 42 °C for 48 h. For wounding treatment, each fully expanded leaf was penetrated with a needleless syringe 10 times. Leaf samples were collected at 0, 2, 6, 12, and 48 h, respectively. Seedlings grown in the chamber with the temperature at 25 °C as the control, and leaf samples were collected at the same times as those of treated seedlings.

Salt and drought treatments: Seedlings were grown in the potting mix were drenched with 200 mM NaCl or 20% PEG6000 solution until leachate appeared from the bottom of pots (about 250 mL solution was used for each treatment). Seedlings were irrigated with 250 mL water as the control treatment. Leaf samples were collected from treated and control seedlings after 0, 2, 6, 12, 48, 96, and 144 h of treatment, respectively.

All the collected leaf samples were immediately frozen in liquid N and stored at −80 °C for RNA extraction.

4.6. RNA Isolation and Expression Analysis

Total RNA was extracted from the collected samples using the RNAprep pure Plant Kit (Tiangen, Beijing, China) according to the manufacturer's instructions. One thousand nanograms of total RNA was used to synthesize first-strand cDNA with EasyScript One-Step gDNA Removal and cDNA Synthesis SuperMix (Transgen, Beijing, China). Quantitative PCR was carried out using TransStart Tip Green qPCR SuperMix (Transgen, Beijing, China) on a CFX96 Real-time System (Bio-Rad, Hercules, CA, USA) according to the manufacturer's protocol. *GAPDH* [77], *Actin*, and *F-box* [78] were used as reference genes to normalize the expression of the investigated genes. Gene specific primers (Additional file 5: Table S4) were designed with the Primer Premier 6 software (Premier Biosoft International, San Francisco, CA, USA). The PCR reaction mixtures were incubated at 95 °C for 30 s, followed by 40 cycles of 95 °C for 5 s and 60 °C for 30 s. The relative expression levels were determined by $2^{-\Delta\Delta Ct}$ method described by Livak and Schmittgen [79], data were analyzed using SPSS 22.0 (IBM Corporation, Somers, NY, USA) statistics software, and mean differences were separated by Tukey's HSD test at $p < 0.05$ level with four biological replicates. The heatmap for the *14-3-3* transcripts with their expression values were generated using HemI tool [66].

4.7. Prediction of Cis-Regulatory Elements

The sequences of 1.5 kb upstream regions from the translation initiation codon of each gene of 14-3-3s were selected and subjected to analysis by the online database PlantCare. Putative developmental and stress-responsive cis-elements in citrus *14-3-3s* were identified.

4.8. Gene Co-Expression Network and Protein–Protein Interaction Prediction

To predict gene interaction on a genome-wide scale, keyword 14-3-3s were searched in publicly accessible tool NICCE (http://citrus.adelaide.edu.au/nicce/home.aspx) to predict potential targets of 14-3-3s [80]. CitrusNet and PPI networks in *C. sinensis* were constructed using ortholog-based and domain-based interaction methods, which contained

8195 proteins and 124,491 interactions [65]. The nine citrus 14-3-3 proteins were used as hub nodes to connect potential target proteins in CitrusNet (http://citrus.hzau.edu.cn/orange/ppi/index.php). All 14-3-3s and client proteins were linked into an interconnected sub-network which was visualized by Cytoscape.

4.9. Analysis of CitGRFs Interactions by Y2H Assay

The full-length coding sequences of 9 citrus 14-3-3 proteins were introduced into the pGBKT7 fusion bait vector and pGADT7 fusion prey vector. The fused pGADT7-CitGF14s and pGBT9-CitGF14s recombinant vectors were then co-transformed into yeast strain Y2H gold by LiAc/SS carrier DNA/PEG method [81]. The transformants were first selected in the SD/-Trp-Leu medium and PCR testing. After that the positive colonies were transferred to the selection medium supplemented with X-α-gal but lacked Trp, Leu, His, and adenine (SD/-Trp-Leu-His-Ade). Aureobasidin A (AbA) was added to the selection plates to suppress the auto-activation of the prey vectors.

5. Conclusions

The present study identified nine *14-3-3 genes* (*CitGF14a* through *CitGF14i*) in *C. sinensis* through genome-wide analysis. All the *CitGF14s* genes were detected in different tissues or organs but varied in abundance. Transcript levels of *CitGF14s* were also analyzed after plants were treated with hormones, extreme temperatures, drought, salinity, wounding, and infected with Xac 29-1 strains. Almost all *CitGF14s* responded to the treatments by variable levels of expression during the experiments, suggesting that *CitGF14s* play important roles in citrus responses to different exogenous and endogenous signals. This study also showed that most gene family members had a functional divergence of 14-3-3 proteins. Y2H assay showed that CitGF14i was the most active and important isoform due to its interaction with six other isoforms and also with itself. Additionally, CitGF14d, CitGF14g, and CitGF14i contribute to their resistance to Xac infection, and CitGF14d could be a negative regulator in response to drought stress. Finally, a citrus 14-3-3 interactome network was constructed by PPI method and microarray gene co-expression. Our study for the first time provides a comprehensive framework about *14-3-3* family genes in *C. sinensis*, which may lead to further investigation of their roles in citrus growth and development as well as in response to abiotic and biotic stresses.

Supplementary Materials: The following are available online at https://www.mdpi.com/1422-0067/22/2/568/s1, Additional file 1: Figure S1. *CitGF14s* sequence alignment and protein homology modeling; Additional file 2: Table S1. *Cis*-acting regulatory elements; Additional file 3: Table S2. 14-3-3s Co-expressed genes. Additional file 4: Table S3. Domain-based 14-3-3s PPI network; Additional file 5: Table S4. Gene specific primers.

Author Contributions: W.S. and J.C. conceived and designed the experiments. S.L. performed the experiments. S.L. and J.C. wrote the paper. G.C. and D.P. provided supervision. All authors have read and agreed to the published version of the manuscript.

Funding: This study was supported in part by the project "Research on Breeding and Quality Formation Mechanism of High-quality and Early-ripening Pomelo Varieties (KSYLC008)".

Institutional Review Board Statement: Not applicable.

Informed Consent Statement: Not applicable.

Data Availability Statement: The data presented in this study are available in article and supplementary materials.

Acknowledgments: The authors thanks Terri A. Mellich at the University of Florida for critical review of this manuscript.

Conflicts of Interest: The authors declare that the research was conducted in the absence of any commercial or financial relationships that could be construed as a potential conflict of interest.

References

1. Chevalier, D.; Morris, E.R.; Walker, J.C. 14-3-3 and fha domains mediate phosphoprotein interactions. *Annu. Rev. Plant Biol.* **2009**, *60*, 67–91. [CrossRef] [PubMed]
2. Ferl, R.J.; Manak, M.S.; Reyes, M.F. The 14-3-3s. *Genome Biol.* **2002**, *3*, REVIEWS3010. [CrossRef] [PubMed]
3. De Boer, A.H.; van Kleeff, P.J.; Gao, J. Plant 14-3-3 proteins as spiders in a web of phosphorylation. *Protoplasma* **2013**, *250*, 425–440. [CrossRef] [PubMed]
4. Moore, B.W.; Perez, V.J. Specific acidic proteins of the nervous system. In *Physiological Biochemical Aspects of Neverous Integration*; Carlson, D., Ed.; Prentice-Hall: New York, NY, USA, 1996; pp. 343–359.
5. Van Heusden, G.P.H. 14-3-3 proteins: Insight from genome-wide studies in yeast. *Genomics* **2009**, *94*, 287–293. [CrossRef] [PubMed]
6. Rosenquist, M.; Alsterfjord, M.; Larsson, C.; Sommarin, M. Data mining the *Arabidopsis* genome reveals fifteen 14-3-3 genes. Expression is demonstrated for two out of five novel genes. *Plant Physiol.* **2001**, *127*, 142–149. [CrossRef]
7. Konagaya, K.I.; Matsushita, Y.; Kasahara, M.; Nyunoya, H. Members of 14-3-3 protein isoforms interacting with the resistance gene product N and the elicitor of Tobacco mosaic virus. *J. Gen. Plant Pathol.* **2004**, *70*, 221–231. [CrossRef]
8. Xu, W.F.; Shi, W.M. Expression profiling of the *14-3-3* gene family in response to salt stress and potassium and iron deficiencies in young tomato (*Solanum lycopersicum*) roots: Analysis by real-time RT-PCR. *Ann. Bot.* **2006**, *98*, 965–974. [CrossRef]
9. Yuan, Y.; Ying, D.; Lin, J.; Jin-Yuan, L. Molecular analysis and expression patterns of the *14-3-3* gene family from *Oryza sativa*. *J. Biochem. Mol. Biol.* **2007**, *40*, 349–357.
10. DeLille, J.M.; Sehnke, P.C.; Ferl, R.J. The arabidopsis 14-3-3 family of signaling regulators. *Plant Physiol.* **2001**, *126*, 35–38. [CrossRef] [PubMed]
11. Chung, H.J.; Sehnke, P.C.; Ferl, R.J. The 14-3-3 proteins: Cellular regulators of plant metabolism. *Trends Plant Sci.* **1999**, *4*, 367–371. [CrossRef]
12. Yaffe, M.B.; Rittinger, K.; Volinia, S.; Caron, P.R.; Aitken, A.; Leffers, H.; Gamblin, S.J.; Smerdon, S.J.; Cantley, L.C. The structural basis for 14-3-3: Phosphopeptide binding specificity. *Cell* **1997**, *91*, 961–971. [CrossRef]
13. Wurtele, M.; Jelich-Ottmann, C.; Wittinghofer, A.; Oecking, C. Structural view of a fungal toxin acting on a 14-3-3 regulatory complex. *EMBO J.* **2003**, *22*, 987–994. [CrossRef] [PubMed]
14. Yang, X.; Lee, W.H.; Sobott, F.; Papagrigoriou, E.; Robinson, C.V.; Grossmann, J.G.; Sundstrom, M.; Doyle, D.A.; Elkins, J.M. Structural basis for protein-protein interactions in the 14-3-3 protein family. *Proc. Natl. Acad. Sci. USA* **2006**, *103*, 17237–17242. [CrossRef] [PubMed]
15. Wilker, E.W.; Grant, R.A.; Artim, S.C.; Yaffe, M.B. A structural basis for 14-3-3sigma functional specificity. *J. Biol. Chem.* **2005**, *280*, 18891–18898. [CrossRef]
16. Paul, A.L.; Denison, F.C.; Schultz, E.R.; Zupanska, A.K.; Ferl, R.J. 14-3-3 phosphoprotein interaction networks—Does isoform diversity present functional interaction specification? *Front. Plant Sci.* **2012**, *3*, 190. [CrossRef]
17. Muslin, A.J.; Tanner, J.W.; Allen, P.M.; Shaw, A.S. Interaction of 14-3-3 with signaling proteins is mediated by the recognition of phosphoserine. *Cell* **1996**, *84*, 889–897. [CrossRef]
18. Ganguly, S.; Weller, J.L.; Ho, A.; Chemineau, P.; Malpaux, B.; Klein, D.C. Melatonin synthesis: 14-3-3-dependent activation and inhibition of arylalkylamine N-acetyltransferase mediated by phosphoserine-205. *Proc. Natl. Acad. Sci. USA* **2005**, *102*, 1222–1227. [CrossRef]
19. Lozano-Duran, R.; Robatzek, S. 14-3-3 proteins in plant-pathogen interactions. *Mol. Plant Microbe Interact.* **2015**, *28*, 511–518. [CrossRef]
20. Petosa, C.; Masters, S.C.; Bankston, L.A.; Pohl, J.; Wang, B.; Fu, H.; Liddington, R.C. 14-3-3zeta binds a phosphorylated Raf peptide and an unphosphorylated peptide via its conserved amphipathic groove. *J. Biol. Chem.* **1998**, *273*, 16305–16310. [CrossRef]
21. Andrews, R.K.; Harris, S.J.; McNally, T.; Berndt, M.C. Binding of purified 14-3-3 zeta signaling protein to discrete amino acid sequences within the cytoplasmic domain of the platelet membrane glycoprotein Ib-IX-V complex. *Biochemistry* **1998**, *37*, 638–647. [CrossRef]
22. Denison, F.C.; Paul, A.L.; Zupanska, A.K.; Ferl, R.J. 14-3-3 proteins in plant physiology. *Semin. Cell Dev. Biol.* **2011**, *22*, 720–727. [CrossRef] [PubMed]
23. Oh, C.-S. Characteristics of 14-3-3 proteins and their role in plant immunity. *Plant Pathol. J.* **2010**, *26*, 1–7. [CrossRef]
24. Darling, D.L.; Yingling, J.; Wynshaw-Boris, A. Role of 14-3-3 proteins in eukaryotic signaling and development. *Curr. Top. Dev. Biol.* **2005**, *68*, 281–315. [PubMed]
25. Zhou, H.; Lin, H.; Chen, S.; Becker, K.; Yang, Y.; Zhao, J.; Kudla, J.; Schumaker, K.S.; Guo, Y. Inhibition of the Arabidopsis salt overly sensitive pathway by 14-3-3 proteins. *Plant Cell* **2014**, *26*, 1166–1182. [CrossRef] [PubMed]
26. Chen, F.; Li, Q.; Sun, L.; He, Z. The rice 14-3-3 gene family and its involvement in responses to biotic and abiotic stress. *DNA Res. Int. J. Rapid Publ. Rep. Genes Genomes* **2006**, *13*, 53–63. [CrossRef] [PubMed]
27. Jarillo, J.A.; Capel, J.; Leyva, A.; Martinez-Zapater, J.M.; Salinas, J. Two related low-temperature-inducible genes of *Arabidopsis* encode proteins showing high homology to 14-3-3 proteins, a family of putative kinase regulators. *Plant Mol. Biol.* **1994**, *25*, 693–704. [CrossRef] [PubMed]
28. Roberts, M.R.; Salinas, J.; Collinge, D.B. 14-3-3 proteins and the response to abiotic and biotic stress. *Plant Mol. Biol.* **2002**, *50*, 1031–1039. [CrossRef]

29. Xu, W.; Shi, W.; Jia, L.; Liang, J.; Zhang, J. TFT6 and TFT7, two different members of tomato 14-3-3 gene family, play distinct roles in plant adaption to low phosphorus stress. *Plant Cell Environ.* **2012**, *35*, 1393–1406. [CrossRef] [PubMed]

30. Yang, J.L.; Chen, W.W.; Chen, L.Q.; Qin, C.; Jin, C.W.; Shi, Y.Z.; Zheng, S.J. The 14-3-3 protein GENERAL REGULATORY FACTOR11 (GRF11) acts downstream of nitric oxide to regulate iron acquisition in *Arabidopsis thaliana*. *New Phytol.* **2013**, *197*, 815–824. [CrossRef]

31. Schoonheim, P.J.; Costa Pereira, D.D.; De Boer, A.H. Dual role for 14-3-3 proteins and ABF transcription factors in gibberellic acid and abscisic acid signalling in barley (*Hordeum vulgare*) aleurone cells. *Plant Cell Environ.* **2009**, *32*, 439–447. [CrossRef]

32. Roberts, M.R.; Bowles, D.J. Fusicoccin, 14-3-3 proteins, and defense responses in tomato plants. *Plant Physiol.* **1999**, *119*, 1243–1250. [CrossRef] [PubMed]

33. Seehaus, K.; Tenhaken, R. Cloning of genes by mRNA differential display induced during the hypersensitive reaction of soybean after inoculation with *Pseudomonas syringae* pv. *glycinea. Plant Mol. Biol.* **1998**, *38*, 1225–1234. [CrossRef] [PubMed]

34. Klink, V.P.; Hosseini, P.; Matsye, P.; Alkharouf, N.W.; Matthews, B.F. A gene expression analysis of syncytia laser microdissected from the roots of the *Glycine max* (soybean) genotype PI 548402 (Peking) undergoing a resistant reaction after infection by Heterodera glycines (soybean cyst nematode). *Plant Mol. Biol.* **2009**, *71*, 525–567. [CrossRef]

35. Hill, M.K.; Lyon, K.; Lyon, B.R. Identification of disease response genes expressed in *Gossypium hirsutum* upon infection with the wilt pathogen *Verticillium dahliae. Plant Mol. Biol.* **1999**, *40*, 289–296. [CrossRef] [PubMed]

36. Li, X.; Dhaubhadel, S. Soybean 14-3-3 gene family: Identification and molecular characterization. *Planta* **2011**, *233*, 569–582. [CrossRef] [PubMed]

37. Li, R.; Jiang, X.; Jin, D.; Dhaubhadel, S.; Bian, S.; Li, X. Identification of 14-3-3 family in common bean and their response to abiotic stress. *PLoS ONE* **2015**, *10*, e0143280. [CrossRef] [PubMed]

38. Tian, F.; Wang, T.; Xie, Y.; Zhang, J.; Hu, J. Genome-wide identification, classification, and expression analysis of 14-3-3 gene family in *Populus. PLoS ONE* **2015**, *10*, e0123225. [CrossRef]

39. Kumar, K.; Muthamilarasan, M.; Bonthala, V.S.; Roy, R.; Prasad, M. Unraveling 14-3-3 proteins in C4 panicoids with emphasis on model plant *Setaria italica* reveals phosphorylation-dependent subcellular localization of RS splicing factor. *PLoS ONE* **2015**, *10*, e0123236. [CrossRef]

40. Yashvardhini, N.; Bhattacharya, S.; Chaudhuri, S.; Sengupta, D.N. Molecular characterization of the 14-3-3 gene family in rice and its expression studies under abiotic stress. *Planta* **2018**, *247*, 229–253. [CrossRef]

41. Guo, J.; Dai, S.; Li, H.; Liu, A.; Liu, C.; Cheng, D.; Cao, X.; Chu, X.; Zhai, S.; Liu, J.; et al. Identification and expression analysis of wheat *TaGF14* genes. *Front. Genet.* **2018**, *9*, 12. [CrossRef]

42. Jaspert, N.; Throm, C.; Oecking, C. *Arabidopsis* 14-3-3 proteins: Fascinating and less fascinating aspects. *Front. Plant Sci.* **2011**, *2*, 96. [CrossRef] [PubMed]

43. Xu, Q.; Chen, L.L.; Ruan, X.; Chen, D.; Zhu, A.; Chen, C.; Bertrand, D.; Jiao, W.B.; Hao, B.H.; Lyon, M.P.; et al. The draft genome of sweet orange (*Citrus sinensis*). *Nat. Genet.* **2013**, *45*, 59–66. [CrossRef]

44. Wang, W.; Shakes, D.C. Molecular evolution of the 14-3-3 protein family. *J. Mol. Evol.* **1996**, *43*, 384–398. [CrossRef]

45. Paul, A.L.; Sehnke, P.C.; Ferl, R.J. Isoform-specific subcellular localization among 14-3-3 proteins in *Arabidopsis* seems to be driven by client interactions. *Mol. Biol. Cell* **2005**, *16*, 1735–1743. [CrossRef] [PubMed]

46. Tzivion, G.; Avruch, J. 14-3-3 proteins: Active cofactors in cellular regulation by serine/threonine phosphorylation. *J. Biol. Chem.* **2002**, *277*, 3061–3064. [CrossRef] [PubMed]

47. Li, X.; Chen, L.; Dhaubhadel, S. 14-3-3 proteins regulate the intracellular localization of the transcriptional activator *GmMYB176* and affect isoflavonoid synthesis in soybean. *Plant J. Cell Mol. Biol.* **2012**, *71*, 239–250. [CrossRef]

48. Su, A.I.; Wiltshire, T.; Batalov, S.; Lapp, H.; Ching, K.A.; Block, D.; Zhang, J.; Soden, R.; Hayakawa, M.; Kreiman, G.; et al. A gene atlas of the mouse and human protein-encoding transcriptomes. *Proc. Natl. Acad. Sci. USA* **2004**, *101*, 6062–6067. [CrossRef] [PubMed]

49. Yang, Y.; Yu, M.D.; Xu, F.X.; Yu, Y.S.; Liu, C.Y.; Li, J.; Wang, X.L. Identification and expression analysis of the *14-3-3* gene family in the mulberry tree. *Plant Mol. Biol. Report.* **2015**, *33*, 1815–1824. [CrossRef]

50. Shi, H.; Wang, X.; Li, D.; Tang, W.; Wang, H.; Xu, W.; Li, X. Molecular characterization of cotton *14-3-3L* gene preferentially expressed during fiber elongation. *J. Genet. Genom. Yi Chuan Xue Bao* **2007**, *34*, 151–159. [CrossRef]

51. Zhang, Z.T.; Zhou, Y.; Li, Y.; Shao, S.Q.; Li, B.Y.; Shi, H.Y.; Li, X.B. Interactome analysis of the six cotton *14-3-3s* that are preferentially expressed in fibres and involved in cell elongation. *J. Exp. Bot.* **2010**, *61*, 3331–3344. [CrossRef]

52. Sorrell, D.A.; Marchbank, A.M.; Chrimes, D.A.; Dickinson, J.R.; Rogers, H.J.; Francis, D.; Grierson, C.S.; Halford, N.G. The Arabidopsis 14-3-3 protein, *GF14omega*, binds to the *Schizosaccharomyces pombe* Cdc25 phosphatase and rescues checkpoint defects in the rad24- mutant. *Planta* **2003**, *218*, 50–57. [CrossRef] [PubMed]

53. Oh, C.S.; Martin, G.B. Tomato 14-3-3 protein TFT7 interacts with a MAP kinase kinase to regulate immunity-associated programmed cell death mediated by diverse disease resistance proteins. *J. Biol. Chem.* **2011**, *286*, 14129–14136. [CrossRef] [PubMed]

54. Teper, D.; Salomon, D.; Sunitha, S.; Kim, J.G.; Mudgett, M.B.; Sessa, G. *Xanthomonas euvesicatoria* type III effector XopQ interacts with tomato and pepper 14-3-3 isoforms to suppress effector-triggered immunity. *Plant J. Cell Mol. Biol.* **2014**, *77*, 297–309. [CrossRef] [PubMed]

55. Aksamit, A.; Korobczak, A.; Skala, J.; Lukaszewicz, M.; Szopa, J. The *14-3-3* gene expression specificity in response to stress is promoter-dependent. *Plant Cell Physiol.* **2005**, *46*, 1635–1645. [CrossRef]

56. Yang, X.; Wang, W.; Coleman, M.; Orgil, U.; Feng, J.; Ma, X.; Ferl, R.; Turner, J.G.; Xiao, S. *Arabidopsis 14-3-3 lambda* is a positive regulator of RPW8-mediated disease resistance. *Plant J. Cell Mol. Biol.* **2009**, *60*, 539–550. [CrossRef]

57. Schoonheim, P.J.; Sinnige, M.P.; Casaretto, J.A.; Veiga, H.; Bunney, T.D.; Quatrano, R.S.; de Boer, A.H. 14-3-3 adaptor proteins are intermediates in ABA signal transduction during barley seed germination. *Plant J. Cell Mol. Biol.* **2007**, *49*, 289–301. [CrossRef]

58. Visconti, S.; D'Ambrosio, C.; Fiorillo, A.; Arena, S.; Muzi, C.; Zottini, M.; Aducci, P.; Marra, M.; Scaloni, A.; Camoni, L. Overexpression of 14-3-3 proteins enhances cold tolerance and increases levels of stress-responsive proteins of *Arabidopsis* plants. *Plant Sci. Int. J. Exp. Plant Biol.* **2019**, *289*, 110215. [CrossRef]

59. Lapointe, G.; Luckevich, M.D.; Cloutier, M.; Seguin, A. 14-3-3 gene family in hybrid poplar and its involvement in tree defence against pathogens. *J. Exp. Bot.* **2001**, *52*, 1331–1338. [CrossRef]

60. Keller, C.K.; Radwan, O. The functional role of 14-3-3 proteins in plant-stress interactions. *I-ACES* **2015**, *1*, 100–110.

61. Sun, X.; Luo, X.; Sun, M.; Chen, C.; Ding, X.; Wang, X.; Yang, S.; Yu, Q.; Jia, B.; Ji, W.; et al. A Glycine soja 14-3-3 protein GsGF14o participates in stomatal and root hair development and drought tolerance in *Arabidopsis thaliana*. *Plant Cell Physiol.* **2014**, *55*, 99–118. [CrossRef]

62. Van Dam, S.; Vosa, U.; van der Graaf, A.; Franke, L.; de Magalhaes, J.P. Gene co-expression analysis for functional classification and gene–disease predictions. *Brief. Bioinform.* **2018**, *19*, 575–592. [CrossRef] [PubMed]

63. Fields, S.; Song, O.-k. A novel genetic system to detect protein–protein interactions. *Nature* **1989**, *340*, 245–246. [CrossRef] [PubMed]

64. Schoonheim, P.J.; Veiga, H.; Pereira Dda, C.; Friso, G.; van Wijk, K.J.; de Boer, A.H. A comprehensive analysis of the 14-3-3 interactome in barley leaves using a complementary proteomics and two-hybrid approach. *Plant Physiol.* **2007**, *143*, 670–683. [CrossRef] [PubMed]

65. Johnson, C.; Tinti, M.; Wood, N.T.; Campbell, D.G.; Toth, R.; Dubois, F.; Geraghty, K.M.; Wong, B.H.; Brown, L.J.; Tyler, J.; et al. Visualization and biochemical analyses of the emerging mammalian 14-3-3-phosphoproteome. *Mol. Cell. Proteom.* **2011**, *10*, M110.005751. [CrossRef]

66. Ding, Y.D.; Chang, J.W.; Guo, J.; Chen, D.; Li, S.; Xu, Q.; Deng, X.X.; Cheng, Y.J.; Chen, L.L. Prediction and functional analysis of the sweet orange protein-protein interaction network. *BMC Plant Biol.* **2014**, *14*, 213. [CrossRef]

67. Deprost, D.; Yao, L.; Sormani, R.; Moreau, M.; Leterreux, G.; Nicolai, M.; Bedu, M.; Robaglia, C.; Meyer, C. The Arabidopsis TOR kinase links plant growth, yield, stress resistance and mRNA translation. *EMBO Rep.* **2007**, *8*, 864–870. [CrossRef]

68. Guo, A.Y.; Zhu, Q.H.; Chen, X.; Luo, J.C. GSDS: A gene structure display server. *Hereditas* **2007**, *29*, 1023–1026. [CrossRef]

69. Larkin, M.A.; Blackshields, G.; Brown, N.P.; Chenna, R.; McGettigan, P.A.; McWilliam, H.; Valentin, F.; Wallace, I.M.; Wilm, A.; Lopez, R.; et al. Clustal W and clustal X version 2.0. *Bioinformatics* **2007**, *23*, 2947–2948. [CrossRef]

70. Tamura, K.; Dudley, J.; Nei, M.; Kumar, S. MEGA4: Molecular Evolutionary Genetics Analysis (MEGA) software version 4.0. *Mol. Biol. Evol.* **2007**, *24*, 1596–1599. [CrossRef]

71. Deng, W.; Wang, Y.; Liu, Z.; Cheng, H.; Xue, Y. HemI: A toolkit for illustrating heatmaps. *PLoS ONE* **2014**, *9*, e111988. [CrossRef]

72. Aritua, V.; Achor, D.; Gmitter, F.G.; Albrigo, G.; Wang, N. Transcriptional and microscopic analyses of citrus stem and root responses to *Candidatus* Liberibacter asiaticus infection. *PLoS ONE* **2013**, *8*, e73742. [CrossRef] [PubMed]

73. Kim, J.S.; Sagaram, U.S.; Burns, J.K.; Li, J.L.; Wang, N. Response of sweet orange (*Citrus sinensis*) to 'Candidatus Liberibacter asiaticus' infection: Microscopy and microarray analyses. *Phytopathology* **2009**, *99*, 50–57. [CrossRef]

74. Cernadas, R.A.; Camillo, L.R.; Benedetti, C.E. Transcriptional analysis of the sweet orange interaction with the citrus canker pathogens *Xanthomonas axonopodis* pv. *citri* and *Xanthomonas axonopodis pv. aurantifolii*. *Mol. Plant Pathol.* **2008**, *9*, 609–631. [CrossRef] [PubMed]

75. Smyth, G.K. Limma: Linear models for microarray data. In *Bioinformatics and Computational Biology Solutions Using R and Bioconductor*; Gentleman, R., Carey, V.J., Huber, W., Irizarry, R.A., Dudoit, S., Eds.; Springer: New York, NY, USA, 2005; pp. 397–420.

76. Gautier, L.; Cope, L.; Bolstad, B.; Irizarrt, R. Affy—Analysis of affymetrix genechip data at the probe level. *Bioinformatics* **2004**, *20*, 307–315. [CrossRef] [PubMed]

77. Wu, J.; Su, S.; Fu, L.; Zhang, Y.; Chai, L.; Yi, H. Selection of reliable reference genes for gene expression studies using quantitative real-time pcr in navel orange fruit development and pummelo floral organs. *Sci. Hortic.* **2014**, *176*, 180–188. [CrossRef]

78. Mafra, V.; Kubo, K.S.; Alves-Ferreira, M.; Ribeiro-Alves, M.; Stuart, R.M.; Boava, L.P.; Rodrigues, C.M.; Machado, M.A. Reference genes for accurate transcript normalization in citrus genotypes under different experimental conditions. *PLoS ONE* **2012**, *7*, e31263. [CrossRef]

79. Livak, K.J.; Schmittgen, T.D. Analysis of relative gene expression data using real-time quantitative PCR and the 2(-Delta Delta C(T)) method. *Methods* **2001**, *25*, 402–408. [CrossRef]

80. Wong, D.C.J.; Sweetman, C.; Ford, C.M. Annotation of gene function in citrus using gene expression information and co-expression networks. *BMC Plant Biol.* **2014**, *14*, 186. [CrossRef]

81. Gietz, R.D.; Schiestl, R.H. High-efficiency yeast transformation using the LiAc/SS carrier DNA/PEG method. *Nat. Protoc.* **2007**, *2*, 31–34. [CrossRef]

International Journal of
Molecular Sciences

Article

Annexin 1 Is a Component of eATP-Induced Cytosolic Calcium Elevation in *Arabidopsis thaliana* Roots

Amirah Mohammad-Sidik [1], Jian Sun [2], Ryoung Shin [3], Zhizhong Song [4], Youzheng Ning [1], Elsa Matthus [1], Katie A. Wilkins [1] and Julia M. Davies [1,*]

1   Department of Plant Sciences, University of Cambridge, Cambridge CB2 3EA, UK; amirahbmsa@gmail.com (A.M.-S.); yn283@cam.ac.uk (Y.N.); ematthus@hotmail.com (E.M.); kaw67@cam.ac.uk (K.A.W.)
2   School of Life Sciences, Jiangsu Normal University, Xuzhou 221116, China; sunjian@jsnu.edu.cn
3   RIKEN Centre for Sustainable Resource Science, Yokohama, Kanagawa 230-0045, Japan; ryoung.shin@riken.jp
4   School of Agriculture, Ludong University, Yantai 264205, China; szhzh2000@163.com
*   Correspondence: jmd32@cam.ac.uk; Tel.: +44-1223-333-939

Abstract: Extracellular ATP (eATP) has long been established in animals as an important signalling molecule but this is less understood in plants. The identification of *Arabidopsis thaliana* DORN1 (Does Not Respond to Nucleotides) as the first plant eATP receptor has shown that it is fundamental to the elevation of cytosolic free Ca^{2+} ($[Ca^{2+}]_{cyt}$) as a possible second messenger. eATP causes other downstream responses such as increase in reactive oxygen species (ROS) and nitric oxide, plus changes in gene expression. The plasma membrane Ca^{2+} influx channels involved in eATP-induced $[Ca^{2+}]_{cyt}$ increase remain unknown at the genetic level. *Arabidopsis thaliana* Annexin 1 has been found to mediate ROS-activated Ca^{2+} influx in root epidermis, consistent with its operating as a transport pathway. In this study, the loss of function Annexin 1 mutant was found to have impaired $[Ca^{2+}]_{cyt}$ elevation in roots in response to eATP or eADP. Additionally, this annexin was implicated in modulating eATP-induced intracellular ROS accumulation in roots as well as expression of eATP-responsive genes.

Keywords: extracellular ATP; ADP; root; *Arabidopsis*; annexin 1; calcium; calcium channel; reactive oxygen species

Citation: Mohammad-Sidik, A.; Sun, J.; Shin, R.; Song, Z.; Ning, Y.; Matthus, E.; Wilkins, K.A.; Davies, J.M. Annexin 1 Is a Component of eATP-Induced Cytosolic Calcium Elevation in *Arabidopsis thaliana* Roots. *Int. J. Mol. Sci.* **2021**, 22, 494. https://doi.org/10.3390/ijms22020494

Received: 14 December 2020
Accepted: 4 January 2021
Published: 6 January 2021

Publisher's Note: MDPI stays neutral with regard to jurisdictional claims in published maps and institutional affiliations.

1. Introduction

Extracellular ATP (eATP) is implicated as an apoplastic signal molecule in the abiotic and biotic stress responses of plants, their cellular viability, growth and stomatal regulation [1–6]. In *Arabidopsis thaliana*, eATP can act as a damage-associated molecular pattern (DAMP) and activates immunity signalling through the plasma membrane purinoreceptor AtDORN1 (Does Not Respond to Nucleotides1, also known as P2K1) [7]. A plasma membrane co-receptor P2K2 has recently been identified and both DORN1/P2K1 and P2K2 are lectin receptor kinases [8]. eATP-dependent but AtDORN1-independent effects have also been reported [9,10], pointing to the presence of other perception mechanisms. eATP perception triggers increase in root and leaf free cytosolic Ca^{2+} ($[Ca^{2+}]_{cyt}$) [5] that can lead to the production of reactive oxygen species (ROS) as further putative signalling agents [11–14]. Nitric oxide (NO) production can also be increased [2,15,16]. Eventually, signalling causes changes in gene expression [7,10,12,17–19]. Many eATP-responsive genes contain the CAM-box motif, which suggests that CAMTAs (Calmodulin-binding Transcription Activators) are important components [18]. Indeed, gene regulation could run through CAMTA3 [18], potentially connecting Ca^{2+} as a second messenger to changes in transcription due to eATP. It may be that the nuclear Ca^{2+} increase that follows eATP-induced $[Ca^{2+}]_{cyt}$ increase [20] activates CAMTA3 through Ca^{2+}-CAM interaction. Understanding how eATP causes $[Ca^{2+}]_{cyt}$ increase is, therefore, relevant to downstream responses.

It is known that eATP-induced $[Ca^{2+}]_{cyt}$ increase in *Arabidopsis* roots requires plasma membrane Ca^{2+} influx channels but the molecular identities of these channels remain unknown, as is also the case for other organs [9,11,14,21–23]. It has been hypothesized that the *Arabidopsis* annexin 1 protein (AtANN1) could be involved in mediating plasma membrane Ca^{2+} influx [5,24] and recently AtANN4 was found to mediate eATP-induced $[Ca^{2+}]_{cyt}$ increase when expressed in *Xenopus* oocytes [25]. AtANN1 is thought to act as a plasma membrane Ca^{2+} channel in the root $[Ca^{2+}]_{cyt}$ response to salinity stress, hyperosmotic stress, and oxidative stress [26–28]. In this study, the possible involvement of AtANN1 in the root and leaf eATP-$[Ca^{2+}]_{cyt}$ signalling pathway has been tested using an *Atann1* loss of function mutant constitutively expressing cytosolic (apo)aequorin as a luminescent $[Ca^{2+}]_{cyt}$ reporter [26–28]. As eADP has also been shown to increase $[Ca^{2+}]_{cyt}$ [7,11,12,14,21], this nucleotide was also tested. The consequences for eATP-induced root intracellular ROS elevation and gene transcription were also investigated. The results show that AtANN1 is required for the normal $[Ca^{2+}]_{cyt}$ response towards both eATP and eADP in roots. It affects the spatial extent of intracellular ROS accumulation in roots and influences their eATP-induced transcriptional response.

2. Results

2.1. AtANN1 Mediates Root $[Ca^{2+}]_{cyt}$ Elevation In Response To eATP and eADP

Previously, eATP-dependent root $[Ca^{2+}]_{cyt}$ elevation was found to be wholly reliant on the AtDORN1 receptor [5]. To assess the role of AtANN1 in this response, seven-day-old whole roots of *Atann1* (loss of function mutant) and Col-0 (expressing cytosolic (apo)aequorin under the 35S CaMV promoter) were excised and assayed individually to measure $[Ca^{2+}]_{cyt}$ in the presence of eATP. Addition of control solution after 35 s of measurement evoked a monophasic $[Ca^{2+}]_{cyt}$ increase in response to mechanical stimulation ("touch response") before returning to the basal level (Figure 1a). The touch response of *Atann1* roots was similar to Col-0 in terms of the amplitude ("touch peak") and the total accumulation of $[Ca^{2+}]_{cyt}$ (estimated as the area under the curve, AUC) ($p > 0.05$; Figure 1b). Measurement of $[Ca^{2+}]_{cyt}$ in response to 1 mM eATP revealed a biphasic increase comprising a first peak and second peak after the initial touch response (Figure 1c). A biphasic response in roots was also observed previously [5,11]. The touch response was similar between genotypes ($p > 0.5$) and although the first eATP peak was lower in *Atann1* it was not significantly different to Col-0 ($p > 0.05$) (Figure 1d). The second eATP-induced $[Ca^{2+}]_{cyt}$ peak response of *Atann1* was significantly lower than Col-0 ($p < 0.0001$) (Figure 1e). The total $[Ca^{2+}]_{cyt}$ accumulated was also significantly lower in *Atann1* ($p < 0.0001$; Figure 1e).

eADP also evokes a biphasic $[Ca^{2+}]_{cyt}$ increase in roots [5,11] that is entirely dependent on AtDORN1 [5]. To assess whether AtANN1 is also required, a similar test with eADP on seven-day-old excised roots was carried out for Col-0 and *Atann1*. As shown in Figure 2a, control treatment elicited a monophasic touch response in both genotypes. No significant differences were evident for either the touch peak ($p > 0.05$) or the overall $[Ca^{2+}]_{cyt}$ ($p > 0.05$) between Col-0 and *Atann1* (Figure 2b). In the presence of 1 mM eADP as shown in Figure 2c, both Col-0 and *Atann1* produced a biphasic $[Ca^{2+}]_{cyt}$ increase following the touch response. No significant difference was found between Col-0 and *Atann1* in the touch peak response ($p > 0.05$) (Figure 2d). Unlike eATP treatment however, *Atann1* showed a significantly impaired ability to produce both a normal first peak in response to 1 mM eADP ($p < 0.0001$) (Figure 2d) and a normal second peak ($p < 0.0001$) (Figure 2e). Overall, the loss of functional AtANN1 protein led to a reduced total accumulation of $[Ca^{2+}]_{cyt}$ compared to Col-0 ($p < 0.0001$) (Figure 2e). Lowered *AtDORN1* expression cannot explain the impairments in *Atann1's* response to extracellular nucleotides as no significant difference in the receptor's expression between Col-0 and *Atann1* roots was found in either control conditions ($p > 0.5$) or in the presence of 1 mM ATP ($p > 0.5$) (Figure 2f). Therefore, the defects in $[Ca^{2+}]_{cyt}$ elevation appear to rest with the lack of AtANN1.

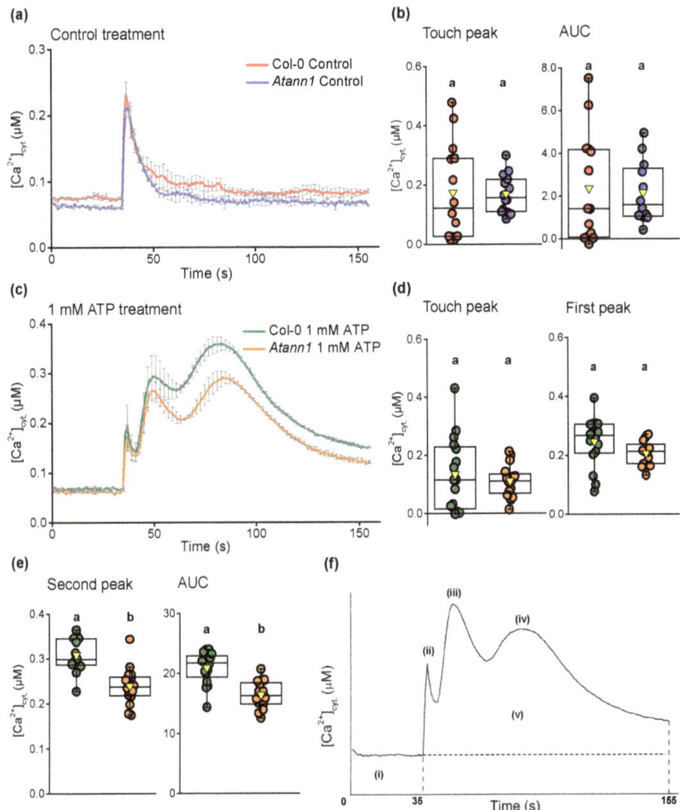

Figure 1. *Arabidopsis* annexin 1 (AtANN1) is needed for normal [Ca²⁺]$_{cyt}$ elevation in a root by eATP. (**a**) Time course of [Ca²⁺]$_{cyt}$ elevation produced by control treatment in three experiments (mean (± SEM): Col-0 in red, *n* = 14 roots in total; *Atann1* loss of function mutant in blue, *n* = 14). (**b**) The [Ca²⁺]$_{cyt}$ touch peak values and area under the curve (AUC) extracted from the control time course (± SEM). Middle line of the boxplot represents the median whereas the inverted triangle represents the mean. (**c**) Time course of [Ca²⁺]$_{cyt}$ elevation with 1 mM eATP treatment in 3 experiments (Col-0 in green, *n* = 16; *Atann1* in orange, *n* = 16). (**d**) The touch peak and the first peak [Ca²⁺]$_{cyt}$ values extracted from the 1 mM eATP time course. (**e**) Second peak and the AUC [Ca²⁺]$_{cyt}$ values. (**f**) Schematic diagram of different time course sections. Each section was calculated with the average baseline value (indicated by (i)) subtracted. Touch peak (ii) was the highest [Ca²⁺]$_{cyt}$ value of the touch response between 35 and 40 s due to mechanical stimulus from solution addition at the 35th second. First peak (iii) and second peak (iv) were the highest [Ca²⁺]$_{cyt}$ value between 40 s and 60 s and 60 s and 155 s, respectively. Total [Ca²⁺]$_{cyt}$ accumulation was obtained from the AUC (v; 35 s–155 s). *p*-values were obtained from analysis of variance (ANOVA) with Tukey's post-hoc test or Kruskal–Wallis test for non-parametric approaches. Different lower-case letters indicate a significant difference between means (*p* < 0.05).

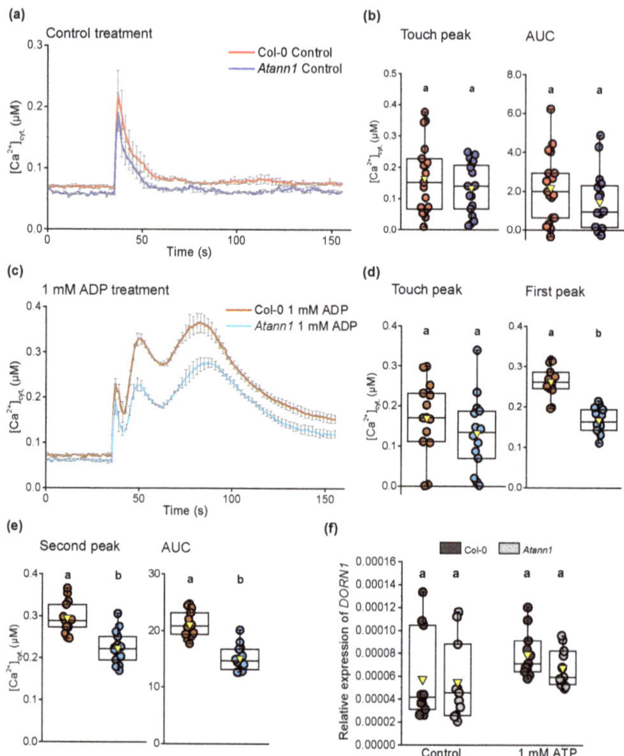

Figure 2. AtANN1 is involved in eADP-induced $[Ca^{2+}]_{cyt}$ elevation in the root. (**a**) Mean ($\pm$ SEM) time course of $[Ca^{2+}]_{cyt}$ increase by control treatment in three experiments (Col-0 in red, $n = 19$; *Atann1* in blue, $n = 18$). (**b**) $[Ca^{2+}]_{cyt}$ touch peak values and AUC extracted from the control time course. Middle line of the boxplot represents the median whereas the inverted triangle represents the mean. (**c**) Mean ($\pm$ SEM) time course of $[Ca^{2+}]_{cyt}$ increase by 1 mM eADP treatment obtained from three experiments (Col-0 in brown, $n = 14$; *Atann1* in light blue, $n = 14$). (**d**) The touch peak and the first peak $[Ca^{2+}]_{cyt}$ values from the 1 mM eADP time course. (**e**) Second peak and the total AUC $[Ca^{2+}]_{cyt}$ values. (**f**) Quantification of *AtDORN1* gene expression in Col-0 and *Atann1* roots after seven days of growth on control medium or 1 mM eATP-containing medium (Col-0 in black, *Atann1* in grey with $n = 11$ for each genotype and treatment) obtained from three experiments. *p*-values were obtained from ANOVA with Tukey's post-hoc test. Different lower-case letters indicate a significant difference between means ($p < 0.05$).

2.2. AtANN1's Involvement in the eATP-Generated First Peak Response Is Concentration-Dependent

A non-hydrolysable ATP analogue (ATPγS; Adenosine 5′-[γ-thio] triphosphate) and ADP analogue (ADPβS; Adenosine 5′-[β-thio] diphosphate) were then used to confirm that the agonists acted as signal molecules rather than as energy sources that drive the $[Ca^{2+}]_{cyt}$ increase. The ATPγS used was a tetralithium salt whereas the ADPβS was a trilithium salt. A LiCl treatment (4 mM for ATPγS and 3 mM for ADPβS) was carried out alongside the non-hydrolysable analogues as a lithium control. Seven-day-old individual whole roots responded with the biphasic $[Ca^{2+}]_{cyt}$ increase when tested with different concentrations of eATPγS, or the eATP/LiCl salt control (Figure 3a). As seen in Figure 3b, the touch peak responses were not significantly different between genotypes ($p > 0.5$). The Col-0 first $[Ca^{2+}]_{cyt}$ peak did not require eATP hydrolysis (no significant difference between 1 mM eATPγS and 1 mM eATP/LiCl) and indeed was already "saturated" at 0.1 mM eATPγS (no significant difference, $p > 0.05$, between 0.1 and 1 mM eATPγS) (Figure 3c).

The first peak responses of *Atann1* also appeared saturated at 0.1 mM eATPγS (Figure 3c). As in the response to 1 mM eATP in Figure 1d, the first peak of *Atann1* in response to 1 mM eATPγS was lower than Col-0 but was not significant (Figure 3c). However, when 1 mM eATP was tested with LiCl as a control for Li$^+$ addition, the difference did become significant. This may be due to the range of Col-0 values in the 1 mM eATPγS test. Importantly, within each genotype there was no evidence for eATP's acting as an energy source at 1 mM. *Atann1* showed a significantly lower first peak [Ca^{2+}]$_{cyt}$ response compared to Col-0 in response to 0.1 mM eATPγS ($p < 0.001$) and 0.2 mM eATPγS ($p < 0.001$) (Figure 3c). These results suggest that the role of AtANN1 in the first peak response relies on the concentration of agonist used. The second [Ca^{2+}]$_{cyt}$ peak of Col-0 showed a significant dependence on the concentration of eATPγS, as did *Atann1* between 0.1 and 1 mM (Figure 3d). AtANN1 proved to be important in generating the second peak as the *Atann1* mutant failed to respond in similar magnitude as the Col-0 over the concentration range ($p < 0.001$) (Figure 3d). This was also evident in the total [Ca^{2+}]$_{cyt}$ accumulated where there were significant differences between Col-0 and *Atann1* ($p < 0.0001$) in the AUC for every concentration tested (Figure 3e).

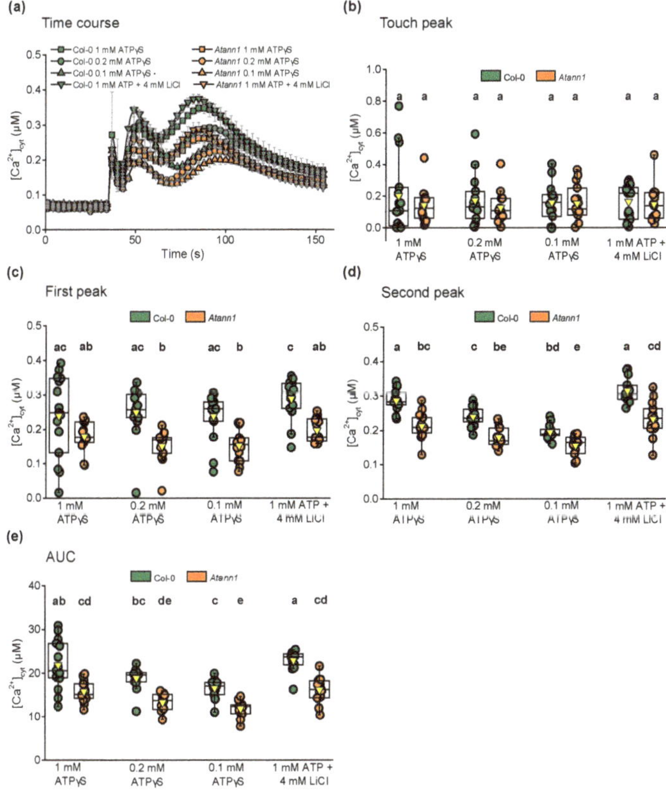

Figure 3. AtANN1 is crucial for the first peak [Ca^{2+}]$_{cyt}$ response at lower eATP concentration. (**a**) Mean (± SEM) [Ca^{2+}]$_{cyt}$ time course in response to different concentrations of eATPγS or eATP with LiCl control from 3 experiments with $n = 14$–15 per genotype and treatment. (**b**) The [Ca^{2+}]$_{cyt}$ touch peak values (± SEM), (**c**) first peak [Ca^{2+}]$_{cyt}$ values (± SEM), (**d**) second peak [Ca^{2+}]$_{cyt}$ values (± SEM) and (**e**) the total [Ca^{2+}]$_{cyt}$ accumulated obtained from AUC (± SEM) for each concentration tested in both Col-0 and *Atann1* extracted from the time course. *p*-values were obtained from ANOVA with Tukey's post-hoc test or Kruskal-Wallis test for non-parametric approach. Different lower case letters indicate significant difference between means ($p < 0.05$).

Both Col-0 and *Atann1* generated the transient biphasic $[Ca^{2+}]_{cyt}$ elevation after the touch peak response when challenged with different concentrations of eADPβS or 1 mM eADP with 3 mM LiCl (Figure 4a). Both Col-0 and *Atann1* produced the same level of touch peak response for each treatment ($p > 0.5$) (Figure 4b). Just like the first peak response to eATPγS, the response of both genotypes to eADPβS appeared saturated at 0.1 mM (Figure 4c). In contrast to the eATPyS test, there were significant differences between Col-0 and *Atann1* in the first peak $[Ca^{2+}]_{cyt}$ response regardless of the concentration of eADPβS tested ($p < 0.001$) (Figure 4c). Consistent with the results in the hydrolysable eADP test, *Atann1* was found to produce lower $[Ca^{2+}]_{cyt}$ responses than the Col-0 in the second peak ($p < 0.001$) (Figure 4d) and in the AUC ($p < 0.001$) (Figure 4e) for all the concentrations tested. There was no evidence for eADP's acting as an energy source. Overall, these results suggest that the involvement of AtANN1 in generating the first peak response is specific to lower concentrations of eATP, with the likelihood of other components participating at higher concentration. AtANN1 is still needed for the first peak response to high concentration of eADP and for the second peak regardless of agonist concentration.

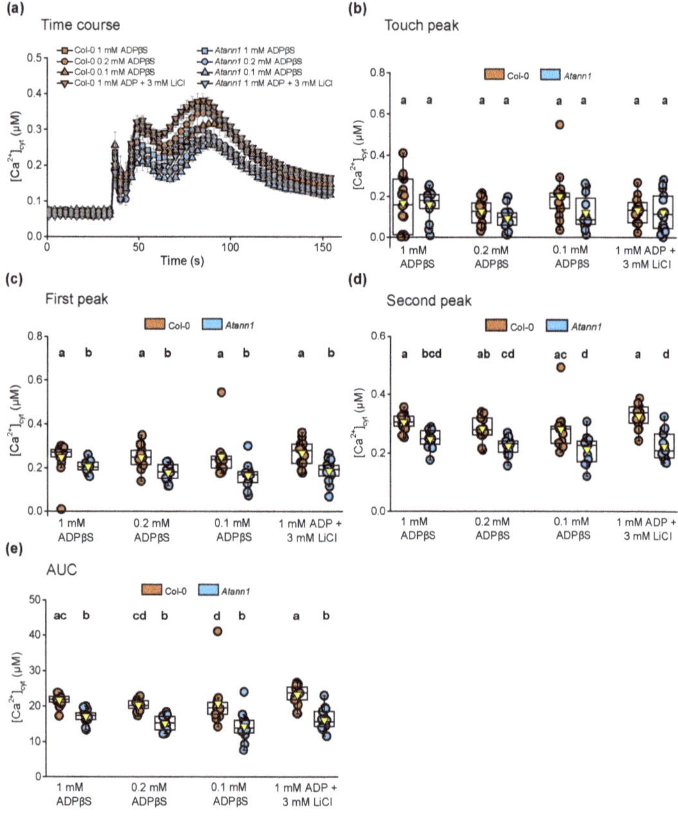

Figure 4. AtANN1 is involved in the eADPβS-induced $[Ca^{2+}]_{cyt}$ response at all concentrations tested. (**a**) Mean ($\pm$ SEM) $[Ca^{2+}]_{cyt}$ time course in response to different concentrations of eADPβS or ADP with LiCl control from 3 experiments ($n = 13$—15 per genotype and treatment). (**b**) The $[Ca^{2+}]_{cyt}$ touch peak values ($\pm$ SEM), (**c**) first peak $[Ca^{2+}]_{cyt}$ values ($\pm$ SEM), (**d**) second peak $[Ca^{2+}]_{cyt}$ values ($\pm$ SEM) and (**e**) the total $[Ca^{2+}]_{cyt}$ accumulated obtained from AUC ($\pm$ SEM) for each concentration tested in both Col-0 and *Atann1* extracted from the time courses. *p*-values were obtained from ANOVA with Tukey's post-hoc test or Kruskal-Wallis test for non-parametric approach. Different lower case letters indicate a significant difference between means ($p < 0.05$).

2.3. *AtANN1 Sets the Spatial Extent of Intracellular ROS in Roots in Response to eATP*

Addition of eATP (but not eADP) causes rapid intracellular accumulation of ROS in *Arabidopsis* roots that requires Ca^{2+} influx and is largely dependent on AtRBOHC activity [14,21]. Whether AtANN1 is involved in the production of intracellular ROS was tested here with ester-loaded CM-H_2DCFDA (5-(and-6-)-chloromethyl-$2',7'$-dichlorodihydrofluorscein diacetate) [14]. Figure 5a shows the baseline ROS detected in control conditions. In the presence of 1 mM eATP (Figure 5b), ROS increase was detectable within 20 s, as reported previously [14]. Signal intensity was higher than the baseline in both Col-0 and *Atann1* with the latter supporting a greater length of ROS production that clearly extended into the mature zone supporting root hairs (Figure 5b). Further statistical analysis carried out confirmed this significant difference between *Atann1* and Col-0 (Figure 5c). In line with previous studies [14,21], 1 mM eADP failed to induce any intracellular ROS accumulation in either genotype (Figure 5d).

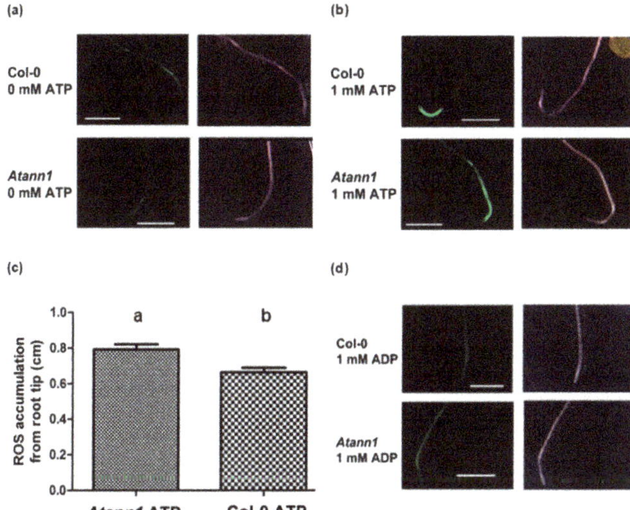

Figure 5. *Atann1* supports a longer zone of intracellular ROS accumulation than Col-0 in response to eATP. (**a**) CM-H2DCFDA fluorescence from a Col-0 or *Atann1* root under control conditions. Corresponding bright field images are also shown. (**b**) Roots after exposure to 1 mM eATP. (**c**) Mean ($\pm$ SEM) length of root from the tip fluorescing after exposure to eATP (Col-0 $n = 48$; *Atann1* $n = 68$; $p = 0.0026$, Student's t-test). (**d**) Roots after exposure to 1 mM eADP. Scale bar = 4 mm. Different lower-case letters indicate a significant difference between means ($p < 0.05$).

The focus of the analysis was then shifted to the root apex to distinguish any differences in signal intensity between Col-0 and *Atann1* (Figure 6a). Based on the mean signal intensity, there was no significant difference in ROS production between genotypes under control conditions. Both Col-0 and *Atann1* treated with 1 mM eATP produced significantly higher ROS than under control conditions but although *Atann1* supported a greater spatial extent of ROS accumulation, the mean signal intensity at the root apex was similar to Col-0 in the presence of eATP. Once again, 1 mM eADP treatment failed to increase ROS in either genotype (Figure 6b). Overall, these data suggest that AtANN1 is involved in controlling the spatial extent of ROS accumulation evoked by 1 mM eATP.

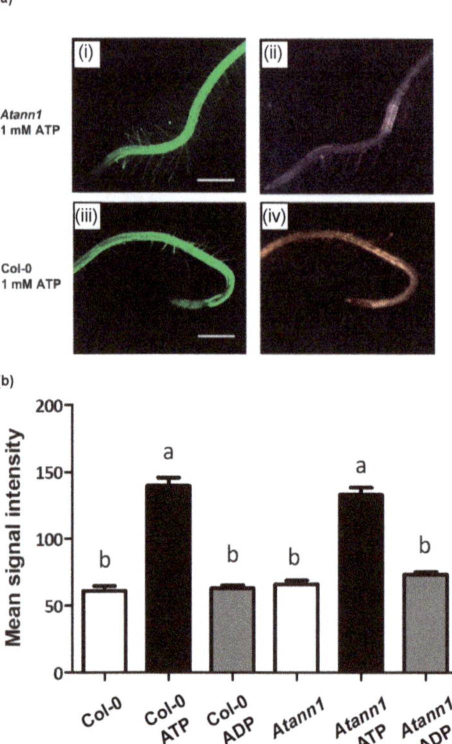

Figure 6. eATP-induced intracellular ROS accumulation at the root apex at higher resolution. (**a**) CM-H$_2$DCFDA fluorescence from (i) a representative *Atann1* root exposed to 1 mM eATP and (ii) corresponding bright field image. (iii) A representative Col-0 root after exposure to 1 mM eATP and (iv) corresponding bright field image. (**b**) Mean ($\pm$ SEM) of fluorescence pixel intensity at root apices under control conditions and after 30 s exposure to 1 mM eATP (Col-0 n = 48; *Atann1 n* = 68) or 1 mM ADP (Col-0 n = 32; *Atann1 n* = 38). Both genotypes responded significantly to eATP but not eADP ($p < 0.001$; ANOVA with Dunnett's post-hoc test). Scale bar = 2 mm. Different lower-case letters indicate a significant difference between means ($p < 0.05$).

2.4. AtANN1 Is Required For eATP-Induced Changes in the Expression of ACS6 and WRKY40

eATP has been shown previously to be able to induce transcription of genes involved in stress responses [7,10,12,18,29–31]. eATP-induced genes *AtRBOHD* (NADPH/Respiratory Burst Oxidase Protein D), *AtWRKY40* (WRKY DNA-Binding Protein 40) and *AtACS6* (1-Aminocyclopropane-1-carboxylic Acid Synthase 6) [7,12,18] were tested for regulation by 1 mM eATP in roots and the possibility of AtANN1's affecting their regulation (Figure 7). Ionic composition of the control solution was identical to that used in the aequorin tests (10 mM CaCl$_2$, 0.1 mM KCl). *AtANN1* transcript was almost completely knocked-down in *Atann1* compared to the Col-0 in control conditions ($p < 0.05$) and after eATP treatment ($p < 0.01$). Transcript level did not increase in Col-0 after eATP treatment (Figure 7a). *AtRBOHD* gene was not up-regulated by either 10 min or 30 min of eATP treatment when compared with the control for both Col-0 ($p > 0.05$) and *Atann1* ($p > 0.05$) (Figure 7b). In contrast, *AtACS6* was significantly upregulated in Col-0 when treated for 10 min with eATP compared to the control treatment ($p < 0.01$) but fell back to control levels after 30 min ($p > 0.05$; Figure 7c). Expression was not significantly upregulated in *Atann1* after 10 min of eATP treatment ($p > 0.5$) and it remained significantly lower than Col-0 at this time point

($p < 0.05$). No significant difference was evident in *AtACS6* expression between 30 min control treatment and eATP treatment for *Atann1* ($p > 0.05$; Figure 7c). These findings suggest a temporal regulation of *AtACS6* by AtANN1 in the presence of eATP.

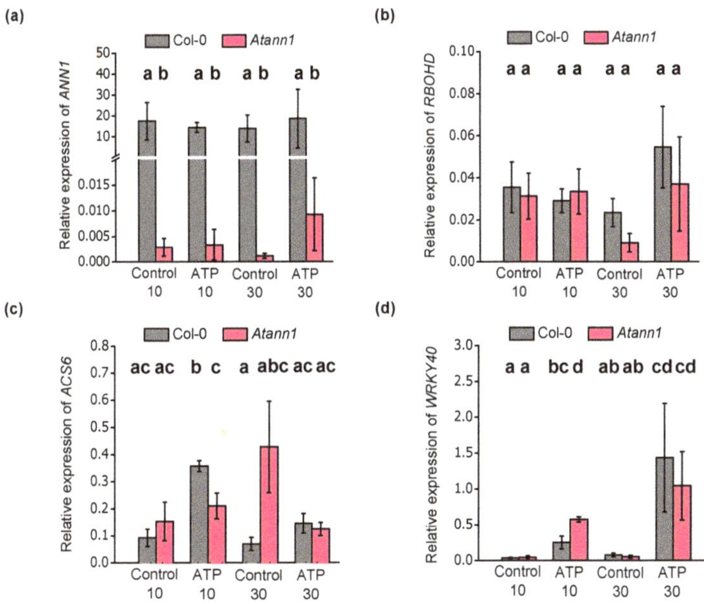

Figure 7. Transcriptional regulation of stress responsive genes by AtANN1. Col-0 and *Atann1* whole roots were treated with control solution or 1 mM eATP for 10 or 30 min. Transcript abundance of (**a**) *AtANN1*, (**b**) *AtRBOHD*, (**c**) *AtACS6* and (**d**) *AtWRKY40* normalised to two housekeeping genes; *AtUBQ10* and *AtTUB4*. Data were from the means (± SEM) of four independent trials. Student's *t*-test and Welch's *t*-test were used to test parametric data whereas Wilcoxon rank sum test was used for non-parametric data. Different lower-case letters indicate a significant difference between means ($p < 0.05$).

Transcript abundance of *AtWRKY40* was upregulated by eATP exposure (Figure 7d). Col-0 and *Atann1* samples treated with eATP for 10 and 30 min had significantly higher *AtWRKY40* transcript abundance compared to their controls (Col-0, $p < 0.05$; *Atann1*, $p < 0.05$). Notably, *Atann1* supported a significantly higher increase in *AtWRKY40* transcript than Col-0 after 10 min of eATP treatment ($p < 0.05$) but over time, no significant difference in transcript abundance between genotypes was found after 30 min of eATP treatment ($p > 0.5$) (Figure 7c), suggesting a temporal effect on the response. Overall, AtANN1 appears important in regulating the eATP-induced changes in the transcription of *AtACS6* and *AtWRKY40*.

2.5. eATP-Induced [Ca^{2+}]$_{cyt}$ Elevation Is Not Mediated by AtANN1 in Cotyledons

Whole seedlings of *A. thaliana* were found previously to elevate [Ca^{2+}]$_{cyt}$ in response to eATP [7]. To assess whether AtANN1 is involved in mediating this response in aerial organs as well as roots, the [Ca^{2+}]$_{cyt}$ elevation by 1 mM eATP of seven-day-old cotyledons of Col-0 and *Atann1* was compared. Consistent with previous results on Col-0 true leaves [5], control treatment of cotyledons caused a monophasic touch response in both Col-0 and *Atann1* (Figure 8a) that was not significantly different between genotypes ($p > 0.05$; Figure 8b). Also in common with previous results on Col-0 true leaves [5], 1 mM eATP treatment of cotyledons caused a prolonged monophasic [Ca^{2+}]$_{cyt}$ increase after the touch response (Figure 8c). No significant differences were found between Col-0 and *Atann1* in either the touch peak ($p > 0.05$), the first peak ($p > 0.05$) or in the total $[Ca^{2+}]_{cyt}$ accumulated ($p > 0.05$; Figure 8d,e). Based on these observations, it is certain that AtANN1's role in eATP signalling does not extend to the cotyledon and that the $[Ca^{2+}]_{cyt}$ signature caused by eATP treatment differs with the type of tissues or organs tested.

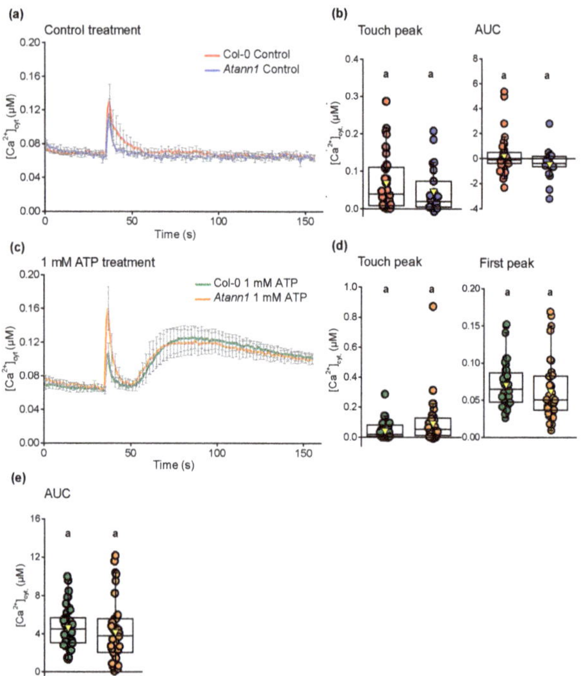

Figure 8. AtANN1 is not needed for eATP-induced $[Ca^{2+}]_{cyt}$ increase in cotyledons. (**a**) Mean ($\pm$ SEM) time course of $[Ca^{2+}]_{cyt}$ increase due to control treatment of seven-day-old individual cotyledons. (**b**) The $[Ca^{2+}]_{cyt}$ touch peak ($\pm$ SEM) and the total $[Ca^{2+}]_{cyt}$ accumulation by the AUC ($\pm$ SEM). Middle line of the boxplot represents the median whereas the inverted triangle represents the mean. (**c**) Mean ($\pm$ SEM) time course of $[Ca^{2+}]_{cyt}$ increase due to 1 mM eATP treatment. (**d**) The $[Ca^{2+}]_{cyt}$ touch peak values ($\pm$ SEM) and the first peak values ($\pm$ SEM). (**e**) Total $[Ca^{2+}]_{cyt}$ accumulated values ($\pm$ SEM) from the AUC. Data were obtained from four experiments with n = 21–39 cotyledons per genotype and treatment. p-values obtained from ANOVA with Tukey's post-hoc test. Identical lower-case letters indicate an insignificant difference between means ($p > 0.05$).

2.6. AtANN1 Is Less Important in the $[Ca^{2+}]_{cyt}$ Response of True Leaves to Extracellular ATP or ADP

AtANN1 expression is evident in true leaves as well as cotyledons [32–34]. To assess whether AtANN1's participation in $[Ca^{2+}]_{cyt}$ elevation is limited to roots, individual 14-days-old leaves were tested with varying concentrations of eATP and eADP. All the eATP concentrations used (0.1 mM, Figure 9a; 0.5 mM, Figure 9b; 1 mM, Figure 9c) generated a prolonged monophasic $[Ca^{2+}]_{cyt}$ response after the touch peak that was similar to both the seven-day-old cotyledon $[Ca^{2+}]_{cyt}$ pattern in the previous test and 14-day-old Col-0 leaves studied previously [5]. Although *Atann1* produced a lower peak response than Col-0, differences between the genotypes were not significant except for 1 mM eATP's causing Col-0 to produce a significantly higher total $[Ca^{2+}]_{cyt}$ than *Atann1* ($p < 0.05$) (Figure 9d).

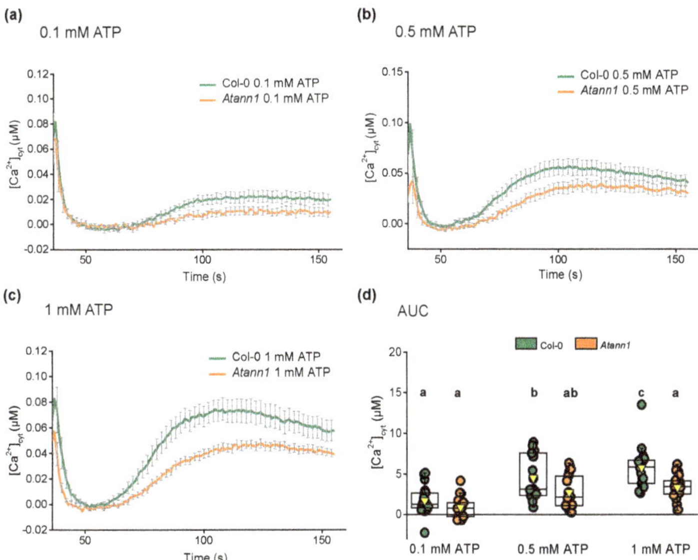

Figure 9. AtANN1 is required for leaf $[Ca^{2+}]_{cyt}$ elevation at higher concentrations of eATP. Mean ($\pm$ SEM) $[Ca^{2+}]_{cyt}$ time course (t = 36 s–155 s) of 14-day-old individual true leaves treated with (**a**) 0.1 mM eATP, (**b**) 0.5 mM eATP or (**c**) 1 mM eATP. (**d**) The total $[Ca^{2+}]_{cyt}$ accumulated over the period of measurement for each concentration given as AUC. Data were obtained from four experiments with n = 17–22 per genotype and treatment. Middle line of the boxplot represents the median whereas the inverted triangle represents the mean. *p*-values were obtained from ANOVA with Tukey's post-hoc test. Different lower-case letters indicate a significant difference between means ($p < 0.05$).

A similar pattern was found in tests of eADP, which evoked a monophasic $[Ca^{2+}]_{cyt}$ response at 0.1 mM (Figure 10a), 0.5 mM (Figure 10b) and 1 mM ADP (Figure 10c). Although *Atann1* leaf samples had lower peak $[Ca^{2+}]_{cyt}$ responses than Col-0, these were not significantly different and no significant differences were found between Col-0 and *Atann1* for the total $[Ca^{2+}]_{cyt}$ accumulated at each concentration (in all cases $p > 0.05$) (Figure 10d). Thus, at this level of resolution, AtANN1 appears only to have an impact in leaves at 1 mM eATP.

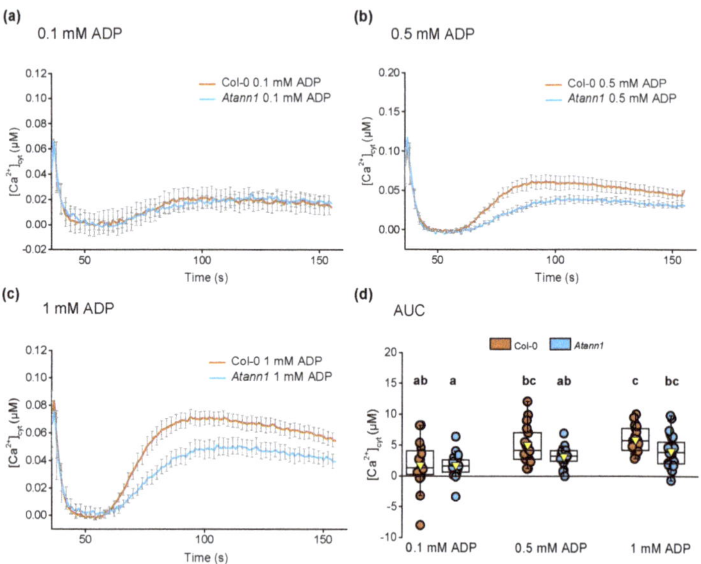

Figure 10. AtANN1 is not involved in mediating eADP-induced [Ca^{2+}]$_{cyt}$ increase in leaves. Mean ($\pm$ SEM) time course of [Ca^{2+}]$_{cyt}$ increase (t = 36s–155s) when 14-day-old individual leaves were treated with (**a**) 0.1 mM eADP, (**b**) 0.5 mM eADP or (**c**) 1 mM eADP agonist. (**d**) The total [Ca^{2+}]$_{cyt}$ accumulated over the period of measurement for each concentration by calculating the AUC. Data were obtained from four experiments (Col-0 *n* = 16–23; *Atann1 n* = 17–24). The middle line of the boxplot represents the median whereas the inverted triangle represents the mean. *p*-values were obtained from ANOVA with Tukey's post-hoc test. Different lower-case letters indicate a significant difference between means (*p* < 0.05).

3. Discussion

Few components of eATP (or eADP) signalling pathways have been identified at the genetic level. In this study, the evidence suggests that AtANN1 is a component of both eATP- and eADP-induced [Ca^{2+}]$_{cyt}$ elevation in roots, with consequences for eATP-induced ROS accumulation and gene expression. Mechanical or "touch" stimulus can cause accumulation of extracellular ATP by *Arabidopsis* root tips but much less so in older regions of the root [35]. Here, the addition of control solution alone caused a touch-induced monophasic increase in [Ca^{2+}]$_{cyt}$ (Figure 1a) but it is not known whether it also caused accumulation of extracellular ATP to trigger all or part of that [Ca^{2+}]$_{cyt}$ increase. There was no significant difference between the touch response of Col-0 and *Atann1*, which may indicate that AtANN1 does not contribute to any [Ca^{2+}]$_{cyt}$ increase that is downstream of any touch-induced eATP increase. There were no indications of further [Ca^{2+}]$_{cyt}$ elevations in response to addition of control solution, suggesting that any extracellular ATP produced by that mechanical stimulus was insufficient to trigger the biphasic [Ca^{2+}]$_{cyt}$ increase seen when ATP is added experimentally. However, it cannot be ruled out that it could affect the root's subsequent response to experimental addition of nucleotides. Previous analysis of the Col-0 root's biphasic [Ca^{2+}]$_{cyt}$ response to experimental additions of eATP or eADP demonstrated that this biphasic pattern is wholly reliant on AtDORN1 as an eATP receptor [5]. For eATP, the first peak originates at the root apex whereas the second peak originates sub-apically in more mature cells, possibly as part of a [Ca^{2+}]$_{cyt}$ "wave" that travels from the apex [5,20,36–38]. Mature regions can also respond to eATP when it is added there specifically rather than to the whole root [5], showing a level of autonomy from the apex. It is assumed that the biphasic eADP response maps to the same areas. AtANN1 is present at the root apex and in more mature cells such as trichoblasts [32,34] so it could contribute to both phases. The near normal first peak $[Ca^{2+}]_{cyt}$ in the *Atann1* mutant with 1 mM eATP (Figure 1d, Figure 3c) but not with 0.2 mM or 0.1 mM eATPγS (Figure 3c) suggests that at high concentration, other components can compensate for the loss of AtANN1 especially in the root apex. No such redundancy was observed with eADP. A far clearer need for AtANN1 was seen in the second eATP- and eADP-induced $[Ca^{2+}]_{cyt}$ increase across the concentration range tested and this could map to mature cells. A study of *Atann1* using a $[Ca^{2+}]_{cyt}$ reporter affording spatial resolution such as YC3.6 (Yellow Cameleon 3.6) or root cell-specific GCaMP3 [20,36–39] is now needed to determine which cells and regions AtANN1 operates in. Challenging specific regions of the root with agonist could also help determine whether an impaired apical/first peak response leads to an impaired sub-apical/second peak response (which could help explain the patterns observed here) and the extent to which the lesion in the *Atann1* second peak is a consequence of a local response to agonist independent of the apex.

In *Arabidopsis* roots, eATP (but not eADP) causes intracellular ROS accumulation, which relies on Ca^{2+} influx and the AtRBOHC NADPH oxidase [14,21]. Another source of ROS production is AtRBOHD, which is a phosphorylation target of AtDORN1 in guard cells [7,40]. It is envisaged that Ca^{2+} influx acts downstream of AtDORN1 and upstream of AtRBOHC/AtRBOHD. It is not known why eADP is ineffective. The pharmacological block that effectively eliminates eATP-activated Ca^{2+} influx across the plasma membrane only inhibits half of the intracellular ROS in roots caused by eATP [14]. The residual $[Ca^{2+}]_{cyt}$ elevation in *Atann1* roots in response to eATP could therefore have been sufficient to cause the observed normal level of ROS accumulation, particularly at the apex (Figure 6b). However, the results show that AtANN1 plays a role in limiting the spatial extent of the ROS increase, limiting its distal spread (Figure 5). AtANN1 can be a cytosolic protein in root cells [34] as well as being a plasma membrane protein. Recombinant AtANN1 has a very low level peroxidase activity in vitro [41], which suggests that loss of its activity would act to increase cytosolic ROS. However, this in vitro activity could have arisen from a co-purified protein [41]. Peroxide treatment of roots suppresses *AtANN1* expression [28], which is inconsistent with a role as a protective peroxidase. Annexin overexpression can protect against oxidative stress by elevating peroxidase, catalase and superoxide dismutase activities [42–44]. One possibility is that the AtANN1-dependent pathway spatially fine-tunes the expression or activity of one or more of the five peroxidases that are regulated by extracellular ATP [45]. Therefore, the *Atann1* mutant would exhibit the observed loss of spatial control of eATP-induced ROS accumulation.

Changes in gene expression were also examined to investigate the effect of *Atann1* in the downstream responses to eATP. eATP induction of *AtRBOHD* expression is *AtDORN1*-dependent in seedlings, suggesting $[Ca^{2+}]_{cyt}$ dependence [7], but here no effect of eATP on its expression was found in Col-0 or *Atann1* roots (Figure 7b). It could be that the effect of eATP treatment on *AtRBOHD* gene expression is more pronounced in leaves compared to the roots. Figure 7c reveals the potential importance of eATP signal regulation on ethylene production. AtACS6 is one of the many isoforms of 1-aminocyclopropane-1-carboxylix acid synthase (ACS) that is an important enzyme in the biosynthesis of ethylene. Its eATP-induced expression is *AtDORN1*-dependent [18]. Here, *AtACS6* expression was transiently upregulated in Col-0 whole roots. Transient up-regulation was also observed in previous studies over a similar time course [12,46]. The *Atann1* mutant failed to upregulate *AtACS6* expression compared to Col-0, indicating an important requirement for AtANN1 in this part of the eATP signalling pathway. A recent study showed that eATP-dependent ethylene production alleviates salinity stress by regulating Na^+ and K^+ homeostasis [47]. Salt stress itself promotes eATP accumulation by *Arabidopsis* roots [48] and *Atann1* is impaired in both the root's transcriptional response to NaCl and the adaptive growth response [26,27]. It is possible, therefore, that a component of the salt stress response involves eATP's acting through AtANN1 to up-regulate *ACS6* expression and ethylene production. Salt stress also up-regulates expression of *AtWRKY40* [49] and its up-regulation by eATP in roots is *AtDORN1*-dependent [7]. Both Col-0 and *Atann1* showed up-regulation of *AtWRKY40* expression by eATP, with *Atann1*'s exhibiting greater expression at 10 min than Col-0 (Figure 7d). This difference in the time course could relate more readily to the higher ROS in the mutant than the lesion in [Ca^{2+}]$_{cyt}$ response. With the demonstration that eATP-induced transcriptional responses can be affected by AtANN1, further studies should expand to the level of RNA$_{seq}$ analysis to determine the extent to which this annexin is involved.

Although expressed in cotyledons and true leaves [32], no evidence was found here for the involvement of AtANN1 in the cotyledon eATP-induced [Ca^{2+}]$_{cyt}$ increase (Figure 8). For true leaves tested with eATP or eADP, only 1 mM eATP supported a significant difference between *Atann1* and Col-0 (Figures 9 and 10). This is might be due to the temporal regulation of AtANN1 function and it seems that, in contrast to the root first [Ca^{2+}]$_{cyt}$ peak, AtANN1 is recruited to the pathway only at higher agonist concentration. It would be worthwhile to repeat these studies with another [Ca^{2+}]$_{cyt}$ reporter to afford spatial resolution and resolve any cell- or tissue-specific involvement of AtANN1.

AtANN1 is now known to be involved in [Ca^{2+}]$_{cyt}$ elevation in response to chitin, salinity stress, heat stress, extracellular hydroxyl radicals and H$_2$O$_2$ [26–28,50–53]. Here, in roots, it most likely operates downstream of AtDORN1, given the dependency of the root eATP- and eADP-induced [Ca^{2+}]$_{cyt}$ response on this receptor [5]. Studies on the root epidermal plasma membrane have indicated that AtANN1 can operate as an extracellular ROS-activated Ca^{2+}-permeable channel and it is likely to act as such in salt stress signalling [26,27]. While AtANN1 function in eATP signalling may well differ from cell type to cell type, we propose that this is the most likely mode of action for eATP signalling in the root epidermis (Figure 11), especially given AtDORN1's ability to activate the AtRBOHD plasma membrane NADPH oxidase that would generate extracellular ROS. This now requires testing. As eATP-and eADP-induced root [Ca^{2+}]$_{cyt}$ increase still occurred in *Atann1*, there remain Ca^{2+} channels downstream of AtDORN1 (and probably upstream of NADPH oxidases) to be discovered.

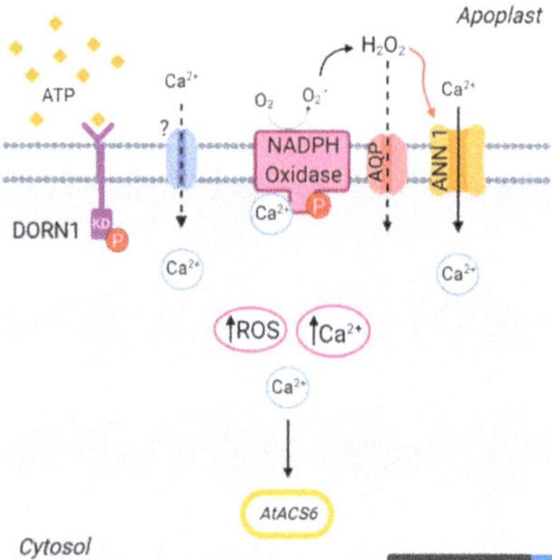

Figure 11. eATP perception by AtDORN1 at the root epidermal plasma membrane could be upstream of AtANN1 via production of extracellular ROS. Perception of eATP by AtDORN1 activates as yet unknown Ca^{2+} influx channels. AtDORN1 could phosphorylate an NADPH oxidase (AtRBOHC and AtRBOHD being the most likely candidates) with its kinase domain (KD) and the NADPH oxidase could also be activated by [Ca^{2+}]$_{cyt}$ at its EF hands. This would result in production of extracellular hydroxyl radicals that could readily be converted to peroxide (H$_2$O$_2$) or hydroxyl radicals [26,27]. These could promote AtANN1-mediated Ca^{2+} influx. Peroxide could enter the cytosol through aquaporins (AQP) [54], which could account for eATP-induced intracellular ROS accumulation. The latter is the clearest point for the divergence between eATP and eADP pathways. Decoding of [Ca^{2+}]$_{cyt}$ as a second messenger leads ultimately to a nuclear transcriptional response that can be impaired in *Atann1*. The position of AtP2K2 in such a pathway has yet to be tested. This original figure was created with BioRender.com.

4. Materials and Methods

4.1. Plant Materials and Growth Conditions

The *annexin 1* loss of function mutant (*Atann1*) was in the Columbia (Col-0) wild type background with a T-DNA insertion in the third exon [26]. The *Atann1* line expressing the (apo)aequorin protein cytosolically under the 35S promoter was as described in [26]. Surface-sterilised seeds were sown on half-strength Murashige-Skoog (MS; Duchefa Biochemie, Haarlem, The Netherlands) medium (0.8% *w/v*) Bactoagar (BD Diagnostics VWR, Sparks, MD, USA); pH 5.6 with 0.1 M KOH +/− 1 mM ATP) in square petri dishes (12 cm × 12 cm, Greiner Bio-One, Frickenhausen, Germany) and incubated in the dark at 4 °C. After two days of stratification, plates were transferred into a growth chamber (PERCIVAL, CLF Plant Climatics, Emersacker, Germany) at 23 °C with a 16 h photoperiod (80 μmol m^{-2} s^{-1}). Plants were grown vertically for excised root experiments or horizontally for leaf experiments.

4.2. Measurement of $[Ca^{2+}]_{cyt}$

A. thaliana (apo)aequorin-expressing samples were incubated in 100 μL of control solution (2 mM Bis-Tris Propane; 10 mM CaCl$_2$; 0.1 mM KCl, pH 5.8 adjusted using 1 M Bis-Tris Propane and 0.5 M MES) containing 10 μL coelanterazine (10 μM; Nanolight Technology, Pinetop, AZ, USA) overnight in the dark at room temperature in a 96-well plate (Greiner Bio-One, Frickenhausen, Germany). All samples were washed with coelanterazine-free control solution before any luminescence measurement was taken and then placed (as individual samples) into a 96-well plate containing 100 μL fresh control solution. Luminescence (as an output of free cytosolic calcium ion, $[Ca^{2+}]_{cyt}$) upon different treatments was measured using a FLUOstar OPTIMA (BMG Labtech, Ortenberg, Germany) plate reader. All the treatment solutions were prepared as additions to the control solution and adjusted to pH 5.8 (1 M Bis-Tris Propane and 0.5 M MES) prior to measurement. After 35 s of initial background measurement, 100 μL of control or treatment solution was added and the luminescence measured every second for 155s. At the end of each measurement, 100 μL discharge solution (10% (*v/v*) ethanol; 1 M CaCl$_2$) was added to quench the total (apo)aequorin luminescence in each sample. $[Ca^{2+}]_{cyt}$ was determined according to the calibration formula given in [55]. ATP (adenosine-5'-triphosphate, disodium salt trihydrate) and ADP (adenosine-5'-diphosphate, disodium salt dihydrate) were purchased from Melford Laboratories Ltd. (Ipswich, UK) whereas non-hydrolysable analogues ATPγS (adenosine 5'-[γ-thio] triphosphate tetralithium salt and ADPβS (Adenosine 5'-[β-thio] diphosphate trilithium salt) were purchased from Sigma-Aldrich (St. Louis, MO, USA).

4.3. Determination of Intracellular ROS

Accumulation of intracellular reactive oxygen species in individual roots was determined using 50 μM CM-H$_2$DCFDA (5-(and-6-)-chloromethyl-2',7'-dichlorodihydrofluorscein diacetate, acetyl ester; Molecular Probes/Invitrogen, Waltham, MA, USA), as described previously [14]. The assay medium comprised LSM medium [56]; when eATP was added, medium was supplemented with CaCl$_2$ to counteract chelation. Images of roots were acquired with a Nikon SMZ 1500 microscope and a QImaging Retiga cooled 12-bit camera (www.qimaging.com).

4.4. Quantification of eATP-Induced Gene Expression

Seven-days-old Col-0 and *Atann1* were acclimatised for 1 h at room temperature in the light in 8 mL control solution (2 mM Bis-Tris Propane; 10 mM CaCl$_2$; 0.1 mM KCl) pH 5.8 in separate small petri dishes (Thermo Scientific, Waltham, MA, USA). Either 8 mL of control solution or 2 mM ATP solution (prepared in control solution to give a final concentration of 1 mM) was added. After 10 or 30 min, seedlings were placed in 10 mL RNAlater solution (25 mM sodium citrate, 10 mM EDTA, 70 g ammonium sulphate/100 mL solution, pH 5.2). The root from each seedling was excised (in the RNAlater solution), dried briefly on filter paper and then frozen in liquid nitrogen. RNA extraction was carried out by using the

RNEasy Plant Mini Kit (QIAGEN, Hilden, Germany) with a DNAse treatment additional step (RNAse free DNAse kit, QIAGEN, Hilden, Germany) and LiCl purification. Complementary DNA (cDNA) was synthesised by using the Quantitect Reverse Transcription kit (QIAGEN, Hilden, Germany) following the manufacturer's protocol. Quantitative Polymerase Chain Reaction (qPCR) was done with a Rotor-Gene 3000 thermocycler with the Rotor-Gene™ SYBR® Green PCR Kit (QIAGEN, Hilden, Germany) according to the manufacturer's protocol. cDNA final concentration was 5 ng with 0.25 µM final primer concentration. The qPCR programme was: Initial stage 95 °C for 5 min, 40 cycles of 95 °C for 5 s, 60 °C for 10 s. Melting curves were used to check for specific amplification (ramping from 55 °C to 95 °C with 1 °C rise each step and 5 s delay between steps). The primer pairs used in the qPCR reaction were as shown in Appendix A (Table 1). Raw data were analysed by using the 'modlist' and the 'getPar' function in R (qpcR package) [57]. Quantification (R value) was performed by using the average of the selected Ct and efficiency values using the formula $E_{sample} = \text{Efficiency}^{(-Ct)}$. Data were normalised with two different housekeeping genes *UBQ10* and *TUB4* using the formula $R_{sample} = E_{sample}/(\text{sqrt}(E_{UBQ10} \times E_{TUB4}))$ [58].

4.5. Statistical Analysis

All the data collected were analysed with the R statistical programme (https://www.r-project.org). ANOVA, Student's *t*-test or Welch two sample *t*-test were used for parametric tests whereas either Wilcoxon rank-sum test or Kruskal–Wallis test was used for non-parametric tests. Further comparison was analysed with Tukey's HSD or Dunnett's post-hoc test. A 95% confidence interval was used for all tests carried out.

5. Conclusions

This study has identified AtANN1 as a component in mediating the root increase of $[Ca^{2+}]_{cyt}$ in response to both eATP and eADP. It is postulated that AtANN1 might operate as an ROS-activated plasma membrane Ca^{2+} channel, downstream of AtDORN1. Since loss of *AtANN1* does not completely abolish the $[Ca^{2+}]_{cyt}$ increase, there are clearly other channels in the pathway. AtANN1 appears to regulate the spatial extent of eATP-induced intracellular ROS in the root. At the gene expression level, AtANN1 is involved in eATP-induced up-regulation of *AtACS6* and *AtWRKY40*, thus potentially directing eATP signalling towards ethylene production and salt stress tolerance.

Author Contributions: Conceptualization: A.M.-S., J.S., R.S. and J.M.D.; methodology: A.M.-S., J.S., E.M., K.A.W. and R.S.; formal analysis: A.M.-S., R.S., Y.N., J.S., and Z.S.; investigation: A.M.-S., J.S., R.S., Y.N., and Z.S.; writing—original draft preparation: A.M.-S. and J.M.D.; writing—review and editing: all authors; visualization: A.M.-S. and R.S. All authors have read and agreed to the published version of the manuscript.

Funding: This research was funded by the UK BBSRC (BB/J014540/1), the University of Cambridge's Broodbank Trust, Commonwealth, European and International Trust, and Tom ap Rees Trust, Jiangsu Normal University, National Key R and D Program of China 2019, Riken CSRS and Yayasan DayaDiri.

Institutional Review Board Statement: Not applicable.

Informed Consent Statement: Not applicable.

Data Availability Statement: Materials are available on request to the corresponding author.

Acknowledgments: We thank Adeeba Dark (University of Cambridge) for technical support.

Conflicts of Interest: The authors declare no conflict of interest.

Abbreviations

ACS	1-Aminocyclopropane-1-carboxylate synthase
ADPβS	Adenosine 5′-[β-thio] diphosphate
ANN1	Annexin1
ANOVA	Analysis of Variance
AQP	Aquaporins
ATPγS	Adenosine 5′-[γ-thio] triphosphate
AUC	Area Under the Curve
$[Ca^{2+}]_{cyt}$	Cytosolic free calcium ion
CAM	Calmodulin
CAMTA	Calmodulin-binding Transcription Activators
CAMV	Cauliflower Mosaic Virus
cDNA	Complementary DNA
CM-H_2DCFDA	5-(and-6-)-Chloromethyl-2′,7′-dichlorodihydrofluorscein diacetate
DAMP	Damage Associated Molecular Pattern
DORN1	Does not Respond to Nucleotides1
eATP	Extracellular ATP
eADP	Extracellular ADP
H_2O_2	Hydrogen peroxide
MES	2-(N-morpholino) ethanesulfonic
MS	Murashige and Skoog nutrient medium
NADPH	Nicotinamide adenine dinucleotide phosphate
NO	Nitric oxide
PA	Phosphatidic Acid
qRT-PCR	Quantitative reverse transcription-polymerase chain reaction
RBOH	Respiratory burst homologue
ROS	Reactive oxygen species
RNAseq	RNA sequencing
SEM	Standard error of mean
T-DNA	Transfer DNA
Tris	Tris base, 2-amino-2-(hydroxymethyl)-1,3-propanediol
TUB4	Tubulin Beta Chain4
UBQ10	Polyubiquitin10
YC3.6	Yellow Cameleon 3.6

Appendix A

Table 1. The forward and reverse sequences of primers used for the gene expression tests.

Gene	Forward (5′–3′)	Reverse (3′ 5′)
AtANN1 (AT1G35720.1)	TGTTCTTCGTTCAGCAATCAAC	GTACTCCTCTCCAATGACCTTC
AtDORN1 (AT5G60300)	ATGGTCACATTGCCTGCAGAAG	TCCCTCTTTACAGGCTGGACTCTC
AtRBOHD (AT5G47910)	ATGATCAAGGTGGCTGTTTACCC	ATCCTTGTGGCTTCGTCATGTG
AtACS6 (AT4G11280)	TATCCAGGGTTTGATAGAGA	TCCACCGTAATCTTGAACC
AtWRKY40 (AT1G80840)	AGCTTCTGACACTACCCTCGTTG	TTGACAGAACAGCTTGGAGCAC
AtUBQ10 (AT4G05320)	CCGACTACAACATTCAGAAGGA	TCAGAACTCTCCACCTCCAAA
AtTUB4 (AT5G44340)	AGGGAAACGAAGACAGCAAG	GCTCGCTAATCCTACCTTTGG

References

1. Chivasa, S.; Ndimba, B.K.; Simon, W.J.; Lindsey, K.; Slabas, A.R. Extracellular ATP functions as an endogenous external metabolite regulating plant cell viability. *Plant Cell* **2005**, *17*, 3019–3034. [CrossRef] [PubMed]
2. Clark, G.; Torres, J.; Herz, N.; Wat, N.; Ogoti, J.; Aranda, G.; Blizard, M.; Wu, M.; Onyirimba, J.; Canales, A.A.; et al. Both the stimulation and inhibition of root hair growth induced by extracellular nucleotides in Arabidopsis are mediated by nitric oxide and reactive oxygen species. *Plant Mol. Biol.* **2010**, *74*, 423–435. [CrossRef] [PubMed]

3. Sun, J.; Zhang, C.L.; Deng, S.R.; Lu, C.F.; Shen, X.; Zhou, X.Y.; Zheng, X.J.; Hu, Z.M.; Chen, S.L. An ATP signalling pathway in plant cells: Extracellular ATP triggers programmed cell death in Populus euphratica. *Plant Cell Environ.* **2012**, *35*, 893–916. [CrossRef] [PubMed]
4. Nizam, S.; Qiang, X.; Wawra, S.; Nostadt, R.; Getzke, F.; Schwanke, F.; Dreyer, I.; Langen, G.; Zuccaro, A. Serendipita indica E5′NT modulates extracellular nucleotide levels in the plant apoplast and affects fungal colonization. *EMBO Rep.* **2019**, *20*, e47430. [CrossRef] [PubMed]
5. Matthus, E.; Sun, J.; Wang, L.; Bhat, M.G.; Sidik, A.B.M.; Davies, J.M. DORN1/P2K1 and purino—Calcium signalling in plants; making waves with extracellular ATP. *Ann. Bot.* **2019**, *124*, 1227–1242. [CrossRef]
6. Hou, Q.; Wang, Y.; Fan, B.; Sun, K.; Liang, J.; Feng, H.; Jia, L. Extracellular ATP affects cell viability, respiratory O_2 uptake, and intracellular ATP production of tobacco cell suspension culture in response to hydrogen peroxide-induced oxidative stress. *Biologia* **2020**, *75*, 1437–1443. [CrossRef]
7. Choi, J.; Tanaka, K.; Cao, Y.; Qi, Y.; Qiu, J.; Liang, Y.; Lee, S.Y.; Stacey, G. Identification of a plant receptor for extracellular ATP. *Science* **2014**, *343*, 290–294. [CrossRef]
8. Pham, A.Q.; Cho, S.-H.; Nguyen, C.T.; Stacey, G. Arabidopsis lectin receptor kinase P_2K_2 is a second plant receptor for extracellular ATP and contributes to innate immunity. *Plant Physiol.* **2020**, *183*, 1364–1375. [CrossRef]
9. Zhu, R.; Dong, X.; Hao, W.; Gao, W.; Zhang, W.; Xia, S.; Liu, T.; Shang, Z. Heterotrimeric G protein-regulated Ca^{2+} influx and PIN2 asymmetric distribution are involved in *Arabidopsis thaliana* roots' avoidance response to extracellular ATP. *Front. Plant Sci.* **2017**, *8*, 1522. [CrossRef]
10. Zhu, R.; Dong, X.; Xue, Y.; Xu, J.; Zhang, A.; Feng, M.; Zhao, Q.; Xia, S.; Yin, Y.; He, S.; et al. Redox-Responsive Transcription Factor 1 (RRFT1) is involved in extracellular ATP-regulated *Arabidopsis thaliana* seedling growth. *Plant Cell Physiol.* **2020**, *61*, 685–698. [CrossRef]
11. Demidchik, V.; Shabala, S.N.; Coutts, K.B.; Tester, M.A.; Davies, J.M. Free oxygen radicals regulate plasma membrane Ca^{2+}-and K^+-permeable channels in plant root cells. *J. Cell Sci.* **2003**, *116*, 81–88. [CrossRef] [PubMed]
12. Jeter, C.R.; Tang, W.; Henaff, E.; Butterfield, T.; Roux, S.J. Evidence of a novel cell signaling role for extracellular adenosine triphosphates and diphosphates in Arabidopsis. *Plant Cell* **2004**, *16*, 2652–2664. [CrossRef] [PubMed]
13. Kim, S.-Y.; Sivaguru, M.; Stacey, G. Extracellular ATP in plants. Visualization, localization, and analysis of physiological significance in growth and signaling. *Plant Physiol.* **2006**, *142*, 984–992. [CrossRef] [PubMed]
14. Demidchik, V.; Shang, Z.; Shin, R.; Thompson, E.; Rubio, L.; Laohavisit, A.; Mortimer, J.C.; Chivasa, S.; Slabas, A.R.; Glover, B.J.; et al. Plant extracellular ATP signalling by plasma membrane NADPH oxidase and Ca^{2+} channels. *Plant J.* **2009**, *58*, 903–913. [CrossRef] [PubMed]
15. Foresi, N.P.; Laxalt, A.M.; Tonón, C.V.; Casalongué, C.A.; Lamattina, L. Extracellular ATP induces nitric oxide production in tomato cell suspensions. *Plant Physiol.* **2007**, *145*, 589–592. [CrossRef] [PubMed]
16. Wu, S.J.; Wu, J.Y. Extracellular ATP-induced NO production and its dependence on membrane Ca^{2+} flux in Salvia miltiorrhiza hairy roots. *J. Exp. Bot.* **2008**, *59*, 4007–4016. [CrossRef] [PubMed]
17. Tripathi, D.; Zhang, T.; Koo, A.J.; Stacey, G.; Tanaka, K. Extracellular ATP acts on jasmonate signaling to reinforce plant defense. *Plant Physiol.* **2017**, *176*, 511–523. [CrossRef]
18. Jewell, J.B.; Sowders, J.M.; He, R.; Willis, M.A.; Gang, D.R.; Tanaka, K. Extracellular ATP shapes a defense-related transcriptome both independently and along with other defense signaling pathways. *Plant Physiol.* **2019**, *179*, 1144–1158. [CrossRef]
19. Dong, X.; Zhu, R.; Kang, E.; Shang, Z. RRFT1 (Redox Responsive Transcription Factor 1) is involved in extracellular ATP-regulated gene expression in *Arabidopsis thaliana* seedlings. *Plant Signal. Behav.* **2020**, *15*, 1748282. [CrossRef]
20. Loro, G.; Drago, I.; Pozzan, T.; Lo Schiavo, F.; Costa, A. Targeting of Cameleons to various subcellular compartments reveals a strict cytoplasmic/mitochondrial Ca^{2+} handling relationships in plant cells. *Plant J.* **2012**, *71*, 1–13. [CrossRef]
21. Demidchik, V.; Shang, Z.; Shin, R.; Colaço, R.; Laohavisit, A.; Shabala, S.; Davies, J.M. Receptor-like activity evoked by extracellular ADP in Arabidopsis root epidermal plasma membrane. *Plant Physiol.* **2011**, *156*, 1375–1385. [CrossRef] [PubMed]
22. Wang, L.; Wilkins, K.A.; Davies, J.M. Arabidopsis DORN1 extracellular ATP receptor; activation of plasma membrane K^+-and Ca^{2+}-permeable conductances. *New Phytol.* **2018**, *218*, 1301–1304. [CrossRef] [PubMed]
23. Wang, L.; Stacey, G.; Leblanc-Fournier, N.; Legué, V.; Moulia, B.; Davies, J.M. Early extracellular ATP signaling in Arabidopsis root epidermis: A multi-conductance process. *Front. Plant Sci.* **2019**, *10*, 1064. [CrossRef] [PubMed]
24. Clark, G.; Roux, S.J. Role of Ca^{2+} in mediating plant responses to extracellular ATP and ADP. *Int. J. Mol. Sci.* **2018**, *19*, 3590. [CrossRef]
25. Ma, L.; Ye, J.; Yang, Y.; Lin, H.; Yue, L.; Luo, J.; Long, Y.; Fu, H.; Liu, X.; Zhang, Y.; et al. The SOS2-SCaBP8 complex generates and fine-tunes an AtANN4-dependent calcium signature under salt stress. *Dev. Cell* **2019**, *48*, 697–709. [CrossRef]
26. Laohavisit, A.; Shang, Z.; Rubio, L.; Cuin, T.A.; Véry, A.A.; Wang, A.; Mortimer, J.C.; Macpherson, N.; Coxon, K.M.; Battey, N.H.; et al. Arabidopsis annexin1 mediates the radical-activated plasma membrane Ca^{2+}-and K^+-permeable conductance in root cells. *Plant Cell* **2012**, *24*, 1522–1533. [CrossRef]
27. Laohavisit, A.; Richards, S.L.; Shabala, L.; Chen, C.; Colaco, R.D.D.R.; Swarbreck, S.M.; Shaw, E.; Dark, A.; Shabala, S.; Shang, Z.; et al. Salinity-induced calcium signaling and root adaptation in Arabidopsis require the calcium regulatory protein Annexin1. *Plant Physiol.* **2013**, *163*, 253–262. [CrossRef]

28. Richards, S.L.; Laohavisit, A.; Mortimer, J.C.; Shabala, L.; Swarbreck, S.M.; Shabala, S.; Davies, J.M. Annexin 1 regulates the H_2O_2-induced calcium signature in *Arabidopsis thaliana* roots. *Plant J.* **2014**, *77*, 136–145. [CrossRef]

29. Chivasa, S.; Murphy, A.M.; Hamilton, J.M.; Lindsey, K.; Carr, J.P.; Slabas, A.R. Extracellular ATP is a regulator of pathogen defence in plants. *Plant J.* **2009**, *60*, 436–448. [CrossRef]

30. Tanaka, K.; Choi, J.; Cao, Y.; Stacey, G. Extracellular ATP acts as a damage-associated molecular pattern (DAMP) signal in plants. *Front. Plant Sci.* **2014**, *5*, 446. [CrossRef]

31. Choi, J.; Tanaka, K.; Liang, Y.; Cao, Y.; Lee, S.Y.; Stacey, G. Extracellular ATP, a danger signal, is recognized by DORN1 in Arabidopsis. *Biochem. J.* **2014**, *463*, 429–437. [CrossRef] [PubMed]

32. Clark, G.B.; Sessions, A.; Eastburn, D.J.; Roux, S.J. Differential expression of members of the annexin multigene family in Arabidopsis. *Plant Physiol.* **2001**, *126*, 1072–1084. [CrossRef] [PubMed]

33. Cantero, A.; Barthakur, S.; Bushart, T.J.; Chou, S.; Morgan, R.O.; Fernandez, M.P.; Clark, G.B.; Roux, S.J. Expression profiling of the Arabidopsis annexin gene family during germination, de-etiolation and abiotic stress. *Plant Physiol. Biochem.* **2006**, *44*, 13–24. [CrossRef] [PubMed]

34. Tichá, M.; Richter, H.; Ovečka, M.; Maghelli, N.; Hrbáčková, M.; Dvořák, P.; Šamaj, J.; Šamajová, O. Advanced microscopy reveals complex developmental and subcellular localization patterns of ANNEXIN 1 in Arabidopsis. *Front. Plant Sci.* **2020**, *11*, 1153. [CrossRef]

35. Weerasinghe, R.R.; Swanson, S.J.; Okada, S.F.; Garrett, M.B.; Kim, S.Y.; Stacey, G.; Boucher, R.C.; Gilroy, S.; Jones, A.M. Touch induces ATP release in Arabidopsis roots that is modulated by the heterotrimeric G-protein complex. *FEBS Lett.* **2009**, *583*, 2521–2526. [CrossRef]

36. Matthus, E.; Wilkins, K.A.; Swarbreck, S.M.; Doddrell, N.H.; Doccula, H.G.; Costa, A.; Davies, J.M. Phosphate starvation alters root calcium signatures. *Plant Physiol.* **2019**, *179*, 1754–1767. [CrossRef]

37. Rincón-Zachary, M.; Teaster, N.D.; Sparks, J.A.; Valster, A.H.; Motes, C.M.; Blancaflor, E.B. Fluorescence resonance energy transfer-sensitized emission of yellow cameleon 3.60 reveals root zone-specific calcium signatures in Arabidopsis in response to aluminum and other trivalent cations. *Plant Physiol.* **2010**, *152*, 1442–1458. [CrossRef]

38. Costa, A.; Candeo, A.; Fieramonti, L.; Valentini, G.; Bassi, A. Calcium dynamics in root cells of *Arabidopsis thaliana* visualized with selective plane illumination microscopy. *PLoS ONE* **2013**, *8*, e75646. [CrossRef]

39. Krogman, W.; Sparks, J.A.; Blancaflor, E.B. Cell type-specific imaging of calcium signaling in Arabidopsis thaliana seedling roots using GCaMP3. *Int. J. Mol. Sci.* **2020**, *21*, 6385. [CrossRef]

40. Chen, D.; Cao, Y.; Li, H.; Kim, D.; Ahsan, N.; Thelen, J.; Stacey, G. Extracellular ATP elicits DORN1-mediated RBOHD phosphorylation to regulate stomatal aperture. *Nat. Commun.* **2017**, *8*, 2265. [CrossRef]

41. Konopka-Postupolska, D.; Clark, G.; Goch, G.; Debski, J.; Floras, K.; Cantero, A.; Fijolek, B.; Roux, S.; Hennig, J. The role of annexin 1 in drought stress in Arabidopsis. *Plant Physiol.* **2009**, *150*, 1394–1410. [CrossRef] [PubMed]

42. Dalal, A.; Kumar, A.; Yadav, D.; Gudla, T.; Viehhauser, A.; Dietz, J.-K.; Kirti, P.B. Alleviation of methyl viologen-mediated oxidative stress by *Brassica juncea* annexin-3 in transgenic Arabidopsis. *Plant Sci.* **2014**, *219*, 9–18. [CrossRef] [PubMed]

43. Qiao, B.; Zhang, Q.; Liu, D.; Wang, H.; Yin, J.; Wang, R.; He, M.; Cui, M.; Shang, Z.; Wang, D.; et al. A calcium-binding protein, rice annexin OsANN1, enhances heat stress tolerance by modulating the production of H_2O_2. *J. Expt. Bot.* **2015**, *66*, 5853–5866. [CrossRef]

44. Zhang, F.; Li, S.F.; Yang, S.M.; Wang, L.K.; Guo, W.Z. Overexpression of a cotton annexin gene, *GhAnn1*, enhances drought and salt stress tolerance in transgenic cotton. *Plant Mol. Biol.* **2015**, *87*, 47–67. [CrossRef] [PubMed]

45. Lim, M.H.; Wu, J.; Yao, J.; Gallardo, I.F.; Dugger, J.W.; Webb, L.J.; Huang, J.; Salmi, M.L.; Song, J.; Clark, G.; et al. Apyrase suppression raises extracellular ATP levels and induces gene expression and cell wall changes characteristic of stress responses. *Plant Physiol.* **2014**, *164*, 2054–2067. [CrossRef]

46. Song, C.J.; Steinebrunner, I.; Wang, X.; Stout, S.C.; Roux, S.J. Extracellular ATP induces the accumulation of superoxide via NADPH oxidases in Arabidopsis. *Plant Physiol.* **2006**, *140*, 1222–1232. [CrossRef] [PubMed]

47. Lang, T.; Deng, C.; Yao, J.; Zhang, H.; Wang, Y.; Deng, S. A salt-signaling network involving ethylene, extracellular ATP, hydrogen peroxide, and calcium mediates K^+/Na^+ homeostasis in Arabidopsis. *Int. J. Mol. Sci.* **2020**, *21*, 8683. [CrossRef]

48. Dark, A.; Demidchik, V.; Richards, S.L.; Shabala, S.; Davies, J.M. Release of extracellular purines from plant roots and effect on ion fluxes. *Plant Signal. Behav.* **2011**, *6*, 1855–1857. [CrossRef]

49. Chen, H.; Lai, Z.; Shi, J.; Xiao, Y.; Chen, Z.; Xu, X. Roles of Arabidopsis WRKY18, WRKY40 and WRKY60 transcription factors in plant responses to abscisic acid and abiotic stress. *BMC Plant Biol.* **2010**, *10*, 281. [CrossRef]

50. Wang, X.; Ma, X.L.; Wang, H.; Li, B.; Clark, G.; Guo, Y.; Roux, S.; Sun, D.; Tang, W. Proteomic study of microsomal proteins reveals a key role for Arabidopsis Annexin1 in mediating heat stress-induced increase in intracellular calcium levels. *Mol. Cell Prot.* **2015**, *14*, 686–694. [CrossRef]

51. Espinoza, C.; Liang, Y.; Stacey, G. Chitin receptor CERK1 links salt stress and chitin-triggered innate immunity in Arabidopsis. *Plant J.* **2017**, *98*, 984–995. [CrossRef] [PubMed]

52. Liao, C.C.; Zheng, Y.; Guo, Y. MYB30 transcription factor regulates oxidative and heat stress responses through annexin-mediated cytosolic calcium signalling in Arabidopsis. *New Phytol.* **2017**, *216*, 163–177. [CrossRef] [PubMed]

53. Zhao, J.; Li, L.; Liu, Q.; Liu, P.; Li, S.; Yang, D.; Chen, Y.; Pagnotta, S.; Favery, B.; Abad, P.; et al. A MIF-like effector suppresses plant immunity and facilitates nematode parasitism by interacting with plant annexins. *J. Exp. Bot.* **2019**, *70*, 5943–5958. [CrossRef] [PubMed]
54. Rodrigues, O.; Reshetnyak, G.; Grondin, A.; Saijo, Y.; Leonhardt, N.; Maurel, C.; Verdoucq, L. Aquaporins facilitate hydrogen peroxide entry into guard cells to mediate ABA- and pathogen-triggered stomatal closure. *Proc. Nat. Acad. Sci. USA* **2017**, *114*, 9200–9205. [CrossRef] [PubMed]
55. Knight, H.; Trewavas, A.J.; Knight, M.R. Calcium signalling in *Arabidopsis thaliana* responding to drought and salinity. *Plant J.* **1997**, *12*, 1067–1078. [CrossRef] [PubMed]
56. Hong, J.P.; Takeshi, Y.; Kondu, Y.; Matsui, M.; Shin, R. Identification and characterization of transcription factors regulating *HAK5*. *Plant Cell Physiol.* **2013**, *54*, 1478–1490. [CrossRef]
57. Andrej-Nikolai Spiess. Package 'qpcR'. 2018. Available online: https://cran.r-project.org/web/packages/qpcR/qpcR.pdf (accessed on 5 June 2020).
58. Swarbreck, S.M.; Guerringue, Y.; Matthus, E.; Jamieson, F.J.C.; Davies, J.M. Impairment in karrikin but not strigolactone sensing enhances root skewing in *Arabidopsis thaliana*. *Plant J.* **2019**, *98*, 607–621. [CrossRef]

Article

OsPP2C09 Is a Bifunctional Regulator in Both ABA-Dependent and Independent Abiotic Stress Signaling Pathways

Myung Ki Min [1], Rigyeong Kim [1], Woo-Jong Hong [2], Ki-Hong Jung [2], Jong-Yeol Lee [1] and Beom-Gi Kim [1,*]

[1]   Division of Metabolic Engineering, National Institute of Agricultural Sciences, RDA, Jeonju-si 54874, Korea; mkmin66@gmail.com (M.K.M.); rigyeong02@gmail.com (R.K.); jy0820@korea.kr (J.-Y.L.)
[2]   Graduate School of Biotechnology & Crop Biotech Institute, Kyung Hee University, Yongin 17104, Korea; hwj0602@khu.ac.kr (W.-J.H.); khjung2010@khu.ac.kr (K.-H.J.)
*   Correspondence: bgkimpeace@gmail.com

Abstract: Clade A Type 2C protein phosphatases (PP2CAs) negatively regulate abscisic acid (ABA) signaling and have diverse functions in plant development and in response to various stresses. In this study, we showed that overexpression of the rice ABA receptor OsPYL/RCAR3 reduces the growth retardation observed in plants exposed to osmotic stress. By contrast, overexpression of the OsPYL/RCAR3-interacting protein OsPP2C09 rendered plant growth more sensitive to osmotic stress. We tested whether OsPP2CAs activate an ABA-independent signaling cascade by transfecting rice protoplasts with luciferase reporters containing the drought-responsive element (DRE) or ABA-responsive element (ABRE). We observed that OsPP2CAs activated gene expression via the cis-acting drought-responsive element. In agreement with this observation, transcriptome analysis of plants overexpressing OsPP2C09 indicated that OsPP2C09 induces the expression of genes whose promoters contain DREs. Further analysis showed that OsPP2C09 interacts with DRE-binding (DREB) transcription factors and activates reporters containing DRE. We conclude that, through activating DRE-containing promoters, OsPP2C09 positively regulates the drought response regulon and activates an ABA-independent signaling pathway.

Keywords: PP2CAS bifunction; dreb regulation; aba-dependent/independent pathway

check for
updates

Citation: Min, M.K.; Kim, R.; Hong, W.-J.; Jung, K.-H.; Lee, J.-Y.; Kim, B.-G. OsPP2C09 Is a Bifunctional Regulator in Both ABA-Dependent and Independent Abiotic Stress Signaling Pathways. *Int. J. Mol. Sci.* **2021**, *22*, 393. https://doi.org/10.3390/ ijms22010393

Received: 19 October 2020
Accepted: 8 December 2020
Published: 1 January 2021

Publisher's Note: MDPI stays neutral with regard to jurisdictional claims in published maps and institutional affiliations.

1. Introduction

As plants inevitably face adverse environmental conditions such as drought, high salt, and extreme temperatures, they have evolved complex signaling networks that respond to environmental cues and balance their resources between promoting growth and mounting tolerance to inauspicious conditions. Abscisic acid (ABA) plays a central role during abiotic stress responses, in addition to influencing various aspects of plant development and growth, such as seed dormancy, leaf abscission, growth inhibition and fruit ripening [1–3].

ABA signal transduction starts with the formation of a complex between the ABA receptor PYRABACTIN RESISTANCE1/PYR-like/REGULATORY COMPONENT OF ABA RECEPTOR (PYR/PYL/RCAR) and Clade A Type 2C protein phosphatases (PP2CAs), resulting in the inactivation of PP2CAs and the activation of SNF1-related protein Kinase 2 (SnRK2). Activated SnRK2s in turn activate ABA-responsive element (ABRE)-binding factors (ABFs). ABA-induced gene expression is mediated through the binding of ABFs to the ABRE in target promoter regions and subsequent transcriptional activation [4–6].

Yamaguchi-Shinozaki et al., (1992) cloned nine genes responsive to desiccation stress in Arabidopsis (*Arabidopsis thaliana*). *RD29A* and *RD29B* are two closely related genes that respond to drought and high salt conditions. The *RD29A* promoter region contains multiple cis-acting elements such as the dehydration-responsive element/C-repeat (DRE/CRT) and ABRE. *RD29A* expression can therefore respond to both ABA-dependent and -independent signaling pathways through the ABRE or DRE/CRT cis-element, respectively. By contrast, the *RD29B* promoter region lacks a DRE/CRT sequence and is activated only by the

ABA-dependent signaling cascade [7]. Dehydration-responsive element binding factor (DREB) transcription factors have been determined to regulate the gene expression in ABA-independent signaling pathways such as cold and drought [8,9].

ABA-dependent signaling is mainly regulated by phosphorylation and dephosphorylation of the signaling components such as SnRK2s and ABFs. However, several studies have shown that DREB degradation is a key regulatory step during ABA-independent signaling. For example, DREB2A, which plays an important role in the ABA-independent pathway in response to drought stress, is marked for degradation by the 26S proteasome via the E3 ubiquitin ligase DREB2A-INTERACTING PROTEIN1 (DRIP1) [10]. In addition, *DREB2A* expression is regulated by GROWTH-REGULATING FACTOR7 (GRF7), which binds to the *DREB2A* promoter and represses transcription [6,11]. The expression of another DREB factor, Arabidopsis *DREB1A*, is influenced by the Myc transcription factor, INDUCER OF CBF EXPRESSION1 (ICE1), which is sumoylated by the SUMO E3 ligase SIZ1, and marked for degradation by the E3 ubiquitin ligase, high expression of osmotically responsive gene1 (HOS1) [8,12,13]. How phosphorylation and dephosphorylation of DREB contributed to its activation was unclear [9] until the recent report that showed phosphorylation at the negative regulatory domain (NRD) destabilizes DREB2A [14].

In this study, we elucidate a new biological role for OsPP2CAs in the ABA-independent pathway and show that they act as a bifunctional regulator of the crosstalk between the ABA-independent and dependent pathways.

2. Results

2.1. Transgenic Rice Overexpressing OsPYL/RCAR3 Are Insensitive to Osmotic Stress

In previous studies, plants overexpressing the cytosolic ABA receptors OsPYL/RCARs showed increased tolerance to abiotic stress and hypersensitivity to ABA and osmotic stress [3,15–18]. We generated transgenic rice lines overexpressing a monomeric ABA receptor, OsPYL/RCAR3, and named the line A30 (Figure S1A,B). As expected for an ABA receptor, the A30 transgenic line exhibited increased tolerance to abiotic stress and hypersensitivity to ABA in terms of young seedling growth compared to the control (empty vector control, named A8). In addition, A30 plants were smaller than A8 control plants even in the absence of exogenous ABA application (Figure S1C–F). A30 seedlings did not show hypersensitivity to an increase in NaCl concentrations (Figure 1C) as measured by seedling growth. However, A30 plants demonstrate hyposensitive phenotypes when mannitol was applied in concentrations over 200 mM (Figure 1A–D).

To characterize the effects of OsPYL/RCAR3 overexpression, we introduced OsPYL/RCAR3-HA into rice protoplasts by transient transfection together with a reporter construct bearing the rice Rab16A promoter driving the expression of firefly luciferase (pRab 16A::fLUC), which is normally induced by ABA and osmotic stress. We then treated protoplasts with ABA or polyethylene glycol (PEG) to mimic osmotic stress conditions (Figure 1E). Normalized firefly luciferase signal clearly increased in protoplasts overexpressing OsPYL/RCAR3 (4.9-fold increase over mock transfection). Treating OsPYL/RCAR3-overexpressing protoplasts with ABA had no effect on the luciferase signal, as it rose 4.7-fold over the mock sample. However, treating OsPYL/RCAR3-overexpressing protoplasts with with 25% PEG resulted in a 53% reduction in luciferase activity relative to mock-treated overexpressing protoplasts (Figure 1E). These results thus raise the possibility that OsPYL/RCAR3 may inhibit ABA-independent osmotic stress signaling.

2.2. Stress-Inducible OsPP2CAs Interact with OsPYL/RCAR3 to Activate Abiotic Stress Responses

OsPP2CAs activate ABA signaling through their ABA-mediated interaction with ABA receptors. OsPYL/RCAR3 belongs to a group of monomeric ABA receptors (Figure S2A). As a first step toward identifying OsPP2CAs that are candidates that interact with OsPYL/RCAR3 under osmotic stress conditions, we profiled *OsPP2CA* expression levels in response to ABA treatment or high concentrations (200 mM) of mannitol. *OsPP2C06*,

OsPP2C09 and *OsPP2C30* showed the highest induction out of all other *OsPP2CA* genes by mannitol and ABA treatment (Figure 2A,B).

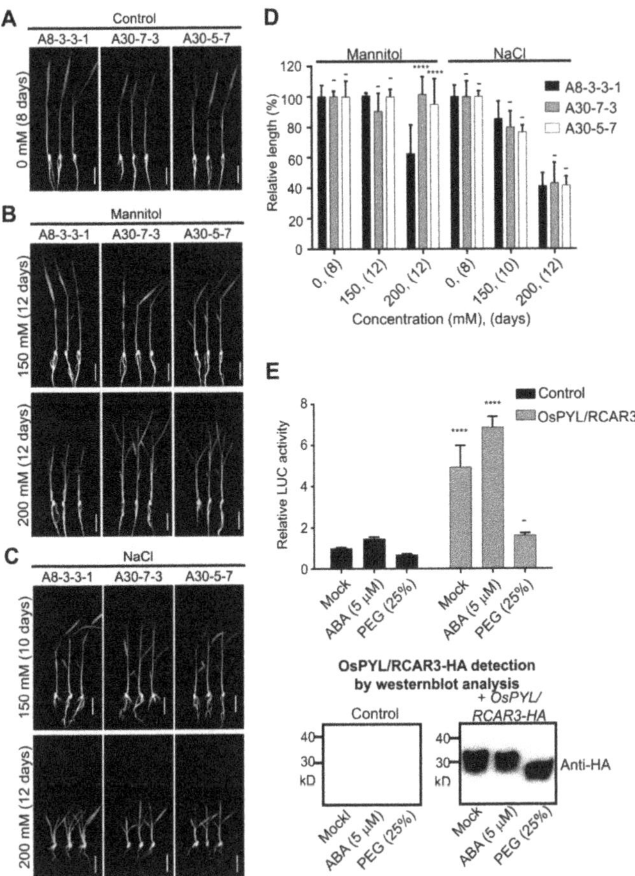

Figure 1. A30 plants, overexpressing OsPYL/RCAR3, have reduced sensitivity to mannitol-imposed osmotic stress. (**A–C**), Seedling growth test. A8 and A30 seedlings were grown in growth medium alone (**A**) or containing mannitol (**B**) or NaCl (**C**). Scale bar: 2 cm. (**D**) Measurement of relative shoot length compared to each mock (0 mM). X axis indicates mannitol or NaCl concentrations, and growth period, (in d), in brackets. Two-way ANOVA was performed with A8 plants as controls. ****: $p < 0.0001$, -: not significant. (**E**) Effect of OsPYL/RCAR3 overexpression on luciferase activity from the *OsRab16a* promoter in rice protoplasts. Control: *pRab16a: LUC* alone. Data presented as mean ± standard deviation (SD), $n = 3$. Two-way ANOVA was performed, comparing with control. ****: $p < 0.0001$, -: not significant. OsPYL/RCAR3-HA expression was confirmed by immunoblot analysis with anti-hemagglutinin (HA) rat antibody.

We then determined the subcellular localization of these three OsPP2CAs by fusing their coding sequences to the *green fluorescent protein* (GFP) gene. OsPP2C06 localized to the cytosol, whereas OsPP2C09 and OsPP2C30 localized to the nucleus (Figure S3A). The interaction between OsPYL/RCAR3 and OsPP2C06, OsPP2C09, or OsPP2C30 was confirmed by yeast two-hybrid assays, even in the absence of ABA in the growth medium (Figure S3B). Bimolecular fluorescence complementation (BiFC) and co-immuno-precipitation assays further corroborated the interaction between the ABA receptor OsPYL/RCAR3 and the three OsPP2C proteins (Figure S3C,D).

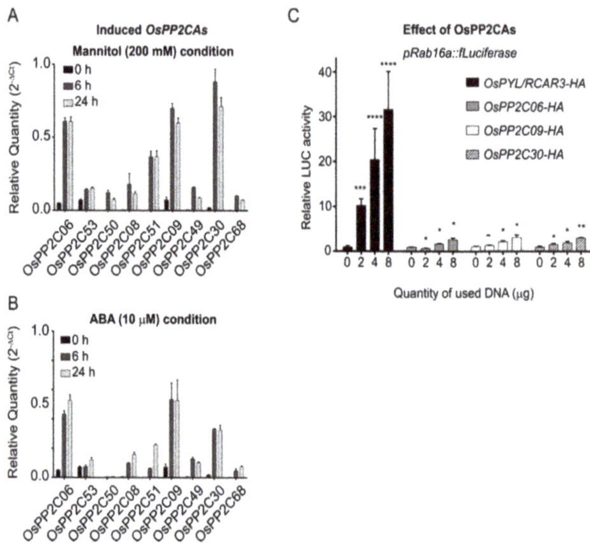

Figure 2. Highly expressed *PP2CAs* in the A30 line activate transcription from the *Rab16A* promoter. (**A,B**), relative transcript levels of 9 *PP2CA* genes after exposure of A30 seedlings to 200 mM Mannitol or 10 μM ABA for 0, 6, or 24 h. All values were determined by RT-qPCR and normalized to *UBIQUTIN5* levels. Error bars are means ± SD (*n* = 3). Similar results were obtained in three iterations. (**C**), Effects associated with overexpression of selected *OsPP2C* genes on *Rab16* promoter activity (*pRab16A:fLUC* and *pUbi:rLUC*) introduced into rice protoplasts by PEG transfection. After a 15 h incubation, luciferase activities were detected. All values were normalized to luciferase activity with 0 μg of *OsPP2C* overexpression vector. Error bars are means ± SD (*n* = 3). One-way ANOVAs were performed with comparing to each 0 μg (****: *p* < 0.0001, ***: *p* < 0.001, **: *p* < 0.01, *: *p* < 0.05, -: not significant).

To examine the effects of *OsPP2CA* overexpression, we introduced the abiotic stress reporter (*pRab16A::fLUC*) together with increasing concentrations of plasmid DNA for the effectors *OsPYL/RCAR3*, *OsPP2C06*, *OsPP2C09*, or *OsPP2C30* into rice protoplasts by transient transfection (Figure 2C). As expected, fLUC activity, reflecting *Rab16A* promoter activation, increased with the amount of *OsPYL/RCAR3* plasmid DNA used during transfection. Unexpectedly, OsPP2Cs also activated transcription from the *Rab16A* promoter in a dose-dependent manner, although measured fLUC activity levels were much lower than with OsPYL/RCAR3. We thus conclude that the three OsPP2Cs OsPP2C06, OsPP2C09, or OsPP2C30 can act as positive regulators of stress signaling.

2.3. OsPP2CAs Induce Gene Expression through the DRE Cis-Element

Rab16A is responsive to both ABA and osmotic stress because its promoter contains both DRE and ABRE elements. To establish whether OsPP2CAs activate *Rab16A* expression through the ABA-dependent or ABA-independent signaling pathway, we constructed two new reporter constructs, in which firefly luciferase was driven by promoter fragments that bore only the DRE (*pDRE::fLUC*) or both the DRE and ABRE elements (*pABRE-DRE::fLUC*); both constructs also contained a minimal promoter consisting a TATA box transcription start site (Figure 3A). DNA for both reporters was introduced into rice protoplasts together with empty vector or with effector plasmid DNA for *OsPYL/RCAR3*, *OsPP2C09*, or *OsPP2C30*. We then examined the level of reporter activation for each effector with and without ABA. The *pABRE-DRE::fLUC* clearly responded to ABA, and fLUC activity rose further when OsPYL/RCAR3 was overexpressed (Figure 3B). By contrast, overexpression of *OsPP2C09* or *OsPP2C30* suppressed the induction of luciferase activity resulting from ABA treatment.

These results demonstrate that OsPYL/RCAR functions as a positive regulator whereas OsPP2C09 and OsPP2C30 are negative regulators of ABA signaling, as previously reported.

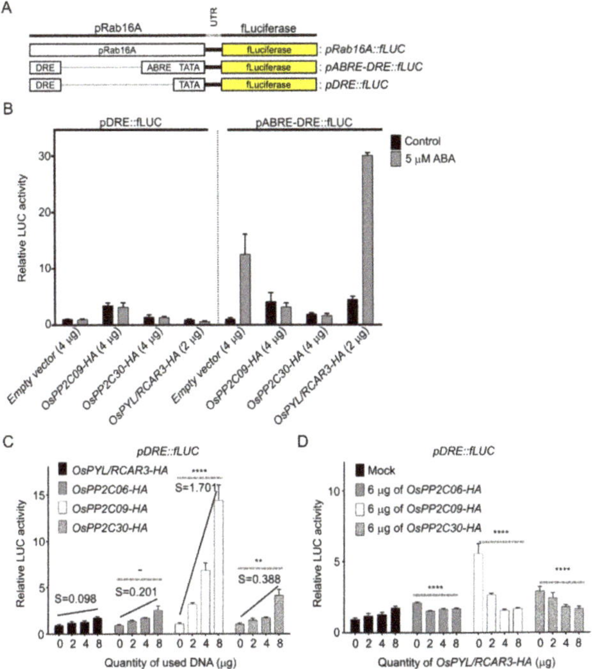

Figure 3. OsPP2Cs activate *Rab16a* transcription via *cis*-DRE. (**A**), Schematics of generated reporter plasmids, *pABRE-DRE::fLUC* and *pDRE::fLUC*. (**B**), Activation test of *pABRE-DRE::fLUC* or *pDRE::fLUC* with effector only or ABA in rice protoplasts. (**C**), Activation of the *pDRE::fLUC* reporter with increasing amounts of the indicated effector DNA during transfection. (**D**), Effects of OsPYL/RCAR3 overexpression on OsPP2C-mediated induction of luciferase activity from the *pDRE::fLUC* reporter. The indicated effector and marker DNAs were introduced into rice protoplasts by PEG-mediated transfection and then incubated for 15 h. Induced luciferase activity was detected with a dual luciferase reporter assay system. All values are means + SD (*n* = 3). S is the slope of the trend line. Two-way ANOVAs (main column effect) were performed, comparing to OsPYL/RCAR3-HA values (****: $p < 0.0001$, **: $p < 0.01$, -: not significant).

In contrast to the *pABRE-DRE::fLUC* reporter, the *pDRE::fLUC* construct did not respond to ABA. Overexpression of *OsPP2C09* or *OsPP2C30* increased fLUC activity in an ABA-independent manner, but *OsPYL/RCAR3* overexpression did not induce fLUC activity (Figure 3B). To validate this result, we examined fLUC activity derived from the *pDRE::fLUC* reporter as a function of increasing effector DNA amounts during transfection. We used *OsPYL/RCAR3* as a control, with a characteristic weakly positive slope between detectable fLUC activity and transfected *OsPYL/RCAR3* DNA amounts (*S* = 0.098). By contrast, transfecting rice protoplasts with increasing amounts of the *OsPP2C09* effector resulted in a 14-fold rise in fLUC activity over background, with a steep slope (*S* = 1.70). The gradual overexpression of either *OsPP2C06* or *OsPP2C30* also induced fLUC activity more than *OsPYL/RCAR3*, with moderately strong slopes (*S* = 0.20 and *S* = 0.39, respectively) (Figure 3C). These results show that overexpression of OsPP2CAs can activate signaling mediated by the *cis*-acting DRE.

In addition, we examined the possible relationship between OsPYL/RCAR3 and OsPP2CAs in the activation of transcription through the *cis*-acting DRE. We measured fLUC activity from the *pDRE::fLUC* reporter when one OsPP2CA and increasing amounts of OsPYL/RCAR3 plasmid effector DNA were co-transfected into rice protoplasts. *Os-PYL/RCAR3* overexpression counteracted the increase in fLUC activity normally associated with *OsPP2CA* overexpression and did so in a dose-dependent manner (Figure 3D). We obtained the same result when we used *pOsRab16A::fLUC* albeit the responses were weaker than with *pDRE::fLUC* (Figure S6). Thus, we conclude that overexpression of *OsPYL/RCAR3* inhibited the action of OsPP2Cs through the ABA-independent *cis*-acting DRE.

2.4. Transgenic Rice Plants Overexpressing OsPP2C09 Are Hypersensitive to Osmotic Stress

To determine the function of OsPP2CAs under osmotic stress, we examined transgenic rice lines (named C18) overexpressing *OsPP2C09* (Figure 4A–C). Growth of C18 seedlings was more sensitive to osmotic stress than the control A8 line, when properly normalized to the mock treatment. Indeed, in the absence of mannitol, C18 seedlings were 10% taller than A8 seedlings. However, C18 seedlings appeared slightly smaller than A8 seedlings when grown in the presence or 150 mM or 200 mM mannitol (Figure 4A), although this difference was not significant. We therefore converted shoot growth to relative lengths by normalizing each value to the 0 mM control growth values in Figure 4C. The two independent transgenic lines C18-1 and C18-2 seedlings thus exhibited a growth deficit of 12.3% p (C18-1) or 24.2% p (C18-2) in the presence of 150 mM mannitol and 13.6% p (C18-1) or 20.5% p (C18-2) in the presence of 200 mM mannitol when compared to the A8 control line. These results suggest that C18 seedlings are more sensitive than A30 seedlings (overexpressing *OsPYL/RCAR3*) to osmotic stress.

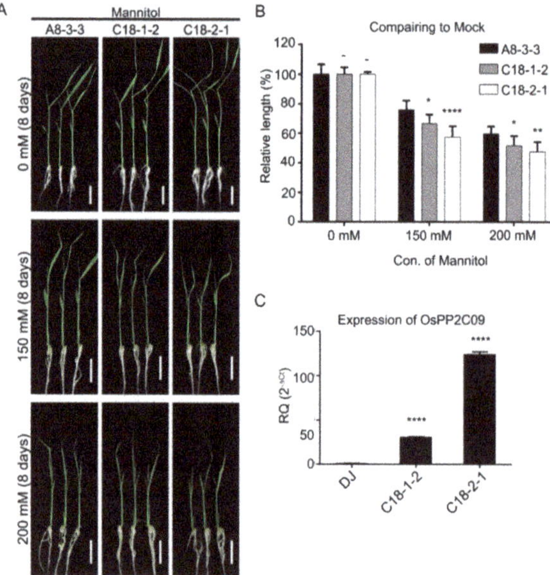

Figure 4. C18 plants, overexpressing *OsPP2C09*, are hypersensitive to osmotic stress. (**A**), A8 and C18 plants transferred after the two-leaf stage to half-strength MS medium containing the indicated mannitol concentrations for 8 d. (**B**), Shoot lengths were measured, with six or more seedlings per replicate. Relative shoot length for A8 and C18 seedlings, normalized to mock control (0 mM mannitol). The values are means ± SD (n = 3). The experiment was repeated three times with similar results. Two-way ANOVA was performed comparing with A8 plants (****: $p < 0.0001$, **: $p < 0.01$, *: $p < 0.05$, -: not significant). (**C**), Relative transcript levels of *OsPP2C09* in C18 seedlings. All values were determined by RT-qPCR and normalized to *UBIQUTIN5* before normalization to relative transcript levels in Dong-Jin (DJ), shown as mean of relative quantity and SD. One-way ANOVA was performed (****: $p < 0.0001$).

2.5. Transcriptomic Analysis of Rice Overexpressing OsPP2CA09 during Osmotic Stress and ABA Treatment

OsPP2CAs can regulate gene expression through activation of *cis*-DREs in promoters but also repress gene expression resulting from ABA signaling. To identify genes regulated by OsPP2C09, we dissected the transcriptome of rice seedlings (C18-2-1) overexpressing OsPP2CA09 and the control rice cultivar Dongjin (DJ) treated with 200 mM mannitol or 10 µM ABA for 24 h using deep sequencing of the transcriptome (RNAseq). Read numbers were normalized to FPKM (fragments per kb of transcripts per million mapped reads) and then compared with those of non-treated control seedlings. We detected 1631 and 1598 genes that were upregulated more than two-fold in response to 200 mM mannitol or 5 µM ABA treatment, respectively (Figure 5A). Of these, 828 genes were upregulated in both conditions, and are given in Table S1.

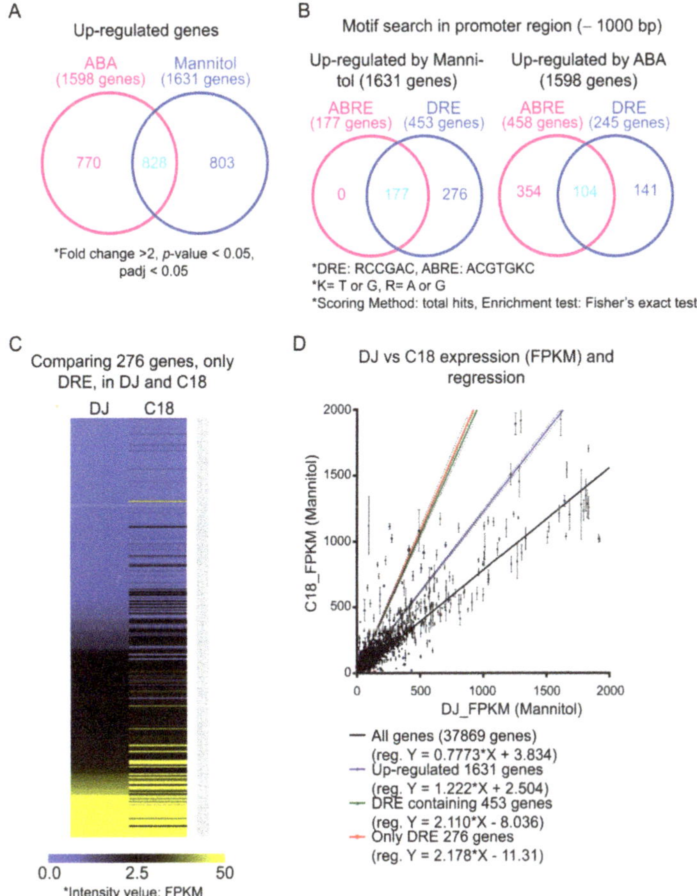

Figure 5. Transcriptome and promoter motif analysis in response to mannitol or ABA treatment. (**A**), Venn diagram of upregulated genes after treatment with 200 mM mannitol or 10 µM ABA. The genes were selected with the indicated criteria (fold-change > 2, *p*-value and padj < 0.05). (**B**), Venn diagrams of genes whose promoters contain DRE and/or ABRE *cis*-acting elements. We performed an AME analysis on the MEME website. Venn diagrams were generated with InteractiVenn (http://interactivenn.net). (**C**), Heatmap of FPKM values for 276 genes containing *cis*-DRE. (**D**), Comparison of FPKMs for the indicated genes between wild-type (DJ) and C18 seedlings treated with 200 mM mannitol. Dotted line: slope of 1. The values are FPKM ± SD (*n* = 3).

Next, we searched for DRE (A/GCCGAC) and ABRE (ACGTGT/GC) *cis*-elements in the 1-kb upstream promoter regions of these upregulated genes using AME (Analysis of Motif Enrichment) on the MEME website (Multiple EM for Motif Elicitation, http://meme-suite.org/index.html) [19]. Among the 1631 genes upregulated by mannitol, 453 genes contained the DRE alone (276 genes) or both the ABRE and the DRE (177 genes) in their promoters. None of the 1631 promoter sequences contained only the ABRE element. By contrast, among the 1598 genes upregulated by ABA, 458 promoter sequences contained *cis*-ABRE element alone (354 genes) or both cis-ABRE element and *cis*-DRE (104 genes). DRE was detected alone in 141 promoters (Figure 5B). The selected genes are listed in Table S2. These results showed that low osmotic stress imposed by mannitol upregulated a limited set of genes, but the *cis*-ABRE element that activates transcription in response to ABA signaling is lacking in their promoters.

Next, we compared the expression levels of the 276 genes whose promoters only contained the *cis*-DRE in C18-2-1 and DJ control seedlings. If one of the functions of OsPP2C09 is to activate gene expression via the *cis*-DRE, we would expect most of the 276 genes to exhibit higher expression in DJ seedlings treated with 200 mM mannitol relative to C18 seedlings grown under the same conditions. For visualization, we generated a heatmap of expression values (in FPKMs) for these 276 genes. Most genes were more strongly expressed in C18-2-1 than in DJ (Figure 5C).

We next compared FPKM values for all genes and all upregulated genes in C18 and DJ treated with mannitol (Figure 5D). The linear regression slope was 0.78, indicating that most genes were generally more highly expressed in the DJ control seedlings than in C18. However, the regression slope between FPKM values of genes upregulated by mannitol was 1.22, in agreement with our heatmap result. Moreover, focusing on FPKM values of genes whose promoters contained either *cis*-DRE and *cis*-ABRE or *cis*-DRE alone, the corresponding regression slopes were 2.11 and 2.18, respectively. These results showed that the presence of *cis*-DREs was associated with higher expression in C18 seedlings overexpressing OsPP2C09 relative to its wild-type parental line. Thus, OsPP2C09 may participate in the activation of gene expression through *cis*-DREs in response to low osmotic stress.

In addition, we examined the expression of several genes in control DJ, A30, or C18 seedlings by RT-qPCR. As previously mentioned, the *Rab16A* promoter contains both a *cis*-DRE and a *cis*-ABRE, whereas the *LIP9* and *OsDREB1C* promoters bear only *cis*-DREs. *Rab16A* transcript levels were highly induced in DJ and C18 seedlings grown in the presence of 250 mM mannitol. *Rab16A* transcript reached higher levels in C18 seedlings subjected to mannitol stress for 24 h compared to that in DJ. By contrast, *Rab16A* transcript levels were greatly induced in A30 seedlings following treatment with 10 µM ABA, and showed a slight and delayed rise in C18, possibly caused by OsPP2C09 (Figure 6A). LIP9 and OsDREB1C expression levels increased to a greater extent in C18 seedlings exposed to 250 mM mannitol than in the DJ control, whereas their transcript levels decreased in A30 following this treatment, likely due to OsPYL/RCAR3-mediated inhibition of OsPP2C. Finally, *LIP9* and *OsDREB1C* transcripts were weakly induced by 10 µM ABA in C18 (Figure 6B,C). Together, these results corroborated our transcriptome analysis above.

2.6. OsPP2CAs Regulate the Transcriptional Activity of OsDREBs

Our combined results thus far have revealed that OsPP2CAs activate transcription via the *cis*-DRE in response to osmotic stress. Thus, we inferred that OsPP2CAs might regulate the activity of DREB transcription factors directly or indirectly. To test this possibility, we examined the interaction potential between OsPP2CAs and OsDREBs. We selected three OsDREBs (OsDREB1A, OsDREB1B, and OsDREB1C), based on their expression in rice leaf blades in microarray datasets (Figure S4). Due to the demonstrated autoactivity of OsDREB and OsPP2Cs, we did not perform a yeast two-hybrid first, but rather conducted a BiFC with OsDREBs and OsPP2Cs. We used OsPYL/RCAR3 as a negative control because it shared the same subcellular localization (nucleus) as OsDREBs, but is known not to interact with these transcription factors (Figure S5). The N-terminus of the fluorescent protein Venus (VN) was fused with OsPP2Cs and OsPYL/RCAR3; the C-terminus of Venus (VC) was fused with OsDREBs. The resulting constructs were then introduced into rice protoplasts in chosen combinations. We also co-transformed a mCherry marker as control for transfection, when no fluorescence signal from Venus was detectable, and normalized Venus fluorescence signal to mCherry fluorescence (Figure 7A,B). All three OsDREBs interacted with OsPP2C09 and OsPP2C30. However, OsPP2C06 failed to complement Venus fluorescence, indicating a lack of interaction between OsDREBs and OsPP2C06. We confirmed the interaction of OsPP2C09 and OsPP2C30 with OsDREBs by a co-immunoprecipitation assay. OsDREB1A showed the strongest interaction, whereas OsDREB1B showed the weakest interaction with OsPPC09 and OsPP2C30 (Figure 7C).

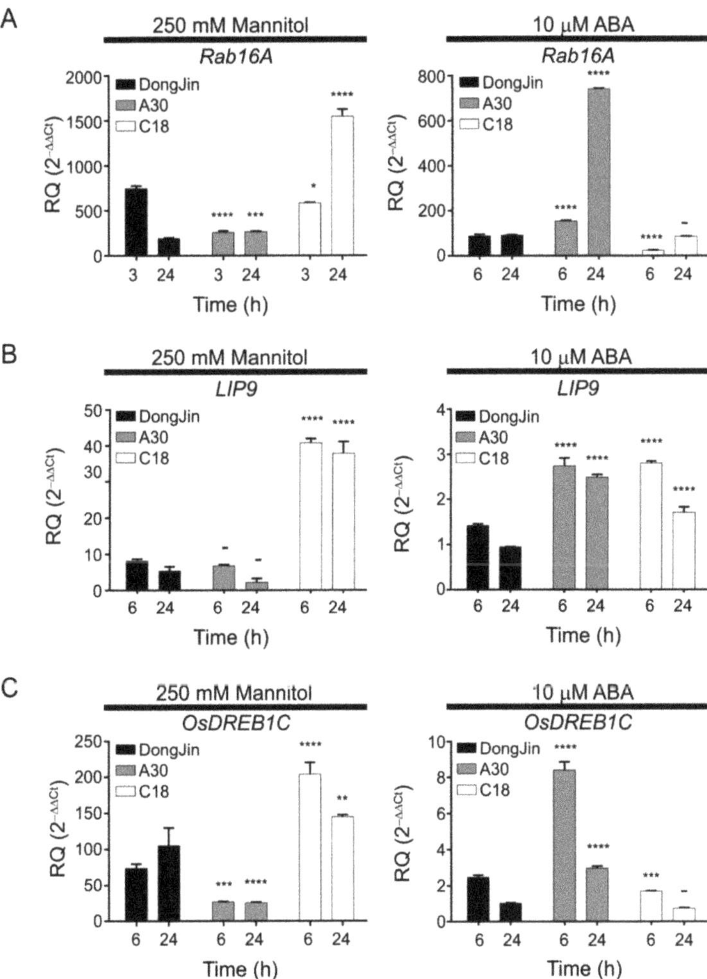

Figure 6. RT-qPCR analysis of three marker genes *Rab16A*, *LIP9*, and *OsDREB1C* in DJ, A30 or C18 seedlings after exposure to 250 mM mannitol or 10 μM ABA. Relative transcript levels of *Rab16A* (**A**), *LIP9* (**B**) or *OsDREB1C* (**C**), determined by RT-qPCR and normalized to *UBIQUITIN5* as internal control. Two way ANOVAs were performed (****: $p < 0.0001$, ***: $p < 0.001$, **: $p < 0.01$, *: $p < 0.05$, -: not significant). Values are means ± SD ($n = 3$). Similar results were obtained in three replicate experiments.

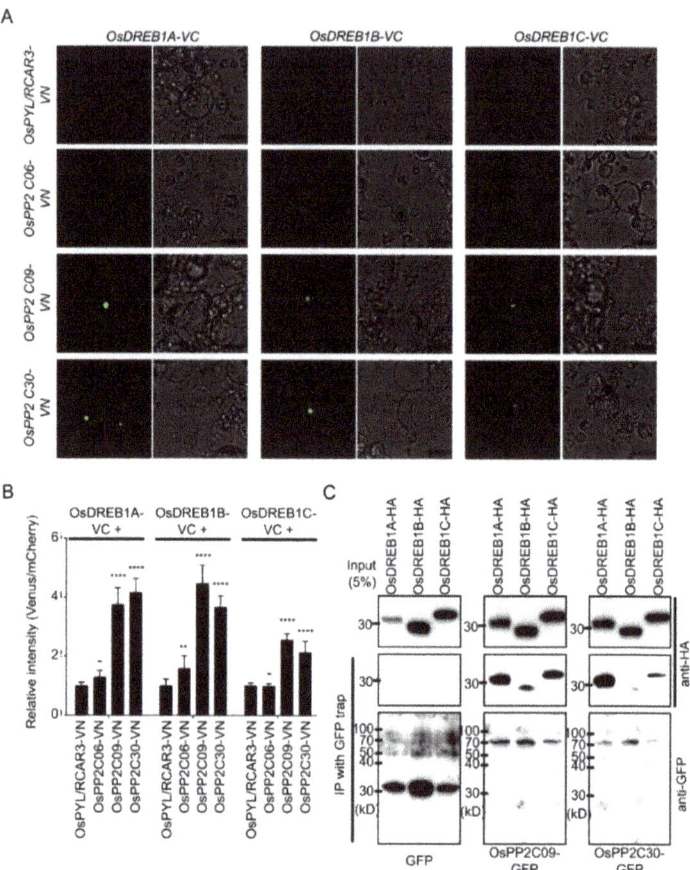

Figure 7. OsPP2CAs interact with OsDREBs. (**A**), Venus *C*-terminus (VC)-tagged OsDREB1A, OsDREB1B and OsDREB1C were examined with Venus *N*-terminus (VN)-tagged OsPYL/RCAR3, OsPP2C06, OsPP2C09, and OsPP2C30 by BiFC analysis. Scale bars: 10 μm. (**B**), BiFC signal intensities normalized to internal control (ER-mCherry). Values are means ±SD (*n* > 20 cells) and normalized to BiFC signal for OsPYL/RCAR3-VN and each OsDREBs-VC. One-way ANOVAs were performed with comparing to signals interacted with OsPYL/RCAR3-VN (****: $p < 0.0001$, **: $p < 0.01$, -: not significant). (**C**), Co-immunoprecipitation analysis. Indicated constructs were introduced into rice protoplasts and GFP-tagged proteins were pull-downed with GFP-trap beads. An immunoblot analysis was performed with anti-GFP rabbit antibodies or anti-HA rat antibodies.

To test whether OsPP2CAs activate OsDREBs, we focused on OsPP2C09 because it showed the strongest interaction with OsDREBs. We tested OsPP2C09-mediated activation of OsDREB1A, OsDREB1B, or OsDREB1C using the *pDRE::fLUC* reporter in transiently-transfected rice protoplasts. First, we established that fLUC activity (as a proxy for activation mediated by the *cis*-DRE) increased when *OsDREBs-HA* effector plasmid DNA was co-introduced into the protoplasts in a dose-dependent manner (Figure 8A). Next, we generated an OsPP2C09 interference line (OsPP2C09-i) in which Gly-139 was replaced by Asn. This point mutant was reported to interfere with ABA signaling as it abrogates phosphatase activity (Sheen et al., 1998). The OsPP2C09-i point mutant also interrupted ABA signaling in our rice protoplast luciferase assay (Figure 8B).

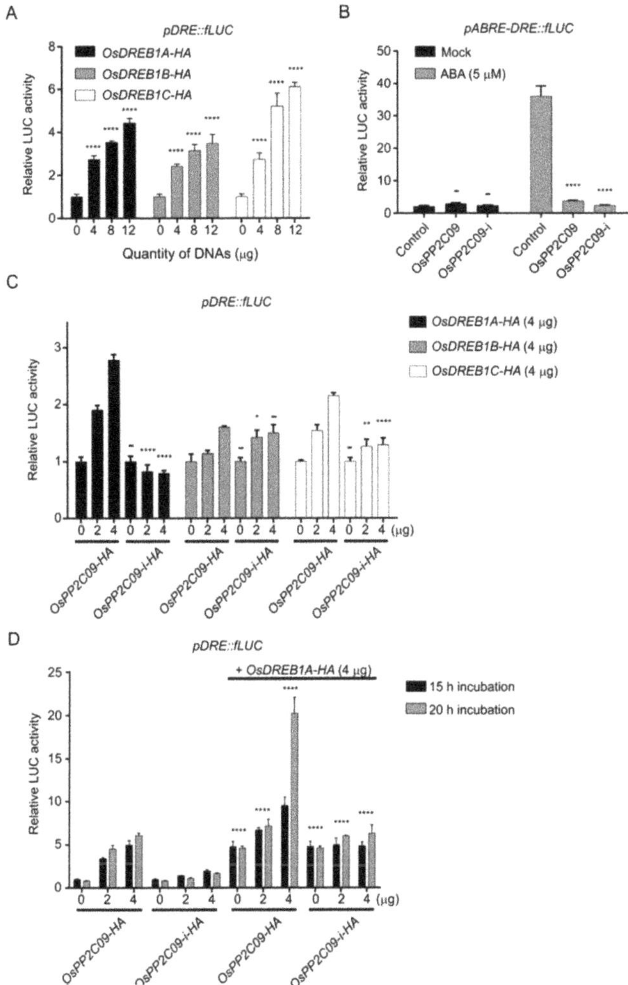

Figure 8. Activation of *pDRE::fLUC* by overexpression of *OsDREB1s*. (**A**), Activation of *pDRE::fLUC* depends on *OsDREB1s* effector plasmid DNA amounts. One-way ANOVAs were performed with comparing to 0 µg of effector DNAs. (**B**), Confirmation of the lack of activation of the *pABRE-DRE::fLUC* reporter by OsPP2C09-i. One-way ANOVAs were performed, comparing to the control. (**C**), Effects of OsDREB1-mediated activation by OsPP2C09 or OsPP2C09-i. Two-way ANOVAs were performed with comparing to WT (OsPP2C09). (**D**), Effects of OsPP2C09 on the activity of OsDREB1A at two time points. Two-way ANOVA was performed, comparing to non-OsDREB1A. The indicated effector and marker DNAs were introduced into rice protoplasts by the PEG-mediated transfection method and then incubated for 15 h or 20 h. Induced LUC activity was detected with a dual-luciferase reporter assay system. All values of panels are means ± SD ($n = 3$). ****: $p < 0.0001$, **: $p < 0.01$, *: $p < 0.05$, -: not significant.

We next monitored the transcriptional activity of *OsDREB1s* co-transfected with *OsPP2C09-i* or *OsPP2C09*. OsDREB1A-mediated activity clearly increased as a function of OsPP2C09 plasmid effector DNA dosage, compared to other OsDREB1s (Figure 8C). However, OsPP2C09i did not show severe effects on luciferase activity (Figure 8C,D). When transfected cells were incubated for up to 20 h, the luciferase activity induced by OsPP2C09 increased sharply, indicating a synergistic response between OsPP2C09 and OsDREB1A.

However, OsPP2C09i did not increase DREB activity under the same conditions (Figure 8D). Instead, OsPP2C09 showed a synergistic effect on the activation of *cis*-DRE by OsDREB1s, especially with OsDREB1A. The OsPP2C09i point mutant, lacking phosphatase activity, did not show activation of *cis*-DRE.

3. Discussion

ABA plays central roles in tolerance to abiotic stress such as drought, cold, high temperatures, and salinity. Other signaling pathways, such as the DREB-mediated pathway, also participate in plant abiotic stress responses. Thus, abiotic stress responses may be largely classified into two groups: ABA-independent and ABA-dependent signaling cascades [6].

PP2CAs are well-known negative regulators of ABA signaling. However, their expression is induced significantly not only during treatment with exogenous ABA, but also during osmotic stress [20]. This phenomenon puzzled scientists for a long time and was just explained as a feedback regulation between ABA signaling and osmotic stress (Figure 2A,B) [21]. However, in this study we suggest that PP2CAs, induced by osmotic stress, can activate drought-responsive regulons from the ABA-independent signaling branch and act as positive regulators of signaling.

3.1. Nucleus-Localized PP2CAs Regulate Diverse Abiotic Stress-Related Transcription Factors

ABA-dependent signal transduction relies heavily on phosphorylation/dephosphorylation mechanisms of the regulatory proteins [22]. Plant genomes harbor several thousands of kinase genes; whose products are responsible for the phosphorylation of very specific substrates [23]. By contrast, plant genomes contain genes encoding only a few hundred phosphatases, which might have broader substrates than kinases. PP2CAs are major negative regulators of the ABA signaling pathway that function by dephosphorylating SnRK2-type kinases [24].

In total, nine rice PP2CAs can be clearly divided into two groups, based on their subcellular localization (nucleus and cytosol) [25]. Cytosol- or membrane-localized OsPP2CAs might regulate the activity of several second messengers and transporters and play diverse functions. Such PP2Cs, as for example Arabidopsis ABA INSENSITIVE1 (ABI1) and ABI2, are the main regulators and initial signal transducers of the ABA signal [26]. However, the six nucleus-localized PP2CAs potentially can modulate transcription factors, directly or indirectly. Bhatnagar et al., 2017 have reported that PP2CAs can interact with OsbZIP and directly regulate this transcription factor independently of SnRKs, suggesting that PP2CAs can directly regulate transcription factors [27]. In addition, Mizoi et al., 2019 reported that phosphorylation of the NRD region makes DREB2A unstable; this result raised the possibility that phosphatases like OsPP2C09 may affect the stability or activity of DREBs [14]. We therefore investigated the abiotic stress responsive transcription factors DREBs during ABA-independent signaling, although their phosphorylation status during such treatment was obscure in this report. We propose here that one of the nucleus-localized PP2CA, OsPP2C09, might regulate DREB1 activity. Thus, OsPP2C09 can modulate ABA-dependent signaling as well as ABA-independent signaling reversibly.

3.2. PP2CAs Might Act as Hubs in ABA-Dependent and -Independent Signaling Crosstalk

ABA, drought, osmotic and cold stress signals all sense environmental cues via different receptors [28]. Thereafter, the receptors transduce these signals to downstream signaling components such as kinases, phosphatases and so on [29]. Transcriptional regulation is finally enacted by transcription factors binding to *cis*-regulatory elements in the promoters of target genes [30]. The receptor(s) for osmotic stress is not known yet, but might initiate a faster response than the ABA signaling cascade. Even though the two signaling pathways have largely distinct signaling components, both rely on PP2CAs, which may target several common substrates and regulate their phosphorylation/dephosphorylation status in response to both signals.

In this and previous studies, PP2CAs are induced quickly by abiotic stresses and can then activate DREBs regulons, which include ABA biosynthesis and growth regulators. Thus, PP2CAs might act as one of the early regulators of abiotic stress responses.

After recognition of the initial osmotic stress, ABA content increases in the cell. This rise in ABA levels may amplify the responses necessary for adaptation to abiotic stresses, and PP2Cs may provide feedback regulation to balance stress tolerance responses (which limit growth) and sustained growth (which limits stress responses) [6,21]. Abiotic stress signaling components are shared by both ABA-independent and -dependent signaling pathways.

We hypothesize that OsPP2CAs might have several substrates in both ABA-dependent and ABA-independent signaling, and antagonistically regulate the two signaling pathways. Thus, PP2CAs might function as a central signaling hub in both abiotic stress-signaling pathways.

4. Materials and Methods

4.1. Plant Materials and Generation of Transgenic Rice

We surface-sterilized rice seed (*Oryza sativa* cv. Dongjin) with 70% ethanol for 30 s and 50% bleach for 40 min, respectively. After 5 washes with distilled water, seeds were sown on half-strength Murashige and Skoog [31] medium pH 5.8 (supplemented with 0.4% phytagel) for 7 days under long-day conditions (16 h light and 8 h darkness) at 28 °C. To generate transgenic rice, we amplified the coding sequence of *OsPYL/RCAR3* (locus number Os02g15640) or *OsPP2C09* (locus number *Os01g62760*) via PCR from first-strand cDNAs with specific primers (5-topo-RCAR3 and 3-RCAR3, or 5-topo-PP2C09 and 3-PP2C09, listed in Table S2) and AccPrime ^{TM} *Pfx* DNA polymerase using the provided protocol. The PCR conditions were 2 min denaturing and 35 cycles (denaturing 95 °C for 30, annealing at 55 °C for 30 s, and extending at 68 °C for 40 s). The PCR products were inserted into pENTR/D-Topo (Invitrogen, Carlsbad, CA, USA) and then transferred to the pGA2897 vector via Gateway LR recombination. The resulting *pUbi:OsPYL/RCAR3* or *pUbi:OsPP2C09* constructs, in which *OsPYL/RCAR3* or *OsPP2C09* was placed under the control of the constitutive maize *UBIQUITIN* promoter, was transformed into Agrobacterium (*Agrobacterium tumefaciens*) strain LBA4404 via electroporation. We generated transgenic rice plants using the Agrobacterium-mediated co-cultivation method and selected the transformants based on hygromycin resistance and before transferring them to the greenhouse (Toki et al., 2006).

4.2. Post-Germination Assays and Stress Treatment

We plated surface-sterilized dehulled seeds on half-strength MS medium supplemented with hygromycin (40 mg/L). Then, 3 days later, we transferred the seedlings to half-strength MS supplemented with 5 μM ABA, 200 mM NaCl, 200 mM Mannitol in square Petri dishes (125 × 125 × 20 mm). Seedling growth was measured 10 days after transfer. This experiment was repeated three times independently.

4.3. RT-PCR and Quantitative PCR

For RT-qPCR analysis, we synthesized first-strand cDNAs from 5 μg total RNA using SuperScript III reverse transcriptase (Invitrogen, Carlsbad, CA, USA) using the provided protocol. A 1:40 dilution of the cDNAs was used for RT-qPCR. The amplification parameters were as follows: 15 min of denaturation and enzyme activation at 95 °C; followed by 40 cycles of 95 °C for 5 s, 60 °C for 15 s, and 72 °C for 30 s; with a final step performed at 65–95 °C (1 °C/s) for melting curve analysis. The amplified signals were detected using a MyiQ real-time PCR system (Bio-Rad Laboratories, Hercules, CA, USA) using SYBR Premix Ex Taq^{TM} (TOPrealTM qPCR 2X PreMix, www.enzynomics.com, Yuseong-gu, Daejeon, Korea). The data were normalized based on the expression of rice *UBIQUITIN5*, and the relative gene expression was analyzed using the $2^{-\Delta\Delta Ct}$ method or the $2^{-\Delta Ct}$ method. Primer sequences used for RT-qPCR analysis are listed in Table S2.

4.4. Luciferase Assay Using Rice Protoplasts

We isolated rice protoplasts using the protocol described by Kim et al., (2015). For the luciferase assay, firefly luciferase (fLUC) and renilla luciferase (rLUC) plasmids and effector plasmids were introduced into purified rice protoplasts as previously described [32]. We introduced all plasmid DNAs (5 μg of *pRab16a:fLUC*, *pABRE-DRE:fLUC*, or *pDRE:fLUC* with 0.5 μg of *pUbi:rLUC* (transformation control) and indicated effector plasmids) into rice protoplasts by the PEG transfection method. After 15–20 h incubation, fLUC and rLUC activity was detected using a dual-luciferase assay kit (Promega, Madison, WI, USA) and a GloMax 96 Microplate Luminometer (Promega, USA), according to the manufacturer's'instructions.

4.5. Subcellular Localization and BiFC

To observe interactions between proteins using bimolecular fluorescence complementation (BiFC), we introduced gene fragments into pENTR-D-TOPO vectors (Invitrogen, Carlsbad, CA, USA) and then transferred them to their destination vectors by LR recombination (Promega, Madison, WI, USA) [25]. pGEM-gw-VC vector was used for fusion of Venus Carboxyl-terminus (VC) with OsDREB1A, OsDREB1B, and OsDREB1C, pGEM-gw-VN vector used for fusion of Venus Amino-terminus (VN) with OsPP2C06, OsPP2C09, osPP2c30, and OsPYL/RCAR3. Generated plasmids were introduced into rice protoplasts as indicated pairs using the PEG-mediated method [32]. The ER-mCherry reporter was used as internal control. Fluorescence signals were captured using a Leica TCS SP8 laser scanning confocal microscope (Leica Microsystems, Wetzlar, Germany). The combination of excitation wavelength/detection range of emission for Venus signals was 488 nm (solid state laser)/ with a bandpass of 505–561. Signal intensities on captured images were analyzed using Leica Application SuiteX provided by the manufacturer (Leica Microsystems, Wetzlar, Germany) with default settings (threshold 30%, background 20%).

4.6. Co-Immunoprecipitation

To perform co-immunoprecipitation experiments, we used GFP-trap (Chromotek, Planegg, Germany) precipitate GFP-fusion proteins. All used genes were PCR-amplified with specific primers (Table S2) and inserted into pENTR/D-topo vectors (Invitrogen, Carlsbad, CA, USA). We then recombined these entry clones with the vectors pGEM-gw-GFP for GFP fusions and pGEM-gw-3xHA for HA tagging [25], using LR recombination (Invitrogen, Carlsbad, CA, USA). The indicated constructs were introduced into rice protoplasts using the PEG-mediated method, and the transformed protoplasts were incubated at 28 °C for 20 h. Cellular extracts from transformed protoplasts extracted in immunoprecipitation buffer (150 mM NaCl, 50 mM Tris-HCl at pH 7.5, 1 mM EDTA, 2 mM EGTA, 2 mM MgCl$_2$, 0.5% NP40, 0.5% Triton X-100, and 1x protease inhibitor cocktail (complete ULTRA tablet, Roche, Indianapolis, IN, USA)) were incubated with pre-cleaned GFP-trap beads at 4 °C for 2 h. After 5 washes with immunoprecipitation buffer, the precipitated proteins, together with GFP-trap, were subjected to SDS-PAGE with mid-range protein marker (Elpis Biotech, Daejeon, Korea) and immunoblot analysis. Precipitated GFP and HA-tagged proteins were detected with anti-GFP rabbit antibody (Life Technologies, Carlsbad, CA, USA) and anti-hemagglutinin (HA) rat antibody (Roche, Indianapolis, IN, USA), respectively.

4.7. Transcriptome Analysis

For the transcriptomic analysis, we prepared the total RNA from 14 day-old plants grown on 1/2 MS medium. C18 (OsPP2C09-overexpressing plant) and its control Dongjin plants were prepared with or without ABA or mannitol treatment for 24 h. Total RNA was extracted from shoots and purified using the RNeasy Mini Kit (Qiagen, Hilden, Germany). Quality control was conducted with the Agilent Technologies 2100 Bioanalyzer (Agilent Technologies, Santa Clara, CA, USA). The libraries for sequencing were prepared using a TruSeqRNA Sample Prep Kit v2 (Illumina, San Diego, CA, USA), following the manufacturer's instructions. The sequencing of the libraries was performed using a HiSeq 4000 system (Illumina, San Diego, CA, USA) generating single-end 101-bp reads. The trimmed reads were mapped to IRGSP (v. 1.0) and assembled into transcripts. The read counts were determined using the StringTig program, and then normalized with the DESeq2 program [33]. The promoter sequences of upregulated genes were obtained from The Rice Annotation Project Database annotated data on OsNipponbare-Reference-IRGSP-1.0. The motifs searches were performed in http://meme-suite.org/tools/ame with indicated conditions [19]. The heatmap image was constructed with MeV program [34]. The graphs were constructed using GraphPad Prism6 program (GraphPad Software, San Diego, CA, USA).

Supplementary Materials: Supplementary materials can be found at https://www.mdpi.com/1422-0067/22/1/393/s1. List of Supplemental Data: Table S1. List of 2fold upregulated genes. Table S2. List of ABRE, DRE *cis*-element hit genes in upregulated genes by ABA or Mannitol. Figure S1. Phylogenetic trees of ABA receptors and phosphatases. Figure S2. A30 plants (overexpressing OsPYL/RCAR3) and their tolerance to abiotic stress. Figure S3. ABA-independent interaction of OsPYL/RCAR3 and OsPP2Cs. Figure S4. Search for OsDREB expression in microarray data base. Figure S5. Subcellular localization of OsPYL/RCAR3 and OsDREB1s.

Author Contributions: M.K.M. and B.-G.K. designed research plans, J.-Y.L. and B.-G.K. supervised the experiments, W.-J.H. and K.-H.J. performed the bioinformatic analysis. M.K.M. and R.K. performed the experiments, M.K.M. wrote the draft manuscript, J.-Y.L. and B.-G.K. edited and completed the writing. All authors have read and agreed to the published version of the manuscript.

Funding: This work was supported by the Agenda program (PJ014838) and Next-Generation Biogreen Program (PJ013676) through the Rural Development Administration.

Conflicts of Interest: The authors declare no conflict of interest. The funders had no role in the design of the study; in the collection, analyses, or interpretation of data; in the writing of the manuscript, or in the decision to publish the results.

Abbreviations

Abfs	Aba-responsive element (abre)-binding factors
Abi1	Aba insensitive1
Abre	Aba-responsive element
Dre/crt	Dehydration-responsive element/c-repeat
Dreb	Dehydration-responsive element binding factor
Drip1	Dreb2a-interacting protein1
Fpkm	Fragments per kilobase million
Gfp	Green fluorescent protein
Grf7	Growth-regulating factor7
Hos1	High expression of osmotically responsive gene1
Ice1	Inducer of cbf expression1
Pp2cas	Clade a type 2c protein phosphatases
Pyr/pyl/rcar	Pyrabactin resistance1/pyr-like/regulatory component of aba receptor
Snrk	Snf1-related serine/threonine-protein kinase

References

1. Wang, W.; Vinocur, B.; Altman, A. Plant responses to drought, salinity and extreme temperatures: Towards genetic engineering for stress tolerance. *Planta* **2003**, *218*, 1–14. [CrossRef] [PubMed]
2. Hasanuzzaman, M.; Nahar, K.; Alam, M.M.; Roychowdhury, R.; Fujita, M. Physiological, biochemical, and molecular mechanisms of heat stress tolerance in plants. *Int. J. Mol. Sci.* **2013**, *14*, 9643–9684. [CrossRef] [PubMed]
3. Sah, S.K.; Reddy, K.R.; Li, J. Abscisic Acid and Abiotic Stress Tolerance in Crop Plants. *Front. Plant Sci.* **2016**, *7*, 571. [CrossRef] [PubMed]
4. Fujita, Y.; Fujita, M.; Satoh, R.; Maruyama, K.; Parvez, M.M.; Seki, M.; Hiratsu, K.; Ohme-Takagi, M.; Shinozaki, K.; Yamaguchi-Shinozaki, K. AREB1 is a transcription activator of novel ABRE-dependent ABA signaling that enhances drought stress tolerance in Arabidopsis. *Plant Cell* **2005**, *17*, 3470–3488. [CrossRef]
5. Kang, J.Y.; Choi, H.I.; Im, M.Y.; Kim, S.Y. Arabidopsis basic leucine zipper proteins that mediate stress-responsive abscisic acid signaling. *Plant Cell* **2002**, *14*, 343–357 [CrossRef]
6. Yoshida, T.; Mogami, J.; Yamaguchi-Shinozaki, K. ABA-dependent and ABA-independent signaling in response to osmotic stress in plants. *Curr. Opin. Plant Biol.* **2014**, *21*, 133–139. [CrossRef]

7. Yamaguchi-Shinozaki, K.; Shinozaki, K. A novel cis-acting element in an Arabidopsis gene is involved in responsiveness to drought, low-temperature, or high-salt stress. *Plant Cell* **1994**, *6*, 251–264.
8. Saibo, N.J.; Lourenco, T.; Oliveira, M.M. Transcription factors and regulation of photosynthetic and related metabolism under environmental stresses. *Ann. Bot.* **2009**, *103*, 609–623. [CrossRef]
9. Lata, C.; Prasad, M. Role of DREBs in regulation of abiotic stress responses in plants. *J. Exp. Bot.* **2011**, *62*, 4731–4748. [CrossRef]
10. Qin, F.; Sakuma, Y.; Tran, L.S.; Maruyama, K.; Kidokoro, S.; Fujita, Y.; Fujita, M.; Umezawa, T.; Sawano, Y.; Miyazono, K.; et al. Arabidopsis DREB2A-interacting proteins function as RING E3 ligases and negatively regulate plant drought stress-responsive gene expression. *Plant Cell* **2008**, *20*, 1693–1707. [CrossRef]
11. Kim, J.S.; Mizoi, J.; Kidokoro, S.; Maruyama, K.; Nakajima, J.; Nakashima, K.; Mitsuda, N.; Takiguchi, Y.; Ohme-Takagi, M.; Kondou, Y.; et al. Arabidopsis growth-regulating factor7 functions as a transcriptional repressor of abscisic acid- and osmotic stress-responsive genes, including DREB2A. *Plant Cell* **2012**, *24*, 3393–3405. [CrossRef]
12. Miura, K.; Jin, J.B.; Lee, J.; Yoo, C.Y.; Stirm, V.; Miura, T.; Ashworth, E.N.; Bressan, R.A.; Yun, D.J.; Hasegawa, P.M. SIZ1-mediated sumoylation of ICE1 controls CBF3/DREB1A expression and freezing tolerance in Arabidopsis. *Plant Cell* **2007**, *19*, 1403–1414. [CrossRef] [PubMed]
13. Chinnusamy, V.; Ohta, M.; Kanrar, S.; Lee, B.H.; Hong, X.; Agarwal, M.; Zhu, J.K. ICE1: A regulator of cold-induced transcriptome and freezing tolerance in Arabidopsis. *Genes Dev.* **2003**, *17*, 1043–1054. [CrossRef]
14. Mizoi, J.; Kanazawa, N.; Kidokoro, S.; Takahashi, F.; Qin, F.; Morimoto, K.; Shinozaki, K.; Yamaguchi-Shinozaki, K. Heat-induced inhibition of phosphorylation of the stress-protective transcription factor DREB2A promotes thermotolerance of Arabidopsis thaliana. *J. Biol. Chem.* **2019**, *294*, 902–917. [CrossRef] [PubMed]
15. Kim, H.; Hwang, H.; Hong, J.W.; Lee, Y.N.; Ahn, I.P.; Yoon, I.S.; Yoo, S.D.; Lee, S.; Lee, S.C.; Kim, B.G. A rice orthologue of the ABA receptor, OsPYL/RCAR5, is a positive regulator of the ABA signal transduction pathway in seed germination and early seedling growth. *J. Exp. Bot.* **2012**, *63*, 1013–1024. [CrossRef] [PubMed]
16. Santiago, J.; Dupeux, F.; Round, A.; Antoni, R.; Park, S.Y.; Jamin, M.; Cutler, S.R.; Rodriguez, P.L.; Marquez, J.A. The abscisic acid receptor PYR1 in complex with abscisic acid. *Nature* **2009**, *462*, 665–668. [CrossRef]
17. Xue, S.; Hu, H.; Ries, A.; Merilo, E.; Kollist, H.; Schroeder, J.I. Central functions of bicarbonate in S-type anion channel activation and OST1 protein kinase in CO2 signal transduction in guard cell. *EMBO J.* **2011**, *30*, 1645–1658. [CrossRef]
18. Tian, X.; Wang, Z.; Li, X.; Lv, T.; Liu, H.; Wang, L.; Niu, H.; Bu, Q. Characterization and Functional Analysis of Pyrabactin Resistance-Like Abscisic Acid Receptor Family in Rice. *Rice* **2015**, *8*, 28. [CrossRef]
19. McLeay, R.C.; Bailey, T.L. Motif Enrichment Analysis: A unified framework and an evaluation on ChIP data. *BMC Bioinform.* **2010**, *11*, 165. [CrossRef]
20. Lu, T.; Zhang, G.; Wang, Y.; He, S.; Sun, L.; Hao, F. Genome-wide characterization and expression analysis of PP2CA family members in response to ABA and osmotic stress in Gossypium. *PeerJ* **2019**, *7*, e7105. [CrossRef]
21. Wang, X.; Guo, C.; Peng, J.; Li, C.; Wan, F.; Zhang, S.; Zhou, Y.; Yan, Y.; Qi, L.; Sun, K.; et al. ABRE-BINDING FACTORS play a role in the feedback regulation of ABA signaling by mediating rapid ABA induction of ABA co-receptor genes. *New Phytol.* **2019**, *221*, 341–355. [CrossRef] [PubMed]
22. Raghavendra, A.S.; Gonugunta, V.K.; Christmann, A.; Grill, E. ABA perception and signalling. *Trends Plant Sci.* **2010**, *15*, 395–401. [CrossRef] [PubMed]
23. Bheri, M.; Mahiwal, S.; Sanyal, S.K.; Pandey, G.K. Plant protein phosphatases: What do we know about their mechanism of action? *FEBS J.* **2020**. [CrossRef] [PubMed]
24. Hauser, F.; Waadt, R.; Schroeder, J.I. Evolution of abscisic acid synthesis and signaling mechanisms. *Curr. Biol.* **2011**, *21*, R346–R355. [CrossRef]
25. Min, M.K.; Choi, E.-H.; Kim, J.-A.; Yoon, I.S.; Han, S.; Lee, Y.; Lee, S.; Kim, B.-G. Two Clade A Phosphatase 2Cs Expressed in Guard Cells Physically Interact with Abscisic Acid Signaling Components to Induce Stomatal Closure in Rice. *Rice* **2019**, *12*, 37. [CrossRef]
26. Wang, H.; Tang, J.; Liu, J.; Hu, J.; Liu, J.; Chen, Y.; Cai, Z.; Wang, X. Abscisic Acid Signaling Inhibits Brassinosteroid Signaling through Dampening the Dephosphorylation of BIN2 by ABI1 and ABI2. *Mol. Plant* **2018**, *11*, 315–325. [CrossRef]
27. Bhatnagar, N.; Min, M.K.; Choi, E.H.; Kim, N.; Moon, S.J.; Yoon, I.; Kwon, T.; Jung, K.H.; Kim, B.G. The protein phosphatase 2C clade A protein OsPP2C51 positively regulates seed germination by directly inactivating OsbZIP10. *Plant Mol. Biol.* **2017**, *93*, 389–401. [CrossRef]
28. Uslu, V.V.; Grossmann, G. The biosensor toolbox for plant developmental biology. *Curr. Opin. Plant Biol.* **2016**, *29*, 138–147. [CrossRef]
29. Durbak, A.; Yao, H.; McSteen, P. Hormone signaling in plant development. *Curr. Opin. Plant Biol.* **2012**, *15*, 92–96. [CrossRef]
30. Nguyen, H.T.K.; Kim, S.Y.; Cho, K.-M.; Hong, J.C.; Shin, J.S.; Kim, H.J. A Transcription Factor γMYB1 Binds to the P1BS cis-Element and Activates PLA 2 -γ Expression with its Co-Activator γMYB2. *Plant Cell Physiol.* **2016**, *57*, 784–797. [CrossRef]
31. Cutler, S.R.; Rodriguez, P.L.; Finkelstein, R.R.; Abrams, S.R. Abscisic acid: Emergence of a core signaling network. *Annu. Rev. Plant Biol.* **2010**, *61*, 651–679. [CrossRef] [PubMed]
32. Kim, N.; Moon, S.J.; Min, M.K.; Choi, E.H.; Kim, J.A.; Koh, E.Y.; Yoon, I.; Byun, M.O.; Yoo, S.D.; Kim, B.G. Functional characterization and reconstitution of ABA signaling components using transient gene expression in rice protoplasts. *Front. Plant Sci.* **2015**, *6*, 614. [CrossRef] [PubMed]

33. Love, M.I.; Huber, W.; Anders, S. Moderated estimation of fold change and dispersion for RNA-seq data with DESeq2. *Genome Biol.* **2014**, *15*, 550. [CrossRef]
34. Saeed, A.I.; Sharov, V.; White, J.; Li, J.; Liang, W.; Bhagabati, N.; Braisted, J.; Klapa, M.; Currier, T.; Thiagarajan, M.; et al. TM4: A Free, Open-Source System for Microarray Data Management and Analysis. *BioTechniques* **2003**, *34*, 374–378. [CrossRef] [PubMed]

International Journal of
Molecular Sciences

Review

Signal Integration by Cyclin-Dependent Kinase 8 (CDK8) Module and Other Mediator Subunits in Biotic and Abiotic Stress Responses

Leelyn Chong [†], Xiaoning Shi [†] and Yingfang Zhu *

State Key Laboratory of Crop Stress Adaptation and Improvement, School of Life Sciences, Henan University, Kaifeng 475001, China; leelyn.chong@yahoo.com (L.C.); shixn555@gmail.com (X.S.)
* Correspondence: zhuyf@henu.edu.cn
† These authors contributed to this work equally.

Abstract: Environmental stresses have driven plants to develop various mechanisms to acclimate in adverse conditions. Extensive studies have demonstrated that a significant reprogramming occurs in the plant transcriptome in response to biotic and abiotic stresses. The highly conserved and large multi-subunit transcriptional co-activator of eukaryotes, known as the Mediator, has been reported to play a substantial role in the regulation of important genes that help plants respond to environmental perturbances. CDK8 module is a relatively new component of the Mediator complex that has been shown to contribute to plants' defense, development, and stress responses. Previous studies reported that CDK8 module predominantly acts as a transcriptional repressor in eukaryotic cells by reversibly associating with core Mediator. However, growing evidence has demonstrated that depending on the type of biotic and abiotic stress, the CDK8 module may perform a contrasting regulatory role. This review will summarize the current knowledge of CDK8 module as well as other previously documented Mediator subunits in plant cell signaling under stress conditions.

Keywords: CDK8 module; Mediator subunits; biotic stress; abiotic stress; cell signaling

Citation: Chong, L.; Shi, X.; Zhu, Y. Signal Integration by Cyclin-Dependent Kinase 8 (CDK8) Module and Other Mediator Subunits in Biotic and Abiotic Stress Responses. *Int. J. Mol. Sci.* **2021**, 22, 354. https://doi.org/10.3390/ijms22010354

Received: 9 December 2020
Accepted: 28 December 2020
Published: 31 December 2020

Publisher's Note: MDPI stays neutral with regard to jurisdictional claims in published maps and institutional affiliations.

1. Introduction

In order to perceive and respond effectively to environmental stresses, plants have developed sophisticated signaling transduction pathways that induce gene expression changes via a complex network of transcription factors (TFs). Phytohormones of abscisic acid (ABA), jasmonic acid (JA), ethylene (ET), and salicylic acid (SA) are known to have extensive crosstalk with other hormones of auxin, brassinosteroid (BR), cytokinins, and gibberellic acid in plant stress response [1,2]. More than 1500 TFs in *Arabidopsis* have been reported to orchestrate the transcriptional control of abiotic stress responses [3]. The regulation of transcription in plant involves RNA polymerase II (RNA Pol II), general TFs, transcriptional activators/repressors, as well as co-regulators such as Mediator. Mediator was initially studied in yeast and subsequently identified to be critical for RNA Pol II-regulated transcription in eukaryotes including mammalian cells and plants through biochemical purification and comparative genomics [4,5].

The Mediator is an evolutionarily conserved large protein complex with multiple components called subunits that transfers upstream regulatory information from activators and repressors to the basal transcriptional machinery in the downstream pathway. A number of Mediator subunits have been identified to play critical roles in plant defense, adaptation, growth and development [6]. More subunits (25–35) are constituted in the Mediator of plant and mammalian cells than in yeast [7]. Mediator subunits of plants have been revealed to involve in various stress-response pathways through genetic analyses. There is also a fourth regulatory kinase module consisting of two Mediator subunits known as MED12 and MED13 as well as a separable kinase unit made up of CDK8 (cyclin-dependent kinase 8) and a C-type cyclin (CycC). CDK8, as its full name implies is a cyclin-dependent kinase and it along with its associated cyclin, CycC, also referred to as the kinase or CDK8-cyclin C module. The whole kinase module is an integral part of the Mediator complex. Studies have indicated that the protein components of the CDK8 kinase module may act as a both positive and negative transcription regulator in cells since CDK8 module was reported to act as a repressor to Mediator when bound to the complex [8,9]. However, there were situations when the CDK8 module worked as an activator on certain genes [10,11]. In plants, the kinase module was initially viewed as a predominant transcriptional repressor in eukaryotic cells by reversibly associating with the core Mediator. Later, evidence that supports the positive regulatory roles of CDK8 module in plants' transcription has also appeared. In fact, the subunits of MED12 and MED13 were reported recently to serve as conditional positive gene regulators in *Arabidopsis* [12–14]. Likewise, the CDK8 subunit is capable of recruiting different TFs to RNA Pol II to regulate multiple signaling pathways in yeast and eukaryotic cells including plant cells [15]. The general working model for the signaling role of CDK8 and other Mediator subunits in plants during biotic and abiotic stresses are illustrated in Figure 1.

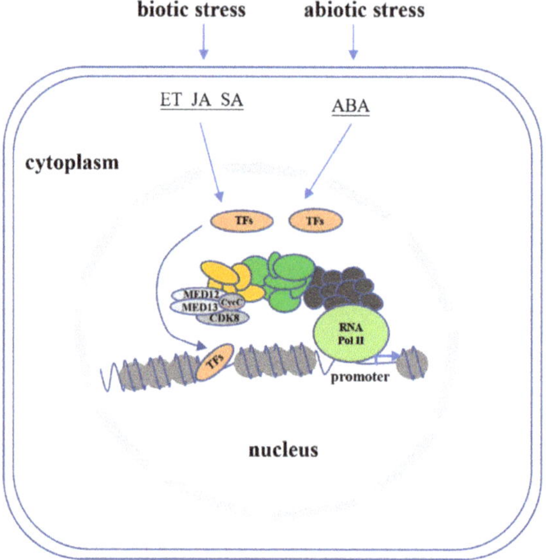

Figure 1. The regulatory functions of Mediator complex in biotic and abiotic stress responses. Biotic and abiotic stresses trigger the plant hormones of ethylene (ET), jasmonic acid (JA), salicylic acid (SA) and abscisic acid (ABA) to activate signaling transduction pathway. A number of transcription factors (TFs) transmit these messages to the transcriptional machinery in the nucleus. Mediator complex functions as a bridge between TFs and RNA polymerase II to precisely regulate the transcription of stress responsive genes.

Arabidopsis CDK8, also named *HUA ENHANCER3* (*HEN3*), was first reported for its regulation of floral organ identity. Thus, it was suggested to contribute to cell differentiation in a multicellular organism [16]. *CDK8* was later found to interact with *Arabidopsis* LEUNIG, a transcription co-repressor [17]. Another name for *Arabidopsis CDK8* is *CDKE1* (cyclin-dependent kinase E1). Studies on the *Arabidopsis regulator of alternative oxidase 1* (*rao1*) mutant which carries a mutation in *CDKE1* further documented that CDKE1 regulates mitochondrial retrograde signaling under H_2O_2 and cold stress [18]. Moreover, *CDK8* has been indicated to contribute in drought stress regulation recently. In terms of plant immune responses, *Arabidopsis CDK8* was reported to perform a contrasting regulatory role [19]. Each subunit of the Mediator complex appears to specifically respond to various environmental conditions to different degrees. The significance of the reported subunits, particularly in the CDK8 module, are being explored next for their signaling involvement in biotic and abiotic stresses [20].

2. The Role of Mediator Subunits in Biotic Stress Regulation

When plants encounter biotic stress such as pathogen infection, they must either activate and/or suppress many genes. It has been documented that more than 620 genes were induced by common necrotrophic fungi such as *Botrytis cinerea* or *Alternaria brassicicola* [21,22]. Undeniably within minutes after infection, significant transcriptional reprogramming occurs [23,24]. In fact, an extensive number of genes such as lignin synthesis for cell wall fortification, defense proteins that attenuate infection, as well as antimicrobial secondary metabolite production, were induced in plants when a pathogen is sensed. TFs including *WRKY33*, *MYC2*, *ethylene response factor1* (*ERF1*), *ethylene-insensitive3* (*EIN3*) along with a few activators are important for defense gene induction as well as resistance to pathogens [25,26].

Since Bäckström et al. isolated the plant Mediator from *Arabidopsis* back in 2007 [27], Mediator has been established as an essential component of plant defense and development. Its subunit of MED21 was determined by Dhawan et al. to have a role in necrotrophic pathogen defense [28]. *MED25*, also called *PHYTOCHROME AND FLOWERING TIME1*, along with *MED8* were further uncovered by Kidd et al. for their involvement in plant defense [29]. Moreover, *MED25* was also identified to regulate JA and SA-induced gene expression. In 2011, Ou et al. identified eight other TFs that have a direct interaction with *Arabidopsis MED25*. Three of the identified eight genes directly recognize the GCC-box of the promoter of *Plant Defensin 1.2* (*PDF1.2*) gene, a gene that involves in plant defense [30]. Other subunits including *MED33a* and *MED33b*, which correspond to the known *Arabidopsis* mutants *REDUCED EPIDERMAL FLUORESCENCE4* (*REF4*) and *REF4-RESEMBLING1* (*RFR1*), have been reported to regulate the phenylpropanoid pathway [31]. The phenylpropanoid pathway generates products of lignin and anthocyanin, which are essential for plant defense as lignin fortification of cell walls curb pathogen growth [32,33] and anthocyanins are plant pigments that possess antimicrobial activity along with free radical scavenging properties. In fact, during pathogen infection, anthocyanins synthesis incorporates compounds of jasmonates, ABA and sugars to support plant defense [34,35]. The accumulation of anthocyanins is affected by *MED25* [29]. As can be seen from these studies, the involvement of Mediator subunits in biotic stress is a sophisticated network of highly interconnected systems. Mediator subunits interact with defense genes to regulate various signaling transduction pathways that involve phytohormones as well as other hormones that enable the strengthening of plants to fight off pathogen. At the time when these subunits were reported for their roles in biotic stress, the involvement of CDK8 module in environmental disturbances was still unspecified, until recently when more evidence began to emerge to show its importance [36].

3. Subunits of the CDK8 Module Play a Role in Regulating Biotic Stress

Along with the subunits of MED8, MED15, MED16, MED21, and MED25, recent evidence about the substantial role of CDK8 module subunits in the resistance against necrotrophic pathogens has appeared [28,37–39]. In response to pathogen infection, plants produce different hormones such as ET, JA and SA [40]. These hormones further activate the expression of defense-related genes. As mentioned previously, very few studies focused on the transcriptional regulation of the CDK8 kinase module in biotic stress. In 2014, Zhu et al. studied the roles of CDK8 subunit in immune responses of *Arabidopsis* [19]. Beside *CDK8*, they further noted that the two other kinase module mutants, *med12* and *med13*, presented signs of disease responses and increased cuticle permeability that were similarly observed in *cdk8* mutant. They proposed that a shared function and structural conservation exist among the kinase module subunits. Since their study was targeting the functional role of CDK8 subunit in biotic stress, they did not mine into the mechanistic roles of *med12* and *med13*. Very recently, *MED12* and *MED13* have been reported for their roles in gene regulation. *MED12* and *MED13* were indicated to participate in the beginning steps of gene transcription and were identified as positive gene regulators under certain conditions. Liu et al. performed mutations of *MED12* and *MED13* and discovered that they suppress *morc1*-reactivated *pSDC:GFP* (*SUPPRESSOR OF DRM1 DRM2 CMT3*) [13] (Figure 2A). *Microrchidia* (*MORC*) is a *GHKL* (gyrase, Hsp90, histidine kinase, MutL)-type ATPase-containing protein that exists in both animal and plant species [41]. It was reported that the silenced DNA methylated genes are reactivated by the mutations of *MORC* proteins in *Arabidopsis* [42]. Apparently, *MED12* and *MED13* are necessary for the expression of genes depleted in active chromatin marks, a chromatin signature shared with *morc1* reactivated loci [43].

As for the subunit of *CDK8*, Zhu et al. have elucidated the functions as well as the underlying molecular and biochemical mechanisms of the Mediator subunit *CDK8* in plant defense. An extensive amount of work has been performed to unfold a contrasting defense function of *CDK8* to two necrotrophic fungi of *A. brassicicola* and *B. cinerea*. Both fungi share similar mechanisms of pathogenesis, virulence, and modes of nutrition. In their investigation of *Arabidopsis'* biotic response to *A. brassicicola*, they pointed out that *CDK8* regulates the expression of defensin genes including *PDF1.2* and several Ethylene Response Transcription Factors (ERFs) which are reported for disease resistance [19]. This finding was consistent with the observation that *ERF1*- and *OCTADECANOID RESPONSIVE ARABIDOPSIS AP2/ERF59 (ORA59)*-dependent activation of *PDF1.2* expression requires *CDK8* [44] (Figure 2B). Since an interaction was found between CDK8 and MED25, it was suggested that *CDK8* regulates plant immunity through a JA-dependent pathway. Simultaneously, *CDK8* contributes to *Arabidopsis'* resistance to *A. brassicicola* through direct regulation of *AGMATINE COUMAROYLTRANSFERASE (AACT1)* transcription. *AACT1* is involved in the biosynthesis of a class of secondary metabolites called hydroxycinnamic acid amides (HCAAs) (Figure 2C), which are known for their functions in fungal resistance [19]. Therefore, *Arabidopsis* would fail to induce critical defense responses without *CDK8*.

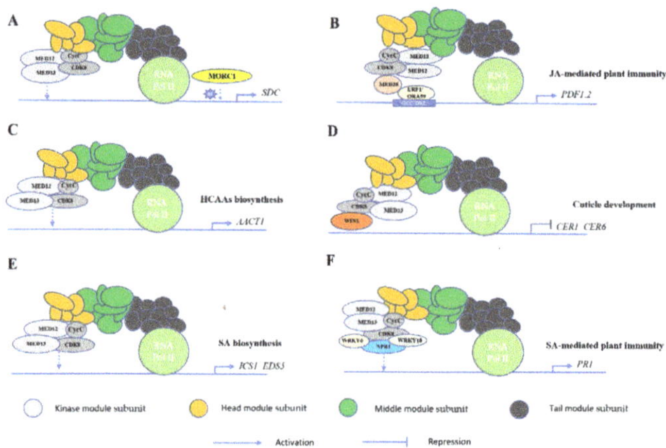

Figure 2. The regulatory functions of CDK8 module in biotic stress response. (**A**) *MED12* and *MED13* reactivate the expression of a silenced DNA methylated *SDC* gene regulated by *MORC1*. (**B**) CDK8, MED12 and MED13 subunits interact with ERF1/ORA59 to regulate the expression of *PDF1.2*. (**C**) *CDK8* is involved in the induced expression of *AACT1* to regulate the biosynthesis of HCAAs for biotic stress response. (**D**) *CDK8* represses the (*ECERIFERUM*) *CER1* and *CER6* gene expression to downregulate cuticle biosynthesis through interacting with WIN1, which binds to the GCC/DRE-box [45]. (**E**) *CDK8* subunit integrates SA signaling through activating the expression of *ICS1* and *EDS5*. (**F**) CDK8, WRKY6, WRKY18, and NPR1 form a complex to regulate the expression of pathogenesis-related (*PR*) genes. Light gray indicates the CDK8 kinase module, yellow indicates the head module, green indicates the middle module, and dark gray indicates the tail module.

Despite their discovery of the positive regulatory role of *CDK8* in plant immunity, they also uncovered another interesting finding that shows *cdk8* mutant exhibits enhanced resistance to *B. cinerea*. Changes in the cuticle structure and permeability of the *cdk8* mutant were noticed [19]. Therefore, they suggested that the increased cuticle permeability and altered cuticle structure are factors that influenced the *cdk8* mutant's biotic resistance as cuticles are related to plants' improved resistance to *B. cinerea* [46,47]. To test this hypothesis, they investigated if *CDK8* could interact with *WAX INDUCER1 (WIN1)* which is an *ERF* family protein known for cuticular wax biosynthesis regulation. In fact, they found that an interaction occurs between CDK8 and WIN1, which indicates that *CDK8* is also involved in cuticle development (Figure 2D). Additionally, they observed that the expression of *CDK8* that was defective in phosphorylation activity failed to rescue the susceptibility of the *cdk8* mutant to *A. brassicicola* while the mutant's resistance to *B. cinerea* was restored to the wild type (WT); indicating two differential functions of *CDK8* in response to different strains of fungi. They also noted a strong interaction of *CDK8* with two CycCs, thereby supporting an evolutionarily conserved structure of the kinase module in plants [19]. Their overall results from their study have revealed that the Mediator subunit CDK8 as well as the CDK8 module possess multiple regulatory roles in plant defense and development.

It has been revealed that *CDK8* mutations diminished the plant's resistance against the necrotrophic fungus *B. cinerea*. CDK8 was also demonstrated to interact with MED25 to play a role in the JA signaling [19]. In 2019, another piece of evidence about the subunit CDK8's role in biotic stress has emerged. Huang et al. discovered *CDK8* through a suppressor screen that used the triple mutant of *camta1/2/3 (calmodulin-binding transcriptional activator)*. Similar to the Mediator complex, CAMTAs are evolutionarily conserved in multicellular eukaryotes in which their roles in transcriptional activity of plants are controversial [48]. They discovered that the mutation of *cdk8* partially suppresses autoimmunity mediated by *camta1/2/3* which includes enhanced resistance against *Hpa Noco2* and SA accumulation levels. Hence, they suggested that *CDK8* positively regulates SA accumulation and systemic acquired resistance (SAR) in *Arabidopsis*. SA is a phytohormone that participates in diverse immune responses including SAR, local defense, and effector-triggered immunity (ETI) whereas SAR is a form of plants' systemic immune response that becomes activated in uninfected distal parts of plants when a local pathogen infection is detected [49]. SA accumulation is reported to be triggered in both infected and distal tissues after an infection occurs. The study further indicated that CDK8 subunit positively regulates these roles through increasing the expression of SA biosynthesis genes of *ICS1 (Isochorismate Synthase 1)* and *EDS5 (Enhanced Disease Susceptibility 5)* as the expression of these two genes was compromised in *cdk8* mutants, preventing the plants to perform immune defense (Figure 2E) [50,51].

Chen et al. later showed that *NONEXPRESSER OF PATHOGENESIS-RELATED GENES (NPR1)* interacts with *CDK8* as well as with WRKY DNA-BINDING PROTEINs such as *WRKY18* to induce the expression of *PATHOGENESIS-RELATED (PR)* genes to promote defense responses in *Arabidopsis* [52]. Additionally, SA was indicated to substantially promote the interactions of these proteins to trigger immune response. *NPR1* has been recognized as a master regulator of SA-mediated local and systemic plant immunity due to its control of the expression of over 2000 genes [53,54]. SAR was found to be included in the *cdk8* and CDK8-associated Mediator mutants. The reduced expression of *NPR1* and *NPR1*-dependent defense genes was observed as well in these mutants compared to WT. *CDK8* positively regulates the expression of *NPR1* through the interaction with both *WRKY6* and *WRKY18* and they are associated with the promoter of *NPR1*. Furthermore, CDK8 interacts with TGACG-Binding (TGA) TFs of *TGA5* and *TGA7*; both are associated with the *PR1* promoter to regulate *PR1* gene expression. Another interesting finding about *CDK8* from their study is that CDK8 recruits RNA Pol II to the promoters and coding regions of *NPR1* and *PR1* to promote their gene expression [52] From their study, CDK8's contribution to plant immunity such as SAR is further established. In fact, the study explained that *CDK8* could essentially be the component that helps clarify the relationship between NPR1 and its activation of *PR1* to initiate plant immune response (Figure 2F). Overall, they have demonstrated that under SA influence, CDK8 links TGA TFs and NPR1 (through interacting with *WRKYs*) with RNA Pol II to facilitate *PR1* gene expression.

4. Signaling Roles of Mediator Subunits in Abiotic Stress

ABA is a phytohormone that contributes significantly to various developmental processes in the plant's life cycle. The hormone plays a vital role in the plant's response to various abiotic stresses including drought, salt, and heat stresses. In addition to biotic stress, Mediator complex, has also been found to serve important roles in the ABA signaling transduction. *MED25* was the first Mediator subunit that was documented to act in response to ABA. *MED25* was found to negatively regulate the ABA signaling pathway as *med25* mutants display an increased sensitivity to ABA during seed germination and early seedling growth. Consistent with its negative role in ABA signaling, *med25* mutant was observed to have an increased expression of ABA-responsive genes in response to ABA treatment compared to WT plants. The transcription of *ABA-INSENSITIVE5* (*ABI5*), a key TF regulating the ABA signaling during seed germination, is induced by ABA, and interestingly, the ABA-induced transcription of *ABI5* was suppressed in *med25* mutants compared to WT. ABI5 protein, however, accumulated at higher abundance in *med25* mutants compared to WT, indicating that *MED25* may negatively regulate *ABI5* at post-transcriptional level. Chromatin immunoprecipitation (ChIP) experiments further indicated that *MED25* was highly enriched at the promoters of *ABI5* downstream genes but this enrichment was decreased upon ABA treatment. A direct interaction was noticed between MED25 and ABI5 and this interaction was attenuated by ABA, which was in accordance with the negative impacts of *MED25* on the *ABI5*-regulated ABA responses [38]. MED25 may act as a critical regulator in hormones crosstalk between JA, ethylene and ABA signaling due to its interaction with MYC2 and several TFs in plants. In addition to *MED25*, the head module subunit *MED18* has also been reported to have a role in ABA signaling. Opposite to *med25*, *med18* mutants are more insensitive to ABA at seed germination and early growth stages, similar to *abi4* and *abi5* mutants. Interestingly, the ABA-induced expressions of *ABI4* and *ABI5* are much lower in *med18* mutants than those in WT, suggesting that the transcription of *ABI4* and *ABI5* are positively regulated by *MED18*. ChIP-qPCR further revealed that *MED18* is recruited to the ABI4 binding site on the *ABI5* promoter under both mock and ABA treatments [55]. The physical interaction between MED18 and TF ABI4 further reinforces that *MED18* regulates ABA response and expression of *ABI5* by interacting with ABI4. *MED12* of the CDK8 module was also found recently to contribute to abiotic stress by blocking transient gene upregulation after light treatment as it has interactions with genes that are responsive to environmental stimuli such as light and radiation [13].

5. Signaling Roles of CDK8 Subunit in Abiotic Stress

In 2013, Ng et al. studied cyclin-dependent kinase E1/Regulator of AOX1a 1 (*CDKE1/RAO1*) to understand how the induction of alternative oxidase (*AOX*) is integrated into the general regulatory context of the cell. To their surprise, *CDKE1/RAO1* did not interact directly with any other cyclin components even though *CDKE1* is a cyclin-dependent kinase. *rao1* mutants in the study did not show an alternation in *ABI4* gene expression under normal conditions but its gene expression was changed in a complex *aox1a* knock-out mutant. In fact, they noticed that only 119 transcripts genome-wide were significantly changed by 2.5-fold greater than the *pAOX1a:LUC* under normal conditions in both *rao1–1* and *rao1–2* [18]. Additionally, KIN10 was found to directly interact with CDK8 to coordinately regulate overlapping target genes in response to mitochondrial stress (Figure 3A). They suggested this change was due to the integrated network regulated by direct phosphorylation events associated with the Mediator complex and not any secondary effect caused by stress responsive TF dysregulation.

They also conducted global transcriptional analyses and found that transcripts relating to both growth and stress are affected in the *rao1* mutant background but in opposite ways. It appears that *CDK8* plays a role in initiating cell wide stress responses, protein metabolism and protein synthesis, which are the essential building blocks for cell division and growth. Photosynthetic components, however, are switched off. Their study demonstrated that *CDKE1/RAO1* has a role as it can integrate cellular responses to environmental signals for cell division or elongation. Furthermore, they have shown that *CDKE1/RAO1* essentially serves as a sensitive relay between specific stress-induced TFs that are bound to the promoter and RNA Pol II, therefore it can directly regulate transcription under stress [18]. In general, the study has indicated that plants can utilize *CDKE1/RAO1* as a mechanism to switch between growth and stress responses when responding to different environmental conditions.

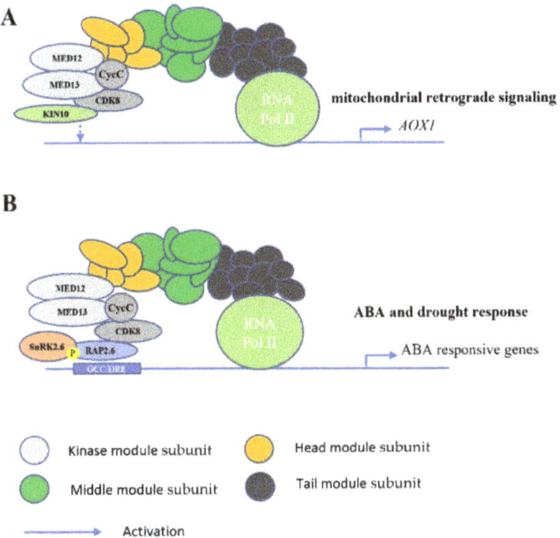

Figure 3. The regulatory functions of CDK8 subunit in abiotic stress response. (**A**) CDK8 cooperates with KIN10 to control the expression of *AOX1* to regulate mitochondrial retrograde signaling. (**B**) CDK8 along with phosphorylated proteins of RAP2.6 and SnRK2.6 bind to the GCC/DRE-box to initiate the transcription of ABA responsive genes in response to ABA and drought stress. Light gray indicates the CDK8 kinase module, yellow indicates the head module, green indicates the middle module and dark gray indicates the tail module.

Drought and cold stresses are some of the environmental challenges that prevent plant growth and development. In order to overcome stress, plants utilize various strategies including regulating signaling transduction pathways to respond to adverse environmental conditions. *MED16, MED25, MED14,* and *MED2* are notable for their roles in cold stress regulation. At the time when these subunits were known for their involvement in abiotic stress responses, the function of *CDK8* in abiotic stress was unidentified. It was not until 2020 that Zhu et al. reported that *CDK8* positively modulates drought response in *Arabidopsis*. Since *CDK8* is known for possessing kinase activity, they thought that it would be interesting to explore its potential in the ABA signaling pathway in which phosphorylation is involved. Through utilizing genetic, transcriptomic, and biochemical approaches, it was solidified that CDK8 associates with RAP2.6 and SnRK2.6 to positively regulate the transcription of ABA-responsive genes (Figure 3B). From their study, they have discovered that *CDK8* mutation in *Arabidopsis* results in higher stomata density, impaired stomatal aperture as well as reduced tolerance to drought. Consistently, over-expression of *CDK8* in *Arabidopsis* enhances drought tolerance. They have also observed improved cuticle permeability and thinner cutin in *cdk8* mutants and, thus, they suggested that *CDK8* may possibly regulate drought response through multiple mechanisms [56]. CDK8 was revealed to have a direct interaction with ERF/AP2 type TFs WIN1 (WAX INDUCER1) and RAP2.6 [19,56]. As mentioned previously, *WIN1* is a key regulator of cuticle wax biosynthesis and *RAP2.6* is an abiotic stress responsive gene. It was very likely that *CDK8* upregulates cutin biosynthesis and wax accumulation through interacting with WIN1. Interestingly, *WIN1* may also participate in abiotic stress response as its expression is significantly induced by various abiotic stresses. Their study suggested that WIN1 can bind to the GCC-box and DRE element sequences to activate several stress-responsive genes, suggesting a potential function of CDK8-WIN1 interaction in drought response. Their findings also uncovered another strategy which *Arabidopsis* utilizes CDK8 to cooperate with RAP2.6-SnRK2.6 complex to facilitate the immediate transcription of stress-responsive genes in drought.

6. Conclusions and Perspectives

When plants encounter biotic stress, they tackle the challenges using their built-in defense mechanisms and/or triggering defense-related signaling pathways. Activation of these responses often leads to the altered expression of various defense genes. Some of these defense genes are responsible for cuticle formation as well as for the biosynthesis and modification of cell wall [53,57]. Other defense genes are related to pathogenesis. Moreover, genes involved in various hormone signaling pathways have also been reported to change when plants respond to biotic stress [58,59]. As an essential player in the transcriptional regulation of gene expression, it was no surprise to find that the subunits of Mediator have connections with stress mitigation. The current reports of the core Mediator subunits along with the CDK8 module in plant cell signaling has revealed that Mediator has contrary roles in gene regulation when responding to stress.

The CDK8 module, consisting of CDK8, MED12, MED13 and C-type cyclins, is an important part of the Mediator complex. This kinase module is dissociable during transcription and thus initially considered to be a negative regulator of transcription [60]. Increasing evidence, however, indicated that the module can also stimulate transcription [61,62], thereby suggesting that *CDK8* can both positively and negatively regulate transcription as indicated in Table 1. It is believed that the positive regulatory roles of *CDK8* in transcription are either performed through the phosphorylation of TFs which would lead to protein degradation in some reported cases or through the promotion of RNA Pol II elongation [63]. *CDK8* was revealed to regulate plant immunity through kinase dependent and independent functions in Arabidopsis. It is worth investigating if CDK8 may potentially interact with Arabidopsis signal responsive1 (AtSR1)/CAMTA3 to regulate plant growth during plant defense against pathogens as *AtSR1* has been indicated to involve in the regulation of auxin- and BRs-related pathways as well as in the suppression of genes elicited by pathogen attack through binding to the "CGCG" containing CG-box in target gene promoters (51). Aside from serving a role in plant immunity, *Arabidopsis CDK8* module also has been shown to play a role in abiotic stresses. Other subunits such as MED12 and MED13 in the CDK8 module are reportedly served as conditional positive regulators in biotic stress. Subunits in the Mediator complex including *MED25* contributes to drought and salt tolerance in *Arabidopsis* by physically interacting with dehydration-responsive element-binding protein 2a (DREB2a), zinc-finger homeodomain protein 1 (ZFHD1), and a MYB TF through the ACID domain [64].

Table 1. Genes regulated by the CDK8 module in *Arabidopsis*.

Subunit	Genes	Regulation	Functions	Reference
CDK8	*ICS1, EDS5*	positive	SA biosynthesis	[12]
MED12, MED13	*SDC*	positive	unknown	[13]
CDK8	*PDF1.2, ERF*	positive	JA-mediated plant immunity	[19]
CDK8	*NPR1, PR1*	positive	SA-mediated plant immunity	[52]
CDK8	*AACT1*	positive	Biosynthesis of defense metabolites HCAAs	[19]
CDK8	*CER1, CER6*	negative	Cuticle development	[19]
CDK8	*AOX1*	positive	Mitochondrial retrograde signaling	[18]
CDK8	*COR15A, RD29B, DREB2A*	positive	ABA and drought responses	[56]

More evidence is emerging regarding the signaling roles of Mediator subunits in biotic and abiotic stresses. The involvement of these subunits in plant defense indicates the capability of the Mediator to accommodate new pathogen resistances in plants. Not only that, recent reports about these subunits in abiotic stresses also indicate the ability of Mediator to assist plants in acclimating in harsh environment [64,65]. The diverse roles depicted about the Mediator, especially its potential with various TFs interaction in regulating abiotic and biotic stress signaling, suggested a unique capacity of plants to recognize new factors to generate effective responses to adverse conditions. Since the subunit(s) of Mediator can interact with various TFs, it is comprehensible to see it serving as a hub for cross-linking many hormones and other signaling pathways to regulate stress responses. It is very possible that the subunits within the Mediator will cooperate with each other when the plants face new environmental challenges. Moreover, the function of Mediator subunits that respond to biotic and abiotic stresses may also be expanded with the possibility of them acquiring other new subunit(s) to recognize new regulatory proteins in plant survival against unfavorable circumstances.

Funding: National Natural Science Foundation of China (NSFC 31900238 and NSFC 32070307) to Y.Z. and 111 Project #D16014.

Institutional Review Board Statement: Not applicable.

Informed Consent Statement: Not applicable.

Conflicts of Interest: The authors declare no conflict of interest.

References

1. Verma, V.; Ravindran, P.; Kumar, P.P. Plant hormone-mediated regulation of stress responses. *BMC Plant Biol.* **2016**, *16*, 1–10. [CrossRef]
2. Yang, J.; Duan, G.; Li, C.; Liu, L.; Han, G.; Zhang, Y.; Wang, C. The Crosstalks Between Jasmonic Acid and Other Plant Hormone Signaling Highlight the Involvement of Jasmonic Acid as a Core Component in Plant Response to Biotic and Abiotic Stresses. *Front. Plant Sci.* **2019**, *10*, 1349. [CrossRef]
3. Qu, L.-J.; Zhu, Y.-X. Transcription factor families in Arabidopsis: Major progress and outstanding issues for future research. *Curr. Opin. Plant Biol.* **2006**, *9*, 544–549. [CrossRef]
4. Sato, S.; Tomomori-Sato, C.; Banks, C.A.S.; Sorokina, I.; Parmely, T.J.; Kong, S.E.; Jin, J.; Cai, Y.; Lane, W.S.; Brower, C.S.; et al. Identification of Mammalian Mediator Subunits with Similarities to Yeast Mediator Subunits Srb5, Srb6, Med11, and Rox3. *J. Biol. Chem.* **2003**, *278*, 15123–15127. [CrossRef]
5. Liu, Y.; Ranish, J.A.; Aebersold, R.; Hahn, S. Yeast nuclear extract contains two major forms of RNA polymerase II mediator complexes. *J. Biol. Chem.* **2001**, *276*, 7169–7175. [CrossRef]
6. Hasan, A.S.M.M.; Schoor, J.K.V.; Hecht, V.; Weller, J.L. The CYCLIN-DEPENDENT KINASE Module of the Mediator Complex Promotes Flowering and Reproductive Development in Pea1[OPEN]. *Plant Physiol.* **2020**, *182*, 1375–1386. [CrossRef]
7. Yang, Y.; Li, L.; Qu, L.-J. Plant Mediator complex and its critical functions in transcription regulation. *J. Integr. Plant Biol.* **2015**, *58*, 106–118. [CrossRef]
8. Akoulitchev, S.; Chuikov, S.; Reinberg, D. TFIIH is negatively regulated by cdk8-containing mediator complexes. *Nat. Cell Biol.* **2000**, *407*, 102–106. [CrossRef]
9. Elmlund, H.; Baraznenok, V.; Lindahl, M.; Samuelsen, C.O.; Koeck, P.J.B.; Holmberg, S.; Hebert, H.; Gustafsson, C.M. The cyclin-dependent kinase 8 module sterically blocks Mediator interactions with RNA polymerase II. *Proc. Natl. Acad. Sci. USA* **2006**, *103*, 15788–15793. [CrossRef]

10. Donner, A.J.; Ebmeier, C.C.; Taatjes, D.J.; Espinosa, J.M. CDK8 is a positive regulator of transcriptional elongation within the serum response network. *Nat. Struct. Mol. Biol.* **2010**, *17*, 194–201. [CrossRef]

11. Donner, A.J.; Szostek, S.; Hoover, J.M.; Espinosa, J.M. CDK8 Is a Stimulus-Specific Positive Coregulator of p53 Target Genes. *Mol. Cell* **2007**, *27*, 121–133. [CrossRef] [PubMed]

12. Huang, J.; Sun, Y.; Orduna, A.R.; Jetter, R.; Li, X. The Mediator kinase module serves as a positive regulator of salicylic acid accumulation and systemic acquired resistance. *Plant J.* **2019**, *98*, 842–852. [CrossRef] [PubMed]

13. Liu, Q.; Bischof, S.; Harris, C.J.; Zhong, Z.; Zhan, L.; Nguyen, C.; Rashoff, A.; Barshop, W.D.; Sun, F.; Feng, S.; et al. The characterization of Mediator 12 and 13 as conditional positive gene regulators in Arabidopsis. *Nat. Commun.* **2020**, *11*, 1–13. [CrossRef]

14. Raya-González, J.; López-Bucio, J.S.; Prado-Rodríguez, J.C.; Ruiz-Herrera, L.F.; Guevara-García, Á.A.; López-Bucio, J. The MEDIATOR genes MED12 and MED13 control Arabidopsis root system configuration influencing sugar and auxin responses. *Plant Mol. Biol.* **2017**, *95*, 141–156. [CrossRef]

15. Galbraith, M.D.; Allen, M.A.; Bensard, C.L.; Wang, X.; Schwinn, M.K.; Qin, B.; Long, H.W.; Daniels, D.L.; Hahn, W.C.; Dowell, R.D.; et al. HIF1A Employs CDK8-Mediator to Stimulate RNAPII Elongation in Response to Hypoxia. *Cell* **2013**, *153*, 1327–1339. [CrossRef]

16. Wang, W.-M.; Chen, X. HUA ENHANCER3 reveals a role for a cyclin-dependent protein kinase in the specification of floral organ identity in Arabidopsis. *Development* **2004**, *131*, 3147–3156. [CrossRef]

17. Gonzalez, D.; Bowen, A.J.; Carroll, T.S.; Conlan, R. The Transcription Corepressor LEUNIG Interacts with the Histone Deacetylase HDA19 and Mediator Components MED14 (SWP) and CDK8 (HEN3) To Repress Transcription. *Mol. Cell. Biol.* **2007**, *27*, 5306–5315. [CrossRef]

18. Ng, S.; Giraud, E.; Duncan, O.; Law, S.R.; Wang, Y.; Xu, L.; Narsai, R.; Carrie, C.; Walker, H.; Day, D.A.; et al. Cyclin-dependent Kinase E1 (CDKE1) Provides a Cellular Switch in Plants between Growth and Stress Responses. *J. Biol. Chem.* **2013**, *288*, 3449–3459. [CrossRef]

19. Zhu, Y.; Schluttenhoffer, C.M.; Wang, P.; Fu, F.; Thimmapuram, J.; Zhu, J.; Lee, S.Y.; Yun, D.; Mengiste, T. CY-CLIN-DEPENDENT KINASE8 Differentially Regulates Plant Immunity to Fungal Pathogens through Kinase-Dependent and -Independent Functions inArabidopsis. *Plant Cell* **2014**, *26*, 4149–4170. [CrossRef]

20. Agrawal, R.; Jiří, F.; Thakur, J.K. The kinase module of the Mediator complex: An important signalling processor for the development and survival of plants. *J. Exp. Bot.* **2020**. [CrossRef]

21. AbuQamar, S.; Chen, X.; Dhawan, R.; Bluhm, B.; Salmeron, J.; Lam, S.; Dietrich, R.A.; Mengiste, T. Expression profiling and mutant analysis reveals complex regulatory networks involved in Arabidopsis response to Botrytis infection. *Plant J.* **2006**, *48*, 28–44. [CrossRef] [PubMed]

22. Van Wees, S.C.M.; Chang, H.-S.; Zhu, T.; Glazebrook, J. Characterization of the Early Response of Arabidopsis to Alternaria brassicicola Infection Using Expression Profiling[w]. *Plant Physiol.* **2003**, *132*, 606–617. [CrossRef] [PubMed]

23. Hahlbrock, K.; Bednarek, P.; Ciolkowski, I.; Hamberger, B.; Heise, A.; Liedgens, H.; Logemann, E.; Nürnberger, T.; Schmelzer, E.; Somssich, I.; et al. Non-self recognition, transcriptional reprogramming, and secondary metabolite accumulation during plant/pathogen interactions. *Proc. Natl. Acad. Sci. USA* **2003**, *100*, 14569–14576. [CrossRef] [PubMed]

24. Cormack, R.S.; Hahlbrock, K.; Somssich, I. Isolation of putative plant transcriptional coactivators using a modified two-hybrid system incorporating a GFP reporter gene. *Plant J.* **1998**, *14*, 685–692. [CrossRef]

25. Zheng, Z.; Qamar, S.A.; Chen, Z.; Mengiste, T. Arabidopsis WRKY33 transcription factor is required for resistance to ne-crotrophic fungal pathogens. *Plant J.* **2006**, *48*, 592–605. [CrossRef]

26. Lorenzo, O.; Chico, J.M.; Sánchez-Serrano, J.J.; Solano, R. JASMONATE-INSENSITIVE1 Encodes a MYC Transcription Factor Essential to Discriminate between Different Jasmonate-Regulated Defense Responses in Arabidopsis. *Plant Cell* **2004**, *16*, 1938–1950. [CrossRef]

27. Bäckström, S.; Elfving, N.; Nilsson, R.; Wingsle, G.; Björklund, S. Purification of a Plant Mediator from Arabidopsis thaliana Identifies PFT1 as the Med25 Subunit. *Mol. Cell* **2007**, *26*, 717–729. [CrossRef]

28. Dhawan, R.; Luo, H.; Foerster, A.M.; Abuqamar, S.; Du, H.N.; Briggs, S.D.; Mittelsten, S.O.; Mengiste, T. HISTONE MON-OUBIQUITINATION1 interacts with a subunit of the mediator complex and regulates defense against necrotrophic fungal pathogens in Arabidopsis. *Plant Cell* **2009**, *21*, 1000–1019. [CrossRef]

29. Kidd, B.N.; Edgar, C.I.; Kumar, K.K.; Aitken, E.A.; Schenk, P.M.; Manners, J.M.; Kazan, K. The Mediator Complex Subunit PFT1 Is a Key Regulator of Jasmonate-Dependent Defense in Arabidopsis. *Plant Cell* **2009**, *21*, 2237–2252. [CrossRef]

30. Ou, B.; Yin, K.Q.; Liu, S.N.; Yang, Y.; Gu, T.; Wing, H.J.; Zhang, L.; Miao, J.; Kondou, Y.; Matsui, M.; et al. A high-throughput screening system for Arabidopsis transcription factors and its application to Med25-dependent transcrip-tional regulation. *Mol. Plant* **2011**, *4*, 546–555. [CrossRef]

31. Stout, J.; Romero-Severson, E.; Ruegger, M.O.; Chapple, C. Semidominant Mutations inReduced Epidermal Fluorescence 4Reduce Phenylpropanoid Content in Arabidopsis. *Genetics* **2008**, *178*, 2237–2251. [CrossRef] [PubMed]

32. Schenke, D.; Bottcher, C.; Scheel, D. Crosstalk between abiotic ultraviolet-B stress and biotic (flg22) stress signalling in Ara-bidopsis prevents flavonol accumulation in favor of pathogen defence compound production. *Plant Cell Environ.* **2011**, *34*, 1849–1864. [CrossRef] [PubMed]

33. Bhuiyan, N.H.; Selvaraj, G.; Wei, Y.; King, J. Role of lignification in plant defense. *Plant Signal. Behav.* **2009**, *4*, 158–159. [CrossRef] [PubMed]

34. Loreti, E.; Povero, G.; Novi, G.; Solfanelli, C.; Alpi, A.; Perata, P. Gibberellins, jasmonate and abscisic acid modulate the su-crose-induced expression of anthocyanin biosynthetic genes in Arabidopsis. *New Phytol.* **2008**, *179*, 1004–1016. [CrossRef]

35. Teng, S.; Keurentjes, J.; Bentsink, L.; Koornneef, M.; Smeekens, S. Sucrose-Specific Induction of Anthocyanin Biosynthesis in Arabidopsis Requires the MYB75/PAP1 Gene. *Plant Physiol.* **2005**, *139*, 1840–1852. [CrossRef]

36. Mao, X.; Kim, J.I.; Wheeler, M.T.; Heintzelman, A.K.; Weake, V.M.; Chapple, C. Mutation of Mediator subunit CDK8 coun-teracts the stunted growth and salicylic acid hyperaccumulation phenotypes of an Arabidopsis MED5 mutant. *New Phytol.* **2019**, *223*, 233–245. [CrossRef]

37. Wathugala, D.L.; Hemsley, P.A.; Moffat, C.S.; Cremelie, P.; Knight, M.R.; Knight, H. The Mediator subunit SFR6/MED16 controls defence gene expression mediated by salicylic acid and jasmonate responsive pathways. *New Phytol.* **2012**, *195*, 217–230. [CrossRef]

38. Chen, R.; Jiang, H.; Li, L.; Zhai, Q.; Qi, L.; Zhou, W.; Liu, X.; Li, H.; Zheng, W.; Sun, J.; et al. The Arabidopsis mediator subunit MED25 differentially regulates jasmonate and abscisic acid signaling through interacting with the MYC2 and ABI5 transcrip-tion factors. *Plant Cell* **2012**, *24*, 2898–2916. [CrossRef]

39. Zhang, X.; Wang, C.; Zhang, Y.; Sun, Y.; Mou, Z. The Arabidopsis mediator complex subunit16 positively regulates salicy-late-mediated systemic acquired resistance and jasmonate/ethylene-induced defense pathways. *Plant Cell* **2012**, *24*, 4294–4309. [CrossRef]

40. Glazebrook, J. Contrasting Mechanisms of Defense Against Biotrophic and Necrotrophic Pathogens. *Annu. Rev. Phytopathol.* **2005**, *43*, 205–227. [CrossRef]

41. Pastor, W.A.; Stroud, H.; Nee, K.; Liu, W.; Pezic, D.; Manakov, S.; Lee, S.A.; Moissiard, G.; Zamudio, N.; Bourc'His, D.; et al. MORC1 represses transposable elements in the mouse male germline. *Nat. Commun.* **2014**, *5*, 5795. [CrossRef] [PubMed]

42. Moissiard, G.; Cokus, S.J.; Cary, J.; Feng, S.; Billi, A.C.; Stroud, H.; Husmann, D.; Zhan, Y.; Lajoie, B.R.; Mccord, R.P.; et al. MORC Family ATPases Required for Heterochromatin Condensation and Gene Silencing. *Science* **2012**, *336*, 1448–1451. [CrossRef] [PubMed]

43. Imura, Y.; Kobayashi, Y.; Yamamoto, S.; Furutani, M.; Tasaka, M.; Abe, M.; Araki, T. CRYPTIC PRECOCIOUS/MED12 is a Novel Flowering Regulator with Multiple Target Steps in Arabidopsis. *Plant Cell Physiol.* **2012**, *53*, 287–303. [CrossRef] [PubMed]

44. Wang, C.; Ding, Y.; Yao, J.; Zhang, Y.; Sun, Y.; Colee, J.; Mou, Z. Arabidopsis Elongator subunit 2 positively contributes to resistance to the necrotrophic fungal pathogens Botrytis cinerea and Alternaria brassicicola. *Plant J.* **2015**, *83*, 1019–1033. [CrossRef]

45. Yang, C.-K.; Huang, B.-H.; Ho, S.-W.; Huang, M.-Y.; Wang, J.-C.; Gao, J.; Liao, P.-C. Molecular genetic and biochemical evidence for adaptive evolution of leaf abaxial epicuticular wax crystals in the genus Lithocarpus (Fagaceae). *BMC Plant Biol.* **2018**, *18*, 196. [CrossRef]

46. Bessire, M.; Chassot, C.; Jacquat, A.-C.; Humphry, M.; Borel, S.; Petétot, J.M.-C.; Métraux, J.-P.; Nawrath, C. A permeable cuticle in Arabidopsis leads to a strong resistance to Botrytis cinerea. *EMBO J.* **2007**, *26*, 2158–2168. [CrossRef]

47. Kurdyukov, S.; Faust, A.; Nawrath, C.; Bar, S.; Voisin, D.; Efremova, N.; Franke, R.; Schreiber, L.; Saedler, H.; Metraux, J.P.; et al. The epidermis-specific extracellular BODYGUARD controls cuticle development and morphogenesis in Ara-bidopsis. *Plant Cell* **2006**, *18*, 321–339. [CrossRef]

48. Finkler, A.; Ashery-Padan, R.; Fromm, H. CAMTAs: Calmodulin-binding transcription activators from plants to human. *FEBS Lett.* **2007**, *581*, 3893–3898. [CrossRef]

49. Vlot, A.C.; Dempsey, D.A.; Klessig, D.F. Salicylic acid, a multifaceted hormone to combat disease. *Annu. Rev. Phytopathol.* **2009**, *47*, 177–206. [CrossRef]

50. Yuan, P.; Yang, T.; Poovaiah, B.W. Calcium Signaling-Mediated Plant Response to Cold Stress. *Int. J. Mol. Sci.* **2018**, *19*, 3896. [CrossRef]

51. Yuan, P.; Du, L.; Poovaiah, B.W. Ca(2+)/Calmodulin-Dependent AtSR1/CAMTA3 Plays Critical Roles in Balancing Plant Growth and Immunity. *Int. J. Mol. Sci.* **2018**, *19*, 1764. [CrossRef]

52. Chen, J.; Mohan, R.; Zhang, Y.; Li, M.; Chen, H.; Palmer, I.A.; Chang, M.; Qi, G.; Spoel, S.H.; Mengiste, T.; et al. NPR1 Promotes Its Own and Target Gene Expression in Plant Defense by Recruiting CDK8. *Plant Physiol.* **2019**, *181*, 289–304. [CrossRef] [PubMed]

53. Fu, Z.Q.; Dong, X. Systemic acquired resistance: Turning local infection into global defense. *Annu. Rev. Plant Biol.* **2013**, *64*, 839–863. [CrossRef] [PubMed]

54. Wang, D.; Amornsiripanitch, N.; Dong, X. A Genomic Approach to Identify Regulatory Nodes in the Transcriptional Network of Systemic Acquired Resistance in Plants. *PLOS Pathog.* **2006**, *2*, e123. [CrossRef] [PubMed]

55. Lai, Z.; Schluttenhofer, C.M.; Bhide, K.; Shreve, J.; Thimmapuram, J.; Lee, S.Y.; Yun, D.-J.; Mengiste, T. MED18 interaction with distinct transcription factors regulates multiple plant functions. *Nat. Commun.* **2014**, *5*, 3064. [CrossRef]

56. Zhu, Y.; Huang, P.; Guo, P.; Chong, L.; Yu, G.; Sun, X.; Hu, T.; Li, Y.; Hsu, C.; Tang, K.; et al. CDK8 is associated with RAP2.6 and SnRK2.6 and positively modulates abscisic acid signaling and drought response in Arabidopsis. *New Phytol.* **2020**. [CrossRef]

57. Boller, T.; Felix, G. A renaissance of elicitors: Perception of microbe-associated molecular patterns and danger signals by pat-tern-recognition receptors. *Annu. Rev. Plant Biol.* **2009**, *60*, 379–406. [CrossRef]

58. Chen, F.; Dixon, R. Lignin modification improves fermentable sugar yields for biofuel production. *Nat. Biotechnol.* **2007**, *25*, 759–761. [CrossRef]

59. Jones, J.D.; Dangl, J.L. The plant immune system. *Nat. Cell Biol.* **2006**, *444*, 323–329. [CrossRef]
60. Gillmor, C.S.; Silva-Ortega, C.O.; Willmann, M.R.; Buendía-Monreal, M.; Poethig, R. The Arabidopsis Mediator CDK8 module genes CCT (MED12) and GCT (MED13) are global regulators of developmental phase transitions. *Development* **2014**, *141*, 4580–4589. [CrossRef]
61. Nemet, J.; Jeličić, B.; Rubelj, I.; Sopta, M. The two faces of Cdk8, a positive/negative regulator of transcription. *Biochimie* **2014**, *97*, 22–27. [CrossRef] [PubMed]
62. Conaway, R.C.; Conaway, J.W. Function and regulation of the Mediator complex. *Curr. Opin. Genet. Dev.* **2011**, *21*, 225–230. [CrossRef] [PubMed]
63. Allen, B.L.; Taatjes, D.J. The Mediator complex: A central integrator of transcription. *Nat. Rev. Mol. Cell Biol.* **2015**, *16*, 155–166. [CrossRef] [PubMed]
64. Elfving, N.; Davoine, C.; Benlloch, R.; Blomberg, J.; Brännström, K.; Müller, D.; Nilsson, A.; Ulfstedt, M.; Ronne, H.; Wingsle, G.; et al. The Arabidopsis thaliana Med25 mediator subunit integrates environmental cues to control plant development. *Proc. Natl. Acad. Sci. USA* **2011**, *108*, 8245–8250. [CrossRef]
65. Crawford, T.; Karamat, F.; Lehotai, N.; Rentoft, M.; Blomberg, J.; Strand, Å.; Björklund, S. Specific functions for Mediator complex subunits from different modules in the transcriptional response of Arabidopsis thaliana to abiotic stress. *Sci. Rep.* **2020**, *10*, 1–18. [CrossRef]

International Journal of
Molecular Sciences

MDPI

Review

ROS and Ions in Cell Signaling during Sexual Plant Reproduction

Maria Breygina * and Ekaterina Klimenko

Department of Plant Physiology, Biological Faculty, Lomonosov Moscow State University, 119991 Moscow, Russia; kleo80@yandex.ru
* Correspondence: breygina@mail.bio.msu.ru; Tel.: +7-499-939-1209

Received: 25 November 2020; Accepted: 10 December 2020; Published: 13 December 2020

Abstract: Pollen grain is a unique haploid organism characterized by two key physiological processes: activation of metabolism upon exiting dormancy and polar tube growth. In gymnosperms and flowering plants, these processes occur in different time frames and exhibit important features; identification of similarities and differences is still in the active phase. In angiosperms, the growth of male gametophyte is directed and controlled by its microenvironment, while in gymnosperms it is relatively autonomous. Recent reviews have detailed aspects of interaction between angiosperm female tissues and pollen such as interactions between peptides and their receptors; however, accumulated evidence suggests low-molecular communication, in particular, through ion exchange and ROS production, equally important for polar growth as well as for pollen germination. Recently, it became clear that ROS and ionic currents form a single regulatory module, since ROS production and the activity of ion transport systems are closely interrelated and form a feedback loop.

Keywords: pollen germination; pollen tube growth; ROS; ions; plant reproduction

---

1. Introduction

Reactive oxygen species (ROS) in plant tissues are a universal regulatory element associated with various signaling systems, such as phospholipids, calcium, and ROP (Rho of plants) GTPases. The participation of ROS in intercellular cross-talk and morphogenesis at the cellular level, for example, during stomatal movements, zygote polarization, and root hair growth, has been convincingly shown [1–4].

Sexual reproduction in plants has been extensively studied since this area has both fundamental and practical significance. Most of the data on male gametophyte germination was obtained in vitro, since pollen cultivation is an accessible and convenient technique that allows one to simplify the experimental system. However, recently, many studies have focused on the interaction (1) between gametophytes and (2) of male gametophyte with female tissues of sporophyte [5–7]. In this case, regulatory factors found in vitro are tested for in vivo efficacy, mainly using genetic approach and improved fluorescence techniques. A lot of attention is paid to the interactions of peptides with their receptors which provide the control of pollen tube growth by female tissues [8]. However, the accumulated evidence suggests also low-molecular communication between sporophyte tissues and pollen, in particular, through ion exchange and production of ROS [6,7]. The ability of pollen grains to respond to changes in ionic environment was discovered around 45 years ago [9], and recently, it became clear that ROS and ionic currents form a single regulatory system, since ROS production and the activity of ion transporters are tightly interrelated [10,11].

2. Pollen Germination

2.1. ROS Production as an Early Event during Germination

When a dry, dormant pollen grain lands on a stigma or on the scales of a female cone, it rehydrates and gradually switches to active metabolism [12]. One of the first physiological changes, apparently, is ROS production, since NBT staining reveals ROS in the aperture area already after 5 minutes of in vitro pollen incubation in kiwi [13]; staining of non-germinated tobacco pollen grains with ROS-sensitive dyes displays their apoplastic and mitochondrial localization [14]. ROS production has also been described in non-germinated pollen grains of Arizona cypress [15]. Mitochondrial ROS production must be tightly controlled, as its excess, for example, in defective pollen leads to cell death and, as a consequence, to cytoplasmic male sterility (CMS) [16]. The involvement of ROS in pollen abortion has been reported for many cytoplasm male sterile crop varieties, for example, cotton [17], pepper [18], and rice [19]. In these cases, excessive ROS production was associated with reduced abundance of superoxide dismutase (SOD), ascorbate peroxidase (APX), catalase [19], and peroxisomal-like protein [17] in mitochondria.

Besides their generation in mitochondria, ROS in pollen are produced on the plasma membrane of vegetative cell and can be accumulated in the apoplast. ROS release from pollen grains in vitro was recorded after 20 minutes in tobacco and kiwi [13,14] and after 10 min in spruce [20]. In these cases, ROS production was associated with NADPH oxidase, since its inhibitor DPI (diphenyl iodonium) suppressed ROS accumulation in the germination medium [14,20] and ROS synthesis in pollen grains [13,15]. In *Arabidopsis* pollen, two isoforms of plasma membrane NADPH oxidase are involved in ROS synthesis: RbohH and RbohJ (respiratory burst oxidase protein homolog) [21]. Cytochemical analysis of pollen grains germinating on stigma showed that double mutants *rbohH,J* lack H_2O_2 accumulation in the apoplast, which is typical for wild-type pollen [22]. SOD and MnTMPP (ROS quencher that mimics the activity of SOD and catalase) severely reduce pollen germination efficiency in blue spruce [20]. A negative effect has been also reported for cypress [15]. In angiosperms, the situation is, apparently, more complicated than in conifers since the balance of ROS production/elimination largely depends on pistil tissues. Thus, in tobacco, low concentrations of ascorbic acid and DPI reduce ROS content in pollen, but stimulated germination as well as MnTMPP [23]. This data indicates that the level of ROS produced by tobacco pollen is excessive relative to optimal. However, no such effect was observed in kiwi; all investigated concentrations of antioxidants blocked pollen germination [13]. Taken together, the facts described indicate the importance of endogenous ROS and, in particular, NADPH oxidase-derived ROS during initial germination stages in both angiosperms and conifers.

During in vivo germination in *Arabidopsis*, ROS synthesis is very important already at the stage of pollen grain hydration [24]. Snf1(sucrose nonfermenting 1)-related kinase1 protein kinase complex belongs to a family of highly conserved serine/threonine kinases and is involved in pollen, embryo, seedling and organ development regulation as well as in sugar, stress and hormonal signaling. It turned out that the mutant lacking subunit $KIN\beta\gamma$ of SnRK1 protein complex simultaneously exhibit a reduced level of endogenous ROS and hydration disorders, which did not appear in vitro (when water was in excess), as well as when water was added on the stigma. The same disturbances were observed upon overexpression of catalase, which drastically reduced the level of endogenous H_2O_2 in pollen [24]. Thus, a relationship was revealed between ROS production at the early germination stage and water flow into dry pollen grains.

2.2. Changes of Ionic Status during Early Germination Stages

In parallel with ROS production, the exact time of which has not been established, since it may differ for different species, active protein synthesis begins, and pollen starts the preparation for polar growth, which includes significant changes in ionic homeostasis. One of the earliest events is the release of anions from pollen grains [25]; in tobacco it begins within 2 minutes of incubation. Other early changes include cytoplasmic pH shift towards alkaline values and hyperpolarization of the plasmalemma [26,27], in tobacco they occur simultaneously. Experiments with proton pump inhibitor and activator demonstrate significance of this enzyme in triggering the early germination stages, apparently due to the cytoplasm alkalization [27–30]. The H^+-ATPase can also take part in plasma membrane hyperpolarization, together with anion channels: suppression of their activity blocks membrane potential (MP) shift in tobacco pollen [25].

When metabolism activation is completed, pollen grain undergoes polarization and segregation of the germination pole (cytoplasmic zone where a pollen tube will form). This domain differs from other zones in the arrangement of organelles: clusters of vesicles and mitochondria, which will provide the delivery of membrane material and energy for pollen tube growth, are accumulated at the germination pole [31]. However, some time before the polar growth is launched and cytoplasm protrusion becomes noticeable, physiological segregation of the pole occurs, for which ionic currents are essential. In early studies, electric currents crossing lily pollen grain were recorded, which, apparently, set the polarity for future germination [9,32]. On *Arabidopsis* pollen, it was shown that 9–15 minutes after in vivo pollination, a local increase in the intracellular calcium concentration begins in the germination pole [33]; it apparently causes NADPH oxidase activation and subsequent local ROS synthesis [21].

2.3. ROS Production on Stigma

ROS synthesis and ion currents at the early germination stage are interrelated and can stimulate each other forming a regulatory feedback loop; however, this system may involve not only endogenous ROS (synthesized by pollen) but also exogenous ROS coming from the sporophyte tissues. A current hypothesis states that ROS synthesis should be considered, inter alia, as a way of activating and/or supporting pollen germination by female tissues.

ROS are synthesized on stigmas of various (if not all) flowering plants; in total, more than 20 species from different families have been studied to date, and this property is common for all studied species [5,34–36]. According to inhibitory experiments, the main ROS on stigma is hydrogen peroxide [35,36]. The presence of various ROS-regulating enzymes, in particular, peroxidases [37,38], on stigma and in stigma exudate has been widely accepted. On the other hand, in vitro pollen grain diffusates caused the inhibition of peroxidase activity [39]. Thus, the final balance between ROS production and elimination during in vivo germination in flowering plants is the result of a complex interaction between sporophyte and male gametophyte, which includes both low molecular weight components and antioxidant enzymes.

2.4. Perception of Exogenous ROS Signal by Pollen

Moderate H_2O_2 concentrations activate pollen germination in tobacco, while high concentrations inhibit [23]; for kiwi, high peroxide concentrations also have an inhibitory effect [13]. In spruce, the latter effect did not appear - the presence of ROS in a wide concentration range (0.1–2 mM) does not reduce the germination efficiency [20]; So far, it can be assumed that during pollen germination in gymnosperms, female tissues (cones) do not produce noticeable amounts of superoxide radical or peroxide, and only endogenous ROS are used to activate ion currents and other physiological effects in pollen. To confirm or disprove this hypothesis, one needs to find out if there are ROS in female cones before and during pollination.

One of the functions performed by ROS in plant cell is the control of cell wall cytomechanics. Moreover, while some ROS loosen the extracellular matrix, others, on the contrary, promote cross-linking of polymers [40]. Wall loosening can occur by non-protein-mediated scission of polysaccharides through •OH attack [41], while H_2O_2 can strengthen polymers via peroxidase-mediated cross-linking of hydroxycinnamates [42]. As shown in tobacco, this function is critical for pollen grains: a shift in ROS balance leads to impaired pollen germination. Hence, pollen grains treated with an excess of •OH were unstable to hypotonic stress and burst, and those treated with high concentrations of H_2O_2 became too hard and could not launch polar growth although remained viable [23]. For gymnosperms, ROS-mediated regulation of cell wall stiffness determines not only the germination efficiency, but also the pattern of pollen tubes appearance: in many species of the Pinaceae family, two tubes can appear from one pollen grain [43,44], which, as it turned out, depends not on the availability of nutrients, but mainly on the properties of the pollen wall regulated by ROS [43].

However, ROS functions are not limited to the cell wall. Exogenous ROS, affecting pollen in flowering plants, can specifically activate ion channels: in protoplasts from lily pollen grains, calcium and potassium currents are stimulated by H_2O_2 (100 µM) [45]; in pear pollen protoplasts 10 mM H_2O_2 activates the calcium current [46]. For tobacco, similar results were obtained in different concentration range: intracellular $[Ca^{2+}]$ was assessed by a fluorescent method, and it clearly reacted to peroxide already at 10 µM. The effect was abolished by calcium channel inhibitor nifedipine [47]. Another important effect was the plasma membrane hyperpolarization in H_2O_2-treated protoplasts (10 µM) [47].

Thus, for angiosperms (although we can speak with confidence only of a few species), ROS, and in particular H_2O_2, are an important product of female sporophyte tissues enhancing germination, causing membrane hyperpolarization, activation of calcium currents, and, possibly, other more delayed effects.

3. Pollen Tube Growth

The pollen tube of angiosperms is characterized by an extremely rapid growth, which is supported by physiological and structural zoning of the cytoplasm [48]. It is generally accepted to distinguish apical, subapical, and distal domains in the pollen tube; the distal, in turn, has its own subdivisions, based on the presence of certain organelles in it. The segregation of cytoplasmic zones is maintained due to a set of regulatory mechanisms, including small GTPases and signaling phospholipids, which are highlighted in a recent review [49]; here, we focus on those that have become the subject of the present review.

Endogenous mechanisms maintaining the polar growth are conveniently studied in vitro, but female sporophyte tissues produce a number of their own molecules that can influence tube growth by enhancing, directing, or blocking it. In particular, the direction of tube growth can be affected by $[Ca^{2+}]$, NO, and ROS in pistil tissues, and the latter can both support growth and stop it if fertilization is undesirable [7,50,51].

3.1. Ionic Homeostasis and ROS Production in Growing Pollen Tube

The uneven distribution and activity of ion transporters cells with polar growth results in gradients of ion concentrations and membrane potential (MP) corresponding to different cell domains (Figure 1). Thus, $[Ca^{2+}]$ is highest in the apical domain [52]; pH—in subapical; concentration of anions—20 to 50 µm from the tip (*Arabidopsis thaliana*) [53]. MP has the lowest (relatively depolarized) value at the tip; further along the tube length, hyperpolarization is observed [20,54,55]. At the moment, the gradient has already been described in pollen tubes of both flowering plants (tobacco, lily) and conifers (spruce). According to inhibitory analysis, at least H^+-ATPase and anion channels take part in its maintenance, and for spruce potassium and calcium channels are also involved. Some of these systems are sensitive to endogenous and exogenous ROS, since the gradient changes shape in the presence of both H_2O_2, antioxidants, and DPI [20,55]. Interestingly, the sensitivity of pollen tubes to hydrogen peroxide is higher in spruce than that in lily: in spruce, depolarization in subapical region was observed already at 100 µM H_2O_2, while in lily the effect was observed at 500 µM, and it was hyperpolarizing. A decrease in the endogenous ROS level in both cases led to hyperpolarization and dissipation of the gradient. Considering these data together with the results obtained on protoplasts from tobacco pollen tubes [47], we can conclude that MP is an indicator with high sensitivity to H_2O_2, which means that systems responsible for maintaining the MP gradient can respond to both endogenous ROS, at least partially generated by NADPH oxidase, and ROS produced by female sporophyte tissues.

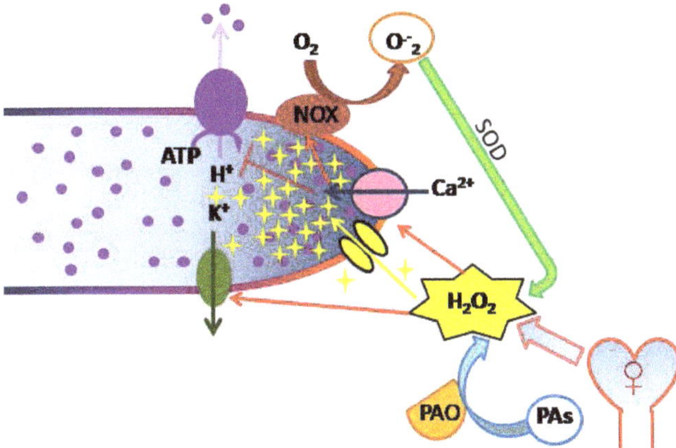

Figure 1. ROS and ion homeostasis in the regulation of pollen tube growth (simplified diagram). Three sources of apoplastic ROS during pollen tube growth are shown: NOX in the plasma membrane produces superoxide radical, SOD dismutates it to hydrogen peroxide; PAO produces H_2O_2 during polyamine oxidation; female tissues of the sporophyte produce ROS that control pollen tube growth in vivo. Hydrogen peroxide enters the pollen tube through aquaporins and forms apical ROS accumulation (yellow stars). The main targets for apoplastic ROS are shown: Ca_{2+} permeable ion channels and K^+-permeable ion channels are activated by H_2O_2, lateral membrane potential is affected (apex is red (depolarized), shank is blue (hyperpolarized)). Apical calcium gradient in the cytoplasm is shown in blue color. High $[Ca^{2+}]$ reduces the proton pump activity, which forms the alkaline band (protons shown as circles). NOX—NADPH-oxidase, SOD—superoxidedismutase, PAO—polyamineoxidase, Pas—polyamines.

In growing pollen tubes, pH gradient is maintained: pH at the tip is acidic, in the subapical zone it rises to alkaline values due to H^+-ATPase activity, and along the tube shank, pH is close to neutral [55–57] (Figure 1). pH gradient is affected by significant alterations in ROS production/elimination balance: in MnTMPP-treated in lily pollen tubes pH shifts towards more alkaline values, wherein the difference between apical and subapical zones is leveled [55]; during short-term exposure to 1 mM H_2O_2 the alkaline band disappears, the gradient is also leveled; lower concentrations do not affect pH [55]. The authors suggest that the observed effect in H_2O_2-treated tubes can be explained by the suppression of H^+-ATPase activity by high $[Ca^{2+}]$, since peroxide stimulates calcium channels [45] (Figure 1). Thus, ROS directly or indirectly regulate the proton pump activity in lily pollen tubes and maintain gradient pH distribution in the cytoplasm.

In all studied pollen tubes, there is a gradient of intracellular calcium concentration: in flowering plants, apical $[Ca^{2+}]$ is two orders of magnitude higher than that in the shank [52,58]; gymnosperms have a flatter gradient [59]. ROS production is important for maintaining normal calcium homeostasis: in lily tubes, MnTMPP, even at low concentrations, caused a decrease in calcium concentration and dissipation of $[Ca^{2+}]$ gradient [55].

Downstream of $[Ca^{2+}]$ and pH is the structure and dynamics of actin cytoskeleton controlled by numerous proteins [60]. Among Ca^{2+}-dependent actin-binding proteins, for example, profilin, LIM domain-containing proteins, ROP-interactive and CRIB motif-containing protein1 (RIC1), and villins should be mentioned [61]. High $[Ca^{2+}]$ in the tube tip supports actin remodeling and ensures its existence in the form of sparse short filaments while low $[Ca^{2+}]$ in the shank area correlates with rather thick and stable actin cables; the main mediators in this relationship are proteins from villin/gelsolin/fragmin superfamily [60,61]; subapical actin fringe is tightly associated with the alkaline band, presumably, via actin depolimerizing factor (ADF) and actin-interacting protein 1 (AIP1) [62,63].

Thus, by provoking Ca^{2+} influx in the pollen tube tip and regulating intracellular pH, ROS can thereby affect the actin cytoskeleton.

Since, as has been revealed in protoplasts, calcium transport is closely related to ROS production, one might assume that ROS are also unevenly distributed in the pollen tube. This has been shown by different methods for *Arabidopsis* [64,65], *Pyrus* [66], cypress [15], and two *Picea* species [20,67]. For spruce, it was shown that H_2O_2 accumulates in the tube apex, apparently coming from the apoplast (where NADPH oxidase and SOD work) (Figure 1), and in amyloplasts, while most of the $O^{\bullet}{}_2{}^-$ is produced in mitochondria, and the localization of the two ROS does not coincide [20].

ROS are produced in pollen tubes by RbohH and RbohJ. Both proteins located on the pollen tube plasma membrane have EF-hands in their structure and are activated upon binding of calcium ion [21]. In tobacco, transfection with NOX-specific antisense oligodeoxynucleotides (ODNs) resulted in pollen tube growth inhibition, which was rescued by exogenous H_2O_2 [68]. In *Arabidopsis*, *rbohH,J* mutants have severe reproductive disorders. The mutant's pollen has inhibited tube growth and impaired calcium homeostasis [21,69]. Interestingly, mutants for calcium channel genes *cngc7,8* (cyclic nucleotide gated, non-selective, Ca_{2+}-permeable ion channels) have phenotypes almost identical to *rbohH,J*, which indicates the feedback regulation of these systems [7]. The features of the *rbohH,J* mutants were revealed in vivo: in wild-type *Arabidopsis* plants, apoplastic ROS production in the area between the pollen tube surface and stigma papilla was detected by histochemistry [22]. In mutants, ROS did not accumulate; growth was impaired. Thus, endogenous ROS produced by NADPH oxidase are essential for pollen tube growth in vitro and in vivo.

However, this enzyme is not the only source of ROS in the male gametophyte: polyamine oxidase (PAO) in pollen can synthesize H_2O_2 from polyamines [70] (Figure 1). In *Oryza sativa* seven PAO isoforms have been identified, and one of these, OsPAO7, is specifically expressed in anthers; OsPAO7 produces H_2O_2 about 100 times more efficiently than other PAO isoforms [71]. In a recent study the relationship between polyamines and H_2O_2 in *Arabidopsis* pollen tubes was reported [65]. Such a relationship had been previously shown for pear: the gradients of total ROS and spermidine in these tubes coincided, and in spermidine-treated tubes (100 µM) cytoplasmic level of these substances increased consistently [66]. It is to be noted that the same exogenously applied polyamines had different effects on the NO/ROS levels pollen grains and tubes. Furthermore, recent studies indicate that PAs regulate pollen germination primarily via regulating the ROS level, while tube elongation primarily influencing the NO level [72].

Polyamines can affect ion homeostasis in plant cells, in many cases, in an indirect manner, with ROS formation as an intermediate stage [70,72,73]. For example, upon treatment with 100 µM spermidine, pear pollen tubes responded with a rapid $[Ca^{2+}]_{cyt}$ increase, pH gradient alterations, and switch of growth pattern [74]. The data on polyamine-induced changes in ion transport in root cells is much more plentiful: H^+-ATPase pumping activity was affected in several species, including both activation (in rice and wheat) and inhibition (in maize) [73]. In some cases, polyamines can cause pollen damage through excessive ROS formation, followed by activation of the antioxidant machinery, degradation of nuclear DNA, and finally, cell death [66].

An ABC transporter carrying polyamines is involved in forming the ROS gradient early during polar growth: short tubes of *atabcg28* (ATP-binding cassette G28) mutant lack the gradient of both total ROS and hydrogen peroxide with maximum at the tip, typical for wild type tubes [65]. It should be noted, however, that total ROS level in this pollen is high, and normal tubes are not formed.

According to colocalization experiments with both fluorescence microscopy and TEM, mitochondria in the pollen tube are also a source of ROS [20,75,76]; however, the role of mitochondria-derived ROS in the regulation of polar growth is still questionable. To date, their reduced production has been associated with loss of mitochondrial function in self-incompatible pear pollen tubes [76].

3.2. ROS Are Involved in Signal Perception and Mediate Pollen Tube Rupture

Signals that determine pollen tube growth direction include sporophyte-derived ROS, NO, polyamines, and peptides [6,77,78]; in many of these cases, ROS are also involved in signal perception [50].

In tobacco pollen tubes, polyamines applied at a low concentration (10 μM) affected pollen elongation differentially [72]: putrescine negatively regulated pollen tube elongation; spermidine enhanced it, spermine had no effect on pollen tube growth. This influence of polyamines correlated well with their effect on ROS and/or NO levels in pollen tube tip: high NO and low ROS levels in the tip region of treated pollen tubes promoted while the opposite inhibited growth [72].

Recently, ROS was found to be involved in a signaling cascade triggered in response to a peptide signal (RALF4—RAPID ALKALINIZATION FACTOR 4), which is the pistil-side control of pollen tube growth. Exposure of tubes growing in vitro to this peptide caused a sharp increase in ROS level, growth stimulation and prevented tube rupture [79]. On the contrary, H_2O_2 quenchers potassium iodide and sodium pyruvate inhibited pollen tube growth and caused rupture. In plant lines that do not produce LORELEI-LIKE GPI-ANCHORED PROTEINS 2/3, involved in the perception of RALF signal, ROS level was significantly reduced, and growth was impaired. The addition of exogenous H_2O_2 partially restored these disturbances [79]. One of the proposed mechanisms was the regulation of the cell wall mechanical properties [7], since the deposition of callose and pectins in pollen tube wall of *llg2,3* mutant was disrupted [79].

Pollen tube rupture upon reaching the embryo sac is a necessary condition for sperm release and, accordingly, for fertilization. This process is tightly controlled: recognition of the pollen tube by synergids and subsequent perception of "permitting" signal by the tube is required [8,80]. Signal peptides (for example, cysteine-rich peptides) and small molecules (for example, ROS), which allow synergid cells to recognize the pollen tube, are accumulated in the filiform apparatus (FA) area. FA is a highly thickened structure of synergids' cell wall at the micropylar end. Pollen tube recognition at this stage is an important barrier to interspecies crossing, since "unrecognized" pollen tube does not stop growth and does not release sperms [5]. Recognition involves receptor kinase FERONIA (FER), a member of the CrRLK1L (*Catharanthus roseus* receptor-like kinase 1-like) subfamily. It regulates NORTIA (NTA) membrane anchoring and interacts with the GPI-anchored protein LORELEI (LRI) on the synergid surface [7]. Lack of recognition and of the following rupture ("overgrowth phenotype") was found in *Arabidopsis* pistils with reduced ROS levels. In the FA area of the synergid, wild-type plants exhibit a ROS production peak, which is absent in *fer* and *lre* mutants, as well as in DPI-treated pistils. In all these cases tubes form the overgrowth phenotype and do not take part in fertilization [81]. Thus, receptor kinase FER, as well as the LRI interacting with it, are responsible for the local ROS production, which provokes tube rupture and sperm release. *abstinence by mutual consent, amc,* also has a phenotype similar to *fer*, but this mutant is self-sterile, that is, the phenotype is observed only when both male and female gamtophytes carry the *amc* allele. AMC encodes a peroxine involved in protein import in peroxisomes, potentially important for ROS production during pollen tube-synergid signaling [82].

4. Conclusions

Thus, ROS are involved in the life of the male gametophyte at all stages, from hydration to the release of sperms. On the one hand, ROS are produced endogenously, on the other, they act as a signal from female tissues. One of the main mechanisms of ROS action is the activation of ionic currents through the plasma membrane. Various ion transport systems exhibit sensitivity to ROS, but the specific pattern of their activity in each of the redox states, through which the male gametophyte passes, remains to be studied.

Author Contributions: Conceptualization, M.B.; writing—original draft preparation, E.K. and M.B.; writing—review and editing, M.B. Both authors have read and agreed to the published version of the manuscript.

Funding: This review was funded by the Russian Science Foundation (19-74-00036).

Acknowledgments: We would like to thank ex-team members participated in the development of the topic of ROS and ion channels in our lab, Nikita Maksimov, Anna Smirnova, and Natalie Matveyeva.

Conflicts of Interest: The authors declare no conflict of interest.

Abbreviations

ROS	reactive oxygen species
ROP	Rho of plants
MnTMPP	Mn-5,10,15,20-tetrakis(1-methyl-4-pyridyl)21H,23H-porphin
CMS	cytoplasmic male sterility
DPI	diphenyl iodonium chloride
NBT	nitro blue tetrazolium
SOD	superoxide dismutase
NOX	NADPH-oxidase
MP	membrane potential
PAO	polyamine oxidase
DPI	diphenyleneiodonium (NOX inhibitor)

References

1. Bell, E.; Takeda, S.; Dolan, L. Reactive oxygen species in growth and development. In *Reactive Oxygen Species in Plant Signaling. Signaling and Communication in Plants*; Springer: Berlin/Heidelberg, Germany, 2009; pp. 43–53.
2. Kwak, J.M.; Mori, I.C.; Pei, Z.-M.; Leonhardt, N.; Torres, M.A.; Dangl, J.L.; Bloom, R.E.; Bodde, S.; Jones, J.D.G.; Schroeder, J.I. NADPH oxidase AtrbohD and AtrbohF genes function in ROS-dependent ABA signaling in Arabidopsis. *EMBO J.* **2003**, *22*, 2623–2633. [CrossRef] [PubMed]
3. Apel, K.; Hirt, H. REACTIVE OXYGEN SPECIES: Metabolism, Oxidative Stress, and Signal Transduction. *Annu. Rev. Plant Biol.* **2004**, *55*, 373–399. [CrossRef] [PubMed]
4. Coelho, S.M.B.; Brownlee, C.; Bothwell, J.H.F. A tip-high, Ca^{2+}-interdependent, reactive oxygen species gradient is associated with polarized growth in Fucus serratus zygotes. *Planta* **2008**, *227*, 1037–1046. [CrossRef] [PubMed]
5. Johnson, M.A.; Harper, J.F.; Palanivelu, R. A Fruitful Journey: Pollen Tube Navigation from Germination to Fertilization. *Annu. Rev. Plant Biol.* **2019**, *70*, 809–837. [CrossRef]
6. Lopes, A.L.; Moreira, D.; Ferreira, M.J.; Pereira, A.M.; Coimbra, S. Insights into secrets along the pollen tube pathway in need to be discovered. *J. Exp. Bot.* **2019**, *70*, 2979–2992. [CrossRef]
7. Zhang, M.J.; Zhang, X.S.; Gao, X.-Q. ROS in the Male–Female Interactions During Pollination: Function and Regulation. *Front. Plant Sci.* **2020**, *11*, 1–8. [CrossRef]
8. Ge, Z.; Cheung, A.Y.; Qu, L.J. Pollen tube integrity regulation in flowering plants: Insights from molecular assemblies on the pollen tube surface. *New Phytol.* **2019**, *222*, 687–693. [CrossRef]
9. Weisenseel, M.H.; Jaffe, L.F. The major growth current through lily pollen tubes enters as K+ and leaves as H+. *Planta* **1976**, *7*, 1–7. [CrossRef]
10. DeFalco, T.A.; Bender, K.W.; Snedden, W.A. Breaking the code: Ca^{2+} sensors in plant signalling. *Biochem. J.* **2010**, *425*, 27–40. [CrossRef]
11. Feijó, J.A.; Wudick, M.M. Calcium is life. *J. Exp. Bot.* **2018**, *69*, 4147–4150. [CrossRef]
12. Taylor, L.P.; Hepler, P.K. Pollen germination and tube growth. *Annu. Rev. Plant Physiol. Plant Mol. Biol.* **1997**, *48*, 461–491. [CrossRef] [PubMed]
13. Speranza, A.; Crinelli, R.; Scoccianti, V.; Geitmann, A. Reactive oxygen species are involved in pollen tube initiation in kiwifruit. *Plant Biol.* **2012**, *14*, 64–76. [CrossRef] [PubMed]
14. Smirnova, A.V.; Matveyeva, N.P.; Polesskaya, O.G.; Yermakov, I.P. Generation of reactive oxygen species during pollen grain germination. *Russ. J. Dev. Biol.* **2009**, *40*, 345–353. [CrossRef]
15. Pasqualini, S.; Cresti, M.; Del Casino, C.; Faleri, C.; Frenguelli, G.; Tedeschini, E.; Ederli, L. Roles for NO and ROS signalling in pollen germination and pollen-tube elongation in Cupressus arizonica. *Biol. Plant.* **2015**, *59*, 735–744. [CrossRef]

16. Huang, L.; Zhang, C.; Fang, M.; Xu, H.; Cao, M. The Cell Death in CMS Plants. *Int. J. Biosci. Biochem. Bioinforma.* **2011**, *1*, 297–301. [CrossRef]

17. Nie, H.; Cheng, C.; Hua, J. Mitochondrial proteomic analysis reveals that proteins relate to oxidoreductase activity play a central role in pollen fertility in cotton. *J. Proteom.* **2020**, *225*, 103861. [CrossRef]

18. Deng, M.-H.; Wen, J.-F.; Huo, J.-L.; Zhu, H.-S.; Dai, X.-Z.; Zhang, Z.-Q.; Zhou, H.; Zou, X.-X. Relationship of metabolism of reactive oxygen species with cytoplasmic male sterility in pepper (Capsicum annuum L.). *Sci. Hortic. (Amsterdam)* **2012**, *134*, 232–236. [CrossRef]

19. Li, S.; Wan, C.; Kong, J.; Zhang, Z.; Li, Y.; Zhu, Y. Programmed cell death during microgenesis in a Honglian CMS line of rice is correlated with oxidative stress in mitochondria. *Funct. Plant Biol.* **2004**, *31*, 369–376. [CrossRef]

20. Maksimov, N.; Evmenyeva, A.; Breygina, M.; Yermakov, I. The role of reactive oxygen species in pollen germination in *Picea pungens* (blue spruce). *Plant Reprod.* **2018**, *18*, 761–767. [CrossRef]

21. Kaya, H.; Nakajima, R.; Iwano, M.; Kanaoka, M.M.; Kimura, S.; Takeda, S.; Kawarazaki, T.; Senzaki, E.; Hamamura, Y.; Higashiyama, T.; et al. Ca2+-activated reactive oxygen species production by Arabidopsis RbohH and RbohJ is essential for proper pollen tube tip growth. *Plant Cell* **2014**, *26*, 1069–1080. [CrossRef]

22. Kaya, H.; Iwano, M.; Takeda, S.; Kanaoka, M.M.; Kimura, S. Apoplastic ROS production upon pollination by RbohH and RbohJ in Arabidopsis. *Plant Signal. Behav.* **2015**. [CrossRef] [PubMed]

23. Smirnova, A.; Matveyeva, N.; Yermakov, I. Reactive oxygen species are involved in regulation of pollen wall cytomechanics. *Plant Biol.* **2013**, *16*, 252–257. [CrossRef] [PubMed]

24. Gao, X.Q.; Liu, C.Z.; Li, D.D.; Zhao, T.T.; Li, F.; Jia, X.N.; Zhao, X.Y.; Zhang, X.S. The Arabidopsis KINβγ Subunit of the SnRK1 Complex Regulates Pollen Hydration on the Stigma by Mediating the Level of Reactive Oxygen Species in Pollen. *PLoS Genet.* **2016**, *12*, 1–25. [CrossRef] [PubMed]

25. Breygina, M.A.; Matveeva, N.P.; Ermakov, I.P. The role of Cl⁻ in pollen germination and tube growth. *Russ. J. Dev. Biol.* **2009**, *40*, 157–164. [CrossRef]

26. Breygina, M.; Smirnova, A.; Matveeva, N.; Yermakov, I. The role of anion channels in pollen germination and tube growth. In *Pollen: Structure, Types and Effects*; Kaiser, B.J., Ed.; Nova Science Publishers, Inc.: Hauppauge, NY, USA, 2010; pp. 1–19. ISBN 9781616686697.

27. Matveeva, N.P.; Andreyuk, D.S.; Voitsekh, O.O.; Ermakov, I.P. Regulatory Changes in the Intracellular pH and Cl– Efflux at Early Stages of Pollen Grain Germination in vitro. *Russ. J. Plant Physiol.* **2003**, *50*, 318–323. [CrossRef]

28. Rodriguez-Rosales, M.P.; Roldan, M.; Belver, A.; Donaire, J.P. Correlation between in vitro germination capacity and proton extrusion in olive pollen. *Plant Physiol. Biochem.* **1989**, *27*, 723–728.

29. Fricker, M.D.; White, N.S.; Obermeyer, G. pH gradients are not associated with tip growth in pollen tubes of Lilium longiflorum. *J. Cell Sci.* **1997**, *110*, 1729–1740.

30. Certal, A.C.; Almeida, R.B.; Carvalho, L.M.; Wong, E.; Moreno, N.; Michard, E.; Carneiro, J.; Rodriguéz-Léon, J.; Wu, H.; Cheung, A.Y.; et al. Exclusion of a proton ATPase from the apical membrane is associated with cell polarity and tip growth in Nicotiana tabacum pollen tubes. *Plant Cell* **2008**, *20*, 614–634. [CrossRef]

31. Mazina, S.E.; Matveyeva, N.P.; Yermakov, I.P. Determination of a functional pore in the pollen grain of tobacco. *Tsitologiya* **2002**, *44*, 33–39.

32. Weisenseel, M.H.; Nuccitelli, R.; Jaffe, L.F. Large electrical currents traverse growing pollen tubes. *J. Cell Biol.* **1975**, *66*, 556–567. [CrossRef]

33. Iwano, M.; Shiba, H.; Miwa, T.; Che, F.-S.; Takayama, S.; Nagai, T.; Miyawaki, A.; Isogai, A. Ca2+ dynamics in a pollen grain and papilla cell during pollination of Arabidopsis. *Plant Physiol.* **2004**, *136*, 3562–3571. [CrossRef] [PubMed]

34. Zafra, A.; Rejón, J.D.; Hiscock, S.J.; Alché, J.D.D. Patterns of ROS accumulation in the stigmas of angiosperms and visions into their multi-functionality in plant reproduction. *Front. Plant Sci.* **2016**, *7*, 1112–1119. [CrossRef] [PubMed]

35. McInnis, S.M.; Desikan, R.; Hancock, J.T.; Hiscock, S.J. Production of reactive oxygen species and reactive nitrogen species by angiosperm stigmas and pollen: Potential signalling crosstalk? *New Phytol.* **2006**, *172*, 221–228. [CrossRef] [PubMed]

36. Hiscock, S.J.; Bright, J.; McInnis, S.M.; Desikan, R.; Hancock, J.T. Signaling on the stigma. Potential new roles for ROS and NO in plant cell signaling. *Plant Signal. Behav.* **2007**, *2*, 23–24. [CrossRef] [PubMed]

37. Bredemeijer, G.M.M. The role of peroxidases in pistil-pollen interactions. *Theor. Appl. Genet.* **1984**, *68*, 193–206. [CrossRef] [PubMed]

38. Shivanna, K.R.; Rangaswamy, N.S. *Pollen Biology: A Laboratory Manual*; Springer Science & Business Media: Berlin, Germany, 2012; ISBN 3642773060.

39. Žárský, V.; Říhová, L.; Tupý, J. Interference of pollen diffusable substances with peroxidase catalyzed reaction. *Plant Sci.* **1987**, *52*, 29–32. [CrossRef]

40. Kärkönen, A.; Kuchitsu, K. Reactive oxygen species in cell wall metabolism and development in plants. *Phytochemistry* **2015**, *112*, 22–32. [CrossRef]

41. Lindsay, S.E.; Fry, S.C. Redox and Wall-Restructuring. In *The Expanding Cell*; Verbelen, J.-P., Vissenberg, K., Eds.; Springer: Berlin/Heidelberg, Germany, 2007; pp. 159–190, ISBN 978-3-540-39116-6.

42. Bunzel, M. Chemistry and occurrence of hydroxycinnamate oligomers. *Phytochem. Rev.* **2010**, *9*, 47–64. [CrossRef]

43. Breygina, M.; Maksimov, N.; Polevova, S.; Evmenyeva, A. Bipolar pollen germination in blue spruce (*Picea pungens*). *Protoplasma* **2019**, *256*, 941–949. [CrossRef]

44. Çetinbaş-Genç, A.; Vardar, F. Effect of methyl jasmonate on in-vitro pollen germination and tube elongation of *Pinus nigra*. *Protoplasma* **2020**, *257*, 1655–1665. [CrossRef]

45. Breygina, M.A.; Abramochkin, D.V.; Maksimov, N.M.; Yermakov, I.P. Hydrogen peroxide affects ion channels in lily pollen grain protoplasts. *Plant Biol.* **2016**, *18*, 761–767. [CrossRef] [PubMed]

46. Wu, J.; Shang, Z.; Wu, J.; Jiang, X.; Moschou, P.N.; Sun, W.; Roubelakis-Angelakis, K.A.; Zhang, S. Spermidine oxidase-derived H2O2 regulates pollen plasma membrane hyperpolarization-activated Ca2+-permeable channels and pollen tube growth. *Plant J.* **2010**, *63*, 1042–1053. [CrossRef] [PubMed]

47. Maksimov, N.M.; Breygina, M.A.; Ermakov, I.P. Regulation of ion transport across the pollen tube plasmalemma by hydrogen peroxide. *Cell Tissue Biol.* **2016**, *10*, 69–75. [CrossRef]

48. Hafidh, S.; Fíla, J.; Honys, D. Male gametophyte development and function in angiosperms: A general concept. *Plant Reprod.* **2016**, *29*, 31–51. [CrossRef]

49. Scholz, P.; Anstatt, J.; Krawczyk, H.E.; Ischebeck, T. Signalling Pinpointed to the Tip: The Complex Regulatory Network That Allows Pollen Tube Growth. *Plants* **2020**, *9*, 1098. [CrossRef]

50. Sankaranarayanan, S.; Ju, Y.; Kessler, S.A. Reactive Oxygen Species as Mediators of Gametophyte Development and Double Fertilization in Flowering Plants. *Front. Plant Sci.* **2020**, *11*, 1199. [CrossRef]

51. Lan, X.; Yang, J.; Abhinandan, K.; Nie, Y.; Li, X.; Li, Y.; Samuel, M.A. Flavonoids and ROS Play Opposing Roles in Mediating Pollination in Ornamental Kale (*Brassica oleracea* var. *acephala*). *Mol. Plant* **2017**, *10*, 1361–1364. [CrossRef]

52. Iwano, M.; Entani, T.; Shiba, H.; Kakita, M.; Nagai, T.; Mizuno, H.; Miyawaki, A.; Shoji, T.; Kubo, K.; Isogai, A.; et al. Fine-tuning of the cytoplasmic Ca2+ concentration is essential for pollen tube growth. *Plant Physiol.* **2009**, *150*, 1322–1334. [CrossRef]

53. Gutermuth, T.; Lassig, R.; Portes, M.; Maierhofer, T.; Romeis, T.; Borst, J.; Hedrich, R.; Feijó, J.A.; Konrad, K.R. Pollen tube growth regulation by free anions depends on the interaction between the anion channel SLAH3 and calcium-dependent protein kinases CPK2 and CPK20. *Plant Cell* **2013**, *25*, 4525–4543. [CrossRef]

54. Breygina, M.; Smirnova, A.; Matveeva, N.; Yermakov, I. Membrane potential changes during pollen germination and tube growth. *Cell Tissue Biol.* **2010**, *3*, 573–582. [CrossRef]

55. Podolyan, A.; Maksimov, N.; Breygina, M. Redox-regulation of ion homeostasis in growing lily pollen tubes. *J. Plant Physiol.* **2019**, *243*, 153050. [CrossRef]

56. Hepler, P.K.; Winship, L.J. The pollen tube clear zone: Clues to the mechanism of polarized growth. *J. Integr. Plant Biol.* **2015**, *57*, 79–92. [CrossRef] [PubMed]

57. Feijó, J.A.; Sainhas, J.; Hackett, G.R.; Kunkel, J.G.; Hepler, P.K. Growing pollen tubes possess a constitutive alkaline band in the clear zone and a growth-dependent acidic tip. *J. Cell Biol.* **1999**, *144*, 483–496. [CrossRef] [PubMed]

58. Michard, E.; Lima, P.T.; Borges, F.; Silva, A.C.; Portes, M.T.; Carvalho, J.E.; Gilliham, M.; Liu, L.-H.; Obermeyer, G.; Feijó, J. a Glutamate receptor-like genes form Ca2+ channels in pollen tubes and are regulated by pistil D-serine. *Science* **2011**, *332*, 434–437. [CrossRef] [PubMed]

59. Fernando, D.D.; Lazzaro, M.D.; Owens, J.N. Growth and development of conifer pollen tubes. *Sex. Plant Reprod.* **2005**, *18*, 149–162. [CrossRef]

60. Cai, G.; Parrotta, L.; Cresti, M. Organelle trafficking, the cytoskeleton, and pollen tube growth. *J. Integr. Plant Biol.* **2015**, *57*, 63–78. [CrossRef]

61. Qian, D.; Xiang, Y. Actin cytoskeleton as actor in upstream and downstream of calcium signaling in plant cells. *Int. J. Mol. Sci.* **2019**, *20*, 1403. [CrossRef]

62. Lovy-Wheeler, A.; Kunkel, J.G.; Allwood, E.G.; Hussey, P.J.; Hepler, P.K. Oscillatory increases in alkalinity anticipate growth and may regulate actin dynamics in pollen tubes of lily. *Plant Cell* **2006**, *18*, 2182–2193. [CrossRef]

63. Chen, C.Y.; Wong, E.I.; Vidali, L.; Estavillo, A.; Hepler, P.K.; Wu, H.M.; Cheung, A.Y. The regulation of actin organization by actin-depolymerizing factor in elongating pollen tubes. *Plant Cell* **2002**, *14*, 2175–2190. [CrossRef]

64. Muhlemann, J.K.; Younts, T.L.B.; Muday, G.K. Flavonols control pollen tube growth and integrity by regulating ROS homeostasis during high-temperature stress. *Proc. Natl. Acad. Sci. USA* **2018**, *115*, E11188–E11197. [CrossRef]

65. Do, T.H.T.; Choi, H.; Palmgren, M.; Martinoia, E.; Hwang, J.U.; Lee, Y. Arabidopsis ABCG28 is required for the apical accumulation of reactive oxygen species in growing pollen tubes. *Proc. Natl. Acad. Sci. USA* **2019**, *116*, 12540–12549. [CrossRef]

66. Aloisi, I.; Cai, G.; Tumiatti, V.; Minarini, A.; Del Duca, S. Del Natural polyamines and synthetic analogues modify the growth and the morphology of *Pyrus communis* pollen tubes affecting ROS levels and causing cell death. *Plant Sci.* **2015**, *239*, 92–105. [CrossRef] [PubMed]

67. Liu, P.; Li, R.L.; Zhang, L.; Wang, Q.L.; Niehaus, K.; Baluška, F.; Šamaj, J.; Lin, J.X. Lipid microdomain polarization is required for NADPH oxidase-dependent ROS signaling in *Picea meyeri* pollen tube tip growth. *Plant J.* **2009**, *60*, 303–313. [CrossRef] [PubMed]

68. Potocký, M.; Jones, M.A.; Bezvoda, R.; Smirnoff, N.; Žárský, V. Reactive oxygen species produced by NADPH oxidase are involved in pollen tube growth. *New Phytol.* **2007**, *174*, 742–751. [CrossRef] [PubMed]

69. Lassig, R.; Gutermuth, T.; Bey, T.D.; Konrad, K.R.; Romeis, T. Pollen tube NAD(P)H oxidases act as a speed control to dampen growth rate oscillations during polarized cell growth. *Plant J.* **2014**, *78*, 94–106. [CrossRef]

70. Aloisi, I.; Cai, G.; Serafini-Fracassini, D.; Del Duca, S. Polyamines in Pollen: From Microsporogenesis to Fertilization. *Front. Plant Sci.* **2016**, *7*, 155. [CrossRef]

71. Liu, T.; Kim, D.W.; Niitsu, M.; Maeda, S.; Watanabe, M.; Kamio, Y.; Berberich, T.; Kusano, T. Polyamine Oxidase 7 is a Terminal Catabolism-Type Enzyme in Oryza sativa and is Specifically Expressed in Anthers. *Plant Cell Physiol.* **2014**, *55*, 1110–1122. [CrossRef]

72. Benkő, P.; Jee, S.; Kaszler, N.; Fehér, A.; Gémes, K. Polyamines treatment during pollen germination and pollen tube elongation in tobacco modulate reactive oxygen species and nitric oxide homeostasis. *J. Plant Physiol.* **2020**, *244*, 153085. [CrossRef]

73. Pottosin, I.; Shabala, S. Polyamines control of cation transport across plant membranes: Implications for ion homeostasis and abiotic stress signaling. *Front. Plant Sci.* **2014**, *5*, 154. [CrossRef]

74. Aloisi, I.; Cai, G.; Faleri, C.; Navazio, L.; Serafini-Fracassini, D.; Del Duca, S. Spermine Regulates Pollen Tube Growth by Modulating Ca2+-Dependent Actin Organization and Cell Wall Structure. *Front. Plant Sci.* **2017**, *8*, 1701. [CrossRef]

75. Cárdenas, L.; McKenna, S.T.; Kunkel, J.G.; Hepler, P.K. NAD(P)H oscillates in pollen tubes and is correlated with tip growth. *Plant Physiol.* **2006**, *142*, 1460–1468. [CrossRef]

76. Wang, C.L.; Wu, J.; Xu, G.H.; Gao, Y.B.; Chen, G.; Wu, J.Y.; Wu, H.Q.; Zhang, S.L. S-RNase disrupts tip-localized reactive oxygen species and induces nuclear DNA degradation in incompatible pollen tubes of *Pyrus pyrifolia*. *J. Cell Sci.* **2010**, *123*, 4301–4309. [CrossRef]

77. Pereira, A.M.; Lopes, A.L.; Coimbra, S. Arabinogalactan proteins as interactors along the crosstalk between the pollen tube and the female tissues. *Front. Plant Sci.* **2016**, *7*, 1–15. [CrossRef] [PubMed]

78. Mizuta, Y.; Higashiyama, T. Chemical signaling for pollen tube guidance at a glance. *J. Cell Sci.* **2018**, *131*. [CrossRef] [PubMed]

79. Feng, H.; Liu, C.; Fu, R.; Zhang, M.; Li, H.; Shen, L.; Wei, Q.; Sun, X.; Xu, L.; Ni, B.; et al. LORELEI-LIKE GPI-ANCHORED PROTEINS 2/3 Regulate Pollen Tube Growth as Chaperones and Coreceptors for ANXUR/BUPS Receptor Kinases in Arabidopsis. *Mol. Plant* **2019**, *12*, 1612–1623. [CrossRef]

80. Li, H.J.; Yang, W.C. Ligands Switch Model for Pollen-Tube Integrity and Burst. *Trends Plant Sci.* **2018**, *23*, 369–372. [CrossRef] [PubMed]

81. Duan, Q.; Kita, D.; Johnson, E.A.; Aggarwal, M.; Gates, L.; Wu, H.-M.; Cheung, A.Y. Reactive oxygen species mediate pollen tube rupture to release sperm for fertilization in Arabidopsis. *Nat. Commun.* **2014**, *5*, 3129. [CrossRef]
82. Boisson-Dernier, A.; Frietsch, S.; Kim, T.-H.; Dizon, M.B.; Schroeder, J.I. The peroxin loss-of-function mutation abstinence by mutual consent disrupts male-female gametophyte recognition. *Curr. Biol.* **2008**, *18*, 63–68. [CrossRef]

 International Journal of
Molecular Sciences

Review

The Effect of Virulence and Resistance Mechanisms on the Interactions between Parasitic Plants and Their Hosts

Luyang Hu [1], Jiansu Wang [1], Chong Yang [2], Faisal Islam [1], Harro J. Bouwmeester [3], Stéphane Muños [4] and Weijun Zhou [1,*]

[1] Institute of Crop Science and Zhejiang Key Lab of Crop Germplasm, Zhejiang University, Hangzhou 310058, China; hu.luyang@foxmail.com (L.H.); 21716129@zju.edu.cn (J.W.); faisalislam@zju.edu.cn (F.I.)
[2] Bioengineering Research Laboratory, Institute of Bioengineering, Guangdong Academy of Sciences, Guangzhou 510316, China; parker815@163.com
[3] Swammerdam Institute for Life Sciences, University of Amsterdam, 1000 BE Amsterdam, The Netherlands; H.J.Bouwmeester@uva.nl
[4] LIPM, Université de Toulouse, INRAE, CNRS, 31326 Castanet-Tolosan, France; stephane.munos@inra.fr
* Correspondence: wjzhou@zju.edu.cn; Tel.: +86-571-88982770

Received: 31 August 2020; Accepted: 31 October 2020; Published: 27 November 2020

Abstract: Parasitic plants have a unique heterotrophic lifestyle based on the extraction of water and nutrients from host plants. Some parasitic plant species, particularly those of the family Orobanchaceae, attack crops and cause substantial yield losses. The breeding of resistant crop varieties is an inexpensive way to control parasitic weeds, but often does not provide a long-lasting solution because the parasites rapidly evolve to overcome resistance. Understanding mechanisms underlying naturally occurring parasitic plant resistance is of great interest and could help to develop methods to control parasitic plants. In this review, we describe the virulence mechanisms of parasitic plants and resistance mechanisms in their hosts, focusing on obligate root parasites of the genera *Orobanche* and *Striga*. We noticed that the resistance (R) genes in the host genome often encode proteins with nucleotide-binding and leucine-rich repeat domains (*NLR* proteins), hence we proposed a mechanism by which host plants use *NLR* proteins to activate downstream resistance gene expression. We speculated how parasitic plants and their hosts co-evolved and discussed what drives the evolution of virulence effectors in parasitic plants by considering concepts from similar studies of plant–microbe interaction. Most previous studies have focused on the host rather than the parasite, so we also provided an updated summary of genomic resources for parasitic plants and parasitic genes for further research to test our hypotheses. Finally, we discussed new approaches such as CRISPR/Cas9-mediated genome editing and RNAi silencing that can provide deeper insight into the intriguing life cycle of parasitic plants and could potentially contribute to the development of novel strategies for controlling parasitic weeds, thereby enhancing crop productivity and food security globally.

Keywords: parasitic plant; host; virulence; race; resistance mechanism; pathogen effector; evolution; *NLR*; *Orobanche*; *Striga*; interaction model

---

1. Introduction

Parasitic plants have a unique heterotrophic lifestyle in which they obtain water and nutrients from their hosts via an invasive root-like organ known as haustorium [1]. Parasitic plants occur in all terrestrial plant communities and ~4500 species have been described, distributed over 28 families, representing 1% of all dicotyledonous angiosperm species [2]. These parasites have independently evolved at least 12 or 13 times [3] and unprecedented horizontal gene transfer (HGT) [4] has contributed to their taxonomic and morphological diversity [1]. Some parasitic plant species attack crops and cause severe damage and yield losses, particularly in the Mediterranean, central and eastern Europe, Africa, and Asia [5,6]. Most research has focused on the genera *Orobanche, Striga, Cuscuta,* and *Viscum* (Figure 1).

Figure 1. Some representative parasitic plant species. (**a**) *Triphysaria versicolor*, a hemiparasite, a photosynthetically competent species that, facultatively, parasitizes roots of neighboring plants; (**b**) *Orobanche cumana*, holoparasite, with absolute nutritional dependence on a host, mainly parasitizes roots of sunflower; (**c**) *Cuscuta pentagona*, holoparasite, also known as dodder, that parasitizes aboveground tissues of both monocot and dicot hosts; (**d**) *Striga gesnerioides*, an obligate hemiparasite that mainly parasitizes roots of cowpea.

Striga and *Orobanche* species are especially difficult to control in the field due to their large seed banks and special parasitism traits [6], as well as the economic limitations in developing countries, where these parasites are most prevalent [3]. The life cycles of *Striga* and *Orobanche* species are similar because they coordinate with the life cycle of the host. The essential steps are germination, radicle growth to the host root, haustorium formation and attachment to the host root, establishment of a xylem–xylem connection, and the production of seeds [7,8]. The host–parasite interaction begins with the secretion of chemical signals by the host roots that induce the germination of parasite seeds and are called germination stimulants [9]. Accordingly, the inhibition of parasite seed germination is a primary target for parasitic weed control [10]. Almost all germination stimulants discovered thus far belong to the carotenoid-derived strigolactone (SL) family [11]. Recent studies have shown that the breeding of crops showing limited exudation of SLs from the root is an effective strategy to achieve resistance to *Orobanche* and *Striga* [12–14]. We, therefore, discussed low SL levels as a natural resistance mechanism in host plants as well as biotechnological strategies to induce this trait. Other practical methods to control parasitic plants have been extensively reviewed but are often unsuccessful in the long term because the parasite evolves faster than the resistant host, leading to the emergence of distinct races or pathotypes with renewed virulence [6,10,15–18].

The existence of host-specific races suggests that parasites have evolved complex mechanisms to overcome potential host resistance, but most reviews overlook this aspect and focus on the host's resistance mechanisms. Here, we considered recent examples of virulence and race evolution in parasitic plants before looking at host resistance mechanisms in the context of canonical resistance genes (*R* genes) encoding proteins with nucleotide-binding and leucine-rich repeat (*LRR*) domains, often termed "*NLR* proteins". We used these to develop a plausible model explaining the molecular basis of host–parasite interactions. We also summarized current genomic resources for parasitic plants and discussed the functions of known virulence genes and their roles in the evolution of host-specific races of parasitic plants.

2. Virulence Evolution in the Family Orobanchaceae

2.1. Definition of Race in Parasitic Plants

In biological taxonomy, race is an informal rank below the level of subspecies that may be defined according to any identifiable characteristic (e.g., chromosomal race, geographical race, or physiological race), but the differences are relative rather than absolute. When we talk about race in the context of parasitic plants it usually refers to physiological race, which means a group of individuals that do not necessarily differ in morphology from other members of the species but have distinct physiology or behavior. In parasitic plants, a race signifies a genotype that has the capacity to parasitize on a certain genotype of host plant. For example, *Orobanche cumana* (*O. cumana*) races are classified according to the resistance/susceptibility of a set of sunflower lines carrying different resistances' genes. A new nomenclature, similar to the one used for downy mildew pathogens, was proposed [19]. A physiological race may be an ecotype (subgroup of a species that has adapted to a different local habitat), perhaps defined by a specific food source. Parasitic plant species tied to no geographic location often have races that are adapted to different hosts, but these are, so far, at least difficult to distinguish genetically.

2.2. History of Race in Parasitic Plants

The family Orobanchaceae is part of the order Lamiales, which comprises annual herbs as well as perennial herbs and shrubs. With the exception of the nonparasitic genera *Lindenbergia*, *Rehmannia*, and *Triaenophora*, members of the Orobanchaceae parasitize the roots of other plants and display all known types of plant parasitism: facultative parasitism, obligate parasitism, hemiparasitism, and holoparasitism. *Striga* and *Orobanche* are widely studied because of their impact on agriculture. For example, *O. cumana* (sunflower broomrape) causes yield losses of up to 80% [20–22]. In sub-Saharan Africa, up to 60% of the arable land used to cultivate cereals and grain legumes is infested with one or more *Striga* species [23]. Within parasitic plant species, races can be distinguished. For example, seven races of *Striga gesnerioides* parasitizing cowpea (*Vigna unguiculata*) have been identified [24] and eight races of *O. cumana* parasitizing sunflower (*Helianthus annus*) [25].

The race evolution history of *O. cumana* has been studied on sunflower and wild species of the Asteraceae, mainly *Artemisia maritime* (sea wormwood) [26]. A mature *O. cumana* plant can produce 50,000–500,000 dust-like seeds, which remain viable in soil for up to a decade [15]. The long viability of *O. cumana* seeds limits sunflower production in contaminated fields, mainly in Eastern Europe and Asia, and is found in Spain, France, Turkey, Russia, Ukraine, Israel, Kazakhstan, and China. The virulence or pathogenicity of *O. cumana* has evolved rapidly with the increasing global production of sunflower, especially in Russia, eastern Europe, and Asia since the 1920s [19]. *O. cumana* races were first discussed by local sunflower breeders in Russia in 1920 [27]: race A in the Saratov and Voronezh regions did not attack local sunflower crops, whereas race B in the Rostov and Krasnodar regions was highly virulent against the same sunflower variety. Since then, *O. cumana* has evolved quickly from race A to race H and has parasitized local sunflower varieties for 100 years [19,27].

The dispersion of *O. cumana* races has been systematically reviewed [19]. However, there is no worldwide consensus on the sunflower lines used to identify *O. cumana* races because different countries and regions favor distinct sunflower lines or hybrids in local breeding practices for identification purposes. Therefore, it is difficult to compare results from experiments carried out in different parts of the world for the characterization of *O. cumana* race structure. Race identification is usually carried out by counting emerged shoots of *O. cumana* on distinct sunflower lines or hybrids.

Figure 2 represents the worldwide occurrence of race structure of *O. cumana*. The races E, F, and G are formerly the most commonly reported all over the world, but races F, G, and H are now the most prevalent in Spain and in several countries around the Black Sea, whereas races A, D, E, F, and G are the most prevalent in China [28,29]. A new race (G_{KE}) responsible for ~80% yield losses was identified in Morocco in 2016 [30]. In 2017, race G was found for the first time in Portugal [31] (Figure 2).

Figure 2. Current distribution and virulence level of *Orobanche cumana* in the world. Distribution range of *O. cumana* in the worldwide regions updated to December 2019. Although different levels of virulence are not comparable horizontally, it is notable that the parasitic plant *O. cumana* evolves fast with the increasing production of sunflower. Letters in the figure refer to distinct virulence levels. Race A: China; Race B: None; Race C: Kazakhstan; Race D: China and Russia; Race E: Bulgaria, China, France, Hungary, Moldova, Romania, Russia, Serbia, Spain, and Ukraine; Race F: Bulgaria, China, France, Hungary, Moldova, Romania, Russia, Spain, Turkey, and Ukraine; Race G: Bulgaria, China, Kazakhstan, Romania, Russia, Spain, Turkey, Ukraine, Morocco, and Portugal; and Race H: Russia.

3. The Mechanisms of Virulence-Specific Resistance in Host Plants

3.1. Phenotypic Aspects

The race or pathotype of parasitic plants is often determined by quantification of the infection level. The sunflower lines 2603 and P-96 are typically used as controls because the first is susceptible and the last is resistant to *O. cumana* race F [20]. Rhizotrons, pots, and field experiments were used to characterize all sunflower recombinant inbred lines for resistance to *O. cumana* race F at three life stages: (1) early attachment of the parasite to the sunflower roots, (2) young tubercle, and (3) shoot emergence [20]. This showed that the number of healthy tubercles at stage 3 is the trait best correlating with the number of emerged broomrape shoots in the field. Other researchers have counted the necrotic tubercles (post-haustorial/secondary resistances) in the resistant line or the number of successfully established radicles allowing the development of root tubercles on the susceptible line [32–34]. Another strategy to determine the successful infection by parasitic plants is to measure host plant parameters such as height, weight/biomass, photosynthesis, leaf CO_2 assimilation rates, transpiration rate, stomatal conductance, vapor pressure deficit, and leaf carbon, nitrogen, potassium, phosphorus, and magnesium levels [35–37]. For example, the effects of *Striga* on susceptible rice (*Oryza sativa*) genotypes included 30–65% stunting of the main stem and the inhibition of photosynthesis and CO_2 assimilation in 30-day-old plants (and even more profoundly in 45-day-old plants), whereas these effects were not evident in resistant genotypes, resulting in high grain yields in the field [35]. Interestingly, the comparison of susceptible and resistant sunflower varieties in response to *O. cumana* revealed no physiological differences between the infected and non-infected cohorts of the resistant cultivar (cv) after 23–51 days of planting, including photosynthesis, transpiration rate, stomatal conductance, vapor pressure deficit, nonphotochemical quenching, and chlorophyll levels [36]. However, significant differences were found in the levels of the macro-elements potassium, phosphorus, magnesium, and sulfur between the infected and non-infected plants during the early stage of parasite development. The mineral and carbon content were higher in the broomrape infected sunflower, as compared to the non-infected ones after 31 days of planting [35]. Sunflower leaf nitrogen content, however, was 42% lower in broomrape-infected plants after 56 days of planting, which can be explained by the reduction in mesophyll cells per area leaf and a delay in leaf senescence [36].

Resistance that occurs in multiple layers is often associated with the accumulation of compounds such as H_2O_2, peroxidases, β-glucanase, and callose in the case of *Orobanchecrenata* vs. pea (*Pisum sativum*), or 7-hydroxylated simple coumarins in the case of *Orobanche cernua* vs. sunflower [38]. To reinforce the cell wall, resistant plants deposit lignin in the endodermis and pericycle cells at the penetration site, as seen in the cases of *O. crenata* vs. vetch (*Vicia* spp.), faba bean (*Vicia faba*), pea, chickpea (*Cicer arietinum*) and lentil (*Lens esculenta*), *O. cumana* vs. sunflower, and *Striga hermonthica* vs. rice [39–41]. Genetic analysis has also confirmed that lignification and secondary wall formation promote resistance during incompatible interactions between cowpea and *S. gesnerioides* [42] and between rice and *S. hermonthica* [41].

3.2. Histological Aspects

To ward off infection by parasitic plants, host plants can deploy several defense mechanisms. The first line of defense is a physical barrier (the cuticle and cell wall) supported by the constitutive production of metabolites that deter the invader. Successful penetration of the root cell layers and the establishment of host–parasite vascular connections are necessary for *Orobanche* and *Striga* to survive [8]. The host can, therefore, block parasite development at the epidermis, in the cortex, at the endodermis, and inside the central cylinder [23,43,44] (Figure 3). In rice cultivar Nipponbare, which shows strong resistance to *S. hermonthica*, parasite development is inhibited at the cortex, suggesting that the host blocks signaling pathways required for the parasite to penetrate between endodermal cells [44,45] (Figure 4). In sunflower, the major resistance gene *HaOr7* prevents the connection of *O. cumana* to the sunflower root vascular system [46]

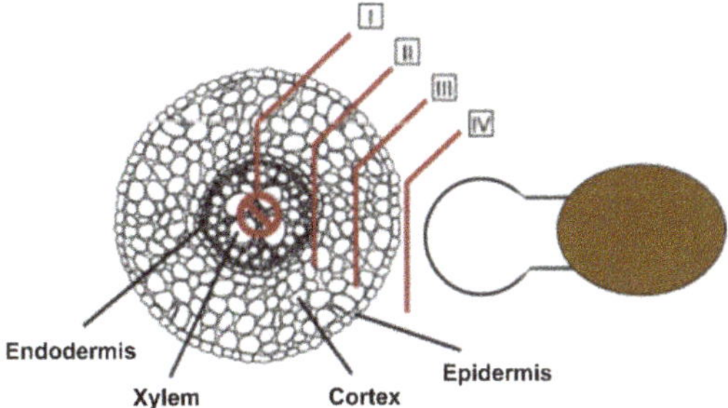

Figure 3. Schematic illustration of the different root cell layers where resistances cans occur. Four layers of incompatibility described in the text are presented. (**I**) Incompatibility expressed after vascular connection, which was observed in *Arabidopsis*, cowpea, and in rice cultivar Nipponbare. (**II**) Endodermis blockage, which is observed in rice cultivar Nipponbare as well as in cultivar Koshihikari. (**III**) Mechanical barrier in the root cortex, observed in interaction with *Lotus japonicus* and occasionally with *Phtheirospermum japonicum*. (**IV**) Incompatibility preventing attachment, observed in interaction with *P. japonicum*. Parasite plant (right) and host plants (left, shown as a transverse section of a root). Adapted from Yoshida et al. [44].

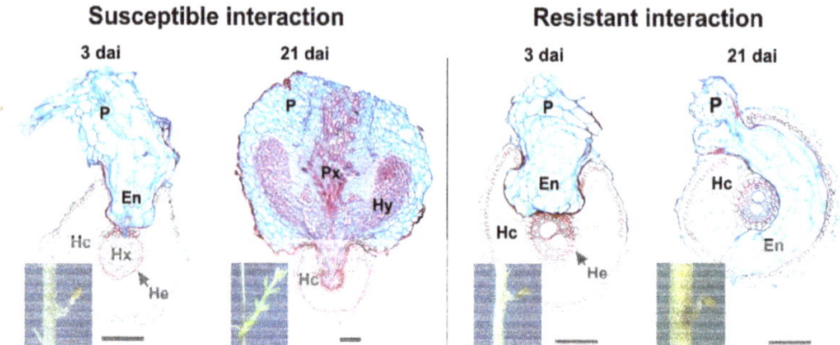

Figure 4. Host resistance to parasite establishment. Transverse sections of embedded tissue of susceptible (cv. Kasalath) and resistant (cv. Nipponbare) rice roots 3 and 21 days after inoculation with *Striga hermonthica*. In the susceptible interaction, the parasite penetrates the cortex and endodermis and connects to the xylem vessels of the host, allowing the haustorium to differentiate. In contrast, in the resistant interaction, although the parasite penetrates the cortex, it is unable to breach the endodermal barrier and grows around the host vascular cylinder. The parasite is unable to access host water and nutrients and the haustorium does not differentiate and the parasite dies. The scale bar represents 0.1 mm. En, endophyte (internal part of haustorium); Hc, host root cortex; He, host endodermis; Hx, host xylem; Hx–Px, host–parasite xylem continuity; Hy, hyaline body; P, parasite haustorium; and Px, parasite xylem vessels. Adapted from Gurney et al. [45].

3.3. Chemical Aspects

The life cycle of parasitic plants begins when seed germination is promoted by root exudates from a presumptive host. The inhibition of seed germination is, therefore, a key target for parasitic weed management [47] and this can be achieved by reducing the amount of SLs exuded by host roots [12]. Early studies identified host germplasm that produces lower levels of *Orobanche* germination stimulants in root exudates, yielding resistant varieties of pea, chickpea and *Lathyrus* spp. [32,48–51], faba bean [52], and sunflower [53–55]. Similarly, host germplasm that produces lower levels of *Striga* germination stimulants were identified in sorghum (*Sorghum bicolor*) [11] and maize (*Zea mays*) [56], and such traits have already been used to breed resistant sorghum varieties [57].

Initially, the nature of the stimulant was unclear. To collect sunflower root exudates, seedlings were transferred to sterile distilled water for 1–5 weeks after germination and cultivated for three days before preconditioned *O. cumana* seeds were incubated in the root exudate solution [54,55]. In these experiments, the germination of *O. cumana* seeds was stimulated using one ppm GR24, an artificial SL analog. The exudates of different sunflower genotypes had different effects on the broomrape seedlings, indicating that chemicals in the exudates have an effect on *O. cumana*. The development of high-performance liquid chromatography connected to tandem mass spectrometry allows the identification of specific compounds that act as germination stimulants, revealing that the abundance of SLs in pea root exudate correlates with resistance [14]. Most germination stimulants discovered thus far are SLs, and this suggests that resistance may involve the reduced secretion of SLs in root exudates [58].

SLs were first found in cotton (*Gossypium* spp.) root exudates as a potent germination stimulant of *Striga lutea* [59]. Subsequent research revealed that SLs stimulate hyphal branching of arbuscular mycorrhizal fungi [60,61], but also function as endogenous hormones to inhibit shoot branching or tillering [62,63]. SL biosynthesis begins with all-trans/9-cis β-carotene isomerase (*DWARF27* or *D27*), which converts all-trans-β-carotene to 9-cis-β-carotene [64,65]. In the next step, carotenoid cleavage dioxygenase 7 (*CCD7*) cleaves 9-cis-β-carotene into the volatile β-ionone and 9-cis-β-apo-10′-carotenal [64,66]. This cis-configured intermediate is the substrate for *CCD8*, which catalyzes a combination of reactions including repeated dioxygenation and intramolecular rearrangements to yield carlactone and the C_8-product ω-OH-(4-CH_3) heptanal [64,67].

Carlactone is the precursor of all SLs, although the later reaction steps are not completely understood [68,69]. In *Arabidopsis thaliana*, carlactone is converted into carlactonoic acid by the cytochrome P450 monooxygenase *MORE AXILLARY GROWTH 1 (MAX1)*, followed by methylation by an unknown enzyme and hydroxylation by lateral branching oxidoreductase into a yet unidentified SL [70,71]. In rice, *MAX1* homologs convert carlactone into 4-deoxyorobanchol and orobanchol [72,73]. The biosynthesis and transport of SLs was reviewed in detail [69].

It is currently unclear whether the germination of parasitic plants involves additive, synergistic, and/or antagonistic effects in response to the usually multiple germination stimulants produced by their hosts. There is evidence that germination stimulants are involved in species-dependent and race-specific effects because the synthetic SL GR24 induces germination up to ~70–90% in *Orobanche ramosa*, *O. cumana*, and *O. minor*, but only up to ~50% in *S. hermonthica* [74,75]. In vitro bioassays with *O. minor* seeds revealed that the activity of 1 µM heliolactone is similar to that of GR24 after 24–72 h (75% germination) but only ~25% of *O. cumana* seeds germinated, suggesting additional compounds may be required for the latter species [76–80]. Moreover, additive or antagonistic effects were observed between two SLs [81]. For example, strigol and orobanchol together germinate 24% of *O. cumana* seeds, while orobanchol alone induces 64% germination of *O. cumana* [81].

More evidence of host-dependent parasite germination comes from the analysis of SL biosynthesis in different hosts. The first step, the formation of carlactone, is common to all hosts, but the next steps are species-dependent. In many species carlactone is converted to carlactonoic acid, for example, carlactonoic acid is converted into methyl carlactonoate and then converted into heliolactone in sunflower [72], while in rice carlactonoic acid is converted to 4-deoxyorobanchol [73]. Carlactonoic acid is the precursor for strigol in moonseed, for sorgomol in sorghum, and for strigol in cotton [72,82].

The analysis of SLs in autotrophic plants such as *Arabidopsis* has revealed that the receptor for endogenous strigolactones in nonparasitic plants is encoded by *DWARF 14 (D14)*, while the strigolactone receptor in root parasitic plants, responsible for the detection of host strigolactones, is encoded by *HYPOSENSITIVE TO LIGHT/KARRIKIN INSENSITIVE2 (HTL/KAI2)* [83–86]. Intriguingly, genes encoding strigolactone biosynthetic enzymes as well as the receptor D14 have also been identified in parasitic plants [87]. Several *HTL/KAI2* strigolactone receptors from parasitic plants have been functionally characterized [88–92]. For example, *ShHTL7* (*Striga hermonthica* HTL protein) is found to be a very sensitive SL receptor that binds with several natural strigolactones [88].

3.4. Genetic and Genomic Aspects

3.4.1. R Genes against *Orobanche cumana*

Genetic studies related to the virulence/race of *O. cumana* have been poorly described. However, an avirulence gene interacting with *Or5* resistance gene in sunflower (see below) was characterized [93] and was mapped [94]. The *O. cumana* genome sequence will help to identify the avirulence genes [95]. The sunflower genes *Or1*, *Or2*, *Or3*, *Or4*, *Or5*, and *Or6* confer resistance to *O. cumana* races A, B, C, D, E, and F, respectively, and are inherited as single dominant alleles [96–98]. Resistance to race F has also been associated with two recessive alleles [91], two partially dominant alleles [99], and multiple quantitative trait loci (QTLs) [19]. Preliminary results suggest that resistance to race G may be conferred by a single dominant allele [100] or a single recessive allele [101]. However, all these studies used traditional crosses to determine whether the resistance gene is transmitted in a dominant or recessive manner, with no indication of the candidate genes or their potential functions. The sequencing of the sunflower genome should help to identify resistance genes [102].

The genomic location of an *O. cumana* resistance gene in sunflower was recently verified by bulk segregant analysis combined with genotyping-by-sequencing technology. Two major QTLs associated with resistance were resolved to sunflower chromosome 3 (*or3.1* and *or3.2*) and the former maps to the same region as *Or5* (conferring resistance to race E) whereas the latter is associated with markers of resistance to race G. Exploration of the first region (31.9–38.48 Mb) revealed 123 candidate genes, including a known disease resistance gene (*HanXRQChr03g0065841*) encoding an oxygen-dependent choline dehydrogenase and *FAD/NAD(P)*-binding domain [103]. The second region (97.13–100.85 Mb) contained 71 candidate genes, including one with an *NLR* domain (*HanXRQChr03g0076321*) that is often found in resistance gene (*R*) products [104], such as the *Arabidopsis* R proteins *RPM1* and *RPS5* [105]. The sunflower orthologs of *PRM1* and *PRS5* are both induced in the *O. cumana*-resistant cultivar JY207 following inoculation with the parasite, but there is no change (or even a slight fall) in the susceptible cultivar TK0409 compared to non-inoculated controls [40]. The *HaOr7* resistance gene to race F from Spain was identified by a map-based cloning approach and encodes a receptor-like kinase [46].

3.4.2. R Genes against *Striga gesnerioides*

Another well-studied example of race evolution in parasitic plants is *S. gesnerioides* (Figure 1d), which has multiple races that significantly affect cowpea production in sub-Saharan Africa [106]. Crossing and backcrossing experiments among resistant and susceptible cowpea cultivars indicated a monogenic resistance locus with a dominant inheritance pattern [107–109]. This led to the proposed designations of *Rsg1*, *Rsg2*, and *Rsg3* for the genes present in cowpea cultivars B 301, IT82D 849, and SUVITA-2, respectively [18]. Later studies in cowpea identified amplified fragment length polymorphism markers tightly linked to different race-specific *S. gesnerioides* resistance genes [110] and the microsatellite/simple sequence repeat marker SSR-1 co-segregating with *S. gesnerioides* race 3 (SG3) resistance [111], which was ultimately identified in a cowpea gene-space sequence read [112]. The gene was named *RSG3-301* (resistance to *S. gesnerioides* race 3 in cowpea cultivar B301) and was shown to encode an R protein with an *NLR* domain [113]. When *RSG3-301* expression is knocked down by virus-induced gene silencing (VIGS) in the multirace-resistant cowpea cultivar B301, *S. gesnerioides* can invade the endodermis and establish xylem–xylem connections with the host vascular system [113], suggesting that *S. gesnerioides* may interfere with the regulation of *NLR* proteins to overcome host plant defenses.

Recently, a transcriptome study focusing directly on parasitic plants rather than their hosts has revealed that candidate haustorium-specific genes in *T. versicolor* and *S. hermonthica* are significantly enriched for aspartyl protease, peroxidase, and *NLR* protein domains [114]. Moreover, a novel decoy effector *SHR4z* was recently identified from the haustorium of *S. gesnerioides* that can suppress the hypersensitive response in host cowpea plants to boost parasite growth. *SHR4z* has significant homology to the short *LRR* domain of somatic embryogenesis receptor kinase (*SERK*) proteins and functions by binding to *VuPOB1*, a positive regulator of the hypersensitive response [115].

3.4.3. Virulence Genes in Parasitic Plants

Parasitic plants are more complex organisms than microbes and pathogens. Parasitic plants could possess specific proteins involved in virulence and they must be considered as pests because they also induce diseases in a host plant [116]. Insight into the distinction between virulence genes (pathogen effectors) and host resistance genes could be gained by genome annotation, transcriptome sequencing, and the functional classification of single nucleotide polymorphisms to determine the roles of specific gene families. The *O. cumana* genome encodes 221 proteins with an *LRR* domain [102]. These genes are also annotated according to the presence of other domains (e.g., L domain, FBD domain, or F-box domain) and according to predicted molecular and cellular functions (e.g., ATP binding, cell wall organization, or oxidoreductase activity). Combined with the *S. gesnerioides* transcriptome analysis discussed above, showing that haustorium genes are also enriched for *LRR* domains [114], we can begin to see the outline of a process in which parasitic plants overcome host resistance by targeting components of signal transduction pathways activated by the *R* genes containing LRR domains to block defense response cascades directly or indirectly. Beside, *LRR* domains, a secretome analysis of

Striga hermonthica revealed a large number of cysteine-rich small proteins associated with protease and cell wall modification activities was also involved in *S. hermonthica*–host plant interaction [117]. To counter the virulence of parasitic plants, hosts also detect and respond to molecular signals secreted by parasitic plant. For example, a surface receptor *Cuscuta Receptor 1* (*CuRe1*) was also identified in tomato plants that responded to *Cuscuta* spp. peptide factor and activates immune response and identify parasitic plants in a manner similar to perception of microbial pathogens [118].

Recently, a high-throughput silencing approach was developed to study *NLR* proteins in *Nicotiana benthamiana* in which 257 VIGS constructs based on tobacco rattle virus were used to target 386 of the 403 identified proteins, providing an efficient strategy to discover new immune receptors [119]. Agrobacterium-mediated transformation, together with transcriptome analysis of differentially expressed genes in *S. gesnerioides*, was used to dissect the involvement of resistance cascades in cowpea that is being attacked by this parasite [115]. Transcriptome assembly to identify genes in *Striga* and *Orobanche* was used to investigate the involvement of virulence proteins on a genome-wide scale [120–122].

3.4.4. Genome and Transcriptome of Parasitic Plants

Genomics, transcriptomics, proteomics, bioinformatics, biochemistry, and cell biology have all played major roles in the identification and functional characterization of pathogen and host proteins involved in plant–pathogen interactions. One of the key challenges when applying such methods to parasitic plants is the need to extract pure nucleic acids or proteins from the parasite during infection, without contaminating host material. The distinction between host and parasite is complicated by the extensive mutual HGT and high substitution rates in parasitic plant genomes, and this also makes it more difficult to construct accurate phylogenetic trees [116,123,124]. New dating approaches have been applied to solve the problem of long branch lengths in gene tree analysis, allowing the absolute divergence time of parasitic plants to be determined more precisely [123]. For example, Kim et al. [125] developed a protocol to study the movement of parasitic mRNA into the host plant.

Thus far, 25 parasitic plant genomes have been sequenced and assembled, including nuclear, plastome, and mitochondriome sequences. Ten nuclear genome sequences have been reported, namely, the holoparasite *Cynomorium* [126], *Viscum scurruloideum* [4], *Viscum album* [4], *Castilleja paramensis* [126], *Cuscuta campestris* [127], *Cuscuta australis* [128], *Hydnora visseri* [129], *Aphyllon epigalium* [130], *Striga asisatica* [131], and *O. cumana* [95]. These resources could promote the analysis of candidate parasite-specific genes and provide evidence for host-to-parasite, parasite-to-host, and bidirectional [116,124,132–134]. For example, Illumina sequencing of the *Cynomorium* plastid and mitochondrial genomes revealed that the plastome contigs assembled into inverted repeat (IR) regions and a large single copy region. All genes involved in photosynthesis (*ndh*, *atp*, *pet*, *psa*, *psb*, and *rbcL*) have been lost as anticipated, but the predominant IR region contains genes encoding an ATP-dependent Clp protease proteolytic subunit (clpP) and metal-resistance protein (*YCF1*) [130]. Evidence of HGT was presented showing the transfer of host mitochondrial genes into the *Cynomorium* mitochondrial and nuclear genomes and the intracellular transfer of *Cynomorium* mitochondrial and plastid genes into the nuclear genome [135].

Transcriptome sequencing allows the functional analysis of parasitic plant genomes, and the Parasitic Plant Genome Project (Available online: http://ppgp.huck.psu.edu/) mainly focuses on the identification of genes related to haustorium initiation and development by applying comparative transcriptomics to multiple stages of parasite growth and development in three species of Orobanchaceae: *T. versicolor* (a facultative hemiparasite) (Figure 1a), *S. hermonthica* (an obligate hemiparasite), and *P. aegyptiaca* (an obligate holoparasite) [3,120]. A core set of "parasitism genes" was identified that are enriched for proteases, cell wall-modifying enzymes, and proteins secreted during haustorium development. Genes encoding transporters (cationic amino acid transporter, major facilitator family protein, NOD26-like intrinsic protein, and an oligopeptide transporter) and regulatory proteins (transcription factors and receptor protein kinases) are co-expressed during the parasitic stages and may be required for haustorium development and function [114]. *NLR* resistance genes are found in all three species and are significantly enriched in *T. versicolor* and *S. hermonthica*, suggesting an underlying important function that may facilitate the future analysis of race/virulence in parasitic plants.

The expression levels of the parasitism genes differed between the hemiparasites and holoparasite at the haustorium stage. The genes encoding cell wall-modifying enzymes (cellulase, Pectate Lyases, glycosyl hydrolases, and pectin methylesterase) and peroxidases were strongly expressed in both hemiparasites (*T. versicolor* and *S. hermonthica*) but not in *P. aegyptiaca*, although expression increased at a later stage of the life cycle [114]. There were also differences in expression between the facultative parasite (*T. versicolor*) and the obligate parasites. In *T. versicolor*, the haustorium initiation genes were primarily Ca^{2+} ATPases, including genes coding for proteins with functions such as Ca^{2+}-binding activity, Ca^{2+}-transporting ATPase activity, Ca^{2+} transmembrane transporter activity, and cation-transporting ATPase activity. In contrast, the *S. hermonthica* haustorium initiation genes were enriched for a distinct set of gene ontology terms, including nucleotide binding and ATP-dependent helicase activity, suggesting that facultative and obligate parasitic plants used different underlying processes for haustorium initiation [114].

Transcriptomic studies have also shown that, once connections between the parasite and host plant are established, the host–parasite relationship relies on the transfer of nutrients and solutes from host to parasite via multiple transporters, including amino acid and sugar transporters [114]. For example, the transcriptomic analysis of *S. hermonthica* infected leaves and flower buds during the parasitism of maize and sorghum hosts identified transporters (primarily carbohydrate and amino acid transporters) as the most common functional class of parasitism genes, followed by cell wall-modifying enzymes [136]. Similarly, the de novo assembly and characterization of the *S. gesnerioides* transcriptome during the pre-haustorium and haustorium stages of infection (Figure 1c) revealed the strong induction of genes encoding cell wall-modifying enzymes and transporters, including sugar transporters, amino acid transporters, ATP-binding cassette-type transporters, ammonium transporters, phosphate transporters, nitrate transporters, and potassium transporters [136].

All candidate haustorium genes are valuable resources for future functional and evolutionary studies, which will help to determine whether they are secreted by the parasite and whether they influence parasite–host interactions. For example, upregulated haustorium genes that encode subtilisin-like serine proteases [114] are similar to those acting as virulence factors in bacterial pathogens [137]. However, serine proteases are often involved in protein degradation and processing, the hypersensitive response, and signal transduction in nonparasitic plants, so their specific role in the parasitic life cycle has yet to be determined [138,139]. Indeed, numerous questions remain concerning the functional role of core parasitism genes identified by genomic and transcriptomic studies. The three species of Orobanchaceae considered by the Parasitic Plant Genome Project feature 84 orthologous groups with no BLAST (basic local alignment search tool) hits against annotated genes in nonparasitic species (178, 180, and 139 unique genes in *T. versicolor*, *S. hermonthica*, and *P. aegyptiaca*, respectively) although a small number match predicted protein sequences in nonparasitic plants but the functions are currently unknown (6, 18, and 13 sequences in *T. versicolor*, *S. hermonthica*, and *P. aegyptiaca*, respectively) [108]. Three focal transcripts of *S. hermonthica* generate no BLASTx hits at all [136]. Genes of unknown function are also found in the *Cynomorium plastome* [135].

4. Models of Interaction and Co-Evolution between Parasitic Plants and Their Hosts

R genes in plants play an important role in the recognition of pathogen virulence factors, which is required to induce resistance. They typically show dominant phenotypes, but recessive resistance genes have also been reported. As discussed above, most *R* gene products contain a nucleotide-binding ATPase domain and an *LRR* domain (Figure 5) and are, thus, described as *NLR* proteins [104]. The *LRR* domain includes individual repeats that recognize pathogen proteins [140]. Several *NLR* proteins have been described, including *MLA10*, *Sr50*, *RPP13*, *RPS4*, *RPS5*, *ZAR1*, and *L6* [141]. This has led to the definition of two subclasses, namely, the Toll and interleukin-1 receptor subclass (TNL) [142] and the coiled coil subclass (CNL). Structural models of both have been constructed based on the Arabidopsis proteins RPS4 (ribosomal protein s4) (TNL) and RPS5 (CNL) using self-consistent mean-field homology modeling in the absence of ADP. This ligand was then added by inference from the *APAF-1–ADP* complex without further refinement of the models to illustrate the position of the nucleotide relative to the conserved motifs [143,144]. The Toll and interleukin-1 receptor, coiled coil, and LRR protein interaction domains have the ability to swap functions, such as the recognition and recruitment of transcription factors or other host proteins [104]. *RPS5* is normally activated when a second host protein (*PBS1*) is cleaved by the pathogen-secreted protease *AvrPphB*. The *AvrPphB* cleavage site within *PBS1* can be replaced with cleavage sites for other pathogen proteases, which then enables *RPS5* to be activated by these proteases, thereby conferring resistance against new pathogens [105].

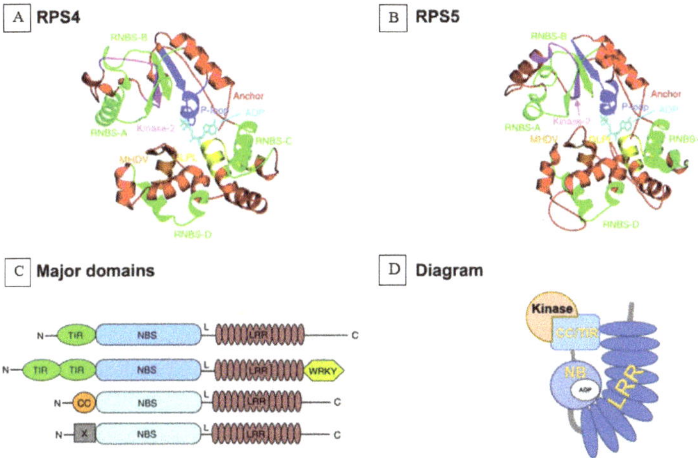

Figure 5. Predicted structures and models of nucleotide-binding site domain and a leucine-rich repeat domain (NB-LLR). (**A**) The structures of the NBS domains (nucleotide-binding site domain and a leucine-rich repeat domain) of TNL and *RPS4*. The protein structures are shown as ribbon diagrams and ADP (adenosine diphosphate) is shown as a stick model. TIR-type and CC-type *NBS* domains are made up of motifs: P-loop (or Walker A site, blue), *RNBS-A* (green), kinase-2 (or Walker B site, magenta), *RNBS-B* (green), *RNBS-C* (green), *GLPL* (yellow), *RNBS-D* (green), and *MHDV* (orange) [131]. Structural models for the NBS domain of *TNL RPS4* and CNL RPS5 of *Arabidopsis* were created by self-consistent mean-field homology modeling technique [132]. (**B**) The structures of the NBS domains of *CNL RPS5*; (**C**) major domains of NBL proteins. N, amino terminus; TIR, Toll/interleukin-1 receptor-like domain; CC, coiled-coil domain; X, domain without obvious CC motif; NBS, nucleotide binding site; L, linker; LRR, leucine-rich repeat domain; WRKY, zinc-finger transcription factor-related domain containing the WRKY sequence; C, carboxyl terminus; (**D**) Diagram of *NLR*.

4.1. Model of Defense Activated by Host NLR Proteins Triggered by the Parasitic Plant

The induction of defense responses by *NLR* proteins proceeds in three stages, as shown in Figure 6. The *LRR* region is an inhibitory domain, associating with the nucleotide-binding domain when there is no infection. The *N*-terminal coiled coil domain associates with a protein kinase such as PBS1 [104,105] (not shown in the model). In the first stage, the inactive *NLR* receptor (blue) perceives specific virulence proteins (pathogen effectors, shown in brown) secreted from the parasitic plant and binds to them (Figure 6, step 1). *NLR* proteins are highly specific, with each *NLR* protein capable of detecting only a limited number of effectors [105]. In some cases, the kinase (also described as a host factor) takes part in indirect recognition, enabling the *NLR* protein's N-terminal domain to bind pathogen effectors via an intermediary kinase. In the second stage, the *NLR* receptor is activated by a conformation change and ATP binding to the nucleotide-binding domain, relieving the latter from LRR repression. The exchange of ADP for ATP at the nucleotide-binding domain may generate an activated, ATP-bound form of *NLR* [133]. Recently, a highly conserved nucleotide-binding domain shared by APAF-1, various R proteins, and *CED-4* (the NB-ARC domain) was proposed to act as a molecular switch, cycling between ADP binding (repressed) and ATP binding (active) [145] (Figure 6, step 2). Finally, the activated *NLR* protein translocates to the nucleus to induce defense-related gene expression and corresponding signaling pathways (Figure 6, step 3). In the presence of the pathogen effector, the activated form of an *NLR* accumulates in the nucleus to initiate defense signaling.

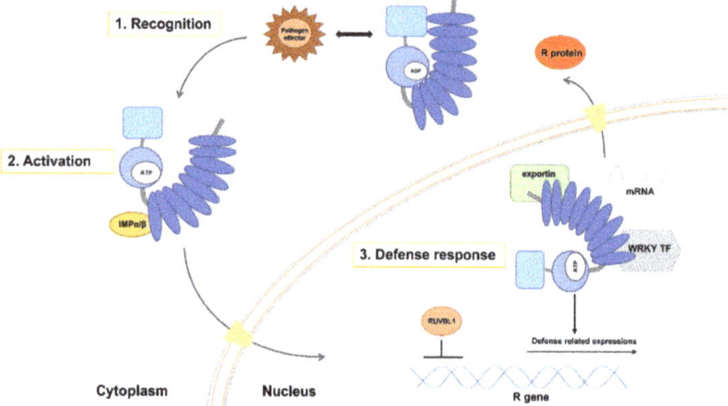

Figure 6. Putative model for defense activation by NLRs in host plant triggered by parasitic plant. Induction of defense responses by NB-LRRs (nucleotide-binding site domain and a leucine-rich repeat domain) proceeds in three stages. In some cases, kinase will take part in indirect recognition. In the first stage, the inactive *NLR* receptor (blue) perceives the presence of specific virulence proteins, called pathogen effector (brown), secreted from parasitic plant, then binds with pathogen effector. In some cases, kinase (also referred as host factor) will take part in indirect recognition. *NB-LRR* could indirectly recognize pathogen effector through *N*-terminal domain (CC or TIR) by an intermediary kinase. In the second stage, NB-LRR receptor is activated by a conformation change and ATP binding to NB domain. A highly conserved nucleotide-binding domain that is shared with apoptotic protease activating factor 1 (*APAF-1*), various R-proteins, and *CED-4* (NB-ARC domain) is proposed to act as a molecular switch, cycling between ADP (repressed) and ATP (active) bound forms [132]. In the third stage, activated NB-LRR work in the nucleus to induce defense-related signaling and gene expression. *NLR* negative regulators of defense such as (*TIP49a*) transcription factor (TF) is inhibited. Alternatively, WRKY transcription factor (TF) may bind to *NLR* to positively regulate and induce defense expression. Chimeric proteins comprise domains typical for both intracellular type-R proteins (NBS–LRR proteins) and WRKY transcription factors [146], suggesting that WRKY TF binds to *NLR* closely. To cross the nuclear pore, *NLRs* with a classical nuclear localization signal will require importin-α and importin-β (light yellow) for import and export [104]. Last, specific defense-related mRNAs or proteins are exported through nuclear pore. *IMPα/β*: importin-α/β; R genes: *RPS5*.

Defense responses include a localized hypersensitive response that serves to prevent spread of infection by triggering cell death. *NLR* negative regulators of defense are inhibited during this response, including the transcription factor *RUVBL1* (*TIP49a*). Alternatively, WRKY transcription factors may bind to *NLR* to induce defense gene expression. *RUVBL1* (*TIP49a*) is a member of the AAA+ ATPase family (ATPases associated with various cellular activities), and *Arabidopsis TIP49a* (*RUVBL1*) can act as a negative regulator of some *R* gene functions [147]. WRKY transcription factors contain a highly conserved, ~60 amino acid domain featuring the consensus sequence *WRKYGQK* and a zinc-finger motif [148,149]. WRKY transcription factors recognize the *cis*-regulatory element (T/A)TGAC(T/A), also known as the W-box, in the promoters of target genes [150,151]. Certain soybean (*Glycine max*) WRKY genes (*GmWRKY154, GmWRKY62, GmWRKY36, GmWRKY28,* and *GmWRKY5*) promote resistance to the soybean cyst nematode (*Heterodera glycines*) [152]. The presence of chimeric proteins featuring the domains of both intracellular R proteins (*NLR* proteins) and WRKY transcription factors suggests that these protein families work closely together [148]. To cross the nuclear pore, *NLRs* with a classical nuclear localization signal require importin-α and importin-β (light yellow) for import and export, respectively [104]. Finally, defense-related mRNAs or proteins are exported through the nuclear pore.

4.2. Model of Antagonistic Host–Parasite Co-Evolution

A remarkable consensus has emerged concerning the genetic basis of virulence and resistance in typical interactions between plants and microbial parasites, and we can build a similar hypothesis for the interaction with parasitic plants based on a co-evolution model (Figure 7). When host plant R proteins win a "match" against the race-specific effectors or virulence proteins of parasitic plants, then the effectors become redefined as avirulence (Avr) proteins. The nature of R–Avr interactions is now well understood [151]. Host plants have receptor proteins (including the *NLR* proteins discussed above) that perceive parasite proteins and trigger responses that confer immunity [104,119] via the activation of WRKY transcription factors [151]. To circumvent host immunity, parasitic plants evolve new virulence proteins that disrupt the host defense pathways. To overcome these virulence effectors, host plants adapt their R proteins to recognize the new virulence proteins [152–154], enabling the reactivation of the downstream response, and so the cycle continues [155].

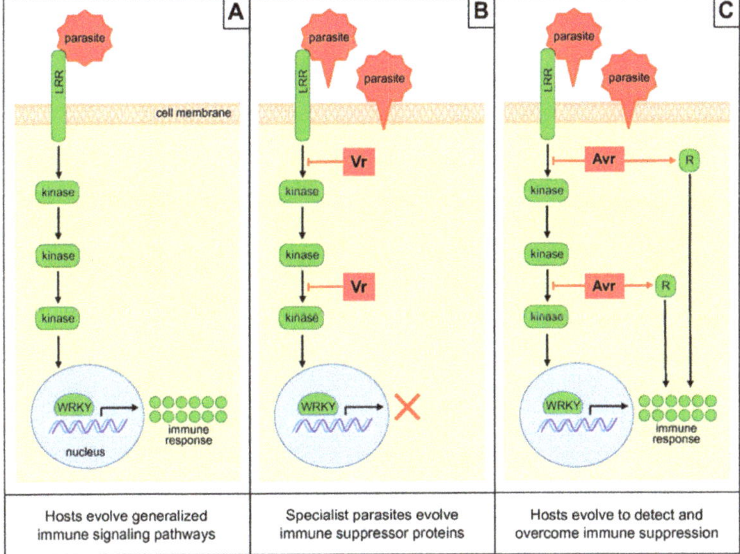

Figure 7. Parasitic plant-host antagonistic co-evolution. In figure (**A**), host plants evolve an antiparasite immune response that protects them from most parasitic plant invasion. New parasitic plant race evolves suppressive virulence mechanisms in figure (**B**), selecting the plant hosts to counter-evolve secondary immune mechanisms in figure (**C**). Figures (**B**) and (**C**) can then repeatedly cycle in an evolutionary "arms race". Figures are adapted from Keebaugh et al. [151].Vr = virulence, Avr = avirulence, *LRR* = leucine-rich repeat, WRKY – Zinc-finger transcription factor related domain containing the WRKY sequence.

4.3. Drivers of Pathogen Effector Evolution

The evolution of pathogen effectors is driven by two forms of selection pressure, adaption to targets in the host and adaption to evade detection by the host (Figure 8). The interplay between these dynamic selection pressures creates an inherently unstable biotic environment for pathogen effectors, accelerating effector evolution [156]. One perfect case to study what drives the evolution of virulence effectors in parasitic plant is natural populations of parasitic plant parasitizing different host species based on gene expression level. A number of differential *S. hermonthica* transcripts were identified depending on whether it grew on maize or sorghum [136]. These differential transcripts including genes are involved in defense mechanisms and pathogenesis, some of which might be parasite effectors that subdue host defense [4]. Pathogen effectors show marked patterns of gene evolution following host jumps, where there is extreme pressure to adapt to new host targets (Figure 8a). Parasitic plants also suppress plant host defense by producing a battery of molecules (effectors), just like bacterial and fungal pathogens [157,158]. Host specialization that leads to evolutionary divergence depends on reciprocal single amino acid changes that tailor the pathogen effector to a specific host protein that is disabled. Thus, small changes in either the host or the pathogen can allow pathogens to jump to another host species [159]. For example, orthologous protease inhibitors from the oomycetes *Phytophthora infestans* and *Phytophthora mirabilis* have adapted to target unique proteases in different hosts, allowing the pathogens to specifically target potato (*Solanum tuberosum*) and the four o'clock flower (*Mirabilis jalapa*), respectively [159].

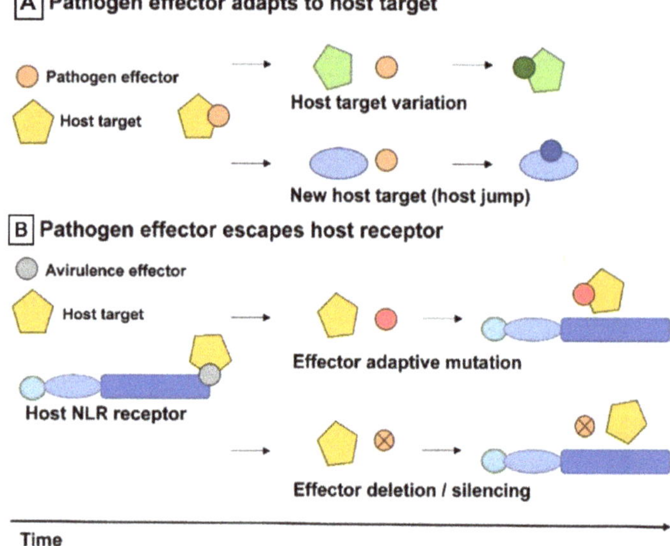

Figure 8. Forces of virulence effector evolution of parasitic plant. (**A**) Parasitic plant virulence effectors adapt to new host targets. Natural variation in a host targets or changes in the biotic environment, for example, a host jump, drive effector adaptation. This results in effectors binding or acting on a variant of the original target or on a totally new host target. (**B**) Effectors evade host immune receptors. Effectors also evolve to evade recognition by host immune receptors, for example, *NLR*. This can occur through adaptive mutations that result in stealthy effectors, which avoid host recognition but retain virulence activity. Alternatively, effector genes can also escape host immunity through pseudogenization, deletion, or gene silencing. Figure adapted from Upson et al. [156].

Avirulence effectors that are detected by plant receptors are prominent examples of rapid evolutionary adaptations. Notably, parasite pathogen effector variants with an excess of nonsynonymous polymorphisms (amino acid replacements) that escape detection by the host plant while retaining virulence (Figure 8b) can carry extreme signatures of adaptive evolution [160]. In a recent case, three avirulence effectors from the rice blast fungus *Magnaporthe oryzae* (*AVR-Pik*, *AVR-Pia*, and *AVR-Pii*), matching three rice resistance genes (*Pik*, *Pia*, and *Pii*), have been validated by comparative genomics. Among these effectors, *AVR-Pik* (in which allelic variants only carry nonsynonymous polymorphisms) binds to an interface of the *NLR* receptor *Pik-1* [161,162]. Effector genes can evolve through a birth and death process via chromosomal rearrangements, resulting in significant levels of presence/absence of polymorphisms within pathogen populations (Figure 8b) [163]. Effector genes can escape host recognition through pseudogenization [164], deletion [165], or gene silencing [166]. Many effector gene loci segregate as presence/absence polymorphisms, leading to a mosaic of effector genes within species such as *Magnaporthe oryzae* [156] and the cereal pathogen *Zymoseptoria tritici* [167].

Interestingly, the analysis of pathogen genomes has revealed that effector genes often arise in highly repetitive but gene-sparse regions rich in transposable elements [164,166,168]. This genome architecture has affected nearly every aspect of effector evolution, including transcriptional control, mutation rates, loss of function, and deletions [151]. Interestingly, the recent availability of parasitic plant genome sequences has shown that these, too, are rich in transposable elements, suggesting that effector genes may evolve in a similar manner [4,113,133].

This arms race probably stands on the foundation of gene-for-gene interaction between plant resistance and pathogen avirulence factors [141]. In theoretical models, the frequencies of resistance and virulence alleles in a population progress in an infinite cycle, sometimes called the boom-and-bust cycle, because of the frequent dramatic rise and fall in the effectiveness of resistance genes against pathogen populations in agriculture. Researchers also offer mathematical models for the co-evolution of host and parasites in terms of genetic diversity to test the gene-for-gene model, as discussed in a comprehensive review [169].

5. The Effect on Parasitic Plants

Clustered regularly interspaced short palindromic repeats/CRISPR associated protein 9 (CRISPR/Cas9)-mediated mutagenesis and RNA interference (RNAi) silencing have both been used to disrupt SL biosynthesis in host plants, aiming to suppress the germination and, thereby, infection of parasitic weeds. CRISPR/Cas9 is a form of adaptive immunity found in bacteria and archaea, which has been engineered as a powerful gene editing tool that has been applied in more than 20 crop species [170]. It has been applied in rice to disrupt the *CCD7* gene, reducing SL biosynthesis and inhibiting the germination of *S. hermonthica* [171] and similarly in tomato (*Solanum lycopersicum*) to disrupt the *CCD8* gene, inhibiting the germination of *P. aegyptiaca* [172]. In both cases, resistance was transmitted as a transgene-free trait by segregating the Cas9 cassette from the induced mutation. The tomato *CCD8* gene has also been targeted by RNA interference (RNAi), in which the expression of double-stranded RNA corresponding to the target gene causes post-transcriptional silencing [173]. Interestingly, targeting the tomato *CCD8* gene by RNAi led to the faster development of parasite tubercles when pregerminated *O. ramosa* seeds were used to infect the plant, suggesting that SLs inhibit parasite development after attachment [173]. It is possible that auxin levels or transport efficiency increased as a result of the decreased *CCD8* expression in the RNAi lines, as previously reported for the *Arabidopsis* SL-deficient mutant *max4* [174].

Eukaryotes have evolved several gene-silencing pathways to defend against viruses, mediated by small interfering RNA (siRNA) molecules 21–24 nt in length [175–177]. Although the natural purpose of these pathways is to recognize and attack of viral nucleic acids invading the cell, the components can be harnessed to recognize specific mRNA molecules, thus suppressing gene expression by either destroying the mRNA or blocking protein synthesis [176,178]. This has been exploited to develop host-induced gene silencing (HIGS) technology to control plant pathogens, in which the plant expresses siRNAs targeting gene expression in the attacking parasite [179]. RNAi or HIGS strategies have been used to try to affect gene expression in parasitic plants such as *Triphysaria versicolor* [180,181], *Cuscuta pentagona* [182], *S. hermonthica* [183], *S. asiatica* [184], *O. aegyptiaca* mboxciteB181-ijms-934400,B185-ijms-934400,B186-ijms-934400, and *Phelipanche ramosa* [173,187]. A β-glucuronidase (GUS) silencing signal can move from the transgenic host to another host and there silence GUS through the parasite *Triphysaria versicolor* as a physiological bridge [180]. Interspecific silencing of a SHOOT MERISTEMLESS-like (*STM*) gene in dodder driven by a vascular promoter in transgenic host plants disrupts dodder growth, demonstrating the efficacy of interspecific small RNA-mediated silencing of parasite genes [182]. Three *O. aegyptiaca* genes were suppressed to induce parasite mortality by virus-induced gene silencing (VIGS) and hairpin silencing on host tomato [186], which impaired expression of essential parasite virulent genes.

The Effect on Host Plants

Although the principal effect of suppressing the SL biosynthesis pathway is to inhibit the parasite, SLs are also required for normal plant development by inhibiting shoot branching/tillering and regulating the growth of primary and lateral roots [63,64,188]. Thus, the rice *ccd7* mutants discussed above exhibited stunting and a striking increase in tillering [171,189,190]. Similarly, tomato *ccd8* mutants were stunted, with increased shoot branching and adventitious root growth, and similar phenotypes were observed in *Arabidopsis*, tobacco (*Nicotiana tabacum*), and kiwifruit (*Actinidia deliciosa*) [170,191,192]. In wild-type tomato plants, *P. ramosa* infection reduces the root biomass, whereas both root and shoot biomass were affected when the same parasite infected tomato plants expressing the *CCD8* RNAi construct [173]. One way to avoid a dwarf phenotype is by grafting. Genome editing could be applied to rootstock already resistant to fungal pathogens, viruses, and nematodes, and this could be grafted to a wild-type scion in order to simultaneously achieve parasite resistance and normal growth [172]. Another solution is the application of the synthetic SL analog GR24, which can reduce the tiller number to wild-type levels in rice *ccd7* and *ccd8* mutants [171,189].

For farmers, the benefits of mutant lines with increased parasite resistance must be balanced against any trade-off against crop quality and yield. Mutation of the *CCD8* gene in tobacco caused a loss of shoot biomass [191] but mutating the same gene in tomato resulted in the production of numerous additional fruits, although these were smaller than wild-type fruits [174]. The depletion of SL in root exudates not only inhibits the germination of parasitic plants but also impairs symbiotic relationships with arbuscular mycorrhizal fungi [71,193]. This is because *CCD7* not only regulates branching, but also arbuscular mycorrhizal symbiosis [194]. More research is required to determine the possibilities to manipulate SLs in order to suppress the growth of parasitic plants while retaining normal growth characteristics and beneficial relationships with symbionts.

6. Conclusions

The intensification of agriculture has led to a surge in the prevalence and dispersal of parasitic plants that utilize crop species as their hosts, resulting in extensive yield losses. Although resistant crop varieties can be engineered or bred, some parasitic plants rapidly evolve as new races to break the resistance and establish infestations. Therefore, it is essential to understand the virulence mechanisms of parasitic plants and the corresponding host plant defensive responses. Phenotypic quantification of the infection level is the primary approach to identify the race or pathotype at the shoot emergence stage. Host plants deploy multiple layers of defenses including physical barriers (the cuticle and cell wall) and constitutively produced metabolites, but one of the major strategies to avoid parasitism is reducing the exudation of germination stimulants, particularly SLs released by the roots. We illustrated the history of parasitic plant race evolution and host resistance using *O. cumana* vs. sunflower and *S. gesnerioides* vs. cowpea as model systems. Genetic, genomic, and transcriptomic studies have shown that the major host resistance components are R genes encoding *NLR* domain proteins that play an important role in host immunity by recognizing parasite virulence factors. The induction of host defense responses by *NLR* proteins' model is proposed in three stages: recognition, activation, and defense response. We also proposed a hypothesis for the virulence effector evolution of parasitic plants based on genetic basis of typical interactions between plants and microbial parasites. The evolution of pathogen effectors is driven by two forms of selection pressure: adaption to targets in the host and adaption to evade detection by the host. Transcriptomic and genomic studies are also beginning to identify the virulence effectors and corresponding signaling pathways in parasitic plants, building into a rich information resource for future studies. Biotechnology-based approaches (CRISPR/Cas9 and RNAi) has resulted in reduced host infection by parasitic plants, but it is important to ensure that the endogenous functions of SLs are not disrupted, as well as preserving the crosstalk with other hormone pathways. In the future, gene co-expression network analysis could be used to select parasite gene candidates for targeted knockdown to develop parasitic weed-resistant crop varieties.

Funding: The authors acknowledge financial support from the National Natural Science Foundation of China (31570434, 31701333), the Inner Mongolia Science and Technology Plan (201802072), the Jiangsu Collaborative Innovation Center for Modern Crop Production, and the Science and Technology Department of Zhejiang Province (2016C02050-8, LGN18C130007).

Conflicts of Interest: The authors declare no conflict of interest.

Abbreviations

HIGS	Host-induced gene silencing
HGT	Horizontal gene transfer
LRR	Leucine-rich repeat
NLR	Nucleotide-binding and leucine-rich repeat domain
PR	Pathogen-related protein gene
QTL	Quantitative trait loci
R	Resistance gene
SiRNA	Small interfering RNA
SL	Strigolactone
HGT	Horizontal gene transfer
GR24	Synthetic analog of strigolactones
WRKY	Zinc-finger transcription factor-related domain containing the WRKY sequence
TF	Transcription factor
Avr	Avirulence
Vr	Virulence
RNAi	RNA interference
Crispr/Cas9	Clustered regularly interspaced short palindromic repeats/CRISPR associated protein 9

References

1. Yoshida, S.; Cui, S.; Ichihashi, Y.; Shirasu, K. The haustorium, a specialized invasive organ in parasitic plants. *Annu. Rev. Plant. Biol.* **2016**, *67*, 643–667. [CrossRef] [PubMed]

2. Heide-Jørgensen, H.S. Introduction: The parasitic syndrome in higher plants. In *Parasitic Orobanchaceae*; Springer: Berlin/Heidelberg, Germany, 2013; pp. 1–18.

3. Westwood, J.H.; Yoder, J.I.; Timko, M.P.; dePamphilis, C.W. The evolution of parasitism in plants. *Trends Plant. Sci.* **2010**, *15*, 227–235. [CrossRef] [PubMed]

4. Skippington, E.; Barkman, T.J.; Rice, D.W.; Palmer, J.D. Comparative mitogenomics indicates respiratory competence in parasitic *Viscum* despite loss of complex I and extreme sequence divergence, and reveals horizontal gene transfer and remarkable variation in genome size. *BMC Plant. Biol.* **2017**, *17*, 49. [CrossRef] [PubMed]

5. Parker, C. Observations on the current status of *Orobanche* and *Striga* problems worldwide. *Pest. Manag. Sci.* **2009**, *65*, 453–459. [CrossRef]

6. Fernández-Aparicio, M.; Reboud, X.; Gibot-Leclerc, S. Broomrape weeds. Underground mechanisms of parasitism and associated strategies for their control: A review. *Front. Plant. Sci.* **2016**, *7*, 135. [CrossRef] [PubMed]

7.	Bouwmeester, H.J.; Matusova, R.; Zhongkui, S.; Beale, M.H. Secondary metabolite signalling in host-parasitic plant interactions. *Curr. Opin. Plant. Biol.* **2003**, *6*, 358–364. [CrossRef]

8.	Zhou, W.J.; Yoneyama, K.; Takeuchi, Y.; Iso, S.; Rungmekarat, S.; Chae, S.H.; Joel, D.M. In vitro infection of host roots by differentiated calli of the parasitic plant *Orobanche*. *J. Exp. Bot.* **2004**, *55*, 899–907. [CrossRef]

9.	Matusova, R.; Rani, K.; Verstappen, F.W.A.; Franssen, M.C.R.; Beale, M.H.; Bouwmeester, H.J. The strigolactone germination stimulants of the plant-parasitic *Striga* and *Orobanche* spp. are derived from the carotenoid pathway. *Plant. Physiol.* **2005**, *139*, 920–934. [CrossRef]

10.	Screpanti, C.; Yoneyama, K.; Bouwmeester, H.J. Strigolactones and parasitic weed management 50 years after the discovery of the first natural strigolactone strigol: Status and outlook. *Pest. Manag. Sci.* **2016**, *72*, 2013–2015. [CrossRef]

11.	Ejeta, G. Breeding for *Striga* resistance in *Sorghum*: Exploitation of an intricate host-parasite biology. *Crop. Sci.* **2007**, *47*, 216–227. [CrossRef]

12.	Fernández-Aparicio, M.; Kisugi, T.; Xie, X.; Rubiales, D.; Yoneyama, K. Low strigolactone root exudation: A novel nechanism of Broomrape (*Orobanche* and *Phelipanche* spp.) resistance available forfaba bean breeding. *J. Agric. Food Chem.* **2014**, *62*, 7063–7071. [CrossRef] [PubMed]

13.	Jamil, M.; Rodenburg, J.; Charnikhova, T.; Bouwmeester, H.J. Pre-attachment *Striga hermonthica* resistance of New Rice for Africa (NERICA) cultivars based on low strigolactone production. *New Phytol.* **2011**, *192*, 964–975. [CrossRef] [PubMed]

14.	Pavan, S.; Schiavulli, A.; Marcotrigiano, A.R.; Bardaro, N.; Bracuto, V.; Ricciardi, F.; Charnikhova, T.; Lotti, C.; Bouwmeester, H.; Ricciardi, L. Characterization of low-strigolactone germplasm in pea (*Pisum sativum* L.) resistant to crenate broomrape (*Orobanche crenata* Forsk.). *Mol. Plant. Microbe. Interact.* **2016**, *29*, 743–749. [CrossRef]

15.	Gevezova, M.; Dekalska, T.; Stoyanov, K.; Hristeva, T.; Kostov, K.; Batchvarova, R.; Denev, I. Recent advances in broomrapes research. *J. Biosci. Biotechnol.* **2012**, 91–105.

16.	Samejima, H.; Sugimoto, Y. Recent research progress in combatting root parasitic weeds. *Biotechnol. Biotechnol. Equip.* **2018**, *32*, 221–240. [CrossRef]

17.	Islam, F.; Wang, J.; Farooq, M.A.; Khan, M.S.; Xu, L.; Zhu, J.; Zhou, W. Potential impact of the herbicide 2,4-dichlorophenoxyacetic acid on human and ecosystems. *Environ. Intl.* **2018**, *111*, 332–351. [CrossRef]

18.	Zhu, J.; Wang, J.; DiTommaso, A.; Zhang, C.; Zheng, G.; Liang, W.; Zhou, W. Weed research status, challenges, and opportunities in China. *Crop. Prot.* **2020**, *134*, 104449. [CrossRef]

19.	Molinero-Ruiz, L.; Delavault, P.; Pérez-Vich, B.; Pacureanu-Joita, M.; Bulos, M.; Altieri, E.; Domínguez, J. History of the race structure of *Orobanche cumana* and the breeding of sunflower for resistance to this parasitic weed: A review. *Spanish J. Agric. Res.* **2015**, *13*, 1–19. [CrossRef]

20.	Louarn, J.; Boniface, M.-C.; Pouilly, N.; Velasco, L.; Pérez-Vich, B.; Vincourt, P.; Muños, S. Sunflower resistance to broomrape (*Orobanche cumana*) is controlled by specific QTLs for different parasitism stages. *Front. Plant. Sci.* **2016**, *7*, 590. [CrossRef]

21.	Alcántara, E.; Morales-García, M.; Díaz-Sánchez, J. Effects of broomrape parasitism on sunflower plants: Growth, development, and mineral nutrition. *J. Plant. Nutr.* **2006**, *29*, 1199–1206. [CrossRef]

22.	Duca, M. Historical aspects of sunflower researches in the Republic of Moldova. *Helia* **2015**, *38*, 79–92. [CrossRef]

23.	Yoder, J.I.; Scholes, J.D. Host plant resistance to parasitic weeds; recent progress and bottlenecks. *Curr. Opin. Plant. Biol.* **2010**, *13*, 478–484. [CrossRef]

24.	Botanga, C.J.; Timko, M.P. Phenetic relationships among different races of *Striga gesnerioides* (Willd.) Vatke from West Africa. *Genome* **2006**, *49*, 1351–1365. [CrossRef] [PubMed]

25.	Antonova, T.S.; Araslanova, N.M.; Strelnikov, E.A.; Ramazanova, S.A.; Guchetl, S.Z.; Chelyustnikova, T.A. Distribution of highly virulent races of sunflower broomrape (*Orobanche cumana* Wallr.) in the southern regions of the Russian Federation. *Russ. Agric. Sci.* **2013**, *39*, 46–50. [CrossRef]

26.	Delipavlov, D. Orobanchaceae. In *Flora Reipublicae Bulgaricae*; Kozuharov, S.I., Kuzmanov, B.A., Eds.; Marin Drinov Academic Publishing House: Sofia, Bulgaria, 1995; Volume 10, pp. 291–325.

27.	Antonova, T.S. The history of interconnected evolution of *Orobanche cumana* Wallr. and sunflower in the Russian Federation and Kazakhstan. *Helia* **2014**, *37*, 215–225. [CrossRef]

28.	Shi, B.; Lei, Z.; Xiang, L.; Zhao, J. Identification of sunflower as physiological species in 4 provinces of China. *Chin. J. Oil Crops Sci.* **2016**, *38*, 116–119. [CrossRef]

29. Shi, B.X.; Chen, G.H.; Zhang, Z.J.; Hao, J.J.; Jing, L.; Zhou, H.Y.; Zhao, J. First report of race composition and distribution of sunflower broomrape, *Orobanche cumana*, in China. *Plant. Dis.* **2015**, *99*, 291. [CrossRef] [PubMed]

30. Nabloussi, A.; Velasco, L.; Assissel, N. First report of sunflower broomrape, *Orobanche cumana* Wallr., in Morocco. *Plant. Dis.* **2018**, *102*, 457. [CrossRef]

31. González-Cantón, E.; Velasco, A.; Velasco, L.; Pérez-Vich, B.; Martín-Sanz, A. First report of sunflower broomrape (*Orobanche cumana*) in Portugal. *Plant. Dis.* **2019**, *8*, 2143. [CrossRef]

32. Pérez de Luque, A.; Jorrín, J.; Cubero, J.I.; Rubiales, D. *Orobanche crenata* resistance and avoidance in pea (Pisum spp.) operate at different developmental stages of the parasite. *Weed Res.* **2005**, *45*, 379–387. [CrossRef]

33. Rubiales, D.; Rojas-Molina, M.M.; Sillero, J.C. Characterization of resistance mechanisms in faba bean (*Vicia faba*) against broomrape species (*Orobanche* and *Phelipanche* spp.). *Front. Plant. Sci.* **2016**, *7*, 1747. [CrossRef] [PubMed]

34. Martín-Sanz, A.; Pérez-Vich, B.; Rueda, S.; Fernández-Martínez, J.M.; Velasco, L. Characterization of post-haustorial resistance to sunflower broomrape. *Crop. Sci.* **2020**, *60*, 1188–1198. [CrossRef]

35. Rodenburg, J.; Cissoko, M.; Kayongo, N.; Dieng, I.; Bisikwa, J.; Irakiza, R.; Masoka, I.; Midega, C.A.O.; Scholes, J.D. Genetic variation and host-parasite specificity of *Striga* resistance and tolerance in rice: The need for predictive breeding. *New Phytol.* **2017**, *214*, 1267–1280. [CrossRef] [PubMed]

36. Cochavi, A.; Rapaport, T.; Gendler, T.; Karnieli, A.; Eizenberg, H.; Rachmilevitch, S.; Ephrath, J.E. Recognition of *Orobanche cumana* below-ground parasitism through physiological and hyper spectral measurements in sunflower (*Helianthus annuus* L.). *Front. Plant. Sci.* **2017**, *8*, 909. [CrossRef] [PubMed]

37. Pincovici, S.; Cochavi, A.; Karnieli, A.; Ephrath, J.; Rachmilevitch, S. Source-sink relations of sunflower plants as affected by a parasite modifies carbon allocations and leaf traits. *Plant. Sci.* **2018**, *271*, 100–107. [CrossRef]

38. Serghini, K.; Pérez de Luque, A.; Castejón-Muñoz, M.; García-Torres, L.; Jorrín, J. V Sunflower (*Helianthus annuus* L.) response to broomrape (*Orobanche cernua* Loefl.) parasitism: Induced synthesis and excretion of 7-hydroxylated simple coumarins. *J. Exp. Bot.* **2001**, *52*, 2227–2234. [CrossRef]

39. Pérez-DE-Luque, A.; Rubiales, D.; Cubero, J.I.; Press, M.C.; Scholes, J.; Yoneyama, K.; Takeuchi, Y.; Plakhine, D.; Joel, D.M. Interaction between *Orobanche crenata* and its host legumes: Unsuccessful haustorial penetration and necrosis of the developing parasite. *Ann. Bot.* **2005**, *95*, 935–942. [CrossRef]

40. Yang, C.; Xu, L.; Zhang, N.; Islam, F.; Song, W.; Hu, L.; Liu, D.; Xie, X.; Zhou, W. iTRAQ-based proteomics of sunflower cultivars differing in resistance to parasitic weed *Orobanche cumana*. *Proteomics* **2017**, *17*, 1700009. [CrossRef]

41. Mutuku, J.M.; Cui, S.; Hori, C.; Takeda, Y.; Tobimatsu, Y.; Nakabayashi, R.; Mori, T.; Saito, K.; Demura, T.; Umezawa, T.; et al. The structural integrity of lignin is crucial for resistance against *Striga hermonthica* parasitism in rice. *Plant. Physiol.* **2019**, *179*, 1796–1809. [CrossRef]

42. Huang, K.; Mellor, K.E.; Paul, S.N.; Lawson, M.J.; Mackey, A.J.; Timko, M.P. Global changes in gene expression during compatible and incompatible interactions of cowpea (*Vigna unguiculata* L.) with the root parasitic angiosperm *Striga gesnerioides*. *BMC Genomics* **2012**, *13*, 1. [CrossRef]

43. Pérez-de-Luque, A.; Moreno, M.T.; Rubiales, D. Host plant resistance against broomrapes (*Orobanche* spp.): Defence reactions and mechanisms of resistance. *Ann. Appl. Biol.* **2008**, *152*, 131–141. [CrossRef]

44. Yoshida, S.; Shirasu, K. Multiple layers of incompatibility to the parasitic witchweed, *Striga hermonthica*. *New Phytol.* **2009**, *183*, 180–189. [CrossRef] [PubMed]

45. Gurney, A.L.; Slate, J.; Press, M.C.; Scholes, J.D. A novel form of resistance in rice to the angiosperm parasite *Striga hermonthica*. *New Phytol.* **2006**, *169*, 199–208. [CrossRef] [PubMed]

46. Duriez, P.; Vautrin, S.; Auriac, M.C.; Bazerque, J.; Boniface, M.C.; Callot, C.; Carrère, S.; Cauet, S.; Chabaud, M.; Gentou, F.; et al. A receptor-like kinase enhances sunflower resistance to *Orobanche cumana*. *Nat. Plants* **2019**, *5*, 1211–1215. [CrossRef]

47. López-Ráez, J.A.; Matusova, R.; Cardoso, C.; Jamil, M.; Charnikhova, T.; Kohlen, W.; Ruyter-Spira, C.; Verstappen, F.; Bouwmeester, H. Strigolactones: Ecological significance and use as a target for parasitic plant control. *Pest. Manag. Sci.* **2009**, *65*, 471–477. [CrossRef]

48. Rubiales, D.; Pérez-de-Luque, A.; Joel, D.M.; Alcántara, C.; Sillero, J.C. Characterization of resistance in chickpea to crenate broomrape (*Orobanche crenata*). *Weed Sci.* **2003**, *51*, 702–707. [CrossRef]

49. Rubiales, D.; Alcántara, C.; Pérez-de-Luque, A.; Gil, J.; Sillero, J.C. Infection of chickpea (*Cicer arietinum*) by crenate broomrape (*Orobanche crenata*) as influenced by sowing date and weather conditions. *Agronomie* **2003**, *23*, 359–362. [CrossRef]

50. Sillero, J.C.; Cubero, J.I.; Fernández-Aparicio, M.; Rubiales, D. Search for resistance to crenate broomrape (*Orobanche crenata*). *Lathyrus Lathyrism Newsl.* **2005**, *4*, 7–9.

51. Sillero, J.C.; Moreno, M.T.; Rubiales, D. Sources of resistance to crenate broomrape in *Vicia* species. *Plant. Dis.* **2005**, *89*, 23–27. [CrossRef]

52. Abbes, Z.; Kharrat, M.; Pouvreau, J.B.; Delavault, P.; Chaibi, W.; Simier, P. The dynamics of faba bean (*Vicia faba* L.) parasitism by *Orobanche foetida*. *Phytopathol. Mediterr.* **2010**, *49*, 239–248.

53. Jorrín, J.; Pérez-de-Luque, A.; Serghini, K.; Cubero, J.I.; Moreno, M.T.; Rubiales, D.; Sillero, J.C. *Resistance to Orobanche: The State of the Art*; Junta de Andalucía: Sevilla, Spain, 1999; Volume 51, pp. 163–177.

54. Labrousse, P. Several mechanisms are involved in resistance of *Helianthus* to *Orobanche cumana* Wallr. *Ann. Bot.* **2001**, *88*, 859–868. [CrossRef]

55. Labrousse, P.; Arnaud, M.C.; Griveau, Y.; Fer, A.; Thalouran, P. Analysis of resistance criteria of sunflower recombined inbred lines against *Orobanche cumana* Wallr. *Crop. Prot.* **2004**, *23*, 407–413. [CrossRef]

56. Pierce, S.; Mbwaga, A.M.; Press, M.C.; Scholes, J.D. Xenognosin production and tolerance to *Striga asiatica* infection of high-yielding maize cultivars. *Weed Res.* **2003**, *43*, 139–145. [CrossRef]

57. Fernández-Aparicio, M.; Moral, A.; Kharrat, M.; Rubiales, D. Resistance against broomrapes (*Orobanche* and *Phelipanche* spp.) in faba bean (*Vicia faba*) based in low induction of broomrape seed germination. *Euphytica* **2012**, *186*, 897–905. [CrossRef]

58. Al-Babili, S.; Bouwmeester, H.J. Strigolactones, a novel carotenoid-derived plant hormone. *Annu. Rev. Plant. Biol.* **2015**, *66*, 161–186. [CrossRef]

59. Cook, C.E.; Whichard, L.P.; Turner, B.; Wall, M.E.; Egley, G.H. Germination of Witchweed (*Striga lutea* Lour.): Isolationand properties of a potent stimulant. *Science* **1966**, *154*, 1189–1190. [CrossRef]

60. Akiyama, K.; Matsuzaki, K.I.; Hayashi, H. Plant sesquiterpenes induce hyphal branching in arbuscular mycorrhizal fungi. *Nature* **2005**, *435*, 824–827. [CrossRef]

61. Besserer, A.; Puech-Pagès, V.; Kiefer, P.; Gomez-Roldan, V.; Jauneau, A.; Roy, S.; Portais, J.C.; Roux, C.; Bécard, G.; Séjalon-Delmas, N. Strigolactones stimulate arbuscular mycorrhizal fungi by activating mitochondria. *PLoS Biol.* **2006**, *4*, 1239–1247. [CrossRef]

62. Gomez-Roldan, V.; Fermas, S.; Brewer, P.B.; Puech-Pagès, V.; Dun, E.A.; Pillot, J.P.; Letisse, F.; Matusova, R.; Danoun, S.; Portais, J.C.; et al. Strigolactone inhibition of shoot branching. *Nature* **2008**, *455*, 189–194. [CrossRef]

63. Umehara, M.; Hanada, A.; Yoshida, S.; Akiyama, K.; Arite, T.; Takeda-Kamiya, N.; Magome, H.; Kamiya, Y.; Shirasu, K.; Yoneyama, K.; et al. Inhibition of shoot branching by new terpenoid plant hormones. *Nature* **2008**, *455*, 195–200. [CrossRef]

64. Alder, A.; Jamil, M.; Marzorati, M.; Bruno, M.; Vermathen, M.; Bigler, P.; Ghisla, S.; Bouwmeester, H.; Beyer, P.; Al-Babili, S. The path from β-Carotene to carlactone, a strigolactone-like plant hormone. *Science* **2012**, *335*, 1348–1351. [CrossRef] [PubMed]

65. Bruno, M.; Al-Babili, S. On the substrate specificity of the rice strigolactone biosynthesis enzyme *DWARF27*. *Planta* **2016**, *243*, 1429–1440. [CrossRef] [PubMed]

66. Bruno, M.; Hofmann, M.; Vermathen, M.; Alder, A.; Beyer, P.; Al-Babili, S. On the substrate- and stereospecificity of the plant carotenoid cleavage dioxygenase 7. *FEBS Lett.* **2014**, *588*, 1802–1807. [CrossRef] [PubMed]

67. Bruno, M.; Vermathen, M.; Alder, A.; Wüst, F.; Schaub, P.; van der Steen, R.; Beyer, P.; Ghisla, S.; Al-Babili, S. Insights into the formation of carlactone from in-depth analysis of the *CCD8*-catalyzed reactions. *FEBS Lett.* **2017**, *591*, 792–800. [CrossRef]

68. Jia, K.-P.; Baz, L.; Al-Babili, S. From carotenoids to strigolactones. *J. Exp. Bot.* **2017**, *69*, 2189–2204. [CrossRef]

69. Waters, M.T.; Gutjahr, C.; Bennett, T.; Nelson, D.C. Strigolactone Signaling and Evolution. *Annu. Rev. Plant. Biol.* **2017**, *68*, 291–322. [CrossRef]

70. Brewer, P.B.; Yoneyama, K.; Filardo, F.; Meyers, E.; Scaffidi, A.; Frickey, T.; Akiyama, K.; Seto, Y.; Dun, E.A.; Cremer, J.E.; et al. *LATERAL BRANCHING OXIDOREDUCTASE* acts in the final stages of strigolactone biosynthesis in *Arabidopsis*. *Proc. Natl. Acad. Sci. USA* **2016**, *113*, 6301–6306. [CrossRef]

71. Abe, S.; Sado, A.; Tanaka, K.; Kisugi, T.; Asami, K.; Ota, S.; Kim, H., II; Yoneyama, K.; Xie, X.; Ohnishi, T.; et al. Carlactone is converted to carlactonoic acid by *MAX1* in *Arabidopsis* and its methyl ester can directly interact with *AtD14* in vitro. *Proc. Natl. Acad. Sci. USA* **2014**, *111*, 18084–18089. [CrossRef]

72. Iseki, M.; Shida, K.; Kuwabara, K.; Wakabayashi, T.; Mizutani, M.; Takikawa, H.; Sugimoto, Y. Evidence for species-dependent biosynthetic pathways for converting carlactone to strigolactones in plants. *J. Exp. Bot.* **2018**, *69*, 2305–2318. [CrossRef]

73. Zhang, Y.; van Dijik, A.D.; Scaffidi, A.; Flematti, G.R.; Hofmann, M.; Charnikhova, T.; Verstappen, F.; Hepworth, J.; van der Krol, S.; Leyser, O.; et al. Rice cytochrome P450 *MAX1* homologs catalyze distinct steps in strigolactone biosynthesis. *Nat. Chem. Biol.* **2014**, *12*, 1028–1033. [CrossRef]

74. Pouvreau, J.B.; Gaudin, Z.; Auger, B.; Lechat, M.M.; Gauthier, M.; Delavault, P.; Simier, P. A high-throughput seed germination assay for root parasitic plants. *Plant. Methods* **2013**, *9*, 32. [CrossRef]

75. Matusova, R.; van Mourik, T.; Bouwmeester, H.J. Changes in the sensitivity of parasitic weed seeds to germination stimulants. *Seed Sci. Res.* **2004**, *14*, 335–344. [CrossRef]

76. Ueno, K.; Furumoto, T.; Umeda, S.; Mizutani, M.; Takikawa, H.; Batchvarova, R.; Sugimoto, Y. Heliolactone, a non-sesquiterpene lactone germination stimulant for root parasitic weeds from sunflower. *Phytochemistry* **2014**, *108*, 122–128. [CrossRef]

77. Li, J.; Yang, C.; Liu, H.; Cao, M.; Yan, G.; Si, P.; Zhou, W.; Xu, L. 5-Aminolevolinic acid enhances sunflower resistance to *Orobanche cumana* (broomrape). *Ind. Crops Prod.* **2019**, *140*, 111467. [CrossRef]

78. Yang, C.; Hu, L.Y.; Ali, B.; Islam, F.; Bai, Q.J.; Yun, X.P.; Zhou, W.J. Seed treatment with salicylic acid invokes defense mechanism of *Helianthus annuus* against *Orobanche cumana*. *Ans. App. Biol.* **2016**, *169*, 408–422. [CrossRef]

79. Song, W.J.; Zhou, W.J.; Jin, Z.L.; Zhang, D.; Yoneyama, K.; Takeuchi, Y.; Joel, D.M. Growth regulators restore germination of *Orobanche* seeds that are conditioned under water stress and suboptimal temperature. *Aus. J. Agri. Res.* **2006**, *57*, 1195–1201. [CrossRef]

80. Song, W.J.; Zhou, W.J.; Jin, Z.L.; Cao, D.D.; Joel, D.M.; Takeuchi, Y.; Yoneyama, K. Germination response of *Orobanche* seeds subjected to conditioning temperature, water potential and growth regulator treatments. *Weed Res.* **2005**, *45*, 467–476. [CrossRef]

81. Yoneyama, K. Recent progress in the chemistry and biochemistry of strigolactones. *J. Pestic. Sci.* **2020**, *45*, 45–53. [CrossRef]

82. Yoneyama, K.; Xie, X.; Yoneyama, K.; Nomura, T.; Takahashi, I.; Asami, T.; Mori, N.; Akiyama, K.; Kusajima, M.; Nakashita, H. Regulation of biosynthesis, perception, and functions of strigolactones for promoting arbuscular mycorrhizal symbiosis and managing root parasitic weeds. *Pest. Manag. Sci.* **2019**, *75*, 2353–2359. [CrossRef]

83. Yao, R.; Ming, Z.; Yan, L.; Li, S.; Wang, F.; Ma, S.; Yu, C.; Yang, M.; Chen, L.; Chen, L.; et al. *DWARF14* is a non-canonical hormone receptor for strigolactone. *Nature* **2016**, *536*, 469–473. [CrossRef]

84. Hamiaux, C.; Drummond, R.S.M.; Janssen, B.J.; Ledger, S.E.; Cooney, J.M.; Newcomb, R.D.; Snowden, K.C. *DAD2* is an α/βhydrolase likely to be involved in the perception of the plant branching hormone, strigolactone. *Curr. Biol.* **2012**, *22*, 2032–2036. [CrossRef]

85. de Saint Germain, A.; Clavé, G.; Badet-Denisot, M.-A.; Pillot, J.-P.; Cornu, D.; Le Caer, J.-P.; Burger, M.; Pelissier, F.; Retailleau, P.; Turnbull, C.; et al. An histidine covalent receptor and butenolide complex mediates strigolactone perception. *Nat. Chem. Biol.* **2016**, *12*, 787–794. [CrossRef]

86. Toh, S.; Holbrook-Smith, D.; Stokes, M.E.; Tsuchiya, Y.; McCourt, P. Detection of parasitic plant suicide germination compounds using a high-throughput *arabidopsishtl/kai2* strigolactone perception system. *Chem. Biol.* **2014**, *21*, 988–998. [CrossRef]

87. Das, M.; Fernández-Aparicio, M.; Yang, Z.; Huang, K.; Wickett, N.J.; Alford, S.; Wafula, E.K.; DePamphilis, C.; Bouwmeester, H.; Timko, M.P.; et al. Parasitic plants *striga* and *phelipanche* dependent upon exogenous strigolactones for germination have retained genes for strigolactone biosynthesis. *Am. J. Plant. Sci.* **2015**, *6*, 1151–1166. [CrossRef]

88. Toh, S.; Holbrook-Smith, D.; Stogios, P.J.; Onopriyenko, O.; Lumba, S.; Tsuchiya, Y.; Savchenko, A.; McCourt, P. Structure-function analysis identifies highly sensitive strigolactone receptors in *Striga*. *Science* **2015**, *350*, 203–207. [CrossRef]

89. Conn, C.E.; Bythell-Douglas, R.; Neumann, D.; Yoshida, S.; Whittington, B.; Westwood, J.H.; Shirasu, K.; Bond, C.S.; Dyer, K.A.; Nelson, D.C. Convergent evolution of strigolactone perception enabled host detection in parasitic plants. *Science* **2015**, *349*, 540–543. [CrossRef]

90. Tsuchiya, Y.; Yoshimura, M.; Sato, Y.; Kuwata, K.; Toh, S.; Holbrook-Smith, D.; Zhang, H.; McCourt, P.; Itami, K.; Kinoshita, T.; et al. Probing strigolactone receptors in *Striga hermonthica* with fluorescence. *Science* **2015**, *349*, 864–868. [CrossRef]

91. Yao, R.; Wang, F.; Ming, Z.; Du, X.; Chen, L.; Wang, Y.; Zhang, W.; Deng, H.; Xie, D. ShHTL7 is a non-canonical receptor for strigolactones in root parasitic weeds. *Cell Res.* **2017**, *27*, 838–841. [CrossRef]

92. Xu, Y.; Miyakawa, T.; Nosaki, S.; Nakamura, A.; Lyu, Y.; Nakamura, H.; Ohto, U.; Ishida, H.; Shimizu, T.; Asami, T.; et al. Structural analysis of *HTL* and *D14* proteins reveals the basis for ligand selectivity in *Striga*. *Nat. Commun.* **2018**, *9*, 1–11. [CrossRef]

93. Rodríguez-Ojeda, M.I.; Rocío Pineda-Martos, L.C.; Alonso, J.; Fernández-Martínez, J.M.; Pérez-Vich, B.; Velasco, L. A dominant avirulence gene in *Orobanche cumana* triggers Or5 resistance in sunflower. *Weed Res.* **2013**, *53*, 322–327.

94. Calderón-González, Á.; Pouilly, N.; Muñoz, S.; Grand, X.; Coque, M.; Velasco, L.; Pérez-Vich, B. A SSR-SNP linkage map of the parasitic weed *Orobanche cumana* Wallr. including a gene for plant pigmentation. *Front. Plant. Sci.* **2019**, *10*, 797. [CrossRef] [PubMed]

95. Gouzy, J.; Pouilly, N.; Hu, L.; Delavault, P.; Simier, P.; Boniface, M.C.; Louarn, J.; Catrice, O.; Carrère, S.; Cottret, L.; et al. The whole genome sequence of the obligate root parasitic plant *Orobanche cumana* (sunflower broomrape). In Proceedings of the Plant and Animal Genome XXVII Conference (PAG), San Diego, CA, USA, 12–16 January 2019.

96. Vranceanu, A.V.; Tudor, V.A.; Stoenescu, F.M.; Pirvu, N. Virulence groups of *Orobanche cumana* Wallr. [root parasite], differential hosts and resistance sources and genes in sunflower. In Proceedings of the 9th International Conference of Sunflower, Malaga, Spain, 8–13 June 1980.

97. Pérez-Vich, B.; Akhtouch, B.; Munoz-Ruz, J.; Fernandez-Martinez, J.M.; Jan, C.C. Inheritance of resistance to a highly virulent race F of *Orobanche cumana* Wallr. in a sunflower line derived from interspecific amphiploids. *Helia* **2002**, *25*, 137–143. [CrossRef]

98. Akhtouch, B.; del Moral, L.; Leon, A.; Velasco, L.; Fernández-Martínez, J.M.; Pérez-Vich, B. Genetic study of recessive broomrape resistance in sunflower. *Euphytica* **2016**, *209*, 419–428. [CrossRef]

99. Velasco, L.; Pérez-Vich, B.; Jan, C.C.; Fernández-Martínez, J.M. Inheritance of resistance to broomrape (*Orobanche cumana* Wallr.) race F in a sunflower line derived from wild sunflower species. *Plant. Breed.* **2007**, *126*, 67–71. [CrossRef]

100. Velasco, L.; Pérez-Vich, B.; Yassein, A.A.M.; Jan, C.C.; Fernández-Martínez, J.M. Inheritance of resistance to sunflower broomrape (*Orobanche cumana* Wallr.) in an interspecific cross between *Helianthus annuus* and *Helianthus debilis* subsp. tardiflorus. *Plant. Breed.* **2012**, *131*, 220–221. [CrossRef]

101. Imerovski, I.; Dimitrijević, A.; Miladinović, D.; Dedić, B.; Jocić, S.; Tubić, N.K.; Cvejić, S. Mapping of a new gene for resistance to broomrape races higher than F. *Euphytica* **2016**, *209*, 281–289. [CrossRef]

102. Hélène, B.; Gouzy, J.; Grassa, C.J.; Murat, F.; Staton, E.; Cottret, L. The sunflower genome provides insights into oil metabolism, flowering and Asterid evolution. *Nature* **2017**, *7656*, 148–152.

103. Imerovski, I.; Dedić, B.; Cvejić, S.; Miladinović, D.; Jocić, S.; Owens, G.L.; Tubić, N.K.; Rieseberg, L.H. BSA-seq mapping reveals major QTL for broomrape resistance in four sunflower lines. *Mol. Breed.* **2019**, *39*, 41. [CrossRef]

104. Caplan, J.; Padmanabhan, M.; Dinesh-Kumar, S.P. Plant NB-LRR immune receptors: From recognition to transcriptional reprogramming. *Cell Host Microbe.* **2008**, *3*, 126–135. [CrossRef]

105. Kim, S.H.; Qi, D.; Ashfield, T.; Helm, M.; Innes, R.W. Using decoys to expand the recognition specificity of a plant disease resistance protein. *Science* **2016**, *351*, 684–687. [CrossRef] [PubMed]

106. Lane, J.A.; Moore, T.H.M.; Child, D.V.; Cardwell, K.F.; Singh, B.B.; Bailey, J.A. Virulence characteristics of a new race of the parasitic angiosperm, *Striga gesnerioides*, from southern Benin on cowpea (*Vigna unguiculata*). *Euphytica* **1994**, *72*, 183–188. [CrossRef]

107. Atokple, I.D.K.; Singh, B.B.; Emechebe, A.M. Independent inheritance of *Striga* and *Alectra* resistance in cowpea genotype B301. *Crop. Sci.* **1993**, *33*, 714–715. [CrossRef]

108. Singh, B.B.; Emechebe, A.M. Inheritance of *Striga* resistance in cowpea genotype B301. *Crop. Sci.* **1990**, *30*, 879–881. [CrossRef]

109. Atokple, I.D.K.; Singh, B.B.; Emechebe, A.M. Genetics of resistance to *Striga* and *Alectra* in cowpea. *J. Hered.* **1995**, *86*, 45–49. [CrossRef]

110. Ouédraogo, J.T.; Maheshwari, V.; Berner, D.K.; St-Pierre, C.-A.; Belzile, F.; Timko, M.P. Identification of AFLP markers linked to resistance of cowpea (*Vigna unguiculata* L.) to parasitism by *Striga gesnerioides*. *Theor. Appl. Genet.* **2001**, *102*, 1029–1036. [CrossRef]

111. Li, J.; Lis, K.E.; Timko, M.P. Molecular genetics of race-specific resistance of cowpea to *Striga gesnerioides* (Willd.). *Pest. Manag. Sci.* **2009**, *65*, 520–527. [CrossRef]

112. Timko, M.P.; Rushton, P.J.; Laudeman, T.W.; Bokowiec, M.T.; Chipumuro, E.; Cheung, F.; Town, C.D.; Chen, X. Sequencing and analysis of the gene-rich space of cowpea. *BMC Genomics* **2008**, *9*, 103. [CrossRef]

113. Li, J.; Timko, M.P. Gene-for-gene resistance in *Striga*-cowpea associations. *Science* **2009**, *325*, 1094. [CrossRef]

114. Yang, Z.; Wafula, E.K.; Honaas, L.A.; Zhang, H.; Das, M.; Fernandez-Aparicio, M.; Huang, K.; Bandaranayake, P.C.G.; Wu, B.; Der, J.P.; et al. Comparative transcriptome analyses reveal core parasitism genes and suggest gene duplication and repurposing as sources of structural novelty. *Mol. Biol. Evol.* **2015**, *32*, 767–790. [CrossRef]

115. Su, C.; Liu, H.; Wafula, E.K.; Honaas, L.; Pamphilis, C.W.; Timko, M.P. *SHR4z*, a novel decoy effector from the haustorium of the parasitic weed *Striga gesnerioides*, suppresses host plant immunity. *New Phytol.* **2020**, *226*, 891–908. [CrossRef]

116. Zhang, D.; Qi, J.; Yue, J.; Huang, J.; Sun, T.; Li, S.; Wen, J.-F.; Hettenhausen, C.; Wu, J.; Wang, L.; et al. Root parasitic plant *Orobanche aegyptiaca* and shoot parasitic plant *Cuscuta australis* obtained Brassicaceae-specific strictosidine synthase-like genes by horizontal gene transfer. *BMC Plant. Biol.* **2014**, *14*, 1–14. [CrossRef]

117. Hegenauer, V.; Fürst, U.; Kaiser, B.; Smoker, M.; Zipfel, C.; Felix, G.; Stahl, M.; Albert, M. Detection of the plant parasite *Cuscuta reflexa* by a tomato cell surface receptor. *Science* **2016**, *353*, 478–481. [CrossRef]

118. Bradley, J.; Qiu, S.; Butlin, R.; Chaudhuri, R.; Scholes, J. The identification of candidate pathogenicity-related genes from the genome of Striga hermonthica. In Proceedings of the 15th World Congress on Parasitic Plants, Amsterdam, The Netherlands, 30 June–5 July 2019.

119. Wang, Y.; Xu, Y.; Sun, Y.; Wang, H.; Qi, J.; Wan, B.; Ye, W.; Lin, Y.; Shao, Y.; Dong, S.; et al. Leucine-rich repeat receptor-like gene screen reveals that Nicotiana *RXEG1* regulates glycoside hydrolase *12 MAMP* detection. *Nat. Commun.* **2018**, *9*, 1–12. [CrossRef]

120. Westwood, J.H.; DePamphilis, C.W.; Das, M.; Fernández-Aparicio, M.; Honaas, L.A.; Timko, M.P.; Wafula, E.K.; Wickett, N.J.; Yoder, J.I. The parasitic plant genome project: New tools for understanding the biology of *Orobanche* and *Striga*. *Weed Sci.* **2012**, *60*, 295–306. [CrossRef]

121. Yang, C.; Fu, F.; Zhang, N.; Wang, J.; Hu, L.; Islam, F.; Zhou, W. Transcriptional profiling of underground interaction of two contrasting sunflower cultivars with the root parasitic weed *Orobanche cumana*. *Plant. Soil* **2020**, *450*, 303–321. [CrossRef]

122. Li, J.; Liu, H.; Yang, C.; Wang, J.; Yan, G.; Si, P.; Bai, Q.; Lu, Z.; Zhou, W.; Xu, L. Genome-wide identification of *MYB* genes and expression analysis under different biotic and abiotic stresses in *Helianthus annuus* L. *Ind. Crops Prod.* **2020**, *143*, 111924. [CrossRef]

123. Bellot, S.; Renner, S.S. Exploring new dating approaches for parasites: The worldwide Apodanthaceae (Cucurbitales) as an example. *Mol. Phylogenet. Evol.* **2014**, *80*, 1–10. [CrossRef] [PubMed]

124. Xi, Z.; Wang, Y.; Bradley, R.K.; Sugumaran, M.; Marx, C.J.; Rest, J.S.; Davis, C.C. Massive mitochondrial gene transfer in a parasitic flowering plant clade. *PLoS Genet.* **2013**, *9*, 1–10. [CrossRef] [PubMed]

125. Kim, G.; LeBlanc, M.L.; Wafula, E.K.; DePamphilis, C.W.; Westwood, J.H. Genomic-scale exchange of mRNA between a parasitic plant and its hosts. *Science* **2014**, *345*, 808–811. [CrossRef]

126. Fan, W.; Zhu, A.; Kozaczek, M.; Shah, N.; Pabón-Mora, N.; González, F.; Mower, J.P. Limited mitogenomic degradation in response to a parasitic lifestyle in *Orobanchaceae*. *Sci. Rep.* **2016**, *6*, 36285. [CrossRef]

127. Vogel, A.; Schwacke, R.; Denton, A.K.; Usadel, B.; Hollmann, J.; Fischer, K.; Bolger, A.; Schmidt, M.H.W.; Bolger, M.E.; Gundlach, H.; et al. Footprints of parasitism in the genome of the parasitic flowering plant *Cuscuta campestris*. *Nat. Commun.* **2018**, *9*, 1–11.

128. Sun, G.; Xu, Y.; Liu, H.; Sun, T.; Zhang, J.; Hettenhausen, C.; Shen, G.; Qi, J.; Qin, Y.; Li, J.; et al. Large-scale gene losses underlie the genome evolution of parasitic plant *Cuscuta australis*. *Nat. Commun.* **2018**, *9*, 1–8.

129. Naumann, J.; Der, J.P.; Wafula, E.K.; Jones, S.S.; Wagner, S.T.; Honaas, L.A.; Ralph, P.E.; Bolin, J.F.; Maass, E.; Neinhuis, C.; et al. Detecting and characterizing the highly divergent plastid genome of the nonphotosynthetic parasitic plant *Hydnora visseri* (Hydnoraceae). *Genome Biol. Evol.* **2016**, *8*, 345–363.

130. Schneider, A.C.; Chun, H.; Stefanović, S.; Baldwin, B.G. Punctuated plastome reduction and host–parasite horizontal gene transfer in the holoparasitic plant genus Aphyllon. *Proc. R. Soc. B Biol. Sci.* **2018**, *285*, 1–8.

131. Yoshida, S.; Kim, S.; Wafula, E.K.; Tanskanen, J.; Kim, Y.M.; Honaas, L.; Yang, Z.; Spallek, T.; Conn, C.E.; Ichihashi, Y.; et al. Genome sequence of *Striga asiatica* provides insight into the evolution of plant parasitism. *Curr. Biol.* **2019**, *29*, 3041–3052.

132. Yang, Z.; Wafula, E.K.; Kim, G.; Shahid, S.; McNeal, J.R.; Ralph, P.E.; Timilsena, P.R.; Yu, W.B.; Kelly, E.A.; Zhang, H.; et al. Convergent horizontal gene transfer and cross-talk of mobile nucleic acids in parasitic plants. *Nat.Plants* **2019**, *5*, 991–1001.

133. Yang, Z.; Zhang, Y.; Wafula, E.K.; Honaas, L.A.; Ralph, P.E.; Jones, S.; Clarke, C.R.; Liu, S.; Su, C.; Zhang, H.; et al. Horizontal gene transfer is more frequent with increased heterotrophy and contributes to parasite adaptation. *Proc. Natl. Acad. Sci. USA* **2016**, *113*, E7010–E7019.

134. Xi, Z.; Bradley, R.K.; Wurdack, K.J.; Wong, K.M.; Sugumaran, M.; Bomblies, K.; Rest, J.S.; Davis, C.C. Horizontal transfer of expressed genes in a parasitic flowering plant. *BMC Genomics* **2012**, *13*, 227.

135. Bellot, S.; Cusimano, N.; Luo, S.; Sun, G.; Zarre, S.; Gröger, A.; Temsch, E.; Renner, S.S. Assembled plastid and mitochondrial genomes, as well as nuclear genes, place the parasite family Cynomoriaceae in the Saxifragales. *Genome Biol. Evol.* **2016**, *8*, 2214–2230. [PubMed]

136. Lopez, L.; Bellis, E.S.; Wafula, E.; Hearne, S.J.; Honaas, L.; Ralph, P.E.; Timko, M.P.; Unachukwu, N.; DePamphilis, C.W.; Lasky, J.R. Transcriptomics of host-specific interactions in natural populations of the parasitic plant purple witchweed (*Striga hermonthica*). *Weed Sci.* **2019**, *67*, 397–411.

137. Ranjan, A.; Ichihashi, Y.; Farhi, M.; Zumstein, K.; Townsley, B.; David-Schwartz, R.; Sinha, N.R. De novo assembly and characterization of the transcriptome of the parasitic weed dodder identifies genes associated with plant parasitism. *Plant. Physiol.* **2014**, *166*, 1186–1199.

138. Kennan, R.M.; Wong, W.; Dhungyel, O.P.; Han, X.; Wong, D.; Parker, D.; Rosado, C.J.; Law, R.H.P.; McGowan, S.; Reeve, S.B.; et al. The subtilisin-like protease Aprv2 is required for virulence and uses a novel disulphide-tethered exosite to bind substrates. *PLoS Pathog.* **2010**, *6*, e1001210.

139. Antão, C.M.; Malcata, F.X. Plant serine proteases: Biochemical, physiological and molecular features. *Plant. Physiol. Biochem.* **2005**, *43*, 637–650.

140. Moffett, P. Mechanisms of recognition in dominant R gene mediated resistance. *Adv. Virus Res.* **2009**, *75*, 1–229. [PubMed]

141. Adachi, H.; Derevnina, L.; Kamoun, S. NLR singletons, pairs, and networks: Evolution, assembly, and regulation of the intracellular immunoreceptor circuitry of plants. *Curr. Opin. Plant. Biol.* **2019**, *50*, 121–131. [PubMed]

142. Wirthmueller, L.; Maqbool, A.; Banfield, M.J. On the front line: Structural insights into plant–pathogen interactions. *Nat. Rev. Microbiol.* **2013**, *11*, 761–776. [PubMed]

143. McHale, L.; Tan, X.; Koehl, P.; Michelmore, R.W. Plant NBS-LRR proteins: Adaptable guards. *Genome Biol.* **2006**, *7*, 1–11.

144. Koehl, P.; Delarue, M. A self consistent mean field approach to simultaneous gap closure and side-chain positioning in homology modelling. *Nat. Struct. Mol. Biol.* **1995**, *2*, 163–170.

145. Steele, J.F.C.; Hughes, R.K.; Banfield, M.J. Structural and biochemical studies of an NB-ARC domain from a plant NLR immune receptor. *PLoS ONE* **2019**, *14*, e0221226.

146. Holt, B.F.; Boyes, D.C.; Ellerström, M.; Siefers, N.; Wiig, A.; Kauffman, S.; Grant, M.R.; Dangl, J.L. An evolutionarily conserved mediator of plant disease resistance gene function is required for normal *Arabidopsis* development. *Dev. Cell* **2002**, *2*, 807–817.

147. Machens, F.; Becker, M.; Umrath, F.; Hehl, R. Identification of a novel type of WRKY transcription factor binding site in elicitor-responsive cis-sequences from *Arabidopsis thaliana*. *Plant. Mol. Biol.* **2014**, *84*, 371–385. [PubMed]

148. Rushton, P.J.; Somssich, I.E.; Ringler, P.; Shen, Q.J. WRKY transcription factors. *Trends Plant. Sci.* **2010**, *15*, 247–258. [PubMed]

149. Yang, Y.; Zhou, Y.; Chi, Y.; Fan, B.; Chen, Z. Characterization of soybean WRKY gene family and identification of soybean WRKY genes that promote resistance to soybean cyst nematode. *Sci. Rep.* **2017**, *7*, 17804. [PubMed]

150. Deslandes, L.; Olivier, J.; Theulieres, F.; Hirsch, J.; Feng, D.X.; Bittner-Eddy, P.; Beynon, J.; Marco, Y. Resistance to *Ralstonia solanacearum* in *Arabidopsis thaliana* is conferred by the recessive *RRS1-R* gene, a member of a novel family of resistance genes. *Proc. Natl. Acad. Sci. USA* **2002**, *99*, 2404–2409. [PubMed]

151. Keebaugh, E.S.; Schlenke, T.A. Insights from natural host–parasite interactions: The *Drosophila* model. *Dev. Comp. Immunol.* **2014**, *42*, 111–123.

152. Dodds, P.N.; Lawrence, G.J.; Catanzariti, A.-M.; Teh, T.; Wang, C.-I.A.; Ayliffe, M.A.; Kobe, B.; Ellis, J.G. Direct protein interaction underlies gene-for-gene specificity and coevolution of the flax resistance genes and flax rust avirulence genes. *Proc. Natl. Acad. Sci. USA* **2006**, *103*, 8888–8893.

153. Jones, J.D.G.; Dangl, J.L. The plant immune system. *Nature* **2006**, *444*, 323–329.

154. DeYoung, B.J.; Innes, R.W. Plant NBS-LRR proteins in pathogen sensing and host defense. *Nat. Immunol.* **2006**, *7*, 1243–1249.

155. Chisholm, S.T.; Coaker, G.; Day, B.; Staskawicz, B.J. Host-microbe interactions: Shaping the evolution of the plant immune response. *Cell* **2006**, *124*, 803–814.

156. Upson, J.L.; Zess, E.K.; Białas, A.; Wu, C.; Kamoun, S. The coming of age of EvoMPMI: Evolutionary molecular plant–microbe interactions across multiple timescales. *Curr. Opin. Plant. Biol.* **2018**, *44*, 108–116.

157. Runo, S.; Kuria, E.K. Habits of a highly successful cereal killer, *Striga*. *PLoS Pathog.* **2018**, *14*, e1006731. [CrossRef] [PubMed]

158. Bellis, E.S.; Kelly, E.A.; Lorts, C.M.; Gao, H.; DeLeo, V.L.; Rouhan, G.; Budden, A.; Bhaskara, G.B.; Hu, Z.; Muscarella, R.; et al. Genomics of sorghum local adaptation to a parasitic plant. *Proc. Natl. Acad. Sci. USA* **2020**, *117*, 4243–4251. [CrossRef] [PubMed]

159. Dong, S.; Stam, R.; Cano, L.M.; Song, J.; Sklenar, J.; Yoshida, K.; Bozkurt, T.O.; Oliva, R.; Liu, Z.; Tian, M.; et al. Effector specialization in a lineage of the Irish potato famine pathogen. *Science* **2014**, *343*, 552–555. [CrossRef] [PubMed]

160. Białas, A.; Zess, E.K.; De la Concepcion, J.C.; Franceschetti, M.; Pennington, H.G.; Yoshida, K.; Upson, J.L.; Chanclud, E.; Wu, C.-H.; Langner, T.; et al. Lessons in effector and NLR biology of plant-microbe systems. *Mol. Plant. Microbe. Interact.* **2018**, *31*, 34–45. [CrossRef]

161. Yoshida, K.; Saitoh, H.; Fujisawa, S.; Kanzaki, H.; Matsumura, H.; Yoshida, K.; Tosa, Y.; Chuma, I.; Takano, Y.; Win, J.; et al. Association genetics reveals three novel avirulence genes from the rice blast fungal pathogen *Magnaporthe oryzae*. *Plant. Cell* **2009**, *21*, 1573–1591. [CrossRef]

162. Maqbool, A.; Saitoh, H.; Franceschetti, M.; Stevenson, C.E.M.; Uemura, A.; Kanzaki, H.; Kamoun, S.; Terauchi, R.; Banfield, M.J. Structural basis of pathogen recognition by an integrated HMA domain in a plant NLR immune receptor. *Elife* **2015**, *4*.

163. Fouché, S.; Plissonneau, C.; Croll, D. The birth and death of effectors in rapidly evolving filamentous pathogen genomes. *Curr. Opin. Microbiol.* **2018**, *46*, 34–42. [CrossRef]

164. Dong, S.; Raffaele, S.; Kamoun, S. The two-speed genomes of filamentous pathogens: Waltz with plants. *Curr. Opin. Genet. Dev.* **2015**, *35*, 57–65. [CrossRef]

165. Dong, Y.; Li, Y.; Zhao, M.; Jing, M.; Liu, X.; Liu, M.; Guo, X.; Zhang, X.; Chen, Y.; Liu, Y.; et al. Global genome and transcriptome analyses of *Magnaporthe oryzae* epidemic isolate 98-06 uncover novel effectors and pathogenicity-related genes, revealing gene gain and lose dynamics in genome evolution. *PLoS Pathog.* **2015**, *11*, e1004801.

166. Raffaele, S.; Kamoun, S. Genome evolution in filamentous plant pathogens: Why bigger can be better. *Nat. Rev. Microbiol.* **2012**, *10*, 417–430. [CrossRef]

167. Hartmann, F.E.; Croll, D. Distinct trajectories of massive recent gene gains and losses in populations of a microbial eukaryotic pathogen. *Mol. Biol. Evol.* **2017**, *34*, 2808–2822. [CrossRef] [PubMed]

168. Möller, M.; Stukenbrock, E.H. Evolution and genome architecture in fungal plant pathogens. *Nat. Rev. Microbiol.* **2017**, *15*, 756–771. [PubMed]

169. Brown, J.K.M.; Tellier, A. Plant-parasite coevolution: Bridging the gap between genetics and ecology. *Annu. Rev. Phytopathol.* **2011**, *49*, 345–367. [CrossRef] [PubMed]

170. Ricroch, A.; Clairand, P.; Harwood, W. Use of CRISPR systems in plant genome editing: Toward new opportunities in agriculture. *Emerg. Top. Life Sci.* **2017**, *1*, 169–182.

171. Butt, H.; Jamil, M.; Wang, J.Y.; Al-Babili, S.; Mahfouz, M. Engineering plant architecture via CRISPR/Cas9-mediated alteration of strigolactone biosynthesis. *BMC Plant. Biol.* **2018**, *18*, 174. [CrossRef]

172. Bari, V.K.; Nassar, J.A.; Kheredin, S.M.; Gal-On, A.; Ron, M.; Britt, A.; Steele, D.; Yoder, J.; Aly, R. CRISPR/Cas9-mediated mutagenesis of *CAROTENOID CLEAVAGE DIOXYGENASE 8* in tomato provides resistance against the parasitic weed *Phelipanche aegyptiaca*. *Sci. Rep.* **2019**, *9*, 11438. [CrossRef]

173. Cheng, X.; Floková, K.; Bouwmeester, H.; Ruyter-Spira, C. The role of endogenous strigolactones and their interaction with aba during the infection process of the parasitic weed phelipanche ramosa in tomato plan. *Front. Plant. Sci.* **2017**, *8*, 392. [CrossRef]

174. Bennett, T.; Sieberer, T.; Willett, B.; Booker, J.; Luschnig, C.; Leyser, O. The arabidopsis Max pathway controls shoot branching by regulating auxin transport. *Curr. Biol.* **2006**, *16*, 553–563. [CrossRef]

175. Harvey, J.J.W.; Lewsey, M.G.; Patel, K.; Westwood, J.; Heimstadt, S.; Carr, J.P.; Baulcombe, D.C. An antiviral defense role of AGO2 in plants. *PLoS ONE* **2011**, *6*, e14639. [CrossRef]

176. Baulcombe, D. RNA silencing in plants. *Nature* **2004**, *431*, 356–363.

177. Qi, T.; Guo, J.; Peng, H.; Liu, P.; Kang, Z.; Guo, J. Host-induced gene silencing: A powerful strategy to control diseases of wheat and barley. *Int. J. Mol. Sci.* **2019**, *20*, 206. [CrossRef] [PubMed]

178. Ding, S.-W. RNA-based antiviral immunity. *Nat. Rev. Immunol.* **2010**, *10*, 632–644. [CrossRef] [PubMed]

179. Huang, G.; Allen, R.; Davis, E.L.; Baum, T.J.; Hussey, R.S. Engineering broad root-knot resistance in transgenic plants by RNAi silencing of a conserved and essential root-knot nematode parasitism gene. *Natl. Acad. Sci.* **2006**, *103*, 14302–14306. [CrossRef]

180. Tomilov, A.A.; Tomilova, N.B.; Wroblewski, T.; Michelmore, R.; Yoder, J.I. Trans-specific gene silencing between host and parasitic plants. *Plant. J.* **2008**, *56*, 389–397. [CrossRef] [PubMed]

181. Westwood, J.H.; Roney, J.K.; Khatibi, P.A.; Stromberg, V.K. RNA translocation between parasitic plants and their hosts. *Pest. Manag. Sci.* **2009**, *65*, 533–539. [CrossRef] [PubMed]

182. Alakonya, A.; Kumar, R.; Koenig, D.; Kimura, S.; Townsley, B.; Runo, S.; Garces, H.M.; Kang, J.; Yanez, A.; David-Schwartz, R.; et al. Interspecific RNA interference of SHOOT MERISTEMLESS-like disrupts *Cuscuta pentagona* plant parasitism. *Plant. Cell* **2012**, *24*, 3153–3166. [CrossRef] [PubMed]

183. Kirigia, D.; Runo, S.; Alakonya, A. A virus-induced gene silencing (VIGS) system for functional genomics in the parasitic plant *Striga hermonthica*. *Plant. Methods* **2014**, *10*, 1–8. [CrossRef] [PubMed]

184. de Framond, A.; Rich, P.J.; McMillan, J.; Ejeta, G. Effects on Striga parasitism of transgenic maize armed with RNAi constructs targeting essential S. asiatica genes. In *Integrating New Technologies for Striga Control*; Gebisa, E., Jonathan, G., Eds.; World Scientific: Singapore, 2007; pp. 185–196.

185. Aly, R.; Cholakh, H.; Joel, D.M.; Leibman, D.; Steinitz, B.; Zelcer, A.; Naglis, A.; Yarden, O.; Gal-On, A. Gene silencing of mannose 6-phosphate reductase in the parasitic weed *Orobanche aegyptiaca* through the production of homologous dsRNA sequences in the host plant. *Plant. Biotechnol. J.* **2009**, *7*, 487–498. [CrossRef]

186. Dubey, N.K.; Eizenberg, H.; Leibman, D.; Wolf, D.; Edelstein, M.; Abu-Nassar, J.; Marzouk, S.; Gal-On, A.; Aly, R. Enhanced host-parasite resistance based on down-regulation of *Phelipanche aegyptiaca* target genes is likely by mobile small RNA. *Front. Plant. Sci.* **2017**, *8*, 1574. [CrossRef]

187. Kohlen, W.; Charnikhova, T.; Lammers, M.; Pollina, T.; Tóth, P.; Haider, I.; Pozo, M.J.; Maagd, R.A.; Ruyter-Spira, C.; Bouwmeester, H.J.; et al. The tomato *CAROTENOID CLEAVAGE DIOXYGENASE 8* (*SlCCD8*) regulates rhizosphere signaling, plant architecture and affects reproductive development through strigolactone biosynthesis. *New Phytol.* **2012**, *196*, 535–547. [CrossRef]

188. Waldie, T.; McCulloch, H.; Leyser, O. Strigolactones and the control of plant development: Lessons from shoot branching. *Plant. J.* **2014**, *79*, 607–622. [CrossRef] [PubMed]

189. Kulkarni, K.P.; Vishwakarma, C.; Sahoo, S.P.; Lima, J.M.; Nath, M.; Dokku, P.; Gacche, R.N.; Mohapatra, T.; Robin, S.; Sarla, N.; et al. A substitution mutation in *OsCCD7* cosegregates with dwarf and increased tillering phenotype in rice. *J. Genet.* **2014**, *93*, 389–401. [CrossRef] [PubMed]

190. Yang, X.; Chen, L.; He, J.; Yu, W. Knocking out of carotenoid catabolic genes in rice fails to boost carotenoid accumulation, but reveals a mutation in strigolactone biosynthesis. *Plant. Cell Rep.* **2017**, *36*, 1533–1545. [CrossRef] [PubMed]

191. Gao, J.; Zhang, T.; Xu, B.; Jia, L.; Xiao, B.; Liu, H.; Liu, L.; Yan, H.; Xia, Q. Crispr/cas9-mediated mutagenesis of carotenoid cleavage dioxygenase 8 (*ccd8*) in tobacco affects shoot and root architecture. *Int. J. Mol. Sci.* **2018**, *19*, 1062. [CrossRef] [PubMed]

192. Ledger, S.E.; Janssen, B.J.; Karunairetnam, S.; Wang, T.; Snowden, K.C. Modified CAROTENOID CLEAVAGE DIOXYGENASE8 expression correlates with altered branching in kiwifruit (*Actinidia chinensis*). *New Phytol.* **2010**, *188*, 803–813. [CrossRef]
193. Tsuchiya, Y.; Yoshimura, M.; Hagihara, S. The dynamics of strigolactone perception in *Striga hermonthica*: A working hypothesis. *J. Exp. Bot.* **2018**, *69*, 2281–2290. [CrossRef]
194. Vogel, J.T.; Walter, M.H.; Giavalisco, P.; Lytovchenko, A.; Kohlen, W.; Charnikhova, T.; Simkin, A.J.; Goulet, C.; Strack, D.; Bouwmeester, H.J.; et al. *SlCCD7* controls strigolactone biosynthesis, shoot branching and mycorrhiza-induced apocarotenoid formation in tomato. *Plant. J.* **2010**, *61*, 300–311. [CrossRef]

International Journal of
Molecular Sciences

Article

Modification of Serine 1040 of SlBRI1 Increases Fruit Yield by Enhancing Tolerance to Heat Stress in Tomato

Shufen Wang [1,2,†], Tixu Hu [1,2,†], Aijuan Tian [1,2], Bote Luo [1,2], Chenxi Du [1,2], Siwei Zhang [1,2], Shuhua Huang [1,2], Fei Zhang [1,2] and Xiaofeng Wang [1,2,*]

1   State Key Laboratory of Crop Stress Biology in Arid Areas, College of Horticulture, Northwest A&F University, Yangling 712100, China; shufenwang@nwafu.edu.cn (S.W.); htx0729@nwsuaf.edu.cn (T.H.); tianaijuan@nwafu.edu.cn (A.T.); robert2018@nwafu.edu.cn (B.L.); cxdu@nwafu.edu.cn (C.D.); Siweizhang@nwafu.edu.cn (S.Z.); shhuang@nwsuaf.edu.cn (S.H.); feizhang@nwafu.edu.cn (F.Z.)
2   Shaanxi Engineering Research Center for Vegetables, Yangling 712100, China
*   Correspondence: wangxff99@nwsuaf.edu.cn
†   These authors contributed equally to this work.

Received: 30 September 2020; Accepted: 15 October 2020; Published: 16 October 2020

Abstract: High temperature is a major environmental factor that adversely affects plant growth and production. SlBRI1 is a critical receptor in brassinosteroid signalling, and its phosphorylation sites have differential functions in plant growth and development. However, the roles of the phosphorylation sites of SlBRI1 in stress tolerance are unknown. In this study, we investigated the biological functions of the phosphorylation site serine 1040 (Ser-1040) of SlBRI1 in tomato. Phenotype analysis indicated that transgenic tomato harbouring SlBRI1 dephosphorylated at Ser-1040 showed increased tolerance to heat stress, exhibiting better plant growth and plant yield under high temperature than transgenic lines expressing SlBRI1 or SlBRI1 phosphorylated at Ser-1040. Biochemical and physiological analyses further showed that antioxidant activity, cell membrane integrity, osmo-protectant accumulation, photosynthesis and transcript levels of heat stress defence genes were all elevated in tomato plants harbouring SlBRI1 dephosphorylated at Ser-1040, and the autophosphorylation level of SlBRI1 was inhibited when SlBRI1 dephosphorylated at Ser-1040. Taken together, our results demonstrate that the phosphorylation site Ser-1040 of SlBRI1 affects heat tolerance, leading to improved plant growth and yield under high-temperature conditions. Our results also indicate the promise of phosphorylation site modification as an approach for protecting crop yields from high-temperature stress.

Keywords: tomato; SlBRI1; phosphorylation site; heating tolerance; yield

---

1. Introduction

Heat stress is a major abiotic stress that threatens crop production by affecting plant growth processes such as seed germination, root growth, hypocotyl elongation, and fertilization [1]. Physiological and physiochemical analyses have further indicated that heat stress affects photosynthesis and induces excessive reactive oxygen species (ROS) accumulation, which subsequently leads to membrane lipid peroxidation and increased membrane permeability of plants [2]. To avoid heat-induced damage, plants upregulate a series of processes involved in osmotic adjustment, ROS removal, photosynthetic reactions, and saturation of membrane-associated lipids. The genes encoding superoxide dismutase (SOD), peroxidase (POD), and catalase (CAT) participate in ROS scavenging. The major role of heat shock proteins (HSPs) is to act as molecular chaperones regulating protein folding, accumulation, location, and degradation to protect cells against damage due to high-temperature stress. Heat shock factors (HSFs) specifically bind to the heat shock element (HSE) of high-temperature-regulated genes and interact with HSPs to regulate the transcription of genes under high-temperature stress, while the trans-acting *WRKY* factors are overexpressed to help plants respond to high-temperature stress [3–7].

Brassinosteroids (BRs) are a group of steroid hormones that play a potential role in crop yield boosting due to their positive roles in plant growth and tolerance to biotic and abiotic stresses [8–10]. BRASSINOSTEROID INSENSITIVE1 (BRI1) is the major BR signalling receptor, and intensive research has confirmed the involvement of BRI1-mediated BR signalling in plant growth and stress responses. In BR signal transduction, the BR first binds to BRI1 to promote its sequential transphosphorylation and heterodimerization with its coreceptor BRI1-ASSOCIATED RECEPTOR KINASE1 (BAK1), which results in the activation of the BR signal [11–13]. In *Arabidopsis*, over 30 mutant *bri1* alleles were reported with characteristic BR-insensitive phenotypes, including short hypocotyls in the dark, dwarf stature, prolonged vegetative phase, and male sterility of varying strength; of these, only *bri1-301* additionally exhibited temperature sensitivity [14]. *BRI1* orthologues in different crops, such as rice and tomato, were also found to regulate plant growth and stress tolerance; either loss of function or suppression of *BRI1* usually resulted in shorter plant height, twisted leaves, compromised BR signalling, and altered environmental stress tolerance in plants, while overexpression of *BRI1* could enhance BR signalling to promote plant germination, flowering, and yield [15–20]. All these results highlight the critical role of BRI1 in plant growth and stress adaption, and further research on its molecular mechanisms is necessary to realize its valuable potential in crop production.

As a receptor kinase, BRI1 activates BR signal transduction through phosphorylation [11]. Our previous study identified the phosphorylation sites of BRI1 and further revealed their different effects on the biological functions of BRI1 [13,21,22]. Phosphorylation sites in the kinase domain of BRI1 exhibited the strongest functions in BR signalling and plant growth, preventing phosphorylation, which caused severe plant growth inhibition. Phosphorylation sites in the juxtamembrane and C-terminal domains influenced BR signalling to varying degrees, indicating their diverse functions in plant development. For example, dephosphorylation of Ser-1168 and Ser-1172 in the C-terminal domain of BRI1 resulted in slight leaf growth inhibition and greatly reduced seed yields, while dephosphorylation of Tyr-831 in the juxtamembrane domain influenced flowering time and leaf growth [22,23]. In tomato, phosphorylation sites of SlBRI1 were also found to be critical for SlBRI1-mediated regulation of BR signal transduction and plant biological development. Dephosphorylation of Thr-1054 in SlBRI1 severely attenuated BR signalling and disturbed plant growth in tomato [24]. Our previous study further found that the tomato SlBRI1 phosphorylation site Thr-1050 could affect tomato yield by regulating BR signalling [25]. These results suggest the potential value of phosphorylation sites of BRI1 in crop agronomic trait improvement, given their precise modification of the biological function of BRI1. However, functional analyses of the phosphorylation sites of SlBRI1 are limited and focus on plant growth, while the functions of these sites in stress tolerance are still unknown.

In this work, we investigated the biological functions of the phosphorylation site serine 1040 (Ser-1040) of SlBRI1 in tomato. The weak SlBRI1 mutant *cu3*^−abs1^ (the result of a His-1012-Tyr (H1012Y) missense mutation in SlBRI1), as well as *cu3*^−abs1^ transformed with SlBRI1 constitutively phosphorylated at Ser-1040, SlBRI1 dephosphorylated at Ser-1040, or wild-type SlBRI1, was used for phenotype comparison. Our results showed that transgenic plants harbouring dephosphorylated Ser-1040 exhibited similar growth phenotypes under normal temperature conditions but better growth and more efficient stress responses than plants harbouring phosphorylated Ser-1040 or SlBRI1 under high-temperature conditions. Furthermore, compared with wild-type SlBRI1 and SlBRI1 phosphorylated at Ser-1040, dephosphorylation of Ser-1040 resulted in a lower autophosphorylation level of SlBRI1 in vitro but a similar BR signal strength in plants. These results suggest that Ser-1040 can modulate BRI1 autophosphorylation to promote plant heat tolerance. Our results provide a theoretical basis for revealing the molecular modulation mechanism of SlBRI1 in stress adaption, as well as coordinating tomato yield with environmental changes through fine-tuning of the phosphorylation site of SlBRI1.

2. Results

2.1. SlBRI1 Ser-1040 Influences Autophosphorylation of SlBRI1

SlBRI1 is a receptor kinase that transmits BR signals through phosphorylation [24]. To determine the function of SlBRI1 Ser-1040 in tomato, we first explored the conservation of Ser-1040. Protein sequence alignment of BRI1 homologues revealed that Ser-1040 was a highly conserved phosphorylation site among the distinct species. This high conservation indicated that Ser-1040 had a crucial role in the biological function of SlBRI1 (Figure 1A).

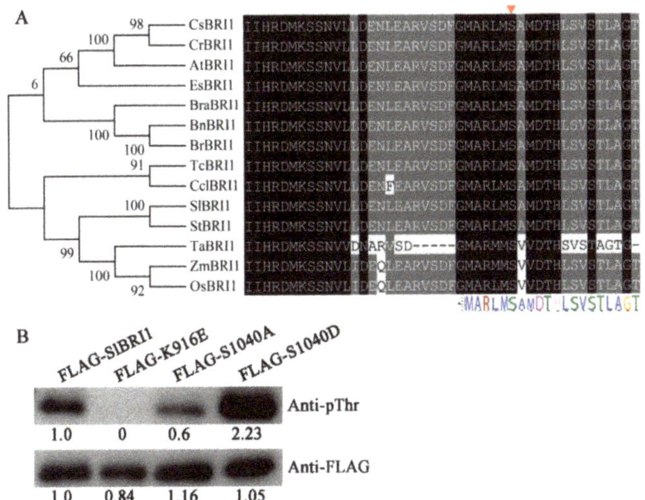

Figure 1. SlBRI1 Ser-1040 influences autophosphorylation of SlBRI1. (**A**) Alignment of the partial kinase domain sequences of BRI1 homologues. Conserved and similar residues were highlighted with black and gray grounds, respectively. The set of Ser-1040 among BRI1 homologues was marked by the red arrow. Each symbol with different colors at the bottom indicated the conservation of residue at each site. SlBRI1 (*Solanum lycopersicum*, NP_001296180.1), CsBRI1 (*Camelina sativa*, XP_010431911.1), CrBRI1 (*Camelina sativa*, XP_010431911.1), AtBRI1 (*Arabidopsis thaliana*, NP_195650.1), EsBRI1 (*Eutrema salsugineum*, XP_006411743.1), BraBRI1 (*Brassica oleracea* var. oleracea, XP_013597742.1), BnBRI1 (*Brassica napus*, NP_001303105.1), BrBRI1 (*Brassica rapa* XP_009101880.2), TcBRI1 (*Theobroma cacao*, XP_017985424.1), CclBRI1 (*Citrus clementina*, XP_006427932.1), StBRI1 (*Solanum tuberosum*, XP_006357355.1), TaBRI1 (*Triticum aestivum*, DQ_655711.1), ZmBRI1 (*Zea mays*, XP_008656807.1) and OsBRI1 (*Oryza sativa*, NP_001044077.1). (**B**) Autophosphorylation level of SlBRI1 in vitro. Autophosphorylation analysis of recombinant FLAG-SlBRI1, FLAG-K916E, FLAG-S1040A, and FLAG-S1040D proteins was detected by anti-pThr antibodies, and anti-FLAG antibodies were used to show the loading levels for western blotting. Intensities of bands were presented as relative values compared with the FLAG-SlBRI1.

Autophosphorylation of SlBRI1 is usually important for its functions in plants, and most of the phosphorylation sites of SlBRI1 positively regulate the autophosphorylation. To determine whether the phosphorylation site Ser-1040 could influence the autophosphorylation of SlBRI1, we compared the autophosphorylation levels of SlBRI1, a kinase-inactive form of SlBRI1 (K916E, in which Lys-916 was replaced with glutamic acid), S1040A (in which Ser-1040 of SlBRI1 was replaced with alanine), and S1040D (in which Ser-1040 was replaced with aspartic acid) in vitro. The results suggested a positive role for Ser-1040 phosphorylation in SlBRI1 autophosphorylation, since the intensity of the phosphorylation band of FLAG-S1040D was strongest, the phosphorylation level of which was 4.1 and

2.1-fold that of FLAG-S1040A and FLAG-SlBRI1, respectively (Figure 1B). Thus, dephosphorylation of Ser-1040 might attenuate kinase active of SlBRI1, and further disturb the biological functions.

2.2. SlBRI1 Ser-1040 Slightly Affects BR Signalling in Tomato

To investigate the biological function of SlBRI1 Ser-1040 in tomato, transgenic plants in the *cu3*[−abs1] background expressing SlBRI1, S1040A, or S1040D were generated for phenotype evaluation (Figure 2A). The transgenic lines P$_{SlBRI1}$::SlBRI1-GFP-1 and P$_{SlBRI1}$::SlBRI1-GFP-3 (SlBRI1-1 and SlBRI1-3 for short) were selected as the positive controls, while the weak *SlBRI1* mutant *cu3*[−abs1] was considered the negative control. Both protein level and transcription level analyses suggested that all transgenic lines used in this study had a higher level of SlBRI1 than *cu3*[−abs1] plants (Figure 2B,C).

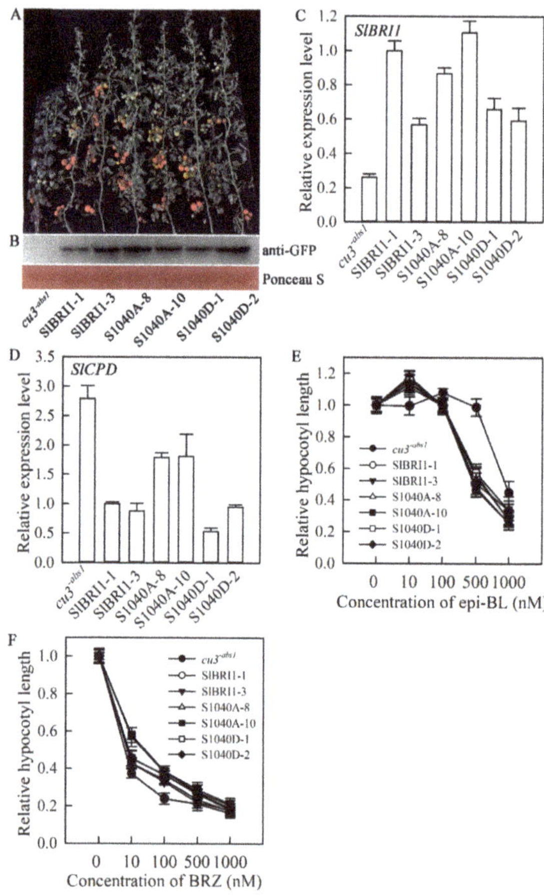

Figure 2. SlBRI1 Ser-1040 slightly affects BR signalling in tomato. (**A**) Phenotypes of plants at the maturation stage. The plants shown from left to right are as follows: *cu3*[−abs1], SlBRI1-1, SlBRI1-3, S1040A-8, S1040A-10, S1040D-1, and S1040D-2. (**B**) Western blot analysis of transgenic protein expression using anti-green fluorescent protein (GFP) antibodies. Ponceau S (Solarbio, P8330, Beijing, China) staining shows loading. (**C**) and (**D**) Relative transcript levels of SlBRI1 (**C**) and the BR signalling marker gene *SlCPD* (**D**) in tomato. (**E**) and (**F**) Dose-response curves of relative hypocotyl lengths of tomato seedlings grown in the dark for 10 days on the surface of media with increasing concentrations of epi-BL (**E**) and BRZ (**F**). The data for (**E**) and (**F**) are the means ± SDs of 15 independent biological samples.

Previous studies have suggested that the BR biosynthesis gene *SlCPD* could be suppressed by BR signalling feedback; thus, the expression of this gene could be considered to be a marker of BR signal strength [26]. Consistent with this feedback regulation, the transcript levels of *SlCPD* in P_{SlBRI1}::SlBRI1-GFP and P_{SlBRI1}::S1040D-GFP (S1040D-1 and S1040D-2 for short) plants were similar and lower than those in *cu3-abs1* plants, while in P_{SlBRI1}::S1040A-GFP (S1040A-8 and S1040A-10 for short) lines, the levels were between those in P_{SlBRI1}::SlBRI1-GFP and *cu3-abs1* plants, which suggested that both wild-type SlBRI1 and SlBRI1 with phosphorylated Ser-1040 could rescue the BR signalling defect in *cu3-abs1* plants, and the BR signal strength in P_{SlBRI1}::S1040A-GFP lines was slightly weaker than that in P_{SlBRI1}::SlBRI1-GFP and P_{SlBRI1}::S1040D-GFP lines (Figure 2D).

The BR sensitivity of the seedlings was also analysed to further quantitatively evaluate the BR signal strength among *cu3-abs1*, P_{SlBRI1}::SlBRI1-GFP, P_{SlBRI1}::S1040A-GFP, and P_{SlBRI1}::S1040D-GFP plants. The hypocotyl lengths of the transgenic lines and *cu3-abs1* were measured, and their relative hypocotyl lengths under five increasing concentrations of exogenous 24-epibrassinolide (epi-BL) or BR inhibitor brassinazole (BRZ) were compared. As shown in Figure 2E and 2F, *cu3-abs1* was insensitive to epi-BL and sensitive to BRZ, since the hypocotyl length this plant was nearly unchanged when the concentration of epi-BL was lower than 500 nM, while a 62.7% decrease was observed under 10 nM BRZ. However, the change rates of the hypocotyl length among the transgenic plants were similar regardless of treatment with epi-BL or BRZ. This result demonstrated that Ser-1040A could rescue BR signal transduction in tomato.

2.3. Dephosphorylation of Ser-1040 Improves Tomato Yield under Heat Stress

To assess the effect of Ser-1040 on tomato yield under normal and heat stress conditions, plant height and stem diameter at the mature stage; flower number, stamen length, pistil length, and flower phenotype of the second and fifth inflorescences; fruit number and fruit setting rate of the second and fifth clusters; single-fruit weight; and early yield and total yield per plant were compared. As shown in Figures 3A and 3B, the flower numbers of the second inflorescences that grew at normal temperature were similar; however, the fifth inflorescences that grew in the late-spring stage with high temperature showed a significant difference among the transgenic lines. The flower numbers of the fifth inflorescences of P_{SlBRI1}::S1040A-GFP lines were the largest, approximately 12.9% and 25.3% larger than those of P_{SlBRI1}::SlBRI1-GFP and P_{SlBRI1}::S1040D-GFP lines, respectively. Fruit numbers per second cluster were similar among the transgenic lines, whereas the fruit number per fifth cluster of P_{SlBRI1}::S1040A-GFP lines were 24.5% and 39.2% higher than those of P_{SlBRI1}::SlBRI1-GFP and P_{SlBRI1}::S1040D-GFP lines (Figure 3C). In terms of fruit setting rate, the second clusters from all transgenic lines were the same, while the fifth clusters were also constant for P_{SlBRI1}::SlBRI1-GFP and P_{SlBRI1}::S1040D-GFP plants; however, P_{SlBRI1}::S1040A-GFP lines showed higher fruit setting rates of the fifth cluster, the values for which were 13.8% and 16.4% greater than those for P_{SlBRI1}::SlBRI1-GFP and P_{SlBRI1}::S1040D-GFP lines (Figure 3D). Furthermore, the early yield per plant for each transgenic line was similar, while P_{SlBRI1}::S1040D-GFP lines exhibited 1.52-fold and 1.59-fold higher total yields per plant than P_{SlBRI1}::SlBRI1-GFP and P_{SlBRI1}::S1040D-GFP lines, respectively (Figure 3A,E). Furthermore, plant height, stem diameter, and single-fruit weight at the mature stage were similar among the transgenic lines (Figures S1A–C). Phenotype analysis of the second and fifth inflorescences showed that there were no significant differences in flower phenotype and stamen length among the transgenic plants under either normal or high-temperature conditions. However, the pistils were nearly the same under normal temperature conditions but showed different extension trends under high temperature, and this extension was most obvious in P_{SlBRI1}::S1040D-GFP lines compared with other transgenic lines (Figures S1D–F). In conclusion, Ser-1040 appears to play an important role in tomato fruiting under high temperature.

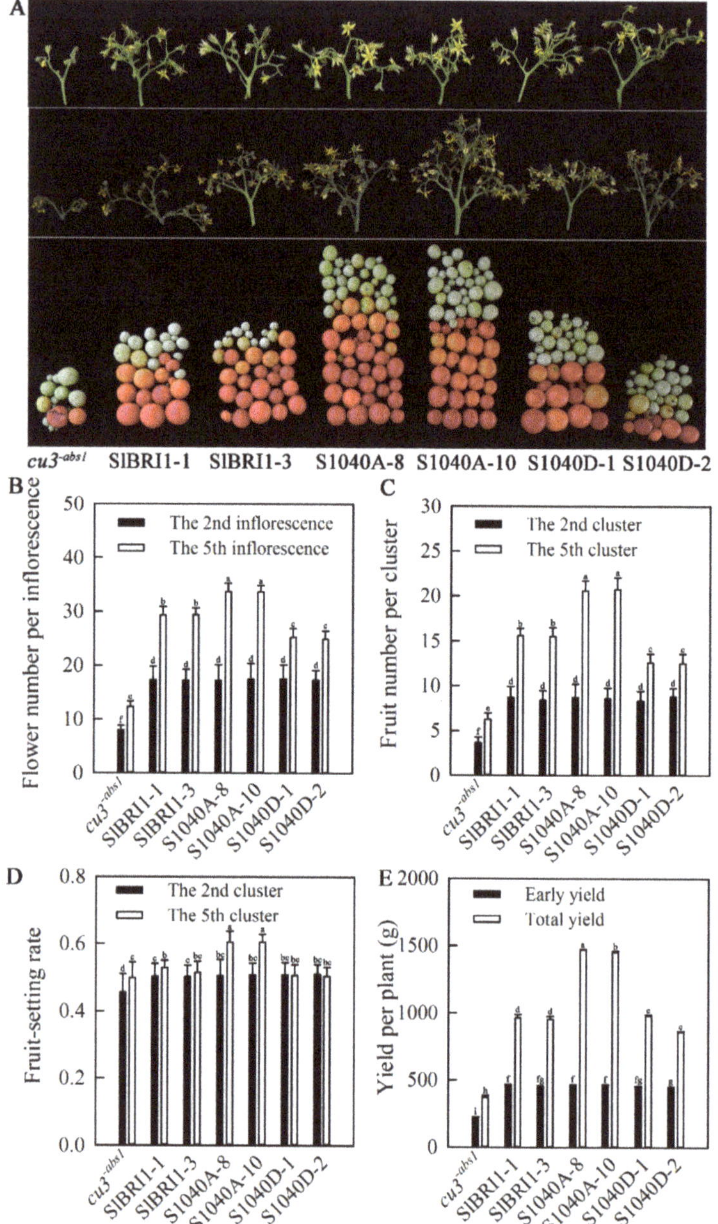

Figure 3. Dephosphorylation of Ser-1040 improves tomato yield under heat stress. (**A**) Phenotypes of the second inflorescence (top), fifth inflorescence (middle) and total yield per plant (bottom). (**B**) Flower numbers of the second and fifth inflorescences. (**C**) Fruit number and (**D**) fruit setting rate of the second and fifth clusters. (**E**) Early yield and total yield per plant. The different letters indicate significant differences at the 0.05 level. The data for (**B**) and (**E**) are the means ± SDs of 10 independent biological samples.

2.4. Dephosphorylation of Ser-1040 Promotes Germination and Seedling Growth under Heat Stress

To determine how Ser-1040 phosphorylation extensively affects seed germination under heat stress, the seed germination phenotypes of transgenic lines and $cu3^{-abs1}$ were analysed. The difference in germination rates among transgenic lines was not obvious at 28 °C; however, the germination rates of P_{SlBRI1}::S1040A-GFP lines were 1.17-fold and 1.28-fold higher than those of P_{SlBRI1}::SlBRI1-GFP and P_{SlBRI1}::S1040D-GFP lines when the temperature increased to 33 °C (Figure 4B). The germination potential of P_{SlBRI1}::SlBRI1-GFP lines decreased most dramatically, by approximately half, while the decreases in germination potentials of P_{SlBRI1}::S1040D-GFP and P_{SlBRI1}::S1040A-GFP lines were 33.0% and 23.4%, respectively, when the temperature increased from 28 °C to 33 °C (Figure 4C). The sensitivity of root growth to heat stress was also analysed. P_{SlBRI1}::S1040A-GFP plants were insensitive to heat stress, and their root lengths decreased by 22.3% at 33 °C, while those of P_{SlBRI1}::SlBRI1-GFP and P_{SlBRI1}::S1040D-GFP plants decreased by 44.5% and 44.1%, respectively (Figure 4D). At 28 °C, the malondialdehyde (MDA) content in P_{SlBRI1}::S1040A-GFP leaves was higher than that in other lines. When the germination temperature was 33 °C, the MDA content in all plants increased to different degrees, and the MDA content in P_{SlBRI1}::S1040A-GFP plants increased slightly and was similar to that in P_{SlBRI1}::SlBRI1-GFP plants, while in P_{SlBRI1}::S1040D-GFP plants, the MDA content increased rapidly by more than 1.6-fold compared with that in other lines (Figure 4E).

Tomato seedlings of transgenic lines and $cu3^{-abs1}$ at the four-leaf stage were exposed to 38 °C/28 °C or 25 °C/25 °C for 9 days, and seedling phenotypes, shoot fresh weights, shoot dry weights, seedling heights and seedling stem diameters were subsequently analysed to determine how Ser-1040 phosphorylation affects tomato seedling growth under heat stress. As shown in Figure 4A, the values for all the transgenic plants were the same and were higher than those for $cu3^{-abs1}$ plants under normal conditions. When the seedlings were exposed to 38 °C/28 °C (day/night) for 9 days, P_{SlBRI1}::S1040D-GFP and P_{SlBRI1}::SlBRI1-GFP plants showed more severe wilting phenotypes than P_{SlBRI1}::S1040A-GFP and $cu3^{-abs1}$ plants. Shoot fresh weights among the transgenic lines were nearly the same and higher than those of $cu3^{-abs1}$ before heat stress. Following heat stress, the decreases in shoot fresh weights of P_{SlBRI1}::S1040A-GFP plants were 21.0% and 23.4% lower than those of P_{SlBRI1}::SlBRI1-GFP and P_{SlBRI1}::S1040D-GFP plants (Figure 4F). The shoot dry weights showed similar changes when the temperature increased (Figure S2A). The seedling heights of P_{SlBRI1}::SlBRI1-GFP and P_{SlBRI1}::S1040D-GFP plants were similar and slightly higher than those of P_{SlBRI1}::S1040A-GFP plants when the temperature was 25 °C. However, P_{SlBRI1}::S1040A-GFP lines showed the second-fastest growth rates after $cu3^{-abs1}$ at high temperature, approximately 38.5% and 43.6% faster than those of P_{SlBRI1}::SlBRI1-GFP and P_{SlBRI1}::S1040D-GFP lines (Figure 4G). In addition, P_{SlBRI1}::S1040A-GFP lines also exhibited the lowest decline rates for the seedling stem diameter compared with P_{SlBRI1}::SlBRI1-GFP and P_{SlBRI1}::S1040D-GFP plants (Figure S2B).

2.5. Dephosphorylation of Ser-1040 Promotes Heat Stress Tolerance of Seedlings

To determine whether Ser-1040 phosphorylation affects heat tolerance in tomato, the free proline accumulation, electrolyte leakage and MDA content, which are indicators of cell membrane damage caused by heat stress, were measured. As shown in Figure 5A, the MDA content increased in all of the seedlings after heat stress and was highest in P_{SlBRI1}::S1040D-GFP lines, approximately 1.8, 2.7 and 2.9 times that in P_{SlBRI1}::SlBRI1-GFP, P_{SlBRI1}::S1040A-GFP, and $cu3^{-abs1}$ plants, respectively, on the ninth day of heat stress. Similarly, the levels of electrolyte leakage in P_{SlBRI1}::S1040D-GFP and P_{SlBRI1}::SlBRI1-GFP leaves were significantly higher than those in P_{SlBRI1}::S1040A-GFP and $cu3^{-abs1}$ plants after heat stress (Figure 5B). In contrast, the proline content was lowest in P_{SlBRI1}::S1040D-GFP lines, approximately 44.6%, 25.8% and 26.2% that in P_{SlBRI1}::SlBRI1-GFP, P_{SlBRI1}::S1040A-GFP, and $cu3^{-abs}$ plants, under treatment with high temperature for nine days (Figure 5C).

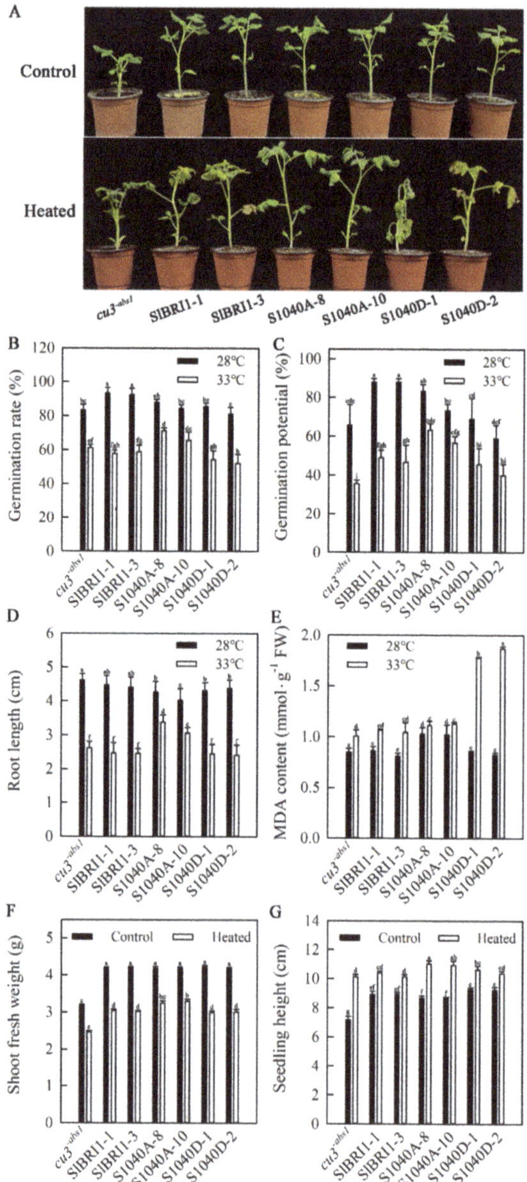

Figure 4. Dephosphorylation of Ser-1040 promotes germination and seedling growth under heat stress. (**A**) Phenotypes of seedlings treated with or without heat stress for 9 days. (**B**) Germination rate of plants on the fourteenth day after seeding at 33 °C or 28 °C. (**C**) Germination potential of plants on the fourth day after seeding at 33 °C or 28 °C. (**D**) Root length and (**E**) MDA content of plants on the fourteenth day after seeding at 33 °C or 28 °C. (**F**) Shoot fresh weights and (**G**) seedling heights of plants at the four-leaf stage treated with or without heat stress (38 °C/28 °C, day/night) for 12 days. The data for (**B**) and (**E**) are the means ± SDs of three replicates, and each replicate had 30 plants. The data for (**F**) and (**G**) are the means ± SDs of three independent biological samples. The different letters indicate significant differences at the 0.05 level.

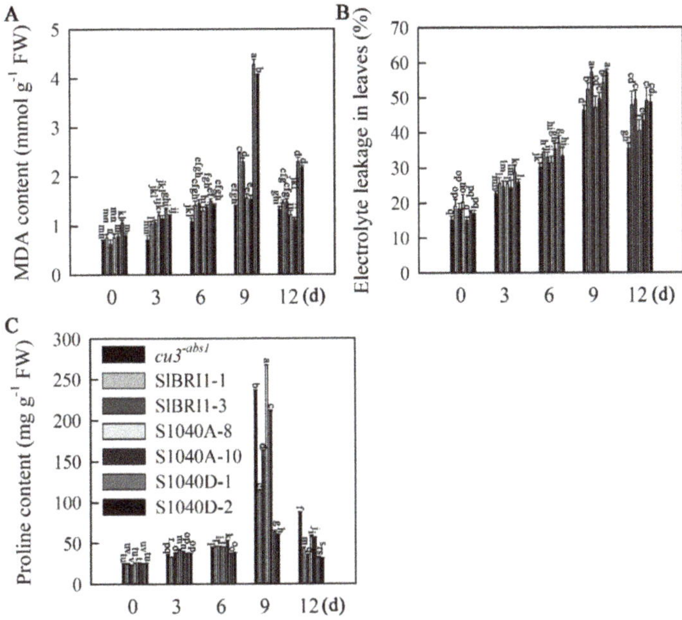

Figure 5. Dephosphorylation of Ser-1040 promotes the heat stress tolerance of seedlings. (**A–C**) Time course of changes in MDA content (**A**), ion leakage (**B**), and proline content (**C**) in tomato seedlings in response to heat stress. Tomato seedlings at the four-leaf stage were placed into a growth chamber set at 38 °C/28 °C (day/night) for 9 days and then transferred to 25 °C for 3 days. The data are the means ± SDs of at least three independent biological samples. The different letters indicate significant differences at the 0.05 level.

2.6. Dephosphorylation of Ser-1040 Promotes ROS Detoxification of Seedlings under Heat Stress

The activity of antioxidants is important in the protection of plants from ROS-induced damage under exposure to heat stress [27]. To investigate whether Ser-1040 phosphorylation affects heat stress tolerance though altered redox status, H_2O_2 accumulation, and antioxidant enzyme activities, the expression levels of antioxidant-related genes in $cu3^{-abs1}$ and transgenic seedlings under normal and high temperature were determined. As shown in Figure 6A, all of the plants accumulated nearly equal amounts of basal H_2O_2 when grown under normal conditions. Under high temperature, the H_2O_2 content in these plants decreased in the following order: P_{SlBRI1}::S1040D-GFP, P_{SlBRI1}::SlBRI1-GFP, P_{SlBRI1}::S1040A-GFP and $cu3^{-abs1}$. In the antioxidant enzyme activity analysis, CAT activity in all plants showed similar peak values during heat stress; however, CAT activity peaked more quickly in P_{SlBRI1}::SlBRI1-GFP and P_{SlBRI1}::S1040D-GFP lines, showing rapid spikes on the sixth day of heat stress, three days earlier than those observed in P_{SlBRI1}::S1040A-GFP and $cu3^{-abs1}$ plants (Figure 6B). SOD activity in P_{SlBRI1}::SlBRI1-GFP and P_{SlBRI1}::S1040D-GFP transgenic lines peaked on the third day of heat stress and then declined rapidly. While SOD activity in P_{SlBRI1}::S1040A-GFP and $cu3^{-abs1}$ plants peaked on the sixth day of heat stress and the decline was relatively slow, SOD activity in P_{SlBRI1}::S1040A-GFP plants was more than 2.5-fold that in P_{SlBRI1}::SlBRI1-GFP and P_{SlBRI1}::S1040D-GFP lines after heat stress (Figure 6C). POD activity in P_{SlBRI1}::SlBRI1-GFP and P_{SlBRI1}::S1040D-GFP plants was higher until the sixth day of heat stress but was then surpassed by that in P_{SlBRI1}::S1040A-GFP and $cu3^{-abs1}$ plants. The peak value of POD activity in P_{SlBRI1}::S1040A-GFP plants was at least 10% higher than those in P_{SlBRI1}::SlBRI1-GFP and P_{SlBRI1}::S1040D-GFP lines (Figure 6D).

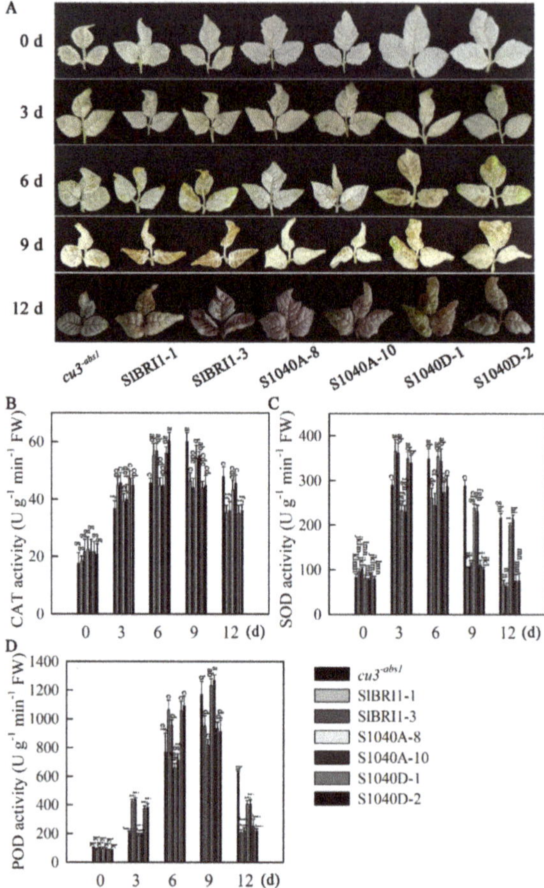

Figure 6. Dephosphorylation of Ser-1040 promotes ROS detoxification in seedlings under heat stress. (**A**) DAB staining for hydrogen peroxide (H_2O_2) in leaves from tomato seedlings during heat stress treatment. (**B–D**) Time course of changes in CAT (**B**), SOD (**C**), and POD (**D**) activities in tomato seedlings in response to heat stress. Tomato seedlings at the four-leaf stage were placed into a growth chamber set at 38 °C/28 °C (day/night) for 9 days and then transferred to 25 °C for 3 days. The data for (**B**) to (**D**) are the means ± SDs of at least three replicates, and each replicate had 10 plants. The different letters indicate significant differences at the 0.05 level.

2.7. Dephosphorylation of Ser-1040 Promotes Photosynthesis in Seedlings under Heat Stress

Photosynthesis in plants is sensitive to high temperature and is usually inhibited under heat stress [28]. To determine whether Ser-1040 phosphorylation affects tomato photosynthesis under heat stress, the total chlorophyll content and CO_2 assimilation rate (Pn) of *cu3^abs1* and transgenic plants under normal and high temperatures were determined. The total chlorophyll content in all the plants decreased under heat stress; however, P$_{SlBRI1}$::S1040D-GFP exhibited the greatest decline at 51.4%. P$_{SlBRI1}$::S1040A-GFP showed the slowest decline, with the value decreasing by 30.8% on the twelfth day of heat stress (Figure 7A). The CO_2 assimilation rates of all the transgenic plants were the same at the flowering stage of the second inflorescence and decreased at the flowering stage of the fifth inflorescence. The decline rates of P$_{SlBRI1}$::S1040A-GFP lines were the lowest and only approximately 61.3% and 69.3% those of P$_{SlBRI1}$::SlBRI1-GFP and P$_{SlBRI1}$::S1040D-GFP lines, respectively (Figure 7B).

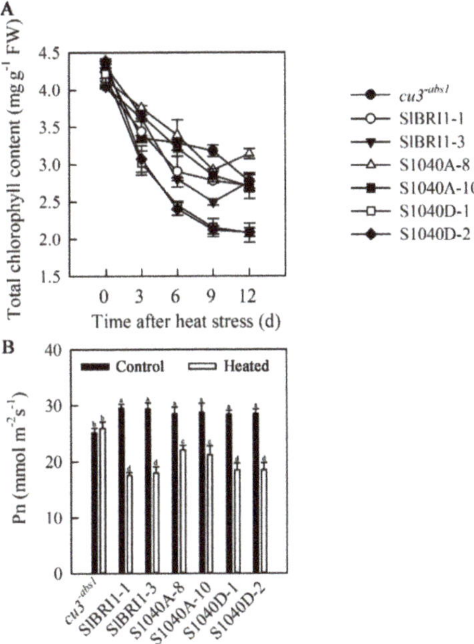

Figure 7. Dephosphorylation of Ser-1040 promotes photosynthesis in seedlings under heat stress. (**A**) Time course of changes in total chlorophyll content in tomato seedlings in response to heat stress. (**B**) CO_2 assimilation rate in tomato at the flowering stages of the second inflorescence (control) and fourth inflorescence (heated). The data are the means ± SDs of at least 3 replicates, and each replicate had 10 plants. The different letters indicate significant differences at the 0.05 level.

2.8. Dephosphorylation of Ser-1040 Enhances the Expression of Tomato Defence-Related Genes under Heat Stress

To reveal the molecular mechanisms underlying the role of Ser-1040 phosphorylation in tomato heat tolerance, the transcription levels of defence-related genes, such as *CAT1*, *Cu/Zn-SOD*, *POD1*, *HSPs*, *HSFs* and *WRKYs*, were analysed at 0, 3, 6, 9, and 12 days under heat stress. The differences in the transcripts of antioxidant-related genes, such as *CAT1*, *Cu/Zn-SOD*, and *POD1*, were nonsignificant when the transgenic plants were grown under normal conditions, and heat stress markedly induced the expression of all the analysed antioxidant-related genes. With prolonged heating, the increase in expression of *Cu/Zn-SOD*, *POD1*, and *RBOH* in P$_{SlBRI1}$::S1040A-GFP plants was significantly higher than that in P$_{SlBRI1}$::SlBRI1-GFP and P$_{SlBRI1}$::S1040D-GFP plants (Figure 8A–D). The transcription levels of *HSFA2* in P$_{SlBRI1}$::SlBRI1-GFP and P$_{SlBRI1}$::S1040D-GFP transgenic lines were unchanged under heat stress; however, in P$_{SlBRI1}$::S1040A-GFP transgenic lines, the transcription increased rapidly and peaked on the ninth day of heat stress (Figure 8E). The expression of *HSFA3* in P$_{SlBRI1}$::S1040D-GFP lines was unchanged under heat stress; however, it increased in P$_{SlBRI1}$::SlBRI1-GFP and P$_{SlBRI1}$::S1040A-GFP lines, and P$_{SlBRI1}$::S1040A-GFP lines showed the greatest increase when exposed to high-temperature conditions (Figure 8F). The expression of *HSP70* and *HSFA90* in all the plants increased and peaked on the sixth day of heat stress, while P$_{SlBRI1}$::S1040A-GFP lines showed the greatest increase compared with P$_{SlBRI1}$::SlBRI1-GFP and P$_{SlBRI1}$::S1040D-GFP lines (Figure 8G,H). The expression of *WRKY1* and *WRKY72* in all the plants increased and peaked on the ninth day of heat stress, and again, P$_{SlBRI1}$::S1040A-GFP lines showed the greatest increase among all the plants, while P$_{SlBRI1}$::S1040D-GFP lines showed the lowest increase (Figure 8I,J)

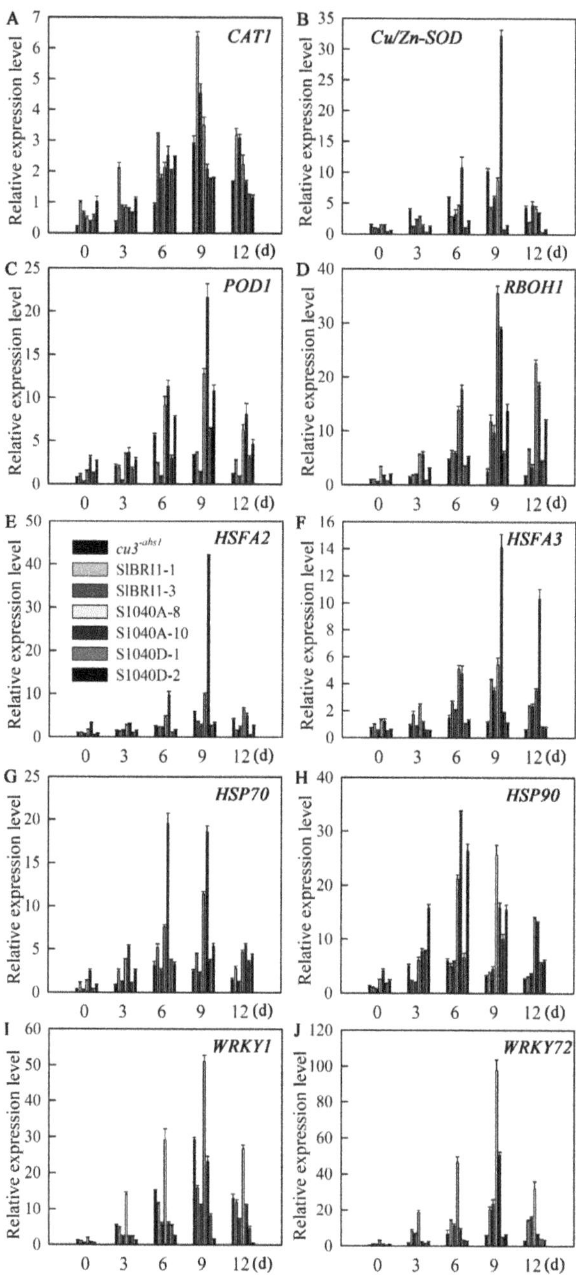

Figure 8. Dephosphorylation of Ser-1040 enhances the expression of tomato defence-related genes under heat stress. (**A–J**) Transcription levels of *CAT1* (**A**), *Cu/Zn-SOD* (**B**), *POD1* (**C**), *RBOH1* (**D**), *HSFA2* (**E**), *HSFA3* (**F**), *HSP70* (**G**), *HSP90* (**H**), *WRKY1* (**I**), and *WRKY72* (**J**) in tomato seedlings in response to heat stress. Tomato seedlings at the four-leaf stage were placed into a growth chamber set at 38 °C/28 °C (day/night) for 9 days and then transferred to 25 °C for 3 days. The data are the means ± SDs of three biological replicates and technical replications.

3. Discussion

High temperature is one of the most serious problems for agricultural production, and BRs are endogenous plant hormones that are involved in the processes of plant growth and environmental stress adaption [8]. As a BR receptor, the functions of BRI1 in plant growth are conserved and well established; however, its functional mechanism for adjusting plant growth and development according to heat stress remains largely unknown. Previous studies in *Arabidopsis* found that increased temperature specifically impacts BRI1 expression to downregulate BR signalling and mediate root elongation [28]. The *Arabidopsis bri1-301* mutant exhibited severe growth defects when the temperature increased from 22 °C to 29 °C [14]. In tomato, the SlBRI1 mutant *cu3-abs1* seedlings showed high basal tolerance to heat stress [29]. In this study, we first found that the phosphorylation site of SlBRI1 could influence its functions on stress tolerance, and dephosphorylation of Ser-1040 significantly increased the capability of tomato to maintain growth parameters and maintained a stable yield under heat stress. Compared with the transgenic line with wild-type SlBRI1 and the line with SlBRI1 phosphorylated at Ser-1040, plants harbouring dephosphorylated Ser-1040 exhibited a higher yield, more stable photosynthesis, a more active ROS-scavenging system, and higher expression levels of stress defence genes. Previous studies have suggested that Ser-1040 is located in the activation loop of SlBRI1, and dephosphorylation of Ser-1040 almost completely abolishes the kinase activity of SlBRI1 in vitro [24]. Consistent with this, our results showed that Ser-1040 acts as a positive regulator of SlBRI1 kinase activity, since dephosphorylated and phosphorylated Ser-1040 showed decreased and increased phosphorylation levels compared with that of wild-type SlBRI1 in vitro (Figure 1). However, the BR signal strengths in P_{SlBRI1}::S1040A-GFP, P_{SlBRI1}::S1040D-GFP and P_{SlBRI1}::SlBRI1-GFP were similar. This raises the question of whether phosphorylation sites of BRI1 influence its functions only through classic BR signal transduction. Previous studies conducted in *Arabidopsis* reported that dephosphorylation of Thr-872 of BRI1 resulted in a dramatic increase in BRI1 phosphorylation levels in vitro; however, transgenic plants with dephosphorylated Thr-872 did not show the strong performance exhibited by plants expressing wild-type BRI1. In addition, phosphorylation sites in the C-terminal domain of BRI1 did not influence BR signalling but could regulate the reproductive development of plants [22]. Thus, phosphorylation sites might influence BRI1 function not only by influencing BR signalling but also by influencing other BR-induced pathways through changing posttranslational modifications, such as protein interactions, folding, localization, and degradation. Furthermore, the molecular mechanism of the tomato BR signal transduction pathway remains unclear, and the relevant indicators are still limited. All these results demonstrated that Ser-1040 regulated tomato heat tolerance mainly by affecting the kinase activity of SlBRI1, and this pattern was distinct from the classic BR signal pathway.

High temperature triggers lipid peroxidation, which could be marked by ROS accumulation. During this process, both respiratory burst oxidase homologue (RBOH) and antioxidant enzymes such as CAT, SOD, and POD play pivotal roles in plants, helping them to adapt to environmental stress by ROS production and scavenging, respectively [30,31]. Studies on BR signalling mutants of *Arabidopsis*, tomato and other crops revealed that BRs play a protective role under high temperature, as they promote antioxidant enzyme activities and elevate *RBOH1* transcription [27,32]. In tomato, overexpression of *BZR1* enhanced the transcription of *RBOH1* and subsequently elevated heat tolerance, and seedling of the SlBRI1 mutant *cu3-abs1* exhibited induction of thermotolerance, showing increased signs of oxidative stress [29,33]. In this study, P_{SlBRI1}::S1040A-GFP lines that were tolerant to heat stress showed higher expression levels of *RBOH1* but lower accumulation of H_2O_2 than P_{SlBRI1}::S1040D-GFP and P_{SlBRI1}::SlBRI1-GFP plants after heat stress (Figures 6 and 8). The reduced accumulation of H_2O_2 in P_{SlBRI1}::S1040A-GFP lines resulted from increases in activities and transcription of ROS-scavenging enzymes, namely, CAT, SOD, and POD, and the increased activities of these ROS-scavenging enzymes were caused by increased expression levels of *Cu/Zn-SOD* and *POD1* (Figures 6 and 8). The cell membrane is the major target of heat injury and is sensitive to lipid peroxidation. P_{SlBRI1}::S1040A-GFP lines, with the highest activities of the ROS-scavenging system, showed less membrane damage that P_{SlBRI1}::S1040D-GFP and P_{SlBRI1}::SlBRI1-GFP plants after heat stress, as the MDA content and electrolyte leakage were lower in P*SlBRI1*::S1040A-GFP lines, while the proline content was higher (Figure 5). Numerous studies have demonstrated that chlorophyll, which is the most important pigment for photosynthesis, is inhibited by heat stress [34]. Under high-temperature conditions, P*SlBRI1*::S1040A-GFP lines showed higher chlorophyll content and CO_2 assimilation rates than P*SlBRI1*::S1040D-GFP and P*SlBRI1*::SlBRI1-GFP plants (Figure 7). These results clearly suggested that dephosphorylation of SlBRI1 Ser-1040 could protect tomato plants from ROS-induced cellular injury and sustain their photosynthetic capacity to enhance their heating tolerance.

High-temperature tolerance in plants has largely been achieved by elevating the expression of *HSP* genes and of trans-acting factors such as HSFs and *WRKYs*. Previous studies demonstrated that HSP70 and HSP90 are abundant in eukaryotic cells and responsible for cell protection under high-temperature conditions [35–37]. The *HvBRI1* mutant of barley was characterized by decreased accumulation of *HSPs* under high temperature [38]. In tomato, overexpressing *HSP21* could protect the photosynthesis system from oxidative stress at high temperatures; however, the BR signalling gene *BZR1* could not influence *HSP* expression during heat stress [33,39]. In soybean and rice, *HSFA2* and *HSFA3* enhance heat tolerance by activating the expression of *HSPs* [40–42]. WRKY transcription factors have been extensively utilized to enhance heat stress tolerance in crops. AtWRKY25 and AtWRKY39 in *Arabidopsis* influence high-temperature stress tolerance by regulating the transcription levels of *HSFs* and *HSP100* [43,44]. Increased expression of *WRKY11* in rice also resulted in enhanced tolerance to heat [45]. Our study also examined the expression profiles of genes associated with *HSPs*, *HSFs*, and *WRKYs*, and the result was basically consistent with those previous studies. All the analysed genes were expressed equally when plants were grown under normal conditions, and the expression increased in response to heat stratification. P*SlBRI1*::S1040A-GFP lines, which were most resistant to high temperature, exhibited the highest accumulation of *HSPs*, *HSFs*, and *WRKYs* (Figure 8). Furthermore, the expression of *SlBRI1* and *SlCPD* showed no obvious regulation during heat stress (data not shown). These data suggested that the modulatory effect of SlBRI1 Ser-1040 on tomato heat tolerance was exerted mainly via changes in the expression levels of *HSPs*, *HSFs*, and *WRKYs*. Notably, in contrast to P*SlBRI1*::S1040A-GFP, the heat-responsive modulatory mechanism in *cu3*$^{-abs1}$ was mostly associated with BR signal strength, since *cu3*$^{-abs1}$ plants with high tolerance to heat stress showed similar expression levels of *HSPs*, *HSFs*, and *WRKYs* as P*SlBRI1*::S1040D-GFP and P*SlBRI1*::SlBRI1-GFP plants, and the expression of *SlBRI1* and *SlCPD* increased and decreased, respectively, in response to heat stress (Figure 8 and Figure S4).

Tomato is a major horticultural crop that is thermophilic but cannot withstand very high temperatures; therefore, tomato plants cannot survive the high temperatures during the summer in most tomato planting regions, which shortens the growing season. Thus, studies on the mechanism of heat tolerance in tomato to improve yield under high temperature are very important in tomato breeding. Previous studies have demonstrated that both plant growth and fertilization, which are determinants of plant yield, are sensitive to ambient temperature and are usually inhibited when the temperature increases [46]. The phytohormone BR is involved in plant promotion and stress adaption, and BR-treated plants exhibit improved photosynthetic activity, total chlorophyll content, and membrane integrity, which together promote tomato yield under heat stress [47]. The BR receptor BRI1 has been found to play an important role in crop yield via regulation of plant architecture. Several *bri1*-defective mutants of many species exhibit a semi-dwarf phenotype and decreased stem-leaf angles, which are recognized as indispensable attributes for intensive agriculture [15,19,48]. We previously reported that SlBRI1-overexpressing plants exhibited increased tomato yields, and dephosphorylated Thr-1050 of SlBRI1 could promote tomato yield by compacting plant architecture, as well as increasing fruit number and fruit weight, under normal conditions [16,25]. In the present investigation, the effects of Ser-1040 of SlBRI1 on tomato yield were distinct from those of Thr-1050. Ser-1040 had no effects on plant development under normal conditions but could improve tomato yield under heat stress. The early yield of individual P*SlBRI1*::S1040A-GFP plants was similar to that of P*SlBRI1*::SlBRI1-GFP and P*SlBRI1*::S1040D-GFP plants; however, the yield of P*SlBRI1*::S1040A-GFP at the late-spring stage with high temperature was significantly higher than that of other lines, mainly due to an increase in number of flowers and in the fruit setting rates (Figure 3). Further investigation indicated that this increase in the fruit setting rate occurred regardless of plant fertilization, since the flower phenotype was unchanged and similar (Figure 3 and Figure S1), and the pollen viability, which is usually sensitive to high-temperature stress. was also the same among all of the transgenic plants. Thus, we concluded that Ser-1040 could increase the yield at high temperatures based on the following aspects. First, the increased yield of P_{SlBRI1}::S1040A-GFP lines might be related to their excellent photosynthetic capabilities under heat stress, as the decline rates of both the total chlorophyll content and CO_2 assimilation rates in P_{SlBRI1}::S1040A-GFP lines were less than those in P_{SlBRI1}::SlBRI1-GFP and P_{SlBRI1}::S1040D-GFP lines under heat stress (Figure 7). Second, the increased activity of the ROS detoxification system under heat stress might increase the yield of P_{SlBRI1}::S1040A-GFP lines. The activities of several enzymes, such as SOD and POD, were higher in P_{SlBRI1}::S1040A-GFP lines than in other transgenic lines, and these increased activities were caused by an increase in *Cu/Zn-SOD* and *POD1* gene expression levels (Figure 6 and 8). Third, compared with the levels in P_{SlBRI1}::SlBRI1-GFP and P_{SlBRI1}::S1040D-GFP lines, the high expression levels of *HSPs*, *HSFs*, and *WRKYs* in P_{SlBRI1}::S1040A-GFP also helped increase the yield of this line under heat stress (Figure 8). Furthermore, to confirm that the higher yield at later stages in P_{SlBRI1}::S1040A-GFP lines was indeed due to their tolerance to high temperature rather than their phenotypes at different developmental stages, we delayed the seeding times of all the plants for two months to ensure that their seedling stages occurred entirely under high-temperature conditions. The results showed that P_{SlBRI1}::S1040A-GFP plants grew stronger and had more flowers in the first two inflorescences than plants of P_{SlBRI1}::SlBRI1-GFP and P_{SlBRI1}::S1040D-GFP lines (Figure S3). Taken together, our results clearly demonstrated that SlBRI1 Ser-1040 could influence the capacities of photosynthesis and ROS detoxification, as well as elevate the expression of heat defense-related genes in tomato, all of which functioned together to maintain stable high yields under high-temperature conditions.

4. Materials and Methods

4.1. Sequence Alignment and Phylogenetic Analysis

The multiple amino acid sequence alignment of BRI1 was performed in ClustalX2 (ftp://ftp.ebi.ac.uk/pub/software/clustalw2/2.1/) with default parameters. The phylogenetic analysis was performed using MEGA7 software (The Pennsylvania State University, State College, Pennsylvania, USA) and the neighbour-joining method and Poisson substitution model with 1000 bootstrap replications. The motif analysis was performed using MEME software (http://meme-suite.org/tools/meme) with the following parameters: (1) the optimum motif width was between 6 wide and 30 wide; (2) the maximum number of motifs was 30; and (3) the other parameters were set to the default values.

4.2. Generation of Transgenic Plants by Site-Directed Mutagenesis

Both the native promoter (2989 bp) of *SlBRI1* (Solyc04g051510) and its coding sequence without a stop codon were amplified from tomato (*Solanum lycopersicum* cv. Moneymaker). The amplified fragments were then recombined into the plant expression vector pBI121 (CLONTECH, Palo Alto, CA, USA) with a GFP tag followed by the CT region to construct the vector encoding wild-type *SlBRI1* (P_{SlBRI1}::SlBRI1-GFP-pBI121). The S1040A or S1040D mutation of SlBRI1 was amplified from P_{SlBRI1}::SlBRI1-GFP-pBI121 using overlap PCR amplification to obtain the plant expression vector P_{SlBRI1}::S1040A-GFP-pBI121 or P_{SlBRI1}::S1040D-GFP-pBI121, respectively. All the constructed vectors were transformed into *Agrobacterium tumefaciens* GV3101 (WEIDI, AC1001S, Shanghai, China) for tomato transformation. Primers designed for plant expression vector construction are listed in Table S1.

For tomato transformation, $cu3^{-abs1}$ was used in accordance with the cotyledon transformation method [49]. The transgenic plants were screened by kanamycin and PCR analysis, while quantitative real-time PCR and western blot analysis were used to confirm that the transgenic proteins were stably expressed in these plants. The T_2 generation of two independent homozygous lines from P_{SlBRI1}::S1040A-GFP-pBI121 (S1040A-8 and S1040A-10), P_{SlBRI1}::S1040D-GFP-pBI121 (S1040D-1 and S1040D-2), and P_{SlBRI1}::SlBRI1-GFP-pBI121 (SlBRI1-1 and SlBRI1-3) were used in this study.

4.3. Agronomic Trait Characterization

For agronomic trait investigations, both transgenic lines and $cu3^{-abs1}$ were planted in early spring and grown in a glasshouse without additional temperature and light controlling equipment. The plant height was measured as the distance from the cotyledon to the top of the plant, while the stem diameter was measured as the diameter of the cotyledon node at the flowering stage of the fourth inflorescence. Single-fruit weight was considered as the individual fruit weight at the RR stage. The early yield was the total weight of the first to the third fruit nodes per plant, while the total yield was the total weight of the first to the sixth fruit nodes per plant. Each agronomic trait was measured for at least 10 independent biological replicates.

4.4. Growth Response of Hypocotyls to Exogenous BRZ and BL

To analyse the growth response of hypocotyls to BR signalling, sterilized seeds of both transgenic lines and $cu3^{-abs1}$ were inoculated in Petri dishes that contained solid 1/2 strength Murashige and Skoog medium (1/2 MS medium) with BRZ (Tokyo Chemical Industry Co., Ltd., Tokyo, Japan) at 0 nM, 10 nM, 100 nM, 500 nM, and 1000 nM or with epi-BL (Shanghai Yuanye Biotechnology Co. Ltd., Shanghai, China) at 0 nM, 10 nM, 100 nM, 500 nM, and 1000 nM. These seeds were maintained at 25 °C in the dark for 10 days. Hypocotyl lengths were subsequently measured, and the relative hypocotyl length represented the changes in the hypocotyl length at different concentrations of BRZ or epi-BL. Each treatment was performed on at least 15 seedlings.

4.5. Germination Analysis of Plants under Heat Stress

For germination assays under heat stress, the seeds of both transgenic lines and $cu3^{-abs1}$ were inoculated in Petri dishes with wet filter paper under 33 °C or 28 °C (control), a 16 h light/8 h dark cycle, and 250 $\mu mol \cdot m^{-2} \cdot s^{-1}$ photosynthetic photon flux density (PPFD) (BN058C LED11/WW L1200 GC OL, Philips, Amsterdam, Holland) as previously reported [50]. The germination rate and germination potential were calculated on the fourteenth day and fourth day, respectively. The root length and MDA content of early-stage seedlings were both detected on the fourteenth day after seeding. There were three replicates for each index, and each replicate had 30 plants.

4.6. Heat Treatment of Seedlings

To analyse the seedling tolerance to heat stress, tomato seeds of both transgenic lines and $cu3^{-abs1}$ were sown in plastic pots (15 × 15 × 15 cm) containing peat (PH 6.0, Pindstrup Plus, Pindstrup Rosebrug A/S Denmark) and vermiculite (8:2, v/v) under the following conditions: 16 h light/8 h dark photoperiod, 25 °C, 70% relative humidity (RH), and 250 $\mu mol \cdot m^{-2} \cdot s^{-1}$ PPFD. For heat shock stress, seedlings at the four-leaf stage were exposed to 38 °C/28 °C (day/night) for 9 days and then transferred to 25 °C for 3 days, with other environmental conditions remaining unchanged. The control seedlings were maintained at 25 °C during the same period. Each index was measured for at least 3 independent biological replicates.

4.7. Growth Performances of Seedlings under Heat Stress

Both the fresh weight and dry weight of the shoots and roots, as well as the seedling height and seedling stem diameter, were measured on the ninth day during heat stress. For seedling dry weight measurement, the seedlings were dried at 105 °C for 15 min and then placed at 75 °C to achieve constant weight. There were 3 replicates, and each replicate had 10 plants.

4.8. Physiological Parameter Measurements

The third fully expanded leaves from the top of the control and stressed seedlings were harvested for physiological parameter measurements at 0, 3, 6, 9, and 12 days during heat stress. The MDA content, electrolyte leakage, and free proline accumulation were determined in accordance with the methods described by Liu [51]. The chlorophyll content and CAT, SOD, and POD activities in leaves were measured as described previously [17]. The CO_2 assimilation rate of third fully expanded leaves from the top was measured by using an infrared gas analyser-based portable photosynthesis system (LI-6800; LI-COR, Lincoln, NE, USA), the measurements were carried out at 25°C, 800 $\mu mol\cdot m^{-2}\cdot s^{-1}$ light intensity, 70% RH, and 400 $\mu mol\ mol^{-1}$ CO_2 concentration, respectively. The chlorophyll content were measured according to the methods described by Kong [52]. Each index had at least 3 replicates, and each replicate had 10 plants.

4.9. DAB Staining

The upper-third fully expanded leaves of the control and stressed seedlings were stained by a solution containing 1 mg ml^{-1} 3,3′-diaminobenzidine (DAB, D12384, Sigma-Aldrich, Saint Louis, MO, USA), pH 5.5 to detect the accumulation of H_2O_2 in leaves. The histochemical analysis of H_2O_2 was performed as described previously [33].

4.10. Autophosphorylation Analysis

For autophosphorylation analyses in vitro, the cytoplasmic domain (824 to 1207 aa) of SlBRI1, S1040A, S1040D, or K916E was amplified from P_{SlBRI1}::SlBRI1-GFP-pBI121, P_{SlBRI1}::S1040A-GFP-pBI121, P_{SlBRI1}::S1040D-GFP-pBI121, or P_{SlBRI1}::SlBRI1-GFP-pBI121 with the specific primers (*SlBRI1*-CD-F and *SlBRI1*-CD-R) listed in Table S1. The amplified fragment was subcloned into the prokaryotic expression vector pFLAG-MAC and then transformed into *E. coli* BL21 (DE3) pLysS (Transgene, CD901-02, Beijing, China). SlBRI1-FLAG-MAC and K916E-FLAG-MAC were used as the positive and negative controls, respectively, and subsequent protein purification and autophosphorylation analyses in vitro were performed according to previously described methods [21,53]. Intensities of bands were quantified by using ImageJ software (NIH, Bethesda, Maryland, USA) and presented as relative values compared with the FLAG-SlBRI1.

4.11. Western Blot Analysis

Target proteins were extracted from 4-week-old transgenic tomato leaves (0.2 g) expressing SlBRI1, S1040A, or S1040D, and the experiment was performed as previously described [21].

4.12. Quantitative Real-time PCR Analysis

The second leaves from the top of the seedlings were selected for transcription analysis at 0, 3, 6, 9, and 12 days during heat stress. Total RNA from the leaves of the seedlings was extracted with an RNAiso Plus Kit (TaKaRa, 9109, Kusatsu, Japan) and transcribed to cDNA with a Transcriptor First Strand cDNA Synthesis Kit (Roche, 4896866001, Mannheim, Germany). qRT-PCR was performed by using a SYBR Green Master Mix Kit (Vazyme, Q121-02, Nanjing, China). The method was performed according to the manufacturer's protocol, and primers for target genes were designed based on previously described methods [33,54]. The tomato housekeeping gene *Actin* was amplified as an internal reference. Each data point had 3 biological replicates and technical replications.

4.13. Statistical Analysis

The data in this study were analysed using a one-way analysis of variance (ANOVA) via SPSS 17.0. (IBM, Armonk, New York, USA). The means and standard errors of each sample were calculated, significant differences between means were examined by LSD test ($p < 0.05$) and represented by different letters.

5. Conclusions

Heat stress is a major abiotic stress influencing agricultural production, and BRI1 is a BR receptor that plays a critical role in plant growth and stress adaption. Previous studies of SlBRI1 phosphorylation sites have suggested that modifying SlBRI1 Thr-1050 in tomato could promote yield through precise control of BR signal strength; however, this functional analysis was limited to Thr-1050, and the associated agronomic traits of other phosphorylation sites, especially their functions in stress tolerance, remained unclear. In this study, we revealed for the first time the biological role of SlBRI1 phosphorylation sites in tomato stress adaption. SlBRI1 Ser-1040 was found to participate in plant responses to heat stress by influencing the autophosphorylation of SlBRI1. Transgenic plants harbouring S1040A exhibited better performance, including faster growth rates, better germination, more active photosynthetic and ROS detoxification systems, and higher expression of heat defence genes, under high temperature, which helped them adapt to and maintain high yields under heat stress. Our results not only shed light on the molecular mechanisms of SlBRI1 in stress adaption regulation but also provide a molecular basis for establishing high-yield tomato lines via fine-tuning of phosphorylation sites of SlBRI1.

Supplementary Materials: Supplementary Materials can be found at http://www.mdpi.com/1422-0067/21/20/7681/s1. Figure S1: Yield trait analysis of *cu3^{-abs1}* and transgenic plants harbouring SlBRI1, SlBRI1 dephosphorylated at Ser-1040, and SlBRI1 phosphorylated at Ser-1040; Figure S2: Dephosphorylation of Ser-1040 promotes seedling growth under heat stress; Figure S3: Dephosphorylation of Ser-1040 improves tomato growth and flower number under heat stress; Figure S4: BR signalling influences the heat tolerance of *cu3^{-abs1}*; Table S1: Primers used in this research.

Author Contributions: Conceptualization, S.W. and X.W.; Data curation, S.W., T.H., A.T. and S.Z.; Funding acquisition, S.W. and X.W.; Methodology, S.W., T.H. and S.H.; Validation, B.L., C.D. and F.Z.; Writing—Original draft, S.W., T.H. and X.W.; Writing—Review & editing, S.W., T.H. and X.W. All authors have read and agreed to the published version of the manuscript.

Funding: This research was funded by the National Natural Science Foundation of China (nos. 31672142 and 31501771), and Major Projects for Industrial, Academic and Research Collaborative Innovation of Yangling Agricultural High-tech Industrial Demonstration Zone (nos. 2018CXY-09).

Conflicts of Interest: The authors declare no conflict of interest.

Abbreviations

BR	Brassinosteroid
BRI1	Brassinosteroid Insensitive1
CPD	Constitutive Photomorphogenesis and Dwarf
epi-BL	24-Epibrassinolide
BRZ	Brassinazole
K916E	Lysine-916-Glutamic acid
GFP	Green fluorescent protein
Ser	Serine
S1040A	Serine-1040-alanine
S1040D	Serine-1040-aspartic acid
Thr	Threonine
Tyr	Tyrosine
ROS	Reactive oxygen species
DAB	3,3′-Diaminobenzidine
CAT	Catalase
SOD	Superoxide dismutase
RBOH1	RESPIRATORY BURST OXIDASE HOMOLOG1
HSF	Heat shock transcription factor
HSP	Heat shock protein
POD	Peroxidase
MDA	Malondialdehyde

Pn   Net photosynthetic rate
qRT-PCR   Quantitative real-time PCR analysis
MS   Murashige and Skoog medium
PPFD   Photosynthetic photon flux density

References

1. Bita, C.E.; Gerats, T. Plant tolerance to high temperature in a changing environment: Scientific fundamentals and production of heat stress-tolerant crops. *Front. Plant. Sci.* **2013**, *4*, 273. [CrossRef]
2. Suzuki, N.; Mittler, R. Reactive oxygen species and temperature stresses: A delicate balance between signaling and destruction. *Physiol. Plant.* **2006**, *126*, 45–51. [CrossRef]
3. Kim, S.Y.; Kim, B.H.; Lim, C.J.; Lim, C.O.; Nam, K.H. Constitutive activation of stress-inducible genes in a *brassinosteroid-insensitive 1 (bri1)* mutant results in higher tolerance to cold. *Physiol. Plant* **2010**, *138*, 191–204. [CrossRef]
4. Xia, X.J.; Wang, Y.J.; Zhou, Y.H.; Mao, W.H.; Shi, K.; Asami, T.; Chen, Z.; Yu, J.Q. Reactive oxygen species are involved in brassinosteroid-induced stress tolerance in cucumber. *Plant Physiol.* **2009**, *150*, 801–814. [CrossRef]
5. Fariduddin, Q.; Yusuf, M.; Ahmad, I.; Ahmad, A. Brassinosteroids and their role in response of plants to abiotic stresses. *Biol. Plantarum* **2014**, *58*, 9–17. [CrossRef]
6. Niu, J.H.; Anjum, S.A.; Wang, R.; Li, J.H.; Liu, M.R.; Song, J.X.; Zohaib, A.; Lv, J.; Wang, S.G.; Zong, X.F. Exogenous application of brassinolide can alter morphological and physiological traits of *Leymus chinensis* (Trin.) Tzvelev under room and high temperatures. *Chil. J. Agr. Res.* **2016**, *76*, 27–33. [CrossRef]
7. Singh, A.; Mittal, D.; Lavania, D.; Agarwal, M.; Mishra, R.C.; Grover, A. OsHsfA2c and OsHsfB4b are involved in the transcriptional regulation of cytoplasmic OsClpB (*Hsp100*) gene in rice (*Oryza sativa* L.). *Cell Stress Chaperones* **2012**, *17*, 243–254. [CrossRef] [PubMed]
8. Planas-Riverola, A.; Gupta, A.; Betegon-Putze, I.; Bosch, N.; Ibanes, M.; Cano-Delgado, A.I. Brassinosteroid signaling in plant development and adaptation to stress. *Development* **2019**, *146*, 151894. [CrossRef]
9. Nawaz, F.; Naeem, M.; Zulfiqar, B.; Akram, A.; Ashraf, M.Y.; Raheel, M.; Shabbir, R.N.; Hussain, R.A.; Anwar, I.; Aurangzaib, M. Understanding brassinosteroid-regulated mechanisms to improve stress tolerance in plants: A critical review. *Environ. Sci. Pollut. Res.* **2017**, *24*, 15959–15975. [CrossRef]
10. Yuan, G.F.; Jia, C.G.; Li, Z.; Sun, B.; Zhang, L.P.; Liu, N.; Wang, Q.M. Effect of brassinosteroids on drought resistance and abscisic acid concentration in tomato under water stress. *Sci. Hortic.* **2010**, *126*, 103–108. [CrossRef]
11. Wang, X.F.; Kota, U.; He, K.; Blackburn, K.; Li, J.; Goshe, M.B.; Huber, S.C.; Clouse, S.D. Sequential transphosphorylation of the BRI1/BAK1 receptor kinase complex impacts early events in brassinosteroid signaling. *Dev. Cell.* **2008**, *15*, 220–235. [CrossRef]
12. Clouse, S.D. Brassinosteroid signal transduction: From receptor kinase activation to transcriptional networks regulating plant development. *Plant Cell* **2011**, *23*, 1219–1230. [CrossRef] [PubMed]
13. Kim, T.W.; Wang, Z.Y. Brassinosteroid signal transduction from receptor kinases to transcription factors. *Annu. Rev. Plant. Biol.* **2010**, *61*, 681–704. [CrossRef]
14. Zhang, X.W.; Zhou, L.Y.; Qin, Y.K.; Chen, Y.W.; Liu, X.L.; Wang, M.Y.; Mao, J.; Zhang, J.J.; He, Z.H.; Liu, L.C.; et al. A temperature-sensitive misfolded bri1-301 receptor requires its kinase activity to promote growth. *Plant Physiol.* **2018**, *178*, 1704–1719. [CrossRef] [PubMed]
15. Gruszka, D.; Szarejko, I.; Maluszynski, M. New allele of *HvBRI1* gene encoding brassinosteroid receptor in barley. *J. Appl. Genetics* **2011**, *52*, 257–268. [CrossRef] [PubMed]
16. Nie, S.M.; Huang, S.H.; Wang, S.F.; Cheng, D.D.; Liu, J.W.; Lv, S.Q.; Li, Q.; Wang, X.F. Enhancing brassinosteroid signaling via overexpression of tomato (*Solanum lycopersicum*) SlBRI1 improves major agronomic traits. *Front. Plant. Sci.* **2017**, *8*, 1386. [CrossRef]
17. Nie, S.M.; Huang, S.H.; Wang, S.F.; Mao, Y.J.; Liu, J.W.; Ma, R.L.; Wang, X.F. Enhanced brassinosteroid signaling intensity via *SlBRI1* overexpression negatively regulates drought resistance in a manner opposite of that via exogenous BR application in tomato. *Plant Physiol. Biochem.* **2019**, *138*, 36–47. [CrossRef] [PubMed]

18. Singh, A.; Breja, P.; Khurana, J.P.; Khurana, P. Wheat *Brassinosteroid-Insensitive1* (*TaBRI1*) interacts with members of *TaSERK* gene family and cause early flowering and seed yield enhancement in *Arabidopsis*. *PLoS ONE* **2016**, *11*, e0153273. [CrossRef]

19. Morinaka, Y.; Sakamoto, T.; Inukai, Y.; Agetsuma, M.; Kitano, H.; Ashikari, M.; Matsuoka, M. Morphological alteration caused by brassinosteroid insensitivity increases the biomass and grain production of rice. *Plant. Physiol.* **2006**, *141*, 924–931. [CrossRef]

20. Zheng, B.W.; Bai, Q.W.; Wu, L.; Liu, H.; Liu, Y.P.; Xu, W.J.; Li, G.S.; Ren, H.Y.; She, X.P.; Wu, G. EMS1 and BRI1 control separate biological processes via extracellular domain diversity and intracellular domain conservation. *Nat. Commun.* **2019**, *10*, 4165. [CrossRef]

21. Wang, X.F.; Goshe, M.B.; Soderblom, E.J.; Phinney, B.S.; Kuchar, J.A.; Li, J.; Asami, T.; Yoshida, S.; Huber, S.C.; Clouse, S.D. Identification and functional analysis of in vivo phosphorylation sites of the *Arabidopsis* BRASSINOSTEROID-INSENSITIVE1 receptor kinase. *Plant Cell* **2005**, *17*, 1685–1703. [CrossRef]

22. Wang, Q.N.; Wang, S.F.; Gan, S.F.; Wang, X.; Liu, J.W.; Wang, X.F. Role of specific phosphorylation sites of *Arabidopsis* Brassinosteroid-Insensitive 1 receptor kinase in plant growth and development. *J. Plant. Growth Regul.* **2016**, *35*, 755–769. [CrossRef]

23. Oh, M.H.; Sun, J.D.; Oh, D.H.; Zielinski, R.E.; Clouse, S.D.; Huber, S.C. Enhancing *Arabidopsis* leaf growth by engineering the BRASSINOSTEROID INSENSITIVE1 receptor kinase. *Plant. Physiol.* **2011**, *157*, 120–131. [CrossRef] [PubMed]

24. Bajwa, V.S.; Wang, X.F.; Blackburn, R.K.; Goshe, M.B.; Mitra, S.K.; Williams, E.L.; Bishop, G.J.; Krasnyanski, S.; Allen, G.; Huber, S.C.; et al. Identification and functional analysis of tomato BRI1 and BAK1 receptor kinase phosphorylation sites. *Plant Physiol.* **2013**, *163*, 30–42. [CrossRef] [PubMed]

25. Wang, S.F.; Liu, J.W.; Zhao, T.; Du, C.X.; Nie, S.M.; Zhang, Y.Y.; Lv, S.Q.; Huang, S.H.; Wang, X.F. Modification of Threonine-1050 of SlBRI1 regulates BR Signalling and increases fruit yield of tomato. *BMC Plant Biol.* **2019**, *19*, 256. [CrossRef] [PubMed]

26. Mathur, J.; Molnar, G.; Fujioka, S.; Takatsuto, S.; Sakurai, A.; Yokota, T.; Adam, G.; Voigt, B.; Nagy, F.; Maas, C.; et al. Transcription of the *Arabidopsis CPD* gene, encoding a steroidogenic cytochrome P450, is negatively controlled by brassinosteroids. *Plant J.* **1998**, *14*, 593–602. [CrossRef]

27. Nie, W.F.; Wang, M.M.; Xia, X.J.; Zhou, Y.H.; Shi, K.; Chen, Z.; Yu, J.Q. Silencing of tomato *RBOH1* and *MPK2* abolishes brassinosteroid-induced H_2O_2 generation and stress tolerance. *Plant Cell Environ.* **2013**, *36*, 789–803. [CrossRef]

28. Hussain, M.; Khan, T.A.; Yusuf, M.; Fariduddin, Q. Silicon-mediated role of 24-epibrassinolide in wheat under high-temperature stress. *Environ. Sci. Pollut. Res. Int.* **2019**, *26*, 17163–17172. [CrossRef]

29. Mazorra, L.M.; Holton, N.; Bishop, G.J.; Nunez, M. Heat shock response in tomato brassinosteroid mutants indicates that thermotolerance is independent of brassinosteroid homeostasis. *Plant Physiol. Biochem.* **2011**, *49*, 1420–1428. [CrossRef]

30. Miller, G.; Schlauch, K.; Tam, R.; Cortes, D.; Torres, M.A.; Shulaev, V.; Dangl, J.L.; Mittler, R. The plant NADPH oxidase RBOHD mediates rapid systemic signaling in response to diverse stimuli. *Sci. Signal.* **2009**, *2*, 45. [CrossRef]

31. Marino, D.; Dunand, C.; Puppo, A.; Pauly, N. A burst of plant NADPH oxidases. *Trends Plant Sci.* **2012**, *17*, 9–15. [CrossRef] [PubMed]

32. Chen, C.T.; Chen, T.H.; Lo, K.F.; Chiu, C.Y. Effects of proline on copper transport in rice seedlings under excess copper stress. *Plant Sci.* **2004**, *166*, 103–111. [CrossRef]

33. Yin, Y.L.; Qin, K.Z.; Song, X.W.; Zhang, Q.H.; Zhou, Y.H.; Xia, X.J.; Yu, J.Q. BZR1 transcription factor regulates heat stress tolerance through FERONIA receptor-like kinase-mediated reactive oxygen species signaling in tomato. *Plant Cell Physiol.* **2018**, *59*, 2239–2254. [CrossRef]

34. Farhad, M.S.; Babak, A.M.; Reza, Z.M.; Hassan, R.S.M.; Afshin, T. Response of proline, soluble sugars, photosynthetic pigments and antioxidant enzymes in potato (*Solanum tuberosum* L.) to different irrigation regimes in greenhouse condition. *Aust. J. Crop. Sci.* **2011**, *5*, 55–60.

35. Al-Whaibi, M.H. Plant heat-shock proteins: A mini review. *J. King Saud Univ. Sci.* **2011**, *23*, 139–150. [CrossRef]

36. Wang, W.X.; Vinocur, B.; Shoseyov, O.; Altman, A. Role of plant heat-shock proteins and molecular chaperones in the abiotic stress response. *Trends Plant Sci.* **2004**, *9*, 244–252. [CrossRef] [PubMed]

37. Usman, M.G.; Rafii, M.Y.; Martini, M.Y.; Yusuff, O.A.; Ismail, M.R.; Miah, G. Molecular analysis of Hsp70 mechanisms in plants and their function in response to stress. *Biotechnol. Genet. Eng. Rev.* **2017**, *33*, 26–39. [CrossRef]

38. Sadura, I.; Libik-Konieczny, M.; Jurczyk, B.; Gruszka, D.; Janeczko, A. HSP transcript and protein accumulation in brassinosteroid barley mutants acclimated to low and high temperatures. *Int. J. Mol. Sci.* **2020**, *21*, 1889. [CrossRef] [PubMed]

39. Neta-Sharir, I.; Isaacson, T.; Lurie, S.; Weiss, D. Dual role for tomato heat shock protein 21: Protecting photosystem II from oxidative stress and promoting color changes during fruit maturation. *Plant Cell* **2005**, *17*, 1829–1838. [CrossRef]

40. Ogawa, D.; Yamaguchi, K.; Nishiuchi, T. High-level overexpression of the *Arabidopsis HsfA2* gene confers not only increased themotolerance but also salt/osmotic stress tolerance and enhanced callus growth. *J. Exp. Bot.* **2007**, *58*, 3373–3383. [CrossRef]

41. Zhu, B.G.; Ye, C.J.; Lu, H.Y.; Chen, X.J.; Chai, G.H.; Chen, J.N.; Wang, C. Identification and characterization of a novel heat shock transcription factor gene, *GmHsfA1*, in soybeans (*Glycine max*). *J. Plant Res.* **2006**, *119*, 247–256. [CrossRef]

42. Liu, J.G.; Qin, Q.L.; Zhang, Z.; Peng, R.H.; Xiong, A.S.; Chen, J.M.; Yao, Q.H. *OsHSF7* gene in rice, *Oryza sativa* L., encodes a transcription factor that functions as a high temperature receptive and responsive factor. *BMB Rep.* **2009**, *42*, 16–21. [CrossRef]

43. Li, S.J.; Fu, Q.T.; Huang, W.D.; Yu, D.Q. Functional analysis of an *Arabidopsis* transcription factor *WRKY25* in heat stress. *Plant Cell Rep.* **2009**, *28*, 683–693. [CrossRef] [PubMed]

44. Li, S.J.; Zhou, X.; Chen, L.G.; Huang, W.D.; Yu, D.Q. Functional characterization of *Arabidopsis thaliana* WRKY39 in heat stress. *Mol. Cells* **2010**, *29*, 475–483. [CrossRef]

45. Wu, X.L.; Shiroto, Y.; Kishitani, S.; Ito, Y.; Toriyama, K. Enhanced heat and drought tolerance in transgenic rice seedlings overexpressing *OsWRKY11* under the control of *HSP101* promoter. *Plant Cell Rep.* **2009**, *28*, 21–30. [CrossRef]

46. Casal, J.J.; Balasubramanian, S. Thermomorphogenesis. *Annu. Rev. Plant Biol.* **2019**, *70*, 321–346. [CrossRef] [PubMed]

47. Khan, A.R.; Hui, C.Z.; Ghazanfar, B.; Khan, M.A.; Ahmad, S.S.; Ahmad, I. Acetyl salicylic acid and 24-epibrassinolide attenuate decline in photosynthesis, chlorophyll contents and membrane thermo-stability in tomato (*Lycopersicon Esculentum* Mill.) under heat stress. *Pak. J. Bot.* **2015**, *47*, 63–70.

48. Li, J.; Chory, J. A putative leucine-rich repeat receptor kinase involved in brassinosteroid signal transduction. *Cell* **1997**, *90*, 929–938. [CrossRef]

49. Park, S.H.; Morris, J.L.; Park, J.E.; Hirschi, K.D.; Smith, R.H. Efficient and genotype-independent *Agrobacterium*-mediated tomato transformation. *J. Plant Physiol.* **2003**, *160*, 1253–1257. [CrossRef]

50. Qian, C.M.; Wu, X.J.; Chen, L.; Jiang, C.H. Effects of high temperature stress on the germination of tomato seed. *Seed* **2002**, *20*, 89.

51. Liu, H.; Yu, C.Y.; Li, H.X.; Ouyang, B.; Wang, T.T.; Zhang, J.H.; Wang, X.; Ye, Z.B. Overexpression of *ShDHN*, a dehydrin gene from *Solanum habrochaites* enhances tolerance to multiple abiotic stresses in tomato. *Plant Sci.* **2015**, *231*, 198–211. [CrossRef] [PubMed]

52. Kong, F.Y.; Deng, Y.S.; Zhou, B.; Wang, G.D.; Wang, Y.; Meng, Q.W. A chloroplast-targeted DnaJ protein contributes to maintenance of photosystem II under chilling stress. *J. Exp. Bot.* **2014**, *65*, 143–158. [CrossRef]

53. Li, J.; Wen, J.Q.; Lease, K.A.; Doke, J.T.; Tax, F.E.; Walker, J.C. BAK1, an *Arabidopsis* LRR receptor-like protein kinase, interacts with BRI1 and modulates brassinosteroid signaling. *Cell* **2002**, *110*, 213–222. [CrossRef]

54. Zhou, J.; Wang, J.; Li, X.; Xia, X.J.; Zhou, Y.H.; Shi, K.; Chen, Z.; Yu, J.Q. H_2O_2 mediates the crosstalk of brassinosteroid and abscisic acid in tomato responses to heat and oxidative stresses. *J. Exp. Bot.* **2014**, *65*, 4371–4383. [CrossRef] [PubMed]

Publisher's Note: MDPI stays neutral with regard to jurisdictional claims in published maps and institutional affiliations.

 International Journal of
Molecular Sciences

Article

Cell Type-Specific Imaging of Calcium Signaling in *Arabidopsis thaliana* Seedling Roots Using GCaMP3

William Krogman, J. Alan Sparks and Elison B. Blancaflor *

Noble Research Institute LLC, 2510 Sam Noble Parkway, Ardmore, OK 73401, USA;
wlkrogman@noble.org (W.K.); jasparks@noble.org (J.A.S.)
* Correspondence: eblancaflor@noble.org

Received: 22 July 2020; Accepted: 28 August 2020; Published: 2 September 2020

Abstract: Cytoplasmic calcium ($[Ca^{2+}]_{cyt}$) is a well-characterized second messenger in eukaryotic cells. An elevation in $[Ca^{2+}]_{cyt}$ levels is one of the earliest responses in plant cells after exposure to a range of environmental stimuli. Advances in understanding the role of $[Ca^{2+}]_{cyt}$ in plant development has been facilitated by the use of genetically-encoded reporters such as GCaMP. Most of these studies have relied on promoters such as *Cauliflower Mosaic Virus* (*35S*) and *Ubiquitin10* (*UBQ10*) to drive expression of *GCaMP* in all cell/tissue types. Plant organs such as roots consist of various cell types that likely exhibit unique $[Ca^{2+}]_{cyt}$ responses to exogenous and endogenous signals. However, few studies have addressed this question. Here, we introduce a set of *Arabidopsis thaliana* lines expressing *GCaMP3* in five root cell types including the columella, endodermis, cortex, epidermis, and trichoblasts. We found similarities and differences in the $[Ca^{2+}]_{cyt}$ signature among these root cell types when exposed to adenosine tri-phosphate (ATP), glutamate, aluminum, and salt, which are known to trigger $[Ca^{2+}]_{cyt}$ increases in root cells. These cell type-targeted GCaMP3 lines provide a new resource that should enable more in depth studies that address how a particular environmental stimulus is linked to specific root developmental pathways via $[Ca^{2+}]_{cyt}$.

Keywords: GCaMP3; Arabidopsis; cell type specificity; calcium response; $[Ca^{2+}]_{cyt}$

---

1. Introduction

Calcium (Ca^{2+}) plays a vital role as a second messenger system in plant development and stress response. Because Ca^{2+} acts as a stress signal, it is one of the first responses plants have to environmental stimuli. This fast response is achieved through an influx of extracellular Ca^{2+} and an efflux of stored Ca^{2+} from the vacuole, endoplasmic reticulum (ER), and mitochondria that is achieved via a concentration gradient between the intracellular Ca^{2+} stores and the cytoplasm [1]. In non-stimulated conditions, the concentration of Ca^{2+} in the cytoplasm rests around 100 nM [1–3]. This low concentration is maintained through H^+/Ca^{2+} antiporters and Ca^{2+}-ATPases that move Ca^{2+} into the apoplast or into intracellular stores [3,4]. Once the response in the cell is initiated, the cell will return to homeostasis by storing excess Ca^{2+} back in the vacuole, ER, or mitochondria through active transport [2], or by pumping it out into extracellular space through gated-ion channels in the plasma membrane.

Once Ca^{2+} enters the cytoplasm in high concentrations, it binds with Ca^{2+}-binding proteins to initiate a second messenger cascade [4,5]. Part of this signaling cascade is the activation of Ca^{2+}-dependent protein kinases and the regulation of transcription factors to increase tolerance to the stress that caused the initial Ca^{2+} response. Much of our understanding about Ca^{2+} as a second messenger has come from the development of technologies that enable monitoring of free cytosolic Ca^{2+} ($[Ca^{2+}]_{cyt}$) changes in living cells. Early on, these technologies consisted of indicator dyes that had to be chemically or physically loaded into the cell, such as Calcium Green or Indo-1. When Ca^{2+} levels in the cell became elevated upon exposing the cell to a specific stimulus, these indicator dyes increased fluorescence [6]. However, use of these indicator dyes in plant cells were often met with technical challenges including uneven dye loading, loss of cell viability after dye loading, and cells failing to take up the dyes [6,7].

Problems with these early indicator dyes were mitigated with the discovery of a suite of fluorescent proteins from jellyfish and other marine organisms [8,9]. Using molecular cloning techniques, selected regions of Ca^{2+}-binding proteins could be fused with these fluorescent proteins to generate genetically-encoded Ca^{2+} sensors that change conformation upon Ca^{2+} binding [8]. Ca^{2+} binding induced changes in the conformation of fluorescent protein-based sensors result in enhanced fluorescence emission. One example of a genetically-encoded Ca^{2+} sensor is GCaMP. GCaMP consists of an enhanced green fluorescent protein (eGFP) fused to calmodulin (CaM) and myosin light-chain kinase (M13) [1,10]. GCaMP3 is a modified form of GCaMP that features a circularly permutated enhanced GFP (eGFP) flanked by CaM and M13 that gives low fluorescence in the absence of Ca^{2+}, which increases upon reversible binding of Ca^{2+} to the CaM domain [11].

To date, expression of CGaMP and other genetically-encoded Ca^{2+} sensors in plants are driven by constitutive promoters such as *Cauliflower Mosaic Virus 35S (35S)* and *Ubiquitin 10 (UBQ10)* [12,13]. Although plant lines constitutively expressing genetically-encoded Ca^{2+} sensors enabled the functional study of $[Ca^{2+}]_{cyt}$ changes across entire plant organs [5,13,14], it is likely that each cell type within the organ is characterized by a unique $[Ca^{2+}]_{cyt}$ signature. While $[Ca^{2+}]_{cyt}$ changes typically observed in the entire plant organ likely originate from more than one cell type, we lack tools to monitor these changes within specific tissues or cells [15]. To address this need, we generated a set of lines in the model plant *Arabidopsis thaliana*, expressing *GCaMP3* targeted to the columella, endodermis, epidermis, cortex, and trichoblast using cell-type specific promoters [16]. We demonstrate the utility of these new GCaMP3 lines in reporting cell-type $[Ca^{2+}]_{cyt}$ changes in roots treated with chemicals previously shown to induce $[Ca^{2+}]_{cyt}$ increases in plants.

2. Results

2.1. Expression of GCaMP3 in Different Root Cell Types of A. thaliana

To the best of our knowledge, plant lines expressing genetically encoded $[Ca^{2+}]_{cyt}$ sensors that are available to the scientific community are driven by constitutive promoters. *A. thaliana* lines expressing the *UBQ10:GCaMP3* construct, which we generated and used previously to study $[Ca^{2+}]_{cyt}$ oscillations in root hairs [17], is one example. In addition to root hairs, GCaMP3 fluorescence was detected in all cell-types of the primary root of *A. thaliana* expressing the *UBQ10:GCaMP3* construct (Figure 1). Using confocal microscopy, strong fluorescence was observed in cells located at the root surface, such as the peripheral root cap and epidermis (Figure 1A,B,D). Moderate levels of fluorescence could also be detected in the underlying cortex (Figure 1C). However, in cells located deeper within the root, such as the columella and endodermis, fluorescence was obscured by the fluorescence from surface cells (Figure 1A,B).

This prompted us to generate a set of constructs that drive *GCaMP3* expression in specific root cell types. For this purpose, we used promoters described in the SWELL promoter collection including *ATHB8* (columella), *SCARECROW* (SCR, endodermis), *Pin-Formed 2 (PIN2, epidermis and cortex)*, *PEP* (cortex), and *Proline rich protein 3(PRP3, trichoblasts)* [16]. For the five constructs, we found that GCaMP3 signal was most prominent in the expected cell types (Figure 2). For example, *ATHB8* and *SCR* promoters drove strong *GCaMP3* expression in the columella and endodermis, respectively (Figure 2A,B). For *ATHB8:GCaMP3*-expressing lines, we also observed weak fluorescence in the peripheral cap cells and the stele. For *SCR:GCaMP3*, fluorescence in the endodermis was observed in all root developmental zones from the meristem and elongation zone, continuing into the maturation zone. Roots of *PEP:GCaMP3-* and *PRP3:GCaMP3*-expressing seedlings also showed fluorescence in the expected cell types (i.e., cortex and trichoblast, respectively) (Figure 2C,D). On the other hand, fluorescence was detected in the epidermis and cortex of *PIN2:GCaMP3*-expressing lines, consistent with observations of Marques-Bueno et al. (2016) [16] (Figure 2E).

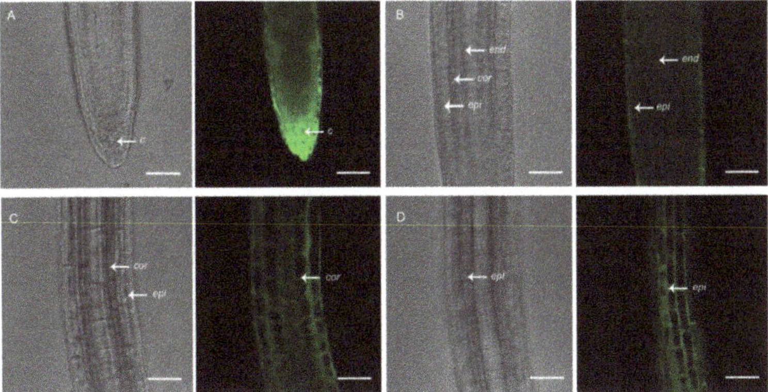

Figure 1. Confocal imaging of primary roots of 5-day-old *A. thaliana* seedlings expressing *UBQ10:GCaMP3*. Representative bright-field and corresponding single-optical section confocal images of the root tip with a focus on the root cap (**A**), meristem (**B**), and elongation zone (**C,D**). *c* = columella; *end* = endodermis; *epi* = epidermis; *cor* = cortex. Scale bar: 50 µm.

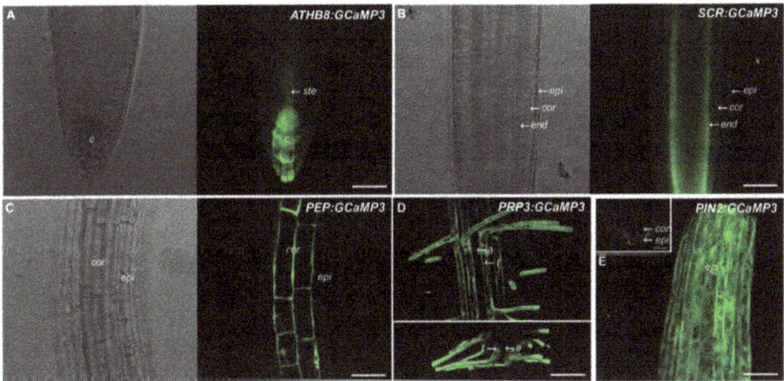

Figure 2. Expression of *GCaMP3* in specific root cell types. Representative bright-field and confocal images of primary roots of 5-day-old *A. thaliana* seedlings expressing *ATHB8:GCaMP3* (**A**), *SCR:GCaMP3* (**B**), and *PEP:GCaMP3* (**C**), respectively. *ATHB8:GCaMP3* is expressed in the columella with some fluorescence visible in the stele. *SCR:GCaMP3* is expressed in the endodermis and *PEP:GCaMP3* is expressed in the cortex. Maximum projection images (top panel in (**D**)) and computer reconstructed transverse section (lower panel in (**D**)) of the root maturation zone of seedlings expressing *PRP3:GCaMP3* show fluorescence confined to the trichoblasts. A maximum projection image of the root elongation zone of seedlings expressing *PIN2:GCaMP3* (**E**). The small box on the upper right corner of panel E shows a computer reconstructed cross-section of the primary root of a *PIN2:GCaMP3*-expressing line. The image was generated from 50 optical sections taken at 0.50 µm intervals. Note that GCaMP3 is predominantly expressed in the epidermis and cortex. *c* = columella; *ste* = stele; *epi* = epidermis; *cor* = cortex; *end* = endodermis; *t* = trichoblast; *a* = atrichoblasts. Scale bar: 50 µm.

2.2. Chemical Treatments of Cell-Type Specific GCaMP3 Constructs

Studies on Ca^{2+} signaling in plants over several years have revealed various stimuli that can trigger a rapid elevation in $[Ca^{2+}]_{cyt}$ including cold, touch, wounding, hormones, reactive oxygen species, cyclic nucleotides, amino acids and nutrient/ionic stress [5,18–23]. We used some of these known [Ca^{2+}]$_{cyt}$ triggers to test the responsiveness of our root cell-specific GCaMP3 lines.

We first analyzed the response of the GCaMP3 lines to adenosine tri-phosphate (ATP) (Figure 3). In mammalian systems, ATP is a neurotransmitter that is perceived by plasma membrane bound purinergic receptors, and its signaling role is facilitated through increases in [Ca^{2+}]$_{cyt}$ [24]. While plants do not have the canonical purinergic receptors found in mammals, they perceive ATP via lectin receptor-like kinases and, similar to mammals, ATP elicits a rapid increase in [Ca^{2+}]$_{cyt}$ [25–27]. Comparable to previous reports, application of ATP to the elongation zone of roots expressing *UBQ10:GCaMP3* elicited a rapid rise in [Ca^{2+}]$_{cyt}$-dependent fluorescence followed by a gradual decline (Figure 3A) [5,25]. To determine if the increase in [Ca^{2+}]$_{cyt}$ was due exclusively to ATP, we applied the solvent control solution, which consisted of 0.5× Murashige Skoog (MS) solution without ATP to roots. We found that applying MS solution triggered a rise in [Ca^{2+}]$_{cyt}$ that was delayed and lower in amplitude compared to the [Ca^{2+}]$_{cyt}$ change caused by ATP (Figure 3A). This observation suggests that under our growing conditions, adding a drop of solvent control solution to the root can induce an increase in [Ca^{2+}]$_{cyt}$, which could be the result of a touch or hypoosmotic response [19,20,27,28]. Therefore, in testing the cell-type GCaMP3 lines, we included solvent control applications in all of the experiments so we could tease apart [Ca^{2+}]$_{cyt}$ changes due to the desired stimulus from those resulting from a touch or hypoosmotic response.

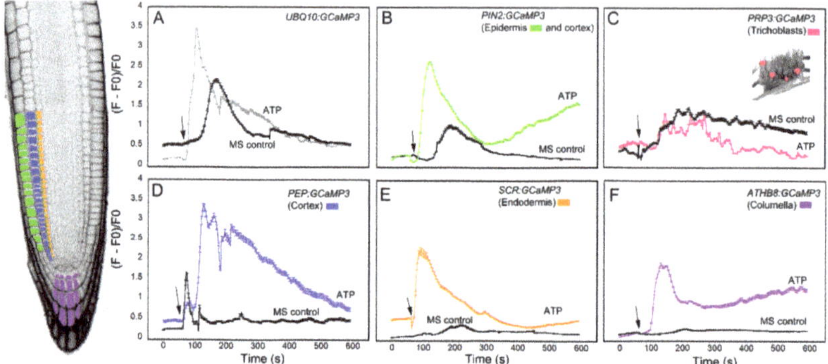

Figure 3. Time course of [Ca^{2+}]$_{cyt}$ changes in *A. thaliana* roots after 1 mM ATP application. The leftmost panel shows an inverted fluorescence image of the terminal 300 µm of an *A. thaliana* primary root to illustrate different cells in which *GCaMP3* was expressed. Cell types are color coded with green = epidermis; blue = cortex; orange = endodermis; purple = columella; and the pink cells in the inset in panel C = trichoblast. (**A–F**) Quantification of [Ca^{2+}]$_{cyt}$ -dependent fluorescence of various lines expressing *GCaMP3* after ATP and solvent control (0.5× MS) application. Black arrows indicate the time of treatment. Plotted values represent the average normalized fluorescence intensity from 3–6 regions of interest (ROI) per line with standard error bars every fifth time point. Fluorescence values were normalized to the lowest fluorescence value.

Upon applying ATP to roots expressing cell-specific GCaMP3 lines, we found similarities and differences in the resulting [Ca^{2+}]$_{cyt}$ signatures. For example, the *PIN2:GCaMP3* lines, which expressed *GCaMP3* in the epidermis and cortex (Figure 2), displayed similar [Ca^{2+}]$_{cyt}$ signatures as *UBQ10:GCaMP3*-expressing lines in response to ATP and the solvent control solution (Figure 3A,B). This is not surprising given that *GCaMP3* is strongly expressed in both the epidermis and cortex under the control of the *UBQ10* and *PIN2* promoters (Figures 1 and 2). For lines expressing *GCaMP3* in the trichoblasts, application of both ATP and MS control solution triggered a small increase in [Ca^{2+}]$_{cyt}$ that had similar patterns (Figure 3C). For lines expressing *GCaMP3* in cells located in the root interior such as the cortex, endodermis, and columella (supplementary video 1), ATP-induced [Ca^{2+}]$_{cyt}$ increases resembled the patterns of *UBQ10:GCaMP3* and *PIN2:GCaMP3*-expressing lines. However, the $[Ca^{2+}]_{cyt}$ increases triggered by application of MS control solution were not as prominent in these lines when compared to those lines in which *GCaMP3* was expressed in the epidermis (Figure 3D–F).

The next chemical we used to test the cell-specific GCaMP3 lines was glutamic acid (Figure 4). Like ATP, glutamic acid (glutamate) is a neurotransmitter in mammalian cells that can trigger a $[Ca^{2+}]_{cyt}$ increase [5,14]. Application of glutamate to the elongation zone of roots expressing the *UBQ10:GCaMP3* construct induced a strong initial $[Ca^{2+}]_{cyt}$ increase with a quick return to equilibrium (Figure 4A). The $[Ca^{2+}]_{cyt}$ spikes induced by application of MS controls were also observed in this dataset. Compared to glutamate, the onset of $[Ca^{2+}]_{cyt}$ increase was delayed in roots treated with MS solution. In this case, however, the $[Ca^{2+}]_{cyt}$ spikes triggered by glutamate and the MS control solution were similar in amplitude (Figure 4A). Like the ATP treatment, $[Ca^{2+}]_{cyt}$ response of roots expressing *PIN2:GCaMP3* to glutamate was similar to roots expressing *UBQ10:GCaMP3*. Furthermore, $[Ca^{2+}]_{cyt}$ changes in the root elongation zone of *PIN2:GCaMP3*-expressing seedlings in response to the MS control solution was delayed, but exhibited a lower amplitude (Figure 4B). Lines expressing *PRP3:GCaMP3* also had a strong initial $[Ca^{2+}]_{cyt}$ peak after glutamate treatment with a rapid return to equilibrium that resembled *UBQ10:GCaMP3* and *PIN2:GCaMP3* lines. Like *UBQ10:GCaMP3* and *PIN2:GCaMP3* lines, application of MS control solution elicited a $[Ca^{2+}]_{cyt}$ response that was delayed when compared to glutamate exposure (Figure 4C). By contrast, *PEP:GCaMP3*-expressing lines showed a dampened response to glutamate compared to lines in which *GCaMP3* was expressed in the root surface (Figure 4D). Close examination of the time course of $[Ca^{2+}]_{cyt}$ signals in the cortex after glutamate application revealed a triphasic response, with two larger peaks and a delayed smaller peak. Surprisingly, the amplitude of glutamate-triggered $[Ca^{2+}]_{cyt}$ increases in the cortex was less than that of MS treated roots (Figure 4D). The lines expressing *SCR:GCaMP3* and *ATHB8:GCaMP3* showed monophasic responses to glutamate with a gradual return to equilibrium, similar to lines in which *GCaMP3* was expressed in root surface cells. Furthermore, unlike lines with *GCaMP* expressed in the cortex and root surface cells, the MS control solution response for endodermis- and columella- targeted GCaMP3 had delayed responses and lower amplitudes (Figure 4E,F).

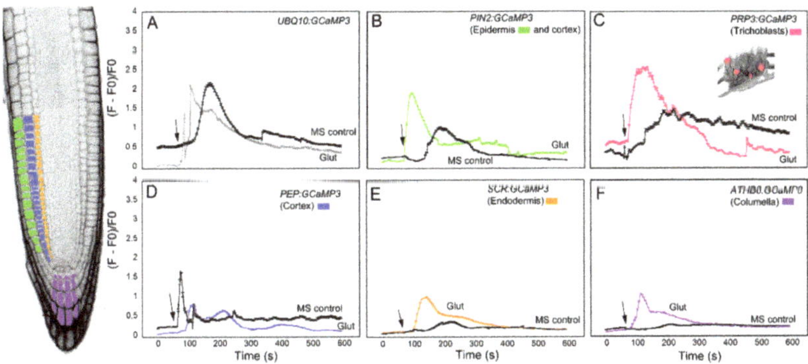

Figure 4. Time course of $[Ca^{2+}]_{cyt}$ changes in *A. thaliana* roots after 1 mM glutamate (Glu) application. The leftmost panel shows an inverted fluorescence image of the terminal 300 μm of an *A. thaliana* primary root to illustrate different cell types in which *GCaMP3* was expressed. Cell types are color coded with green = epidermis; blue = cortex; orange = endodermis; purple = columella; and the pink cells in the inset in panel C = trichoblast. (**A–F**) Quantification of $[Ca^{2+}]_{cyt}$-dependent fluorescence of various lines expressing *GCaMP3* after Glu and solvent control (0.5× MS) application. Black arrows indicate the time of treatment. Plotted values represent the average normalized fluorescence intensity from 3–4 regions of interest (ROI) per line with standard error bars every fifth time point. Fluorescence values were normalized to the lowest fluorescence value.

The third chemical we tested against the cell-type specific GCaMP3 lines was Al^{3+} in the form of aluminum chloride (Figure 5). Al^{3+} is a toxic tri-valent cation that competes with both Mg^{2+} and Ca^{2+} to increase the activity of Ca^{2+}-ATPases while disrupting transport of Ca^{2+} through the plasma membrane [5,29]. Similar to previous treatments, the root elongation zone of seedlings expressing *UBQ10:GCaMP3* showed a monophasic $[Ca^{2+}]_{cyt}$ response to Al^{3+} (Figure 5A). The onset of the $[Ca^{2+}]_{cyt}$ increase after Al^{3+} treatment was delayed and lower in amplitude compared to ATP, glutamate and solvent control application (Figure 3A, Figure 4A and Figure 5A) [5]. In roots expressing *PIN2:GCaMP3*, Al^{3+} application elicited a biphasic $[Ca^{2+}]_{cyt}$ response that consisted of an initial small peak that was quickly followed by a second larger peak, which was broader than the first peak. The onset of the $[Ca^{2+}]_{cyt}$ response in *PIN2:GCaMP3*-expressing roots after Al^{3+} treatment occurred earlier than the $[Ca^{2+}]_{cyt}$ response triggered by solvent controls (Figure 5B). In roots expressing *PRP3:GCaMP3*, Al^{3+}-induced $[Ca^{2+}]_{cyt}$ increase was larger in amplitude and had a broader peak compared to those observed in *UBQ10:GCaMP3*- and *PIN2:GCaMP3*-expressing roots (Figure 5C). The $[Ca^{2+}]_{cyt}$ increases in roots expressing *GCaMP3* in interior root tissues were less than roots expressing *GCaMP3* in the surface tissues. Among the three lines in which *GCaMP3* was expressed in the interior cell types, the columella-specific GCaMP3 line did not show an Al^{3+}-induced $[Ca^{2+}]_{cyt}$ peak (Figure 5F). The *PEP:GCaMP3*-expressing lines showed a biphasic $[Ca^{2+}]_{cyt}$ response with a distinct first peak followed by a second peak with lower amplitude. Although the amplitude of the $[Ca^{2+}]_{cyt}$ peak induced by application of MS solution was similar to the amplitude of the first Al^{3+}-triggered $[Ca^{2+}]_{cyt}$ peak, the onset of the latter was delayed (Figure 5D). *SCR:GCaMP3*-expressing roots exhibited an Al^{3+}-induced $[Ca^{2+}]_{cyt}$ response characterized by a broad peak and with a lower amplitude than peaks observed in the root surface- and cortex *GCaMP3*-expressing lines (Figure 5E).

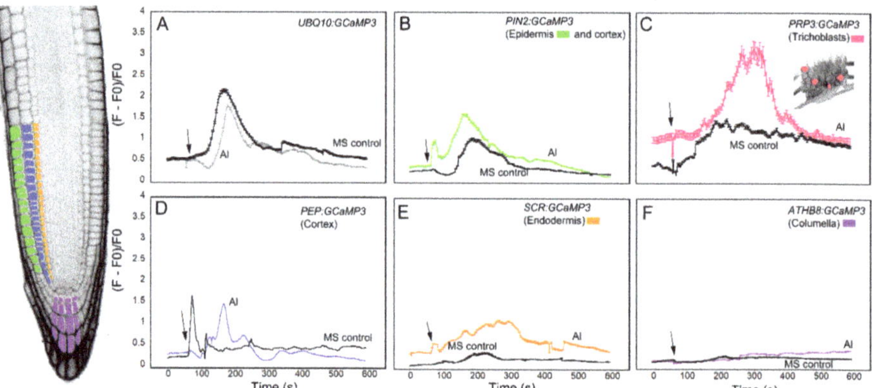

Figure 5. Time course of $[Ca^{2+}]_{cyt}$ changes in *A. thaliana* roots after 1 mM aluminum (Al^{3+}) chloride application. The leftmost panel shows an inverted fluorescence image of the terminal 300 μm of an *A. thaliana* primary root to illustrate different cell types in which *GCaMP3* was expressed. Cell types are color coded with green = epidermis; blue = cortex; orange = endodermis; purple = columella; and the pink cells in the inset in panel C = trichoblast. (**A–F**) Quantification of $[Ca^{2+}]_{cyt}$-dependent fluorescence of various lines expressing *GCaMP3* after Al^{3+} and solvent control (0.5× MS) application. Black arrows indicate the time of treatment. Plotted values represent the average normalized fluorescence intensity from 3–7 regions of interest (ROI) per line with standard error bars every fifth time point. Fluorescence values were normalized to the lowest fluorescence value.

The final stimulus we used to test the cell-type specific GCaMP3 lines was salt in the form of NaCl. Na^+ is used to regulate the voltage equilibrium in cells, but also has direct effects on the efflux of K^+ into the cytosol [30]. Ca^{2+} is a regulator for the K^+ efflux channels [30], so disruption of K^+ into the cytosol may cause variation in the resting concentration of Ca^{2+}. In roots expressing the fluorescent-based

[Ca^{2+}]$_{cyt}$ sensor Yellow Cameleon 3.60 (YC3.60), NaCl treatment resulted in cell-specific [Ca^{2+}]$_{cyt}$ transients in the early elongation zone [31]. Consistent with the results of Feng et al. (2018) [31], roots expressing *UBQ10:GCaMP3* exhibited late-onset [Ca^{2+}]$_{cyt}$ transients (i.e., >10 min) in cells of the elongation zone (Figure 6A; supplementary video 2). These NaCl-triggered [Ca^{2+}]$_{cyt}$ transients were observed in the cell-targeted GCaMP3 lines except for the endodermis-localized GCaMP3 line (Figure 6A–D). In certain lines, the onset of [Ca^{2+}]$_{cyt}$ transients occurred much earlier after NaCl application when compared to *UBQ10:GCaMP3*-expressing lines. The onset of NaCl-induced [Ca^{2+}]$_{cyt}$ transients was fastest in roots of *PRP3:GCaMP3*-expressing lines followed by *PEP:GCaMP3*-expressing lines. The [Ca^{2+}]$_{cyt}$ transients in the columella were only observed several minutes after NaCl application (i.e., 12 min after NaCl application; Figure 6C).

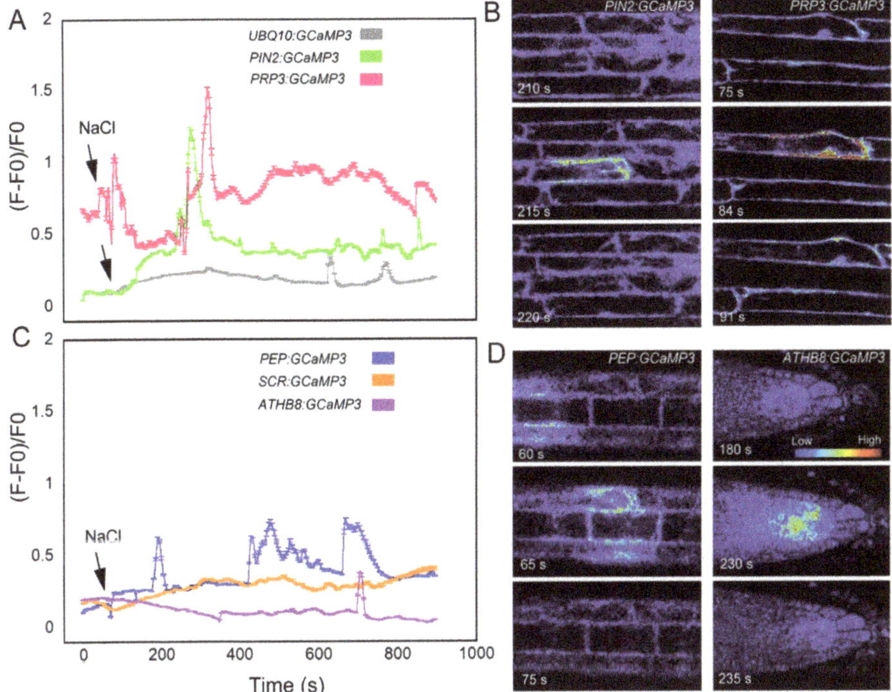

Figure 6. NaCl-triggered [Ca^{2+}]$_{cyt}$ transients in cell type-specific GCaMP3 lines. Time course (**A**) and representative heat maps (**B**) of [Ca^{2+}]$_{cyt}$ transients in roots expressing *UBQ10:GCaMP3* and root surface cell-targeted GCaMP3 lines. Time course (**C**) and representative heat maps (**D**) of [Ca^{2+}]$_{cyt}$ transients in roots where *GCaMP3* is expressed in interior root cell types. Colors of the line graphs in A and C correspond to the cell types shown in preceding figures. The grey line shows the time course of *UBQ10:GCaMP3*-expressing lines. The arrows indicate time of application of sodium chloride. Intensity bar in panel D correspond to high (red) and low (blue) fluorescence with red indicating elevated [Ca^{2+}]$_{cyt}$.

2.3. Modification of Growth Conditions and Chemical Application Methods to Mitigate Solvent Control-Induced [Ca^{2+}]$_{cyt}$ Transients

As shown in the preceding sections, the 0.5X MS control solution induced [Ca^{2+}]$_{cyt}$ changes in *UBQ10:CCaMP3* and lines in which *GCaMP3* was expressed in cells on the root surface and cortex (Figures 3–5). Although the MS-induced [Ca^{2+}]$_{cyt}$ responses were clearly distinct from ATP, glutamate, and Al^{3+} treatments, we were concerned that the growth conditions in which these first experiments were conducted made roots more sensitive to osmotic changes and/or mechanical perturbation due to the process of adding the solutions. We argued that MS-induced $[Ca^{2+}]_{cyt}$ transients could be dampened relative to the chemical applications by: (1) using freshly prepared MS plates for planting seeds and (2) pretreating roots two times with MS solution prior to application of the actual chemical solution. These modifications in our treatment protocols were tested using ATP which typically generated peaks with the highest amplitude.

Compared to the datasets shown in Figures 3–5, the $[Ca^{2+}]_{cyt}$ transients triggered by the MS solution were significantly dampened when fresh plates were used for the experiments (Figure 7). The dampened MS-induced $[Ca^{2+}]_{cyt}$ transients were observed in the two sequential MS applications. Pretreatment with MS solution had the added benefit of allowing roots to adapt prior to ATP application (adapted ATP = ATP:MS). Under these modified growth and pretreatment conditions, the ATP-induced $[Ca^{2+}]_{cyt}$ increases were lower in amplitude when compared to the $[Ca^{2+}]_{cyt}$ increases shown in Figure 3. Despite the lower ATP-induced $[Ca^{2+}]_{cyt}$ transients, these $[Ca^{2+}]_{cyt}$ responses were better separated from the $[Ca^{2+}]_{cyt}$ responses induced by MS solution, particularly in lines expressing *UBQ10:GCaMP3* and those in which *GCaMP3* was expressed in root surface cell types and the cortex (Figure 7). In addition to the dampened response, $[Ca^{2+}]_{cyt}$ response to ATP:MS for *UBQ10:GCaMP3* and *PIN2:GCaMP3* lines had broader peaks than those shown in Figure 3. For *PRP3:GCaMP3*-expressing lines, the $[Ca^{2+}]_{cyt}$ response was markedly different from the initial ATP treatment (Figures 3C and 7C). Application of ATP after 0.5X MS adaptation caused a strong, monophasic response earlier than either *UBQ10:GCaMP3* or *PIN2:GCaMP3* lines. Lines expressing *PEP:GCaMP3* showed a continued elevated $[Ca^{2+}]_{cyt}$ response that oscillated throughout the 10 min time course compared to either *UBQ10:GCaMP3* lines or the previous ATP treatment (Figures 3D and 7D). For lines expressing *SCR:GCaMP3* or *ATHB8:GCaMP3*, the response was similar to previous ATP treatments, but without the strong transient peak (Figures 3E,F and 7E,F). After the cell-type specific lines were adapted to the 0.5× MS control solution, the response time for ATP:MS was initiated later than the previous ATP treatment (Table 1).

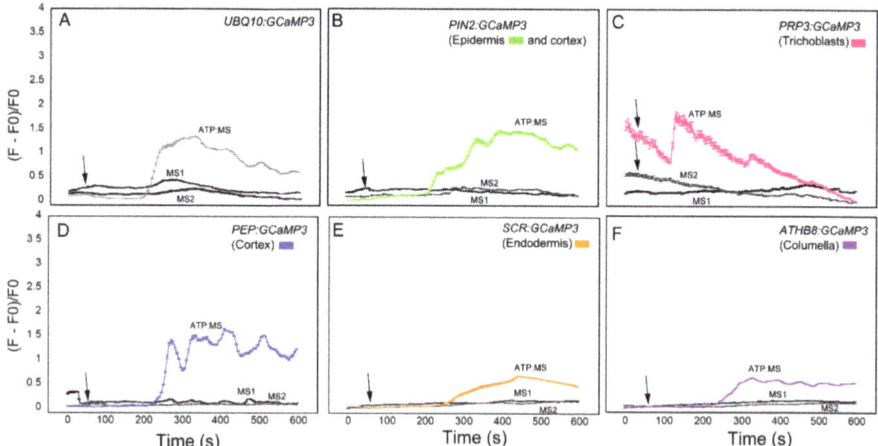

Figure 7. ATP-triggered $[Ca^{2+}]_{cyt}$ responses in cell type-specific GCaMP3 lines planted on fresh MS plates and pre-treated twice with MS. MS pretreatment 1 (MS1) time course (black) of $[Ca^{2+}]_{cyt}$ in roots show construct treatment with MS control solution. MS pretreatment 2 (MS2) time course (grey) shows construct treatment with a second dose of MS control solution 15 min after first treatment. ATP:MS time course of $[Ca^{2+}]_{cyt}$ in roots show cell-type specific response 10 min after MS1 and MS2 treatment. Colors of the line graphs in (**A–F**) correspond to the cell types shown in preceding figures with standard error bars every fifth time point. Black arrows indicate the time of application of ATP.

To further tease out the differences among *GCaMP3*-expressing lines, we obtained the time elapsed from chemical application to maximum GCaMP3 fluorescence (referred to as reaction time), and analyzed the relationship between chemicals using Analysis of Variance (ANOVA) with Tukey Post-Hoc test (Table 1). Our analysis showed that reaction time was delayed in lines where GCaMP3 was expressed in inner cell types. When comparing the different chemical treatments, ATP, glutamate, and Al^{3+} showed similar responses (except *ATHB8:GCaMP3* to Al^{3+}) while adapted ATP:MS showed significantly slower reaction times to both ATP and glutamate (Table 1). However, with such a small sample size, statistical significance may indicate a trend rather than actual significance. NaCl reaction time was not included due to the large variation in responses across each line.

Table 1. Reaction time analysis of $[Ca^{2+}]_{cyt}$ response where reaction time is the time elapsed from chemical application to maximum GCaMP3 fluorescence (reaction time $\pm$ SE; n = sample size; reaction time is in seconds). Statistical analysis using Analysis of Variance (ANOVA) with Tukey Post-Hoc test ($p \leq 0.05$ for reported significance) was performed for differences in chemical treatments only due to the lack of comparable ROI sections between cell-type lines. Due to limited sample size, significance may indicate a general trend rather than actual significance. ATP = adenosine tri-phosphate; Glu = glutamate; Al^{3+} = aluminum; MS = control solution; ATP:MS = adapted ATP.

Treatment	UBQ10:GCaMP3	PIN2:GCaMP3	PRP3:GCaMP3	PEP:GCaMP3	SCR:GCaMP3	ATHB8:GCaMP3
ATP [a,b]	99.3 ± 4.67 (n = 3)	120.3 ± 4.3 (n = 3)	210.3 ± 26.5 (n = 3)	159.0 ± 16.3 (n = 6)	141.2 ± 22.6 (n = 5)	134.8 ± 8.0 (n = 4)
Glu [a,b]	100.0 ± 7.1 (n = 3)	94.0 ± 4.9 (n = 3)	114.0 ± 16.4 (n = 3)	145.7 ± 28.0 (n = 3)	128.7 ± 17.1 (n = 3)	114.7 ± 3.3 (n = 3)
Al^{3+} [a,c]	196.4 ± 16.5 (n = 5)	178.3 ± 37.8 (n = 7)	219.0 ± 27.0 (n = 6)	154.7 ± 8.0 (n = 3)	325.0 ± 94.4 (n = 3)	569.0 ± 12.8 (n = 3)
MS [c]	174.3 ± 5.6 (n = 8)	201.7 ± 33.5 (n = 6)	266.2 ± 27.5 (n = 5)	217.0 ± 46.6 (n = 6)	350.5 ± 77.7 (n = 6)	373.8 ± 88.2 (n = 6)
ATP:MS [c]	315.3 ± 40.4 (n = 3)	404.0 ± 33.7 (n = 3)	329.0 ± 95.9 (n = 3)	366.0 ± 18.9 (n = 3)	424.7 ± 32.9 (n = 3)	311.7 ± 9.3 (n = 3)

[a,b,c] indicates significant differences between chemical treatments.

2.4. Evaluation of $[Ca^{2+}]_{cyt}$ Signatures from Similar Root Development Regions and Cell Types

In comparing *UBQ10:GCaMP3* lines with the other cell-type specific GCaMP3 lines, a ROI was selected and the average of normalized fluorescence intensity was plotted. For the most part, the ROI in *UBQ10:GCaMP3* lines was from the root epidermis and the average normalized fluorescence intensity from this ROI was compared with that of individual or groups of cells from the cell-type specific GCaMP3 lines. This process did not enable direct comparisons of $[Ca^{2+}]_{cyt}$ signatures between the same cell types or root developmental regions. To address this issue, we conducted another set of imaging experiments using ATP-treated *UBQ10:GCaMP3* and *PIN2:GCaMP3* lines. For these experiments, a more direct comparison was conducted by drawing ROIs in epidermal cells from similar positions along the root longitudinal axis encompassing the meristem, distal elongation zone, and central elongation zone (Figure 8A). In doing so, we found similarities and differences in $[Ca^{2+}]_{cyt}$ signatures between *UBQ10:GCaMP3* and *PIN2:GCaMP3* lines. For example, the ATP-induced $[Ca^{2+}]_{cyt}$ signatures of *UBQ10:GCaMP3* in all root developmental regions were identical to those in the distal and central elongation zone of *PIN2:GCaMP3* lines. In *PIN2:GCaMP3* lines, however, the ATP-induced $[Ca^{2+}]_{cyt}$ signature had a much greater amplitude in the meristem (Figure 8A,B).

We extended our analysis by drawing ROIs in presumptive columella or endodermis of *UBQ10:GCaMP3* and comparing the resulting $[Ca^{2+}]_{cyt}$ signatures from these ROIs with those of *ATHB8:GCaMP3* and *SCR:GCaMP3*. For the columella, we found that *ATHB8:GCaMP3* lines displayed an ATP-triggered $[Ca^{2+}]_{cyt}$ signature with a sustained peak throughout the 10 min time course. On the other hand, the ATP-induced $[Ca^{2+}]_{cyt}$ signature from the presumptive columella region of *UBQ10:GCaMP3* lines was transient (Figure 8C). For the endodermis, the peak of ATP-induced $[Ca^{2+}]_{cyt}$ increase was higher in *UBQ10:GCaMP3* when compared to *SCR:GCaMP3* (Figure 8D).

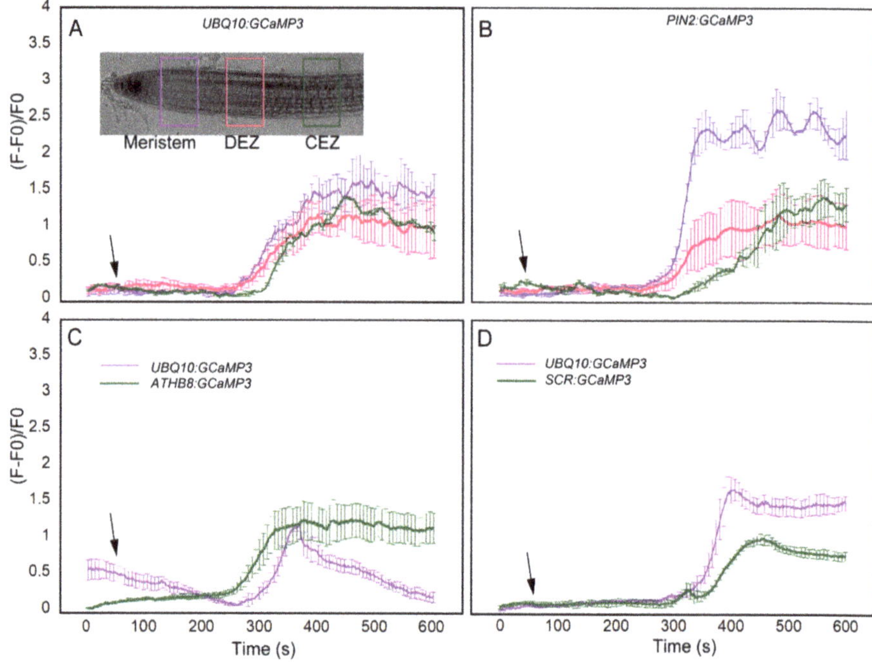

Figure 8. Direct comparison of ATP-induced $[Ca^{2+}]_{cyt}$ signatures in equivalent root developmental regions between *UBQ10:GCaMP3* and cell-type specific GCaMP3 lines. Regions of interest (ROIs) were drawn in the meristem, distal elongation zone (DEZ), and central elongation zone (CEZ) in *UBQ10:GCaMP3* (**A**) and *PIN2:GCaMP3* (**B**) roots (inset in A). Comparison of ATP-induced $[Ca^{2+}]_{cyt}$ signatures in presumptive columella and endodermis of *UBQ10:GCaMP3* lines with *ATHB8:GCaMP3* (**C**) and *SCR:GCaMP3* (**D**). Black arrows indicate the time of treatment. Plotted values represent the average normalized fluorescence intensity from 3-4 ROI per line with standard error bars every fifth time point. Fluorescence values were normalized to the lowest fluorescence value.

3. Discussion

In this paper, we introduced a set of *A. thaliana* lines expressing the intensiometric $[Ca^{2+}]_{cyt}$ reporter GCaMP3 targeted to specific root cell types. Roots of these cell type specific GCaMP3 lines responded to chemical treatments such as ATP, glutamate, Al^{3+}, and NaCl by exhibiting increases in $[Ca^{2+}]_{cyt}$-dependent CGaMP3 fluorescence. Imaging $[Ca^{2+}]_{cyt}$ changes in cell types in response to some of these chemicals using lines in which genetically-encoded $[Ca^{2+}]_{cyt}$ reporters are expressed ubiquitously have been attempted previously. This was done by drawing a region of interest corresponding to a particular tissue/cell type. For example, it was shown that Al^{3+}-induced biphasic $[Ca^{2+}]_{cyt}$ signatures in the root elongation zone occurred predominantly in the cortex when compared with the epidermis [5]. While this has proven effective for marking developmental regions along the longitudinal axis of the root [5,27], specifying a region of interest to designate root tissue/cell types along the radial axis for fluorescence-based measurements of $[Ca^{2+}]_{cyt}$ can be problematic because this does not guarantee complete separation of signals from adjacent cell types. This problem becomes even more challenging for cell types located in the innermost regions of the root such as the columella and endodermis. Moreover, the ability to consistently draw a region of interest to outline the root cortex requires that the root grew in a particular orientation during the imaging experiments [5]. Here, we documented some of the advantages of the cell type specific GCaMP3 lines by comparing $[Ca^{2+}]_{cyt}$ signatures from ROIs marking presumptive columella and endodermis in *UBQ10:GCaMP3* lines with those of *ATHB8:GCaMP3* and *SCR:GCaMP3*. This analysis revealed stark differences in both the amplitude and shape of $[Ca^{2+}]_{cyt}$ signatures. Differences were also uncovered when equivalent root developmental regions were compared, particularly with regard to the amplitude of $[Ca^{2+}]_{cyt}$ signatures in the meristem (Figure 8). The differences are likely due to contaminating signals from other cell types in *UBQ10:GCaMP3* lines. The use of cell type specific GCaMP3 lines described here could potentially mitigate such problems.

The $[Ca^{2+}]_{cyt}$ transients in the cell type specific lines induced by these various chemicals had various patterns when compared with lines in which *GCaMP3* was expressed ubiquitously. A similarity in patterns occurred within the root surface cells' responses to the different chemical treatments and again within the root inner cell types. Lines in which GCaMP3 was expressed in innermost root cell types such as the columella and endodermis generally had slower response times when compared to lines with GCaMP3 expressed in root surface cells. The delayed response times in lines with GCaMP3 expressed in the columella and endodermis was observed in all chemical treatments (Table 1; Figure 6). One could argue that because of the innermost location of the columella and endodermis, the delay in the response times might be due to the longer time it takes for the chemical to make contact with the target cells. The fact that columella and endodermis exhibit minimal or no $[Ca^{2+}]_{cyt}$ increase after application of 0.5X MS supports this possibility. An alternative hypothesis is that the delayed $[Ca^{2+}]_{cyt}$ response in the innermost cell types is due to an inward moving $[Ca^{2+}]_{cyt}$ wave, which is not easily revealed when GCaMP3 is expressed ubiquitously or on the root surface. From our data, these inward-directed $[Ca^{2+}]_{cyt}$ waves appear to be most prominent in ATP- and glutamate-treated roots (Figures 3, 4 and 7). The GCaMP3 lines described here could present a significant tool for better understanding the radial direction of $[Ca^{2+}]_{cyt}$ waves in roots. A meaningful set of experiments with GCaMP3 lines expressed in innermost cell types will be to locally apply a small volume of the chemical stimulus to the root epidermis.

The line that was most similar to *UBQ10:GCaMP3* with regard to chemical-induced $[Ca^{2+}]_{cyt}$ transients was *PIN2:GCaMP3*. This was not surprising given that the *PIN2* promoter drives *GCaMP3* expression in root epidermal cells, which is the cell type that is readily imaged in *UBQ10:GCaMP3* lines due to its location on the root surface. With the *PEP:GCaMP3* lines described here, we were able observe biphasic, triphasic and oscillatory $[Ca^{2+}]_{cyt}$ signatures after Al^{3+} treatment that were reminiscent of some of the observations made by Rincon-Zachary et al. (2010) [5] (Figure 5). The cortex- and trichoblast-targeted GCaMP3 lines also revealed an earlier onset of NaCl-induced $[Ca^{2+}]_{cyt}$ transients when compared to *UBQ10:GCaMP3* lines (Figure 6). It has been shown that $[Ca^{2+}]_{cyt}$ transients in roots exposed to NaCl is part of the cellular machinery that prevents root cells from rupturing under salinity stress. The plasma membrane-localized receptor-like kinase, Feronia (FER), is required for root growth recovery after encountering high salinity conditions [31]. Because *fer* mutants displayed a high incidence of cortical cell rupture when compared to wild type after exposure to NaCl, it is tempting to speculate that earlier onset of $[Ca^{2+}]_{cyt}$ transients in the cortex revealed by the *PEP:GCaMP3*-expressing lines could be a mechanism to maintain root growth integrity. The cell type GCaMP3 lines described here should be useful in testing this hypothesis.

While testing the various cell type specific GCaMP3 lines, we encountered a number of technical challenges brought about by our seedling growth conditions. In several experiments, we found that applying the 0.5X MS control solution alone could trigger a rise in $[Ca^{2+}]_{cyt}$. The MS only-induced $[Ca^{2+}]_{cyt}$ transients were most prominent in the *UBQ10:GCaMP3*-expressing lines and those in which *GCaMP3* was expressed in root surface cell types (e.g., *PIN2:GCaMP3* and *PRP3:GCaMP3*). The $[Ca^{2+}]_{cyt}$ transients observed after MS treatment were reminiscent of touch-induced $[Ca^{2+}]_{cyt}$ responses [20], and recent studies on ATP-induced $[Ca^{2+}]_{cyt}$ signaling [27], suggesting that we might have been eliciting a touch response as the various solutions were added to the root. In support of this hypothesis, our method for root $[Ca^{2+}]_{cyt}$ imaging involved planting seed of the reporter lines on a thin layer of agarose to secure the root while at the same time allowing for the root cells to be accessible to the various chemical treatments ([5]; see Methods). It is possible that some of the $[Ca^{2+}]_{cyt}$ transients observed in response to MS application resulted from roots that were not securely anchored to the agarose medium. Alternatively, the prevalence of MS-induced $[Ca^{2+}]_{cyt}$ increases in some of our experiments could be due to osmotic changes when the solution was applied. Because agar for imaging roots was poured in advance, it was possible that liquid in some of the plates may have evaporated during storage. This could have led to some plates having an MS solution that was more concentrated than the treatment solution.

To test the possibility that MS-induced $[Ca^{2+}]_{cyt}$ changes were due to touch or hypoosmotic shock, we repeated the ATP application experiments using roots that emerged from seeds planted on plates with freshly prepared MS-supplemented agarose. Furthermore, roots were pretreated with MS solution prior to ATP treatment. Under these modified growth and treatment conditions, we found that MS-induced $[Ca^{2+}]_{cyt}$ increases were significantly dampened (Figures 3 and 7). Given that $[Ca^{2+}]_{cyt}$ changes are responses that can be elicited by a wide range of stimuli, the results described here highlight the need to carefully consider the growth and treatment conditions, and to include appropriate controls when conducting $[Ca^{2+}]_{cyt}$ imaging experiments. Based on these observations, it is imperative that fresh plates be prepared prior to planting the reporter lines and that roots be pretreated with the solvent control solution for a few hours to allow seedlings to adapt.

Some studies have shown $[Ca^{2+}]_{cyt}$ signals propagate along the longitudinal axis of roots [5,26]. Meanwhile, another study has indicated that $[Ca^{2+}]_{cyt}$ signal propagation from the apical into the sub-apical can also occur [27]. The cell type specific GCaMP3 lines described here, particularly those lines in which *GCaMP3* is expressed in inner root cells such as the cortex-, endodermis- and columella-specific lines could shed new insights into the direction of $[Ca^{2+}]_{cyt}$ signal propagation within the roots. The MS-induced $[Ca^{2+}]_{cyt}$ responses in lines in which *GCaMP3* was expressed in the endodermis and columella were not observed. However, these cell types exhibited $[Ca^{2+}]_{cyt}$ increases in response to some of the chemicals, indicating that these $[Ca^{2+}]_{cyt}$ changes can be attributed to the chemical itself. In this regard, it is noteworthy that columella-targeted GCaMP3 lines did not exhibit a $[Ca^{2+}]_{cyt}$ increase after Al^{3+} treatment while those lines with *GCaMP3* expressed in the endodermis lacked a $[Ca^{2+}]_{cyt}$ response to NaCl. With the establishment of response patterns from the cell type specific GCaMP3 lines when exposed to various chemical treatments, investigation into unique aspects of different root tissues/cells, such as signal propagation direction and response intensity, can be more readily explored.

4. Materials and Methods

4.1. Generation of Cell-Type Specific Promoter GCaMP Lines

Each promoter was amplified from genomic DNA extracted from wild-type *Arabidopsis thaliana* seedlings using the Plant DNAzol Reagent (Invitrogen, Carlsbad, CA, USA) (Table S1). The sequence preceding the start codon for each gene was used in each instance. The resulting fragments were digested with PstI and SalI (New England Biolabs, Ipswich, MA, USA; http://neb.com) and cloned into a modified pCAMBIA1390 vector [32]. *GCaMP3* was amplified from plasmid DNA and digested with XmaI and BstEII (New England Biolabs, Ipswich, MA, USA; http://neb.com) and then cloned behind each promoter construct in pCAMBIA1390. To generate *UBQ10:GCaMP3*, a similar approach was taken. *GCaMP3* was again amplified from plasmid DNA and then digested with EcoRI and SpeI (New England Biolabs, Ipswich, MA, USA; http://neb.com) and cloned behind the *Ubiquitin 10* promoter in a previously described pCAMIA1390 vector [33]. Plasmid containing *GCaMP3* was a gift from Loren Looger (Addgene plasmid # 22692; http://n2t.net/addgene:22692; RRID:Addgene_22692; [34]). *Agrobacterium tumefaciens*-mediated transformation using the floral dip method was used to generate Arabidopsis ecotype Col-0 plants expressing the GCaMP3 constructs [35]. All primers used in this study are listed in Table S2.

4.2. Preparation of A. thaliana Seedlings for Imaging

Seeds of the various *A. thaliana* GCaMP3 lines were planted on coverslips coated with a thin layer of MS-supplemented low-melting agarose. This set-up was prepared by pouring 3 mL of autoclaved 0.5% low-melting agarose in 0.5X MS solution (pH 5.7) on 48 × 60 mm No. 1 coverslips (Thermo Scientific, gold seal cover glass reorder no. 3334, Waltham, MA, USA; www.thermoscientific.com) as described in Rincon-Zachary et al. (2010) [5]. Coverslips with polymerized agarose were placed in 9 cm diameter round Petri dishes and stored at 4 °C prior to planting seeds. In another set of experiments, seeds were planted as soon as the agarose polymerized. Petri dishes and coverslips with the planted seeds were kept in 4 °C for 2 days and transferred to a Conviron growth chamber (Controlled Environments Ltd., Winnipeg, MB, Canada) set to 24 °C with 14 h/10 h day/night cycle. Petri dishes with the coverslips were kept vertical to enable the roots to grow down and straight along the surface of the agarose. Seedlings were imaged when primary roots were about 3–4 cm long, which was 5–6 days after transfer to the Conviron.

4.3. Validation of GCaMP3 Expression in Root Cell Types

Validation of GCaMP3 expression in the target root cell types was conducted with an inverted Leica TCS SP8-X Confocal Laser Scanning Microscope (Leica Microsystems, Wetzlar, Germany; http://leica-microsystems.com). Seedlings from the Conviron were secured in a horizontal orientation on the stage of the microscope. Seedling roots were illuminated with the 488 nm line of the Argon laser using a 40× (numerical aperture 1.1) water immersion objective and emitted light detected at 510 nm. For some lines, single optical images were acquired at a pixel resolution of 1024 × 1024. For other lines, a Z-series was acquired by capturing 81 images at 0.4 µm intervals and 3-D images were generated using the LAS visualization software of the Leica confocal microscope.

4.4. Chemical Treatments and Measurement of Ca^{2+}-Dependent GCaMP3 Fluorescence

Stock solutions of 1 M ATP (Sigma-Aldrich, St. Louis, MO, USA; A7699-1G), glutamate (Sigma-Aldrich, St. Louis, MO, USA; G1149-100G), aluminum chloride (Sigma-Aldrich, St. Louis, MO, USA; Cas. No. 7784-13-6) and sodium chloride (J.T. Baker, Phillipsburg, NJ, USA; 3624-19) were made with deionized water and stored at 4 °C prior to use. Working solutions of 1 mM Al^{3+}, 1 mM ATP, 1 mM Glu, and 150 mM NaCl were made by adding the appropriate volume of the stock solution in 0.5X MS at pH 5.7. Working solutions were made fresh prior to the imaging experiments.

To measure $[Ca^{2+}]_{cyt}$-dependent GCaMP3 fluorescence, images of growing roots were acquired every 1 s for 10 to 15 min using the Leica SP8-X confocal microscope. Images were captured at a scanning speed of 600 MHz and pixel resolution of 512 × 300. Baseline GCaMP3 fluorescence of the roots was first acquired for 1 min after placing seedlings on the stage of the microscope. Twenty µL of the treatment solution was then added on top of the agarose medium where the root was growing using an adjustable volume pipette while imaging continued. From the collected images, the average fluorescence intensity was acquired by marking a rectangular region of interest corresponding to the specific cell type using the rectangular selection tool of the SPX-8 LAS software. GCaMP3 fluorescence (F) values were normalized using the formula $I = \frac{(F-F0)}{F0}$ where F is the fluorescence intensity at some time point and F0 is the lowest fluorescence intensity point for the data set. Normalized fluorescence values were reported as the mean ± SE of at least three independent seedlings per GCaMP3 construct [36]. Another round of intensity video imaging was taken for the 0.5× MS control solution to generate the MS control fluorescence curves.

Supplementary Materials: The following are available online at http://www.mdpi.com/1422-0067/21/17/6385/s1, Supplementary video 1 Time lapse video of $[Ca^{2+}]_{cyt}$ changes in the columella cells of *ATHB8:GCaMP3* lines in response to ATP. Red indicates elevated $[Ca^{2+}]_{cyt}$. Total elapsed time is 5 min with images captured every 1 s. Video was sped up to 30 frames per second. Supplementary video 2 Time lapse video of $[Ca^{2+}]_{cyt}$ changes in the cortex of *PEP:GCaMP3* lines in response to salt. Red indicates elevated $[Ca^{2+}]_{cyt}$. Total elapsed time is 10 min with images captured every 1 s. Video was sped up to 60 frames per second. Table S1 Promoters used for development of *GCaMP3* constructs that target to specific cell types in *Arabidopsis thaliana* roots. Table S2 Primer sequences used to generate the *GCaMP3* constructs in *Arabidopsis thaliana*.

Author Contributions: All authors conceived and designed the research. W.K. and J.A.S. performed confocal microscopy of GCaMP lines. W.K., J.A.S. and E.B.B. analyzed the data and wrote the manuscript. J.A.S. made the GCaMP constructs and performed the floral dip transformations to generate the transgenic Arabidopsis lines. All authors have read and agreed to the published version of the manuscript.

Funding: This work was supported by the National Aeronautics and Space Administration (NASA grants 80NSSC18K1462 and 80NSSC19KO129) and the Noble Research Institute LLC.

Conflicts of Interest: The authors declare no conflict of interest.

References

1. Luo, J.; Chen, L.; Huang, F.; Gao, P.; Zhao, H.; Wang, Y.; Han, S. Intraorganellar calcium imaging in Arabidopsis seedling roots using the GCaMP variants GCaMP6m and R-CEPIA1er. *J. Plant Physiol.* **2020**, *246*, 153127. [CrossRef] [PubMed]

2. Qian, D.; Xiang, Y. Actin cytoskeleton as actor in upstream and downstream of calcium signaling in plant cells. *Int. J. Mol. Sci.* **2019**, *20*, 1403. [CrossRef] [PubMed]

3. Thor, K. Calcium—Nutrient and messenger. *Front. Plant Sci.* **2019**, *10*, 440. [CrossRef] [PubMed]

4. Tuteja, N.; Mahajan, S. Calcium signaling network in plants. *Plant Signal. Behav.* **2007**, *2*, 79–85. [CrossRef] [PubMed]

5. Rincon-Zachary, M.; Teaster, N.D.; Sparks, J.A.; Valaster, A.H.; Motes, C.M.; Blancaflor, E.B. Fluorescence resonance energy transfer—Sensitized emission of yellow cameleon 3.60 reveals root zone—Specific calcium signatures in Arabidopsis in response to aluminum and other trivalent cations. *Plant Physiol.* **2010**, *152*, 1442–1458. [CrossRef] [PubMed]

6. Rudd, J.J.; Franklin-Tong, V.E. Calcium signaling in plants. *Cell. Mol. Life Sci.* **1999**, *55*, 214–232. [CrossRef]

7. Plieth, C. Plant calcium signaling and monitoring: Pros and cons and recent experimental approaches. *Protoplasma* **2001**, *218*, 1–23. [CrossRef]

8. Robert, V.; Pinton, P.; Tosello, V.; Rizzuto, R.; Pozzan, T. Recombinant aequorin as tool for monitoring calcium concentration in subcellular compartments. *Method Enzymol.* **2000**, *327*, 400–456.

9. Shimomura, O.; Johnson, F.H. Properties of the bioluminescent protein aequorin. *Biochemistry* **1969**, *8*, 3991–3997. [CrossRef]

10. Nakai, J.; Ohkura, M.; Imoto, K. A high signal-to-noise Ca2+ probe composed of a single green fluorescent protein. *Nat. Biotechnol.* **2001**, *19*, 137–141. [CrossRef]

11. Defalco, T.A.; Toyota, M.; Phan, V.; Karia, P.; Moeder, W.; Gilroy, S.; Yoshioka, K. Using GCaMP3 to study Ca2+ signaling in Nicotiana species. *Plant Cell Physiol.* **2017**, *58*, 1173–1184. [CrossRef] [PubMed]

12. Grefen, C.; Donald, N.; Hashimoto, K.; Kudla, J.; Schumacher, K.; Blatt, M.R. A ubiquitin-10 promoter-based vector set for fluorescent protein tagging facilitates temporal stability and native protein distribution in transient and stable expression studies. *Plant J.* **2010**, *64*, 355–365. [CrossRef] [PubMed]

13. Vincent, T.R.; Canham, J.; Toyota, M.; Avramova, M.; Mugford, S.T.; Gilroy, S.; Miller, A.J.; Hogenhout, S.; Sanders, D. Real-time In Vivo recording of Arabidopsis calcium signals during insect feeding using a fluorescent biosensor. *J. Vis. Exp.* **2017**, *126*, 56142. [CrossRef] [PubMed]

14. Toyota, M.; Spencer, D.; Sawai-Toyota, S.; Jiaqi, W.; Zhang, T.; Koo, A.J.; Howe, G.A.; Gilroy, S. Glutamate triggers long-distance, calcium-based plant defense signaling. *Science* **2018**, *361*, 1112–1115. [CrossRef]

15. Vigani, G.; Costa, A. Harnessing the new emerging imaging technologies to uncover the role of Ca2+ signaling in plant nutrient homeostasis. *Plant Cell Environ.* **2019**, *42*, 2885–2901. [CrossRef]

16. Marques-Bueno, M.M.; Karina Morao, A.; Cayrel, A.; Pierre Platre, M.; Barberon, M.; Caillieux, E.; Colot, V.; Jaillais, Y.; Roudier, F.; Vert, G. A versatile Multisite Gateway-compatible promoter and transgenic line collection for cell type-specific functional genomics in Arabidopsis. *Plant J.* **2016**, *85*, 320–333. [CrossRef]

17. Kwon, T.; Sparks, J.A.; Liao, F.; Blancaflor, E.B. ERULUS is a plasma membrane-localized receptor-like kinase that specifies root hair growth by maintaining tip-focused cytoplasmic calcium oscillations. *Plant Cell.* **2018**, *30*, 1173–1177. [CrossRef]

18. Plieth, C.; Hansen, U.P.; Knight, H.; Knight, M.R. Temperature sensing by plants: The primary characteristics of signal perception and calcium response. *Plant J.* **1999**, *18*, 491–497. [CrossRef]

19. Legue, V.; Blancaflor, E.; Wymer, C.; Perbal, G.; Fantin, D.; Gilroy, S. Cytoplasmic free Ca^{2+} in Arabidopsis roots changes in response to touch but not gravity. *Plant Physiol.* **1997**, *114*, 789–800. [CrossRef]
20. Fasano, J.M.; Massa, G.D.; Gilroy, S. Ionic signaling in plant responses to gravity and touch. *J. Plant Growth Regul.* **2002**, *21*, 71–88. [CrossRef]
21. Li, T.; Yan, A.; Bhatia, N.; Altinok, A.; Afik, E.; Durand-Smet, P.; Tarr, P.T.; Schroeder, J.I.; Heisler, M.G.; Meyerowitz, E.M. Calcium signals are necessary to establish auxin transporter polarity in a plant stem cell niche. *Nat. Commun.* **2019**, *10*, 726. [CrossRef] [PubMed]
22. Sierla, M.; Waszczak, C.; Vahisalu, T.; Kangasjarvi, J. Reactive oxygen species in the regulation of stomatal movements. *Plant Physiol.* **2016**, *171*, 1569–1580. [CrossRef] [PubMed]
23. Wilkins, K.A.; Matthus, E.; Swarbreck, S.M.; Davies, J.M. Calcium-mediated abiotic stress signaling in roots. *Front. Plant Sci.* **2016**, *7*, 1296. [CrossRef] [PubMed]
24. Dubyak, G.R.; El-Moatassim, C. Signal transduction via P_2-purinergic receptors for extracellular ATP and other nucleotides. *Am. J. Physiol.* **1993**, *265*, 577–606. [CrossRef] [PubMed]
25. Tanaka, K.; Gilroy, S.; Jones, A.M.; Stacey, G. Extracellular ATP signaling in plants. *Trends Cell Biol.* **2010**, *20*, 601–608. [CrossRef]
26. Choi, W.; Toyota, M.; Kim, S.; Hilleary, R.; Gilroy, S. Salt stress-induced Ca^{2+} waves are associated with rapid, long distance root-to-shoot signaling in plants. *Proc. Natl. Acad. Sci. USA* **2014**, *111*, 6497–6502. [CrossRef]
27. Matthus, E.; Sun, J.; Wang, L.; Bhat, M.G.; Mohammad-Sidik, A.B.; Wilkins, K.A.; Leblanc-Fournier, N.; Legue, V.; Moulia, B.; Stacey, G.; et al. DORN1/P2K1 and purino-calcium signalling in plants: Making waves with extracellular ATP. *Ann. Bot.* **2020**, *124*, 1227–1242. [CrossRef]
28. Erickson, G.R.; Northrup, D.L.; Guilak, F. Hypo-osmotic stress induces calcium-dependent actin reorganization in articular chondrocytes. *Osteoarthr. Cartil.* **2003**, *11*, 187–197. [CrossRef]
29. Mundy, W.R.; Kodavanti, P.R.S.; Dulchinos, V.F.; Tilson, H.A. Aluminum alters calcium transport in plasma membrane and endoplasmic reticulum from rat brain. *J. Biochem. Toxic.* **1994**, *9*, 17–23. [CrossRef]
30. Shabala, S.; Demidchik, V.; Shabala, L.; Cuin, T.A.; Smith, S.J.; Miller, A.J.; Davies, J.M.; Newman, I.A. Extracellular Ca^{2+} ameliorates NaCl-induced K^+ loss from Arabidopsis root and leaf cells by controlling plasma membrane K^+-permeable channels. *Plant Physiol.* **2006**, *141*, 1653–1665. [CrossRef]
31. Feng, W.; Kita, D.; Peaucelle, A.; Cartwright, H.N.; Doan, V.; Duan, Q.; Liu, M.; Maman, J.; Steinhorst, L.; Schmitz-Thom, I.; et al. The FERONIA receptor kinase maintains cell-wall integrity during salt stress through Ca^{2+} signaling. *Curr. Biol.* **2018**, *28*, 666–675. [CrossRef] [PubMed]
32. Wang, Y.S.; Yoo, C.M.; Blancaflor, E.B. Improved imaging of actin filaments in transgenic Arabidopsis plants expressing a green fluorescent protein fusion to the C- and N-termini of the fimbrin actin-binding domain 2. *New Phytol.* **2008**, *177*, 525–536. [CrossRef] [PubMed]
33. Dyachok, J.; Sparks, J.A.; Liao, F.; Wang, Y.S.; Blancaflor, E.B. Fluorescent protein-based reporters of the actin cytoskeleton in living plant cells: Fluorophore variant, actin binding domain, and promoter considerations. *Cytoskeleton* **2014**, *71*, 311–327. [CrossRef] [PubMed]
34. Tian, L.; Hires, S.A.; Mao, T.; Huber, D.; Chiappe, M.E.; Chalasani, S.H.; Petreanu, L.; Akerboom, J.; McKinney, S.A.; Schreiter, E.R.; et al. Imaging neural activity in worms, flies and mice with improved GCaMP calcium indicators. *Nat. Methods* **2009**, *6*, 875–881. [CrossRef]
35. Clough, S.J.; Bent, A.F. Floral dip: A simplified method for Agrobacterium-mediated transformation of Arabidopsis thaliana. *Plant J.* **1998**, *16*, 735–743. [CrossRef]
36. Leitao, N.; Dangeville, P.; Carter, R.; Charpentier, M. Nuclear calcium signatures are associated with root development. *Nat. Commun.* **2019**, *10*, 4865. [CrossRef]

International Journal of
Molecular Sciences

Review

Retrograde Signaling: Understanding the Communication between Organelles

Jakub Mielecki, Piotr Gawroński and Stanisław Karpiński *

Department of Plant Genetics, Breeding and Biotechnology, Institute of Biology,
Warsaw University of Life Sciences, 02-787 Warsaw, Poland; jakub_mielecki@sggw.edu.pl (J.M.);
piotr_gawronski@sggw.edu.pl (P.G.)
* Correspondence: stanislaw_karpinski@sggw.edu.pl

Received: 14 July 2020; Accepted: 20 August 2020; Published: 26 August 2020

Abstract: Understanding how cell organelles and compartments communicate with each other has always been an important field of knowledge widely explored by many researchers. However, despite years of investigations, one point—and perhaps the only point that many agree on—is that our knowledge about cellular-signaling pathways still requires expanding. Chloroplasts and mitochondria (because of their primary functions in energy conversion) are important cellular sensors of environmental fluctuations and feedback they provide back to the nucleus is important for acclimatory responses. Under stressful conditions, it is important to manage cellular resources more efficiently in order to maintain a proper balance between development, growth and stress responses. For example, it can be achieved through regulation of nuclear and organellar gene expression. If plants are unable to adapt to stressful conditions, they will be unable to efficiently produce energy for growth and development—and ultimately die. In this review, we show the importance of retrograde signaling in stress responses, including the induction of cell death and in organelle biogenesis. The complexity of these pathways demonstrates how challenging it is to expand the existing knowledge. However, understanding this sophisticated communication may be important to develop new strategies of how to improve adaptability of plants in rapidly changing environments.

Keywords: retrograde signaling; biogenic control; operational control; stress response; cell death

1. Introduction

Oxygenic photosynthesis is an ancient process that likely evolved over 3.7 billion years ago in free-living bacteria. According to the endosymbiosis theory, the ancestor of the current alphaproteobacteria from which mitochondria are derived were incorporated into prokaryotic Archaean cells. Some of these newly formed eukaryotic cells underwent another endosymbiosis event, incorporating a photosynthetically active ancestor of cyanobacteria into their cells. These two events—followed by a long period of evolution—resulted in the emergence of modern eukaryotic plant cells [1]. During the evolution of eukaryotic plant cells, the genomes of the original endosymbionts evolved and rearranged in such way that many genes were transferred from organellar genomes to nucleus. This process was aimed at securing endosymbionts in eukaryotic cell and simplifying metabolic pathways to allow eukaryotic cells to manage their resources in a more efficient manner. However, these genetic rearmaments during the evolution of the modern plant cell also required the evolution of a communication between the prokaryotic precursors of organelles and the eukaryotic nucleus.

During evolution, the majority of genes encoded by the genome of free-living cyanobacteria were transferred to the nuclear genome of the newly formed symbiotic cell [2,3]. Around 18% of the *Arabidopsis thaliana* nuclear genes are derived from a plastid ancestor. Surprisingly, the majority of proteins encoded by those genes are not transferred back to the chloroplasts. Even in nonphotosynthetic organisms like *Plasmodium* or *Trypanosoma*—which evolved from red algae—we still find some functional genes of cyanobacteria origin [4]. This demonstrate that the benefits from endosymbiosis can be various and the evolution of eukaryotes took many different paths in order to gain advantage over prokaryotic cells. We cannot clearly distinguish which pathway was more beneficial, but it is obvious that the emergence of the eukaryotic cell led to the possibility for the evolution of plant, fungi and animal modern cells, which was a major breakthrough in evolution of oxygenic life on earth. To make this happen, the original symbionts had to develop a highly coordinated two—or three-way communication system. It is obvious that the expression of many genes encoded by the nuclear genome depends on organellar signals, in a mechanism called retrograde signaling. In this review, we present role of the signals derived from dysfunctional organelles such as chloroplasts, mitochondria and peroxisomes in regulation of nuclear gene expression (NGE). The proper folding and assembly of many plastid protein complexes (e.g., photosystems) requires the highly coordinated coupled expression of photosynthesis-associated nuclear genes (PhANGs) and photosynthesis-associated plastid genes (PhAPGs). On the other hand, organellar gene expression (OGE) is controlled by nuclear-encoded factors; this kind of regulation is called anterograde signaling. Disturbances in retrograde—as well as anterograde—signaling may lead to their uncoupled expression, which has harmful effects on plant cells by impairing the proper functioning of chloroplasts [5]. Under photooxidative stress, chloroplasts are not able to carry out the biosynthesis of carbohydrates efficiently, which limits the energy supply of cells. Instead, excess excitation energy (EEE) results in a rapid foliar temperature increase due to nonphotochemical quenching (NPQ) and formation of reactive oxygen species (ROS) which—apart from their destructive capabilities—are also able to act as signaling molecules, allowing plants to adapt to suboptimal environmental conditions. The inhibition of mitochondrial electron transport and the tricarboxylic acid cycle requires retrograde induction of the alternative oxidase pathways in order to reduce oxygen and maintain energy production. Alternative respiratory pathways have a lower ATP yield than the main pathway. However, in stressful conditions, it may be the only way to produce energy. Severe oxidative stress can surpass the abilities of the antioxidative system, and ultimately results in the induction of programmed cell death (PCD). Plants induce PCD in order to survive in extreme environmental conditions, or to prevent spread of pathogens which eventually could lead to death of the whole plant organism. Retrograde signals originated from peroxisomes are connected with the inhibition of catalases. Recently, the role of peroxisome-derived H_2O_2 in the induction of PCD was established; we will discuss this later in this review. On the other hand—apart from discussing the role of each mentioned organelle in retrograde signaling—we also present several putative signal integrators, indicating that under some conditions, changes in the NGE may require a coordinated signal from more than one type of organelle.

Crosstalk between the nucleus and other organelles is crucial, not only for their development, but also to trigger stress responses or to acclimate to a constantly changing environment. Understanding of those complex signaling pathways may allow us to modify them in the future in order to improve crop productivity or to respond to several combined abiotic and biotic stresses [6–9].

Along with climate change and the overheating of the Earth which we are experiencing nowadays, extreme conditions such as water deficit, extreme temperatures, storms, deforestation, desert expansion and high soil salinity occurs more often [10,11]. The Pace of those changes raises at least a few questions: Will plants be able to cope with such rapid climate change? What about diseases and pests that will migrate to more favorable habitats? This is currently happening, for example in the pandemic wheat disease—stem rust (TTTTF)—which is migrating from Africa to Europe [12]. This is why we should think about these challenges more carefully in coming years. If not, we will be not prepared for what may come [13]. One strategy to resolve this problem is to enhance natural plant defense and acclimation mechanisms in crops with genes from their wild ancestors. Especially helpful in this case may be some knowledge about retrograde-signaling pathways. However, acclimation to stressful conditions requires energetic investment which could be utilized for growth and development. Finding the proper balance between those two processes may be crucial [14].

Even widely described, retrograde-signaling pathways still has some unknowns (mostly how signals are transduced from organelles to nucleus). Recently, some scientists working in this field question that even genes that has been considered to be key parts of some retrograde-signaling pathways for last few years. In this review, we try to describe different classes of retrograde signals, already known pathways related to organelles biogenesis or stress response and also discuss all the controversy that have recently arisen around some of the putative retrograde candidates.

2. Plurality of Putative Retrograde Signals

Retrograde signals can be classified according to the nature of the signal. We can distinguish few major types of the signals which are transduced by different biomolecules such as: RNA, ROS, proteins and other metabolites. In pioneering work of chloroplast-to-nucleus signaling, it was proposed that RNA from plastids can regulate protein synthesis in cytoplasm [15]. Indeed, recently was identified link between retrograde signaling and RNA editing in chloroplasts [16]. Noteworthy, recent studies show that RNA metabolism processes such as alternative splicing and micro RNA synthesis are not only affected by retrograde signaling but can also trigger it as well. One hypothesis indicates that it could be achieved through alterations in RNA editing levels for transcripts encoding subunits of RNA polymerase such as *rpoC1* and *rpoB* [17–20]. Even without evidence of RNA being signaling molecule for retrograde signaling itself, it is clear that RNA metabolism (both nuclear and plastid) plays major role in transduction of retrograde signals.

ROS are products of several metabolic pathways; it was described that their accumulation can affect NGE [21]. ROS can be produced in apoplast, chloroplasts, mitochondria, glyoxysomes, peroxisomes, endoplasmic reticulum (ER) and cytosol [22]. Photosynthesis and photorespiration are well known metabolic processes that generates ROS. NADPH oxidases (NOXs) activity results in ROS formation in apoplast. Respiratory burst oxidase homologs (RBOHs) are plant NADPH oxidases located in the plasma membrane. Apoplastic ROS are scavenged in the same way as it happens intracellular, but by a specific extracellular isoforms of antioxidative enzymes [23,24]. H_2O_2 formed in apoplast can be transported into the cells via aquaporins (AQPs) (Figure 1). Plasma membrane intrinsic protein 1.4 (AtPIP1.4) was described as the AQP that can facility intracellular H_2O_2 transport through plasma membrane from apoplast [25]. This mechanism may allow apoplastic communication between plant cells and allow induction of systemic responses. Redox state of the apoplast is important in acclimation of photosynthesis to variable light intensity [26]. To further support this hypothesis recently apoplastic H_2O_2 sensor—HPCA1—was identified [27]. Cells developed ROS scavenging mechanisms to maintain homeostasis, it is achieved mainly by enzymatic and nonenzymatic compounds [28]. Glutathione (GSH) and especially ascorbate (AsA) are main nonenzymatic antioxidants in plants [29]. However, those nonenzymatic compounds depend on their recycling by specific set of enzymes. For example, DEHYDROASCORBATE REDUCTASE2 (DHER2) is an important enzyme taking part in ascorbate recycling [30]. There are also other nonenzymatic antioxidative compounds such as phenolic compounds, carotenoids, flavonoids, tocopherols and alkaloids [31]. Mentioned compounds play role in regulating redox state of plant cells. Under abiotic or biotic stress ROS production surpass abilities of antioxidative systems eventually changing redox state of chloroplasts which is considered to be important in triggering retrograde signaling. Because some of ROS like singlet oxygen ($^1O_2$) have short lifespan it is clear that it rather takes part as one of the signaling molecules than migrate from plastids to nucleus itself to modulate NGE. It is still unclear how ROS escape chloroplasts. However, in cytoplasm there are several redox sensitive proteins such as Mitogen-activated protein kinases (MAPKs) which can trigger signal cascades that eventually lead to changes in NGE [32]. ROS stability, apart from their short half-time is also linked with their reactivity. More reactive species like hydroxyl radical (OH•) are chemically unstable and they can only oxidize compounds in their close vicinity. In contrast to hydroxyl radical, H_2O_2 is the most stable ROS. However, short-lived ROS like superoxide anion radical (O_2•$^-$) generated by photosystem I (PSI) can be dismutated to H_2O_2 enzymatically by superoxide dismutase (SOD) or spontaneously [33]. It was reported that hydrogen peroxide can oxidase Calvin cycle enzymes containing thiol group [34]. Oxidation of cysteine thiol group to sulfenic acid (Cys–SOH) is well established reversible post translational modification that regulates protein activity. It depends on the redox state of cell. Sulfenylation of important antioxidant enzyme DHER2 prevents it from irreversible overoxidation [30]. Endosymbiosis events also contributed to evolution of redox sensitive proteins containing cysteine thiols in order to sense and regulate redox state of cell [35]. Many hypotheses about role of H_2O_2 in plant signaling under light illumination were thoughtfully discussed in literature [36,37]. H_2O_2 derived from photosynthesis is also able to diffuse out of the chloroplasts during illumination [32,38,39]. However, authors suggested that it cannot pass chloroplast membrane by simple diffusion. It is more likely that it is transported out of chloroplasts through AQPs [40]. Some researchers hypothesize that it may be delivered from chloroplast to nucleus via stromules [41]. They are tubules filled with stroma formed by all plastid types discovered in vascular plants and it was described that they can emanate after exposure to ROS [42]. Recently, research group of Phillip Mullineaux suggested that photosynthesis derived H_2O_2 from chloroplasts associated nearby nuclei may affect NGE omitting transport by cytosol [43].

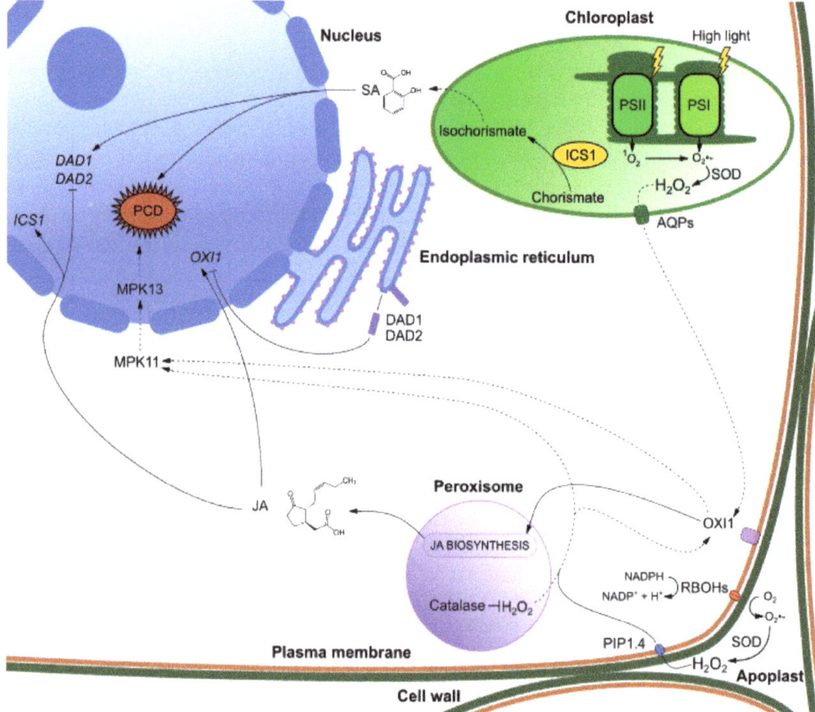

Figure 1. Role of oxidative signal inducible 1 (OXI1) in retrograde signaling in mature mesophyll cells. OXI1 in response to reactive oxygen species (ROS) induces biosynthesis of jasmonic acid (JA) and induces programmed cell death (PCD) through MPK11/MPK13-mediated pathway. JA induces expression of *ICS1*. Crosstalk between SA and JA along with endoplasmic reticulum (ER)-associated proteins DEFENDER AGAINST CELL DEATH (DAD1)/DAD2 regulates expression of *OXI1*. Dotted lines represent pathways which lack experimental evidence or exact nature of indicated regulation is still unknown.

Hydrogen peroxide in chloroplasts is detoxified by peroxiredoxin (PrxR), but mostly by ascorbate–glutathione cycle in which ascorbate peroxidase (APX) is key enzyme [44,45]. There are different enzymes that can also scavenge ROS, including catalase (CAT), superoxide dismutase

(SOD) and glutathione peroxidase (GPX). Because of their broad substrate specificity usually each ROS can be utilized by at least two different enzymes [46,47]. It is also worth mentioning that each cell compartment has specific protein isoforms and contain more than one type of antioxidative enzymes. For example, we can distinguish APX isoenzymes located in stromal (sAPX) and thylakoid (tAPX) membrane of chloroplasts (in mature mesophyll cells), cytosol APX (cAPX) and microbody APX (mAPX) bound to membrane of peroxisome and glyoxysome [48]. APX change their activity under high light illumination [49–51]. *APXs* expression is regulated by redox status of plastoquinone pool [50]. Existence of several APX isoenzymes allow plant cells to regulate ROS content in each compartment independently [52]. Cells may purposely not scavenge all ROS in efficient manner and in every cell compartment so the signal can be transduced and proper response for high light intensity induced. It is still unclear if response to H_2O_2 depends on in which organelle it was produced or rather signal is integrated regardless of the source [53]. Signaling pathways originating from peroxisomes were often discovered analyzing catalase mutants (*cat*) which accumulate H_2O_2 inside peroxisomes in differentiated cells. Its accumulation lead to the induction of pathogenesis related genes and ultimately to the cell death [54,55]. *cat* mutants differentially express many nuclear encoded genes compared to the wild-type plants under high light intensity stress [56]. Transcriptome analysis of *cat2* mutant shows higher expression of many genes related to protein repair [57]. More severe changes are observed in transcriptome of *cat1 cat2 cat3* triple mutant exhibiting deregulation in expression pattern of several receptor-like kinases and transcription factors (TFs) [58]. Among those genes *oxidative signal inducible 1* (*OXI1*) encoding serine/threonine kinase, *MPK11* and *MPK13* were especially interesting considering redox sensitivity of some MAPKs. OXI1 belongs to AGC family of plasma membrane bound kinases [59]. OXI1 was described as activator of MPK3 and MPK6 in response to H_2O_2 and pathogens [60]. However, no changes in expression pattern of those kinases were observed in triple catalase mutant [58]. This may suggest existence of separate H_2O_2 related signaling pathways for peroxisomes and chloroplasts. H_2O_2 accumulation in peroxisomes promote transduction of the signals through OXI1/MPK11/MPK13 (Figure 1). To support this statement, it was observed that *MPK3* and *MPK6* expression levels are not affected in *OXI1* overexpressing lines [61]. Recently, two PCD inhibitors: DEFENDER AGAINST CELL DEATH (DAD1) and its homolog DAD2 were described as regulators of OXI1 induced cell death in response to high light stress [61]. *DAD1* and *DAD2* overexpression lines exhibit lower expression of *OXI1* than wild-type and *OXI1* is upregulated in *dad1* and *dad2* mutants. This suggest antagonistic role of DAD1/DAD2 and OXI1. Induction of OXI1 dependent cell death is coregulated by two plant hormones salicylic acid (SA) and jasmonic acid (JA). SA apart from its role in pathogen defense signaling also regulates redox homeostasis and light acclimation [62]. Wild-type plants treated with JA exhibit increased expression of *OXI1* and decreased expression of *DAD1* and *DAD2*. *oxi1* mutants have lower content of JA than wild-type plants and high light treatment does not induce JA accumulation like it does in the wild-type. This suggest that OXI1 regulates JA biosynthesis in peroxisomes. In contrast, treatment with SA induces *DAD1* and *DAD2* expression while *OXI1* is expressed at same level. Treatment with JA induce expression of SA biosynthesis genes: *ENHANCED DISEASE SUSCEPTIBILITY1* (*EDS1*) and *ISOCHORISMATE SYNTHASE1* (*ICS1*) in the wild-type which leads to increased isochorismate biosynthesis in chloroplasts. ICS1 converts chorismate to isochorismate which is a precursor for SA biosynthesis in cytosol. In contrast treatment with SA does not affect expression of JA biosynthesis genes (*AOS* and *OPR3*) in the wild-type [61]. Regulation of PCD by OXI1 in response to highlight and H_2O_2 accumulation in peroxisomes are shown in Figure 1.

Another interesting aspect in singlet-oxygen–mediated signaling is the executer pathway. EXECUTER1 (EX1) and EXECUTER2 (EX2) are proteins associated with thylakoid membranes and were identified using *fluorescent* (*flu*) mutant suppressor screen [63,64]. Seedling of *Arabidopsis thaliana flu* mutant in the dark accumulate protochlorophyllide which is an intermediate in the biosynthesis of chlorophyll *a*. Transitioning mutants to light results in burst production of singlet oxygen which affect NGE and eventually leads to the inhibition of growth, chlorosis and ultimately cell death [65]. EX1 and EX2 are part of singlet-oxygen-dependent retrograde signaling under low light conditions [63]. In excess light conditions retrograde signals are transduced independently of EX1 and EX2 through generation of β-cyclocitral [66,67]. EX1 is localized to the grana margins in chloroplasts. Chlorophyll synthesis, disassemble and reassemble of damaged PSII take place in a close vicinity of EX1. Singlet-oxygen–mediated retrograde signaling depends on degradation of EX1 protein by ATP-dependent zinc metalloprotease FtsH. The FtsH also catalyzes cleavage of D1 protein in the reaction center of the damaged PSII. Based on that, it is suggested that EX1 dependent signaling is connected with repair of PSII [68]. EX1 degradation depends on oxidation of tryptophan at position 643 by singlet oxygen. The substitution of this amino acid with leucine or alanine (which also are singlet oxygen sensitive amino acids) inhibits EX1 degradation by FtsH2 [69]. Recently, another novel singlet oxygen induced retrograde singling pathway was discovered by ethyl methanesulfonate (EMS) mutagenization in *flu ex1* double mutant. SAFEGUARD1 (SAFE1) is localized in the stroma of chloroplasts and is degraded by the release of singlet oxygen. Plants lacking functional SAFE1 protein were more susceptible to singlet-oxygen-induced damage of thylakoids grana margins [70]. Singlet-oxygen-mediated retrograde-signaling pathways are demonstrated in Figure 2.

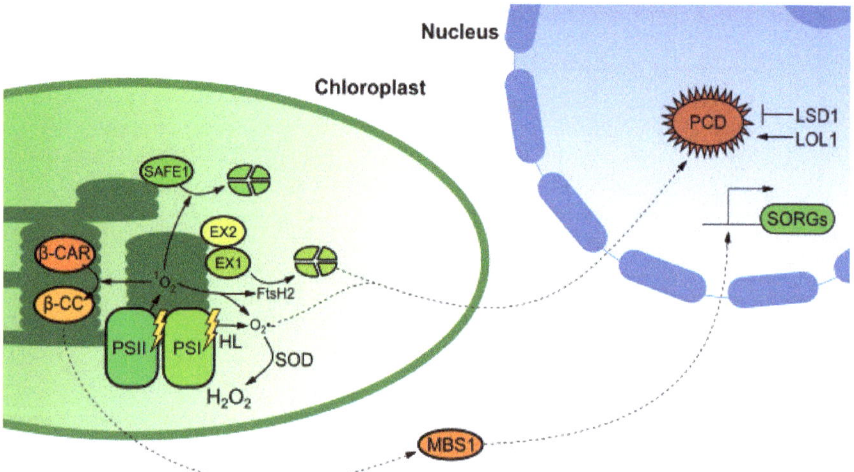

Figure 2. Retrograde-signaling pathways mediated by singlet oxygen in mature mesophyll cell. High light stress induces ROS production in photosystem I (PSI) and PSII. Depending on its concentration singlet oxygen induce different pathways. At low concentrations it promotes degradation of EXECUTER1 (EX1) by FtsH2 protease and induce EX1-mediated PCD. It is negatively regulated by LESION STIMULATING DISEASE1 (LSD1) and positively by its homolog LOL1. SAFEGUARD1 (SAFE1) similarly, to EX1 protects grana margins from oxidative damage by singlet oxygen. At high concentration singlet oxygen oxidizes β-carotene to β-cyclocitral and induces expression of many singlet oxygen related genes (SORGs) through METHYLENE BLUE SENSITIVITY 1 (MBS1). Dotted lines represent pathways which lack experimental evidence or exact nature of indicated regulation is still unknown.

Retrograde signals can also be carried out by different classes of proteins in which transcription factors are worth mentioning. There are few well described proteins such as Whirly1, Plant homeodomain transcription factor with transmembrane domains (PTM) or ABSCISIC ACID-INSENSITIVE4 (ABI4) [71–73], but we will describe those widely in other section of this review.

One of most interesting and well known retrograde-signaling pathway is connected with tetrapyrrole biosynthesis. Under norflurazon (NF) treatment—which inhibits biosynthesis of carotenoids—several nuclear genes are downregulated [73,74]. Almost three decades ago research group of Joanne Chory identified several *genomes uncoupled* (*gun*) mutants under NF treatment [75].

Arabidopsis thaliana mutants exhibiting *gun* phenotype are not able to downregulate PhANGs (e.g., *LIGHT HARVESTING CHLOROPHYLL A/B BINDING PROTEIN1.2 (LHCB1.2)*) like wild type plants after chloroplast damage caused by NF treatment [76]. Thus, it was concluded that GUNs are part of retrograde-signaling pathway regulating expression of PhANGs. All *GUNs* except *GUN1* encode proteins involved in tetrapyrrole biosynthesis pathway (TBP) in chloroplasts. GUN1 is pentatricopeptide repeat (PPR) protein localized in chloroplasts containing a small MutS related (SMR) domain. Analysis of *gun2-5* mutants initially led to conclusion that Mg–protoporphyrin (MgProto) is a retrograde metabolite able to move between the nucleus and chloroplasts [77]. However, further studies did not support this hypothesis because no correlation was found between level of MgProto and *LHCB1.2* expression [78]. In contrary some researchers suggested that this may be due to difficulties with identification of tetrapyrroles by HPLC while their content is low [79]. Recent study confirmed accumulation of MgProto in first two days after newly germinated seedlings were treated with NF. The accumulation of MgProto was also correlated with repression of *LHCB1.2* expression suggesting that accumulation of MgProto is a retrograde signal [80].

The identification of *gun6* mutant exhibiting higher activity of ferrochelatase 1 (FC1) suggested that heme synthetized by FC1 may be precursor or retrograde signal itself [81]. Overexpression of *FC1* targeted to chloroplasts rescued nuclear gene expression after NF treatment and increased expression levels of *CA1*, *LHCB2.1* and *GUN4* even without this treatment. However, targeting FC1 to mitochondria did not affect NGE [82]. These results further support role of heme synthetized by FC1 in the chloroplast-to-nucleus retrograde signaling. On the other hand, a link between ROS and tetrapyrroles pathway may be a GUN4 protein. GUN4 and Protoporphyrin IX form $^1O_2$ generating complex which can initiate retrograde signaling [83].

An exception is a GUN1 protein which is not involved in tetrapyrrole synthesis pathway. There are many different hypotheses about exact role of GUN1, but most of them are related to plastid protein homeostasis [84–87]. When proper functioning of chloroplasts is disturbed GUN1 can also regulate chloroplast RNA editing by physical interaction with MULTIPLE ORGANELLAR RNA EDITING FACTOR 2 (MORF2) to affect maturation of many transcripts among which are subunits of plastid encoded RNA polymerase. MORF2 interacts with ORGANELLE TRANSCRIPT PROCESSING 81 (OTP81), ORGANELLE TRANSCRIPT PROCESSING 84 (OTP84) and YELLOW SEEDLINGS 1 (YS1). *otp81*, *otp84* and *ys1* mutants exhibit weak *gun* phenotype which is enhanced in double and triple mutant in those genes. Overexpression of *MORF2* results in strong *gun* phenotype similar to *gun1* mutant suggesting that plastid RNA editing and retrograde signaling are functionally connected [16]. Another GUN1 interacting protein is FUG1 which functions as translation initiation factor in chloroplasts. Both functional proteins are required to maintain plastid protein homeostasis [88]. GUN1 probably does not affect plastid gene expression per *se*, but it interacts with chaperone cpHSC70-1 in order to regulate nuclear encoded protein import to chloroplast. This allows to maintain protein homeostasis (proteostasis) in the chloroplasts. Mutation in gene encoding *cpHSC70-1* leads to a *gun* phenotype. However, *gun1* mutant grown under normal conditions does not have affected protein import capacity [89]. On the other hand, it was recently demonstrated that GUN1 can directly bind to heme as well as other porphyrins, increases FC1 activity and also limits heme and protochlorophyllide synthesis [90]. Based on that, GUN1 is linked to the tetrapyrroles which are considered to be retrograde signaling molecules [76]. Complex role of GUN1 and its interactors were described in Figure 3. Role of transcription factors connected with GUN pathways is described further in this manuscript in a section dedicated to biogenic control.

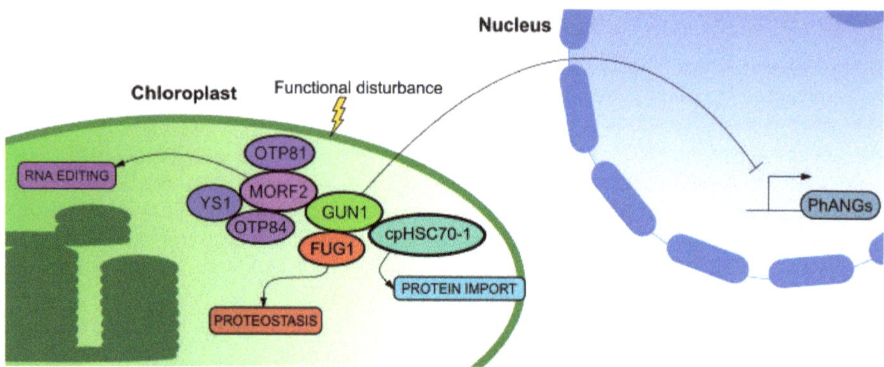

Figure 3. Retrograde signals mediated by GENOMES UNCOUPLED 1 (GUN1). In response to functional disturbance of chloroplast GUN1 inhibits expression of photosynthesis-associated nuclear genes (PhANGs), regulates protein import through interaction with cpHSC70-1, maintains proteostasis through interaction with FUG1 and modulates RNA editing through interaction with MULTIPLE ORGANELLAR RNA EDITING FACTOR 2 (MORF2).

An interesting group of metabolites involved in retrograde signaling and mentioned above are carotenoids. They are main scavengers of $^1O_2$ in chloroplasts and products of their oxidation: β-carotene (β-CAR) and its oxidation product β-cyclocitral (β-CC) can induce changes in NGE [91]. However, transduction of the signals through this pathway depends on METHYLENE BLUE SENSITIVITY 1 (MBS1) PROTEIN. MBS1 is a zinc finger protein located in the nucleus and in the cytosol. Lack of functional protein in *msb1* mutant caused increased susceptibility to $^1O_2$ generated during high light stress [92]. Increased GFP fluorescence observed in plants expressing MBS1:GFP under native promoter treated with β-CC lead to conclusion that MBS1 is involved in β-CC retrograde-signaling pathway. It is hypothesized that MBS1 can induce expression of singlet oxygen related genes (SORGs) in order to cope with high light intensity stress [93].

Another metabolite linked to retrograde signaling is methylerythritol cyclodiphosphate (MEcPP) [94]. It is a precursor of isoprenoids and its accumulation is corelated with changes in NGE. Increased levels of MEcPP induce unfolded protein response in the ER [95]. Similar to ROS, MEcPP accumulates after abiotic stresses such as wounding or high light [94]. Hydroxymethylbutenyl diphosphate synthase (GcpE) is an enzyme responsible for reducing MEcPP to hydroxymethylbutenyl diphosphate (HMBPP). GcpE is encoded by *CEH1* gene. This enzyme is a redox sensitive protein which could explain MEcPP accumulation during photooxidative stress [96]. Recently, two new photoreceptor phytochrome B (phyB) mutant alleles that are able to revert phenotype of *constitutively expressing HPL* (*ceh1*) mutant were described. *ceh1* mutant was identified in a screen for regulators of stress induced *hydroperoxide lyase* (*HPL*) gene [94]. *ceh1* mutant exhibits a dwarf phenotype, has high concentration of SA and accumulates MEcPP [97,98].

3′-phosphoadenosine 5′-phosphate (PAP) is another metabolite which can function as a retrograde signaling molecule in response to drought and highlight stress by altering expression of *APX2*, *ELIP2 ZAT10* and *DREB2A* genes [99]. Abscisic acid (ABA) is one of hormones playing major role in response to those stresses and it is synthetized in chloroplasts. In *Arabidopsis thaliana* PAP acts as secondary messenger in ABA regulated stomatal closure and germination [100]. PAP is synthetized from 3′-phosphoadenosine 5′-phosphosulfate (PAPS) by sulfotransferases [101]. It was considered to be a byproduct without function in plants however it can alter RNA catabolism in yeast (*Saccharomyces cerevisiae*) by inhibiting two 5′ → 3′ exoribonucleases (XRNs) [102]. PAP degradation in chloroplasts is mediated by inositol polyphosphate 1-phosphatase (SAL1) which function as nucleotide phosphatase [99]. Point mutation as well as T-DNA insertion in *SAL1* gene result in greater drought tolerance which leads to conclusion that SAL1 is a negative regulator of drought tolerance and it is connected with PAP accumulation [103]. Additionally, functional SAL1–PAP pathway is important for biotic stress responses since mutations in *SAL1* gene lead to higher susceptibility to *Pseudomonas syringae* pv. *tomato* DC3000 and *Pectobacterium carotovorum* subsp. *carotovorum* EC1 [104]. SAL1–PAP retrograde-signaling pathway is well conserved in all land plants [105]. It is also interesting that SAL1 is localized in cytosol [106], nucleus [107], chloroplasts [108] and mitochondria [99]. One of most recent reports shows SAL1–PAP retrograde-signaling pathway involvement in iron homeostasis [109].

3. Retrograde Signaling in Regulation of Organelles Biogenesis

Retrograde-pathway-transducing signals from plastids to the nucleus in order to regulate chloroplast biogenesis are often called "biogenic control" [110]. Main purpose of this type of retrograde signaling is to modulate NGE so proteins encoded by several of PhANGs can be produced and transported to chloroplasts during their development from proplastids [111]. The coupled expression of PhANGs and PhAPGs allows proper folding and assembly of photosystem complexes [112]. Disturbances in photosystem stoichiometry leads to photoinhibition because amount of energy absorbed from photons exceeds photochemical efficiency of PSII. Photoinhibition eventually leads to ROS formation which can be lethal for developing seedlings.

Signals conditioning proper plastid biogenesis are connected with tetrapyrrole biosynthesis pathway, changes in plastid gene expression (PGE) and activity of the photosynthetic electron transport (PET). ABI4 was discovered during a screen for ABA-insensitive (*abi*) mutants which are able to germinate in presence of ABA [113]. Different *abi4* alleles were also discovered in independent screens for mutants with altered responses to glucose and other sugars [114–117]. ABI4 is a TF classified to APETALA2/ethylene-responsive factor (AP2/ERF) family. Genome of *Arabidopsis thaliana* encodes 147 members of AP2/ERF family and many members of this family are involved in signaling pathways including responses to abiotic and biotic stresses [118,119]. ABI4 takes part in mitochondrial retrograde signaling regulating expression of *ALTERNATIVE OXIDASE1a* (*AOX*) and chloroplast retrograde signaling [73,120]. Higher expression of nuclear-encoded *RbcS* in *abi4* mutant compared to the wild-type after NF treatment allowed to conclude that ABI4 is involved in chloroplast retrograde signaling [121]. In addition, it was reported that *abi4* mutants were able to rescue expression of *LHCB1.2* after lincomycin (Lin) treatment [73]. Lincomycin is plastid translation inhibitor and similar to NF it is often used to screen for *gun* phenotype. Both of those treatments cause damage to chloroplast resulting in photobleached phenotype and drastically reduced expression of most PhANGs [73,74,122]. Activation of ABI4 depends on phosphorylation by MPK3/MPK6 [123]. Based on that knowledge ABI4 was established as one of key proteins involved in plastid development [71,123,124]. It was considered to be a nuclear target of GUN1-dependent retrograde-signaling pathway [71,73,76]. Although there were also studies in which researchers were not able to observe *gun* phenotype in *abi4* mutant when quantifying expression of *CARBONIC ANHYDRASE1* (*CA1*), *GOLDEN2-LIKE1* (*GLK2*) and *LIGHT HARVESTING CHLOROPHYLL A/B BINDING PROTEIN1.1* (*LHCB1.1*) [125–127]. In contrast to *gun1* mutant, crossing *abi4* mutant with *plastid protein import2* (*ppi2*) mutant did not rescue loss of NGE [128]. These results suggest that ABI4 may act independently from GUN1. Recently, an independent study performed on four different alleles of *abi4* did not support a role of ABI4 in biogenic retrograde signaling and researchers were unable to obtain strong or consistent *gun* phenotype in tested *abi4* mutant alleles [129].

It is interesting that this is not first *gun* 'dismantled' by this research group. Before, focusing on ABI4 they decided to investigate role of PTM in biogenic retrograde signaling [130]. PTM is a plant homeodomain (PHD) transcription factor bound to chloroplast envelope [71]. It was proposed that PTM can be cleaved off chloroplast membrane after changes in plastid metabolism and its N terminal domain can affect NGE after its accumulation in nucleus [71]. *ABI4* is one of genes which expression is induced by PTM [71]. Treatments with Lin and NF lead to conclusion that *ptm* mutant exhibit *gun* phenotype because it was able to rescue expression of *LHCB* after treatment with these chemicals. Since its discovery, PTM has been included in many models describing biogenic control pathways [112,131–133]. PTM was also further described in literature by the same research group which provided the first report as regulator of flowering after exposure to high light and it takes part in integration of the signals during de-etiolation [134,135]. However, observation of *gun* phenotype in *ptm* mutant was still to be confirmed by another research group. Because of potential important role of PTM in retrograde signaling Matthew Terry group decided to further examine its role under NF and Lin treatment. In conducted experiments they were unable to confirm *gun* phenotype after both Lin and NF treatment. Expression of selected PhANGs in *ptm* mutant after treatment were not elevated compared to the wild-type plants. Authors concluded that PTM should be excluded from existing models describing plastid signaling or at least its role in it is not as important as we thought before [130].

There are also other interesting mutants in genes encoding transcription factors that exhibit *gun* or *gun*-like phenotype such as: *hy5* and *glk1glk2* [136,137]. In contrast to the *gun* phenotype, there are also number of mutants called *happy on norflurazon* (*hon*) that can tolerate higher concentration of NF in comparison to the wild-type plants. One of the identified *hon* mutants had mutation in ClpR4 (HON5) subunit of Clp protease complex which localizes in chloroplasts. Another example is a *hon23* mutant, which has mutation in putative chloroplast translation elongation factor, and it clearly shows that these mutations interfere with chloroplast protein homeostasis [138]. It is also worth mentioning that not only mutants in genes encoding transcription factors, but also other proteins such as *cry1* (encoding blue light photoreceptor) or *coe1* (encoding mitochondrial transcription termination factor 4) can exhibit *gun* phenotype and eventually take part in biogenic control [137,139].

4. Retrograde Signaling in Stress Response and Acclimation

Retrograde-pathway-transducing signals from plastids to the nucleus in order to cope with environmental stresses and acclimate to them are often called "operational control". Since plants are in general immobile and unable to avoid many unfavorable environmental conditions, they had to evolve sophisticated mechanisms in order to survive and effectively reproduce.

Chloroplasts are sensors of visible light and crop yield is often corelated with the efficiency of photosynthesis. Second, but equally important metabolic process that provides energy for plant cell is aerobic respiration. This processes however are sensitive to changes in plant growth environment [140]. Abiotic stresses can result in photoinhibition of PSII and inhibition of carbon assimilation enzymes [141]. One of the ultimate responses to severe abiotic and biotic stresses is PCD. Apart from its major contribution during tissue development, PCD is also a mechanism that allows plants to prevent pathogens from reproducing and spreading to uninfected cells. During abiotic stresses PCD promotes dismantling of a limited number of affected cells to prevent severe systemic damage to the whole organism [142]. Cells undergoing PCD exhibit extensive chromatin condensation and developmental PCD can occur only in specific cell types [143]. One of the most interesting regulators of PCD is LESION STIMULATING DISEASE1 (LSD1). LSD1 acts as a transcription regulator and condition dependent scaffold protein [7,144]. LSD1 is negative regulator of cell death and defense responses. Along with its homolog LOL1 (which exhibits an antagonistic function) they cooperate in order to induce adequate response through PCD. Mutation in *lsd1* results in a runaway cell death (RCD) in a light dependent manner [145,146]. It was proved that reduction of the PSII antenna size, thus reduction of light absorption and EEE pressure by crossing *lsd1* mutant with *cao1* mutant caused an increase of NPQ and reversion of the RCD phenotype in *lsd1* [147]. The reduction of plastoquinone (PQ) pool induced by EEE results in burst of ROS which lead to induction of SA and ethylene dependent signaling pathway through EDS1 and PHYTOALEXIN DEFICIENT4 (PAD4) which lie on the same pathway as LSD1 because RCD phenotype of *lsd1* mutant is abolished by inactivation of ROS, SA and ethylene signaling components [6,147,148]. The overexpression of bacterial salicylate hydroxylase (NahG) fused with chloroplast transit peptide from RbcS in *lsd1* mutant background also abolished RCD indicating correlation between *lsd1* RCD and SA accumulation. Mutation in *lsd1* causes uncoupled expression of

PhANGs and PhAPGs before induction of RCD; it was hypothesized that LSD1 can act downstream of GUN-mediated pathway. However, *lsd1* treatment with Lin does not rescue expression of *LHCBs* which indicates that it is not a *gun* mutant. Based on that knowledge LSD1 is more likely involved in operational than biogenic control. Uncoupled expression of photosynthesis-associated genes in *lsd1* cause accumulation of singlet oxygen which result in induction of EX1-mediated PCD. *lsd1;ex1* double mutant partially reverts RCD phenotype [5]. Based on that knowledge it is suggested that LSD1 plays a role in regulation of PCD and responses towards biotic and abiotic stresses, it also may be an integrator of SA, ROS (including singlet oxygen), ethylene and other hormones (e.g., IAA) mediated pathways [6]. It is also worth mentioning that LSD1, EDS1 and PAD4 are conditional dependent PCD regulators. For example, in optimal laboratory conditions *lsd1* mutant displays deregulation of over 2000 genes while in the suboptimal field conditions it has 62 deregulated genes and only 43 of those genes were commonly deregulated in both of these conditions [9].

In this manuscript, we mainly focused on retrograde pathways related to functioning of chloroplasts. On the other hand, mitochondria also depend on retrograde signaling during their biogenesis and stress responses. However, it needs further investigation whether chloroplasts and mitochondria induce separate signaling pathways or they converge into the same pathways [149]. Knowing that functioning of both is strongly connected through the energy, metabolism and redox status makes second hypothesis a viable one [150–152]. However, it is considered that mitochondrial retrograde signaling plays greater role in nonphotosynthetic tissues. To support this hypothesis, it was observed that overexpression of TF ANAC013 in *Arabidopsis thaliana* resulted in enhanced tolerance of chloroplasts during oxidative stress [153]. It is achieved mostly by dissipating excess of reducing equivalents [154–156]. Regulators of mitochondrial retrograde signaling are often identified in genetic screens for TFs that can regulate the expression of nuclear genes encoding mitochondrial proteins related to alternative respiration or stress response. Promoters of *AOX1a*, *UPREGULATED BY OXIDATIVE STRESS (UPOX)*, *NAD(P)H-UBIQUINONE OXIDOREDUCTASE B2 (NDB2)* and *CYTOCHROME BC1 SYNTHASE1* (AtBCS1) are often used in such screens. After high light and antimycin A (electron transport chain blocker similar to cyanide) treatment, AtWRKY40 downregulated and AtWRKY63 upregulated expression of *AOX1a*, *UPOX*, *NDB2* and AtBCS1 suggesting their antagonistic function [157]. Interestingly among identified *AOX* regulators there are many components involved in auxin signaling [158]. Another interesting regulator of mitochondrial retrograde signaling is OM66. *OM66* encodes mitochondrial outer membrane protein and its promoter is highly induced by SA in contrary to promoter of *AOX1a* which is responsive to H_2O2 and rotenone [159]. Because *PATHOGENESIS-RELATED1 (PR1)* is downregulated in *OM66* mutant and *OM66* overexpressing lines have higher content of SA it was proposed that *OM66* is regulated in a SA dependent manner [157]. SA inhibits both cytochrome and alternative respiratory pathways [160]. It also inhibits alpha-ketoglutarate dehydrogenase (α-kGDH) a tricarboxylic acid cycle (TCA) enzyme [161]. This information along with well-established SA interactions with chloroplasts metabolism suggests that SA may be a link between chloroplast and mitochondrial retrograde signaling [162].

Retrograde signals transduced from mitochondria as well as chloroplasts can change NGE in a similar manner. The convergence of those two retrograde-signaling pathways is often linked to CYCLIN DEPENDENT KINASE E1 (CDKE1). It is encoded by *REGULATOR OF ALTERNATIVE OXIDASE1* (*RAO1*) gene. It was first described as an important component of mitochondrial retrograde signaling in response to inhibitors. Functional kinase is needed to regulate *AOX1a* in response to oxidative (H_2O_2) and cold stress [163]. It was also demonstrated that CDKE1 regulates expression of *AOX1a* and *Lhcb2.4* in response to photosynthesis inhibitors such as 2,5-dibromo-3-methyl-6-isopropyl-benzoquinone (DBMIB) and 3-(3,4-dichlorophenyl)-1,1-dimethylurea (DCMU). *cdke1*-mutants under high light stress exhibited *gun*-like phenotype [164]. Based on that it was proposed that CDKE1 can integrate retrograde signals from mitochondria and chloroplasts. KIN10 is a subunit of SnRK1 kinase complex conditioning its catalytic activity and it was proposed as an integrator of stress and energy signaling [165,166]. The interaction between KIN10 and CDKE1 in nucleus was also reported [163]. CDKE1 is a part of mediatory complex regulating RNA polymerase II (RNAP II) dependent transcription. Such complexes are considered to integrate stress signals from organelles and initiate proper transcriptional response [167]. Another similarity can be observed in how mitochondrial retrograde signaling is triggered by stress or dysfunction affecting respiratory electron transport chain or TCA [168]. In order to maintain these processes and their energy production, plants need to change their metabolism. That is why proper communication between organelles and nucleus is required [14].

Transcriptome meta-analyses demonstrate that 10% to 20% from differentially regulated genes during abiotic stress responses encode proteins localized to the chloroplasts [169]. We briefly described the role of ROS in retrograde signaling in previous sections of this manuscript. However, it is worth mentioning that ROS signaling is crucial to respond to several abiotic stresses such as drought, variable high light intensity, salinity and heat [170]. Exposure to abiotic and biotic stresses can induce unfolding protein response in ER. As was mentioned before it is connected with MEcPP accumulation [171]. Response to drought stress is usually connected to the SAL1–PAP pathway. Some researchers hypothesize that chloroplasts and mitochondrial retrograde signals converge through TF called ANAC017 (encoded by *RAO2*) to regulate PCD as response to severe organellar stress [172]. NAC family members are ER bound TFs which upon activation are cleaved and relocated to nucleus where they can affect NGE [173]. RADICAL-INDUCED CELL DEATH1 (RCD1) is another putative integrator of chloroplastic and mitochondrial ROS signaling pathways. It was identified in a screen for sensitivity to ozone [174]. *rcd1* mutant is resistant to methyl viologen (MV) and UV-B which suggests that RCD1 may be a ROS sensitive protein [175]. In the *rcd1* mutant more than 400 genes are differently expressed under standard growth conditions. Among those genes there are those encoding mitochondrial AOXs as well as chloroplast 2-Cys peroxiredoxin (2CP) [176–179]. RCD1 interacts with several TFs such as ANAC017 and DREB2A [176,180]. Cleavage of ANAC017 from ER is probably dependent on elevated H_2O_2 levels [181]. Recently, RCD1 was proposed to act as negative regulator of ANAC013 and ANAC017 and thus integrator of NAC and PAP retrograde-signaling pathways [182]. Putative integrators of retrograde-signaling pathways and their interactors were demonstrated on Figure 4. Recently, β-cyclocitral induced protein SCARECROW LIKE14 (SCL14) was described along with TF ANAC102 and xenobiotic detoxification enzymes in lowering levels of toxic carbonyls and peroxides in order to limit damage to the intracellular components caused by photooxidative stress [183].

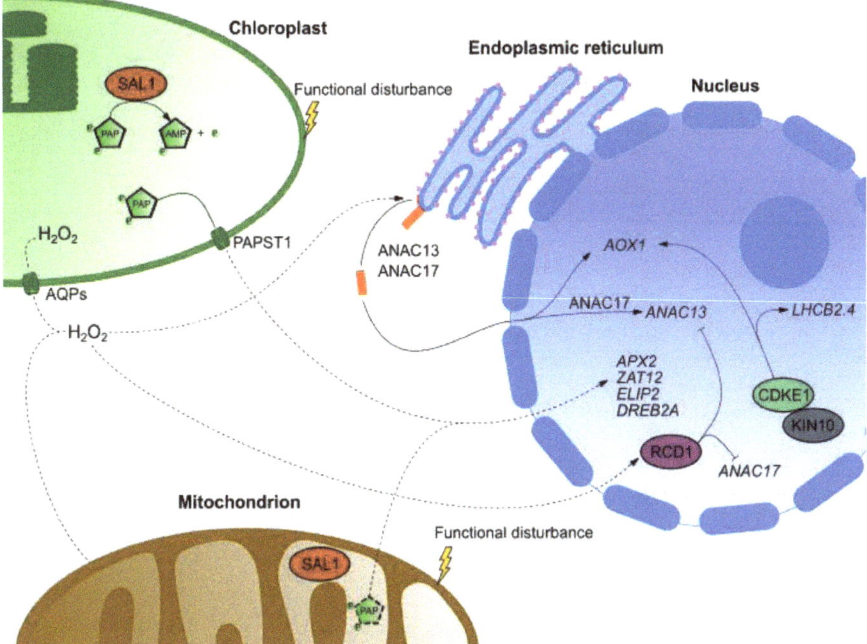

Figure 4. Putative integrators of retrograde-signaling pathways. Inhibition of inositol polyphosphate 1-phosphatase (SAL1) results in accumulation of 3′-phosphoadenosine 5′-phosphate (PAP) which induces expression of many drought and high light related genes such as *APX2, ZAT12, ELIP2* and *DREB2A*. RCD1 after induction with ROS inhibits expression of *ANAC13* and *ANAC17* which regulates *AOX1* expression. CDKE1 induces expression of *AOX1* and *LHCB2.4* and many other genes as part of mediatory complex regulating RNAPII dependent transcription. Dotted lines represent pathways which lack experimental evidence or exact nature of indicated regulation is still unknown.

5. Conclusions

We briefly demonstrated complexity and different nature of retrograde-signaling pathways. It is commonly considered that in each retrograde pathway there are at least two different types of biomolecules involved. It is worth remembering that the signaling pathways we described are not universal for every cell type and in every epigenetic status. While some of them occur only during development, others such as singlet-oxygen–mediated pathways, can occur only in differentiated, photosynthetically active mesophyll cells. Many existing, even well understood pathways still have unknowns and expanding knowledge about them often brings up more new questions than answers. However, even if we are not yet close to understanding retrograde signaling, benefits from it could be worth the effort. Recently, few interesting strategies to improve crop production under field conditions were demonstrated. First aimed at improvement of photorespiration by implementing synthetic glycolate metabolic pathways into tobacco chloroplasts [184]. Others improved crop production and water-use efficiency by accelerating recovery from photoprotection. It was achieved by combined overexpression of *PsbS* and genes encoding xanthophyll cycle enzymes [185,186]. In the future studies we shall also consider other physical retrograde signaling pathways, for example, direct heat radiation and vibration of organelles, electrical and calcium wave signaling from chloroplasts and mitochondria.

Author Contributions: Conceptualization, J.M., P.G. and S.K.; writing—original draft preparation, J.M.; writing—review and editing, J.M., P.G. and S.K.; visualization, J.M.; supervision, P.G. and S.K.; project administration, S.K.; funding acquisition, S.K. All authors have read and agreed to the published version of the manuscript.

Funding: This research and the APC was funded by National Science Center in Poland, grant number UMO-2018/29/B/NZ3/01198.

Acknowledgments: In this section you can acknowledge any support given which is not covered by the author contribution or funding sections. This may include administrative and technical support or donations in kind (e.g., materials used for experiments).

Conflicts of Interest: The authors declare no conflict of interest.

Abbreviations

$^1O_2$	Singlet oxygen
ABA	Abscisic acid
ABI4	ABSCISIC ACID-INSENSITIVE4
AOX	*Alternative oxidase 1a* gene
APX	Ascorbate peroxidase
AQPs	Aquaporins
CAT	Catalase
CDKE1	CYCLIN DEPENDENT KINASE E1
DAD	DEFENDER AGAINST CELL DEATH
EDS1	ENHANCED DISEASE SUSCEPTIBILITY1
EEE	Excess excitation energy
EMS	Ethyl methanesulfonate
ER	Endoplasmic reticulum
EX1	EXECUTER1
EX2	EXECUTER2
FC1	Ferrochelatase 1
GcpE	Hydroxymethylbutenyl diphosphate synthase
GPX	Glutathione peroxidase
GSH	Glutathione
GUN	Genomes uncoupled
H_2O_2	Hydrogen peroxide
IAA	Indole-3-acetic acid
ICS1	ISOCHORISMATE SYNTHASE1
JA	Jasmonic acid
LHCB	*LIGHT HARVESTING CHLOROPHYLL A/B BINDING PROTEIN* gene
Lin	Lincomycin
LSD1	LESION STIMULATING DISEASE1
MAPK	Mitogen-activated protein kinase
MBS1	METHYLENE BLUE SENSITIVITY 1
MEcPP	Methylerythritol cyclodiphosphate
MgProto	Mg–protoporphyrin
MORF2	Multiple organellar RNA editing factor 2
NF	Norflurazon
NGE	Nuclear gene expression
NOXs	NADPH oxidases
NPQ	Nonphotochemical quenching
$O_2{}^{\bullet-}$	Superoxide anion radical
OGE	Organellar gene expression
OH$\bullet$	Hydroxyl radical
OXI1	Oxidative signal inducible 1
PAD4	Phytoalexin deficient4
PAP	3′-phosphoadenosine 5′-phosphate
PCD	Programmed cell death
PET	Photosynthetic electron transport

PGE	Plastid gene expression
PhANGs	Photosynthesis-associated nuclear genes
PhAPGs	Photosynthesis-associated plastid genes
PrxR	Peroxiredoxin
PSI	Photosystem I
PSII	Photosystem II
PTM	Plant homeodomain transcription factor with transmembrane domains
RbcS	*RIBULOSE-1,5-BISPHOSPHATE CARBOXYLASE/OXYGENASE SMALL SUBUNIT* gene
RBOH	Respiratory burst oxidase homologs
RCD1	Radical-induced cell death1
RNAP II	RNA polymerase II
ROS	Reactive oxygen species
SA	Salicylic acid
SAFE1	Safeguard1
SAL1	Inositol polyphosphate 1-phosphatase
SOD	Superoxide dismutase
SORGs	Singlet oxygen related genes
TBP	Tetrapyrrole biosynthesis pathway
TCA	Tricarboxylic acid cycle
TF	Transcription factor
β-CAR	β-carotene
β-CC	β-cyclocitral

References

1. Gray, M.W. Lynn Margulis and the endosymbiont hypothesis: 50 years later. *Mol. Biol. Cell* **2017**, *28*, 1285–1287. [CrossRef] [PubMed]
2. Stiller, J.W. Plastid endosymbiosis, genome evolution and the origin of green plants. *Trends Plant Sci.* **2007**, *12*, 391–396. [CrossRef] [PubMed]
3. Archibald, J.M. Endosymbiosis and eukaryotic cell evolution. *Curr. Biol.* **2015**, *25*, R911–R921. [CrossRef]
4. Raven, J.A.; Allen, J.F. Genomics and chloroplast evolution: What did cyanobacteria do for plants? *Genome Biol.* **2003**, *4*, 209. [CrossRef]
5. Lv, R.; Li, Z.; Li, M.; Dogra, V.; Lv, S.; Liu, R.; Lee, K.P.; Kim, C. Uncoupled expression of nuclear and plastid photosynthesis-associated genes contributes to cell death in a lesion mimic mutant. *Plant Cell* **2019**, *31*, 210–230. [CrossRef] [PubMed]
6. Mühlenbock, P.; Szechyńska-Hebda, M.; Płaszczyca, M.; Baudo, M.; Mullineaux, P.M.; Parker, J.E.; Karpińska, B.; Karpińskie, S. Chloroplast signaling and lesion simulating disease1 regulate crosstalk between light acclimation and immunity in Arabidopsis. *Plant Cell* **2008**, *20*, 2339–2356. [CrossRef] [PubMed]
7. Wituszyńska, W.; Szechyńska-Hebda, M.; Sobczak, M.; Rusaczonek, A.; Kozlowska-Makulska, A.; Witoń, D.; Karpiński, S. LESION SIMULATING DISEASE 1 and enhanced disease susceptibility 1 differentially regulate UV-C-induced photooxidative stress signalling and programmed cell death in Arabidopsis thaliana. *Plant Cell Environ.* **2015**, *38*, 315–330. [CrossRef]
8. Szechyńska-Hebda, M.; Kruk, J.; Górecka, M.; Karpińska, B.; Karpiński, S. Evidence for light wavelength-specific photoelectrophysiological signaling and memory of excess light episodes in Arabidopsis. *Plant Cell* **2010**, *22*, 2201–2218. [CrossRef]
9. Wituszyńska, W.; Ślesak, I.; Vanderauwera, S.; Szechyńska-Hebda, M.; Kornaś, A.; Van Der Kelen, K.; Mühlenbock, P.; Karpińska, B.; Maćkowski, S.; Van Breusegem, F.; et al. Lesion simulating disease1, enhanced disease susceptibility1, and phytoalexin deficient4 conditionally regulate cellular signaling homeostasis, photosynthesis, water use efficiency, and seed yield in Arabidopsis. *Plant Physiol.* **2013**, *161*, 1795–1805. [CrossRef]
10. Sterling, S.M.; Ducharne, A.; Polcher, J. The impact of global land-cover change on the terrestrial water cycle. *Nat. Clim. Chang.* **2013**, *3*, 385–390. [CrossRef]

11. Fedoroff, N.V.; Battisti, D.S.; Beachy, R.N.; Cooper, P.J.M.; Fischhoff, D.A.; Hodges, C.N.; Knauf, V.C.; Lobell, D.; Mazur, B.J.; Molden, D.; et al. Radically rethinking agriculture for the 21st century. *Science* **2010**, *327*, 833–834. [CrossRef] [PubMed]

12. Bhattacharya, S. Deadly new wheat disease threatens Europe's crops. *Nature* **2017**, *542*, 145–146. [CrossRef]

13. Franks, S.J.; Weber, J.J.; Aitken, S.N. Evolutionary and plastic responses to climate change in terrestrial plant populations. *Evol. Appl.* **2014**, *7*, 123–139. [CrossRef]

14. Baena-González, E. Energy signaling in the regulation of gene expression during stress. *Mol. Plant* **2010**, *3*, 300–313. [CrossRef] [PubMed]

15. Bradbeer, J.W.; Atkinson, Y.E.; Börner, T.; Hagemann, R. Cytoplasmic synthesis of plastid polypeptides may be controlled by plastid-synthesised RNA. *Nature* **1979**, *279*, 816–817. [CrossRef]

16. Zhao, X.; Huang, J.; Chory, J. GUN1 interacts with MORF2 to regulate plastid RNA editing during retrograde signaling. *Proc. Natl. Acad. Sci. USA* **2019**, *116*, 10162–10167. [CrossRef]

17. Godoy Herz, M.A.; Kubaczka, M.G.; Brzyżek, G.; Servi, L.; Krzyszton, M.; Simpson, C.; Brown, J.; Swiezewski, S.; Petrillo, E.; Kornblihtt, A.R. Light Regulates Plant Alternative Splicing through the Control of Transcriptional Elongation. *Mol. Cell* **2019**, *73*, 1066–1074.e3. [CrossRef]

18. Fang, X.; Zhao, G.; Zhang, S.; Li, Y.; Gu, H.; Li, Y.; Zhao, Q.; Qi, Y. Chloroplast-to-Nucleus Signaling Regulates MicroRNA Biogenesis in Arabidopsis. *Dev. Cell* **2019**, *48*, 371–382.e4. [CrossRef]

19. Zhao, X.; Huang, J.; Chory, J. Unraveling the Linkage between Retrograde Signaling and RNA Metabolism in Plants. *Trends Plant Sci.* **2020**, *25*, 141–147. [CrossRef]

20. Petrillo, E.; Godoy Herz, M.A.; Fuchs, A.; Reifer, D.; Fuller, J.; Yanovsky, M.J.; Simpson, C.; Brown, J.W.S.; Barta, A.; Kalyna, M.; et al. A chloroplast retrograde signal regulates nuclear alternative splicing. *Science* **2014**, *344*, 427–430. [CrossRef]

21. Apel, K.; Hirt, H. Reactive oxygen species: Metabolism, Oxidative Stress, and Signal Transduction. *Annu. Rev. Plant Biol.* **2004**, *55*, 373–399. [CrossRef] [PubMed]

22. Mignolet-Spruyt, L.; Xu, E.; Idänheimo, N.; Hoeberichts, F.A.; Mühlenbock, P.; Brosche, M.; Van Breusegem, F.; Kangasjärvi, J. Spreading the news: Subcellular and organellar reactive oxygen species production and signalling. *J. Exp. Bot.* **2016**, *67*, 3831–3844. [CrossRef] [PubMed]

23. Podgórska, A.; Burian, M.; Szal, B. Extra-cellular but extra-ordinarily important for cells: Apoplastic reactive oxygen species metabolism. *Front. Plant Sci.* **2017**, *8*, 1353. [CrossRef] [PubMed]

24. Sagi, M.; Fluhr, R. Production of reactive oxygen species by plant NADPH oxidases. *Plant Physiol.* **2006**, *141*, 336–340. [CrossRef]

25. Tian, S.; Wang, X.; Li, P.; Wang, H.; Ji, H.; Xie, J.; Qiu, Q.; Shen, D.; Dong, H. Plant aquaporin AtPIP1;4 links apoplastic H_2O_2 induction to disease immunity pathways. *Plant Physiol.* **2016**, *171*, 1635–1650. [CrossRef]

26. Karpinska, B.; Zhang, K.; Rasool, B.; Pastok, D.; Morris, J.; Verrall, S.R.; Hedley, P.E.; Hancock, R.D.; Foyer, C.H. The redox state of the apoplast influences the acclimation of photosynthesis and leaf metabolism to changing irradiance. *Plant Cell Environ.* **2018**, *41*, 1083–1097. [CrossRef]

27. Wu, F.; Chi, Y.; Jiang, Z.; Xu, Y.; Xie, L.; Huang, F.; Wan, D.; Ni, J.; Yuan, F.; Wu, X.; et al. Hydrogen peroxide sensor HPCA1 is an LRR receptor kinase in Arabidopsis. *Nature* **2020**, *578*, 577–581. [CrossRef]

28. Scandalios, J.G. Oxidative stress: Molecular perception and transduction of signals triggering antioxidant gene defenses. *Braz. J. Med. Biol. Res.* **2005**, *38*, 995–1014. [CrossRef]

29. Asada, K. Production and scavenging of reactive oxygen species in chloroplasts and their functions. *Plant Physiol.* **2006**, *141*, 391–396. [CrossRef]

30. Waszczak, C.; Akter, S.; Eeckhout, D.; Persiau, G.; Wahni, K.; Bodra, N.; Van Molle, I.; De Smet, B.; Vertommen, D.; Gevaert, K.; et al. Sulfenome mining in Arabidopsis thaliana. *Proc. Natl. Acad. Sci. USA* **2014**, *111*, 11545–11550. [CrossRef]

31. Gratão, P.L.; Polle, A.; Lea, P.J.; Azevedo, R.A. Making the life of heavy metal-stressed plants a little easier. *Funct. Plant Biol.* **2005**, *32*, 481. [CrossRef]

32. Dietz, K.J.; Turkan, I.; Krieger-Liszkay, A. Redox-and reactive oxygen species-dependent signaling into and out of the photosynthesizing chloroplast. *Plant Physiol.* **2016**, *171*, 1541–1550. [CrossRef]

33. Laloi, C.; Stachowiak, M.; Pers-Kamczyc, E.; Warzych, E.; Murgia, I.; Apel, K. Cross-talk between singlet oxygen-and hydrogen peroxide-dependent signaling of stress responses in Arabidopsis thaliana. *Proc. Natl. Acad. Sci. USA* **2007**, *104*, 672–677. [CrossRef]

34. Kaiser, W.M. Reversible inhibition of the calvin cycle and activation of oxidative pentose phosphate cycle in isolated intact chloroplasts by hydrogen peroxide. *Planta* **1979**, *145*, 377–382. [CrossRef]

35. Woehle, C.; Dagan, T.; Landan, G.; Vardi, A.; Rosenwasser, S. Expansion of the redox-sensitive proteome coincides with the plastid endosymbiosis. *Nat. Plants* **2017**, *3*, 17066. [CrossRef]

36. Mullineaux, P.; Karpinski, S. Signal transduction in response to excess light: Getting out of the chloroplast. *Curr. Opin. Plant Biol.* **2002**, *5*, 43–48. [CrossRef]

37. Mullineaux, P.M.; Karpinski, S.; Baker, N.R. Spatial dependence for hydrogen peroxide-directed signaling in light-stressed plants. *Plant Physiol.* **2006**, *141*, 346–350. [CrossRef]

38. Mubarakshina, M.M.; Ivanov, B.N.; Naydov, I.A.; Hillier, W.; Badger, M.R.; Krieger-Liszkay, A. Production and diffusion of chloroplastic H_2O_2 and its implication to signalling. *J. Exp. Bot.* **2010**, *61*, 3577–3587. [CrossRef]

39. Dietz, K.J.; Mittler, R.; Noctor, G. Recent progress in understanding the role of reactive oxygen species in plant cell signaling. *Plant Physiol.* **2016**, *171*, 1535–1539. [CrossRef]

40. Bienert, G.P.; Møller, A.L.B.; Kristiansen, K.A.; Schulz, A.; Møller, I.M.; Schjoerring, J.K.; Jahn, T.P. Specific aquaporins facilitate the diffusion of hydrogen peroxide across membranes. *J. Biol. Chem.* **2007**, *282*, 1183–1192. [CrossRef]

41. Caplan, J.L.; Kumar, A.S.; Park, E.; Padmanabhan, M.S.; Hoban, K.; Modla, S.; Czymmek, K.; Dinesh-Kumar, S.P. Chloroplast Stromules Function during Innate Immunity. *Dev. Cell* **2015**, *34*, 45–57. [CrossRef]

42. Hanson, M.R.; Hines, K.M. Stromules: Probing formation and function. *Plant Physiol.* **2018**, *176*, 128–137. [CrossRef]

43. Exposito-Rodriguez, M.; Laissue, P.P.; Yvon-Durocher, G.; Smirnoff, N.; Mullineaux, P.M. Photosynthesis-dependent H_2O_2 transfer from chloroplasts to nuclei provides a high-light signalling mechanism. *Nat. Commun.* **2017**, *8*, 49. [CrossRef]

44. Asada, K. Ascorbate peroxidase—A hydrogen peroxide-scavenging enzyme in plants. *Physiol. Plant* **1992**, *85*, 235–241. [CrossRef]

45. Dietz, K.J. Thiol-based peroxidases and ascorbate peroxidases: Why plants rely on multiple peroxidase systems in the photosynthesizing chloroplast? *Mol. Cells* **2016**, *39*, 20–25. [CrossRef]

46. Mittler, R. Oxidative stress, antioxidants and stress tolerance. *Trends Plant Sci.* **2002**. [CrossRef]

47. Mittler, R.; Vanderauwera, S.; Gollery, M.; Van Breusegem, F. Reactive oxygen gene network of plants. *Trends Plant Sci.* **2004**, *7*, 405–410. [CrossRef]

48. Shigeoka, S.; Ishikawa, T.; Tamoi, M.; Miyagawa, Y.; Takeda, T.; Yabuta, Y.; Yoshimura, K. Regulation and function of ascorbate peroxidase isoenzymes. *J. Exp. Bot.* **2002**, *53*, 1305–1319. [CrossRef]

49. Mano, J.; Ohno, C.; Domae, Y.; Asada, K. Chloroplastic ascorbate peroxidase is the primary target of methylviologen-induced photooxidative stress in spinach leaves: Its relevance to monodehydroascorbate radical detected with in vivo ESR. *Biochim. Biophys. Acta-Bioenerg.* **2001**, *1504*, 275–287. [CrossRef]

50. Karpinski, S.; Escobar, C.; Karpinska, B.; Creissen, G.; Mullineaux, P.M. Photosynthetic electron transport regulates the expression of cytosolic ascorbate peroxidase genes in arabidopsis during excess light stress. *Plant Cell* **1997**, *9*, 627–640. [CrossRef]

51. Karpinski, S.; Reynolds, H.; Karpinska, B.; Wingsle, G.; Creissen, G.; Mullineaux, P. Systemic signaling and acclimation in response to excess excitation energy in Arabidopsis. *Science* **1999**, *284*, 654–657. [CrossRef]

52. Janku, M.; Luhová, L.; Petrivalský, M. On the origin and fate of reactive oxygen species in plant cell compartments. *Antioxidants* **2019**, *8*, 105. [CrossRef]

53. Sewelam, N.; Jaspert, N.; Van Der Kelen, K.; Tognetti, V.B.; Schmitz, J.; Frerigmann, H.; Stahl, E.; Zeier, J.; Van Breusegem, F.; Maurino, V.G. Spatial H_2O_2 signaling specificity: H_2O_2 from chloroplasts and peroxisomes modulates the plant transcriptome differentially. *Mol. Plant* **2014**, *7*, 1191–1210. [CrossRef]

54. Takahashi, H.; Chen, Z.; Du, H.; Liu, Y.; Klessig, D.F. Development of necrosis and activation of disease resistance in transgenic tobacco plants with severely reduced catalase levels. *Plant J.* **1997**, *11*, 993–1005. [CrossRef]

55. Chaouch, S.; Queval, G.; Vanderauwera, S.; Mhamdi, A.; Vandorpe, M.; Langlois-Meurinne, M.; van Breusegem, F.; Saindrenan, P.; Noctor, G. Peroxisomal hydrogen peroxide is coupled to biotic defense responses by ISOCHORISMATE SYNTHASE1 in a daylength-related manner. *Plant Physiol.* **2010**, *153*, 1692–1705. [CrossRef]

56. Vandenabeele, S.; Vanderauwera, S.; Vuylsteke, M.; Rombauts, S.; Langebartels, C.; Seidlitz, H.K.; Zabeau, M.; Van Montagu, M.; Inzé, D.; Van Breusegem, F. Catalase deficiency drastically affects gene expression induced by high light in Arabidopsis thaliana. *Plant J.* **2004**, *39*, 45–58. [CrossRef]

57. Queval, G.; Issakidis-Bourguet, E.; Hoeberichts, F.A.; Vandorpe, M.; Gakière, B.; Vanacker, H.; Miginiac-Maslow, M.; Van Breusegem, F.; Noctor, G. Conditional oxidative stress responses in the Arabidopsis photorespiratory mutant cat2 demonstrate that redox state is a key modulator of daylength-dependent gene expression, and define photoperiod as a crucial factor in the regulation of H_2O_2-induced cel. *Plant J.* **2007**, *52*, 640–657. [CrossRef]

58. Su, T.; Wang, P.; Li, H.; Zhao, Y.; Lu, Y.; Dai, P.; Ren, T.; Wang, X.; Li, X.; Shao, Q.; et al. The Arabidopsis catalase triple mutant reveals important roles of catalases and peroxisome-derived signaling in plant development. *J. Integr. Plant Biol.* **2018**, *60*, 591–607. [CrossRef]

59. Shumbe, L.; Chevalier, A.; Legeret, B.; Taconnat, L.; Monnet, F.; Havaux, M. Singlet oxygen-induced cell death in arabidopsis under high-light stress is controlled by OXI1 kinase. *Plant Physiol.* **2016**, *170*, 1757–1771. [CrossRef]

60. Rentel, M.C.; Lecourieux, D.; Ouaked, F.; Usher, S.L.; Petersen, L.; Okamoto, H.; Knight, H.; Peck, S.C.; Grierson, C.S.; Hirt, H.; et al. OXI1 kinase is necessary for oxidative burst-mediated signalling in Arabidopsis. *Nature* **2004**, *427*, 858–861. [CrossRef]

61. Beaugelin, I.; Chevalier, A.; D'Alessandro, S.; Ksas, B.; Novák, O.; Strnad, M.; Forzani, C.; Hirt, H.; Havaux, M.; Monnet, F. OXI1 and DAD regulate light-induced cell death antagonistically through jasmonate and salicylate levels. *Plant Physiol.* **2019**, *180*, 1691–1708. [CrossRef] [PubMed]

62. Mateo, A.; Funck, D.; Mühlenbock, P.; Kular, B.; Mullineaux, P.M.; Karpinski, S. Controlled levels of salicylic acid are required for optimal photosynthesis and redox homeostasis. *J. Exp. Bot.* **2006**, *57*, 1795–1807. [CrossRef] [PubMed]

63. Keun, P.L.; Kim, C.; Landgraf, F.; Apel, K. EXECUTER1-and EXECUTER2-dependent transfer of stress-related signals from the plastid to the nucleus of Arabidopsis thaliana. *Proc. Natl. Acad. Sci. USA* **2007**, *104*, 10270–10275. [CrossRef]

64. Wagner, D.; Przybyla, D.; Op Den Camp, R.; Kim, C.; Landgraf, F.; Keun, P.L.; Würsch, M.; Laloi, C.; Nater, M.; Hideg, E.; et al. The genetic basis of singlet oxygen-induced stress response of Arabidopsis thaliana. *Science* **2004**, *306*, 1183–1185. [CrossRef] [PubMed]

65. Meskauskiene, R.; Nater, M.; Goslings, D.; Kessler, F.; Op den Camp, R.; Apel, K. FLU: A negative regulator of chlorophyll biosynthesis in arabidopsis thaliana. *Proc. Natl. Acad. Sci. USA* **2001**, *98*, 12826–12831. [CrossRef] [PubMed]

66. Zhang, S.; Apel, K.; Kim, C. Singlet oxygen-mediated and EXECUTER dependent signalling and acclimation of Arabidopsis thaliana exposed to light stress. *Philos. Trans. R. Soc. B Biol. Sci.* **2014**, *369*, 20130227. [CrossRef] [PubMed]

67. Dogra, V.; Duan, J.; Lee, K.P.; Lv, S.; Liu, R.; Kim, C. FtsH2-dependent proteolysis of EXECUTER1 is essential in mediating singlet oxygen-triggered retrograde signaling in Arabidopsis thaliana. *Front. Plant Sci.* **2017**, *8*. [CrossRef]

68. Wang, L.; Kim, C.; Xu, X.; Piskurewicz, U.; Dogra, V.; Singh, S.; Mahler, H.; Apel, K. Singlet oxygen-and EXECUTER1-mediated signaling is initiated in grana margins and depends on the protease FtsH2. *Proc. Natl. Acad. Sci. USA* **2016**, *113*, E3792–E3800. [CrossRef]

69. Dogra, V.; Li, M.; Singh, S.; Li, M.; Kim, C. Oxidative post-translational modification of EXECUTER1 is required for singlet oxygen sensing in plastids. *Nat. Commun.* **2019**, *10*. [CrossRef]

70. Wang, L.; Leister, D.; Guan, L.; Zheng, Y.; Schneider, K.; Lehmann, M.; Apel, K.; Kleine, T. The Arabidopsis SAFEGUARD1 suppresses singlet oxygen-induced stress responses by protecting grana margins. *Proc. Natl. Acad. Sci. USA* **2020**, *117*, 6918–6927. [CrossRef]

71. Sun, X.; Feng, P.; Xu, X.; Guo, H.; Ma, J.; Chi, W.; Lin, R.; Lu, C.; Zhang, L. A chloroplast envelope-bound PHD transcription factor mediates chloroplast signals to the nucleus. *Nat. Commun.* **2011**, *2*, 477. [CrossRef] [PubMed]

72. Isemer, R.; Krause, K.; Grabe, N.; Kitahata, N.; Asami, T.; Krupinska, K. Plastid located WHIRLY1 enhances the responsiveness of Arabidopsis seedlings toward abscisic acid. *Front. Plant Sci.* **2012**, *3*. [CrossRef]

73. Koussevitzky, S.; Nott, A.; Mockler, T.C.; Hong, F.; Sachetto-Martins, G.; Surpin, M.; Lim, J.; Mittler, R.; Chory, J. Signals from chloroplasts converge to regulate nuclear gene expression. *Science* **2007**, *316*. [CrossRef]

74. Page, M.T.; McCormac, A.C.; Smith, A.G.; Terry, M.J. Singlet oxygen initiates a plastid signal controlling photosynthetic gene expression. *New Phytol.* **2017**, *213*, 1168–1180. [CrossRef] [PubMed]

75. Susek, R.E.; Ausubel, F.M.; Chory, J. Signal transduction mutants of arabidopsis uncouple nuclear CAB and RBCS gene expression from chloroplast development. *Cell* **1993**, *74*, 787–799. [CrossRef]

76. Nott, A.; Jung, H.-S.; Koussevitzky, S.; Chory, J. Plastid-to-nucleus retrograde signaling. *Annu. Rev. Plant Biol.* **2006**, *57*, 739–759. [CrossRef]

77. Strand, Å.; Asami, T.; Alonso, J.; Ecker, J.R.; Chory, J. Chloroplast to nucleus communication triggered by accumulation of Mg-protoporphyrinix. *Nature* **2003**, *421*, 79–83. [CrossRef] [PubMed]

78. Mochizuki, N.; Tanaka, R.; Tanaka, A.; Masuda, T.; Nagatani, A. The steady-state level of Mg-protoporphyrin IX is not a determinant of plastid-to-nucleus signaling in Arabidopsis. *Proc. Natl. Acad. Sci. USA* **2008**, *105*, 15184–15189. [CrossRef] [PubMed]

79. Moulin, M.; McCormac, A.C.; Terry, M.J.; Smith, A.G. Tetrapyrrole profiling in Arabidopsis seedlings reveals that retrograde plastid nuclear signaling is not due to Mg-protoporphyrin IX accumulation. *Proc. Natl. Acad. Sci. USA* **2008**, *105*, 15178–15183. [CrossRef] [PubMed]

80. Zhang, Z.W.; Yuan, S.; Feng, H.; Xu, F.; Cheng, J.; Shang, J.; Zhang, D.W.; Lin, H.H. Transient accumulation of Mg-protoporphyrin IX regulates expression of PhANGs-New evidence for the signaling role of tetrapyrroles in mature Arabidopsis plants. *J. Plant Physiol.* **2011**, *168*, 714–721. [CrossRef]

81. Woodson, J.D.; Perez-Ruiz, J.M.; Chory, J. Heme synthesis by plastid ferrochelatase i regulates nuclear gene expression in plants. *Curr. Biol.* **2011**, *21*, 897–903. [CrossRef] [PubMed]

82. Page, M.T.; Garcia-Becerra, T.; Smith, A.G.; Terry, M.J. Overexpression of chloroplast-targeted ferrochelatase 1 results in a genomes uncoupled chloroplast-to-nucleus retrograde signalling phenotype. *Philos. Trans. R. Soc. B Biol. Sci.* **2020**, *375*, 20190401. [CrossRef] [PubMed]

83. Tabrizi, S.T.; Sawicki, A.; Zhou, S.; Luo, M.; Willows, R.D. GUN4-Protoporphyrin IX Is a singlet oxygen generator with consequences for plastid retrograde signaling. *J. Biol. Chem.* **2016**, *291*, 8978–8984. [CrossRef]

84. Colombo, M.; Tadini, L.; Peracchio, C.; Ferrari, R.; Pesaresi, P. GUN1, a jack-of-all-trades in chloroplast protein homeostasis and signaling. *Front. Plant Sci.* **2016**, *7*. [CrossRef] [PubMed]

85. Tadini, L.; Pesaresi, P.; Kleine, T.; Rossi, F.; Guljamow, A.; Sommer, F.; Mühlhaus, T.; Schroda, M.; Masiero, S.; Pribil, M.; et al. Gun1 controls accumulation of the plastid ribosomal protein S1 at the protein level and interacts with proteins involved in plastid protein homeostasis. *Plant Physiol.* **2016**, *170*, 1817–1830. [CrossRef] [PubMed]

86. Llamas, E.; Pulido, P.; Rodriguez-Concepcion, M. Interference with plastome gene expression and Clp protease activity in Arabidopsis triggers a chloroplast unfolded protein response to restore protein homeostasis. *PLoS Genet.* **2017**, *13*, e1007022. [CrossRef]

87. Wu, G.Z.; Chalvin, C.; Hoelscher, M.; Meyer, E.H.; Wu, X.N.; Bock, R. Control of retrograde signaling by rapid turnover of Genomes Uncoupled1. *Plant Physiol.* **2018**, *176*, 2472–2495. [CrossRef]

88. Marino, G.; Naranjo, B.; Wang, J.; Penzler, J.F.; Kleine, T.; Leister, D. Relationship of GUN1 to FUG1 in chloroplast protein homeostasis. *Plant J.* **2019**, *99*, 521–535. [CrossRef]

89. Wu, G.Z.; Meyer, E.H.; Richter, A.S.; Schuster, M.; Ling, Q.; Schöttler, M.A.; Walther, D.; Zoschke, R.; Grimm, B.; Jarvis, R.P.; et al. Control of retrograde signalling by protein import and cytosolic folding stress. *Nat. Plants* **2019**, *5*, 525–538. [CrossRef]

90. Shimizu, T.; Kacprzak, S.M.; Mochizuki, N.; Nagatani, A.; Watanabe, S.; Shimada, T.; Tanaka, K.; Hayashi, Y.; Arai, M.; Leister, D.; et al. The retrograde signaling protein GUN1 regulates tetrapyrrole biosynthesis. *Proc. Natl. Acad. Sci. USA* **2019**, *116*, 24900–24906. [CrossRef]

91. Ramel, F.; Birtic, S.; Ginies, C.; Soubigou-Taconnat, L.; Triantaphylidès, C.; Havaux, M. Carotenoid oxidation products are stress signals that mediate gene responses to singlet oxygen in plants. *Proc. Natl. Acad. Sci. USA* **2012**, *109*, 5535–5540. [CrossRef] [PubMed]

92. Shao, N.; Duan, G.Y.; Bock, R. A mediator of singlet oxygen responses in chlamydomonas reinhardtii and arabidopsis identified by a luciferase-based genetic screen in algal cells. *Plant Cell* **2013**, *25*, 4209–4226. [CrossRef] [PubMed]

93. Shumbe, L.; D'Alessandro, S.; Shao, N.; Chevalier, A.; Ksas, B.; Bock, R.; Havaux, M. Methylene Blue Sensitivity 1 (MBS1) is required for acclimation of Arabidopsis to singlet oxygen and acts downstream of β-cyclocitral. *Plant Cell Environ.* **2017**, *40*, 216–226. [CrossRef]

94. Xiao, Y.; Savchenko, T.; Baidoo, E.E.K.; Chehab, W.E.; Hayden, D.M.; Tolstikov, V.; Corwin, J.A.; Kliebenstein, D.J.; Keasling, J.D.; Dehesh, K. Retrograde signaling by the plastidial metabolite MEcPP regulates expression of nuclear stress-response genes. *Cell* **2012**, *149*, 1525–1535. [CrossRef]

95. Walley, J.; Xiao, Y.; Wang, J.Z.; Baidoo, E.E.; Keasling, J.D.; Shen, Z.; Briggs, S.P.; Dehesh, K. Plastid-produced interorgannellar stress signal MEcPP potentiates induction of the unfolded protein response in endoplasmic reticulum. *Proc. Natl. Acad. Sci. USA* **2015**, *112*, 6212–6217. [CrossRef] [PubMed]

96. Seemann, M.; Wegner, P.; Schünemann, V.; Bui, B.T.S.; Wolff, M.; Marquet, A.; Trautwein, A.X.; Rohmer, M. Isoprenoid biosynthesis in chloroplasts via the methylerythritol phosphate pathway: The (E)-4-hydroxy-3-methylbut-2-enyl diphosphate synthase (GcpE) from Arabidopsis thaliana is a [4Fe-4S] protein. *J. Biol. Inorg. Chem.* **2005**, *10*, 131–137. [CrossRef] [PubMed]

97. Jiang, J.; Zeng, L.; Ke, H.; De La Cruz, B.; Dehesh, K. Orthogonal regulation of phytochrome B abundance by stress-specific plastidial retrograde signaling metabolite. *Nat. Commun.* **2019**, *10*, 2904. [CrossRef] [PubMed]

98. Jiang, J.; Xiao, Y.; Chen, H.; Hu, W.; Zeng, L.; Ke, H.; Ditengou, F.A.; Devisetty, U.K.; Palme, K.; Maloof, J.N.; et al. Retrograde induction of phyB orchestrates ethylene-auxin hierarchy to regulate growth. *Plant Physiol.* **2020**, *183*, 1268–1280. [CrossRef]

99. Estavillo, G.M.; Crisp, P.A.; Pornsiriwong, W.; Wirtz, M.; Collinge, D.; Carrie, C.; Giraud, E.; Whelan, J.; David, P.; Javot, H.; et al. Evidence for a SAL1-PAP chloroplast retrograde pathway that functions in drought and high light signaling in Arabidopsis. *Plant Cell* **2011**, *23*, 3992–4012. [CrossRef]

100. Pornsiriwong, W.; Estavillo, G.M.; Chan, K.X.; Tee, E.E.; Ganguly, D.; Crisp, P.A.; Phua, S.Y.; Zhao, C.; Qiu, J.; Park, J.; et al. A chloroplast retrograde signal, 3'phosphoadenosine 5'-phosphate, acts as a secondary messenger in abscisic acid signaling in stomatal closure and germination. *eLife* **2017**, *6*. [CrossRef]

101. Klein, M.; Papenbrock, J. The multi-protein family of Arabidopsis sulphotransferases and their relatives in other plant species. *J. Exp. Bot.* **2004**, *55*, 1809–1820. [CrossRef] [PubMed]

102. Dichtl, B. Lithium toxicity in yeast is due to the inhibition of RNA processing enzymes. *EMBO J.* **1997**, *16*, 7184–7195. [CrossRef] [PubMed]

103. Wilson, P.B.; Estavillo, G.M.; Field, K.J.; Pornsiriwong, W.; Carroll, A.J.; Howell, K.A.; Woo, N.S.; Lake, J.A.; Smith, S.M.; Harvey Millar, A.; et al. The nucleotidase/phosphatase SAL1 is a negative regulator of drought tolerance in Arabidopsis. *Plant J.* **2009**, *58*, 299–317. [CrossRef] [PubMed]

104. Ishiga, Y.; Watanabe, M.; Ishiga, T.; Tohge, T.; Matsuura, T.; Ikeda, Y.; Hoefgen, R.; Fernie, A.R.; Mysore, K.S. The SAL-PAP chloroplast retrograde pathway contributes to plant immunity by regulating glucosinolate pathway and phytohormone signaling. *Mol. Plant-Microbe Interact.* **2017**, *30*, 829–841. [CrossRef]

105. Zhao, C.; Wang, Y.; Chan, K.X.; Marchant, D.B.; Franks, P.J.; Randall, D.; Tee, E.E.; Chen, G.; Ramesh, S.; Phua, S.Y.; et al. Evolution of chloroplast retrograde signaling facilitates green plant adaptation to land. *Proc. Natl. Acad. Sci. USA* **2019**, *116*, 5015–5020. [CrossRef]

106. Zhang, J.; Vanneste, S.; Brewer, P.B.; Michniewicz, M.; Grones, P.; Kleine-Vehn, J.; Löfke, C.; Teichmann, T.; Bielach, A.; Cannoot, B.; et al. Inositol trisphosphate-induced Ca^{2+} signaling modulates auxin transport and pin polarity. *Dev. Cell* **2011**, *20*, 855–866. [CrossRef]

107. Kim, B.H.; Von Arnim, A.G. FIERY1 regulates light-mediated repression of cell elongation and flowering time via its 3'(2'),5'-bisphosphate nucleotidase activity. *Plant J.* **2009**, *58*, 208–219. [CrossRef]

108. Rodríguez, V.M.; Chételat, A.; Majcherczyk, P.; Farmer, E.E. Chloroplastic phosphoadenosine phosphosulfate metabolism regulates basal levels of the prohormone jasmonic acid in Arabidopsis leaves. *Plant Physiol.* **2010**, *152*, 1335–1345. [CrossRef]

109. Balparda, M.; Armas, A.M.; Estavillo, G.M.; Roschzttardtz, H.; Pagani, M.A.; Gomez-Casati, D.F. The PAP/SAL1 retrograde signaling pathway is involved in iron homeostasis. *Plant Mol. Biol.* **2020**, *102*, 323–337. [CrossRef]

110. Pogson, B.J.; Woo, N.S.; Förster, B.; Small, I.D. Plastid signalling to the nucleus and beyond. *Trends Plant Sci.* **2008**, *13*, 602–609. [CrossRef]

111. Jarvis, P.; López-Juez, E. Biogenesis and homeostasis of chloroplasts and other plastids. *Nat. Rev. Mol. Cell Biol.* **2013**, *14*, 787–802. [CrossRef] [PubMed]

112. Chan, K.X.; Phua, S.Y.; Crisp, P.; McQuinn, R.; Pogson, B.J. Learning the languages of the chloroplast: Retrograde signaling and beyond. *Annu. Rev. Plant Biol.* **2016**, *67*, 25–53. [CrossRef] [PubMed]

113. Finkelstein, R.R. Mutations at two new Arabidopsis ABA response loci are similar to the abi3 mutations. *Plant J.* **1994**, *5*, 765–771. [CrossRef]

114. Arenas-Huertero, F.; Arroyo, A.; Zhou, L.; Sheen, J.; León, P. Analysis of Arabidopsis glucose insensitive mutants, gin5 and gin6, reveals a central role of the plant hormone ABA in the regulation of plant vegetative development by sugar. *Genes Dev.* **2000**, *14*, 2085–2096. [CrossRef]

115. Huijser, C.; Kortstee, A.; Pego, J.; Weisbeek, P.; Wisman, E.; Smeekens, S. The Arabidopsis SUCROSE UNCOUPLED-6 gene is identical to abscisic acid Insensitive-4: Involvement of abscisic acid in sugar responses. *Plant J.* **2000**, *23*, 577–585. [CrossRef]

116. Laby, R.J.; Kim, D.; Gibson, S.I. The ram1 mutant of Arabidopsis exhibits severely decreased β-amylase activity. *Plant Physiol.* **2001**, *127*, 1798–1807. [CrossRef]

117. Rook, F.; Corke, F.; Card, R.; Munz, G.; Smith, C.; Bevan, M.W. Impaired sucrose-induction mutants reveal the modulation of sugar-induced starch biosynthetic gene expression by abscisic acid signalling. *Plant J.* **2001**, *26*, 421–433. [CrossRef]

118. Nakano, T.; Suzuki, K.; Fujimura, T.; Shinshi, H. Genome-wide analysis of the ERF gene family in arabidopsis and rice. *Plant Physiol.* **2006**, *140*, 411–432. [CrossRef]

119. Mizoi, J.; Shinozaki, K.; Yamaguchi-Shinozaki, K. AP2/ERF family transcription factors in plant abiotic stress responses. *Biochim. Biophys. Acta-Gene Regul. Mech.* **2012**, *1819*, 86–96. [CrossRef]

120. Giraud, E.; van Aken, O.; Ho, L.H.M.; Whelan, J. The transcription factor ABI4 is a regulator of mitochondrial retrograde expression of Alternative Oxidase1a. *Plant Physiol.* **2009**, *150*, 1286–1296. [CrossRef]

121. Acevedo-Hernández, G.J.; León, P.; Herrera-Estrella, L.R. Sugar and ABA responsiveness of a minimal RBCS light-responsive unit is mediated by direct binding of ABI4. *Plant J.* **2005**, *43*, 506–519. [CrossRef] [PubMed]

122. Woodson, J.D.; Perez-Ruiz, J.M.; Schmitz, R.J.; Ecker, J.R.; Chory, J. Sigma factor-mediated plastid retrograde signals control nuclear gene expression. *Plant J.* **2013**, *73*, 1–13. [CrossRef] [PubMed]

123. Guo, H.; Feng, P.; Chi, W.; Sun, X.; Xu, X.; Li, Y.; Ren, D.; Lu, C.; David Rochaix, J.; Leister, D.; et al. Plastid-nucleus communication involves calcium-modulated MAPK signalling. *Nat. Commun.* **2016**, *7*, 12173. [CrossRef] [PubMed]

124. Zhang, Z.W.; Feng, L.Y.; Cheng, J.; Tang, H.; Xu, F.; Zhu, F.; Zhao, Z.Y.; Yuan, M.; Chen, Y.E.; Wang, J.H.; et al. The roles of two transcription factors, ABI4 and CBFA, in ABA and plastid signalling and stress responses. *Plant Mol. Biol.* **2013**, *83*, 445–458. [CrossRef]

125. Cottage, A.; Gray, J.C. Timing the switch to phototrophic growth: A possible role of GUN1. *Plant Signal. Behav.* **2011**, *6*, 578–582. [CrossRef]

126. Kerchev, P.I.; Pellny, T.K.; Vivancos, P.D.; Kiddle, G.; Hedden, P.; Driscoll, S.; Vanacker, H.; Verrier, P.; Hancock, R.D.; Foyer, C.H. The transcription factor ABi4 Is required for the ascorbic acid-dependent regulation of growth and regulation of jasmonate-dependent defence signaling pathways in arabidopsis. *Plant Cell* **2011**, *23*, 3319–3334. [CrossRef]

127. Martin, G.; Leivar, P.; Ludevid, D.; Tepperman, J.M.; Quail, P.H.; Monte, E. Phytochrome and retrograde signalling pathways converge to antagonistically regulate a light-induced transcriptional network. *Nat. Commun.* **2016**, *7*, 11431. [CrossRef]

128. Kakizaki, T.; Matsumura, H.; Nakayama, K.; Che, F.S.; Terauchi, R.; Inaba, T. Coordination of plastid protein import and nuclear gene expression by plastid-to-nucleus retrograde signaling. *Plant Physiol.* **2009**, *151*, 1339–1353. [CrossRef]

129. Kacprzak, S.M.; Mochizuki, N.; Naranjo, B.; Xu, D.; Leister, D.; Kleine, T.; Okamoto, H.; Terry, M.J. Plastid-to-nucleus retrograde signalling during chloroplast biogenesis does not require ABI4. *Plant Physiol.* **2019**, *179*, 18–23. [CrossRef]

130. Page, M.T.; Kacprzak, S.M.; Mochizuki, N.; Okamoto, H.; Smith, A.G.; Terry, M.J. Seedlings lacking the PTM protein do not show a genomes uncoupled (Gun) mutant phenotype. *Plant Physiol.* **2017**, *174*, 21–26. [CrossRef]

131. Terry, M.J.; Smith, A.G. A model for tetrapyrrole synthesis as the primary mechanism for plastid-to-nucleus signaling during chloroplast biogenesis. *Front. Plant Sci.* **2013**, *4*, 14. [CrossRef] [PubMed]

132. Bobik, K.; Burch-Smith, T.M. Chloroplast signaling within, between and beyond cells. *Front. Plant Sci.* **2015**, *6*. [CrossRef] [PubMed]

133. De Dios Barajas-López, J.; Blanco, N.E.; Strand, Å. Plastid-to-nucleus communication, signals controlling the running of the plant cell. *Biochim. Biophys. Acta-Mol. Cell Res.* **2013**, *1833*, 425–437.

134. Feng, P.; Guo, H.; Chi, W.; Chai, X.; Sun, X.; Xu, X.; Ma, J.; Rochaix, J.D.; Leister, D.; Wang, H.; et al. Chloroplast retrograde signal regulates flowering. *Proc. Natl. Acad. Sci. USA* **2016**, *113*, 10708–10713. [CrossRef]

135. Xu, X.; Chi, W.; Sun, X.; Feng, P.; Guo, H.; Li, J.; Lin, R.; Lu, C.; Wang, H.; Leister, D.; et al. Convergence of light and chloroplast signals for de-etiolation through ABI4-HY5 and COP1. *Nat. Plants* **2016**, *2*, 16066. [CrossRef]

136. Waters, M.T.; Wang, P.; Korkaric, M.; Capper, R.G.; Saunders, N.J.; Langdale, J.A. GLK transcription factors coordinate expression of the photosynthetic apparatus in Arabidopsis. *Plant Cell* **2009**, *21*, 1109–1128. [CrossRef]

137. Ruckle, M.E.; DeMarco, S.M.; Larkin, R.M. Plastid signals remodel light signaling networks and are essential for efficient chloroplast biogenesis in Arabidopsis. *Plant Cell* **2007**, *19*, 3944–3960. [CrossRef]

138. Saini, G.; Meskauskiene, R.; Pijacka, W.; Roszak, P.; Sjögren, L.L.E.; Clarke, A.K.; Straus, M.; Apel, K. Happy on norflurazon (hon) mutations implicate perturbance of plastid homeostasis with activating stress acclimatization and changing nuclear gene expression in norflurazon-treated seedlings. *Plant J.* **2011**, *65*, 690–702. [CrossRef]

139. Sun, X.; Xu, D.; Liu, Z.; Kleine, T.; Leister, D. Functional relationship between mTERF4 and GUN1 in retrograde signaling. *J. Exp. Bot.* **2016**, *67*, 3909–3924. [CrossRef]

140. Bode, R.; Ivanov, A.G.; Hüner, N.P.A. Global transcriptome analyses provide evidence that chloroplast redox state contributes to intracellular as well as long-distance signalling in response to stress and acclimation in Arabidopsis. *Photosynth. Res.* **2016**, *128*, 287–312. [CrossRef]

141. Murata, N.; Takahashi, S.; Nishiyama, Y.; Allakhverdiev, S.I. Photoinhibition of photosystem II under environmental stress. *Biochim. Biophys. Acta-Bioenerg.* **2007**, *1767*, 414–421. [CrossRef] [PubMed]

142. Thanthrige, N.; Jain, S.; Bhowmik, S.D.; Ferguson, B.J.; Kabbage, M.; Mundree, S.; Williams, B. Centrality of BAGs in Plant PCD, Stress Responses, and Host Defense. *Trends Plant Sci.* **2020**. [CrossRef] [PubMed]

143. Latrasse, D.; Benhamed, M.; Bergounioux, C.; Raynaud, C.; Delarue, M. Plant programmed cell death from a chromatin point of view. *J. Exp. Bot.* **2016**, *67*, 5887–5900. [CrossRef] [PubMed]

144. Czarnocka, W.; Van Der Kelen, K.; Willems, P.; Szechyńska-Hebda, M.; Shahnejat-Bushehri, S.; Balazadeh, S.; Rusaczonek, A.; Mueller-Roeber, B.; Van Breusegem, F.; Karpiński, S. The dual role of LESION SIMULATING DISEASE 1 as a condition-dependent scaffold protein and transcription regulator. *Plant Cell Environ.* **2017**, *40*, 2644–2662. [CrossRef]

145. Dietrich, R.A.; Delaney, T.P.; Uknes, S.J.; Ward, E.R.; Ryals, J.A.; Dangl, J.L. Arabidopsis mutants simulating disease resistance response. *Cell* **1994**, *77*, 565–577. [CrossRef]

146. Dietrich, R.A.; Richberg, M.H.; Schmidt, R.; Dean, C.; Dangl, J.L. A novel zinc finger protein is encoded by the Arabidopsis LSD1 gene and functions as a negative regulator of plant cell death. *Cell* **1997**, *88*, 685–694. [CrossRef]

147. Mateo, A.; Mühlenbock, P.; Rustérucci, C.; Chang, C.C.C.; Miszalski, Z.; Karpinska, B.; Parker, J.E.; Mullineaux, P.M.; Karpinski, S. LESION SIMULATING DISEASE 1 is required for acclimation to conditions that promote excess excitation energy. *Plant Physiol.* **2004**, *136*, 2818–2830. [CrossRef]

148. Karpiński, S.; Szechyńska-Hebda, M.; Wituszyńska, W.; Burdiak, P. Light acclimation, retrograde signalling, cell death and immune defences in plants. *Plant Cell Environ.* **2013**, *36*, 736–744. [CrossRef]

149. Woodson, J.D.; Chory, J. Coordination of gene expression between organellar and nuclear genomes. *Nat. Rev. Genet.* **2008**, *9*, 383–395. [CrossRef]

150. Raghavendra, A.S.; Padmasree, K. Beneficial interactions of mitochondrial metabolism with photosynthetic carbon assimilation. *Trends Plant Sci.* **2003**, *8*, 546–553. [CrossRef]

151. Van Lis, R.; Atteia, A. Control of mitochondrial function via photosynthetic redox signals. *Photosynth. Res.* **2004**, *79*, 133–148. [CrossRef] [PubMed]

152. Noguchi, K.; Yoshida, K. Interaction between photosynthesis and respiration in illuminated leaves. *Mitochondrion* **2008**, *8*, 87–99. [CrossRef] [PubMed]

153. De Clercq, I.; Vermeirssen, V.; Van Aken, O.; Vandepoele, K.; Murcha, M.W.; Law, S.R.; Inzé, A.; Ng, S.; Ivanova, A.; Rombaut, D.; et al. The membrane-bound NAC transcription factor ANAC013 functions in mitochondrial retrograde regulation of the oxidative stress response in Arabidopsis. *Plant Cell* **2013**, *25*, 3472–3490. [CrossRef] [PubMed]

154. Yoshida, K.; Terashima, I.; Noguchi, K. Distinct roles of the cytochrome pathway and alternative oxidase in leaf photosynthesis. *Plant Cell Physiol.* **2006**, *47*, 22–31. [CrossRef] [PubMed]

155. Yoshida, K.; Terashima, I.; Noguchi, K. Up-regulation of mitochondrial alternative oxidase concomitant with chloroplast over-reduction by excess light. *Plant Cell Physiol.* **2007**, *48*, 606–614. [CrossRef]

156. Yoshida, K.; Noguchi, K. Differential gene expression profiles of the mitochondrial respiratory components in illuminated arabidopsis leaves. *Plant Cell Physiol.* **2009**, *50*, 1449–1462. [CrossRef]

157. Van Aken, O.; Zhang, B.; Law, S.; Narsai, R.; Whelan, J. AtWRKY40 and AtWRKY63 modulate the expression of stress-responsive nuclear genes encoding mitochondrial and chloroplast proteins. *Plant Physiol.* **2013**, *162*, 254–271. [CrossRef]

158. Ivanova, A.; Law, S.R.; Narsai, R.; Duncan, O.; Lee, J.H.; Zhang, B.; Van Aken, O.; Radomiljac, J.D.; van der Merwe, M.; Yi, K.K.; et al. A functional antagonistic relationship between auxin and mitochondrial retrograde signaling regulates alternative oxidase1a expression in arabidopsis. *Plant Physiol.* **2014**, *165*, 1233–1254. [CrossRef]

159. Ho, L.H.M.; Giraud, E.; Uggalla, V.; Lister, R.; Clifton, R.; Glen, A.; Thirkettle-Watts, D.; Van Aken, O.; Whelan, J. Identification of regulatory pathways controlling gene expression of stress-responsive mitochondrial proteins in arabidopsis. *Plant Physiol.* **2008**, *147*, 1858–1873. [CrossRef]

160. Norman, C.; Howell, K.A.; Millar, A.H.; Whelan, J.M.; Day, D.A. Salicylic Acid Is an Uncoupler and Inhibitor of Mitochondrial Electron Transport. *Plant Physiol.* **2004**, *134*, 492–501. [CrossRef]

161. Liao, Y.; Tian, M.; Zhang, H.; Li, X.; Wang, Y.; Xia, X.; Zhou, J.; Zhou, Y.; Yu, J.; Shi, K.; et al. Salicylic acid binding of mitochondrial alpha-ketoglutarate dehydrogenase E2 affects mitochondrial oxidative phosphorylation and electron transport chain components and plays a role in basal defense against tobacco mosaic virus in tomato. *New Phytol.* **2015**, *205*, 1296–1307. [CrossRef] [PubMed]

162. Duan, J.; Lee, K.P.; Dogra, V.; Zhang, S.; Liu, K.; Caceres-Moreno, C.; Lv, S.; Xing, W.; Kato, Y.; Sakamoto, W.; et al. Impaired psii proteostasis promotes retrograde signaling via salicylic acid1. *Plant Physiol.* **2019**, *180*, 2182–2197. [CrossRef] [PubMed]

163. Ng, S.; Giraud, E.; Duncan, O.; Law, S.R.; Wang, Y.; Xu, L.; Narsai, R.; Carrie, C.; Walker, H.; Day, D.A.; et al. Cyclin-dependent kinase E1 (CDKE1) provides a cellular switch in plants between growth and stress responses. *J. Biol. Chem.* **2013**, *288*, 3449–3459. [CrossRef]

164. Blanco, N.E.; Guinea-Díaz, M.; Whelan, J.; Strand, Å. Interaction between plastid and mitochondrial retrograde signalling pathways during changes to plastid redox status. *Philos. Trans. R. Soc. B Biol. Sci.* **2014**, *369*, 20130231. [CrossRef]

165. Baena-González, E.; Rolland, F.; Thevelein, J.M.; Sheen, J. A central integrator of transcription networks in plant stress and energy signalling. *Nature* **2007**, *448*, 938–942. [CrossRef] [PubMed]

166. Baena-González, E.; Sheen, J. Convergent energy and stress signaling. *Trends Plant Sci.* **2008**, *13*, 474–482. [CrossRef] [PubMed]

167. Yang, Y.; Li, L.; Qu, L.J. Plant Mediator complex and its critical functions in transcription regulation. *J. Integr. Plant Biol.* **2016**, *58*, 106–118. [CrossRef]

168. Van Aken, O.; Whelan, J. Comparison of transcriptional changes to chloroplast and mitochondrial perturbations reveals common and specific responses in Arabidopsi. *Front. Plant Sci.* **2012**, *3*. [CrossRef]

169. Kmiecik, P.; Leonardelli, M.; Teige, M. Novel connections in plant organellar signalling link different stress responses and signalling pathways. *J. Exp. Bot.* **2016**, *67*, 3793–3807. [CrossRef]

170. Choudhury, F.K.; Rivero, R.M.; Blumwald, E.; Mittler, R. Reactive oxygen species, abiotic stress and stress combination. *Plant J.* **2017**, *90*, 856–867. [CrossRef]

171. Liu, J.X.; Howell, S.H. Managing the protein folding demands in the endoplasmic reticulum of plants. *New Phytol.* **2016**, *211*, 418–428. [CrossRef] [PubMed]

172. Van Aken, O.; Pogson, B.J. Convergence of mitochondrial and chloroplastic ANAC017/PAP-dependent retrograde signalling pathways and suppression of programmed cell death. *Cell Death Differ.* **2017**, *24*, 955–960. [CrossRef] [PubMed]

173. Ng, S.; Ivanova, A.; Duncan, O.; Law, S.R.; Van Aken, O.; De Clercq, I.; Wang, Y.; Carrie, C.; Xu, L.; Kmiec, B.; et al. A membrane-bound NAC transcription factor, ANAC017, mediates mitochondrial retrograde signaling in Arabidopsis. *Plant Cell* **2013**, *25*, 3450–3471. [CrossRef] [PubMed]

174. Overmyer, K.; Tuominen, H.; Kettunen, R.; Betz, C.; Langebartels, C.; Sandermann, H.; Kangasjarvi, J. Ozone-Sensitive arabidopsis rcd1 mutant reveals opposite roles for ethylene and jasmonate signaling pathways in regulating superoxide-dependent cell death. *Plant Cell* **2000**, *12*, 1849. [CrossRef] [PubMed]

175. Fujibe, T.; Saji, H.; Arakawa, K.; Yabe, N.; Takeuchi, Y.; Yamamoto, K.T. A Methyl Viologen-Resistant Mutant of Arabidopsis, Which Is Allelic to Ozone-Sensitive rcd1, Is Tolerant to Supplemental Ultraviolet-B Irradiation. *Plant Physiol.* **2004**, *134*, 275–285. [CrossRef]

176. Jaspers, P.; Blomster, T.; Brosché, M.; Salojärvi, J.; Ahlfors, R.; Vainonen, J.P.; Reddy, R.A.; Immink, R.; Angenent, G.; Turck, F.; et al. Unequally redundant RCD1 and SRO1 mediate stress and developmental responses and interact with transcription factors. *Plant J.* **2009**, *60*, 268–279. [CrossRef]

177. Brosché, M.; Blomster, T.; Salojärvi, J.; Cui, F.; Sipari, N.; Leppälä, J.; Lamminmäki, A.; Tomai, G.; Narayanasamy, S.; Reddy, R.A.; et al. Transcriptomics and Functional Genomics of ROS-Induced Cell Death Regulation by RADICAL-INDUCED CELL DEATH1. *PLoS Genet.* **2014**, *10*, e1004112. [CrossRef]

178. Heiber, I.; Ströher, E.; Raatz, B.; Busse, I.; Kahmann, U.; Bevan, M.W.; Dietz, K.J.; Baier, M. The redox imbalanced mutants of Arabidopsis differentiate signaling pathways for redox regulation of chloroplast antioxidant enzymes. *Plant Physiol.* **2007**, *143*, 1774–1788. [CrossRef]

179. Hiltscher, H.; Rudnik, R.; Shaikhali, J.; Heiber, I.; Mellenthin, M.; Meirelles Duarte, I.; Schuster, G.; Kahmann, U.; Baier, M. The radical induced cell death protein 1 (RCD1) supports transcriptional activation of genes for chloroplast antioxidant enzymes. *Front. Plant Sci.* **2014**, *5*. [CrossRef]

180. Vainonen, J.P.; Jaspers, P.; Wrzaczek, M.; Lamminmäki, A.; Reddy, R.A.; Vaahtera, L.; Brosché, M.; Kangasjärvi, J. RCD1-DREB2A interaction in leaf senescence and stress responses in Arabidopsis thaliana. *Biochem. J.* **2012**, *442*, 573–581. [CrossRef]

181. Inzé, A.; Vanderauwera, S.; Hoeberichts, F.A.; Vandorpe, M.; van Gaever, T.; van Breusegem, F. A subcellular localization compendium of hydrogen peroxide-induced proteins. *Plant Cell Environ.* **2012**, *35*, 308–320. [CrossRef] [PubMed]

182. Shapiguzov, A.; Vainonen, J.P.; Hunter, K.; Tossavainen, H.; Tiwari, A.; Järvi, S.; Hellman, M.; Aarabi, F.; Alseekh, S.; Wybouw, B.; et al. Arabidopsis RCD1 coordinates chloroplast and mitochondrial functions through interaction with ANAC transcription factors. *eLife* **2019**, *8*. [CrossRef]

183. D'alessandro, S.; Ksas, B.; Havaux, M. Decoding β-cyclocitral-mediated retrograde signaling reveals the role of a detoxification response in plant tolerance to photooxidative stress. *Plant Cell* **2018**, *30*, 2495–2511. [CrossRef] [PubMed]

184. South, P.F.; Cavanagh, A.P.; Liu, H.W.; Ort, D.R. Synthetic glycolate metabolism pathways stimulate crop growth and productivity in the field. *Science* **2019**, *363*, eaat9077. [CrossRef] [PubMed]

185. Kromdijk, J.; Głowacka, K.; Leonelli, L.; Gabilly, S.T.; Iwai, M.; Niyogi, K.K.; Long, S.P. Improving photosynthesis and crop productivity by accelerating recovery from photoprotection. *Science* **2016**, *354*, 857–861. [CrossRef]

186. Głowacka, K.; Kromdijk, J.; Kucera, K.; Xie, J.; Cavanagh, A.P.; Leonelli, L.; Leakey, A.D.B.; Ort, D.R.; Niyogi, K.K.; Long, S.P. Photosystem II Subunit S overexpression increases the efficiency of water use in a field-grown crop. *Nat. Commun.* **2018**, *9*, 1–9. [CrossRef]

International Journal of
Molecular Sciences

Article

Dephosphorylation of LjMPK6 by Phosphatase LjPP2C is Involved in Regulating Nodule Organogenesis in *Lotus japonicus*

Zhongyuan Yan [1,†], Jingjing Cao [1,†], Qiuling Fan [2], Hongmin Chao [1], Xiaomin Guan [1], Zhongming Zhang [1] and Deqiang Duanmu [1,*]

[1] State Key Laboratory of Agricultural Microbiology, College of Life Science and Technology, Huazhong Agricultural University, Wuhan 430070, China; yanzy@huas.edu.cn (Z.Y.); cjj123@webmail.hzau.edu.cn (J.C.); chaohongmin2014@163.com (H.C.); guximihubei@163.com (X.G.); zmzhang@mail.hzau.edu.cn (Z.Z.)
[2] College of Life Science and Technology, Huazhong Agricultural University, Wuhan 430070, China; qlfan@mail.hzau.edu.cn
* Correspondence: dmdq2008@gmail.com; Tel.: +86-27-8728-2101
† These authors contributed equally to this work.

Received: 7 July 2020; Accepted: 1 August 2020; Published: 3 August 2020

Abstract: The mitogen-activated protein kinase (MAPK) LjMPK6 is a phosphorylation target of SIP2, a MAPK kinase that interacts with SymRK (symbiosis receptor-like kinase) for regulation of legume-rhizobia symbiosis. Both LjMPK6 and SIP2 are required for nodulation in *Lotus japonicus*. However, the dephosphorylation of LjMPK6 and its regulatory components in nodule development remains unexplored. By yeast two-hybrid screening, we identified a type 2C protein phosphatase, LjPP2C, that specifically interacts with and dephosphorylates LjMPK6 in vitro. Physiological and biochemical assays further suggested that LjPP2C phosphatase is required for dephosphorylation of LjMPK6 in vivo and for fine-tuning nodule development after rhizobial inoculation. A non-phosphorylatable mutant variant LjMPK6 (T224A Y226F) could mimic LjPP2C functioning in MAPK dephosphorylation required for nodule development in hairy root transformed plants. Collectively, our study demonstrates that interaction with LjPP2C phosphatase is required for dephosphorylation of LjMPK6 to fine tune nodule development in *L. japonicus*.

Keywords: LjPP2C; LjMPK6; *Lotus japonicus*; MAPK dephosphorylation; nodule development

---

1. Introduction

Root nodule symbiosis (RNS) between legumes and rhizobia allows conversion of atmospheric nitrogen into ammonia absorbed by plants. The establishment of RNS begins with mutualistic dialogue between two partners. Host plants secrete flavonoids, inducing rhizobia to synthesize and secrete Nod factors (NFs), a type of lipo-chitooligosaccharide molecules [1,2]. Two plant LysM-type serine/threonine receptor kinases, NFR1 and NFR5 in *Lotus japonicus* (Nod Factor Receptor 1 and 5) [3–5], cooperating with the leucine-rich repeat (LRR) receptor-like kinase SymRK (Symbiosis Receptor-like Kinase) [6], recognize NFs and initiate the NF signaling pathway [7–10]. The core component of the NF signaling pathway is a calcium oscillation which is decoded by the nucleus-localized CCaMK (calcium/calmodulin-dependent protein kinase) [11,12]. CCaMK phosphorylates and activates CYCLOPS [13,14], which binds to and activates the nodule inception (NIN) gene essential for rhizobial colonization, nodule organogenesis and ultimate nitrogen fixation in mature nodules [15,16].

Mitogen-activated protein kinase (MAPK) cascades play central roles in various intracellular signal transduction processes via sequential phosphorylation of three kinases, a MAPK kinase kinase (MAPKKK), a MAPK kinase (MAPKK) and a MAPK [17], culminating in phosphorylation of both threonine and tyrosine residues of the terminal MAPK components (e.g., MPK3/4/6) within their TXY consensus sequence [18,19]. Reversible phosphorylation and dephosphorylation affects MAPK protein structure and its activity toward downstream targets responsible for efficient cellular signal transduction. This process is achieved via two biochemical reactions performed by a specific pair of kinase and phosphatase, the latter being called MKP (MAPK phosphatase) [20]. Only a few MKP proteins have been identified so far. For example, the dual-specificity (DSP) phosphatase OsMKP1 dephosphorylates both phospho-threonine (pT) and phospho-tyrosine (pY) to negatively regulate the OsMKKK10-OsMKK4-OsMPK6 cascade regulating panicle architecture in rice [21,22]. Meanwhile, the type 2C protein phosphatase (PP2C)-type Ser/Thr phosphatase AP2C1 dephosphorylates pT in the 'pTEpY' loop of MPK3/6 to downregulate basal resistance and defense responses to *Pseudomonas syringae* in *Arabidopsis thaliana* [23].

The legume SIP2 is a MAPKK that directly interacts with SymRK. SIP2/SymRK interaction inhibits the kinase activity of SIP2 on the substrate LjMPK6 in *Lotus japonicus* [24]. Both SIP2 and LjMPK6 are required for efficient nodulation [25]. However, the role of LjMPK6 dephosphorylation in RNS and its regulatory components remain elusive. Here, we showed that LjPP2C, a PP2C-type phosphatase, specifically interacts with and dephosphorylates LjMPK6 in vitro. Moreover, our molecular and phenotypic data suggest that LjPP2C contributes to the dephosphorylation of LjMPK6 in vivo, which is required for regulating nodule development in *L. japonicus*.

2. Results

2.1. LjPP2C Interacts with and Dephosphorylates LjMPK6 In Vitro

To study proteins interacting with the MAPK kinase LjMPK6, we identified several protein candidates from a Lotus cDNA library by yeast two-hybrid (Y2H) screening. One of these candidates is a type 2C protein phosphatase named LjPP2C (Lj2g3v2292680.1). The pairwise Y2H assay showed that LjPP2C interacted only with LjMPK6 and not with other MAP kinases such as LjMPK3 or LjMPK4, indicating that the interaction between LjPP2C and LjMPK6 is specific (Figure 1A). To further verify the direct physical interaction between two proteins, we purified recombinant MBP-tagged LjPP2C and GST-tagged LjMPK6 proteins from *E. coli* for in vitro pull-down assay. Our results confirmed that LjPP2C interacts with LjMPK6 directly (Figure 1B).

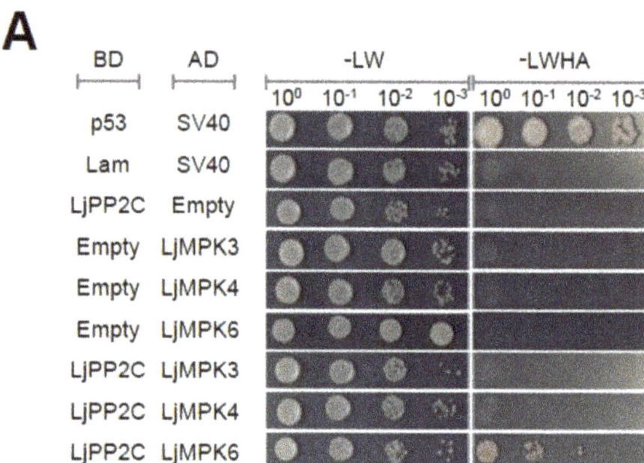

Figure 1. *Cont.*

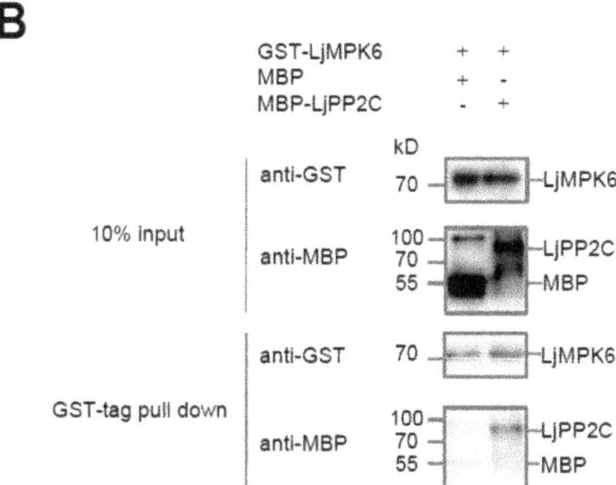

Figure 1. LjPP2C interacts with LjMPK6 in vitro. (**A**) Yeast two-hybrid assays shows interaction between LjPP2C and LjMPK6. Yeast cells were transformed with AD-LjMPK3/4/6 or BD-LjPP2C vectors. Serially diluted yeast cells were grown on SD-LW (lacking leucine and tryptophan) or SD-LWHA medium (lacking leucine, tryptophan, histidine and adenine). Interactions between p53/SV40 and Lam/SV40 were used as positive and negative controls, respectively. (**B**) GST pull-down of interaction between LjPP2C and LjMPK6. Positions of MBP-LjPP2C, MBP and GST-LjMPK6 are indicated. MBP-LiPP2C or MBP itself was incubated with GST-LjMPK6 and glutathione resin. After washing, retained proteins were analyzed by SDS-PAGE and immunoblotted with GST and MBP antibodies. +/-, with (+) or without (-) corresponding protein.

To determine whether LjPP2C can use LjMPK6 as a dephosphorylation substrate, we performed in vitro phosphatase assay by incubating phosphorylated GST-LjMPK6 with MBP-LjPP2C for half an hour at room temperature. Immunoblot analysis using an antibody recognizing specifically the phosphorylated form of MPK6 showed that LjMPK6 phosphorylation level was reduced in a dosage-dependent manner with increasing amount of LjPP2C protein, corroborating the dephosphorylation activity of LjPP2C on LjMPK6 in vitro (Figure 2).

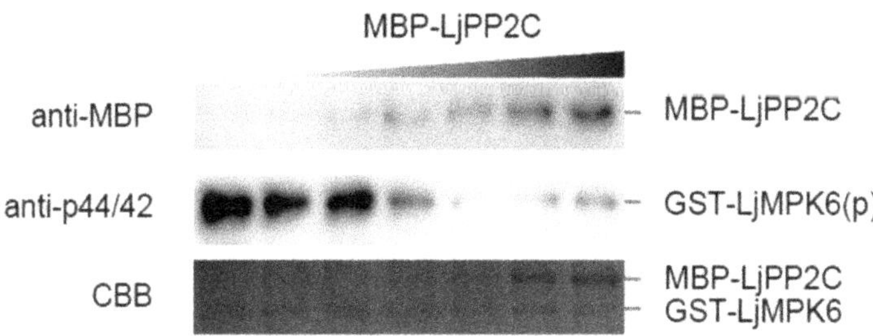

Figure 2. LjPP2C dephosphorylates LjMPK6 in vitro. Phosphorylation of LjMPK6 by SIP2 was performed in advance to prepare phosphorylated GST-tagged LjMPK6. An increasing amount of MBP-LjPP2C was then added to the reaction mixture. MBP-LjPP2C protein was detected by western blotting using anti-MBP antibody. Phosphorylated form of GST-LjMPK6 was detected by anti-p44/42 antibody. CBB, Coomassie Brilliant Blue staining.

2.2. Expression Pattern of LjPP2C in Both Non-Symbiotic and Symbiotic Tissues

To investigate the expression pattern of *LjPP2C* in nodules, we constructed a fusion reporter gene containing the *LjPP2C* promoter and tYFPnls, which carries a nuclear localization signal, and introduced the plasmid into wild-type *L. japonicus* MG20 by stable transformation. We selected several positive transgenic plants for analysis and results from one representative plant was shown. Fluorescence microscopy analysis showed that the *LjPP2C* promoter was actively expressed in root cortex cells (Figure 3A–D). In addition, our qRT-PCR results showed that the relative expression level of *LjPP2C* is also higher in non-symbiotic tissues such as root, shoot and leaf, but to a much lower extent in flower and pod (Figure 3E). Overall, these results suggest that *LjPP2C* is not specifically expressed in symbiotic tissues.

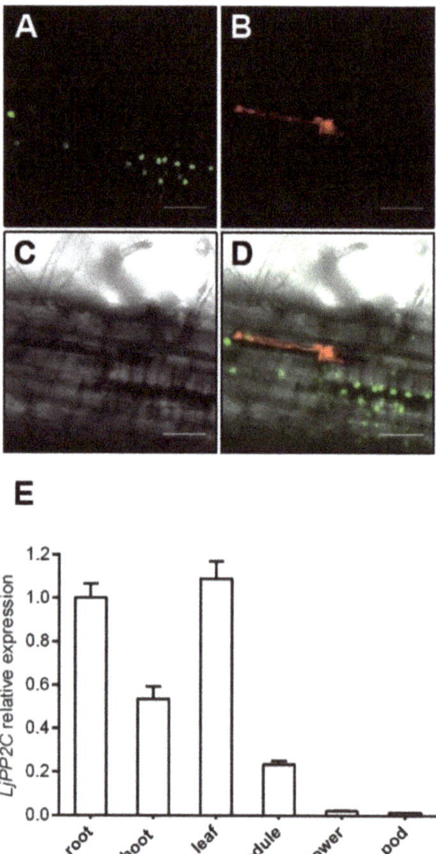

Figure 3. Analysis of *LjPP2C* expression pattern. (**A–D**) Images of *LjPP2C* promoter analysis in nodule primordium stage. MG20 plants were transformed with pUB-Hyg-LjPP2Cpro::tYFPnls construct. Positive hairy roots were inoculated with *M. loti* MAFF303099 (expressing mCherry). Transformed roots were visualized by GFP fluorescence (**A**), mCherry fluorescence (**B**), Brightfield (**C**) and Merge (**D**). Bars: 50 μm (**A–D**). (**E**) Expression of *LjPP2C* in different tissues. Roots were harvested at 14 days post inoculation (pdi) with MAFF303099. Nodules were harvested at 21 dpi. Total RNA was isolated and used for real-time PCR to quantify the expression levels of *LjPP2C* mRNA. The ATPase gene (AW719841) was used as the internal control. Error bars indicate SE of three technical replicates.

2.3. LjPP2C is Required for Nodule Development

To further explore the regulatory role of *LjPP2C* in nodulation, we introduced a CRISPR/Cas9-mediated *LjPP2C* gene knockout construct into Lotus using a stable transformation procedure [26]. In parallel, to generate stable overexpression transgenic plants, we also introduced a *LjPP2C* overexpression construct in a similar way. Two independent knockout mutants of *LjPP2C* (*ljpp2c*#8-3, *ljpp2c*#22-5) were identified by PCR amplification of the target allele and DNA sequencing (Figure 4A).

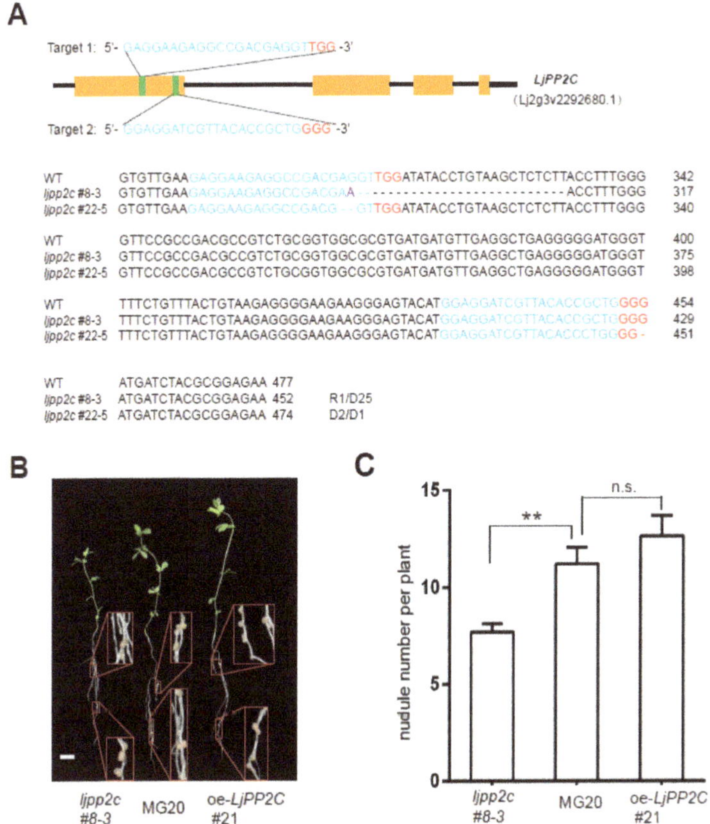

Figure 4. *LjPP2C* mutation affects nodule formation. (**A**) CRISPR/Cas9-mediated *LjPP2C* knockout in stable transgenic MG20 plants. Above: Genomic DNA structure of the *LjPP2C* gene. The PAM sequence of the *LjPP2C* sgRNA target is colored in red and the sgRNA target is in blue. Below: Indel mutations of two stable lines. Both lines have mutations leading to early translational termination. WT, wide-type control. D1/2/25, 1 bp/2 bp/25 bp DNA deletion; R1, 1 bp replacement. (**B**,**C**) Growth phenotype (**B**) and nodule numbers (**C**) of *ljpp2c*#8-3, MG20 and oe-*LjPP2C*#21 at 4 weeks post-inoculation (4 wpi). Bar in (**B**), 1 cm. Error bars in (**C**) indicate SE of 27~29 plants analyzed for each genotype. Student's t-test was used for statistical comparisons. ** $p < 0.01$, n.s., not significant.

We next performed nodulation assays under nitrogen-deficient conditions. The *ljpp2c*#8-3 mutant produced fewer nodules than wild-type MG20, whereas *LjPP2C* overexpression line (oe-*LjPP2C*#21) showed marginally, but insignificantly increased nodule numbers, indicating that *LjPP2C* plays an important role in nodule formation (Figure 4B,C).

2.4. LjPP2C is Required for MAPK Dephosphorylation In Vivo

To further ascertain whether or not dephosphorylation of LjMPK6 affects nodule development, we constructed a non-phosphorylatable version of LjMPK6, LjMPK6$^{T224A;\,Y226F}$ and introduced into wild-type MG20 or the *ljpp2c-KO* mutant by hairy root transformation (Figure 5A). Compared with the empty vector control, overexpression of LjMPK6$^{T224A;\,Y226F}$ reduced the phosphorylation of MAPKs in Lotus plants, especially in the *ljpp2c-KO* mutant background (Figure 5C). Interestingly, compared to empty vector control, nodule numbers were significantly enhanced in LjMPK6$^{T224A;\,Y226F}$ overexpression hairy roots of both MG20 and *ljpp2c*#8-3 plants, suggesting that LjMPK6 phosphorylation/dephosphorylation homeostasis indeed plays an important role for nodule organogenesis (Figure 5B).

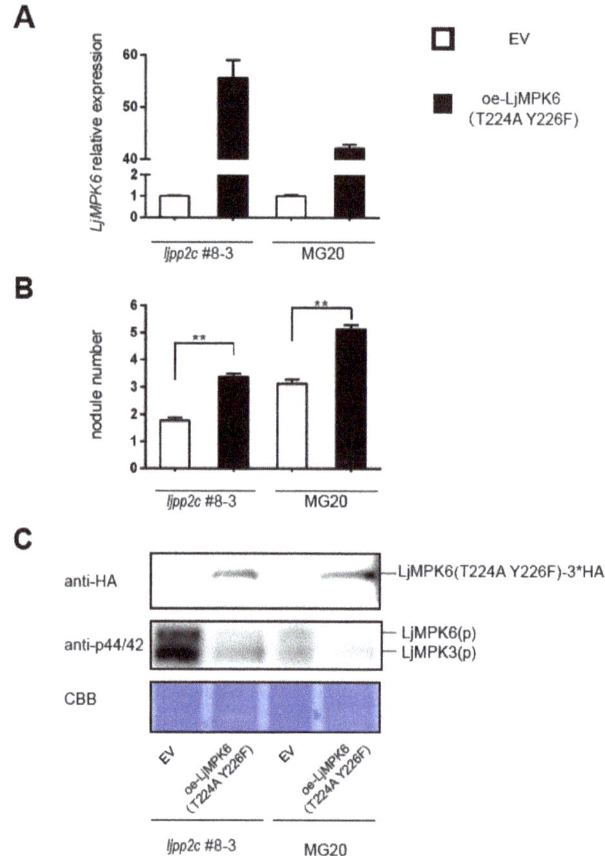

Figure 5. Overexpression of non-phosphorylatable LjMPK6 increased nodule formation in WT and *ljpp2c* mutant plants. MG20 and *ljpp2c*#8-3 were transformed with empty vector and oe-LjMPK6 (T224A Y226F) (non-phosphorylated form of LjMPK6) via hairy root transformation. (**A**) Analysis of transcript abundance of *LjMPK6* in different hairy roots. Error bars represent SE of experimental values from three technical replicates. (**B**) Nodule numbers per plant at 3 weeks post inoculation (3 wpi) in transformed hairy roots. Error bars represent SE of ~13 plants analyzed for each genotype. $^{**}p < 0.01$, Student's *t*-test. (**C**) Immunoblot analysis of protein level of overexpressed non-phosphorylatable LjMPK6 (anti-HA) and endogenous phosphorylated LjMPK6 (anti-p44/42) from those hairy roots. Equal loading of total protein was demonstrated by Coomassie Brilliant Blue (CBB) staining.

3. Discussion

SymRK plays a key role in the NF signaling pathway during legume-rhizobia symbiotic interaction. SymRK-interacting protein SIP2 is required for early symbiotic signal transduction and MAPK signaling during nodule symbiosis [24]. As one of the phosphorylation substrates of SIP2, LjMPK6 is essential for nodulation [25]. However, downstream nodulation events and regulatory components of LjMPK6 signaling are largely unknown. Here, we demonstrated that LjPP2C dephosphorylates LjMPK6 in vitro and LjPP2C is required for MAPK dephosphorylation and nodule organogenesis in response to rhizobial inoculation.

Our phosphatase assay demonstrated that LjPP2C is a genuine protein phosphatase acting on the substrate LjMPK6, in accordance with its pivotal role in cellular signal transduction as reported for other plant PP2Cs in response to various environmental and developmental stimuli. For instance, a phosphatase 2C-1 (PP2C-1) allele from soybean dephosphorylates transcription factor GmBZR1 involved in brassinosteroid signaling [27]. Arabidopsis PP2C38 is an active phosphatase that negatively regulates cytoplasmic kinase BIK1-mediated immune signaling [28]. Rice XB15 harbors PP2C activity and negatively regulates the receptor kinase XA21-mediated innate immune response [29].

Only a few PP2C phosphatases have been identified in *Lotus japonicus*, largely due to the poorly annotated genome and their complicated biological functions. *LjNPP2C1*, encoding a Mg^{2+}- or Mn^{2+}-dependent PP2C, is specifically induced in *L. japonicus* nodules and functions at both early and late stages of nodule development, whereas the *LjPP2C2* gene was expressed at a similar level in nodules and roots [30]. In contrast, LjPP2C reported in this study appears to play an essential role in regulating nodule development, likely downstream of infection thread formation.

The *L. japonicus* 2C-type protein phosphatase LjPP2C interacts with the previously identified LjMPK6, which was shown to play an essential regulatory role in nodule development [25]. As a typical MAPK phosphatase, LjPP2C mainly functions through its dephosphorylation activity on LjMPK6. Compared to wild type MG20, a reduced nodule number and a higher level of LjMPK6 phosphorylation is observed in *ljpp2c* mutant plants, suggesting that a tight regulation of LjMPK6 phosphorylation is required for fine-tuning nodule organogenesis. In consideration of the immune signaling role of numerous plant kinase-PP2C phosphatase pairs previously reported [31], our work has extended our understanding of MAPK signaling in legume-rhizobia symbiosis and provides clues for further research on nodule developmental regulation mediated by protein phosphatases.

4. Materials and Methods

4.1. Plant Materials and Growth Conditions

Lotus japonicus seedlings were grown in pots containing perlite:vermiculite mixture (1.2, v/v) supplied with a 1/2 B&D nitrogen-free nutrient solution. *Nicotiana benthamiana* seedlings were grown in pots filled with perlite:vermiculite:nutrient soil mixture (1:1:1,v/v/v). All plants were cultivated in a growth chamber under a 16-h light/8-h dark cycle at 22 °C and 60% relative humidity.

4.2. Plasmid Construction

To create a *ljpp2c* mutant in *L. japonicus*, the web tool CRISPR-P 2.0 (http://cbi.hzau.edu.cn/crispr/) was used to design appropriate guide RNAs for a CRISPR/Cas9-mediated gene knockout approach. Two guide RNAs were cloned into one construct to improve gene-editing efficiency [26]. To create *LjPP2C* overexpression construct, full-length cDNA of *LjPP2C* was PCR-amplified from wild type *Lotus japonicus* MG20 genomic DNA and was inserted into pUB-Hyg vector [32]. All constructs were validated by DNA sequencing and were then transformed into *Agrobacterium tumefaciens* strain EHA105 for stable transformation into *L. japonicus* MG20 as described previously [33]. Transgenic plants were screened for hygromycin resistance and confirmed by PCR genotyping.

To create AD fusion constructs for pairwise yeast two-hybrid (Y2H) assay, full-length cDNA sequences of *LjMPK3*, *LjMPK4*, and *LjMPK6* were amplified by PCR and inserted into vector pGADT7

(Takara Bio, Beijing, China) to generate AD-LjMPK3, AD-LjMPK4, and AD-LjMPK6, respectively. The full-length cDNA sequence of *LjPP2C* and *LjMPK6* were amplified and inserted into pGBKT7 (Takara Bio) vector to generate BD-LjPP2C and BD-LjMPK6, respectively. All constructs were cloned using Phusion DNA polymerase (New England Biolabs, Ipswich, MA, USA) according to the one-step enzymatic assembly method protocol.

For promoter analysis, a ~3 kb DNA fragment of *LjPP2C* promoter was amplified using PCR and introduced into the pUB-Hyg plasmid to replace the Ubiquitin promoter and fused to tYFPnls (nuclear-localized triple-YFP) reporter [34].

To create plasmids for in vitro pull-down assay, full-length cDNA sequences of *LjMPK6*, *LjPP2C* were PCR amplified and cloned into pGEX-6p-1 and pMAL-c2x, respectively, to obtain the corresponding GST-tagged and MBP-tagged fusion proteins. To create overexpression vector for hairy root transformation in *L. japonicus*, 3*HA tagged full-length cDNA sequence of *LjMPK6* with two introduced point mutations (T224A and Y226F) was cloned into the empty vector pUB-GFP [32]. Sequences of corresponding primers used for plasmids construction are listed in Table S1.

4.3. Yeast Two-Hybrid Assay

Yeast two-hybrid (Y2H) screening was performed according to the manufacturer's instruction (Takara Bio, Beijing, China), using *Saccharomyces cerevisiae* Y187 strain harboring the BD-LjMPK6 fusion construct as bait and *S. cerevisiae* AH109 strain harboring AD fusion construct of Lotus cDNA library derived from root and nodule tissues as prey. A total of 1×10^7 transformants were screened, and positive colonies growing on SD/-Leu-Trp-His-Ade (SD/-4) medium were selected for plasmid extraction and sequence analysis. The candidates were retransformed into yeast for pairwise Y2H assay to validate the interaction results. Cells of yeast strain Y187 transformed with AD fusion constructs and cells of yeast strain AH109 transformed with the BD fusion construct were respectively spreaded onto SD/-Leu or SD/-Trp agar plates using a LiAc-mediated yeast transformation protocol and then mated overnight in 2×YPDA. Aliquots (10 μl) of diploid yeast cells were spotted onto SD/-Leu-Trp (SD/-2) and SD/-Leu-Trp-His-Ade (SD/-4) medium to verify protein-protein interactions. Interactions between p53 (or lamin, Lam) and SV40 were used as positive or negative controls, respectively. Yeast growth was monitored for up to 5 days at 30 °C.

4.4. In Vitro Pull-Down Assay

Protein expression of GST-tagged LjMPK6 and MBP-tagged LjPP2C was induced in *Escherichia coli* strain BL21 (DE3) by 0.1~0.5 mM isopropylthio-β-galactoside (IPTG) for 4 h at 30 °C and purified with glutathione resin (GenScript, Nanjing, China) and Amylose resin (New England Biolabs), respectively. Purified proteins were then incubated with glutathione resin in PBS buffer (2 mM KH_2PO_4, 8 mM Na_2HPO_4, 136 mM NaCl, and 2.6 mM KCl, pH 7.4) for 30 min at 4 °C and separated by centrifugation. The supernatant was discarded, and the glutathione resin was washed at least three times with PBS buffer. Proteins retained with the glutathione resin were boiled in SDS loading buffer (50 mM Tris-HCl, pH 6.8, 2% SDS (w/v), 0.1% bromophenol blue, 10% glycerol, 1% β-mercaptoethanol) and separated by SDS-PAGE electrophoresis. Corresponding antibodies against GST or MBP (PhytoAB, San Jose, CA, USA) were used for immunoblot analysis.

4.5. Phosphatase Assay

The kinase assay between SIP2 and LjMPK6 was performed in advance to prepare phosphorylated GST-tagged LjMPK6, which was then purified by glutathione resin. The in vitro phosphatase assay of LjPP2C was performed by incubating phosphorylated LjMPK6 with LjPP2C in a buffer containing 50 mM Tris-HCl, pH 7.5, 10 mM $MgCl_2$, 0.1% Triton-X100, and 1 mM DTT at 30 °C for 30 min. Reactions were stopped by adding 2×SDS loading buffer and boiling for 5 minutes. The remaining phosphorylated LjMPK6 was identified by immunoblotting with anti-phospho-p44/42 (pThr-X-pTyr) MAPK antibody (Cell Signaling Technology, Danvers, MA, USA).

To detect the phosphatase activity of LjPP2C on LjMPK6 in *L. japonicus* in vivo, roots of MG20 and ljpp2c#8-3 were transformed with empty vector or oe-LjMPK6 (T224A Y226F) (non-phosphorylatable form of LjMPK6). Transgenic hairy roots of above materials were grounded into fine powder in liquid nitrogen. Total protein was extracted using sample buffer containing 50 mM Tris-HCl, pH 6.8, 5% (w/v) SDS, 10 mM DTT, 100 μM PMSF (phenylmethylsulfonyl fluoride), and 1 mM sodium pyrophosphate. Phosphorylated LjMPK6 was identified by immunoblotting with anti-phospho-p44/42 (pThr-X-pTyr) MAPK antibody.

4.6. RNA Extraction and Reverse-Transcription Quantitative PCR (qRT-PCR)

Plant total RNA was isolated using TRIzol reagent (Thermo Fisher Scientific, Hampton, NH, USA) and treated with DNase (Promega, Madison, WI, USA) to eliminate genomic DNA contamination. Less than 1 μg total RNA was used to synthesize first-strand cDNA using oligo(dT) primer according to the instructions of the cDNA synthesis kit (TransGen Biotech, Beijing, China). RT-qPCR was performed on a Roche Light Cycler (thermal cycle: 95 °C for 10 s, 40 cycles of 95 °C for 5 s and 60 °C for 30 s, followed by a melting curve stage at 95 °C for 15 s and a temperature gradient from 65 °C to 95 °C at a rate of 1 °C s^{-1}) based on the instruction of the one-step SYBR Prime Script RT-PCR Kit II (Takara Bio). The housekeeping genes *LjATPase* (AW719841) or *LjUBQ* served as reference genes for relative fold expression changes using the $2^{-\Delta\Delta Ct}$ method. All reactions were performed with three technical replicates.

4.7. Hairy Root Transformation and Nodulation Assays

L. japonicus hairy roots were obtained by transformation with *Agrobacterium rhizogenes* LBA1334 containing the overexpression construct pUB-GFP-LjUBQ1pro::LjMPK6(T224A;Y226F)-3*HA as described previously [35]. Plants transformed with the empty vector (EV) pUB-GFP-3*HA were used as control. Regenerated hairy roots were screened for GFP (green fluorescent protein) fluorescence. Only one GFP-positive root was retained for each plant, and all other roots were removed.

One-week-old seedlings were transferred from Petri dishes to pots containing perlite:vermiculite (1:2, v/v). Plants were watered with 1/2 B&D nitrogen-free nutrient solution for 1 week and were then inoculated with 1 mL of *Mesorhizobium loti* MAFF303099 (OD600 = 0.1) expressing the red fluorescent marker mCherry (pQDN03-ptrp::mCherry) [36,37]. At 3 weeks post inoculation (3 wpi), roots were collected for counting nodules, and root samples were harvested for protein immunoblot analysis.

4.8. Accession Numbers

Sequence data for the following proteins can be found in the Lotus database (v3.0, https://lotus.au.dk). *LjPP2C*, Lj2g3v2292680.1; *LjMPK3*, Lj3g3v3087330.1; *LjMPK4*, Lj4g3v2989020.1; *LjMPK6*, Lj4g3v0510090.1; *SIP2*, Lj3g3v2040150.1.

Supplementary Materials: Supplementary materials can be found at http://www.mdpi.com/1422-0067/21/15/5565/s1.

Author Contributions: D.D. and Q.F. designed the experiments. Z.Y., H.C., J.C. and X.G. created plant transgenic lines, made constructs and performed all experiments. Z.Z. reviewed the article. Z.Y., Q.F. and J.C. analyzed the data and wrote the manuscript. All authors had read and approved the manuscript.

Funding: This work was supported by grants from National Natural Science Foundation of China (project no. 31870220).

Acknowledgments: We are grateful to Chao Wang (University of California, Berkeley) for comments on the manuscript. We also thank Tao Chen and Jun Yin for providing vectors pGEX-6P-1-SIP2 and pCAMBIA1300-35s-LjMPK6-mCherry-Flag, respectively.

Conflicts of Interest: The authors declare no conflict of interest.

Abbreviations

CRISPR	clustered regularly interspaced short palindromic repeats
GFP	Green Fluorescent Protein
GST	Glutathione S-transferase
MAPK	mitogen-activated protein kinase
MBP	maltose-binding protein
NF	nod factor
NIN	nodule inception gene
PP2C	type 2C protein phosphatase
RLK	receptor-like kinase
Y2H	yeast two-hybrid

References

1. Denarie, J.; Debelle, F.; Prome, J.C. Rhizobium lipo-chitooligosaccharide nodulation factors: Signaling molecules mediating recognition and morphogenesis. *Annu. Rev. Biochem.* **1996**, *65*, 503–535. [CrossRef] [PubMed]
2. Broughton, W.J.; Jabbouri, S.; Perret, X. Keys to symbiotic harmony. *J. Bacteriol.* **2000**, *182*, 5641–5652. [CrossRef] [PubMed]
3. Radutoiu, S.; Madsen, L.H.; Madsen, E.B.; Felle, H.H.; Umehara, Y.; Gronlund, M.; Sato, S.; Nakamura, Y.; Tabata, S.; Sandal, N.; et al. Plant recognition of symbiotic bacteria requires two LysM receptor-like kinases. *Nature* **2003**, *425*, 585–592. [CrossRef] [PubMed]
4. Madsen, E.B.; Madsen, L.H.; Radutoiu, S.; Olbryt, M.; Rakwalska, M.; Szczyglowski, K.; Sato, S.; Kaneko, T.; Tabata, S.; Sandal, N.; et al. A receptor kinase gene of the LysM type is involved in legume perception of rhizobial signals. *Nature* **2003**, *425*, 637–640. [CrossRef]
5. Tsikou, D.; Ramirez, E.E.; Psarrakou, I.S.; Wong, J.E.; Jensen, D.B.; Isono, E.; Radutoiu, S.; Papadopoulou, K.K. A Lotus japonicus E3 ligase interacts with the Nod Factor Receptor 5 and positively regulates nodulation. *BMC Plant Biol.* **2018**, *18*, 217. [CrossRef]
6. Stracke, S.; Kistner, C.; Yoshida, S.; Mulder, L.; Sato, S.; Kaneko, T.; Tabata, S.; Sandal, N.; Stougaard, J.; Szczyglowski, K.; et al. A plant receptor-like kinase required for both bacterial and fungal symbiosis. *Nature* **2002**, *417*, 959–962. [CrossRef]
7. Riely, B.K.; Ane, J.M.; Penmetsa, R.V.; Cook, D.R. Genetic and genomic analysis in model legumes bring Nod-factor signaling to center stage. *Curr. Opin. Plant Biol.* **2004**, *7*, 408–413. [CrossRef]
8. Ried, M.K.; Antolin-Llovera, M.; Parniske, M. Spontaneous symbiotic reprogramming of plant roots triggered by receptor-like kinases. *Elife* **2014**, e03891. [CrossRef]
9. Holsters, M. SYMRK, an enigmatic receptor guarding and guiding microbial endosymbioses with plant roots. *Proc. Natl. Acad. Sci. USA* **2008**, *105*, 4537–4538. [CrossRef]
10. Oldroyd, G.E. Speak, friend, and enter: Signalling systems that promote beneficial symbiotic associations in plants. *Nat. Rev. Microbiol.* **2013**, *11*, 252–263. [CrossRef]
11. Miller, J.B.; Pratap, A.; Miyahara, A.; Zhou, L.; Bornemann, S.; Morris, R.J.; Oldroyd, G.E. Calcium/Calmodulin-dependent protein kinase is negatively and positively regulated by calcium, providing a mechanism for decoding calcium responses during symbiosis signaling. *Plant Cell* **2013**, *25*, 5053–5066. [CrossRef] [PubMed]
12. Mitra, R.M.; Gleason, C.A.; Edwards, A.; Hadfield, J.; Downie, J.A.; Oldroyd, G.E.; Long, S.R. A Ca2+/calmodulin-dependent protein kinase required for symbiotic nodule development: Gene identification by transcript-based cloning. *Proc. Natl. Acad. Sci. USA* **2004**, *101*, 4701–4705. [CrossRef] [PubMed]
13. Singh, S.; Katzer, K.; Lambert, J.; Cerri, M.; Parniske, M. CYCLOPS, a DNA-binding transcriptional activator, orchestrates symbiotic root nodule development. *Cell Host Microbe.* **2014**, *15*, 139–152. [CrossRef] [PubMed]
14. Yano, K.; Yoshida, S.; Muller, J.; Singh, S.; Banba, M.; Vickers, K.; Markmann, K.; White, C.; Schuller, B.; Sato, S.; et al. CYCLOPS, a mediator of symbiotic intracellular accommodation. *Proc. Natl. Acad. Sci. USA* **2008**, *105*, 20540–20545. [CrossRef]

15. Schauser, L.; Roussis, A.; Stiller, J.; Stougaard, J. A plant regulator controlling development of symbiotic root nodules. *Nature* **1999**, *402*, 191–195. [CrossRef]

16. Soyano, T.; Kouchi, H.; Hirota, A.; Hayashi, M. Nodule inception directly targets NF-Y subunit genes to regulate essential processes of root nodule development in Lotus japonicus. *PLoS Genet.* **2013**, *9*, e1003352. [CrossRef]

17. Wang, C.; Wang, G.; Zhang, C.; Zhu, P.; Dai, H.; Yu, N.; He, Z.; Xu, L.; Wang, E. OsCERK1-Mediated Chitin Perception and Immune Signaling Requires Receptor-like Cytoplasmic Kinase 185 to Activate an MAPK Cascade in Rice. *Mol Plant.* **2017**, *10*, 619–633. [CrossRef]

18. Pitzschke, A. Modes of MAPK substrate recognition and control. *Trends Plant Sci.* **2015**, *20*, 49–55. [CrossRef]

19. Berriri, S.; Garcia, A.V.; Frei dit Frey, N.; Rozhon, W.; Pateyron, S.; Leonhardt, N.; Montillet, J.L.; Leung, J.; Hirt, H.; Colcombet, J. Constitutively active mitogen-activated protein kinase versions reveal functions of Arabidopsis MPK4 in pathogen defense signaling. *Plant Cell* **2012**, *24*, 4281–4293. [CrossRef]

20. Sarma, U.; Ghosh, I. Different designs of kinase-phosphatase interactions and phosphatase sequestration shapes the robustness and signal flow in the MAPK cascade. *BMC Syst. Biol.* **2012**, *6*, 82. [CrossRef]

21. Xu, R.; Duan, P.; Yu, H.; Zhou, Z.; Zhang, B.; Wang, R.; Li, J.; Zhang, G.; Zhuang, S.; Lyu, J.; et al. Control of Grain Size and Weight by the OsMKKK10-OsMKK4-OsMAPK6 Signaling Pathway in Rice. *Mol. Plant.* **2018**, *11*, 860–873. [CrossRef] [PubMed]

22. Liu, S.; Hua, L.; Dong, S.; Chen, H.; Zhu, X.; Jiang, J.; Zhang, F.; Li, Y.; Fang, X.; Chen, F. OsMAPK6, a mitogen-activated protein kinase, influences rice grain size and biomass production. *Plant J.* **2015**, *84*, 672–681. [CrossRef] [PubMed]

23. Shubchynskyy, V.; Boniecka, J.; Schweighofer, A.; Simulis, J.; Kvederaviciute, K.; Stumpe, M.; Mauch, F.; Balazadeh, S.; Mueller-Roeber, B.; Boutrot, F.; et al. Protein phosphatase AP2C1 negatively regulates basal resistance and defense responses to Pseudomonas syringae. *J. Exp. Bot.* **2017**, *68*, 1169–1183. [PubMed]

24. Chen, T.; Zhu, H.; Ke, D.; Cai, K.; Wang, C.; Gou, H.; Hong, Z.; Zhang, Z. A MAP kinase kinase interacts with SymRK and regulates nodule organogenesis in Lotus japonicus. *Plant Cell* **2012**, *24*, 823–838. [CrossRef]

25. Yin, J.; Guan, X.; Zhang, H.; Wang, L.; Li, H.; Zhang, Q.; Chen, T.; Xu, Z.; Hong, Z.; Cao, Y.; et al. An MAP kinase interacts with LHK1 and regulates nodule organogenesis in Lotus japonicus. *Sci. China Life Sci.* **2019**, *62*, 1203–1217. [CrossRef]

26. Wang, L.; Tan, Q.; Fan, Q.; Zhu, H.; Hong, Z.; Zhang, Z.; Duanmu, D. Efficient Inactivation of Symbiotic Nitrogen Fixation Related Genes in Lotus japonicus Using CRISPR-Cas9. *Front. Plant Sci.* **2016**, *7*, 1333. [CrossRef]

27. Lu, X.; Xiong, Q.; Cheng, T.; Li, Q.T.; Liu, X.L.; Bi, Y.D.; Li, W.; Zhang, W.K.; Ma, B.; Lai, Y.C.; et al. A PP2C-1 Allele Underlying a Quantitative Trait Locus Enhances Soybean 100-Seed Weight. *Mol. Plant.* **2017**, *10*, 670–684. [CrossRef]

28. Couto, D.; Niebergall, R.; Liang, X.; Bucherl, C.A.; Sklenar, J.; Macho, A.P.; Ntoukakis, V.; Derbyshire, P.; Altenbach, D.; Maclean, D.; et al. The Arabidopsis Protein Phosphatase PP2C38 Negatively Regulates the Central Immune Kinase BIK1. *PLoS Pathog.* **2016**, *12*, e1005811. [CrossRef]

29. Park, C.J.; Peng, Y.; Chen, X.; Dardick, C.; Ruan, D.; Bart, R.; Canlas, P.E.; Ronald, P.C. Rice XB15, a protein phosphatase 2C, negatively regulates cell death and XA21-mediated innate immunity. *PLOS Biol.* **2008**, *6*, e231.

30. Kapranov, P.; Jensen, T.J.; Poulsen, C.; de Bruijn, F.J.; Szczyglowski, K. A protein phosphatase 2C gene, LjNPP2C1, from Lotus japonicus induced during root nodule development. *Proc. Natl. Acad. Sci. USA* **1999**, *96*, 1738–1743. [CrossRef]

31. Schweighofer, A.; Kazanaviciute, V.; Scheikl, E.; Teige, M.; Doczi, R.; Hirt, H.; Schwanninger, M.; Kant, M.; Schuurink, R.; Mauch, F.; et al. The PP2C-type phosphatase AP2C1, which negatively regulates MPK4 and MPK6, modulates innate immunity, jasmonic acid, and ethylene levels in Arabidopsis. *Plant Cell* **2007**, *19*, 2213–2224. [CrossRef] [PubMed]

32. Maekawa, T.; Kusakabe, M.; Shimoda, Y.; Sato, S.; Tabata, S.; Murooka, Y.; Hayashi, M. Polyubiquitin promoter-based binary vectors for overexpression and gene silencing in Lotus japonicus. *Mol. Plant Microbe Interact.* **2008**, *21*, 375–382. [CrossRef] [PubMed]

33. Tirichine, L.; Herrera-Cervera, J.A.; Stougaard, J. Transformation-regeneration procedure for Lotus japonicus. In *Lotus Japonicus Handbook*; Springer: Dordrecht, The Netherlands, 2005; pp. 279–284.

34. Reid, D.E.; Heckmann, A.B.; Novak, O.; Kelly, S.; Stougaard, J. CYTOKININ OXIDASE/DEHYDROGENASE3 Maintains Cytokinin Homeostasis during Root and Nodule Development in Lotus japonicus. *Plant Physiol.* **2016**, *170*, 1060–1074. [CrossRef] [PubMed]
35. Diaz, C.L.; Gronlund, M.; Schlaman, H.R.M.; Spaink, H.P. Induction of hairy roots for symbiotic gene expression studies. In *Lotus Japonicus Handbook*; Springer: Dordrecht, The Netherlands, 2005; pp. 261–277.
36. Gage, D.J. Analysis of infection thread development using Gfp- and DsRed-expressing Sinorhizobium meliloti. *J. Bacteriol.* **2002**, *184*, 7042–7046. [CrossRef]
37. Haney, C.H.; Riely, B.K.; Tricoli, D.M.; Cook, D.R.; Ehrhardt, D.W.; Long, S.R. Symbiotic rhizobia bacteria trigger a change in localization and dynamics of the Medicago truncatula receptor kinase LYK3. *Plant Cell* **2011**, *23*, 2774–2787. [CrossRef]

International Journal of
Molecular Sciences

Article

MiR1885 Regulates Disease Tolerance Genes in *Brassica rapa* during Early Infection with *Plasmodiophora brassicae*

Parameswari Paul [1,†], Sushil Satish Chhapekar [1,†], Jana Jeevan Rameneni [1], Sang Heon Oh [1], Vignesh Dhandapani [1,2], Saminathan Subburaj [1], Sang-Yoon Shin [3,4], Nirala Ramchiary [5], Chanseok Shin [3,6,7], Su Ryun Choi [1,*] and Yong Pyo Lim [1,*]

[1] Department of Horticulture, College of Agriculture and Life Science, Chungnam National University, Daejeon 34134, Korea; parameswaripaul@gmail.com (P.P.); sushilchhapekar@gmail.com (S.S.C.); saijeevan7@gmail.com (J.J.R.); rederaser64@gmail.com (S.H.O.); vicky.bioinfo@gmail.com (V.D.); sami_plantbio86@yahoo.co.in (S.S.)
[2] School of Biosciences, University of Birmingham, Birmingham B15 2TT, UK
[3] Department of Agricultural Biotechnology, Seoul National University, Seoul 08826, Korea; nogarded@snu.ac.kr (S.-Y.S.); cshin@snu.ac.kr (C.S.)
[4] Interdisciplinary Program in Agricultural Genomics, Seoul National University, Seoul 08826, Korea
[5] School of Life Sciences, Jawaharlal Nehru University, New Delhi 110067, India; nrudsc@gmail.com
[6] Research Institute of Agriculture and Life Sciences, Seoul National University, Seoul 08826, Korea
[7] Plant Genomics and Breeding Institute, Seoul National University, Seoul 08826, Korea
[*] Correspondence: srchoi@cnu.kr.in (S.R.C.); yplim@cnu.ac.kr (Y.P.L.); Tel.: +82-42-821-8846 (S.R.C.); +82-42-821-5739 (Y.P.L.); Fax: +82-42-821-8847 (Y.P.L.)
[†] These authors contributed equally and share first authorship.

Citation: Paul, P.; Chhapekar, S.S.; Rameneni, J.J.; Oh, S.H.; Dhandapani, V.; Subburaj, S.; Shin, S.-Y.; Ramchiary, N.; Shin, C.; Choi, S.R.; et al. MiR1885 Regulates Disease Tolerance Genes in *Brassica rapa* during Early Infection with *Plasmodiophora brassicae. Int. J. Mol. Sci.* 2021, 22, 9433. https://doi.org/10.3390/ijms22179433

Academic Editor: Anna M. Mastrangelo

Received: 16 July 2021
Accepted: 23 August 2021
Published: 30 August 2021

Abstract: Clubroot caused by *Plasmodiophora brassicae* is a severe disease of cruciferous crops that decreases crop quality and productivity. Several clubroot resistance-related quantitative trait loci and candidate genes have been identified. However, the underlying regulatory mechanism, the interrelationships among genes, and how genes are regulated remain unexplored. MicroRNAs (miRNAs) are attracting attention as regulators of gene expression, including during biotic stress responses. The main objective of this study was to understand how miRNAs regulate clubroot resistance-related genes in *P. brassicae*-infected *Brassica rapa*. Two *Brassica* miRNAs, Bra-miR1885a and Bra-miR1885b, were revealed to target TIR-NBS genes. In non-infected plants, both miRNAs were expressed at low levels to maintain the balance between plant development and basal immunity. However, their expression levels increased in *P. brassicae*-infected plants. Both miRNAs down-regulated the expression of the TIR-NBS genes *Bra019412* and *Bra019410*, which are located at a clubroot resistance-related quantitative trait locus. The Bra-miR1885-mediated down-regulation of both genes was detected for up to 15 days post-inoculation in the clubroot-resistant line CR Shinki and in the clubroot-susceptible line 94SK. A qRT-PCR analysis revealed *Bra019412* expression was negatively regulated by miR1885. Both *Bra019412* and *Bra019410* were more highly expressed in CR Shinki than in 94SK; the same expression pattern was detected in multiple clubroot-resistant and clubroot-susceptible inbred lines. A 5' rapid amplification of cDNA ends analysis confirmed the cleavage of *Bra019412* by Bra-miR1885b. Thus, miR1885s potentially regulate TIR-NBS gene expression during *P. brassicae* infections of *B. rapa*.

Keywords: MicroRNA; TIR-NBS genes; QTL; R gene; *Brassica*; *Plasmodiophora brassicae*; disease resistance; clubroot; *B. rapa*

1. Introduction

Clubroot (CR), which is caused by the soil-borne pathogen *Plasmodiophora brassicae*, is a severe disease of crops in the family Brassicaceae. As a part of sustainable agriculture under deteriorating growth conditions due to long-term cultivation and climate change, the improvement of crops by introducing disease resistance traits is an important goal

339

for plant breeders [1]. Traditionally, the introgression of new traits for crop improvement has been performed via interspecific and intraspecific hybridizations, and these methods have resulted in the successful generation of crop resources with resistance traits [2]. Because CR disease is a major source of economic losses [3], the CR resistance trait has been introduced into *Brassica* species using various resistance resources (primarily turnip, but other subspecies as well).

To fully utilize resistance traits, the mechanisms underlying their genetic regulation must be characterized. In the last two decades, genetic analyses of CR resistance have been conducted using a variety of resistant *Brassica rapa* germplasm. Additionally, the A genome of Brassicaceae species has been extensively investigated. Comparative analyses have revealed the high collinearity between the resistance loci in the genomes of other amphidiploids (with AB and AC genomes) and those in the A genome. Previous studies have identified the following 23 CR resistance loci distributed on seven chromosomes (A01, A02, A03, A05, A06, A07, and A08) in *B. rapa*: *CRa* [4,5], *CRaki* [6,7], *CRb* [8,9], *CRc*, *CRk* [10], *CRd* [11], *Crr1*, *Crr2* [12]), *Crr3* [13], *Crr4* [14], *CrrA05* [15], *CRs* [16], *PbBp3.1*, *PbBp3.3* [9], *qBrCR38-1*, *qBrCR38-2* [17], *Rcr1* [18], *Rcr2* [19], *Rcr4*, *Rcr8*, *Rcr9* [20], *Rcr3*, and *Rcr9^{wa}* [21]. Chromosome A03 harbors at least 12 CR resistance loci (*CRa*, *CRaki*, *CRb*, *CRd*, *CRk*, *PbBp3.1*, *PbBp3.3*, *Crr3*, *Rcr1*, *Rcr2*, *Rcr4*, and *Rcr5*) effective against diverse pathotypes. After *CRa* was first identified at the *CR* locus in *B. rapa*, *CRb*, *CRk*, and *Crr3* were identified using diverse resistant materials and pathotypes. Recent advances in next-generation sequencing technology have enabled the easy, fast, and accurate identification of genomic regions related to qualitative and quantitative resistance traits [22].

Despite many genetic studies on resistance, relatively little is known about the candidate genes and the mechanisms controlling the CR resistance trait. Two loci related to CR resistance (*Crr1* and *CRa*) were discovered by map-based cloning, and TIR-NBS-LRR (TNL) genes were identified as candidate genes involved in resistance, based on gain-of-function analyses [5,23]. In the last decade, several loci (*CRd*, *Rcr1*, *Rcr2*, *Rcr4*, and *Rcr5*) were detected based on the identification of genome-wide variants through bulked segregant RNA sequencing and genotyping-by-sequencing [11,20,24,25]. These earlier studies identified candidate genes on the basis of abundant sequence variations between resistant and susceptible lines, as determined by high-throughput sequencing and analyses of functional similarity. Although different resistant materials and pathogens were used in these studies, the identified loci were localized on chromosome A03 and in genomic regions with a similar set of candidate genes.

Recent genomic, transcriptomic, and proteomic analyses have indicated that hormonal regulation, cell wall structure, secondary metabolites, and resistance genes (R genes) are involved in the CR resistance trait [26,27]. However, to characterize the mechanism regulating CR resistance, future studies should focus on the interrelationships among genes and how genes are regulated.

MicroRNAs (miRNAs) are small non-coding RNAs that are attracting attention as essential regulators of gene expression. They play a major role in the regulation of developmental and physiological processes [28,29] and in the expression of genes responsive to abiotic and biotic stresses, including R genes [30]. On the basis of the presence of specific domains, R genes are generally classified into five functionally distinct classes. The first class contains nucleotide-binding site–leucine-rich repeat (NBS-LRR) genes, which are further classified as Toll/interleukin 1 receptor (TIR)-NBS-LRR (TNL) and coiled-coil (CC)-NBS-LRR (CNL) genes. The second class comprises genes encoding transmembrane proteins, including receptor-like transmembrane proteins. The genes in the third class encode kinases, including serine–threonine kinases, whereas those in the fourth class encode kinases with receptor-like functions (e.g., receptor-like kinases). The fifth class contains atypical R genes [31]. At the molecular level, various R genes contribute to the direct or indirect recognition of pathogen-derived effectors that induce effector-triggered immunity, which frequently involves a series of responses, including the hypersensitive response (i.e., a type of programmed cell death) [32].

In recent decades, several studies have demonstrated the importance of miRNA-regulated RNA silencing for plant innate immunity. Initially, miR472 and miR482 were identified and experimentally confirmed to target various NLR genes [33,34]. In *Glycine max*, most R genes (178/290 CNLs and 171/235 TNLs) are directly regulated by miRNAs [35]. Additionally, miRNA target decoys (i.e., endogenous RNAs that can negatively regulate miRNA activity) have also been identified [36]. Subsequent studies on several plant species detected numerous miRNA families whose members target multiple R genes and are responsible for post-transcriptional gene silencing [18,37–45]. Such miRNAs are generally conserved in identical/similar species and their target sequences encode conserved R protein motifs [45]. For example, members of the miR482 superfamily specifically target a conserved P loop motif in NLR proteins that is crucial for function [46]. In tobacco, nta-miR6019 and nta-miR6020 are involved in the silencing of the N gene, encoding a TNL protein [42]. In *Medicago truncatula*, miR1510 is differentially expressed between the roots and nodules [47,48]. In a soybean, miR1510 cleaves TNL gene transcripts, thereby activating phased secondary small interfering RNA (phasiRNA) synthesis [37]. Similarly, miR482 and miR2118 initiate phasiRNA production from their NLR targets and function as the principal regulators of various genes encoding NLRs [42,43,45,49]. Interestingly, an invading pathogen can alter both the expression of R genes and miRNA-mediated R gene turnover [43,50]. In *B. rapa*, miR1885 regulates both an immune receptor gene (*BraTNL1*) and a growth-related gene (*BraCP24*) via the production of trans-acting small interfering RNAs [18].

The mechanisms underlying the interplay between a pathogen infection and the miRNA-mediated expression of R genes, especially those related to CR disease in *Brassica* species, are still poorly understood. Recent studies explored the role of long non-coding RNAs and mRNAs in *Brassica napus* [51,52] and *Brassica campestris* [17] responses to *P. brassicae*, but there are no reports describing the roles of miRNAs in the response of *B. rapa* to *P. brassicae*.

In this study, we investigated miRNAs that are structurally associated with a group of candidate genes related to CR resistance on chromosome A03. These genes have been consistently identified as being associated with stress resistance [19,20,25,53]. We predicted the targets of miR1885a and miR1885b in *B. rapa*. Further analyses revealed that both miR1885a and miR1885b are differentially expressed between resistant and susceptible genotypes, and these miRNAs negatively regulate R genes in response to pathogens. Under natural conditions, miR1885 expression is maintained at low levels to allow for normal plant development and basal immunity. Its expression levels peak after pathogen infections, indicative of the reallocation of energy between activities related to growth and immunity. Our findings provide insights into how gene expression is precisely controlled during the complex interaction between a pathogen and its host.

2. Results

2.1. Differences in the Responses of Resistant (CR Shinki) and Susceptible (94SK) B. rapa Inbred Lines to P. brassicae Pathotypes

The responses of Chinese cabbage (*B. rapa*) lines to different *P. brassicae* pathotypes were compared on the basis of the disease index (DI). More specifically, the responses of CR Shinki and 94SK Chinese cabbage inbred lines to three well-known *P. brassicae* pathotypes (Race 4, Uiryeong, and Banglim) were analyzed [8,54] (Figure 1). In response to Race 4, the DI was significantly higher for the susceptible line (94SK) than for the resistant line (CR Shinki) (Figure 1A). In contrast to CR Shinki roots, the 94SK roots were severely damaged by CR. Both lines were similarly susceptible to pathotypes Uiryeong and Banglim (i.e., no significant difference in the DI) (Figure 1B,C). This suggests that CR Shinki exhibits pathotype-specific resistance, that is, it is resistant to only Race 4. Plants were photographed at 8 weeks post-inoculation. The DI values confirmed the resistance, and susceptibility, of the investigated Chinese cabbage lines to different *P. brassicae* pathotypes.

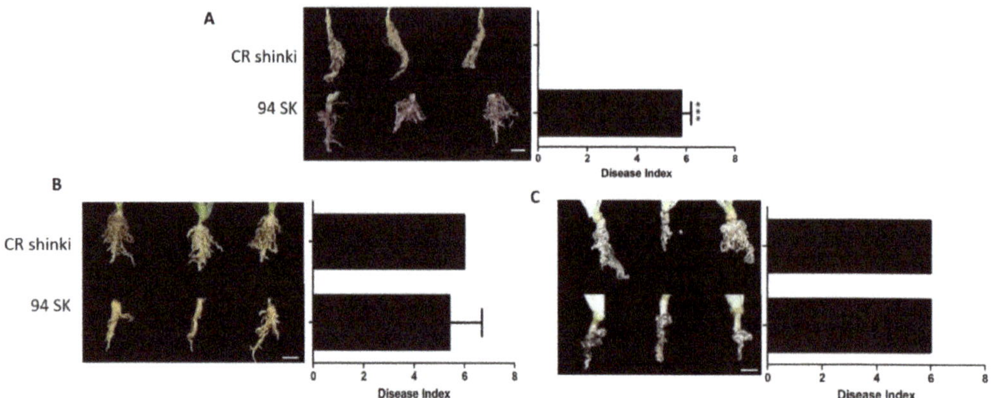

Figure 1. Differences in the responses of CR Shinki (resistant) and 94SK (susceptible) *Brassica rapa* inbred lines to three *Plasmodiophora brassicae* pathotypes. Disease symptoms on *B. rapa* roots infected with *P. brassicae* pathotypes (**A**) Race 4 [8], (**B**) Uiryeong, and (**C**) Banglim [54]. Disease severity was assessed according to the degree of root morphological modifications using the following rating scale: 0 = no symptoms; 1 = very slight swelling, usually confined to the lateral roots; 2 = a few small, separate globular clubs on the lateral roots; 3 = moderate clubbing on the lateral roots; 4 = large clubs on the lateral roots and slight swelling of the main root; 5 = larger clubs on the main root than on the lateral roots; 6 = severe clubbing on all roots. The mean disease index (n = 10) is presented to the right of the photos; the bar represents the standard error of three replicates. Asterisks indicate a significant difference at $p < 0.001$.

2.2. P. brassicae Infection Induces miR1885a and miR1885b Expression

A 22-nucleotide miRNA, Bra-miR1885a, was initially identified as a *B. rapa* R gene-derived novel miRNA during a response to turnip mosaic virus (TuMV) [55]. It was subsequently revealed to be responsive to heat stress [56]. A recent study confirmed that Bra-miR1885 is induced in *B. rapa* and *B. napus* by TuMV, but not by any other pathogen [18]. These previous studies demonstrated that Bra-miR1885a precisely regulates plant growth and immunity in *Brassica* species, suggesting that this *Brassica*-specific miRNA may also play roles in responses to biotic stresses. In this study, miR1885 was induced in *B. rapa* infected with *P. brassicae* (Figure 2).

According to the miRBase database [57], the *Brassica* miR1885 family has the following two members: miR1885a (on chromosome 06: 25285096–25285118 [+]) and miR1885b (on chromosome 06: 25285163–25285185 [+]). The pre-miR1885 structure is presented in Supplementary Figure S1. The expression of both miR1885a and miR1885b in the root tissue of the resistant (CR Shinki) and susceptible (94SK) inbred lines before the inoculation with *P. brassicae* (0 h) as well as at 1.5 h, 3 h, 6 h, 12 h, 24 h, 48 h, 72 h, 96 h, and 15 days post-inoculation (Figure 2A,B) were investigated. At 1.5 h post-inoculation, the miR1885a and miR1885b expression levels increased significantly in CR Shinki, but decreased markedly in 94SK. In CR Shinki and 94SK, the expression levels of both miRNAs decreased at 3 h and 6 h post-inoculation but then increased at 12 h and 24 h. The miR1885a expression level remained steady in CR Shinki at 48 h and 72 h post-inoculation, but it decreased at these time-points in 94SK. At 96 h and 15 days post-inoculation, the miR1885a expression level remained steady in CR Shinki, whereas it increased in 94SK (Figure 2A). Regarding miR1885b, a slightly different expression pattern was observed from 48 h post-inoculation. Its expression level was higher in CR Shinki than in 94SK at 48 h, and it peaked in CR Shinki at 72 h at a level much higher than that in 94SK (Figure 2B). The miR1885b expression level was markedly lower in CR Shinki than in 94SK at 96 h, whereas there were no significant differences between the inbred lines at 15 days post-inoculation. Overall, the miR1885a and miR1885b expression levels were generally higher in CR Shinki than in 94SK following the inoculation with *P. brassicae*. Considered together, these findings suggest that Bra-miR1885 might contribute to the *B. rapa* response to *P. brassicae*.

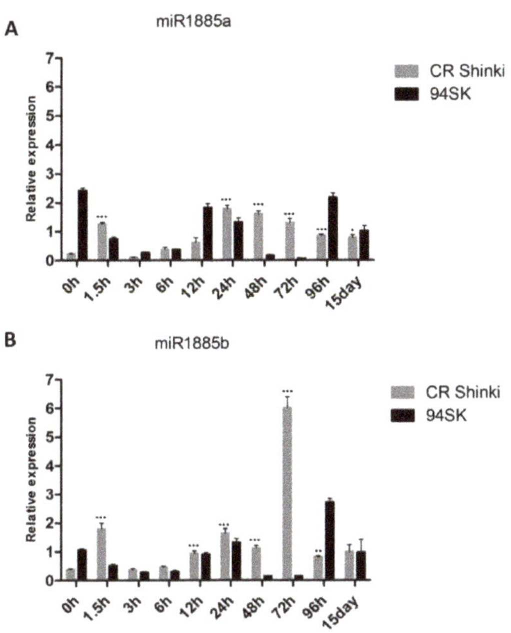

Figure 2. Expression of (**A**) miR1885a and (**B**) miR1885b in the roots of inbred lines CR Shinki (resistant) and 94SK (susceptible) inoculated with *P. brassicae*. The expression of both miRNAs increased significantly in CR Shinki post-inoculation (* $p < 0.01$, ** $p < 0.05$, and *** $p < 0.001$).

2.3. Both miR1885a and miR1885b Target TIR-NBS Genes

The miRNA targets were predicted on the basis of several parameters, including complementarities with miR1885a and miR1885b sequences and the unpaired energy (UPE) required to unfold the target site. Using stringent criteria for target prediction (expectation value of up to 2 and UPE value of up to 25), 11 targets were predicted for miR1885b and five targets were predicted for miR1885a (Supplementary Table S1). Interestingly, gene ontology analyses indicated that most of the target genes were on chromosome A03 and encoded disease resistance proteins with TNL and TX structures. Twelve loci (*CRa*, *CRb*, *CRaki*, *PbBp3.1*, *PbBp3.2*, *CRk*, *CRd*, *Crr3*, *Rcr1*, *Rcr2*, *Rcr4*, and *Rcr5*) related to disease resistance are located on chromosome A03 [4,6,8–11,20,58,59]. The *CRb* resistance-related quantitative trait locus (QTL) region in CR Shinki is located at the basal end of chromosome A03, and several gene loci are located nearby. Even though these earlier studies were performed using different resistant resources/germplasm (Chinese cabbage, pakchoi, and turnip) for introgressing resistance loci as well as different pathotypes, they all consistently detected certain genes (*Bra019409*, *Bra019410*, *Bra019412*, and *Bra019413*) in this QTL region as candidates for the CR resistance trait [19,20,24,25,53,59]. Among the candidate genes within the QTL, *Bra019410*, *Bra019412*, and *Bra020936* had an expectation value of 2, which was calculated on the basis of the affinity of miRNAs for their targets. Interestingly, *Bra020936*, which is located in a different QTL region on chromosome A08 [20], was also predicted as a miR1885b target, but it was excluded from further analyses because CR Shinki lacks the corresponding resistance locus.

Our results indicated that miR1885b targets both *Bra019412* and *Bra019410* with strict complementarity (expectation value approximately 2) and high free energy values, whereas miR1885a only targets *Bra019412* with relatively low complementarity (expectation value approximately 5) (Supplementary Table S1). The prediction analysis indicated that miR1885a inhibits target genes via the 'translation' process, whereas miR1885b inhibits target genes via the 'cleavage' process. These findings suggest that miR1885b is mainly involved in the inhibition of target genes.

2.4. Structure of Proteins Encoded by Bra019412 and Bra019410

Next, the conserved domains of the putative encoded proteins and gene structures of *Bra019412* and *Bra019410* (Figure 3) were identified. A Simple Modular Architecture Research Tool (SMART) analysis of the Bra019410 protein revealed the following two major domains: the TIR domain (amino acids 81 to 218) and the AAA domain (amino acids 276 to 417) (Figure 3B). A SMART analysis of the Bra019412 protein revealed only the TIR domain (amino acids 74 to 211) (Figure 3A). The exon–intron distribution of both genes was examined by comparing the coding and genomic sequences using the GSDS online software (Figure 3). This analysis detected three introns in *Bra019410*, but no introns in *Bra019412*. The characteristics of the putative proteins were determined using the ExPASy ProtParam online tool. The predicted molecular weights of Bra019410 and Bra019412 were 130.82 kDa and 247.88 kDa, respectively. The aliphatic index of Bra019410 was 93.35 and the grand average of hydropathicity (GRAVY) value was −0.22, which confirmed its hydrophilic nature. Similarly, the aliphatic index of Bra019412 was 81.67 and its GRAVY value was −0.228, indicating that it is also hydrophilic in nature. The isoelectric point of Bra019410 was 6.33, indicating it is an acidic protein, whereas that of Bra019412 was 8.74, indicating it is an alkaline protein.

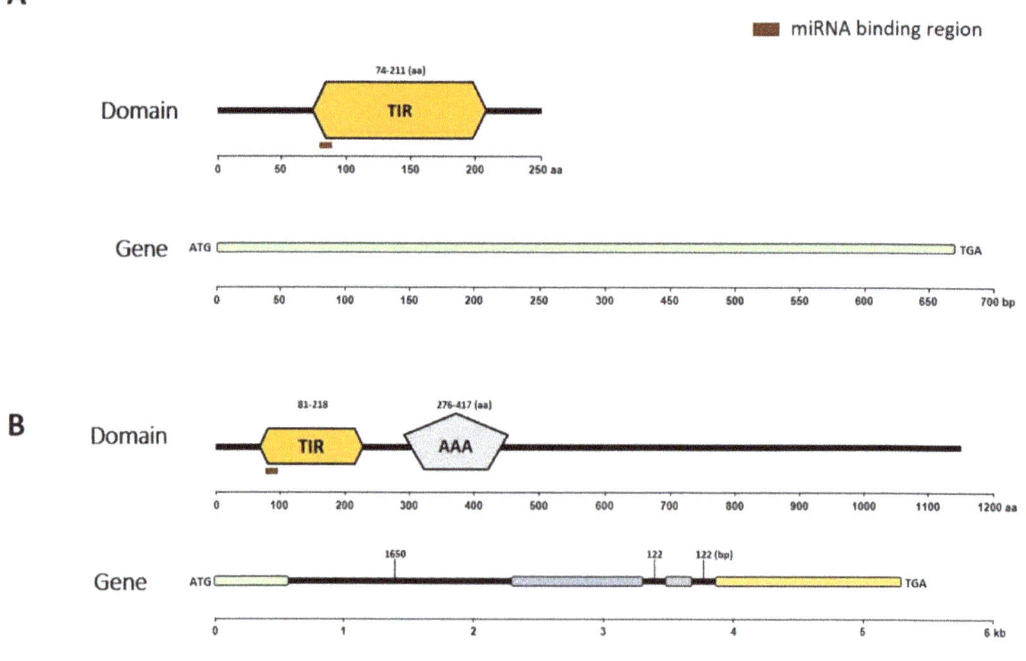

Figure 3. Analyses of conserved domains in putative proteins as well as gene structures. (**A**) *Bra019412* encodes only the TIR domain, whereas (**B**) *Bra019410* encodes the TIR and AAA domains.

2.5. Both miR1885a and miR1885b Negatively Regulate TIR-NBS Genes

The correlation between the expression levels of miRNAs and the transcript levels of both target genes (*Bra019412* and *Bra019410*) was investigated. As the miR1885b expression level increased from 1.5 h onwards, it down-regulated *Bra019412* expression until 15 days post-inoculation, indicating that miR1885b acts as a negative regulator of *Bra019412* in the resistant CR Shinki line (Figure 4A). In the susceptible 94SK line, miR1885b also appears to function as a negative regulator, although the *Bra019412* transcript level was lower in 94SK than in CR Shinki (Figure 4B). The miR1885a expression level in CR Shinki also increased at 1.5 h post-inoculation, but it was not as high as the miR1885b expression level and it was not correlated with the down-regulation of *Bra019412*, implying that miR1885a might be a partial negative regulator (Figure 4C). Additionally, miR1885a down-regulated the expression of *Bra019412* in 94SK, but only at the early post-inoculation time-points (up to 12 h). The expression pattern of miR1885a was dissimilar to that of miR1885b (Figure 4D).

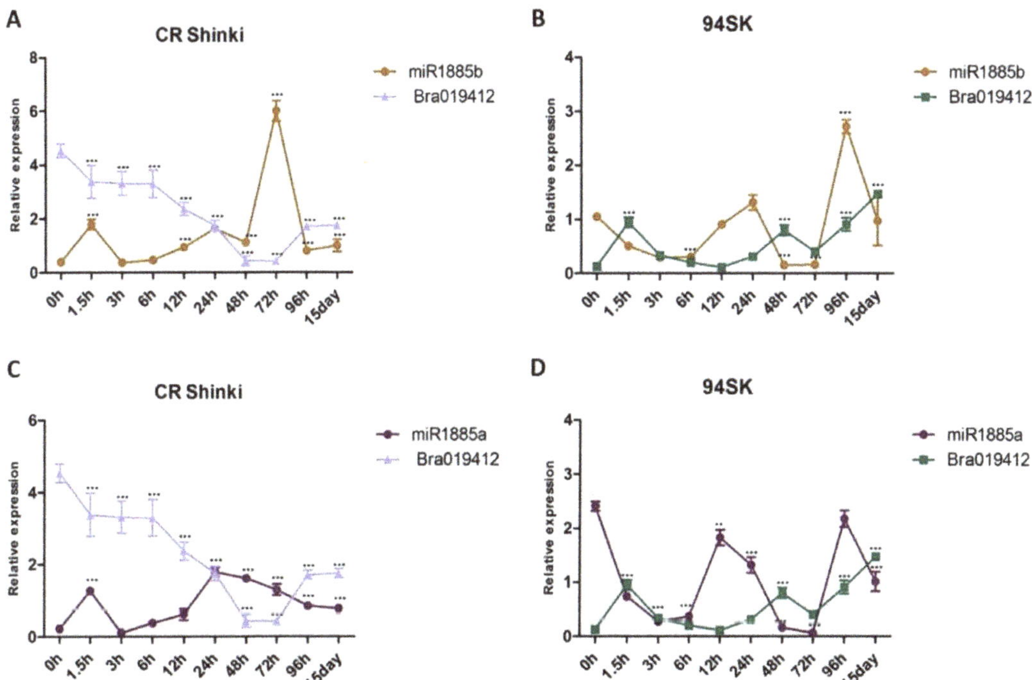

Figure 4. Analysis of the expression of *Bra019412* with miR1885b or miR1885a in the root tissues of *B. rapa* infected with *P. brassicae*. Correlation of the expression between (**A**) miR1885b and *Bra019412* in CR Shinki, (**B**) miR1885b and *Bra019412* in 94SK, (**C**) miR1885a and *Bra019412* in CR Shinki, and (**D**) miR1885a and *Bra019412* in 94SK as determined by qRT-PCR. The transcript level of the *B. rapa* actin gene served as the internal control for target gene expression, whereas the U6 snRNA level served as the internal control for mature miRNA expression. The miRNA and gene expression levels over time in inoculated plant samples were normalized against the corresponding expression levels in untreated plants. For each time-point, more than six individual plants were pooled ($n > 6$) and independent experiments were replicated three times. Standard error bars are presented for each time-point. Asterisks indicate a significant difference at *** $p < 0.001$.

The target analysis with stringent settings predicted that miR1885b targets *Bra019410*. However, the expression analysis demonstrated that miR1885b affects *Bra019410* expression only at post-inoculation time-points after 3 h (Figure 5). In CR Shinki, the *Bra019410* transcript level was stable until 3 h post-inoculation, even though miR1885b expression was up-regulated (Figure 5A,C). However, at the later time-points, there was a negative correlation between miR1885b levels and the *Bra019410* transcript levels. This result suggests that miR1885b only partially controls *Bra019410* expression (Figure 5A,C). In 94SK, there were no correlations between miR1885b or miR1885a levels and *Bra019410* transcript levels (Figure 5B,D), suggesting that both miRNAs may only partially control *Bra019410* expression.

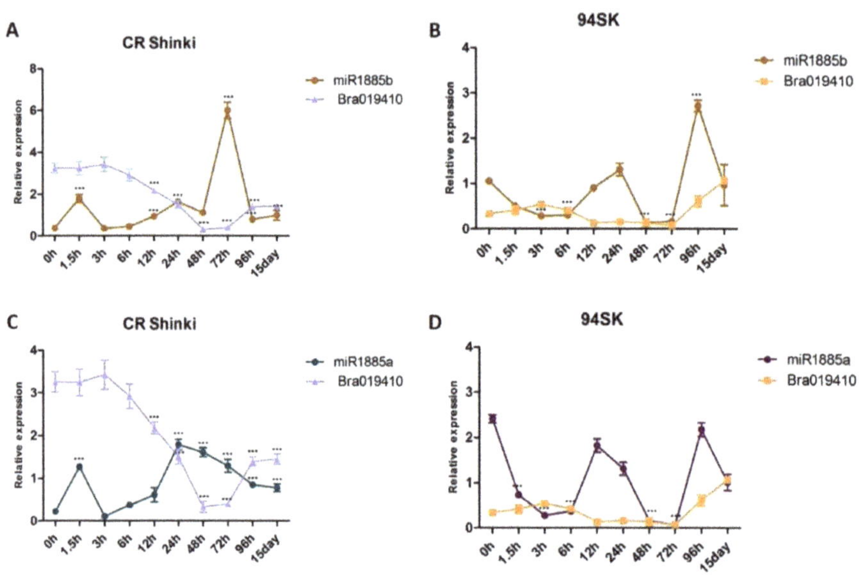

Figure 5. Analysis of the expression of *Bra019410* and miR1885b or miR1885a in the root tissues of *B. rapa* infected with *P. brassicae*. Correlation of the expression between (**A**) miR1885b and *Bra019410* in CR Shinki, (**B**) miR1885b and *Bra019410* in 94SK, (**C**) miR1885a and *Bra019410* in CR Shinki, and (**D**) miR1885a and *Bra019410* in 94SK, as determined by qRT-PCR. The transcript level of the *B. rapa* actin gene served as the internal control for target gene expression, whereas the U6 snRNA level served as the internal control for mature miRNA expression. The miRNA and gene expression levels, over time, in inoculated plant samples were normalized against the corresponding expression levels in untreated plants. For each time-point, more than six individual plants were pooled ($n > 6$) and independent experiments were replicated three times. Standard error bars are presented for each time-point. Asterisks indicate a significant difference at *** $p < 0.001$.

2.6. Expression Patterns of miR1885b, Bra019412, and Bra019410 in Multiple Resistant and Susceptible Lines

To confirm the miR1885b, *Bra019412*, and *Bra019410* expression patterns in CR-resistant and CR-susceptible genotypes, their transcript levels in multiple genotypes with contrasting responses to CR were investigated. On the basis of a DI analysis, three resistant genotypes (CR Shinki, M7, and C-20) and three susceptible genotypes (94SK, Chiifu, and HKC-002) (Figure 6A) were selected. After the inoculation with *P. brassicae*, miR1885b expression increased significantly in all resistant lines (particularly at 48 h) (Figure 6B). However, there were no significant changes in expression in the susceptible lines, suggesting miR1885b may be involved in the regulation of disease development in the resistant lines (Figure 6B). Overall, the root *Bra019412* and *Bra019410* transcript levels were significantly higher in the three resistant lines than in the susceptible lines (Figure 6C,D). The *Bra019412* transcriptional pattern was similar in CR Shinki and M7; the transcript levels gradually decreased from 0 h to 48 h post-inoculation (except at 12 h in M7), but then increased at 96 h and 15 days post-inoculation (Figure 6C). The *Bra019412* transcript levels in C-20 also generally decreased over the first 48 h post-inoculation (except at 24 h), but then increased at 96 h and 15 days post-inoculation, which was consistent with the expression pattern in the other two resistant lines (CR Shinki and M7). The expression profiles indicate that 24–96 h post-inoculation is a critical period for the regulation of resistance. The *Bra019412* transcript levels were lower in the susceptible lines than in the resistant lines at all time-points, with the exception of 24 h in HKC-002 (Figure 6C).

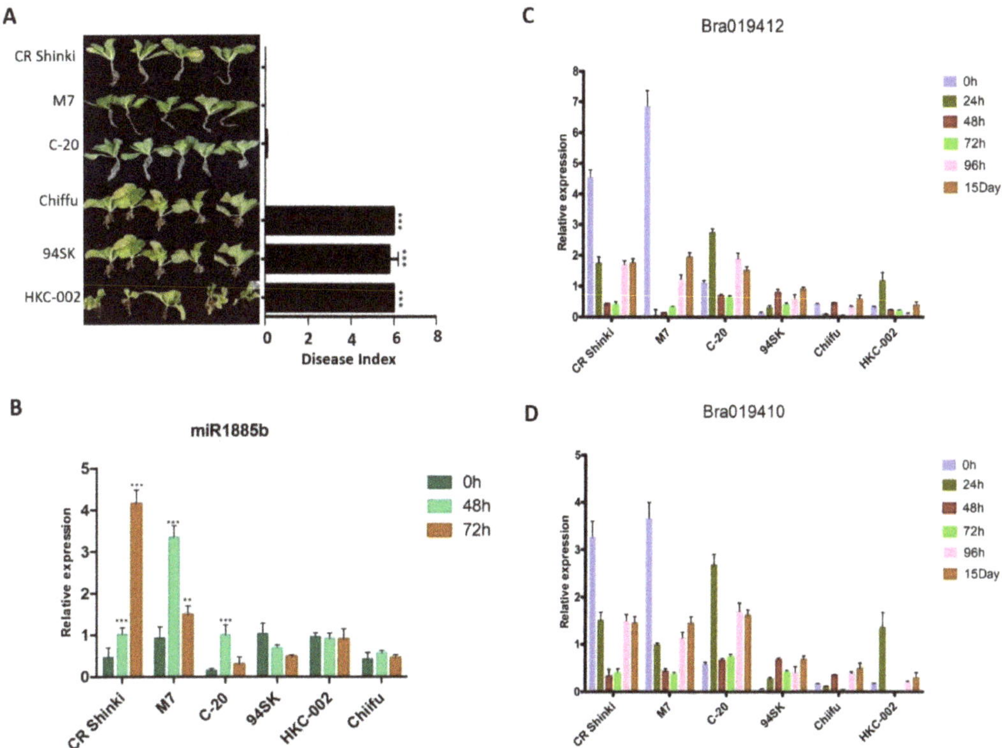

Figure 6. Expression patterns of miR1885b, *Bra019412*, and *Bra019410* in multiple resistant and susceptible lines. (**A**) Effects of the *P. brassicae* Race 4 infection of resistant (CR Shinki, M7, and C-20) and susceptible (94SK, HKC-002, and Chiifu) genotypes with the same alleles. Disease symptoms were examined at 8 weeks post-inoculation. For each line, the disease index, calculated according to the disease severity, is provided to the right of the photos. Bars represent the standard error. Significant differences are indicated with asterisks (** $p < 0.05$, and *** $p < 0.001$). Expression of (**B**) miR1885b, (**C**) *Bra019412*, and (**D**) *Bra019410* in resistant (CR Shinki, M7, and C-20) and susceptible (94SK, HKC-002, and Chiifu) genotypes. For each time-point, the expression levels were calculated on the basis of three independent replicates.

The *Bra019410* transcriptional patterns in the resistant and susceptible lines were similar to those of *Bra019412* (Figure 6D). In general, the *Bra019410* transcript levels post-inoculation were higher in the resistant lines than in the susceptible lines. The *Bra019410* transcript level gradually decreased from 0 h to 48 h post-inoculation in CR Shinki (except at 3 h) and M7 (Figure 6D, Figure 4A), but then increased at 96 h and 15 days post-inoculation. The *Bra019410* and *Bra019412* transcriptional patterns were identical in C-20 (Figure 6C). The *Bra019410* transcript levels were lower in the susceptible lines than in the resistant lines, except at 24 h (i.e., HKC-002).

2.7. Confirmation of miR1885b-Mediated Cleavage of Bra019412 by 5′ Rapid Amplification of cDNA Ends (RACE)

To confirm whether the decrease in the target gene transcript level was due to miR1885b/miR1885a-guided cleavage, the 5′ end of the cleaved *Bra019412* sequences was analyzed by 5′ RACE (Figure 7). The RNA extracted from the roots and leaves was subjected to a 5′ RACE analysis to identify the cleavage products of the target gene. In the root samples, a cleavage site (12th nucleotide) was detected in the miRNA-binding region of the target gene, confirming that *Bra019412* is a target of miR1885b (Figure 7).

Figure 7. 5′ RACE analysis revealing the cleavage site in regions targeted by miR1885b in the roots of infected CR Shinki plants. The miR1885b and *Bra019412* sequences are aligned and the predicted cleavage site is indicated with an arrow and an asterisk. The numbers above the sequence indicate the identified cleavage site in independent clones.

A second cleavage site 47 bases upstream of the predicted cleavage site was also detected. Regarding the RNA extracted from the leaves, no cleavage was detected at the predicted cleavage site of the target gene. The 5′ RACE analysis confirmed that cleavage products were derived from the interaction between the target gene and miR1885b, but not miR1885a. Thus, only miR1885b contributes to mRNA degradation and CR resistance.

3. Discussion

Clubroot is a devastating disease of *Brassica* crops worldwide. The CR resistance trait is reportedly controlled by various genes at both dominant (mostly in *B. rapa*) and recessive loci. Although numerous CR resistance-related loci/genes have been identified over the last few decades, the associated gene expression and regulation remain relatively uncharacterized. MicroRNAs are vital post-transcriptional regulators of R genes, including those related to disease resistance. Therefore, in the current study, the regulation of R gene expression by miRNAs, in Chinese cabbage lines infected with *P. brassicae*, was investigated. In the absence of pathogen infections, the detection of self-antigens by R proteins typically leads to autoimmunity and adversely affects plant survival. Accordingly, maintaining a balance between defense responses and plant development is critical. In plants, to minimize fitness costs, it is essential that R protein activities are stringently regulated [38,60,61]. The miRNA-mediated regulation (or turnover) of R genes allows plants to control the potential fitness consequences in the absence of pathogen infections. Thus, coordinating disease resistance and yield has become a key objective in plant breeding. Several miRNAs targeting R genes have been identified in various plant species in the last decade [35,41–43]. However, the roles of these miRNAs in plant defense responses and how they are regulated during pathogen infections are still unexplored in plants, in general, but especially in *Brassica* crops. Previously, we conducted a genome-wide investigation of miRNAs from *B. rapa* [62]. Additionally, Kim et al. [56] investigated heat stress-responsive miRNAs in *B. rapa*. To the best of our knowledge, the present study is the first to analyze the miRNA roles related to CR resistance in *B. rapa*.

Clubroot decreases the productivity of *Brassica* crops worldwide. A previous study on *B. napus* revealed the differential expression of miRNAs in plants infected with *P. brassicae* as well as several miRNAs targeting genes related to auxin signaling and transcription factors for hormone homeostasis during disease development [51]. Recently, Zhu et al. [17] analyzed long non-coding RNA and mRNA profiles after a *P. brassicae* infection of pakchoi (susceptible line) to elucidate the molecular basis of pathogenesis. The long non-coding RNAs responsive to *P. brassicae* in both resistant and susceptible genotypes were reported by Summanwar et al. [52], and several *P. brassicae*-responsive long non-coding RNAs that mainly target genes on chromosome A08 were identified.

In this study, miR1885, which is associated with CR resistance, was characterized. Initially, miR1885 was identified as an R gene-derived novel miRNA in *B. rapa* by He et al. [55] and Kim et al. [56]. It has since been confirmed to contribute to the regulation of the resistance to TuMV [18]. On the basis of sequence complementarity, miRNAs interact with their target genes to initiate their regulatory functions. The target genes of miR1885 were predicted using a plant-specific psRNA target prediction program [63]. Further analyses confirmed that miR1885 binds to the predicted targets in the region encoding the TIR domain. The TIR domain plays a vital role in the recognition of pathogen effectors [32,64]. It is a signaling domain involved in inducing cell death and it plays a major role in basal defense mechanisms [65]. Nandety et al. [66] proved that the TIR domain alone is sufficient for pathogen recognition.

Genes at the *CRb* locus were predicted to be specific targets of miR1885b. Various studies conducted worldwide using different genetic materials and without shared resources identified several gene loci at *CRb* and adjacent regions [19,20,25,53,58,59]. These studies revealed the importance of this locus for CR resistance and consistently identified a common set of genes, including *Bra019409, Bra019410, Bra019412,* and *Bra019413.*

Although these earlier studies identified various candidate R genes, the underlying regulatory mechanisms were not elucidated. Shivaprasad et al. [43] demonstrated that the miRNA-mediated silencing of disease resistance-related mRNAs in plants infected with viruses and bacteria ultimately leads to the pathogen-induced expression of genes encoding NBS-LRR defense-related proteins. They also identified a member of the miR483/2118 superfamily as the master regulator of disease resistance in tomato. Similarly, in this study, the role of miR1885, which targets *Bra019412*, in disease resistance was investigated. A structural analysis indicated *Bra019412* consists of a single exon and encodes a protein with only a TIR domain. MiR1885 also targets *Bra019410*, but its expression was not perfectly negatively correlated with *Bra019410* transcript levels. Consequently, further analyses were focused on the control of *Bra019412* expression. To understand the differences in defense mechanisms between inbred lines resistant and susceptible to CR, *B. rapa* plants were infected with *P. brassicae* and then gene expression levels at different time-points were analyzed. Consistent with the findings of an earlier transcriptome analysis of a segregating F_1 population containing the *Rcr1* locus [25], *Bra019412* was differentially expressed between the CR-resistant (overall up-regulation) and CR-susceptible (down-regulation) lines. Our results indicate that the observed diversity in the expression patterns was at least partly due to miRNAs.

In the current study, miR1885 was expressed at low levels under control conditions to maintain basal active immunity and *Bra019412* was stably transcribed to facilitate normal plant development (Figure 8A). The miR1885 levels increased only during the pathogen infection (Figure 8B). These results imply that a low miR1885 level is required for basal defense. Similar to our findings, Zou et al. [5] reported that Arabidopsis miR172b, which increases in abundance during seedling development, indirectly promotes the transcription of the gene encoding the immune receptor FLS2 through the post-transcriptional silencing of *TOE1* and *TOE2*, which encode suppressors of *FLS2* transcription. In another study on *B. rapa* infected by TuMV, an increase in miR1885 levels promoted precursor processing, suggesting that a TuMV infection increases the accumulation of miRNA-processing proteins [18].

In the present study, target gene expression increased substantially only in response to the Race 4 pathogen, consistent with the idea that each resistance locus has a different effect on CR development [9,20]. Shivaprasad et al. [43] noted that many genes encoding NBS-LRR proteins are associated with race-specific effector-triggered immunity. Additionally, *Bra019412* may interact with these genes and activate the associated disease resistance genes/clusters during pathogen infections, ultimately leading to quantitative resistance [67–70]. The TIR domain of Bra019412 likely acts as a link between the pathogen and the signaling function of the R protein [65,66,71]. Our results demonstrate that *Bra019412* plays a major role only in *CRb*-mediated CR resistance, which is consistent with the findings of previous studies, in which *Bra019412* was identified as a defense response-related candidate gene at a CR resistance locus in *B. rapa* [20,25,59]. Our observations suggest that the miRNA-mediated active regulation of immune receptors plays dynamic roles in the modulation of plant immunity.

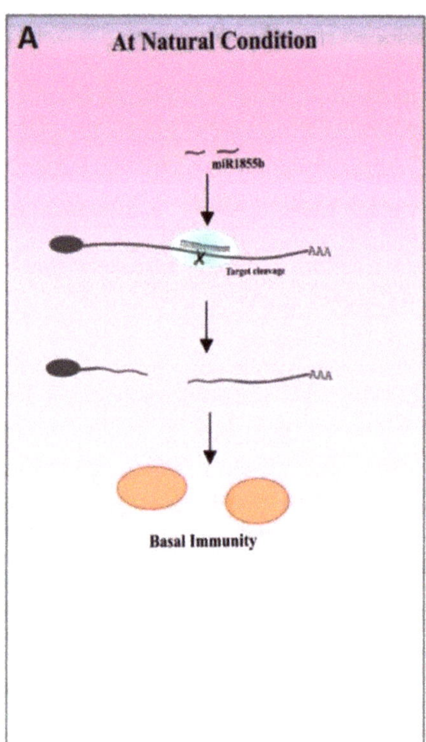
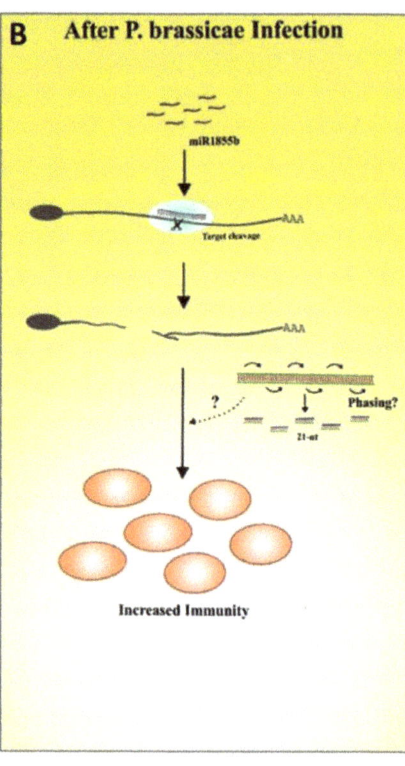

Figure 8. Proposed hypothesis based on the results of this study. (**A**) Under natural conditions, miR1885b is expressed at low levels, which is conducive to basal immunity. (**B**) During pathogen infections, miR1885b levels increase, which may enhance immunity through phasiRNAs.

MicroRNAs targeting NBS-LRR genes have distinct characteristics enabling them to target highly conserved motifs. Thus, they may regulate the expression of multiple members of gene families. This involves evolutionarily conserved interactions between small RNAs and their targets [55,72]. In the present study, miR1885b specifically targeted a gene encoding a protein with a TIR domain (Figure 3). The expression of TIR/TIR-NBS genes may be regulated by miRNAs in various ways. For example, the target sites may be expanded or lost and/or there may be feedback regulation between miRNAs and their target genes (Park and Shin, 2018). Recently, Cui et al. (2020) demonstrated that miR1885 targets NBS-LRR genes to activate phasiRNA generation when there is an excess of NBS-LRR proteins, indicative of self-regulation. This type of gene regulation prevents the undesirable production of TIR-NBS-LRR proteins. Similarly, we revealed that miR1885 regulates R gene turnover, thereby affecting disease resistance possibly through trans-acting RNA silencing [18]. Other studies confirmed the involvement of miR1885a in heat stress responses [56] and TuMV resistance [18]. This is the first report describing the role of miR1885b in a biotic stress response (i.e., CR disease).

4. Materials and Methods

4.1. Plant Materials and Pathogen Inoculation

Brassica rapa CR Shinki (CR resistant) and 94SK (CR susceptible) plants were grown in a growth chamber for 4 weeks and then inoculated with three *P. brassicae* pathotypes (Race 4, Uiryeong, and Banglim). Inbred line CR Shinki is resistant to the Race 4 pathotype [8], but susceptible to the Uiryeong (unpublished data) and Banglim [54] pathotypes. Plant samples were collected at two different infection stages. Samples for the initial infection stage were collected at 1.5, 3, 6, 12, 24, 48, 72, and 96 h post-inoculation, whereas samples for the late infection stage were collected at 15 days post-inoculation. The collected samples were immediately frozen in liquid nitrogen and then stored frozen until used for the gene expression analysis.

A *P. brassicae* suspension (2×10^6 CFU mL^{-1}) was used to infect plants via irrigation (i.e., injected into the soil). Briefly, a 5 mL aliquot of the spore suspension was added to the soil around 4-week-old plants. The differences in gene expression were verified in additional CR-resistant inbred lines (M7 and C-20) and CR-susceptible inbred lines (Chiifu and HKC002). The CR symptoms on plants were evaluated at 5 weeks post-inoculation as described by Choi et al. [54]. The DI was calculated according to a 0–6 scale, with 0 indicating the absence of disease symptoms and 6 indicating severe gall formation all over the roots. Ten plants per line were inoculated, and the experiments were performed in triplicate.

4.2. Prediction of the Target Genes of miR1885a and miR1885b

The target genes of miR1885a and miR1885b were predicted using psRNA Target [63]. This software predicts targets based on the reverse complementarity between miRNAs and target transcripts using the proven scoring scheme. It also evaluates the target site accessibility by calculating the UPE required for unwinding the secondary structure around the miRNA target site on the mRNA. The data were collected and then manually checked to remove repeated or irrelevant information. Finally, targets with an E-value < 4, a UPE < 25, and with significant matches in the seed region (7/8 for the second to eighth bases and 3/5 for the 12th to 16th bases from the 5' end of the miRNA) were selected.

4.3. Validation of miRNA and Target Gene Expression by a qRT-PCR Analysis

Total RNA was isolated from the leaves and roots of plants at different growth stages using the Plant RNeasy kit (Qiagen, Hilden, Germany). The RNA quality and quantity were estimated by gel electrophoresis and by using a NanoDrop spectrophotometer (Agilent, Santa Clara, CA, USA). The miRNA expression levels were verified by stem-loop qRT-PCR. The forward and reverse stem-loop primers for each miRNA were designed and synthesized (Bioneer, Daejeon, Korea) (Table S1). First-strand cDNA was synthesized from 1 μg RNA using the SuperScript III First-Strand Synthesis System (Invitrogen, Carlsbad, CA, USA); the manufacturer-recommended procedure was slightly modified. To increase the efficiency of the reverse transcription, a pulsed reverse transcription reaction was completed using the following PCR program: 16 °C for 30 min, 60 cycles of 30 °C for 30 s, 42 °C for 30 s, and 50 °C for 1 s (Varkonyi-Gasic et al. 2007). The reverse transcriptase was inactivated by incubating the mixture at 85 °C for 5 min. The qRT-PCR analysis was performed using the QuantiSpeed SYBR Kit (PhileKorea, Seoul, Korea) and the CFX96 Touch Real-Time PCR Detection System (Bio-Rad, Berkeley, CA, USA). The PCR program was as follows: 95 °C for 3 min, 40 cycles of 95 °C for 15 s, 58 °C for 20 s, and 72 °C for 15 s. Immediately after the final PCR cycle, a melting curve analysis was conducted (increase from 65 °C to 95 °C in increments of 0.2 °C) to check the PCR product specificity. The reactions were completed in triplicate, and the experiment was repeated at least twice. A control reaction without the template and reverse transcriptase was included for each miRNA. The *B. rapa* U6 snRNA gene served as the internal reference control.

To analyze target gene expression, cDNA was synthesized from 1 μg RNA using the Topscript RT DryMIX kit (Enzynomics, Daejeon, Korea). The qRT-PCR analysis was performed in triplicate using the QuantiSpeed SYBR Kit (PhileKorea, Seoul, Korea) and the CFX96 Touch Real-Time PCR Detection System (Bio-Rad). The PCR program was as follows: 95 °C for 10 min, 40 cycles of 95 °C for 15 s, and 60 °C for 1 min. Immediately after the final PCR cycle, a melting curve analysis was conducted (increase from 65 °C to 95 °C in increments of 0.2 °C). Data were acquired during the annealing/extension step and were analyzed using the CFX manager software (version 2.1) (Bio-Rad, CA, USA). The

B. rapa actin gene served as the internal reference control. The primers used to amplify the miRNA and gene sequences are listed in Supplementary Table S2. The comparative Ct method ($2^{-\Delta\Delta Ct}$) was used to quantify the relative miRNA and target gene expression levels (Livak and Schmittgen 2001). Significant differences were detected using Student's *t*-test. The miRNA and target gene expression levels were quantified using three replicates. Mean miRNA and target gene expression levels were normalized against the U6 snRNA and actin gene expression levels, respectively.

4.4. Target Gene Validation by 5′ RACE

The predicted targets were validated by 5′ RACE using a SMARTer RACE Kit (Clontech Laboratories, Takara Korea Biomedical Inc., Seoul, Korea). Briefly, 1 µg total RNA samples extracted from the roots and leaves of CR Shinki and 94SK plants infected with *P. brassicae* as well as non-infected (control) plants were used for the 5′ RACE assay. For the infected samples, the RNA samples extracted from plants at different infection stages were pooled. These total RNA samples were ligated with SMARTer II A oligonucleotides from the SMARTer RACE Kit and then the RNA was reverse transcribed using the supplied CDS Primer. Next, a PCR amplification was performed twice, using the long universal primer mix (UPM) and gene-specific primers in the first reaction and the short UPM and nested gene-specific primers in the second reaction. The PCR amplification was performed using 2× Advantage Taq Pre-mix. Amplified PCR products were purified and ligated into pRACE vectors using an In-Fusion HD Cloning Kit (Clontech, Laboratories, Takara Korea Biomedical Inc., Seoul, Korea). An AccuPrep® Gel Purification Kit (Bioneer, Daejeon, Korea) was used to purify the target PCR products. The ligated products were inserted into competent cells. Plasmids were isolated from individual transformed clones using an AccuPrep® Plasmid Mini Extraction Kit and then sequenced (Bioneer, Daejeon, Korea). The sequences were analyzed using BioEdit software version 7.2.

5. Conclusions

The current study demonstrated that miR1885a and miR1885b are differentially expressed between CR-resistant and CR-susceptible *B. rapa* genotypes. These miRNAs target TIR-NBS genes, specifically *Bra019412* and *Bra019410*. Overall, the target genes were expressed at higher levels in the resistant plants than in the susceptible plants. A negative correlation between miR1885b and *Bra019412* expression was detected, and a 5′ RACE analysis confirmed the cleavage of *Bra019412* by miR1885b. These findings revealed that miR1885b is critical for CR tolerance/resistance because it regulates *Bra019412* expression, especially in response to a *P. brassicae* infection. Future research should clarify how miR1885b regulates gene expression in plants infected with *P. brassicae*. The findings of this study are relevant for future investigations of miRNA-based regulation of CR development in *B. rapa* and related species. Additionally, the data presented herein will form the basis of the future functional characterization of the miRNAs, controlling the expression of disease resistance genes, in important vegetable and oilseed crops.

Supplementary Materials: The following are available online at https://www.mdpi.com/article/10.3390/ijms22179433/s1: Figure S1: Bra-miR1885 pri-miRNA structure, Table S1: Predicted targets of miR1885a and miR1885b, and Table S2: List of the primers used in this study.

Author Contributions: P.P., S.R.C. and Y.P.L. designed the experiment. P.P., S.S.C., J.J.R. carried out the experiments, generate data. S.S. participates in experiments. P.P., S.S.C., J.J.R., V.D. analyzed data. P.P., S.S.C., J.J.R., S.H.O., V.D., and S.R.C. prepared the samples and investigate phenotypic symptom. P.P., S.S.C., S.R.C. drafted the manuscript. S.-Y.S., C.S., N.R. participated in interpretation of data as advisor, and modification of manuscript. S.R.C. and Y.P.L. participated as a director, modified the manuscript and finalized the manuscript. All authors have read and agreed to the published version of the manuscript.

Funding: This work was supported by Korea Institute of Planning and Evaluation for Technology in Food, Agriculture, and Forestry (IPET) through Golden Seed Project (project number: 213006-05-5-SBD30, project number: 213006-05-5-SB110), funded by Ministry of Agriculture, Food and Rural Affairs (MAFRA), Ministry of Oceans and Fisheries (MOF), Rural Development Administration (RDA) and Korea Forest Services (KFS).

Institutional Review Board Statement: Not applicable.

Informed Consent Statement: Not applicable.

Data Availability Statement: All the necessary data generated are provided in the form of figures, tables and supplementary information.

Acknowledgments: We thank Edanz (https://www.edanz.com/ac, accessed on 16 July 2021) for editing the English text of a draft of this manuscript as well as Sonam Singh for technical support regarding figure preparation.

Conflicts of Interest: The authors declare no conflict of interest.

References

1. Kulwal, P.; Thudi, M.; Varshney, R. Crop Breeding for Sustainable Agriculture, Genomics Interventions in. In *Sustainable Food Production*; Springer: New York, NY, USA, 2013.
2. Kuginuki, Y.; Ajisaka, H.; Yui, M.; Yoshikawa, H.; Hida, K.I.; Hirai, M. RAPD markers linked to a clubroot-resistance locus in *Brassica rapa* L. *Euphytica* **1997**, *98*, 149–154. [CrossRef]
3. Dixon, G.R. The Occurrence and Economic Impact of *Plasmodiophora brassicae* and Clubroot Disease. *J. Plant Growth Regul.* **2009**, *28*, 194–202. [CrossRef]
4. Matsumoto, E.; Yasui, C.; Ohi, M.; Tsukada, M. Linkage analysis of RFLP markers for clubroot resistance and pigmentation in Chinese cabbage (*Brassica rapa* ssp. pekinensis). *Euphytica* **1998**, *104*, 79. [CrossRef]
5. Zou, Y.; Wang, S.; Zhou, Y.; Bai, J.; Huang, G.; Liu, X.; Zhang, Y.; Tang, D.; Lu, D. Transcriptional regulation of the immune receptor FLS2 controls the ontogeny of plant innate immunity. *Plant Cell* **2018**, *30*, 2779–2794. [CrossRef]
6. Kato, T.; Hatakeyama, K.; Fukino, N.; Matsumoto, S. Fine mapping of the clubroot resistance gene *CRb* and development of a useful selectable marker in *Brassica rapa*. *Breed. Sci.* **2013**, *63*, 116–124. [CrossRef] [PubMed]
7. Kato, T.; Hatakeyama, K.; Fukino, N.; Matsumoto, S. Identification of a clubroot resistance locus conferring resistance to a *Plasmodiophora brassicae* classified into pathotype group 3 in Chinese cabbage (*Brassica rapa* L.). *Breed. Sci.* **2012**, *62*, 282–287. [CrossRef]
8. Piao, Z.; Deng, Y.; Choi, S.; Park, Y.; Lim, Y. SCAR and CAPS mapping of *CRb*, a gene conferring resistance to *Plasmodiophora brassicae* in Chinese cabbage (*Brassica rapa* ssp. pekinensis). *Theor. Appl. Genet.* **2004**, *108*, 1458–1465. [CrossRef]
9. Chen, J.; Jing, J.; Zhan, Z.; Zhang, T.; Zhang, C.; Piao, Z. Identification of novel QTLs for isolate-specific partial resistance to *Plasmodiophora brassicae* in *Brassica rapa*. *PLoS ONE* **2013**, *8*, e85307. [CrossRef] [PubMed]
10. Sakamoto, K.; Saito, A.; Hayashida, N.; Taguchi, G.; Matsumoto, E. Mapping of isolate-specific QTLs for clubroot resistance in Chinese cabbage (*Brassica rapa* L. ssp. pekinensis). *Theor. Appl. Genet.* **2008**, *117*, 759–767. [CrossRef] [PubMed]
11. Pang, W.; Fu, P.; Li, X.; Zhan, Z.; Yu, S.; Piao, Z. Identification and mapping of the clubroot resistance gene *CRd* in Chinese cabbage (*Brassica rapa* ssp. pekinensis). *Front. Plant Sci.* **2018**, *9*, 653. [CrossRef]
12. Suwabe, K.; Tsukazaki, H.; Iketani, H.; Hatakeyama, K.; Fujimura, M.; Nunome, T.; Fukuoka, H.; Matsumoto, S.; Hirai, M. Identification of two loci for resistance to clubroot (*Plasmodiophora brassicae* Woronin) in *Brassica rapa* L. *Theor. Appl. Genet.* **2003**, *107*, 997–1002. [CrossRef]
13. Hirai, M.; Harada, T.; Kubo, N.; Tsukada, M.; Suwabe, K.; Matsumoto, S. A novel locus for clubroot resistance in *Brassica rapa* and its linkage markers. *Theor. Appl. Genet.* **2004**, *108*, 639–643. [CrossRef]
14. Suwabe, K.; Tsukazaki, H.; Iketani, H.; Hatakeyama, K.; Kondo, M.; Fujimura, M.; Nunome, T.; Fukuoka, H.; Hirai, M.; Matsumoto, S. Simple sequence repeat-based comparative genomics between *Brassica rapa* and *Arabidopsis thaliana*: The genetic origin of clubroot resistance. *Genetics* **2006**, *173*, 309–319. [CrossRef]
15. Nguyen, M.; Monakhos, G.; Komakhin, R.; Monakhos, S. The new Clubroot resistance locus is located on chromosome A05 in Chinese cabbage (*Brassica rapa* L.). *Russ. J. Genet.* **2018**, *54*, 296–304. [CrossRef]
16. Laila, R.; Park, J.I.; Robin, A.H.K.; Natarajan, S.; Vijayakumar, H.; Shirasawa, K.; Isobe, S.; Kim, H.T.; Nou, I.S. Mapping of a novel clubroot resistance QTL using ddRAD-seq in Chinese cabbage (*Brassica rapa* L.). *BMC Plant Biol.* **2019**, *19*, 13. [CrossRef] [PubMed]
17. Zhu, H.; Zhai, W.; Li, X.; Zhu, Y. Two QTLs controlling clubroot resistance identified from bulked segregant sequencing in pakchoi (*Brassica campestris* ssp. chinensis Makino). *Sci. Rep.* **2019**, *9*, 1–9. [CrossRef] [PubMed]
18. Cui, C.; Wang, J.J.; Zhao, J.H.; Fang, Y.Y.; He, X.F.; Guo, H.S.; Duan, C.G. A *Brassica* miRNA Regulates Plant Growth and Immunity through Distinct Modes of Action. *Mol. Plant* **2020**, *13*, 231–245. [CrossRef] [PubMed]

19. Huang, Z.; Peng, G.; Liu, X.; Deora, A.; Falk, K.C.; Gossen, B.D.; McDonald, M.R.; Yu, F. Fine mapping of a clubroot resistance gene in Chinese cabbage using SNP markers identified from bulked segregant RNA sequencing. *Front. Plant Sci.* **2017**, *8*, 1448. [CrossRef]

20. Yu, F.; Zhang, X.; Peng, G.; Falk, K.C.; Strelkov, S.E.; Gossen, B.D. Genotyping-by-sequencing reveals three QTL for clubroot resistance to six pathotypes of *Plasmodiophora brassicae* in *Brassica rapa*. *Sci. Rep.* **2017**, *7*, 1–11. [CrossRef]

21. Karim, M.; Dakouri, A.; Zhang, Y.; Chen, Q.; Peng, G.; Strelkov, S.E.; Gossen, B.D.; Yu, F. Two Clubroot-resistance genes, *Rcr3* and *Rcr9^wa^*, mapped in *Brassica rapa* using bulk segregant RNA sequencing. *Int. J. Mol. Sci.* **2020**, *21*, 5033. [CrossRef]

22. Lv, H.; Fang, Z.; Yang, L.; Zhang, Y.; Wang, Y. An update on the arsenal: Mining resistance genes for disease management of *Brassica* crops in the genomic era. *Horticulture Res.* **2020**, *7*, 1–18. [CrossRef] [PubMed]

23. Hatakeyama, K.; Suwabe, K.; Tomita, R.N.; Kato, T.; Nunome, T.; Fukuoka, H.; Matsumoto, S. Identification and characterization of *Crr1a*, a gene for resistance to clubroot disease (*Plasmodiophora brassicae* Woronin) in *Brassica rapa* L. *PLoS ONE* **2013**, *8*, 1. [CrossRef]

24. Huang, Z.; Peng, G.; Gossen, B.D.; Yu, F. Fine mapping of a clubroot resistance gene from turnip using SNP markers identified from bulked segregant RNA-Seq. *Mol. Breed.* **2019**, *39*, 1–10. [CrossRef]

25. Chu, M.; Song, T.; Falk, K.C.; Zhang, X.; Liu, X.; Chang, A.; Lahlali, R.; McGregor, L.; Gossen, B.D.; Yu, F.; et al. Fine mapping of Rcr1 and analyses of its effect on transcriptome patterns during infection by *Plasmodiophora brassicae*. *BMC Genom.* **2014**, *15*, 1166. [CrossRef] [PubMed]

26. Mehraj, H.; Akter, A.; Miyaji, N.; Miyazaki, J.; Shea, D.J.; Fujimoto, R.; Doullah, M. Genetics of clubroot and fusarium wilt disease resistance in Brassica vegetables: The application of marker assisted breeding for disease resistance. *Plants* **2020**, *9*, 726. [CrossRef]

27. Lan, M.; Li, G.; Hu, J.; Yang, H.; Zhang, L.; Xu, X.; Liu, J.; He, J.; Sun, R. iTRAQ-based quantitative analysis reveals proteomic changes in Chinese cabbage (*Brassica rapa* L.) in response to *Plasmodiophora brassicae* infection. *Sci. Rep.* **2019**, *9*, 1–13. [CrossRef] [PubMed]

28. Baulcombe, D. RNA silencing in plants. *Nature* **2004**, *431*, 356–363. [CrossRef]

29. Mallory, A.C.; Vaucheret, H. Functions of microRNAs and related small RNAs in plants. *Nat. Genet.* **2006**, *38*, S31–S36. [CrossRef]

30. Stokes, T.L.; Kunkel, B.N.; Richards, E.J. Epigenetic variation in *Arabidopsis* disease resistance. *Genes Dev.* **2002**, *16*, 171–182. [CrossRef]

31. Zhong, Y.; Cheng, Z.M.M. A unique RPW8-encoding class of genes that originated in early land plants and evolved through domain fission, fusion, and duplication. *Sci. Rep.* **2016**, *6*, 1–13. [CrossRef]

32. Jones, J.D.; Dangl, J.L. The plant immune system. *Nature* **2006**, *444*, 323–329. [CrossRef]

33. Lu, C.; Kulkarni, K.; Souret, F.F.; MuthuValliappan, R.; Tej, S.S.; Poethig, R.S.; Henderson, I.R.; Jacobsen, S.E.; Wang, W.; Green, P.J.; et al. microRNAs and other small RNAs enriched in the Arabidopsis RNA-dependent RNA polymerase-2 mutant. *Genome Res.* **2006**, *16*, 1276–1288. [CrossRef] [PubMed]

34. Lu, S.; Sun, Y.H.; Shi, R.; Clark, C.; Li, L.; Chiang, V.L. Novel and mechanical stress-responsive microRNAs in *Populus trichocarpa* that are absent from *Arabidopsis*. *Plant Cell* **2005**, *17*, 2186–2203. [CrossRef] [PubMed]

35. Zhao, M.; Cai, C.; Zhai, J.; Lin, F.; Li, L.; Shreve, J.; Thimmapuram, J.; Hughes, T.J.; Meyers, B.C.; Ma, J. Coordination of microRNAs, phasiRNAs, and NB-LRR genes in response to a plant pathogen: Insights from analyses of a set of soybean *Rps* gene near-isogenic lines. *Plant Gen.* **2015**, *8*. [CrossRef] [PubMed]

36. Ivashuta, S.; Banks, I.R.; Wiggins, B.E.; Zhang, Y.; Ziegler, T.E.; Roberts, J.K.; Heck, G.R. Regulation of gene expression in plants through miRNA inactivation. *PLoS ONE* **2011**, *6*, 6. [CrossRef] [PubMed]

37. Arikit, S.; Xia, R.; Kakrana, A.; Huang, K.; Zhai, J.; Yan, Z.; Valdés-López, O.; Prince, S.; Musket, T.A.; Nguyen, H.T.; et al. An atlas of soybean small RNAs identifies phased siRNAs from hundreds of coding genes. *Plant Cell* **2014**, *26*, 4584–4601. [CrossRef] [PubMed]

38. Deng, Y.; Zhai, K.; Xie, Z.; Yang, D.; Zhu, X.; Liu, J.; Wang, X.; Qin, P.; Yang, Y.; Zhang, G.; et al. Epigenetic regulation of antagonistic receptors confers rice blast resistance with yield balance. *Science* **2017**, *355*, 962–965. [CrossRef] [PubMed]

39. Deng, Y.; Liu, M.; Li, X.; Li, F. microRNA-mediated R gene regulation: Molecular scabbards for double-edged swords. *Sci. China Life Sci.* **2018**, *61*, 138–147. [CrossRef] [PubMed]

40. Deng, Y.; Wang, J.; Tung, J.; Liu, D.; Zhou, Y.; He, S.; Du, Y.; Baker, B.; Li, F. A role for small RNA in regulating innate immunity during plant growth. *PLoS Pathog.* **2018**, *14*, e1006756. [CrossRef] [PubMed]

41. Gonzalez, V.M.; Muller, S.; Baulcombe, D.; Puigdomenech, P. Evolution of NBS-LRR gene copies among dicot plants and its regulation by members of the miR482/2118 superfamily of miRNAs. *Mol. Plant.* **2015**, *8*, 329–331. [CrossRef]

42. Li, F.; Pignatta, D.; Bendix, C.; Brunkard, J.O.; Cohn, M.M.; Tung, J.; Sun, H.; Kumar, P.; Baker, B. MicroRNA regulation of plant innate immune receptors. *Proc. Natl. Acad. Sci. USA* **2012**, *109*, 1790–1795. [CrossRef]

43. Shivaprasad, P.V.; Chen, H.M.; Patel, K.; Bond, D.M.; Santos, B.A.; Baulcombe, D.C. A microRNA superfamily regulates nucleotide binding site-leucine-rich repeats and other mRNAs. *Plant Cell* **2012**, *24*, 859–874. [CrossRef]

44. Vasudevan, S.; Steitz, J.A. AU-rich-element-mediated upregulation of translation by *FXR1* and *Argonaute 2*. *Cell* **2007**, *128*, 1105–1118. [CrossRef]

45. Zhai, J.; Jeong, D.H.; De Paoli, E.; Park, S.; Rosen, B.D.; Li, Y.; Gonzalez, A.J.; Yan, Z.; Kitto, S.L.; Grusak, M.A.; et al. MicroRNAs as master regulators of the plant NB-LRR defense gene family via the production of phased, trans-acting siRNAs. *Genes Dev.* **2011**, *25*, 2540–2553. [CrossRef]

46. Zhao, M.; Meyers, B.C.; Cai, C.; Xu, W.; Ma, J. Evolutionary patterns and coevolutionary consequences of MIRNA genes and microRNA targets triggered by multiple mechanisms of genomic duplications in soybean. *Plant Cell* **2015**, *27*, 546–562. [CrossRef]

47. Lelandais-Brière, C.; Naya, L.; Sallet, E.; Calenge, F.; Frugier, F.; Hartmann, C.; Gouzy, J.; Crespi, M. Genome-wide medicago truncatula small RNA analysis revealed novel microRNAs and isoforms differentially regulated in roots and nodules. *Plant Cell* **2009**, *21*, 2780–2796. [CrossRef] [PubMed]

48. Subramanian, S.; Fu, Y.; Sunkar, R.; Barbazuk, W.B.; Zhu, J.K.; Yu, O. Novel and nodulation-regulated microRNAs in soybean roots. *BMC Genom.* **2008**, *9*, 160. [CrossRef]

49. Jagadeeswaran, G.; Zheng, Y.; Li, Y.F.; Shukla, L.I.; Matts, J.; Hoyt, P.; Macmil, S.L.; Wiley, G.B.; Roe, B.A.; Zhang, W.; et al. Cloning and characterization of small RNAs from *Medicago truncatula* reveals four novel legume-specific microRNA families. *New Phytol.* **2009**, *184*, 85–98. [CrossRef] [PubMed]

50. Boccara, M.; Sarazin, A.; Thiebeauld, O.; Jay, F.; Voinnet, O.; Navarro, L.; Colot, V. The Arabidopsis miR472-RDR6 silencing pathway modulates PAMP-and effector-triggered immunity through the post-transcriptional control of disease resistance genes. *PLoS Pathog.* **2014**, *10*, e1003883. [CrossRef] [PubMed]

51. Verma, S.S.; Rahman, M.H.; Deyholos, M.K.; Basu, U.; Kav, N.N. Differential expression of miRNAs in *Brassica napus* root following infection with *Plasmodiophora brassicae. PLoS ONE* **2014**, *9*, e86648. [CrossRef]

52. Summanwar, A.; Basu, U.; Rahman, H.; Kav, N. Identification of lncRNAs responsive to infection by *plasmodiophora brassicae* in clubroot-susceptible and-resistant *Brassica napus* lines carrying resistance introgressed from rutabaga. *Mol. Plant Microbe Interact.* **2019**, *32*, 1360–1377. [CrossRef]

53. Hatakeyama, K.; Niwa, T.; Kato, T.; Ohara, T.; Kakizaki, T.; Matsumoto, S. The tandem repeated organization of NB-LRR genes in the clubroot-resistant *CRb* locus in *Brassica rapa* L. *Mol. Genet. Genom.* **2017**, *292*, 397–405. [CrossRef]

54. Choi, S.R.; Oh, S.H.; Chhapekar, S.S.; Dhandapani, V.; Lee, C.Y.; Rameneni, J.J.; Ma, Y.; Choi, G.J.; Lee, S.S.; Lim, Y.P. Quantitative trait locus mapping of clubroot resistance and *Plasmodiophora brassicae* pathotype Banglim-specific marker development in *Brassica rapa. Int. J. Mol. Sci.* **2020**, *21*, 4157. [CrossRef]

55. He, X.F.; Fang, Y.Y.; Feng, L.; Guo, H.S. Characterization of conserved and novel microRNAs and their targets, including a TuMV-induced TIR–NBS–LRR class R gene-derived novel miRNA in *Brassica. FEBS Lett.* **2008**, *582*, 2445–2452. [CrossRef]

56. Kim, B.; Yu, H.J.; Park, S.G.; Shin, J.Y.; Oh, M.; Kim, N.; Mun, J.H. Identification and profiling of novel microRNAs in the *Brassica rapa* genome based on small RNA deep sequencing. *BMC Plant Biol.* **2012**, *12*, 1–14. [CrossRef] [PubMed]

57. Kozomara, A.; Birgaoanu, M.; Griffiths-Jones, S. miRBase: From microRNA sequences to function. *Nucleic Acids Res.* **2019**, *47*, D155–D162. [CrossRef] [PubMed]

58. Ueno, H.; Matsumoto, E.; Aruga, D.; Kitagawa, S.; Matsumura, H.; Hayashida, N. Molecular characterization of the *CRa* gene conferring clubroot resistance in *Brassica rapa. Plant Mol. Biol.* **2012**, *80*, 621–629. [CrossRef] [PubMed]

59. Yu, F.; Zhang, X.; Huang, Z.; Chu, M.; Song, T.; Falk, K.C.; Deora, A.; Chen, Q.; Zhang, Y.; McGregor, L.; et al. Identification of genome-wide variants and discovery of variants associated with *Brassica rapa* clubroot resistance gene *Rcr1* through bulked segregant RNA sequencing. *PLoS ONE* **2016**, *11*, e0153218. [CrossRef]

60. Oldroyd, G.E.; Staskawicz, B.J. Genetically engineered broad-spectrum disease resistance in tomato. *Proc. Natl. Acad. Sci. USA* **1998**, *95*, 10300–10305. [CrossRef]

61. Tian, D.; Traw, M.B.; Chen, J.Q.; Kreitman, M.; Bergelson, J. Fitness costs of R-gene-mediated resistance in *Arabidopsis thaliana. Nature* **2003**, *423*, 74–77. [CrossRef]

62. Dhandapani, V.; Ramchiary, N.; Paul, P.; Kim, J.; Choi, S.H.; Lee, J.; Hur, Y.; Lim, Y.P. Identification of potential microRNAs and their targets in *Brassica rapa* L. *Mol. Cells* **2011**, *32*, 21–37. [CrossRef] [PubMed]

63. Dai, X.; Zhuang, Z.; Zhao, P.X. psRNATarget: A plant small RNA target analysis server (2017 release). *Nucleic Acids Res.* **2018**, *46*, W49–W54. [CrossRef] [PubMed]

64. Collier, S.M.; Moffett, P. NB-LRRs work a "bait and switch" on pathogens. *Trends. Plant. Sci.* **2009**, *14*, 521–529. [CrossRef] [PubMed]

65. Swiderski, M.R.; Birker, D.; Jones, J.D. The TIR domain of TIR-NB-LRR resistance proteins is a signaling domain involved in cell death induction. *Mol. Plant Microbe Interact.* **2009**, *22*, 157–165. [CrossRef] [PubMed]

66. Nandety, R.S.; Caplan, J.L.; Cavanaugh, K.; Perroud, B.; Wroblewski, T.; Michelmore, R.W.; Meyers, B.C. The role of TIR-NBS and TIR-X proteins in plant basal defense responses. *Plant Physiol.* **2013**, *162*, 1459–1472. [CrossRef] [PubMed]

67. Gebhardt, C.; Valkonen, J.P. Organization of genes controlling disease resistance in the potato genome. *Annu. Rev. Phytopath.* **2001**, *2001*. *39*, 79–102. [CrossRef]

68. Ramalingam, J.; Vera Cruz, C.M.; Kukreja, K.; Chittoor, J.M.; Wu, J.L.; Lee, S.W.; Baraoidan, M.; George, M.L.; Cohen, M.B.; Hulbert, S.H.; et al. Candidate defense genes from rice, barley, and maize and their association with qualitative and quantitative resistance in rice. *Mol. Plant Microbe Interact.* **2003**, *16*, 14–24. [CrossRef] [PubMed]

69. Trognitz, F.; Manosalva, P.; Gysin, R.; Niño-Liu, D.; Simon, R.; del Rosario Herrera, M.; Trognitz, B.; Ghislain, M.; Nelson, R. Plant defense genes associated with quantitative resistance to potato late blight in Solanum phureja× dihaploid S. tuberosum hybrids. *Mol. Plant Microbe Interact.* **2002**, *15*, 587–597. [CrossRef] [PubMed]

70. Wisser, R.J.; Sun, Q.; Hulbert, S.H.; Kresovich, S.; Nelson, R.J. Identification and characterization of regions of the rice genome associated with broad-spectrum, quantitative disease resistance. *Genetics* **2005**, *169*, 2277–2293. [CrossRef]

71. Burch-Smith, T.M.; Schiff, M.; Caplan, J.L.; Tsao, J.; Czymmek, K.; Dinesh-Kumar, S.P. A novel role for the TIR domain in association with pathogen-derived elicitors. *PLoS Biol.* **2007**, *5*, e68. [CrossRef]
72. Park, J.H.; Shin, C. The role of plant small RNAs in NB-LRR regulation. *Brief. Funct. Genom.* **2015**, *14*, 268–274. [CrossRef] [PubMed]

MDPI

St. Alban-Anlage 66

4052 Basel

Switzerland

Tel. +41 61 683 77 34

Fax +41 61 302 89 18

www.mdpi.com

International Journal of Molecular Sciences Editorial Office

E-mail: ijms@mdpi.com

www.mdpi.com/journal/ijms